AF581910

Risk-Informed Sustainable Development in the Rural Tropics

Risk-Informed Sustainable Development in the Rural Tropics

Editors

Maurizio Tiepolo
Vieri Tarchiani
Alessandro Pezzoli

MDPI • Basel • Beijing • Wuhan • Barcelona • Belgrade • Manchester • Tokyo • Cluj • Tianjin

Editors
Maurizio Tiepolo
DIST
Politecnico and University of Turin
Turin
Italy

Vieri Tarchiani
Institute of Bio Economy
National Research Council
Florence
Italy

Alessandro Pezzoli
DIST
Politecnico and University of Turin
Turin
Italy

Editorial Office
MDPI
St. Alban-Anlage 66
4052 Basel, Switzerland

This is a reprint of articles from the Special Issue published online in the open access journal *Sustainability* (ISSN 2071-1050) (available at: www.mdpi.com/journal/sustainability/special_issues/Risk_Informed_Sustainable).

For citation purposes, cite each article independently as indicated on the article page online and as indicated below:

LastName, A.A.; LastName, B.B.; LastName, C.C. Article Title. *Journal Name* **Year**, *Volume Number*, Page Range.

ISBN 978-3-0365-1370-6 (Hbk)
ISBN 978-3-0365-1369-0 (PDF)

Contents

About the Editors

Maurizio Tiepolo

Maurizio Tiepolo (associate professor of urban and regional planning) investigates how to improve risk management through the integration of local and scientific knowledge and public participation in the planning process. He has coordinated ten hydro-climatic risk reduction plans at the local scale. As a principal investigator, he has led several action–research and action–training projects in the Global South and has provided expertise to official development assistance.

Vieri Tarchiani

Vieri Tarchiani (researcher) investigates the impacts of climate change on the water–soil–vegetation nexus, with particular attention to semiarid environments to identify adaptation and climate risk reduction strategies and solutions. He coordinated several training and research for development projects in West Africa and collaborates with the World Meteorological Organization for the implementation and evaluation of international initiatives on climate services.

Alessandro Pezzoli

Alessandro Pezzoli (senior lecturer in meteo-hydrology risk assessment and weather risk management) has a large expertise in applied meteorology and applied climatology. As an investigator on adaptation strategies to natural hazards generated by extreme weather, he participated to several research projects in Brazil, Ecuador, Ethiopia, Kenya, Niger, and Paraguay. He is a fellow of the Royal Meteorological Society and of the Royal Geographical Society.

Editorial

Risk-Informed Sustainable Development in the Rural Tropics

Maurizio Tiepolo [1,*], Vieri Tarchiani [2] and Alessandro Pezzoli [1]

[1] DIST—Interuniversity Department of Regional and Urban Studies and Planning, Politecnico and University of Turin, 10125 Turin, Italy; alessandro.pezzoli@polito.it

[2] Institute of Bio Economy, National Research Council of Italy, 50019 Sesto Fiorentino, Italy; vieri.tarchiani@ibe.cnr.it

* Correspondence: maurizio.tiepolo@polito.it; Tel.: +39-11-0907491

Citation: Tiepolo, M.; Tarchiani, V.; Pezzoli, A. Risk-Informed Sustainable Development in the Rural Tropics. *Sustainability* **2021**, *13*, 4179. https://doi.org/10.3390/su13084179

Received: 30 March 2021
Accepted: 7 April 2021
Published: 9 April 2021

Publisher's Note: MDPI stays neutral with regard to jurisdictional claims in published maps and institutional affiliations.

1. Overview

In the tropics, rural areas are still the place where many people live, despite ongoing urbanization. In tropical Africa, most of the population is still rural and will be so for at least another generation [1,2]. The development of the rural tropics is not merely a contribution to the growth of individual countries. It can be a way of reducing poverty [3,4] and inequalities in access to water [5], health care [6], and education [7] that the process of urbanization is unable to alleviate. However, it can also be a way to reduce greenhouse gas emissions that drive climate change if rural development is pursued in a sustainable way. This means stopping deforestation [8]. Then, reducing livestock-related emissions, which now account for 56%, 83%, and 87% of the greenhouse gases produced in Asia, Latin America, and Africa, respectively, according to the Food and Agricultural Organization's latest estimates [9].

Efforts to achieve sustainable rural development are often thwarted by hydro-climatic disasters (droughts, flooding, heavy rains, typhoons) which local communities are little prepared to tackle. Understanding these disasters, improving preparation, and strengthening governance have become equal priorities of the Sendai Framework for Disaster Risk Reduction [10]. However, the implementation of disaster risk reduction (DRR) at local scale, to achieve the objectives of the Sendai framework, has come across innumerable obstacles. It is often the case that agricultural practices and local planning are not very risk-informed. Climatic information is absent or not accessible locally [11]. Early warning systems and climate services are habitually not constructed with and for the rural communities [12,13]. Exposure and vulnerability are frequently considered as static determinants of risk [14]. Finally, the frequency and quality of DRR mainstreaming in local development plans are low [15,16] or simply unknown [17,18]. These deficiencies are particularly acute in the Tropics, where many Least Developed Countries are located, and where there is, however, great potential for agricultural development [19,20].

This Special Issue aims to investigate ways of increasing local knowledge of risks to contribute to rural development. It also aims to ascertain the status of essential resources, such as water and soil, and identify what undermines their integrity. Finally, it seeks to identify local policies for risk reduction and adaptation. The 22 articles collected here cover case studies from 12 countries. More than half of the articles concern Africa, as the subcontinent contains most of the Earth's surface in the tropical zone. The 94 authors mobilized cover a wide range of disciplines, such as agronomy, architecture, civil engineering, climatology, earth sciences, ecology, economic policy, environmental engineering, geography, geology, geomatics, hydraulics, materials science, oceanography and atmospheric physics, remote sensing, and spatial and regional planning.

2. A Short Review of the Contributions

Eleven articles are devoted to the knowledge of risk. Two of them are dedicated to hazard knowledge. Vigna et al. consider the best datasets among the Climate Hazards

Group InfraRed Precipitation with Station, the Global Precipitation Climatology Center, and the Kenyan Meteorological Dataset to observe monthly rainfall trends in the North Horr subcounty in northern Kenya between 1983 and 2014. As a result, the Kenyan Meteorological Dataset corrected with a procedure based on the Global Precipitation Climatology Center monthly dataset performs better in terms of resolution and response to local scale precipitation differences.

Baldi et al. consider the severe thunderstorms in Sinai (Egypt) and their future trend. This hazard impacts the arid region as flash floods, which can be a resource if captured by water harvesting works.

Flood exposure is analyzed by Galligari et al. in a 135 km^2 of the most densely populated wetlands in Niger: The Maouri temporary creek in Guéchémé. The dynamics of the built-up area in the flood zone are observed over the last ten years. Human settlements appear to be expanding by 52% in flood prone zones. House consolidation with corrugated sheet metal roofs is stronger in that zone than outside it.

Caselle et al. present a dataset to appreciate the vulnerability of local communities to drought and other threats in the Hodh Chargui region of Mauritania. The dataset is useful for drafting local development plans for the 31 municipalities that make up this vast jurisdiction.

Tiepolo et al. present a pluvial and fluvial risk assessment in four rural settlements along the Sirba River, in western Niger. The assessment is conceived to support planning risk reduction. Set up in a participatory manner, it employs innovative techniques in the region (images captured by unmanned aerial vehicles, hydraulic modelling), integrated with local knowledge. The result is a support to informed decision-making in prioritising and implementing risk reduction policies that local assessments rarely provide.

With another article, Tiepolo et al. offer a similar assessment considering meteorological, hydrological, and agricultural drought and flood risk for 13 rural communities in the Hodh Chargui, Mauritania. A large variability of risk level emerges within a relatively limited geographical area, determined by the risk of heavy rains and agricultural drought. These results are useful to identify risk reduction actions, which are very different from those usually proposed by the literature.

Frontuto et al. propose an assessment of flood impacts in Duran district, Ecuador, that considers income distribution and risk adversity instead of standard monetary damage. This influences damage compensation and the identification of priority areas for intervention.

Risk perception is the subject of the article by Gomez Zapata et al. in the case of the Cotopaxi volcano, Ecuador. The use of modelling for exposed areas identification facilitates discussion with local communities and awareness. Furthermore, it allows us to understand where exposure perceived by communities does not coincide with that calculated with the model.

Finally, three articles deal with the early warning and forecasting of extreme events. Tarchiani et al. present a locally and community-based flood early warning system designed with, and implemented for, the riverine communities along the Sirba River in western Niger. The main result is the demonstration that an early warning system can be set up operationally, involving the beneficiary communities through observation and preparedness.

Bacci et al. analyze the meteorological services delivered by the National Directorate of Meteorology to rural communities from eight municipalities of the Dosso and Tillaberi regions to reduce drought risk in agriculture. Feedback from users demonstrate the positive perception of such services and the willingness to use them, despite the intrinsic incertitude.

Ebhuoma et al. investigate the use of climate services by three rural communities in the Niger delta (Nigeria). Authors find a local preference for using indigenous knowledge rather than climate services due to the lack of local weather stations, the precedent of wrong weather predictions, and the misuse of local communication channels. A way out could be to develop pilot projects with farmers who are willing to use climate services.

Nine articles are dedicated to the state of water and soil, and the threats to these two key natural resources for rural development. Bonetto et al. consider the status of water resources in three districts of the Ethiopian Rift Valley. The study observes trends in fluoride presence, pH, and electrical conductivity values in the wells. The information obtained is useful for increasing access to drinking water in this semi-arid region.

Bertone et al. consider monitoring the presence of fertilizers on waterways in tropical Australia. The use of a mobile station for real-time monitoring of water quality, especially nitrate detection, proves advantageous over traditional laboratory sampling analysis. Nitrates have fluctuations in concentration in a short time that mobile stations can detect.

Water pollution by fecal matter is the subject of the article by Bigi et al. on North-Horr subcounty in Kenya. The presence of nitrates in water sources, and measures foreseen by local government to reduce it, highlight the greater vulnerability of the northern part of the subcounty to this threat.

Baratta et al. focus on actions to ensure greater access to water in the Kayes region of Mali. In particular, the reconstruction of damaged mini-dams with the participation of beneficiary communities is described. The restored dams increase the development of micro businesses.

Lasagna et al. consider the availability of water in Gumbo, east of Juba, South Sudan. The results of the study demonstrated the peculiarity of surface and groundwater and the critical aspects to consider for water use, particularly due to the exceeding of limits suggested by the WHO and national regulations. This research represents a first step for the improvement of water knowledge and management, for sustainable economic development, and for social progress in this African region.

Acciarri et al. consider the best option for producing drinking water at Magoodhoo island, Maldives, whose lens freshwater was damaged by the 2004 tsunami. The result is that a reverse osmosis desalination plant, powered by a photovoltaic plant with batteries, is economically and environmentally more advantageous than using a diesel desalination plant and bottled water supply.

Figueroa et al. consider the impacts of converting forests to grassland in tropical Mexico. The focus is on the dynamics of carbon, nitrogen and phosphorous balances in soil. The study observes a carbon and nutrient loss due to land use change.

Watene et al. observe soil erosion between 1990 and 2015 in the Great Rift Valley region of Kenya. Agriculture with poor soil and water conservation measures in Lake Nakuru and Bogoria–Baringo lake watersheds drive the highest erosion rates. Conservation tillage, curbing deforestation and overgrazing are recommended.

Drextler investigates climate smart agriculture adaptation in Belize. The author finds mulching, soil nutrient enrichment, and cover practices typical of Mayan farming tradition to have positive influences. However, these practices are made unsustainable by poverty, population growth and deforestation. This calls for a renewed commitment of agriculture extension services.

Two articles deal with adaptation and risk reduction at the local scale but observed in a wider context: Nigeria and tropical Africa as a whole. Ogunpaimo et al. identifies the relationship between adaptation and food security and what characteristics of rural households and farms facilitate it. The author finds that adaptation increases food security and that it is linked to access to credit and extension services.

Finally, Tiepolo et al. investigate the state of disaster risk reduction mainstreaming in local development plans for 198 rural jurisdictions over tropical Africa. Emphasis is placed on the quality of the plans rather than their number, as is done in the monitoring of the Sendai framework for DRR. Lack of climate characterization, little DRR, and low participation characterize these plans, which remain anchored in providing basic services such as as electricity, water, sanitation, and hygiene.

In the rural tropics, local communities are exposed to climate-related hazards, as well as to an unsustainable use of land and water resources. Their role in the economy and society is too important to be obscured by urban-centric policies. Support for local risk reduction should be more concerned with informing rural communities, building shared responses to face threats, and the quality of policies implemented, instead of merely considering their quantity.

The guest editors would like to thank the Italian Agency for Development Cooperation (AICS) and the ANADIA 2.0 project for their support in producing this Special Issue.

References

1. Mercandalli, S.; Losch, B. (Eds.) Rural Africa in Motion. In *Dynamics and Drivers of Migration South of the Sahara*; FAO; CIRAD: Rome, Italy, 2017.
2. UNDESA-United Nations Department of Economic and Social Affairs. *World Urbanization Prospects. The 2018 Revision*; United Nations: New York, NY, USA, 2018.
3. Diao, X.; Hazell, P.; Thurlow, J. The Role of Agriculture in African Development. *World Dev.* **2010**, *38*, 1375–1383. [CrossRef]
4. Barrett, C.B.; Christiaensen, L.; Sheahan, M.; Shimeles, A. On the Structural Transformation of Rural Africa. *J. Afr. Econ.* **2017**, *26*, i11–i35. [CrossRef]
5. Agrawal, T. Educational inequality in rural and urban India. *Int. J. Educ. Dev.* **2014**, *34*, 11–19. [CrossRef]
6. Adams, E.A.; Smiley, S.L. Urban-rural water access inequalities in Malawi: Implications for monitoring the Sustainable Development Goals. *Nat. Resour. Forum* **2018**, *42*, 217–226. [CrossRef]
7. Yaya, S.; Uthman, O.A.; Okonofua, F.; Bishwajit, G. Decomposing the rural-urban gap in the factors of under-five mortality in sub-Saharan Africa? Evidence from 35 countries. *BMC Public Health* **2019**, *19*, 616. [CrossRef] [PubMed]
8. Pearson, T.R.H.; Brown, S.; Murray, L.; Sidman, G. Greenhouse gas emissions from tropical forest degradation: An underestimated source. *Carbon Balance Manag.* **2017**, *12*, 3. [CrossRef] [PubMed]
9. FAO. *Greenhouse Gas Emissions from Agriculture, Forestry and Other Land Use*; FAO: Rome, Italy, 2016.
10. UNISDR. *Sendai Framework for Disaster Risk Reduction 2015–2030*; UNISDR: Geneva, Switzerland, 2015.
11. Brasseur, G.P.; Gallardo, L. Climate services: Lessons learned and future prospects. *Earth's Future* **2016**, *4*, 79–89. [CrossRef]
12. Gautam, D.K.; Phaiju, A.G. Community Based Approach to Flood Early Warning in West Rapti River Basin of Nepal. *J. Integr. Disaster Risk Manag.* **2013**, *3*, 155–169. [CrossRef]
13. AsimZia, A.; Wagner, C.H. Mainstreaming Early Warning Systems in Development and Planning Processes: Multilevel Implementation of Sendai Framework in Indus and Sahel. *Int. J. Disaster Risk Sci.* **2015**, *6*, 189–199. [CrossRef]
14. Jurgilevich, A.; Räsänen, A.; Groundstroem, F.; Juhola, S. A systematic review of dynamics in climate risk and vulnerability assessments. *Environ. Res. Lett.* **2017**, *12*, 013002. [CrossRef]
15. Lyles, W.; Berke, P.; Smith, G. A comparison of local hazard mitigation plan quality in six states, USA. *Landsc. Urban Plan.* **2014**, *122*, 89–99. [CrossRef]
16. Horney, J.; Nguyen, M.; Salvesen, D.; Dwyer, C.; Cooper, J.; Berke, P. Assessing the Quality of Rural Hazard Mitigation Plans in the Southeastern United States. *J. Plan. Educ. Res.* **2016**, *37*, 56–65. [CrossRef]
17. UNDRR. *Global Assessment Report on Disaster Risk Reduction*; UNDRR: Geneva, Switzerland, 2019.
18. UNDRR. *Monitoring the Implementation of Sendai Framework for Disaster Risk Reduction 2015–2030: A Snapshot of Reporting for 2018*; UNDRR: Bonn, Germany, 2020.
19. Altieri, M.A.; Nicholls, C.I. The adaptation and mitigation potential of traditional agriculture in a changing climate. *Clim. Chang.* **2013**, *140*, 33–45. [CrossRef]
20. Meijer, S.S.; Catacutan, D.; Ajayi, O.C.; Sileshi, G.W.; Nieuwenhuis, M. The role of knowledge, attitudes and perceptions in the uptake of agricultural and agroforestry innovations among smallholder farmers in sub-Saharan Africa. *Int. J. Agric. Sustain.* **2015**, *13*, 40–54. [CrossRef]

Article

Comparison and Bias-Correction of Satellite-Derived Precipitation Datasets at Local Level in Northern Kenya

Ingrid Vigna *,†, Velia Bigi *,†, Alessandro Pezzoli and Angelo Besana

Interuniversity Department of Regional and Urban Studies and Planning (DIST), Politecnico di Torino & Università di Torino, 10125 Torino, Italy; alessandro.pezzoli@polito.it (A.P.); angelo.besana@unito.it (A.B.)

* Correspondence: ingrid.vigna@polito.it (I.V.); velia.bigi@polito.it (V.B.)

† Both authors contributed equally to this work.

Received: 27 February 2020; Accepted: 30 March 2020; Published: 5 April 2020

Abstract: Understanding ongoing trends at local level is fundamental in research on climate change. However, in the Global South it is hampered by a lack of data. The scarcity of land-based observed data can be overcome through satellite-derived datasets, although performance varies according to the region. The purpose of this study is to compute the normal monthly values of precipitation for the eight main inhabited areas of North Horr Sub-County, in northern Kenya. The official decadal precipitation dataset from the Kenyan Meteorological Department (KMD), the Global Precipitation Climatology Centre (GPCC) monthly dataset and the Climate Hazards Group Infrared Precipitation with Stations (CHIRPS) monthly dataset are compared with the historical observed data by means of the most common statistical indices. The GPCC showed the best fit for the study area. The Quantile Mapping correction is applied to combine the high resolution of the KMD dataset with the high performance of the GPCC set. A new and more reliable bias-corrected monthly precipitation time series for 1983–2014 results for each location. This dataset allows a detailed description of the precipitation distribution through the year, which can be applied in the climate change adaptation and tailored territorial planning.

Keywords: dataset validation; precipitation; Kenya; local climate; ASALs; Quantile Mapping

1. Introduction

Over the past decades, research on climate change has become of primary concern for different disciplines at a global level. However, the understanding of the climate at a local level is key to interpreting undergoing changes. Although there is an abundance of data in the Global North, the countries of the Global South are struggling to fill the gap. More specifically, land-based meteorological stations in African countries are still around half the optimal number required, unevenly distributed and poorly equipped [1–3].

In Kenya, there are thirty-two land-based meteorological stations, distributed mainly in the south and on the coast, which are the most developed and geared towards tourism [4]. To improve the livelihoods of communities, enhance and protect property [5], the Kenyan government is promoting the country's research and development in climate information.

In North Horr Sub-County, situated in Marsabit County in northern Kenya, there are no land-based meteorological stations to provide past climate observations. At a distance of 250 km, there are three weather stations, two located in the highlands and one near lake Turkana. However, they are not close enough to describe the peculiarities of the local climate of North Horr.

The area investigated is mainly inhabited by semi-nomadic pastoral communities which rely on livestock production. They move around the area during the year looking for pasture and water according to the changing season [6]. Climate, therefore, plays an important role in their life, particularly with regard to precipitation, and extreme events such as floods, flash floods and droughts can have a catastrophic impact.

Three complex phenomena and their interaction mainly influence the climate of the Country: the Intertropical Convergence Zone (ITCZ), El Niño Southern Oscillation (ENSO) and the Indian Ocean Dipole (IOD).

Depending on the season, the periodic shift of the ITCZ north and south is mainly responsible for the bimodal rainfall pattern in Kenya. The first rainy season, known as the "long rains season", lasts approximately from March to May (MAM), and the second, the "short rains season", from October to December (OND), with some variation across the country. The ENSO and IOD can affect and alter the onset and duration of the rainy and dry seasons triggering events such as droughts and flooding [7–9]. In previous decades, changes in the amount of rainfall have been recorded. The northern Arid and Semi-Arid Lands (ASALs) region of Kenya, including Marsabit County, showed a decreasing trend. In particular, the period of 1991–2013 was generally drier than the period 1961–1990 with the MAM season having the highest, yet statistically insignificant, decline in seasonal rainfall amounts [10].

Until recently, climate reference literature for the area consisted of outdated studies [6,11,12]. However, climate change effects on climate at a local scale have increased interest in research studies of the area [13–16]. This research shows that agro-pastoralists have an awareness of climate change and that the increasing rainfall variability combined with other environmental, social and political pressures negatively affects their resilience [17–20]. However, although local knowledge is important, if it is not confirmed by official climate information, it could be unusable and ultimately useless [21,22].

The lack of land-based meteorological stations in the area requires the use of satellite-derived data and climatic models for further analysis. The relationship between large-scale weather systems and local climate varies from region to region, making necessary to evaluate and correct them at local scale [23,24], but the scarcity of land surface observation is one of the greatest difficulties in assessing dataset performances [25]. Previous studies have tried to assess the performance of satellite-derived and model-derived datasets in East Africa [26–32], in particular in Kenya [33–35], in order to address the lack of data from land-based meteorological stations. However, these studies have a more regional perspective rather than a local focus, and further investigation on their use at local scale is needed.

This study aims to contribute to precipitation data gap filling in northern Kenya through the design of an innovative methodology for the identification of the normal monthly precipitation values for the main inhabited areas of North Horr Sub-County.

Therefore, it has been necessary to assess and compare the performance of different precipitation datasets with a local scale perspective and to apply a bias correction method, leading to the creation of a new best-fit dataset for the area. Using a direct, point-to-pixel and validation through statistical indices [26,27] approach, this research compares data from models with historical data obtained from land-based meteorological stations in order to assess how well their properties fit the study area characterized by a relatively simple topography. This method was preferred to others due to its adaptability to the available data and to the study area. The no hierarchical k-means clustering method [28] was discarded because of its subjectivity. While it reduces the shortcomings caused by the differences in spatial coverage of the datasets, it requires a subjective choice on the number of clusters of pixels based on the similarity of the annual rainfall cycle. Even an analysis based on the ability of the datasets to detect rainfall events [29] was not suitable because it would have required daily precipitation data instead of monthly data.

Therefore, three model-derived precipitation datasets were selected and compared with the historical series of the nearby land-based meteorological stations of Lodwar, Marsabit and Moyale.

The precipitation datasets used were:

- The decadal dataset from the Kenyan Meteorological Department (KMD), with a resolution of 0.0375°, hereinafter referred to as the KMD dataset [36] available at http://kmddl.meteo.go.ke:8081/SOURCES/.KMD/;

- The Global Precipitation Climatology Centre (GPCC) monthly precipitation dataset, with a 0.5° resolution, hereinafter referred to as the GPCC dataset [37] available at https://www.esrl.noaa.gov/psd/;
- The Climate Hazards group Infrared Precipitation with Stations (CHIRPS) monthly dataset, with a 0.05° resolution, hereinafter referred to as the CHIRPS dataset [38] available at https://iridl.ldeo.columbia.edu/SOURCES/.UCSB/.CHIRPS/.

The most commonly used statistical indices were calculated: Bias, Mean Absolute Error, Mean Squared Deviation, Root Mean Squared Deviation, Correlation Coefficient and standard deviation [26–28,39,40]. The Taylor diagram was used as a graphical evaluation instrument [41].

The comparative analysis highlighted the relatively high performance of the GPCC dataset and the low performance of the KMD dataset. The GPCC gauge-based dataset selected was used to rectify the KMD dataset at local level on sampled reference points—the main inhabited areas, usually cited in policy planning [42,43]—using the Quantile Mapping [44] bias correction algorithm. Specific normal monthly precipitation values were identified for the reference points.

The new normal monthly precipitation values can be used in future studies for local purposes while the experimented methodology can be applied in other scant data contexts.

In Section 2, the study area is described along with the precipitation datasets that were analyzed. The steps of the methodology adopted are also detailed. In Section 3, the main results are presented. Finally, in Section 4 the conclusions are discussed with particular attention to the limits and to the possible future perspectives of the research.

2. Materials and Methods

2.1. Study Area

The study aims to define the best-fit precipitation dataset for North Horr Sub-County, which is situated in Marsabit County, northern Kenya (Figure 1). The area is considered to be part of the ASALs, with an evaporation rate that exceeds rainfall by more than ten times. However, there are some peculiarities due to the influence of the altitude on the precipitation, which makes Mt. Marsabit (1865 m above sea level), Mt. Kulal (2235 m above sea level), Hurry Hills (1685 m above sea level) and the Moyale-Sololo escarpment (up to 1400 m above sea level) quite wet areas. By contrast, the Chalbi Desert, a large salted depression lying between 435 m and 500 m above sea level, is the dryer feature of the area [45].

There are no land-based meteorological stations in the Sub-County. Therefore, an area within a 250 km radius from North Horr, the main village, has been defined and the meteorological stations located inside this area have been selected. These land-based meteorological stations are situated in Lodwar, Moyale and Marsabit.

The main inhabited areas—i.e., reference points—besides North Horr are Balesa, Dukana, El Gadhe, El-Hadi, Gus, Kalacha and Malabot.

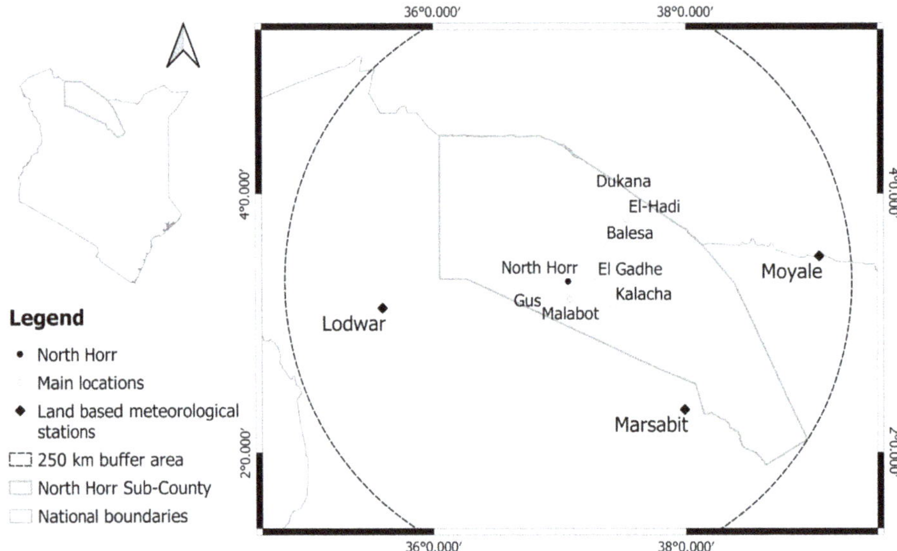

Figure 1. Map of the study area: North Horr sub-County is highlighted along with the main reference points. Within the 250 km radius from the reference point of North Horr, three land-based meteorological stations are identified (Lodwar, Marsabit and Moyale).

2.2. Precipitation Datasets

Previous studies have assessed the performance of different gridded precipitation products over East Africa [32] and Kenya [34,35]. The comparison of GPCC, CHIRPS, TRMM 3B42 (Tropical Rainfall Measuring Mission) and MERRA Modern-Era Retrospective Analysis for Research and Application) based on eight major agro-ecological zones demonstrated that GPCC and CHIRPS achieved improved results in ASALs [35]. In fact, the GPCC dataset best estimates precipitation in tropical warm semiarid areas while CHIRPS best estimates precipitation in tropical warm arid areas. Similarly, the comparison of CHIRPS, TRMM 3B42, PERSIANN-CDR (Precipitation Estimation from Remotely Sensed Information using Artificial Neural Networks Climate Data Record) and ARC2 (African Rainfall Climatology version 2.0) showed that CHIRPS have excellent performance in ASAL regions (high correlation, low RMSE, and low standard deviation) [34]. At regional level (East Africa), GPCC and CHIRPS have similar consistent results [32]. The other principal gridded precipitation products were evaluated. TRMM 3B42—as well as TRMM 3B43—has a reduced temporal (1998–present) [46] coverage compared to the aim of this research (1983–2013). PERSIANN-CDR underestimates rainfall in different topographical features and climatic conditions [34,47]. MERRA has a coarse resolution (0.5°), best estimates rugged mountainous zones and inaccurately predicts the rainfall amounts in relatively low-lying areas [35]. Following these considerations, three precipitation datasets have been selected and compared (Figure 2). The reference dataset is the KMD dataset provided by the National Meteorological Service. The other two datasets where selected on the base of previous studies results. They are highly reliable because they are provided by the World Meteorological Organization and the Climate Hazard Center funded by the U.S. Agency for International Development (USAID), the National Aeronautics and Space Administration (NASA) and the National Oceanic and Atmospheric Administration (NOAA).

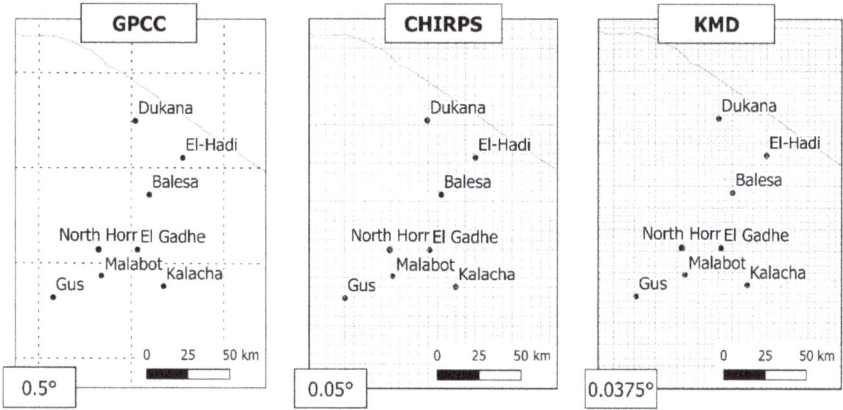

Figure 2. Comparison of the different resolutions of the gridded datasets. The Kenyan Meteorological Department (KMD) dataset has the highest resolution (0.0375°), while the Global Precipitation Climatology Centre (GPCC) has the lowest (0.5°). A precipitation value is available at each intersection point on the grid.

The KMD dataset is a decadal precipitation dataset, part of the Enhancing National Climate Services (ENACTS) project for development in Africa, which focuses on the creation of reliable climate data for national and local decision making. It has been produced by combining quality-controlled data from the national observation network with satellite estimates from the European Meteorological Satellites (METEOSAT). The data processing was performed using the Climate Data Tool software package developed by the International Research Institute (IRI) [36]. The dataset was directly furnished by the KMD but is also available at http://kmddl.meteo.go.ke:8081/SOURCES/.KMD/. It has a spatial resolution of 0.0375° and refers to the period 1983–2014.

The GPCC dataset, a monthly precipitation dataset, was developed by the Global Precipitation Climatology Project in support of the WMO World Climate Research Programme (WCRP) and the Global Energy and Water Cycle Experiment (GEWEX). It is a gauge-only product based on observations from rain gauge stations only available at a coarser resolution of 0.5° and a temporal coverage from 1901 to 2013 [37]. Version 7 (DOI: 10.5676/DWD_GPCC/FD_M_V7_050), available at NOAA/OAR/ESRL PSD website at https://www.esrl.noaa.gov/psd/, has been used for this study.

The CHIRPS dataset was developed to support the USAID Famine Early Warning Systems Network (FEWS NET). It builds on an high resolution and long recording period of precipitation estimates based on infrared Cold Cloud Duration (CCD) observations and on a station blending procedure based on a modified inverse distance weighting algorithm [48]. Several studies ascertain the effectiveness of this dataset in East Africa [26,27,38]. The monthly v2p0 version has been used, which is accessible through the IRI Data Library at https://iridl.ldeo.columbia.edu/SOURCES/.UCSB/.CHIRPS/. It has a spatial resolution of 0.05° and ranges from 1981 to near-present.

Finally, the monthly observed historical series from 1960 to 2016 from Marsabit, Moyale and Lodwar meteorological stations have been used as benchmark for the comparison analysis. They have been directly furnished by the KMD. Their characteristics are summarized in Table 1.

Table 1. Schematic summary of the meteorological stations' characteristics.

Station ID	Station	District	Lat	Long	Elevation
63612	Lodwar	Turkana	3.1°	35.6°	523 m
63641	Marsabit	Marsabit	2.3°	37.9°	1345 m
63619	Moyale	Moyale	3.53°	39.1°	1097 m

2.3. Methodology

The methodology followed in this study was structured in three steps (Figure 3):

- Section 2.3.1 presents the comparison of the datasets with the observed historical series from the selected land-based meteorological stations. This step lead to the identification and selection of the "reference dataset" (D_1) and "dataset to correct" (D_2).
- In Section 2.3.2, the correction method is detailed. The series from D_2 for each reference point are corrected with D_1 through the Quantile Mapping bias correction algorithm. This procedure leads to the definition of a group of five series for each reference point, each series resulting from a different correction method. The comparison of the five series with D_1 identifies the most appropriate correction method. Consequently, one series for each station and reference point has been selected.
- Section 2.3.3 refers to the extraction and computation of the precipitation normals for each reference point.

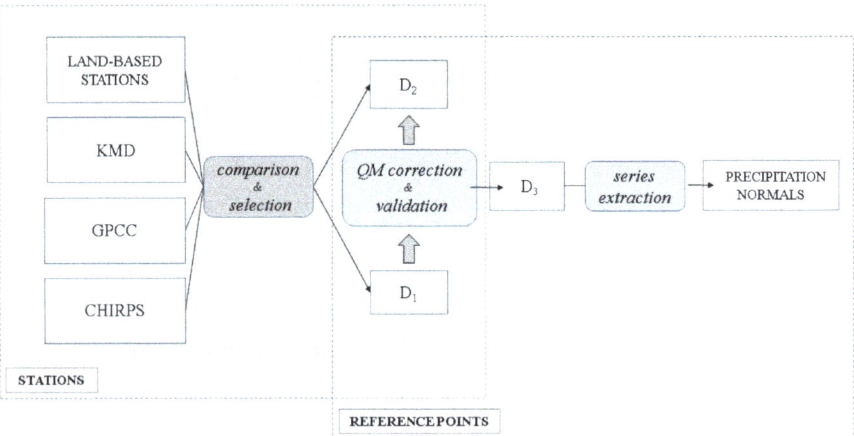

Figure 3. Methodological scheme.

2.3.1. Comparison of Dataset Performance at Meteorological Station Level

The performance of KMD, GPCC and CHIRPS datasets have been evaluated by means of a pixel to station comparison with the historical series of the meteorological stations at Marsabit, Lodwar and Moyale over a selected common period from 1983 to 2013. Five statistical indices have been computed: the bias, the Mean Absolute Error (MAE), the Mean Squared Deviation (MSD), the Root Mean Squared Deviation (RMSD) and the Correlation Coefficient (CC) [49]. A comparison of standard deviations (σ) was also performed in order to assess the dispersion of the values with regard to historical values.

A Taylor Diagram has also been created to provide an easier visual interpretation of the results. In the same graph, the CC, the RMSD, and the σ are shown for each series analyzed [41]. The MATLAB SkillMetrics toolbox developed by Peter Rochford has been used to create the diagram [50].

2.3.2. Correction through the Quantile Mapping Method

The KMD dataset, besides being provided by the official National Meteorological Service, has the higher resolution and therefore it was chosen as the dataset to be corrected (D_2).

The GPCC dataset was preferred to the three land-based meteorological stations as reference dataset (D_1). The stations in a 250 km radius, in fact, are not representative of the ASALs, while the GPCC dataset offers reliable interpolated gauge-derived information. This approach, justified by the scarcity of observed data in the region, is in line with previous studies, which aimed to overcome this obstacle by resorting to gauge-derived datasets for the validation of satellite or reanalysis precipitation datasets [51,52]. The correction procedure aims at merging the information derived by the two datasets, namely integrating the satellite-derived data, which is found to have too low level performances, with the gauge-derived data [53,54]. Findings demonstrate that Quantile Mapping can cause inflation problems (same temporal structure and variability of the coarser grid) when applied to datasets of different resolution [55,56]. However, the procedure here used is opposite to the common downscaling procedure; in fact, it aims at correcting a high-resolution satellite-derived dataset with a coarser grid reference dataset.

The KMD dataset was therefore corrected using the Quantile Mapping bias correction algorithm technique, which has been widely used for correction of precipitation datasets [57–59] and has demonstrated high performances in arid and semi-arid areas [33,60]. In particular, the work of Ringard et al. demonstrated its usefulness for satellite-derived datasets correction in scarce observed data contexts [61]. Moreover, the Quantile Mapping correction method performs very well concerning the reproduction of the precipitation annual cycle and of the wet and dry periods length [62]. This characteristic is fundamental with reference to the monthly normal identification and for future climate analysis of the area.

The Quantile Mapping technique is based on statistical transformation which attempts to adjust the distribution of modelled data such that it closely resembles the observed climatology solved using a theoretical distribution The 'qmap' package developed by Lukas Gudmundsson for R software was used for the computation [63]. The procedure was carried out for all the reference locations using a pixel to pixel approach. The 'qmap' package supports five different analytical methods, both parametric and non-parametric transformations. These methods use different functions to transform the distribution of the modelled data to match the distribution of the observations. The five functions performed are parametric transformations (PTF), distribution derived transformations (DIST), non-parametric quantile mapping using empirical quantiles (QUANT), non-parametric quantile mapping using robust empirical quantiles (RQUANT) and quantile mapping using a smoothing spline (SSPLIN)(for further details see the documentation at the following link https://www.rdocumentation.org/packages/CSTools/versions/2.0.0/topics/CST_QuantileMapping).

Five precipitation series were created for the three stations and for the eight reference locations.

The results of the Quantile Mapping correction were compared with the GPCC series for each reference location through the statistical indices (Section 2.3.1). The most appropriate method was identified, leading to the selection of a best-fit precipitation series for each reference location.

2.3.3. Reference Values Computation

New reference values were computed on the new precipitation series by averaging the monthly precipitation amount for the entire period (1983–2013).

3. Results and Discussion

3.1. Comparison of Dataset Performance at Meteorological Station Level

The comparison of the precipitation datasets with the observed series led to an important first conclusion. The KMD dataset does not feature the best indices values for all the stations. Results from the first step of the analysis conducted indicated that the GPCC dataset was a better choice as the reference series.

According to the statistical indices (Tables 2 and 3) and to the Taylor diagrams (Figure 4), the GPCC dataset fits better for the stations of Marsabit and Moyale, while the KMD dataset fits better for Lodwar. However, for reasons of homogeneity and consistency, the GPCC dataset was also chosen as the reference dataset for Lodwar station since its statistical values are close to the values obtained for the KMD dataset.

Table 2. Comparison based on statistical indices (BIAS, MAE, MSD, RMSD and CC) of the precipitation datasets with the observed historical series from the selected land-based meteorological stations (Lodwar, Marsabit and Moyale) for the period 1983–2013. For the CC index, "*" corresponds to a p-value < 0.01, "**" corresponds to a p-value < 0.001 and "***" corresponds to a p-value < 0.0001. Values in bold correspond to the best value of the index for each station.

	Lodwar			Marsabit			Moyale		
	KMD	GPCC	CHIRPS	KMD	GPCC	CHIRPS	KMD	GPCC	CHIRPS
BIAS	**0.99**	2.52	−5.32	−8.21	**2.35**	13.40	24.7	5.04	**2.16**
MAE	**5.4**	5.74	10.72	14.83	**12.7**	25.32	29.1	**14.9**	18.45
MSD	**237**	260	414	1498	**933**	1528	2797	**966**	1087
RMSD	**15.4**	16.1	20.35	38.71	**30.5**	39.09	52.9	**31.1**	32.97
CC	0.83 ***	**0.85 *****	0.71 ***	0.91 ***	**0.95 *****	0.92 ***	0.82 ***	**0.91 *****	0.90 ***

Table 3. Comparison based on the standard deviation of the precipitation datasets with the observed historical series from the selected land-based meteorological stations (Lodwar, Marsabit and Moyale) for the period 1983–2013.

Standard Deviation											
Lodwar				Marsabit				Moyale			
STAT. [1]	KMD	GPCC	CHIRPS	STAT. [1]	KMD	GPCC	CHIRPS	STAT. [1]	KMD	GPCC	CHIRPS
27.2	24.1	29.9	14.8	91.4	81.2	79.9	93.4	73.5	40.4	67.0	71.6

[1] STAT. = STATION.

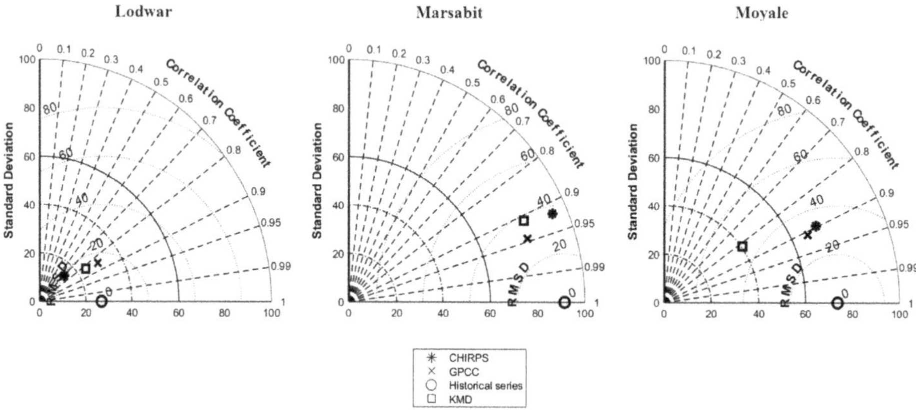

Figure 4. Taylor diagrams showing the agreement between the observed historical series and the precipitation datasets for the selected land-based meteorological stations (Lodwar, Marsabit and Moyale) for the period 1983–2013. The standard deviation of each series (as reported in Table 3) is proportional to the distance from the origin of the diagram. The Correlation Coefficient (in Table 2) between each series and the observed historical series is expressed by the azimuthal angle. Finally, the Root Mean Squared Deviation (in Table 2) between each series and the observed historical series is proportional to the distance from the point representing the observed historical series. Points closer to the historical series' marker, that is, with similar standard deviation, lower RMSD and higher Correlation Coefficient, correspond to the best-fit datasets.

3.2. Correction through the Quantile Mapping Method

As showed in the previous section, the GPCC dataset fits better than the other two datasets compared with the historical series. However, the GPCC dataset has a lower resolution (0.5°) compared to the KMD dataset and CHIRPS dataset (0.0375° and 0.05°, respectively), and the differences in local topography may be biased. Therefore, it was necessary to apply a bias correction method to overcome these two problems. The strategy adopted was to correct the KMD dataset (D_2), which is issued by the official National Meteorological Service and has the highest resolution, with the GPCC dataset (D_1) which performs better on ASALs.

The bias correction method is performed using the Quantile Mapping method from the 'qmap' R package. Five different bias-corrected series are obtained based on the five different transformations applied: Parametric Transformations (PTF), Distribution Derived Transformations (DIST), Robust Empirical Quantiles (RQUANT), Empirical Quantiles (QUANT), Smoothing Spline (SSPLIN).

From comparison analysis, the Parametric Transformations method, which fits a parametric transformation to the quantile-quantile relation of observed and modelled values, provided the best results (see Appendices A and B). Hereinafter, the Bias-Corrected KMD dataset will be referred to as the BCKMD dataset (D_3).

3.2.1. Quantile Mapping Validation at Station Level

The performance of the new BCKMD dataset is assessed by means of the statistical indices mentioned previously (see Section 3.1). The indices have been calculated in relation to the historical series of the land-based meteorological stations, then compared with the same indices calculated for the KMD dataset.

As shown in Table 4, the BCKMD dataset fits the observed historical series better than the KMD dataset, apart from Lodwar station. This may be due to a higher performance of the KMD dataset—before correction—at Lodwar station compared to the GPCC concerning BIAS, MAE, MSD and RMSD. However, the errors obtained are still acceptably low. In fact, the standard deviation values and the relatively low values of the error's indices, even for Lodwar, justify the selection of the BCKMD dataset for the study area.

Table 4. Comparison based on statistical indices (BIAS, MAE, MSD, RMSD, CC and σ) of the KMD and of the Bias-Corrected KMD (BCKMD) datasets with the observed historical series from the selected land-based meteorological stations (Lodwar, Marsabit and Moyale) for the period 1983–2013. For the CC index, "*" corresponds to a p-value < 0.01, "**" corresponds to a p-value < 0.001 and "***" corresponds to a p-value < 0.0001. Values in bold correspond to the best value of the index for each station.

	Lodwar		**Marsabit**		**Moyale**	
	KMD	BCKMD	KMD	BCKMD	KMD	BCKMD
BIAS	**−0.99**	2.29	−8.21	**−4.28**	−24.74	**−5.96**
MAE	**5.41**	7.71	**14.83**	16.78	29.10	**26.02**
MSD	**237.3**	295.32	1498.21	**1427.56**	2796.62	**1881.38**
RMSD	**15.40**	17.18	38.71	**37.78**	52.88	**43.38**
CC	0.83 ***	0.83 ***	0.91 ***	0.91 ***	0.82 ***	0.82 ***

Standard Deviation

Lodwar			**Marsabit**			**Moyale**		
STATION	KMD	BCKMD	STATION	KMD	BCKMD	STATION	KMD	BCKMD
27.15	24.09	**29.93**	91.44	81.21	**81.24**	73.54	40.40	**67.07**

3.2.2. Quantile Mapping at Reference Point Level

The Quantile Mapping correction on the base of the GPCC dataset was also applied to the KMD dataset at the reference points. The Parametric Transformations method has been used in accordance with the validation carried out at the stations level. The result was a best-fit precipitation dataset for the eight locations.

3.3. *Calculating Normal Values of Precipitation at Station Level*

The normal values were calculated on the precipitation series obtained for the three stations, by averaging the monthly precipitation amount for the entire period (1983–2013) (reported in Table 5).

Long rains amount, short rains amount and total annual amount were also calculated. Figure 5 compares the distribution of the precipitation through the year according to the observed series and to the new BCKMD dataset.

Table 5. Comparison of normal values of precipitation (in mm) obtained from the historical time series and the new precipitation dataset BCKMD for the three land-based meteorological stations (Lodwar, Marsabit and Moyale). The table reports monthly cumulative amounts, long rains cumulative amount, short rains cumulative amount and total precipitation amount.

	Lodwar		Marsabit		Moyale	
	Historical	Bckmd	Historical	Bckmd	Historical	Bckmd
Jan	6.5	10.7	30.2	31.3	17.4	18.4
Feb	4.2	4.4	19.2	20.8	19.4	21.1
Mar	30.0	33.2	50.2	44.5	53.2	36.0
Apr	38.3	50.9	212.4	195.4	164.9	193.0
May	28.3	37.8	75.1	77.4	100.1	75.3
June	10.6	10.6	10.6	6.7	18.0	9.8
July	12.1	12.4	9.0	3.7	10.2	3.8
Aug	14.5	16.2	8.1	5.1	10.1	7.8
Sep	10.2	13.9	5.3	5.5	15.5	12.9
Oct	11.2	13.9	88.4	87.1	106.6	75.9
Nov	20.4	17.5	139.3	127.3	92.9	79.9
Dec	15.1	7.1	57.5	48.9	28.7	31.6
Long rains	96.6	121.9	337.6	317.4	318.3	304.2
Short rains	46.7	38.5	285.3	263.3	228.1	187.4
Total	**201.4**	**228.6**	**705.2**	**653.8**	**637.0**	**565.5**

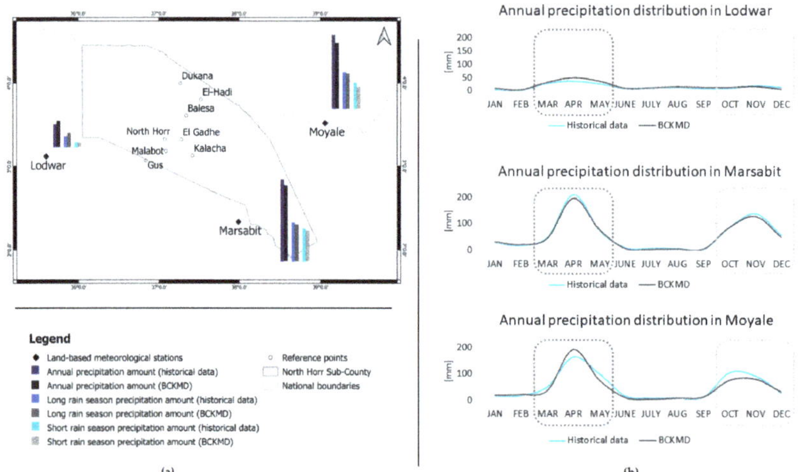

Figure 5. (**a**) Representation at local scale of the historical time series and of new bias-corrected monthly precipitation time series for each land-based meteorological station divided into three cumulative amounts: total annual amount (black and very dark blue bars), long rain season precipitation amount (dark grey and dark blue bars) and short rain season precipitation amount (light grey and light blue bars). (**b**) Comparison of the annual precipitation distribution for each station according to the observed series and to the new bias-corrected monthly precipitation dataset.

3.4. Calculating Normal Values of Precipitation at Reference Point Level

The normal values were calculated on the precipitation series obtained for each reference point, by averaging the monthly precipitation amount for the entire period (1983–2013). Moreover, long rains amount, short rains amount and total amount were calculated.

The normal values for the eight reference points are shown in Table 6. A visual representation of the precipitation distribution at local scale is pictured in Figure 6.

The understanding of climate differences at local scale is crucial for an effective territorial planning against negative impact of climate change. This study succeeded in obtaining normal values of precipitation for each reference point despite the lack of land-based meteorological stations in the area and high-resolution and fitting satellite-derived precipitation time series. Differences in rainfall regime are evident in Figure 6, which shows higher precipitation amounts in the northern part of the Sub-County then in the southern reference points.

The new precipitation time series can be used for the evaluation of drought indices as well as for water security assessment. More specifically, the monthly normal values can be used as reference values for comparing measured or forecasted data in order to evaluate drought or wet periods.

Moreover, it has been possible to calculate the normal values for the entire long rain season and short rain season, by cumulating monthly values for March, April and May and for October, November and December, respectively. Knowing the distribution of the precipitation throughout the year and the possible deviation from normal values is fundamental. This is at the base of the community organization for the local semi-nomadic pastoral population.

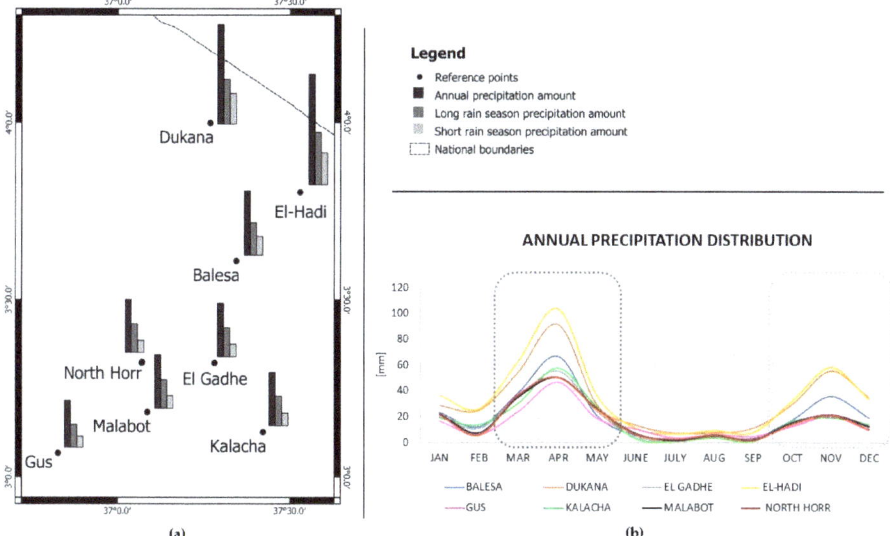

Figure 6. (**a**) Representation at local scale of the new bias-corrected monthly precipitation time series for each reference point divided into three cumulative amounts: total annual amount (black bar), long rain season precipitation amount (dark grey bar) and short rain season precipitation amount (light grey bar). (**b**) Annual precipitation distribution for each reference points based on the new bias-corrected monthly precipitation time series.

Table 6. Normal values of precipitation (in mm) for the eight reference points in North Horr Sub-County. The table reports monthly cumulative amounts, long rains cumulative amount, short rains cumulative amount and total precipitation amount for each reference point.

	Balesa	Dukana	El Gadhe	El-Hadi	Gus	Kalacha	Malabot	North Horr
Jan	23.3	28.6	19.8	37.0	16.8	19.0	22.1	22.0
Feb	12.3	25.0	10.9	26.4	6.6	13.9	7.6	6.1
Mar	38.3	54.2	35.4	63.1	24.0	30.3	35.0	36.8
Apr	66.8	91.5	55.1	104.2	46.6	57.7	50.3	50.7
May	21.1	29.3	24.4	37.1	19.0	27.5	26.4	26.3
June	6.0	13.8	4.5	11.7	10.9	3.2	7.4	7.0
July	2.7	7.6	1.7	7.2	4.1	1.0	1.9	2.9
Aug	4.5	7.8	5.4	9.8	6.4	3.7	6.0	6.0
Sep	3.5	11.4	2.3	8.3	5.3	1.1	2.4	2.5
Oct	17.7	29.5	14.3	32.7	12.8	16.4	15.4	14.7
Nov	35.9	55.4	21.6	58.8	20.7	19.2	21.7	21.9
Dec	19.3	35.0	13.6	33.9	10.2	14.4	12.6	10.7
Long rains	126.2	175	114.9	204.4	89.6	115.5	111.7	113.8
Short rains	72.9	119.9	49.5	125.4	43.7	50	49.7	47.3
Total	**251.4**	**389.1**	**209**	**430.2**	**183.4**	**207.4**	**208.8**	**207.6**

4. Conclusions

The aim of this study was to obtain the normal values of the monthly amount of precipitation for the main inhabited areas in North Horr Sub-County, in order to provide a benchmark for understanding the ongoing changes in the local climate. Therefore, it was necessary to identify an appropriate historical precipitation series. The comparison between the GPCC, KMD and CHIRPS datasets highlighted the lower performance of the KMD dataset compared to the others, despite it being the dataset officially issued and used by the Kenyan Meteorological Department for the whole country. Previous studies on East Africa indicated the CHIRPS dataset to be a reliable global dataset for the region [26,27,32]. The relatively high performance of the GPCC dataset in northern arid Kenya is in line with the results of previous studies, which indicated it as a good fit for the ASALs [35], but with a low capacity in representing complex terrain [28]. However, the need to highlight local differences in the annual trend of precipitation led to the use of the KMD dataset after a correction procedure based on the GPCC dataset. This approach aimed to integrate the higher resolution of the KMD dataset—namely, its ability to detect differences in the precipitation trend at a local scale—with the higher ability of the GPCC dataset to represent the real historical values in the area. The methodology adopted created a new bias-corrected monthly precipitation time series for each reference point, from which the local normal values were extracted.

Since the need for high-resolution precipitation data covering the Global South is becoming urgent for any discipline that must consider the role of climate, this study represents an attempt to provide a solution to the scarcity of observed data. The absence of land-based meteorological stations in the area, however, cannot be ignored and constitutes a limit in the study. Future research should be directed to test the methodology proposed here in other contexts, where the availability of observed data could provide a yardstick for its usefulness and accuracy.

Author Contributions: Conceptualization, A.P., V.B. and I.V.; methodology, V.B. and I.V.; formal analysis, V.B. and I.V.; writing—original draft preparation, V.B. and I.V.; writing—review and editing A.P., V.B., I.V. and A.B.; supervision, A.P. and A.B. All authors have read and agreed to the published version of the manuscript.

Acknowledgments: This study was conducted within the framework of the International Cooperation Project "ONE HEALTH: Multidisciplinary approach to promote the health and resilience of shepherds" communities in North Kenya" funded by the Italian Agency for Development Cooperation (AICS). The authors would like to thank the project coordinator (CCM) and project partners (TRIM and VSF-Germany) and the Kenyan Meteorological Department.

Conflicts of Interest: The authors declare no conflict of interest.

Appendix A

Table A1. Comparison based on statistical indices (BIAS, MAE, MSD, RMSD, CC and σ) of the GPCC dataset with the five bias-corrected series for the period 1983–2013. For the CC index, "*" corresponds to a p-value < 0.01, "**" corresponds to a p-value < 0.001 and "***" corresponds to a p-value < 0.0001. Values in bold correspond to the best value of the index for each reference point.

			Balesa		
	PTF	DIST	RQUANT	QUANT	SSPLIN
BIAS	**−0.09**	0.02	0.14	0.10	0.29
MAE	14.84	**14.68**	14.95	14.96	14.97
MSD	632.88	**609.68**	652.85	652.73	659.10
RMSD	25,16	**24.69**	25.55	25.55	25.67
CC	**0.71** ***	**0.71** ***	0.70 ***	0.70 ***	0.70 ***
Standard Deviation					
GPCC	PT	DIST	RQUANT	QUANT	SSPLIN
32.94	32.83	32.29	32.84	**32.86**	33.18
			Dukana		
	PTF	DIST	RQUANT	QUANT	SSPLIN
BIAS	**−0.23**	29.97	28.68	28.59	28.87
MAE	**19.31**	30.00	29.99	29.96	30.18
MSD	**875.66**	2727.20	2409.16	2398.54	2504.03
RMSD	**29.59**	52.22	49.08	48.97	50.04
CC	**0.73** ***	0.72 ***	0.72 ***	**0.73** ***	0.72 ***
Standard Deviation					
GPCC	PTF	DIST	RQUANT	QUANT	SSPLIN
40.14	**40.09**	42.76	39.83	39.76	40.87
			Elgade		
	PTF	DIST	RQUANT	QUANT	SSPLIN
BIAS	**1.39**	16.51	17.03	17.02	17.20
MAE	**14.47**	16.62	17.26	17.27	17.41
MSD	**856.77**	1343.05	1456.21	1455.56	1520.29
RMSD	**29.27**	36.65	38.16	38.15	38.99
CC	**0.65** ***	**0.65** ***	**0.65** ***	**0.65** ***	**0.65** ***
Standard Deviation					
GPCC	PTF	DIST	RQUANT	QUANT	SSPLIN
34.19	34.10	32.72	**34.15**	**34.15**	34.99
			El Hadi		
	PTF	DIST	RQUANT	QUANT	SSPLIN
BIAS	**1.12**	25.44	24.55	24.48	24.59
MAE	**24.12**	29.13	30.63	30.60	30.57
MSD	**1324.09**	2473.67	2578.99	2578.16	2583.05
RMSD	**36.39**	49.74	50.78	50.78	50.82
CC	**0.71** ***	**0.71** ***	0.70 ***	0.70 ***	0.70 ***
Standard Deviation					
GPCC	PT	DIST	RQUANT	QUANT	SSPLIN
45.00	**45.17**	42.74	44.45	44.49	44.48
			Gas		
	PTF	DIST	RQUANT	QUANT	SSPLIN
BIAS	**1.46**	14.22	14.74	14.81	14.86
MAE	**11.22**	14.30	15.05	15.11	15.12
MSD	**458.36**	832.97	989.23	1016.25	1024.82
RMSD	**21.41**	28.86	31.45	31.88	32.01
CC	**0.69** ***	**0.69** ***	**0.69** ***	**0.69** ***	**0.69** ***

Table A1. *Cont.*

Gas					
Standard Deviation					
GPCC	PTF	DIST	RQUANT	QUANT	SSPLIN
27.80	27.78	25.12	**27.79**	28.23	28.36
Kalacha					
	PT	DIST	RQUANT	QUANT	SSPLIN
BIAS	**1.83**	15.65	15.95	15.92	16.57
MAE	**14.07**	15.65	16.03	16.01	16.61
MSD	**875.82**	1345.08	1404.63	1405.52	1744.06
RMSD	**29.59**	36.68	37.48	37.49	41.76
CC	**0.57** ***	0.56 ***	**0.57** ***	**0.57** ***	0.54 ***
Standard Deviation					
GPCC	PTF	DIST	RQUANT	QUANT	SSPLIN
34.19	34.11	33.33	**34.19**	34.21	38.55
Malabot					
	PTF	DIST	RQUANT	QUANT	SSPLIN
BIAS	**1.37**	16.90	17.11	17.00	17.14
MAE	**13.75**	16.90	17.29	17.19	17.34
MSD	**824.68**	1404.62	1466.47	1443.40	1505.37
RMSD	**28.72**	37.48	38.29	37.99	38.80
CC	**0.68** ***	**0.68** ***	**0.68** ***	**0.68** ***	0.67 ***
Standard Deviation					
GPCC	PTF	DIST	RQUANT	QUANT	SSPLIN
34.19	**34.11**	33.45	34.26	33.98	34.81
North Horr					
	PTF	DIST	RQUANT	QUANT	SSPLIN
BIAS	**1.28**	16.08	17.08	16.95	17.84
MAE	**13.97**	16.18	17.31	17.20	18.05
MSD	**848.22**	1218.58	1472.02	1445.25	1881.88
RMSD	**29.12**	34.91	38.37	38.02	43.38
CC	**0.67** ***	**0.67** ***	**0.67** ***	**0.67** ***	0.64 ***
Standard Deviation					
GPCC	PTF	DIST	RQUANT	QUANT	SSPLIN
34.19	34.01	30.99	34.36	**34.03**	39.54

Appendix B

Table A2. Comparison based on statistical indices (BIAS, MAE, MSD, RMSD, CC and σ) of the historical values with the five bias-corrected series (PTf, DIST, RQUANT, QUANT, SSPLIN) for the period 1983–2013. For the CC index, "*" corresponds to a p-value < 0.01, "**" corresponds to a p-value < 0.001 and "***" corresponds to a p-value < 0.0001. Values in bold correspond to the best value of the index for each station.

Lodwar					
	PTF	DIST	RQUANT	QUANT	SSPLIN
BIAS	**2.30**	2.92	2.92	2.61	2.64
MAE	**7.73**	8.12	8.12	7.84	7.88
MSD	296.12	**294.70**	**294.70**	305.80	307.05
RMSD	17.21	**17.17**	**17.17**	17.49	17.52
CC	**0.83** ***	0.82 ***	0.82 ***	0.82 ***	0.82 ***
Standard Deviation					
STATION	PTF	DIST	RQUANT	QUANT	SSPLIN
27.15	29.93	**29.07**	**29.07**	29.91	29.94

Table A2. *Cont.*

			Marsabit		
	PTF	DIST	RQUANT	QUANT	SSPLIN
BIAS	−4.28	**−0.34**	−2.12	−2.08	−1.38
MAE	**16.78**	17.96	17.65	17.69	18.16
MSD	**1427.56**	1443.34	1430.87	1437.01	1504.62
RMSD	**37.78**	37.99	37.83	37.91	38.79
CC	**0.91 *****	**0.91 *****	**0.91 *****	**0.91 *****	**0.91 *****
Standard Deviation					
STATION	PTF	DIST	RQUANT	QUANT	SSPLIN
91.44	81.24	**84.23**	79.82	80.22	83.47
			Moyale		
	PTF	DIST	RQUANT	QUANT	SSPLIN
BIAS	−5.96	**−2.04**	−4.48	−4.85	−4.24
MAE	26.02	**25.92**	26.32	26.18	26.67
MSD	**1881.38**	1954.03	2206.42	2157.35	2423.31
RMSD	**43.37**	44.20	46.97	46.45	49.23
CC	**0.82 *****	0.81 ***	0.78 ***	0.79 ***	0.77 ***
Standard Deviation					
STATION	PTF	DIST	RQUANT	QUANT	SSPLIN
73.54	67.07	**69.04**	66.99	65.36	69.01

References

1. Hellmuth, M.E.; Moorhead, A.; Thomson, M.C.; Williams, J. (Eds.) *Climate Risk Management in Africa: Learning from Practice*; International Research Institute for Climate and Society (IRI), Columbia University: New York, NY, USA, 2007.
2. Dinku, T.; Block, P.; Sharoff, J.; Hailemariam, K.; Osgood, D.; del Corral, J.; Cousin, R.; Thomson, M.C. Bridging critical gaps in climate services and applications in africa. *Earth Perspect.* **2014**, *1*, 15.
3. LTS International. Acclimatise Technical Report 8: Availability and Accessibility to Climate Data for Kenya. 2012, pp. 1–47. Available online: https://www.kccap.info/index_option_com_phocadownload_view_category_id_37_Itemid_54.html (accessed on 20 January 2020).
4. Kenya Meteorological Department. Available online: https://www.meteo.go.ke/ (accessed on 5 January 2020).
5. Kenya Meteorological Department. *Meteorology Policy 2019*; Kenya Meteorological Department: Nairobi, Kenya, 2019.
6. MoALF Climate Risk Profile for Marsabit. *County Kenya County Climate Risk Profile Series 2017*; Government of Kenya: Nairobi, Kenya, 2017.
7. Orindi, V.A.; Ochieng, A. Case Study 5: Kenya Seed Fairs as a Drought Recovery Strategy in Kenya. *IDS Bull.* **2005**, *36*, 87–102.
8. Karanja, F.; Mutua Nairobi, F. Reducing the impact of environmental emergencies through Early Warning and preparedness-the case of el Niño-Southern Oscillation (ENSO); Nairobi: UNFIP/UNEP/NCAR/WMO/DNDR/UNU. 2000. Available online: https://profiles.uonbi.ac.ke/coludhe/publications/reducing-impacts-environmental-emergencies-through-early-warning-and-preparedne (accessed on 7 January 2020).
9. Ogalo, L.; Owiti, Z.; Mutemi, J. Linkages between the Indian Ocean Dipole and East African Rainfall Anomalies. *J. Kenya Meteorol. Soc.* **2008**, *2*, 3–17.
10. Ouma, J.O.; Olang, L.O.; Ouma, G.O.; Oludhe, C.; Ogallo, L.; Artan, G. Magnitudes of Climate Variability and Changes over the Arid and Semi-Arid Lands of Kenya between 1961 and 2013 Period. *Am. J. Clim. Chang.* **2018**, *7*, 27–39.
11. Edwards, K.A.; Field, C.R.; Hogg, I.G.G. *A Preliminary Analysis of Climatological Data from the Marsabit District of Northern Kenya*; UNEP-MAB Integrated Project in Arid Lands: Nairobi, Kenya, 1979.
12. Schwartz, H.; Shaabani, S.; Walther, D. *Range Management Handbook of Kenya, Marsabit District*; Ministry of Livestock Development, Republic of Kenya: Nairobi, Kenya, 1991.
13. Ongoma, V.; Chen, H.; Omony, G.W. Variability of extreme weather events over the equatorial East Africa, a case study of rainfall in Kenya and Uganda. *Theor. Appl. Climatol.* **2018**, *131*, 295–308.

14. Schmocker, J.; Liniger, H.P.; Ngeru, J.N.; Brugnara, Y.; Auchmann, R.; Brönnimann, S. Trends in mean and extreme precipitation in the Mount Kenya region from observations and reanalyses. *Int. J. Climatol.* **2016**, *36*, 1500–1514.
15. Shongwe, M.E.; van Oldenborgh, G.J.; van den Hurk, B.; van Aalst, M. Projected Changes in Mean and Extreme Precipitation in Africa under Global Warming. Part II: East Africa. *J. Clim.* **2011**, *24*, 3718–3733.
16. Omondi, P.A.; Awange, J.L.; Forootan, E.; Ogallo, L.A.; Barakiza, R.; Girmaw, G.B.; Fesseha, I.; Kululetera, V.; Kilembe, C.; Mbati, M.M.; et al. Changes in temperature and precipitation extremes over the Greater Horn of Africa region from 1961 to 2010. *Int. J. Climatol.* **2014**, *34*, 1262–1277.
17. Opiyo, F.; Wasonga, O.V.; Nyangito, M.M.; Mureithi, S.M.; Obando, J.; Munang, R. Determinants of perceptions of climate change and adaptation among Turkana pastoralists in northwestern Kenya. *Clim. Dev.* **2016**, *8*, 179–189.
18. Herrero, M.; Addison, J.; Bedelian, C.; Carabine, E.; Havlík, P.; Henderson, B.; Van De Steeg, J.; Thornton, P.K. Climate change and pastoralism: Impacts, consequences and adaptation. *Rev. Sci. Tech.* **2016**, *35*, 417–433.
19. Ochieng, J.; Kirimi, L.; Mathenge, M. Effects of climate variability and change on agricultural production: The case of small scale farmers in Kenya. *NJAS Wageningen J. Life Sci.* **2016**, *77*, 71–78. [CrossRef]
20. Judith, C.; Mbogoh, S.G.; Chris, A.-O.; Patrick, I. Smallholder Farmers' Perceptions and Responses to Climate Change in Multi-stressor Environments: The Case of Maasai Agro-pastoralists in Kenya's Rangelands. *Am. J. Rural Dev.* **2017**, *5*, 110–116.
21. Mbaisi, C.N.; Kipkorir, E.C.; Omondi, P. The Perception of Farmers on Climate Change and Variability Patterns in the Nzoia River Basin, Kenya. *Perception* **2016**, *6*. Available online: https://www.semanticscholar.org/paper/The-Perception-of-Farmers-on-Climate-Change-and-in-Mbaisi-Kipkorir/40c0f5ee8576faad730e79b0125087c754c7d0c2 (accessed on 7 January 2020).
22. Deressa, T.T.; Hassan, R.M.; Ringler, C.; Alemu, T.; Yesuf, M. Determinants of farmers' choice of adaptation methods to climate change in the Nile Basin of Ethiopia. *Glob. Environ. Chang.* **2009**, *19*, 248–255. [CrossRef]
23. Hewitson, B.C.; Crane, R.G. Consensus between GCM climate change projections with empirical downscaling: Precipitation downscaling over South Africa. *Int. J. Climatol.* **2006**, *26*, 1315–1337. [CrossRef]
24. Ning, L.; Riddle, E.E.; Bradley, R.S. Projected changes in climate extremes over the northeastern United States. *J. Clim.* **2015**, *28*, 3289–3310. [CrossRef]
25. Wood, A.W.; Maurer, E.P.; Kumar, A.; Lettenmaier, D.P. Long-range experimental hydrologic forecasting for the eastern United States. *J. Geophys. Res. D Atmos.* **2002**, *107*. [CrossRef]
26. Gebrechorkos, S.H.; Hülsmann, S.; Bernhofer, C. Evaluation of multiple climate data sources for managing environmental resources in East Africa. *Hydrol. Earth Syst. Sci.* **2018**, *22*, 4547–4564. [CrossRef]
27. Bayissa, Y.; Tadesse, T.; Demisse, G.; Shiferaw, A. Evaluation of satellite-based rainfall estimates and application to monitor meteorological drought for the Upper Blue Nile Basin, Ethiopia. *Remote Sens.* **2017**, *9*, 669. [CrossRef]
28. Cattani, E.; Merino, A.; Levizzani, V. Evaluation of monthly satellite-derived precipitation products over East Africa. *J. Hydrometeorol.* **2016**, *17*, 2555–2573. [CrossRef]
29. Sahlu, D.; Moges, S.A.; Nikolopoulos, E.I.; Anagnostou, E.N.; Hailu, D. Evaluation of High-Resolution Multisatellite and Reanalysis Rainfall Products over East Africa. *Adv. Meteorol.* **2017**, *2017*, 4957960. [CrossRef]
30. Kimani, M.W.; Hoedjes, J.C.B.; Su, Z. An Assessment of Satellite-Derived Rainfall Products Relative to Ground Observations over East Africa. *Remote Sens.* **2017**, *9*, 430. [CrossRef]
31. Hirpa, F.A.; Gebremichael, M.; Hopson, T. Evaluation of high-resolution satellite precipitation products over very complex terrain in Ethiopia. *J. Appl. Meteorol. Climatol.* **2010**, *49*, 1044–1051. [CrossRef]
32. Agutu, N.O.; Awange, J.L.; Zerihun, A.; Ndehedehe, C.E.; Kuhn, M.; Fukuda, Y. Assessing multi-satellite remote sensing, reanalysis, and land surface models' products in characterizing agricultural drought in East Africa. *Remote Sens. Environ.* **2017**, *194*, 287–302. [CrossRef]
33. Ayugi, B.; Tan, G.; Niu, R.; Babaousmail, H.; Ojara, M.; Wido, H.; Mumo, L.; Nooni, I.K.; Ongoma, V. Quantile Mapping Bias Correction on Rossby Centre Regional Climate Models for Precipitation Analysis over Kenya, East Africa. *Water* **2020**, *12*, 801. [CrossRef]
34. Ayugi, B.; Tan, G.; Ullah, W.; Boiyo, R.; Ongoma, V. Inter-comparison of remotely sensed precipitation datasets over Kenya during 1998–2016. *Atmos. Res.* **2019**, *225*, 96–109. [CrossRef]

35. Macharia, M.J.; Ngetich, F.K.; Shisanya, C.A. Comparison of satellite remote sensing derived precipitation estimates and observed data in Kenya. *Agric. For. Meteorol.* **2020**, *284*, 107875. [CrossRef]
36. Dinku, T.; Thomson, M.C.; Cousin, R.; del Corral, J.; Ceccato, P.; Hansen, J.; Connor, S.J. Enhancing National Climate Services (ENACTS) for development in Africa. *Clim. Dev.* **2018**, *10*, 664–672. [CrossRef]
37. Becker, A.; Finger, P.; Meyer-Christoffer, A.; Rudolf, B.; Schamm, K.; Schneider, U.; Ziese, M. A description of the global land-surface precipitation data products of the Global Precipitation Climatology Centre with sample applications including centennial (trend) analysis from 1901-present. *Earth Syst. Sci. Data* **2013**, *5*, 71–99. [CrossRef]
38. Dinku, T.; Funk, C.; Peterson, P.; Maidment, R.; Tadesse, T.; Gadain, H.; Ceccato, P. Validation of the CHIRPS satellite rainfall estimates over eastern Africa. *Q. J. R. Meteorol. Soc.* **2018**, *144*, 292–312. [CrossRef]
39. Cohen Liechti, T.; Matos, J.P.; Boillat, J.L.; Schleiss, A.J. Comparison and evaluation of satellite derived precipitation products for hydrological modeling of the Zambezi River Basin. *Hydrol. Earth Syst. Sci.* **2012**, *16*, 489–500. [CrossRef]
40. Moazami, S.; Golian, S.; Kavianpour, M.R.; Hong, Y. Comparison of PERSIANN and V7 TRMM multi-satellite precipitation analysis (TMPA) products with rain gauge data over Iran. *Int. J. Remote Sens.* **2013**, *34*, 8156–8171. [CrossRef]
41. Taylor, K.E. Summarizing multiple aspects of model performance in a single diagram. *J. Geophys. Res. Atmos.* **2001**, *106*, 7183–7192. [CrossRef]
42. *County Government of Marsabit Second County Integrated Development Plan 2018–2022*; County Government of Marsabit: Marsabit, Kenya, 2018.
43. Ministry of Health Kenya Master Health Facility List: E-Health Kenya. Available online: http://www.ehealth.or.ke/facilities/ (accessed on 29 January 2020).
44. Gudmundsson, L.; Bremnes, J.B.; Haugen, J.E.; Engen-Skaugen, T. Hydrology and Earth System Sciences Technical Note: Downscaling RCM precipitation to the station scale using statistical transformations-a comparison of methods. *Hydrol. Earth Syst. Sci* **2012**, *16*, 3383–3390. [CrossRef]
45. *County Government of Marsabit Climate Change Mainstreaming Guidelines Agriculture, Livestock and Fisheries Sector*; County Government of Marsabit: Marsabit, Kenya, 2018.
46. Sun, Q.; Miao, C.; Duan, Q.; Ashouri, H.; Sorooshian, S.; Hsu, K.L. A Review of Global Precipitation Data Sets: Data Sources, Estimation, and Intercomparisons. *Rev. Geophys.* **2018**, *56*, 79–107. [CrossRef]
47. Katiraie-Boroujerdy, P.S.; Akbari Asanjan, A.; Hsu, K.L.; Sorooshian, S. Intercomparison of PERSIANN-CDR and TRMM-3B42V7 precipitation estimates at monthly and daily time scales. *Atmos. Res.* **2017**, *193*, 36–49. [CrossRef]
48. Funk, C.; Peterson, P.; Landsfeld, M.; Pedreros, D.; Verdin, J.; Shukla, S.; Husak, G.; Rowland, J.; Harrison, L.; Hoell, A.; et al. The climate hazards infrared precipitation with stations—A new environmental record for monitoring extremes. *Sci. Data* **2015**, *2*, 150066. [CrossRef]
49. Wilks, D.L. *Statistical Methods in the Atmospheric Sciences*, 2nd ed.; Academic Press: Cambridge, MA, USA, 2006.
50. Rochford. PeterRochford/SkillMetricsToolbox, GitHub. Available online: https://www.github.com/PeterRochford/SkillMetricsToolbox (accessed on 24 January 2020).
51. Yatagai, A.; Xie, P.; Kitoh, A. Utilization of a New Gauge-based Daily Precipitation Dataset over Monsoon Asia for Validation of the Daily Precipitation Climatology Simulated by the MRI/JMA 20-km-mesh AGCM. *Sola* **2005**, *1*, 193–196. [CrossRef]
52. Zhao, T.; Yatagai, A. Evaluation of TRMM 3B42 product using a new gauge-based analysis of daily precipitation over China. *Int. J. Climatol.* **2014**, *34*, 2749–2762. [CrossRef]
53. Boushaki, F.I.; Hsu, K.L.; Sorooshian, S.; Park, G.H.; Mahani, S.; Shi, W. Bias adjustment of satellite precipitation estimation using ground-based measurement: A case study evaluation over the southwestern United States. *J. Hydrometeorol.* **2009**, *10*, 1231–1242. [CrossRef]
54. Tesfagiorgis, K.; Mahani, S.E.; Krakauer, N.Y.; Khanbilvardi, R. Bias correction of satellite rainfall estimates using a radar-gauge product – a case study in Oklahoma (USA). *Hydrol. Earth Syst. Sci.* **2011**, *15*, 2631–2647. [CrossRef]
55. Maraun, D. Bias Correction, Quantile Mapping, and Downscaling: Revisiting the Inflation Issue. *J. Clim.* **2013**, *26*, 2137–2143. [CrossRef]
56. Eden, J.M.; Widmann, M.; Grawe, D.; Rast, S. Skill, Correction, and Downscaling of GCM-Simulated Precipitation. *J. Clim.* **2012**, *25*, 3970–3984. [CrossRef]

57. Heo, J.-H.; Ahn, H.; Shin, J.-Y.; Kjeldsen, T.R.; Jeong, C. Probability Distributions for a Quantile Mapping Technique for a Bias Correction of Precipitation Data: A Case Study to Precipitation Data Under Climate Change. *Water* **2019**, *11*, 1475. [CrossRef]
58. Yang, Z.; Hsu, K.; Sorooshian, S.; Xu, X.; Braithwaite, D.; Verbist, K.M.J. Bias adjustment of satellite-based precipitation estimation using gauge observations: A case study in Chile. *J. Geophys. Res.* **2016**, *121*, 3790–3806. [CrossRef]
59. Jakob Themeßl, M.; Gobiet, A.; Leuprecht, A. Empirical-statistical downscaling and error correction of daily precipitation from regional climate models. *Int. J. Climatol.* **2011**, *31*, 1530–1544. [CrossRef]
60. Ajaaj, A.A.; Mishra, A.K.; Khan, A.A. Comparison of BIAS correction techniques for GPCC rainfall data in semi-arid climate. *Stoch. Environ. Res. Risk Assess.* **2016**, *30*, 1659–1675. [CrossRef]
61. Ringard, J.; Seyler, F.; Linguet, L. A Quantile Mapping Bias Correction Method Based on Hydroclimatic Classification of the Guiana Shield. *Sensors* **2017**, *17*, 1413. [CrossRef]
62. Rajczak, J.; Kotlarski, S.; Schär, C. Does quantile mapping of simulated precipitation correct for biases in transition probabilities and spell lengths? *J. Clim.* **2016**, *29*, 1605–1615. [CrossRef]
63. Maintainer, G.; Gudmundsson, L. Software Package. qmap: Statistical Transformations for Post-Processing Climate Model Output. 2016. Available online: https://cran.r-project.org/web/packages/qmap/index.html (accessed on 7 January 2020).

Article

Climatology and Dynamical Evolution of Extreme Rainfall Events in the Sinai Peninsula—Egypt

Marina Baldi [1,*], Doaa Amin [2], Islam Sabry Al Zayed [3] and Giovannangelo Dalu [1,4]

1. CNR-IBE, 00185 Rome, Italy; giovannangelo.dalu@ibe.cnr.it
2. Water Resources Research Institute (WRRI), National Water Research Center (NWRC), Cairo 13621, Egypt; doaa_amin74@yahoo.com
3. Technical Office, Headquarter, National Water Research Center (NWRC), Cairo 13411, Egypt; islam_alzayed@nwrc.gov.eg
4. Accademia dei Georgofili, 50122 Firenze, Italy

* Correspondence: marina.baldi@ibe.cnr.it

Received: 7 June 2020; Accepted: 28 July 2020; Published: 31 July 2020

Abstract: The whole Mediterranean is suffering today because of climate changes, with projections of more severe impacts predicted for the coming decades. Egypt, on the southeastern flank of the Mediterranean Sea, is facing many challenges for water and food security, further exacerbated by the arid climate conditions. The Nile River represents the largest freshwater resource for the country, with a minor contribution coming from rainfall and from non-renewable groundwater aquifers. In more recent years, another important source is represented by non-conventional sources, such as treated wastewater reuse and desalination; these water resources are increasingly becoming valuable additional contributors to water availability. Moreover, although rainfall is scarce in Egypt, studies have shown that rainfall and flash floods can become an additional available source of water in the future. While presently rare, heavy rainfalls and flash floods are responsible for huge losses of lives and infrastructure especially in parts of the country, such as in the Sinai Peninsula. Despite the harsh climate, water from these events, when opportunely conveyed and treated, can represent a precious source of freshwater for small communities of Bedouins. In this work, rainfall climatology and flash flood events are presented, together with a discussion about the dynamics of some selected episodes and indications about future climate scenarios. Results can be used to evaluate the water harvesting potential in a region where water is scarce, also providing indications for improving the weather forecast. Basic information needed for identifying possible risks for population and infrastructures, when fed into hydrological models, could help to evaluate the flash flood water volumes at the outlets of the effective watershed(s). This valuable information will help policymakers and local governments to define strategies and measures for water harvesting and/or protection works.

Keywords: Sinai Peninsula; flash flood; climate change; CORDEX; water harvesting

1. Introduction

In recent decades, as in past times, heavy rainfall resulting in flash floods has affected not only the Egyptian coastal areas along the Mediterranean Sea and the Red Sea but also arid and semi-arid areas such as Upper Egypt (e.g., Luxor, Aswan, and Assiut) and the Sinai Peninsula [1–5].

Under changing climate conditions, the frequency of extreme rainfall episodes is also changing. These episodes, although rare, can be catastrophic in regions characterized by very low annual precipitation, with large impacts on lives, infrastructures, properties, and last but not least, to the great cultural heritage of Egypt [6]. Recent heavy rainfall events in the densely populated region of Cairo Governorate have forced authorities to close schools, offices, and highways connecting Cairo to other provinces, producing electricity disruption and floods in large areas of the town. These floods have also trapped people in their cars for several hours, as happened during the recent episodes that occurred in April 2018, October 2019, and March 2020 [7–10].

Scientific literature [11] has illustrated how the heavy rainfall and subsequent flash floods can be harmful in the Sinai Peninsula and how important the production of risk maps and implementing the flash flood protection works for the main watersheds of the Peninsula is, taking into consideration the complex morphological parameters [12,13]. Concerning the heavy rainfall episodes that have occurred in the Middle East Region and specifically in Egypt, many authors, e.g., [14–18], have focused their analysis on the evolution of extreme rainfall episodes and on the atmospheric pattern precursors of the evolution of these events. Due to the scale of the rare phenomenon, the forecast of a small or, at its largest, mesoscale severe thunderstorm, bringing heavy rainfall and causing flash flood, is not an easy task. Only the use of an appropriate tool like a mesoscale model can help to produce a timely and detailed forecast of the episode, which can be used as the major "ingredient", together with data-driven weather-runoff forecast models, for an Early Warning System (EWS) to be adopted in order to avoid or, at least, minimize major impacts and save lives [19–21]. In addition, if, in this flood-prone region, it is important on one hand to minimize damages and avoid disastrous events deriving from severe thunderstorms, on the other hand, it is also important to favor the use of rainfall as a precious source of fresh water. For this purpose, it is essential that the development and use of EWS [19–21] present high performances.

In order to decrease the uncertainties in the EWS, it is necessary to work on the improvement of Weather Prediction Tools (WPT) in parallel with the adoption of a sophisticated data-driven weather-runoff model; however, this improvement can come only through a thorough validation of the WPT. Due to their randomness, low frequency of occurrence, rapidity in evolution, and small mesoscale dimension, observations related to these thunderstorms are scarce in the Region and do not permit a robust validation of WPT and therefore of EWS. In this respect, the aim of the present study is to contribute to an increase of the knowledge about the present and future occurrence of extreme rainfall events in the Sinai Peninsula and on their dynamics.

To this aim, the study starts with a general overview of the climatic trends in the Sinai Peninsula, followed by a discussion on the climatology of extreme rainfall events in Sinai and by the analysis of a selection of heavy rainfall episodes. The second part of the study presents a general overview of future climate scenarios that might affect the frequency and intensity of extreme events in this region. This part of the work not only gives some general information, but also results can represent a necessary basis for more detailed climate scenario studies, and it will possibly encourage the development of specific regional projections at a higher resolution. More detailed scenario analysis can be used for the implementation of more sophisticated risk maps for future decades and can foster the elaboration and adoption of adaptation plans for population, agriculture, and water resources management and, finally, might help the decision-makers to plan the construction of flood water harvesting structures and flash flood protection works.

2. Materials and Methods

2.1. Data and Methods

The climatological analysis of the current climate and its changes in the last decades is based on the National Centers for Environmental Modeling/National Center for Atmospheric Research global analyses of atmospheric fields, NCEP/NCAR reanalysis, provided by National Oceanic and Atmospheric Administration (NOAA) [22] as well as the global atmospheric reanalysis ERA-Interim provided by the European Centre for Medium-Range Weather Forecasts (ECMWF) available at a spatial resolution of approximately 80 km [23]. The annual rainfall trend over Sinai Peninsula is analyzed using the Global Precipitation Climate Center data (GPCC) provided by the NOAA Physical Sciences Laboratory (PSL), Boulder, Colorado, USA (https://psl.noaa.gov/). The study of the spatio-temporal evolution of the selected episodes of heavy rainfall is based on the analysis of meteorological data from ground stations in Sinai and rainfall estimates derived by satellite images from Tropical Rainfall Measuring Mission (TRMM) and the Hydrologic Data and Information System (HyDIS). In addition, NCEP-NCAR reanalysis is used to highlight the atmospheric patterns before, during, and after the selected episodes in order to individuate (i) the main atmospheric patterns contributing to the evolution of the events, (ii) possible driving mechanisms, and (iii) similarities among the events. Finally, climate scenarios at the regional scale were obtained from the outputs available through the Coordinated Regional Downscaling Experiment (CORDEX).

2.2. Study Area

Egypt is located on the northeastern side of Africa, covering nearly 3% of the total area of the continent. This country is bordered to the north by the Mediterranean Sea and to the east by the Red Sea, by the Gulf of Aqaba and Palestine, to the south by the Republic of Sudan, and to the west by the Republic of Libya. A few geographical sub-regions of the country can be identified: the Nile valley and delta, covering about 4% of the total area; the Eastern desert, covering about 22%; the Western desert, covering about 68%; and the Sinai Peninsula, covering about 6%.

Hot and dry conditions characterize the general climate of Egypt with an average annual rainfall over the whole country of about 10 mm. Even along the narrow northern strip of the Mediterranean coastal land, where most of the rainfall occurs, the average annual rainfall is usually less than 200 mm, and this amount very rapidly decreases proceeding further inland. The only regular supply of water is from the Nile. This river, which has its main source in the highlands of East Africa, crosses Egypt from the south to north end with a large delta in the Mediterranean Sea. In Egypt, the Nile water is artificially channeled on both sides. The Nile supports the life along its length, and allowed the growth of a great civilization in peace and stability [24]. The importance of the river was very clear since ancient times; in fact, Herodotus (484–425 BC) stated that "Egypt is the Gift of the Nile".

The study area of the present work is Sinai, a triangular-shaped peninsula with an area of about 60,000 km^2; this peninsula is bounded by the Mediterranean Sea to the north, the Red Sea to the south, and by the Gulfs of Suez and Aqaba to the west and to the east, respectively (Figure 1). The territory of the region is quite rough with a complex orography and elevated mountains, which reach up to and above 2400 m ASL (above sea level); Mount Catherine, Egypt's highest mountain, reaches 2642 m ASL. The central area of Sinai consists of two plateaus, Al-Tih and Al-Ajmah, both deeply indented and dipping northward towards Wadi El Arish. Towards the Mediterranean Sea, the region is characterized by a plateau, by a system of a number of dome-shaped hills, and by a belt of parallel dunes, which can be as high as 100 m.

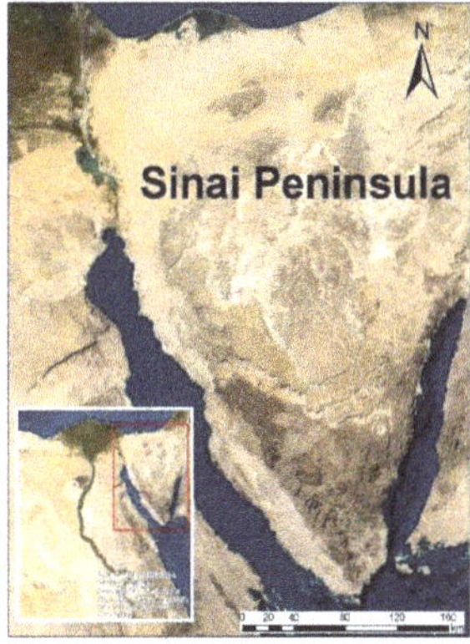

Figure 1. The location of the study area showing the Sinai Peninsula and its orography. Credits: ESRI World Imagery.

The whole territory of the Peninsula is characterized by the presence of numerous watersheds, which discharge rainwater on three sides: the Mediterranean Sea, the Gulf of Aqaba, and the Gulf of Suez (Figure 2). Rivers in these watersheds are short but fierce because of the combined effect of steep slopes and rare episodes of heavy rainfall. The main characteristics of the watersheds in the Sinai Peninsula are summarized in Table 1, as in [25].

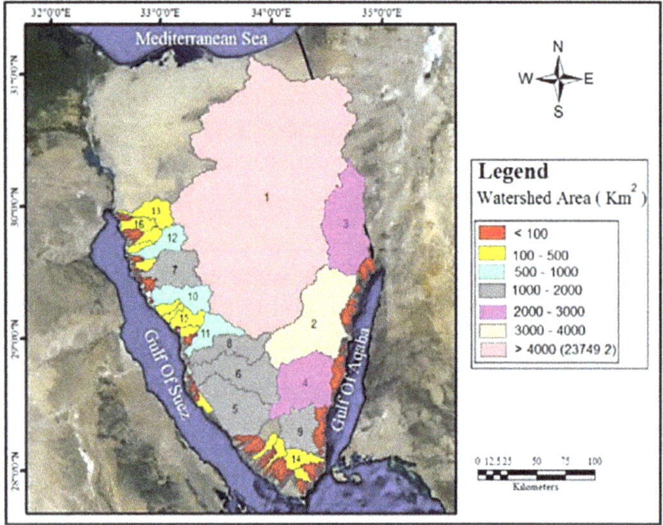

Figure 2. Watersheds of study area (Source: Elsayed et al., 2013).

Table 1. The geomorphology characteristics of the watersheds of Sinai (Source: [25]).

ID	Basin Name	Area (km^2)	Basin Length (m)	Basin Slope (m/m)	Perimeter (m)	Max. Stream Length (m)	Max. Stream Slope (m/m)	Mean Basib Elevation (m)
1	Al Arish	23,784.9	238,040	0.0469	1,213,900	354,250	0.0037	499.23
2	Watier	3521	76,110	0.1446	483,620	118,450	0.0122	912
3	Al Grafee	2610	84,713	0.0334	412,010	114,110	0.0051	640
4	Dhab	2070	56,686	0.2217	353,930	93,238	0.02	1062.9
5	Al Awag	1941	55,070	0.1967	278,530	75,529	0.0082	549.1
6	Fieran	1780	80,341	0.1949	420,050	132,660	0.0177	1015
7	Wardan	1180	59,345	0.1025	273,720	86,222	0.0113	523
8	Ghrandal	1074.1	78,119	0.1536	340,680	105,380	0.0134	801.73
9	Kied	1045.9	47,766	0.3465	258,500	679.14	0.0206	928.8
10	Sidry	868.6	56,292	0.1438	243,580	82,224	0.0107	570.24
11	Paapaa	712.7	54,971	0.1386	232,620	71,005	0.0154	600.24
12	Sidr	623.45	54,196	0.0862	205,660	79,078	0.0073	417
13	Al Rahaa	452	42,466	0.105	181,690	51,580	0.0134	488
14	Om Adwy	367.6	35,744	0.2589	151,320	47,503	0.0282	659.75
15	Tieba	333.9	43,354	0.303	158,240	51,471	0.0311	999.5
16	Lehataa	276.44	31,331	0.0571	114,790	43,528	0.0142	214.64

Sinai is economically important for Egypt because it represents one of its largest mining areas. This peninsula has also a great potential for agriculture and industrial development. Although there are some limitation: the region is prone to flash flood events resulting from heavy, sudden, and short duration rainfall events, which represent a risk for the population, infrastructures, properties, and economic sectors like industry and agriculture itself.

On the other hand, flash floods caused by heavy rainfall events in Sinai and southern/southeastern Egypt represent a potential source of non-conventional freshwater resources. In particular, the water, which usually drains into the Gulf of Suez and the Gulf of Aqaba, if wisely harvested, could fulfill a non-negligible amount of water demand and/or recharge the shallow groundwater aquifers, becoming a precious fresh water source for local people of the region and their agriculture.

3. Results

3.1. Climate Variability in the Sinai Peninsula

The temperature and precipitation monthly mean over the country are shown in Figure 3, evaluated using ECMWF ERA-Interim reanalysis for the period 1981–2010. In the northern part, the Sinai Peninsula is characterized by a Mediterranean climate, whilst in the southern part, the climate is semi-desert to desert. Thus, in most of the Peninsula the climate is hot to very hot. However, sub-regions along the Mediterranean coast in the North and over the mountains are more temperate. In winter, the temperatures are surely lower, with temperatures that can drop to 0 °C at night over the mountains (Figure 3). Most of the precipitation occurs during the winter and then in spring and autumn, while during summer rainfall is almost totally absent over the Sinai (Figure 3).

Using ECMWF ERA-Interim reanalysis, an increase of about 1 °C in mean temperature has been evaluated over Sinai, as shown in Figure 4. This figure shows the anomalies of air temperature at 850 hPa during the last decade over the African Continent at large compared to the baseline period (1979–2000). In addition, the analysis performed using ECMWF ERA-Interim reanalysis for the period 1979 to 2018 over the domain (27°–32° N, 32°–35° E) shows a clear tendency towards increasing average temperatures (Figure 5) and decreasing total rainfall (Figure 6).

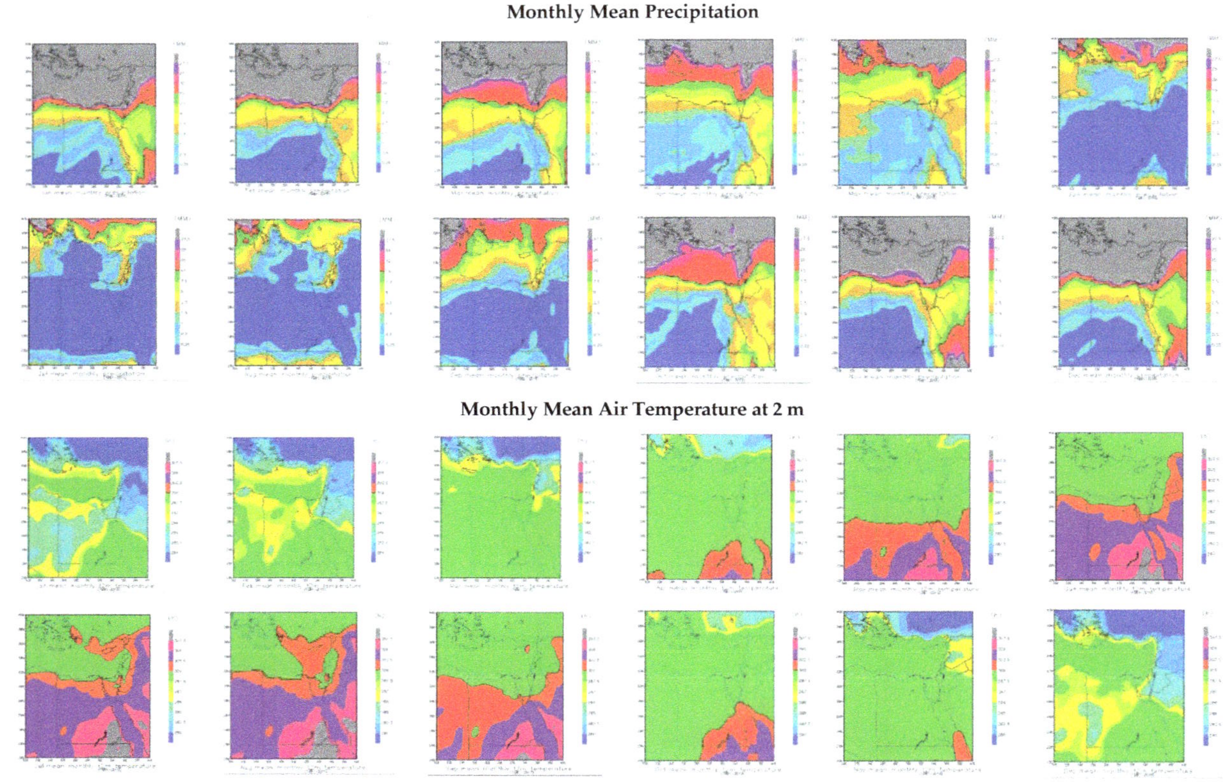

Figure 3. Climatology of Egypt, including the Sinai Peninsula, evaluated over the period 1981–2010. Monthly mean precipitation (in mm) and air temperature (in °K) at 2 m. (Data Source: European Centre for Medium-Range Weather Forecasts (ECMWF) ERA-Interim reanalysis).

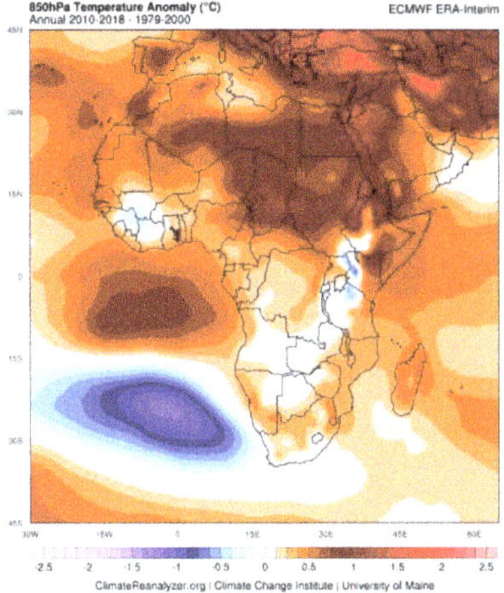

Figure 4. Mean air Temperature at 850 hPa: anomaly over the period 2010–2018 relative to the base period 1979–2000 (Data Source: ECMWF ERA-Interim reanalysis, available at: https://climatereanalyzer.org).

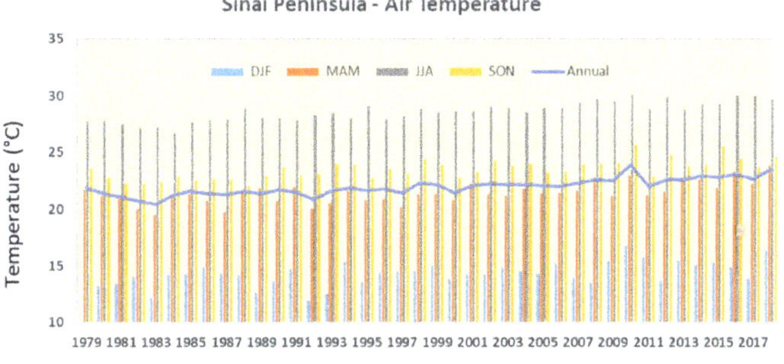

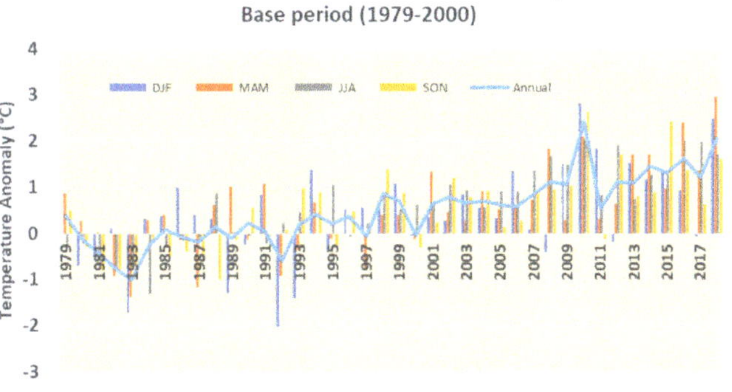

Figure 5. Annual and seasonal mean (top panel) and anomalies (bottom panel) of air temperature at 2 m over the Sinai Peninsula for the period 1979–2018, evaluated in the domain (27°–32° N, 32°–35° E). Base period for anomalies (1979–2000). (Data Source: ECMWF ERA-Interim reanalysis, available at: https://ClimateReanalyzer.org).

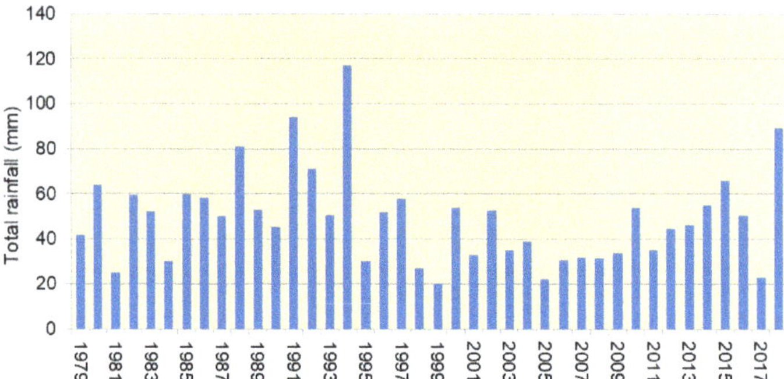

Figure 6. Annual Total Precipitation over the Sinai Peninsula for the period 1979–2018, evaluated in the domain (27°–32° N, 32°–35° E). (Data Source: ECMWF ERA-Interim reanalysis, available at: https://ClimateReanalyzer.org).

The analysis of the GPCC historical data for the period 1901–2013, shows a tendency to decrease precipitation in different sub-regions of the Sinai Peninsula: in the North, at the border with the Mediterranean Sea, in the Centre, and in the South (Figure 7). In general, the total rainfall distribution is quite different along the Peninsula, and it decreases from the northeast towards the southwest, with a maximum in Rafah. The southern part of the Peninsula is characterized by much lower rainfall totals in coastal areas, along the Gulfs of Suez and Aqaba.

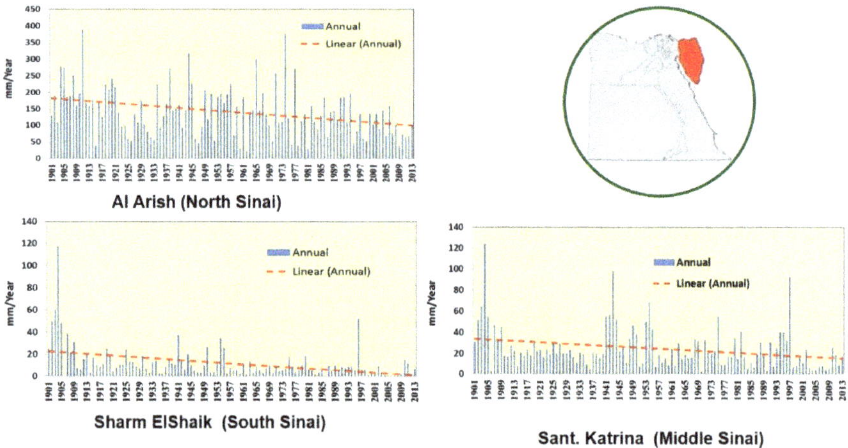

Figure 7. Annual Total Precipitation in North, Middle, and South Sinai Peninsula with linear decreasing trend superimposed. (Data Source: Global Precipitation Climate Center data (GPCC) —https://www.esrl.noaa.gov/psd/data/gridded/data.gpcc.html).

3.2. Extreme Rainfall Events in Sinai

The analysis of rainfall data is the most important element in hydrological studies because it can be used in the EWS development as input to determine the flood discharges. Timing, duration, and extent of rainfall episodes are especially important in Sinai because they represent the major potential source of renewable water. Unfortunately, these flash floods are very difficult to forecast; in addition, their fresh water is difficult to collect because of the steepness of the slopes of the terrain in the peninsula. Moreover, in this semiarid-to-arid region, rainfall is usually characterized by extremely high spatial and temporal variability [26], with intense episodes of very short duration, spatially scattered over the territory. These characteristics make the collection of this water very difficult without the construction of important infrastructures. An additional source of fresh water is the non-renewable groundwater, as in the case of the Nubian Aquifer.

As already mentioned, in recent decades Sinai has been affected by significant changes in the climate, with drier and hottest conditions relative to the baseline period (Figures 4–6), and with more intense storms with associated short, but intense rainfall events. This trend caused severe and long dry periods suddenly interrupted by sporadic intense rainfall with increased space-time variability; this has increased the forecast difficulties in recent years. If this tendency continues, the impacts on the natural environment and resources, including renewable water, as well as on population, infrastructure, and properties, will be severe.

The major storms which affected North and South Sinai in the past years are listed in Table 2 for South Sinai and in Table 3 for North Sinai. The lists show how the number of events since the year 2000 is quite significant.

Table 2. List of heavy rainfall events in the South Sinai Peninsula at different locations.

Flash Flood Events in South Sinai [1] 1990–2015			
Ras Sedr (32°43′00″ E, 29°35′00″ N)	**Ras Sedr (Elmelha) (32°59′54″ E, 29°44′23″ N)**	**Saint Katherin (33°56′58″ E, 28°33′43″ N)**	**Newabaa (34°41′13″ E, 28°58′57″ N)**
02/03/1997	07/12/2000	22/03/1991	17/01/2010
07/02/1999	04/12/2001	17/10/1993	27/01/2013
07/12/2000	06/01/2003	20/10/1993	09/03/2014
10/01/2002	14/12/2003	01/01/1994	08/05/2014
16/12/2003	22/01/2004	18/01/2010	12/09/2015
14/01/2004	05/02/2004	08/01/2013	26/10/2015
04/02/2004	04/04/2011	25/01/2013	
18/01/2010	31/12/2013	27/01/2013	
25/02/2010	09/03/2014	09/03/2014	
09/01/2013		08/05/2014	
01/02/2013		12/09/2015	
09/03/2014		25/10/2015	
07/05/2014			

[1] Other events occurred in South Sinai more recently, including a 27–29 October 2016 episode when the storm also affected the Red Sea, Ismailia, Beni Suef, Qena, Assiut, and Suhag provinces (https://reliefweb.int/disaster/fl-2016-000114-egy).

Table 3. List of heavy rainfall events in the North Sinai Peninsula at different locations.

Flash Flood Events in North Sinai 1990–2015	
Godirat (34°24′35″ E, 30°38′28″ N)	**Maghara (34°32′ E, 30°23′ N)**
22/03/1991	27/10/2005
05/01/2001	21/10/2007
29/01/2004	18/01/2010
17/04/2006	09/01/2013
10/12/2009	09/05/2014
17/01/2010	29/10/2015
09/03/2014	
09/05/2014	
25/10/2015	
29/10/2015	

Among all the events that occurred in the period 2000–2015, the following episodes that affected the whole Peninsula, although with different intensity, have been selected and analyzed: 8 January 2013 and 9 March 2014. The 2013 event started on January 6th with light rainfall and reached a peak on January 7th. On January 8th, the storm continued only along the north coast of North Sinai, covering mainly the north Sinai and parts of south Sinai, and the maximum daily rainfall (29.8 mm) was observed at Rafah station. The 2014 event lasted over Egypt for three days from March 7th until March 9th. This event affected all Sinai with a maximum daily rainfall of 30 mm measured at Dahab station. Table 4 summarizes the total amount of precipitation that occurred in Upper Egypt, the Red Sea, and Sinai Peninsula during the storm (7–9 March 2014), while on the 10th it moved rapidly towards East. The evolution of the 2014 storm as captured by TRMM satellite images is shown in Figure 8, where it is clear how the storm, which started to form in Upper Egypt on the 7th, then rapidly moved toward the North-East and strongly affected Sinai Peninsula in the following days (8th and 9th of March).

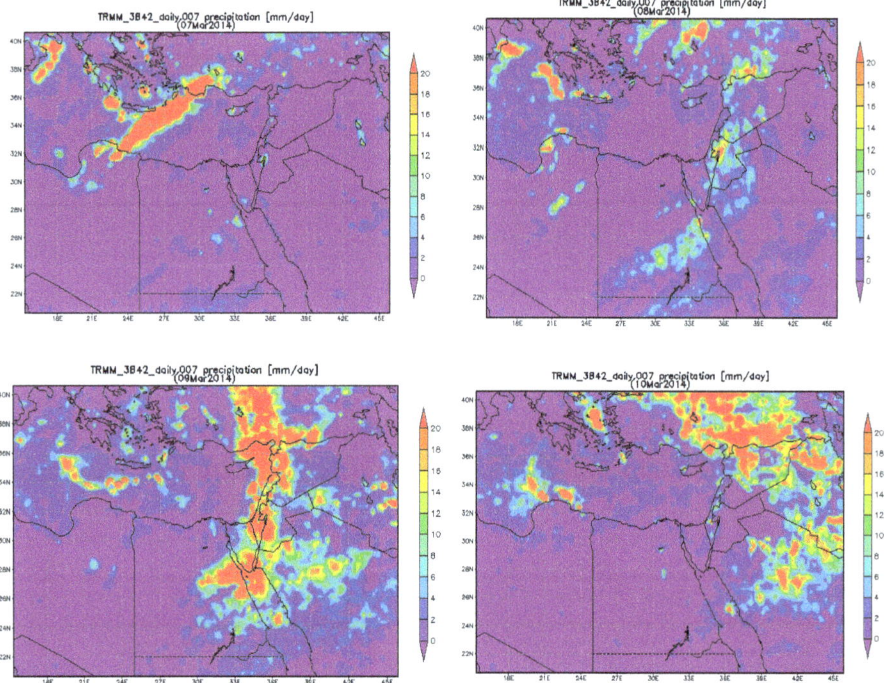

Figure 8. Episode 7–10 March 2014. Storm evolution over Sinai and surrounding region. (Source: TRMM satellite data).

Results of the analysis of the large-scale configuration during these episodes show some similarities, with heavy rainfall usually resulting from mesoscale convective systems embedded in synoptic-scale disturbances interacting with the complex terrain of the Peninsula. The evolution of the storm (not shown) is similar in the selected cases (as an example, Figure 9 shows the spatial distribution of the geopotential at 700 hPa during the selected episodes with a low pressure system over the Eastern Mediterranean bringing a disturbance to the region of interest, which then moves eastward), and a correlation is also noticed with the atmospheric circulation over the Arabic Peninsula and with low-pressure systems over the Eastern Mediterranean. In addition, the distribution of the rainfall between the Northern and Southern parts of the Peninsula is strongly affected by the complexity of the terrain and by the position of the mountain ranges, which is strongly influencing the areas where the majority of rain is observed.

Table 4. Total rainfall in Upper Egypt, the Red Sea, and the Sinai Peninsula during the 7–9 March 2014 event.

Amount of Precipitation Observed in the Upper Egypt, the Red Sea, and Sinai Peninsula during the 7–9 March 2014 Event		
	Station	**Precipitation mm**
UPPER EGYPT	Aswan	29
	Luxor	29
	Asyut	9
RED SEA	Hurgada	21
SINAI	Sharm El-Sheikh	27
	Dahab	30
	Saint Cathrene	29
	El Tor	17
	Nuwaibaa	8
	Nekhel	5

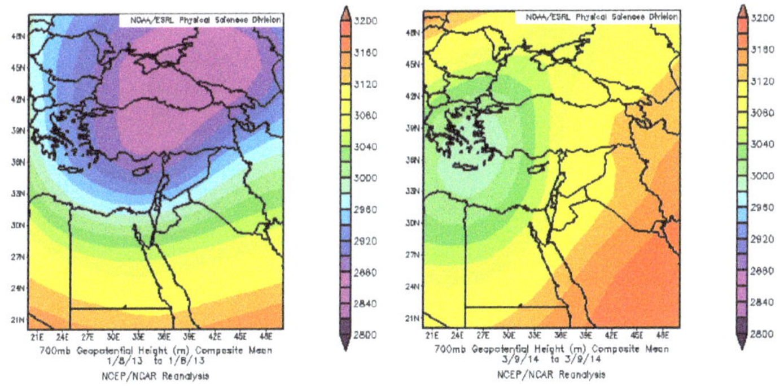

Figure 9. Geopotential height at 700 hPa for the selected episodes: 8 January 2013 and 9 March 2014. (Source: NCEP-NCAR reanalysis).

The mechanism leading to storms with torrential rainfall, thunder, and lightning has been studied by some authors [21] and references therein [18,27,28]; in these works, it is emphasized that they are phenomena of very limited spatial and time extent. Although this part of the study is not yet fully completed, some general remarks are outlined here. Climatically the whole Levant is under the influence of the westerly winds, i.e., on average, the winds blow from the west towards the east and their intensity increases with altitude, and the weather is usually characterized by the presence of a Red Sea Through (RST), i.e., a low-level pressure extending from the African Monsoon over Equatorial Africa, northward over the Red Sea region toward the Eastern Mediterranean (EM). The RST, generated and supported by thermal forcing and supported by the presence of the local complex topography, strongly influences the weather in the whole Levant, usually bringing dry and hot conditions.

The size of storms moving in the eastern part of the Mediterranean basin is of the order of a few km, and their lifetime is only few days. Many storms in the East Mediterranean originate in the region around Cyprus; these storms travel eastwards. Therefore, they do not usually hit Egypt. However, if the conditions are favorable in the Red Sea, they are diverted southeastwards, reaching North Egypt; if their lifetime is sufficiently long, they can hit Central Egypt. These storms are usually associated with the presence of an Active Red Sea Trough. While a (non-active) Red Sea Trough (RST) is a low-level trough extending northward from East Africa over the Red Sea region towards the Levant that brings dry, hot weather, an Active Red Sea Trough (ARST) is an RST accompanied by an upper trough which brings severe weather, heavy precipitation, and possible flash floods as in the case studies analyzed.

In a very simplified scheme, several factors can contribute to transform an RST into the active Red Sea storm; the presence of these factors enhances the tropospheric instability and the synoptic-scale forcing and inducing ascending motions with the formation of an ARST. Usually, the ARST configurations are more active in autumn, when there are two coinciding favorable elements: the position of the African Monsoon and Subtropical Jet Stream (STJ). These elements are absent in winter and spring, when the Middle East (ME) can be affected by extreme rainfall resulting from tropical-extratropical interactions. In the beginning, the systems, which started over Egypt, presented characteristics very similar to a Mesoscale Convective System (MCS), with rainfall particularly intense during its first stage when the convection was dominant. After this first phase, the intensity of rain, as well the role of convection, diminished considerably while the storms move eastwards. The kind storms weather affects not only Egypt but also the Middle East.

3.3. Expected Future Climate Changes

Considering the fact that regional climate information is needed for decision-making on societal issues such as vulnerability and adaptation to a changing climate with weather/water extremes, the Coordinated Regional Climate Downscaling Experiment (CORDEX: https://cordex.org/) has been developed. CORDEX is a World Climate Research Project (WCRP) framework to evaluate regional climate model performance through a set of experiments aimed to the production of regional climate projections [29]. The CORDEX vision is to advance and coordinate the science and application of regional climate downscaling through global partnerships, and its main goals can be summarized as follows:

- To better understand relevant regional/local climate phenomena, their variability, and changes, through downscaling.
- To evaluate and improve regional climate downscaling models and techniques.
- To produce coordinated sets of regional downscaled projections worldwide.
- To foster communication and knowledge exchange with users of regional climate information.

Thanks to CORDEX, a set of experiments are available for different domains, including the Middle East-North Africa (MENA) domain, which contains the region of interest of the present study [30,31], and few selected results are briefly discussed here. Figures 10 and 11 show, respectively, the expected changes in temperature and precipitation over a domain including Egypt and the Arabian Peninsula for the control period 1971–2000 and for three future time periods: 2011–2040, 2041–2070, and 2071–2100 for an intermediate stabilization Representative Concentration Pathway (RCP4.5). These figures show the result of an ensemble of nine different Global Circulation Models (GCMs), in which a general increase in the mean temperature for the mid-to-end of the century is evident. This will occur all over the country except for a narrow region along the Mediterranean Sea, and there will be a decrease in the mean annual precipitation for the same periods; this pattern is more evident in the Eastern Desert and Sinai Peninsula.

Focusing on the Sinai Peninsula, several studies [32–34] showed that several parameters affected the occurrence of the flash floods over the Sinai Peninsula, namely urban expansion and extreme rainfall events. Three ensembles from three Global Circulation Models (GCMs)—CNRM-CM5, EC-EARTH and MPI-ESM-LR—downscaled by the Regional Climate Model (RCM) were used to present the future change in temperature and rainfall over Sinai through the mid (2041–2070) and the end of the century (2071–2100), where both periods are compared with the baseline period (1971–2000). All three ensembles are obtained from CORDEX data, with two ensembles (CNRM-CM5, EC-EARTH) from MENA Domain (created by RICCAR Project: https://riccar.org/) and one ensemble (MPI-ESM-LR) from Africa Domain. The ensembles analyzed are for one emission scenario (RCP4.5) with resolution 50 km × 50 km. Figures 12 and 13 show the future change in temperature over Sinai during the periods of 2050s and 2080s, respectively, while Figures 14 and 15 show the future change in precipitation over Sinai during the same periods. The results show increase in temperature for all three ensembles, while the change of temperature will increase highly during the end of the century (2071–2100). In addition, winter will become warmer, especially in mid and south Sinai, with increases in temperature between (0.2 °C to 3 °C) for both future periods, 2050s and 2080s. It is not clear from the results which model give the worst result, but all of them agree with the increase in temperature. The precipitation will be affected: Where the results show that the rainfall will be shifted to the summer, sudden storms will be expected to occur in summer during the mid and end periods (2041–2100). The results show increases in amount of the precipitation may reach to 300% in some local areas. There is no change expected during winter and autumn; however, south Sinai will suffer by decreasing rainfall during autumn. The expected distribution of the rainfall shows the localization, which indicates increases in damage (see Figures 14 and 15).

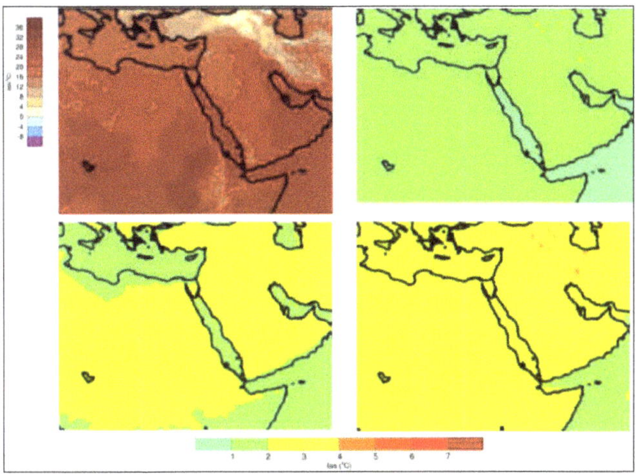

Figure 10. Temperature. Results from an ensemble of nine GCM, for RCP4.5 scenario. Ensemble mean (mm/day) for the control period (1971–2000) (upper left panel-color scale on the upper left). Change in Ensemble mean (%-color scale at the bottom), for 2011–2040 (upper right), 2041–2070 (bottom left) and 2071–2100 (bottom right) compared with 1971–2000. (Source: https://cordex.org/).

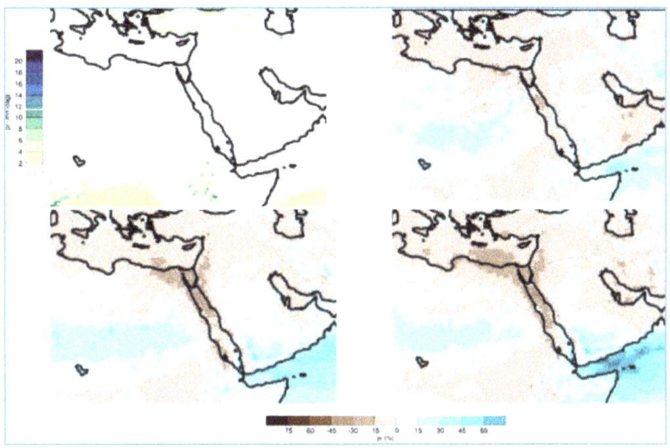

Figure 11. Precipitation. Results from an ensemble of nine Global Circulation Models (GCMs), for RCP4.5 scenario. Ensemble mean (mm/day) for the control period (1971–2000) (upper left panel-color scale on the upper left). Change in Ensemble mean (%-color scale at the bottom), for 2011–2040 (upper right), 2041–2070 (bottom left) and 2071–2100 (bottom right) compared with 1971–2000. (Source: https://cordex.org/).

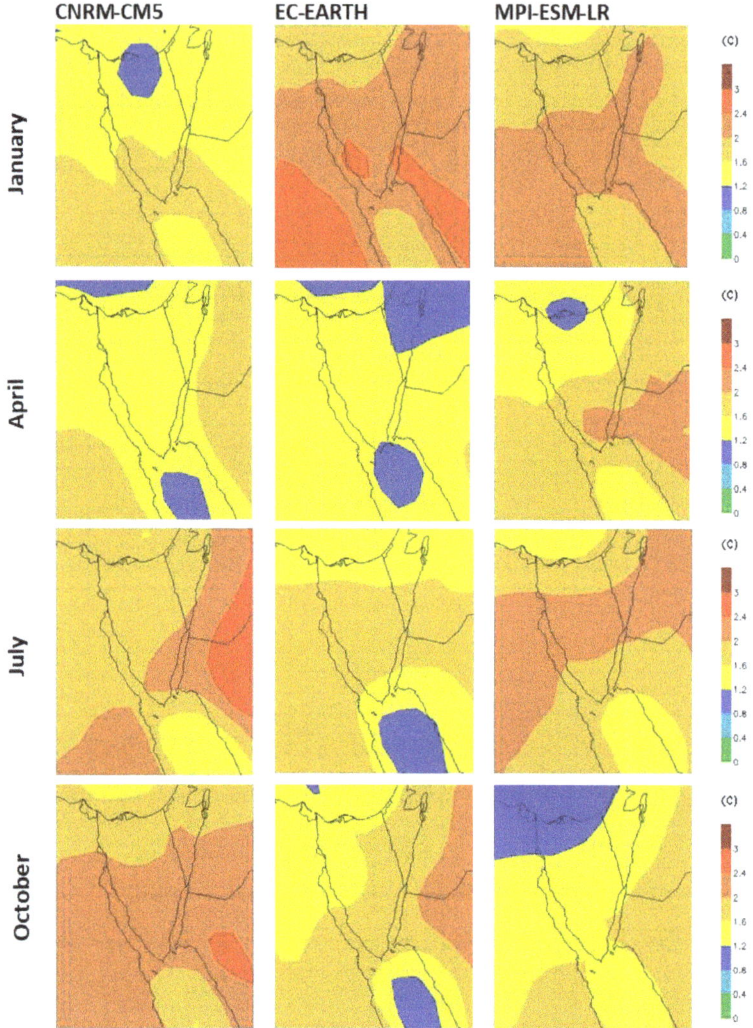

Figure 12. Temperature. Results from an ensemble of three GCMs for RCP4.5 scenario. Change in Ensemble mean monthly (C) for the future period 2041–2070 (2050s) compared with 1971–2000. January, April, July, and October are examples presenting winter, spring, summer, and autumn, respectively.

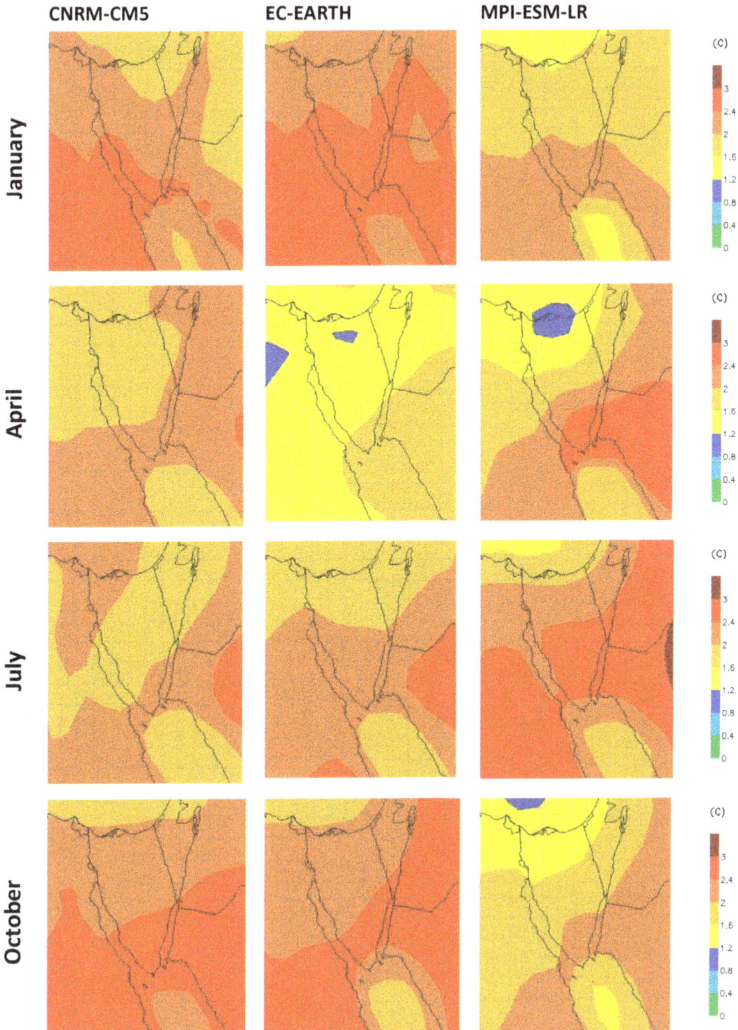

Figure 13. Temperature. Results from an ensemble of three GCMs for RCP4.5 scenario. Change in Ensemble mean monthly (C) for the future period 2071–2100 (2080s) compared with 1971–2000. January, April, July, and October are examples presenting winter, spring, summer, and autumn, respectively.

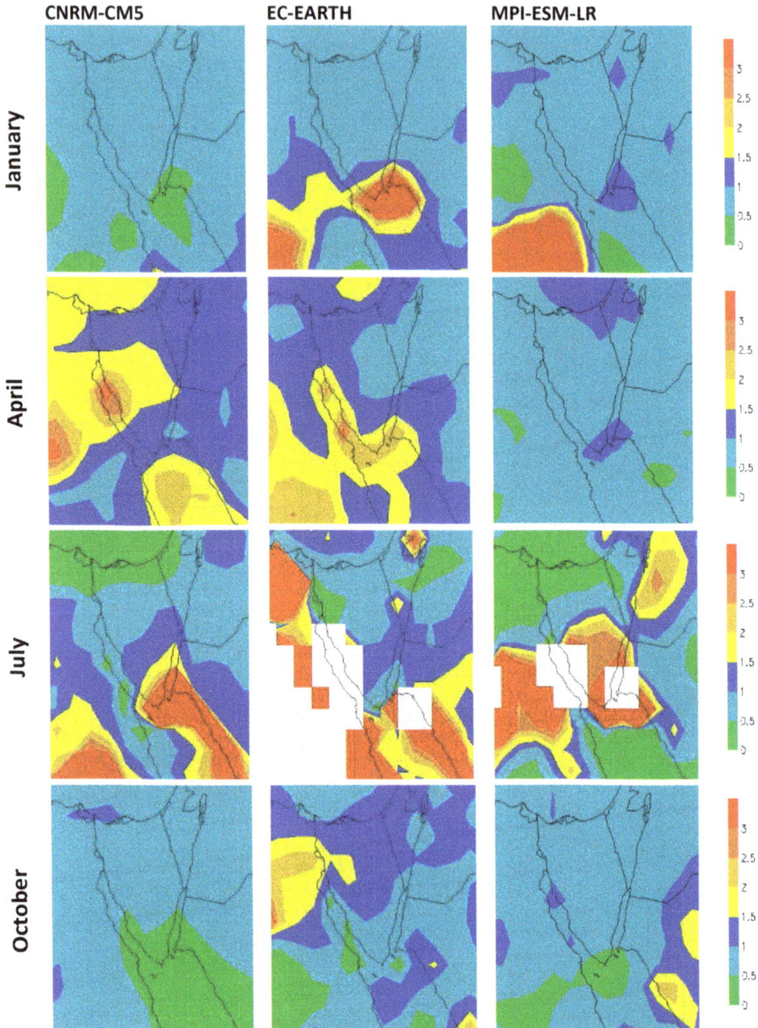

Figure 14. Precipitation. Results from an ensemble of three GCMs for RCP4.5 scenario. Change in ensemble mean monthly (no sign where the change is produced by dividing future by baseline) for the future period 2041–2070 (2050s) compared with baseline period 1971–2000. January, April, July, and October are examples present winter, spring, summer, and autumn, respectively. (White color indicates that baseline value equals zero).

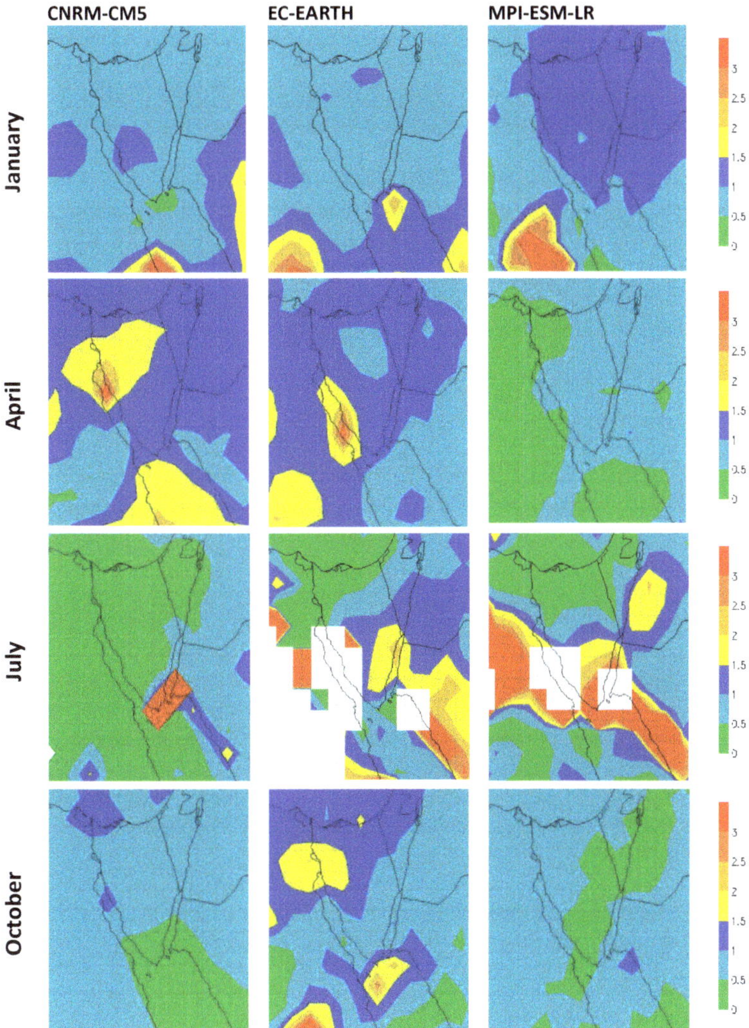

Figure 15. Precipitation. Results from an ensemble of three GCMs for RCP4.5 scenario. Change in ensemble mean monthly (no sign where the change is produced by dividing future by baseline) for the future period 2071–2100 (2080s) compared with baseline period 1971–2000. January, April, July, and October are examples present winter, spring, summer, and autumn, respectively. (White color indicates that baseline value equals zero).

4. Conclusions

The Sinai Peninsula is a region characterized by severe data limitations and by the occurrence of thunderstorms very limited in time and space; therefore, the improvement of Weather Prediction Systems feeding Early Warning Systems is difficult. The lack of reliable WPS and appropriate EWS impair not only the design and adoption of adaptation and mitigation measures but also the development of emergency plans to save lives as well as the harvesting of water precious for the local Bedouin Communities, which is otherwise lost.

The scientific problem addressed in the paper is to assemble information about severe thunderstorms leading to flash flood in the two sub-regions, North and South Sinai, and to present a first attempt to describe, for the same region, the expected evolution of climate in the future. The study presented here is a step forward in order to respond to the need of a deeper knowledge of extreme rainfall in Sinai Peninsula that is driving flash floods affecting population and infrastructure with damage and elevated costs.

The results presented are derived from the Academy of Scientific Research and Technology of Egypt (ASRT) and the National Research Council of Italy (CNR) joint project and show, in particular, the intensity of the current climate change in terms of increase in temperature, decrease in precipitation, and increase (in number and intensity) of heavy rainfall episodes in the interested region. Some results were also discussed showing an increase in the severity of the changes in temperature and precipitation under future climate scenario for an intermediate stabilization Representative Concentration Pathway, RCP4.5.

Further analysis of the heavy rainfall episodes generating flash floods in both the North and South part of the Peninsula is not only needed but also particularly useful to increase the knowledge about the generation and evolution of these short-lived and patchy storms, focusing, in particular, on the atmospheric factors driving their formation. A deeper knowledge of the mechanism could also provide some indication for improving the weather forecast systems over the region, while results will contribute to the development of regional climate scenarios at higher resolution to be used for the implementation of more sophisticated risk maps for future decades. Results can also encourage and sustain the elaboration of adaptation plans for population, agriculture, and water resources management and help decision-makers to plan the construction of water harvesting structures [35].

In conclusion, the knowledge of extreme rain events in Sinai is presently scarce; thus, this study intends to give a contribution to the improvement of their understanding. A possible future study to address this could be a comparative analysis of episodes occurring in the Sinai Peninsula and in neighboring countries, because thunderstorms affecting the Peninsula then move eastward and largely impact the whole Levant.

Further specific studies on the Sinai's present and future climate that can help to understand better if there will be an evolution of heavy rainfall episodes driving flash floods are needed. Sophisticated risk maps for future decades are also needed, together with adoption of adaptation plans for population, agriculture, and water resources management. Finally, results from this and future studies will be largely used by decision-makers in their design and planning of the construction of water harvesting structures in the near future for the good of the Sinai people.

Author Contributions: Individual contributions to the research is as follows: Conceptualization, M.B. and D.A.; methodology, M.B. and D.A.; formal analysis and investigation, M.B., D.A., I.S.A.Z., and G.D.; writing—original draft preparation and review and editing, M.B.; writing—review and editing, D.A., I.S.A.Z., and G.D. All authors have read and agreed to the published version of the manuscript.

Funding: This research was funded by ARST and CNR in the framework of the Agreement on Scientific Cooperation between the Academy of Scientific Research and Technology of Egypt (ASRT) and the National Research Council of Italy (CNR), who provided support for scientists' mobility.

Acknowledgments: The study has been funded in the framework of the Agreement on Scientific Cooperation between the Academy of Scientific Research and Technology of Egypt (ASRT) and the National Research Council of Italy (CNR). The two Institutes participating to the project were, respectively: the Water Resources Research Institute (WRRI) of the National Water Research Center (NWRC), and the Institute of Biometeorology (CNR-Ibimet, now CNR-IBE). M.B and G.D. are grateful to the research group at the WRRI-NWRC for their kind hospitality and useful discussions and to the Italian Scientific Attaché Office in Cairo for support. D.A. and I.S.A. are also thankful to the research staff of the CNR-IBE for supporting the research stay in Italy that enabled the completion of this work.

Conflicts of Interest: The authors declare no conflict of interest. The funders had no role in the design of the study; in the collection, analyses, or interpretation of data; in the writing of the manuscript, or in the decision to publish the results.

References

1. Al Zayed, I.S.; Ribbe, L.; Al Salhi, A. Water harvesting and flashflood mitigation-wadi watier case study (South Sinai, Egypt). *Int. J. Water Resour. Arid Environ.* **2013**, *2*, 102–109.
2. Idowu, O.; Gawusu, S.; Mugo, V.; Bilal, A.; Buumba, M.; Kamga, M.A.; Komassi, A.; Awovi, S.; Al Zayed, I.S.; Khalifa, M.; et al. *GEO-6 for Youth Africa: A Wealth of Green Opportunities—Global Environment Outlook*; United Nations Environment Programme (UNEP): Nairobi, Kenya, 2019.
3. Baldi, M.; Amin, D.; Al Zayed, I.S.; Dalu, G.A. Extreme rainfall events in the Sinai Peninsula. In Proceedings of the EGU General Assembly Conference Abstracts 2017, Vienna, Austria, 23–28 April 2017; Volume 19, p. 13971.
4. Al Zayed, I.S. Observing flash flood in arid and semi-arid regions from space: Wadi Watier of Egypt as a case study. In Proceedings of the Second International Symposium on Flash Floods in Wadi Systems, El Gouna Campus, Hurghada, Egypt, 25–27 October 2016.
5. Attia, K.; Al Zayed, I.S. Water harvesting in Sinai Peninsula, Egypt. In Proceedings of the International Forum on Water Scarcity and Harvesting in the Arid and Semiarid Areas (IFWSH-ASA-2015), Hurghada, Egypt, 23–26 November 2015.
6. Baldi, M. Climate Change, Water availability, and Cultural Heritage in Egypt. In *Sciences and Technologies Applied to Cultural Heritage—STACH 1*; Baldi, M., Vittozzi, G.C., Eds.; Cnr Edizioni: Roma, Italy, 2019; ISBN 978 88 8080 347 8. Available online: https://iiccairo.esteri.it/iic_ilcairo/it/istituto/centro-archeologico/link-utili (accessed on 25 July 2020).
7. FloodList. Eastern Mediterranean—Deadly Flash Floods After Heavy Rain. Available online: http://floodlist.com/asia/eastern-mediterranean-egypt-israel-floods-april-2018 (accessed on 25 July 2020).
8. FloodList. Egypt—Heavy Rain Causes Flood Chaos in Cairo. Available online: http://floodlist.com/africa/egypt-cairo-floods-october-2019 (accessed on 25 July 2020).
9. FloodList. Egypt—5 Dead After Storms Trigger Floods. Available online: http://floodlist.com/africa/egypt-storm-floods-march-2020 (accessed on 25 July 2020).
10. BBC New. Egypt Experiences Worst Flooding in 35 Years. 2020. Available online: https://www.bbc.co.uk/programmes/w172wn3b49zcw4h (accessed on 25 July 2020).
11. El-Rakaiby, M.L. Drainage Basins and Flash Flood Hazard in Selected Parts of Egypt. *Egypt. J. Geol.* **1989**, *33*, 307–323.
12. Elmoustafa, A.M.; Mohamed, M.M. Flash Flood Risk Assessment Using Morphological Parameters in Sinai Peninsula. *Open J. Mod. Hydrol.* **2013**, *3*, 122–129. [CrossRef]
13. Prama, M.; Omran, A.; Schröder, D.; Abouelmagd, A. Vulnerability assessment of flash floods in Wadi Dahab Basin, Egypt. *Environ. Earth Sci.* **2020**, *79*, 1–17. [CrossRef]
14. Dayan, U.; Ziv, B.; Margalit, A.; Morin, E.; Sharon, D. A severe autumn storm over the Middle-East: Synoptic and mesoscale convection analysis. *Theor. Appl. Climatol.* **2001**, *69*, 103–122. [CrossRef]
15. Dayan, U.; Nissen, K.; Ulbrich, U. Review article: Atmospheric conditions inducing extreme precipitation over the eastern and western Mediterranean. *Nat. Hazards Earth Syst. Sci.* **2015**, *15*, 2525–2544. [CrossRef]
16. De Vries, A.J.; Tyrlis, E.; Edry, D.; Krichak, S.O.; Steil, B.; Lelieveld, J. Extreme precipitation events in the Middle East: Dynamics of the Active Red Sea Trough. *J. Geophys. Res. Atmos.* **2013**, *118*, 7087–7108. [CrossRef]
17. Kahana, R.; Ziv, B.; Dayan, U.; Enzel, Y. Atmospheric predictors for major floods in the Negev Desert, Israel. *Int. J. Climatol.* **2004**, *24*, 1137–1147. [CrossRef]
18. Ziv, B.; Dayan, U.; Sharon, D. A mid-winter, tropical extreme flood producing storm in southern Israel: Synoptic scale analysis. *Meteorol. Atmos. Phys.* **2005**, *88*, 53–63. [CrossRef]
19. Cools, J.; Vanderkimpen, P.; El Afandi, G. An early warning system for flash floods in hyper-arid Egypt. *Nat. Hazards Earth Syst. Sci.* **2012**, *12*, 443–457. [CrossRef]
20. El Afandi, G.; Morsy, M.; El Hussieny, F. Heavy Rainfall Simulation over Sinai Peninsula Using the Weather Research and Forecasting Model. *Int. J. Atmos. Sci.* **2013**, *2013*, 11. [CrossRef]
21. El Afandi, G.; Morsy, M. Developing an Early Warning System for Flash Flood in Egypt: Case Study Sinai Peninsula. In *Flash Floods in Egypt*; Negm, A.M., Ed.; Springer: Cham, Switzerland, 2020; pp. 45–60.

22. Kalnay, E.; Kanamitsu, M.; Kistier, R.; Collins, W.; Deaven, D.; Gandin, L.; Iredell, M.; Saha, S.; White, G.; Woollen, J.; et al. The NCEP/NCAR Reanalysis 40-year Project. *Bull. Am. Meteor. Soc.* **1996**, *77*, 437–471. [CrossRef]
23. Dee, D.P.; UPPALA, S.M.; Simmons, A.J.; Berrisford, P.; Poli, P.; Kobayashi, S.; Andrae, U.; Balmaseda, M.A.; Balsamo, G.; Bauer, P. The ERA-Interim reanalysis: Configuration and performance of the data assimilation system. *Q. J. R. Meteorol. Soc.* **2011**, *137*, 553–597. [CrossRef]
24. Zahran, M.A.; Willis, A.J. Egypt: The Gift of the Nile. In *The Vegetation of Egypt*; Springer: Dordrecht, The Netherlands, 2009; Volume 2, pp. 1–3.
25. Elsayad, M.A.; Sanad, A.M.; Kotb, G.; Eltahan, A.H. Flood Hazard Mapping in Sinai Region. In *Global Climate Change, Biodiversity, and Sustainability Conference*; Arab Academy for Science, Technology and Maritime Transport: Cairo, Egypt, 2013; Volume 15.
26. Sharon, D. The spottiness of rainfall in a desert area. *J. Hydrol.* **1972**, *17*, 161–175. [CrossRef]
27. Baldi, M. Climate of Egypt between dryness and torrential rainfall: Patterns of climate variability. In *AHMES-Egyptian Curses Vol. 2—A Research on Ancient Catastrophes*; Capriotti Vittozzi, C., Ed.; Cnr Edizioni: Roma, Italy, 2015; pp. 255–264. ISBN 978 88 8080 140 5.
28. Krichak, S.O.; Breitgand, J.S.; Feldstein, S.B. A conceptual model for the identification of the Active Red Sea Trough Synoptic events over the southeastern Mediterranean. *J. App. Meteorol. Climatol.* **2012**, *51*, 962–971. [CrossRef]
29. Giorgi, F.; Jones, C.; Asrar, G.R. Addressing climate information needs at the regional level: The CORDEX framework. *WMO Bull.* **2009**, *58*, 175–183.
30. Bucchignani, E.; Mercogliano, P.; Panitz, H.-J.; Montesarchio, M. Climate change projections for the Middle East—North Africa domain with COSMO-CLM at different spatial resolutions. *Adv. Clim. Chang. Res.* **2018**, *9*. [CrossRef]
31. Bucchignani, E.; Cattaneo, L.; Panitz, H.-J.; Mercogliano, P. Sensitivity analysis with the regional climate model COSMO-CLM over the CORDEX-MENA domain. *Meteorol. Atmos. Phys.* **2016**, *128*, 73–95. [CrossRef]
32. Abuzied, M.; Mansour, B.M. Geospatial hazard modeling for the delineation of flash flood-prone zones in Wadi Dahab basin, Egypt. *J. Hydroinformatics* **2019**, *21*, 180–206. [CrossRef]
33. Saber, M.; Abdrabo, K.I.; Habiba, O.M.; Kantosh, S.A.; Sumi, T. Impacts of Triple Factors on Flash Flood Vulnerability in Egypt: Urban Growth, Extreme Climate, and Mismanagement. *Geosciences* **2020**, *10*, 24. [CrossRef]
34. El Osta, M.M.; El Sabri, M.S.; Masoud, M.H. Estimation of flash flood using surface water model and GIS technique in Wadi El Azariq, East Sinai, Egypt. *Nat. Hazards Earth Syst. Sci. Discuss.* **2016**, 1–51. [CrossRef]
35. Al Zayed, I.S.; Keshta, E.; Attia, K. Rainwater Harvesting as an Adaptation Measure to Climate Change: A Case Study of Mamora Beach Area, Egypt. *Int. Water Technol. J.* **2018**, *8*, 72–77.

Article

Floodplain Settlement Dynamics in the Maouri Dallol at Guéchémé, Niger: A Multidisciplinary Approach

Andrea Galligari [1,*], Fabio Giulio Tonolo [2] and Giovanni Massazza [1]

[1] DIST, Politecnico and University of Turin, Viale Mattioli 39, 10125 Turin, Italy; giovanni.massazza@polito.it
[2] DAD, Politecnico of Turin, Viale Mattioli 39, 10125 Turin, Italy; fabio.giuliotonolo@polito.it
* Correspondence: andrea.galligari@polito.it

Received: 12 June 2020; Accepted: 9 July 2020; Published: 13 July 2020

Abstract: In Sahelian Africa, rural centers have been hit by catastrophic floods for many years. In order to prevent the impact of flooding, the flood-prone areas and the settlement dynamics within them must be identified. The aim of this study is to ascertain the floodplain settlement dynamics in the Maouri valley (135 km^2) in the municipality of Guéchémé, Niger. Through hydraulic modeling, the analysis identified the flood-prone areas according to three return periods. The dynamics of the settlements in these areas between 2009 and 2019 were identified through the photointerpretation of high-resolution satellite images and compared with those in the adjacent non-flood-prone areas. Spatial planning was applied to extract the main dynamics. The synergic application of these disciplines in a rural context represents a novelty in the research field. Since 2009, the results have shown a 52% increase of the built-up area and a 12% increase in the number of buildings, though the increase was higher in the flood-prone areas. The factors that transform floods into catastrophes were identified through perceptions gathered from the local communities. Three dynamics of the expansion and consolidation of buildings were observed. Specific flood risk prevention and preparation actions are proposed for each type of dynamic.

Keywords: building consolidation; extreme precipitations; flood exposure; flood risk; satellite remote sensing; settlement dynamics; sustainable rural development; vulnerability

1. Introduction

In recent years, floods in Sahelian Africa have become more frequent and catastrophic [1–3]. Over the last decade, flood-related issues have begun in this area as a result of the unusual magnitude of climatic events [4–6]. Extreme rainfall and its effect on an increasingly altered land surface area are the main flood-driving factors usually reported in the literature [7,8]. Arboreal and shrub cover has been greatly reduced to satisfy the incessant need for arable land [9] and firewood for a continuously growing population. The change of land cover has exposed the soils to degradation processes [10] that increase the runoff [11,12] and lower the rainfall threshold, which, in turn, causes related damage.

The impact of hydrometeorological events is aggravated by insufficient rainwater management [13–17] and water-resistant homes. Anthropic pressure in rural areas [18] has led many people to settle within the flood-prone area (FPA) in the absence of regulation and monitoring by the authorities [19]. Flood damage further exacerbates poverty and risks nullifying investments in the primary sector. This sector in the Sahel is closely linked to the reduction of poverty, food insecurity, and inequalities [20].

In rural Sahel, prevention and preparation for flooding through local plans is struggling to take hold due to a lack of financial resources [21], human resources [22], and systematic and detailed knowledge on the exposed settlements [23]. The recourse to low resolution remote sensing and digital elevation models (DEM), photointerpretation with semi-automatic classification algorithms [24], or with urban growth simulation models [25] does not achieve the level of detail on the assets and exposed population that is needed on a local scale. The use of hydraulic numerical models is still occasional [26] and reveals different application fields and different precision levels [27–29]. FPA precision depends on the availability and quality of topographic and hydrological observations [30]. The application field ranges from flood assessment to early warning systems [31], hydraulic constructions [32], and eco-hydraulics [33]. As a result, the literature still offers little on floodplain settlement dynamics and the expansion of settlements is often not quantified [34,35]. This is the first gap that must be addressed in order to improve flood disaster prevention. The second gap concerns the durability of houses in the event of flooding. Currently, the focus is mainly on the precariousness of urban construction [36–39], but less on consolidation [40] in the rural context, which increases the capital exposed to flooding. Once again, here, it is essential to know the characteristics of the housing stock in order to identify the retrofitting measures that can make buildings more resistant to water [41].

The objective of this study is to ascertain the changes of settlements and housing stock in the FPA over the past decade by answering two questions: first, does demographic growth in a rural context increase the exposure of settlements to flooding? Second, does the improvement of building materials increase the flood resistance of homes? Knowledge of these dynamics could help local administrations to identify and localize flood prevention and preparation measures [42].

The synergic contribution of three different disciplines (remote sensing, hydraulic, and spatial planning) and their application in a rural context represents a novelty of the research. To simplify the interpretation, the acronyms adopted in this study are listed in Table 1.

Table 1. Glossary of acronyms used in the text.

Acronyms	Full Names
FPA	Flood-prone areas
NFPA	Non-flood-prone areas
DEM	Digital elevation model
DMN	Direction de la Météorologie Nationale du Niger
BDINA	Base de Données sur les Inondations Niger ANADIA
HydroSHEDS	Hydrological data and maps based on shuttle elevation derivates at multiple scales
PDF	Probability density functions
RP	Return period
HEC-HMS	Hydrologic Engineering Center - hydrologic modeling system
HEC-RAS	Hydrologic Engineering Center - river analysis system
CN	Curve number
SAM	Spectral angle mapper
NDBI	Normalized difference built-up index
NDVI	Normalized difference vegetation index

2. Materials and Methods

2.1. Study Area

For this study, a rural area exposed to floods in one of the countries with the highest rural population growth in the Sahel was considered: the municipality of Guéchémé (109,000 inhabitants in 2012) in Niger. Between 2001 and 2012, the municipality's population grew by 23%. The municipal territory is in a watered region (annual average rainfall of 560 mm) of a semi-arid country [43,44]. The studied area (135 km^2) is the Maouri Dallol (meaning "valley" in the Fula language), a floodplain that is densely cultivated thanks to the presence of water, but partially exposed to a flood hazard. The floodplain accommodates 87 settlements consisting of 23,000 inhabitants, with one of the country's highest densities (167 inhabitants/km^2) [45] (Figure 1).

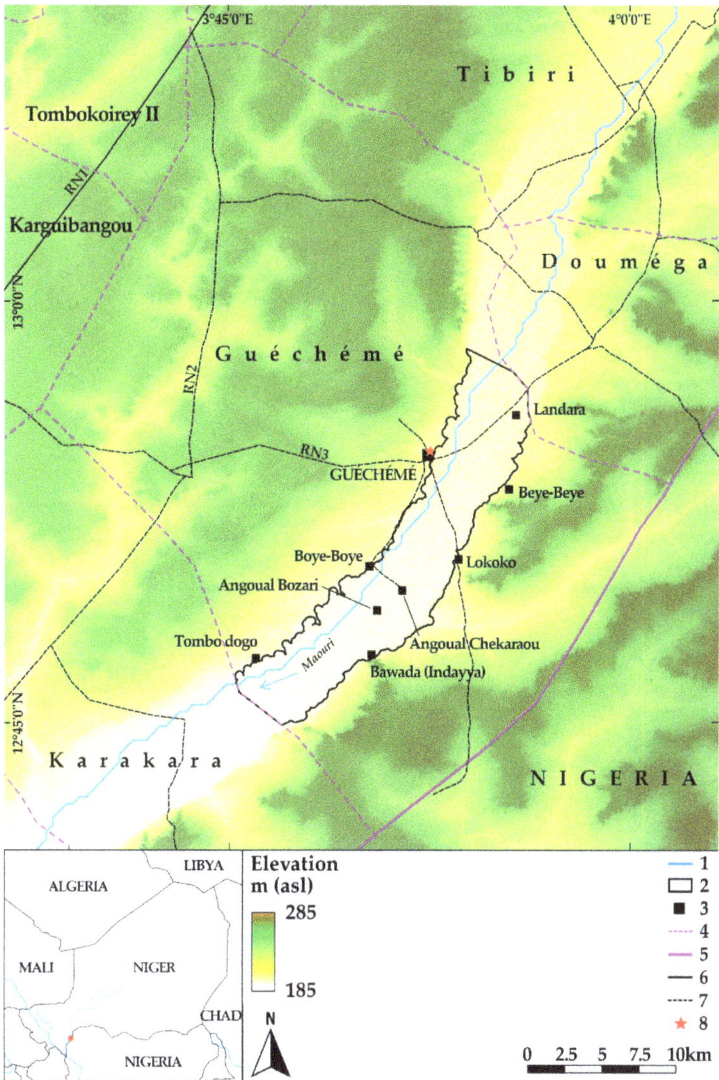

Figure 1. Territorial framework of the Maouri Dallol in the municipality of Guéchémé: Dallol river channel (1), study area (2), main settlements (3), municipal border (4), national border (5), national road no. 1 (6), unpaved tracks (7), and weather stations (8).

Primarily, the study determined the past trend of rainfall and its impact and identified the FPA according to three return period scenarios. Subsequently, it analyzed the settlement dynamics between 2009 and 2019 by identifying the built-up areas and type of buildings based on high-resolution satellite images (Figure 2).

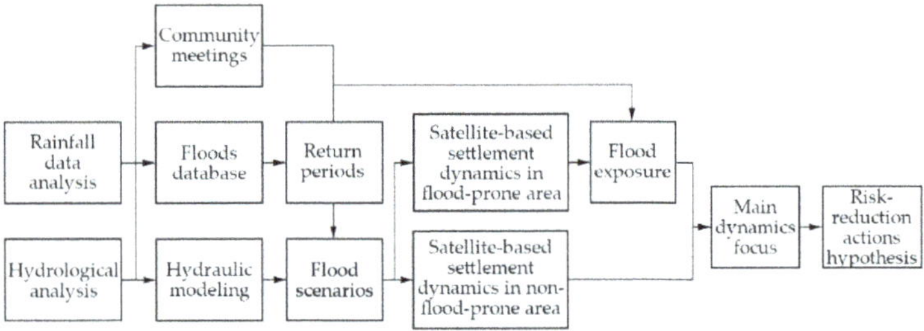

Figure 2. Methodology flowchart.

2.2. *Rainfall Data Analysis*

The analysis of the settlement dynamics was based upon the synergic contribution of different disciplines. The first phase used elements of climatology for the pluviometric characterization of the territory. A series of 38 years of daily rainfall (1981–2018) registered at two weather stations of the National Meteorology Directorate (DMN, according to the French acronym) was used to track the local rainfall profile (annual accumulation) and the intense rainfall within the 95th percentile (trend) [46]. That trend was compared with the database of flood damage recorded and made accessible by the Niger flood database (BDINA, according to the French acronym) [47]. In addition, in 2018–2019, two meetings were organized with the municipality and with the community of Guéchémé to ascertain the local perception of heavy rainfalls, their consequences, and the factors that transformed them into a catastrophic event.

2.3. *Hydrological Analysis and Hydraulic Modeling*

The second phase applied hydraulic methods to identify the FPA. The upstream hydrographic basin of the study area was defined through the DEM, the hydrological data and maps based on shuttle elevation derivates at multiple scales (HydroSHEDS) hydrographic grid [48,49], and GIS software. The morphology was derived from a 90 m DEM, which was resampled and merged with a 10 m DEM for the North part of the study area [50]. The concentration time was computed through Giandotti's formula according to the size and characteristics of the catchment area and the river channel [51–53]. Using the rainfall data, the maximum rainfall, meaning the maximum precipitation during the concentration time—i.e., the period that generated the maximum runoff in the outlet point of the considered basin—was calculated.

The statistical analysis of the annual maximum rainfall was conducted through the probability density functions (PDF) commonly applied to hydrology, such as log-normal, exponential, Gumbel, and generalized extreme value distributions [54,55]. Anderson–Daring and Pearson tests were used to identify the distributional adequacy of PDF and to ascertain the best fit with the dataset [56]. The statistical analysis facilitated the determination of the critical precipitation according to the annual probability of occurrence (P) and the computation of the relative return periods (R_P), which were equal to the inverse of the annual probability (P):

$$R_P = 1/P \quad (1)$$

The R_P values chosen for this case study were 2, 20, and 200 years, corresponding to events with high, medium, and low probability of occurrence.

These return periods were adopted to ensure the maximum variability and a clear distinction of the three scenarios, which were quite close due to the flat morphology of the large floodplain and the DEM inaccuracy [57].

Hydrographs and maximum discharges were computed through a hydrological model covering the upstream basin. The model was developed on the Hydrologic Engineering Center—hydrologic modeling system (HEC-HMS) software according to the curve number (CN) method, the meteorological data recorded in the Guéchémé gauge station, and the endorheic configuration of the catchment area [58–60]. The CN was defined according to the land cover identified from the Copernicus dataset [61–63]: the watershed was almost completely covered by herbaceous vegetation (54%), sparse vegetation (24%), cropland (13%), and shrubs (8%). No discharge data were available for calibration. A validation was conducted based on the literature values of the runoff coefficient for the Sahelian area [64].

The determination of the FPA was carried out through a monodimensional hydraulic numerical model realized along 50 km of the Maouri Dallol floodplain [65]. The model was created with the Hydrologic Engineering Center—river analysis system (HEC-RAS) software [66] and was opportunely extended upstream and downstream to avoid anomalies related to boundary conditions. The hydraulic model was based on the composed 90 m and 10 m DEM morphology, the typical roughness of the vegetation present in the rainy season, and the discharge resulting from the hydrological analysis [67]. No surface water level observations were available for calibration. A validation of the flooded areas was conducted with the in-situ observations taken during the survey in July, 2018 and the ground-water levels of the piezometers of the study area.

2.4. Satellite-Based Settlement Dynamics

The third phase consisted of identifying the dynamics of the built-up area and the type of buildings (durable, non-durable) in the FPA and in the surrounding areas in 2009 and 2019 through high resolution satellite image analysis. This phase was developed in three main steps.

First, the opportunity of extracting single buildings through semi-automatic elaborations of high-resolution satellite images was verified, testing different typologies of classification algorithms. Images were acquired between 6 and 14 October, 2009 [68] and between 1 and 6 September, 2019 [69], as detailed in Table 2.

Table 2. Satellite imagery specifications.

Image Dataset	Year	Acquisition Date	Sensor	Spectral Resolution	Geometric Resolution	Format
1	2009	6–14 October	GeoEye-1	4 bands	0.5 m	Bundle [1]
2	2019	1–6 September	WorldView-2	8 bands	0.5 m	Bundle

[1] Multispectral and panchromatic dataset.

The images were subsequently pre-processed radiometrically (pansharpening [70] and radiometric calibration [71]) and geometrically (mosaicking and orthorectification [72]). The classification tests with pixel-based algorithms [73] (spectral angle mapper (SAM)), object-oriented algorithms [74] (image segmentation), and spectral indices (normalized difference built-up index (NDBI)/normalized difference vegetation index (NDVI)) [75] highlighted the difficulty in distinguishing, in a semi-automatic way, the roofs from the roads due to the very similar material used (respectively, earth and laterite) with overlapping radiometric responses.

Consequently, the second step consisted of photointerpretation by an image analyst. The inhabited areas of the riverbed were determined, and the buildings were subsequently digitalized in 2009 and 2019 through computer aided photo interpretation (CAPI). The use of a GIS (geographic information system) environment made it possible to generate a database containing not only the building geometries, but also the ancillary information, such as the material used for the roofs (straw, mud, and corrugated iron sheets, which constitute as many stages of the consolidation of constructions). The delimitation of the built-up areas took into account contiguous building lots (identifiable with walls or hedges) and non-contiguous building lots less than 50 m away from the contiguous area. Twelve settlements with significant building expansion in the FPA, and as many in the non-flood-prone areas (NFPA), making a total of 24, were selected as case studies. The settlements toponyms were extracted from the national repertoire of inhabited locations.

The third step superimposed the built-up area with the FPA according to three return period scenarios, allowing for an automatic multi-temporal analysis of the changes occurring between 2009 and 2019 in terms of the share of built-up area compared to the total area of each settlement and the share of buildings with roofs made from corrugated iron sheets compared to the total buildings of each settlement.

The lack of precise data for the calibration of the model was the main limitation of this FPA characterization: it ensured a precision that was higher than GIS flood hazard mapping [76] but lower than state-of-the-art hydraulic models [77]. The low accuracy achieved with standard semi-automatic building extraction methods from satellite images was also a limitation, but was bypassed by means of manual digitization.

3. Results

In the past, the Maouri Dallol was crossed by a stream that originated in Northern Niger and flowed into the Niger River 910 km downstream. Today, its regime is intermittent and its endorheic behavior feeds a wetland with an aquifer close to the surface [78]. Heavy rains flood the valley both due to surface flow and water table rise. In Guéchémé, the Dallol crosses one of the most densely populated zones of Niger. Of the 87 settlements identified in 2019, one is large (Angoual Chekaraou), six are medium, 21 are small, and 60 consist of a few houses (Table 3).

Table 3. Settlements in the Maouri Dallol of Guéchémé by size in 2019.

Size Classes ha	Settlements n.
≥50	1
10.00–49.99	6
2.00–9.99	21
<2	59
All	87

3.1. *Climatic and Hydraulic Characterization*

In the 1980s, Guéchémé had a dry period, followed by a wet decade and then an alternation of wet (with a declining trend) and dry years. The accumulated rainfall remained above 400 mm per year, a critical threshold for dry farming [79–81], except in the years 1984, 2016, and 2017 (Figure 3). One of the most significant changes in rainfall over the last forty years has been the increase in value (expressed in mm) of rainfall falling in the 95th percentile, or extreme rainfall. These precipitations increase from 40–50 mm/day (in the nineties) to today's 60–70 mm/day (Figure 4).

The hydrologic analysis revealed the following results: (1) the concentration time of the Dallol Maouri watershed was 210 h (about 9 days), (2) the only PDF that passed the adequacy distribution tests was the log-normal one, and (3) the cumulated rainfall, based on the concentration time of 9 days for the computed return period (RP) ranged from 133 to 287 mm. The flood hydrographs, computed with the hydrological model, agreed with the hydrology of this endorheic area.

The FPA extended between 25 and 67 km^2, depending on the flood scenario considered (Table 3). The FPA covered a considerable surface area due to the high transverse extent of the floodplain at Guéchémé (approximately 4 km), characterized by a considerable number of counter slopes. The NFPA surrounding the RP200 flood limit, considered as a reference to compare the settlement dynamics, extended over 68 km^2 (Figures 5 and 6).

The hydraulic model also enabled the determination of flow velocity and water depth, which could be valuable results for evaluating the effects on the assets concerned. The flow velocity was quite low due to the low longitudinal slope (0.04% = 40 cm/km) of the riverbed, with a maximum value of 0.85 m/s in the RP200 scenario. By contrast, the water depth was quite considerable, reaching a maximum value of 1.78 m in the RP200 scenario (Table 4).

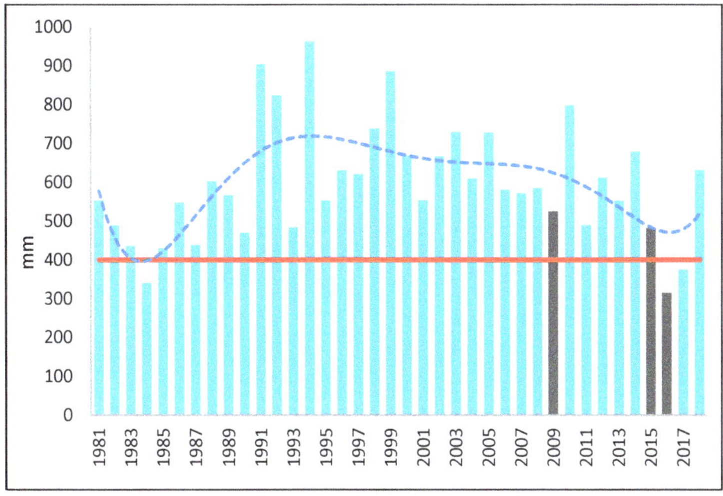

Figure 3. Annual accumulated rainfall (blue) (Guéchémé and Guéchémé Centre de Santé stations) between 1981 and 2018, critical rainfall threshold of 400 mm/year (red), incomplete data (grey).

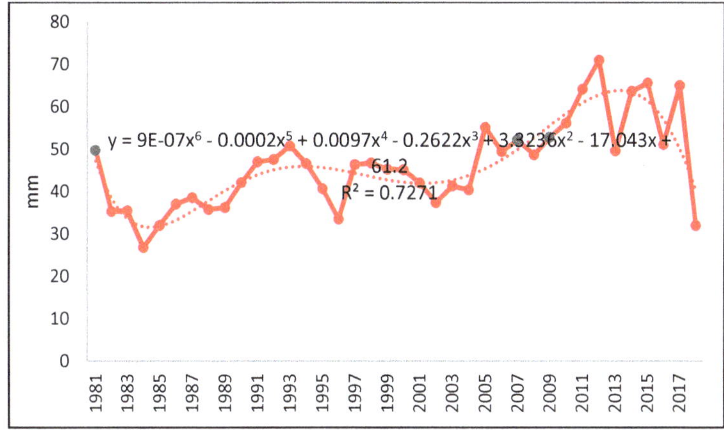

Figure 4. Trend of precipitations in the 95th percentile (red) (Guéchémé and Guéchémé Centre de Santé stations) between 1981 and 2018 and incomplete data (grey).

Table 4. Extension and characteristics of the flood-prone areas in the Maouri Dallol based upon the return periods (RP): velocity (V) and depth (D) referred to the urban area distribution.

RP Years	Rainfall mm	Q_{MAX} l/s·km^2	Flood-Prone Area		V_{MEAN} m/s	V_{MAX} m/s	D_{MEAN} m	D_{MAX} m
			km^2	%				
2	133	2.73	25.2	12.5	0.14	0.43	0.63	1.34
20	217	9.02	48.1	23.9	0.26	0.73	0.84	1.67
200	287	18.05	66.7	33.2	0.33	0.85	0.97	1.78

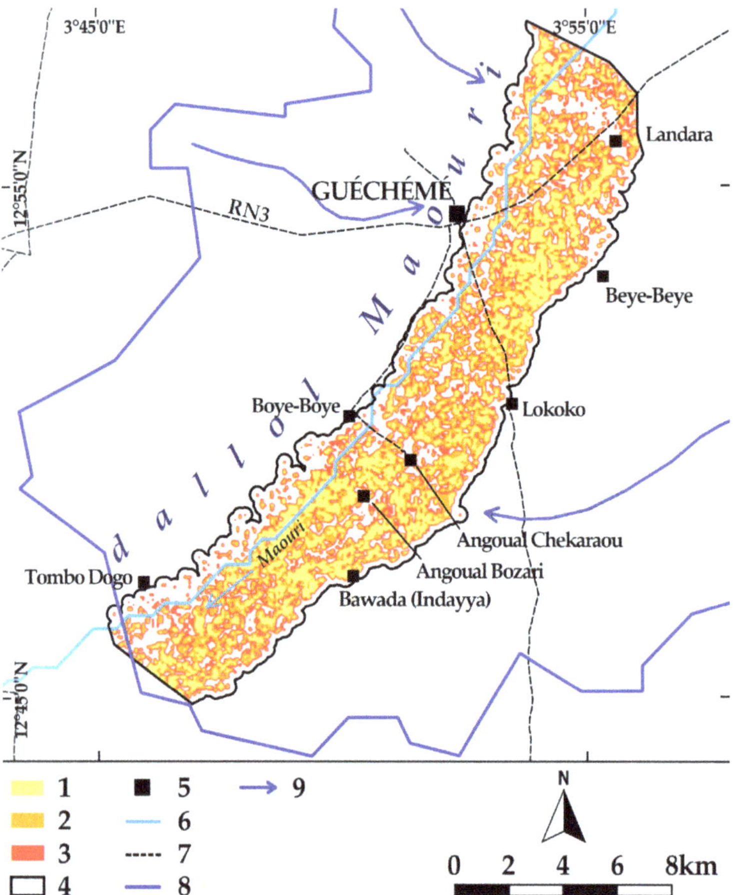

Figure 5. Flood scenarios in the Guéchémé Dallol with probability of occurrence: high RP2 (1), medium RP20 (2), low RP200 (3), study area (4), main settlements (5), Maouri Dallol river channel (6), unpaved tracks (7), hydrographic basin limit (8), and creek flow direction (9).

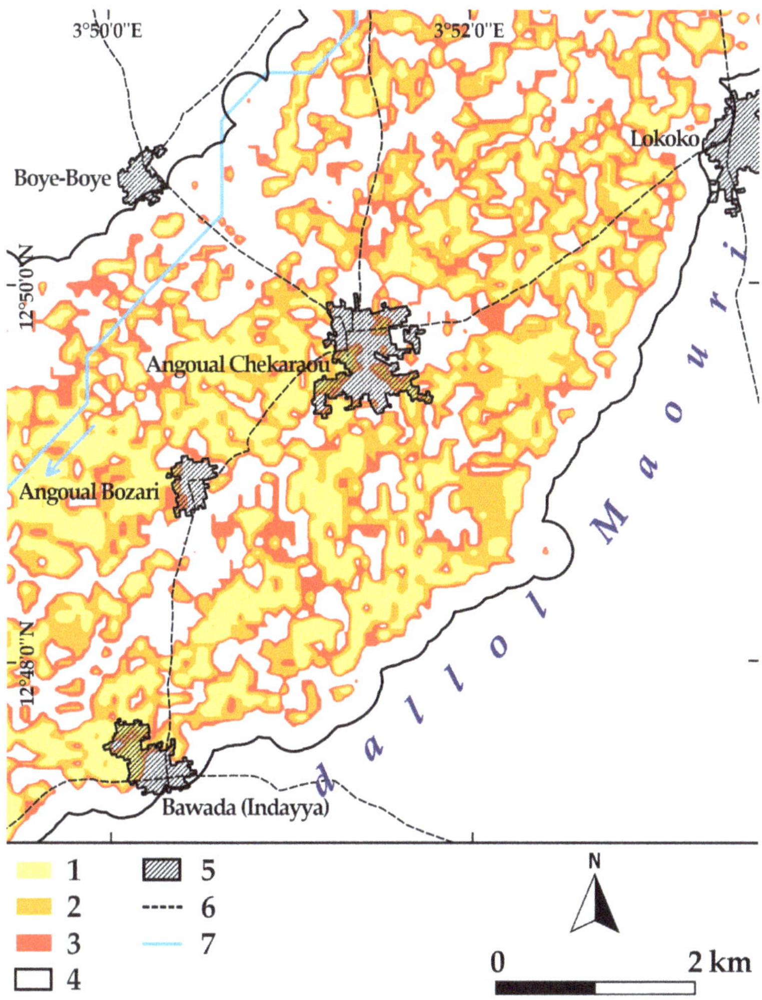

Figure 6. Close-up view of the flood scenarios in the Guéchémé Dallol with probability of occurrence: high RP2 (1), medium RP20 (2), low RP200 (3), study area (4), main settlements (5), unpaved tracks (6), and Maouri Dallol river channel (7).

3.2. Local Perception on Floods

The meetings with the local authorities and the community of Guéchémé identified heavy rainfall, the reduction of vegetation, and the degradation of the soil as drivers of the increased runoff. These, together with the water table rise, caused floods in the Maouri floodplain. These floods—particularly the memorable ones of 1994, 2012, 2015, and 2016—led to the collapse of many buildings. These events are substantially reflected in those listed in the BDINA database relating to the period 2007–2016. In 2012, 2015, and 2016, many collapsed buildings were reported, particularly in 2013 (Table 5). These events also caused extensive damages to cultures, but this aspect was not considered in this research.

Table 5. Catastrophic flood comparison between community remembrance and the Niger flood (BDINA) database between 2012 and 2016.

Year	Catastrophic Floods	
	Community Remembrance	BDINA
2012	X	X
2013		X
2015	X	X
2016	X	X

3.3. Settlement Dynamics

In ten years, the number of settlements in the study area has increased by 7%: 13% in the NFPA and stable in the FPA; in 2009, 45 settlements were in the NFPA and 36 partially or entirely in the RP200 FPA; ten years later, 51 settlements were in the NFPA and 36 settlements in the RP200 FPA (Table 6, Figure 7).

Table 6. Evolution of the number of settlements in the study area between 2009 and 2019.

Settlements in	2009	2019	$\Delta_{2009–2019}$
	no.	no.	%
Non-flood-prone area	45	51	+13
Flood-prone area	36	36	+0
All	81	87	+7

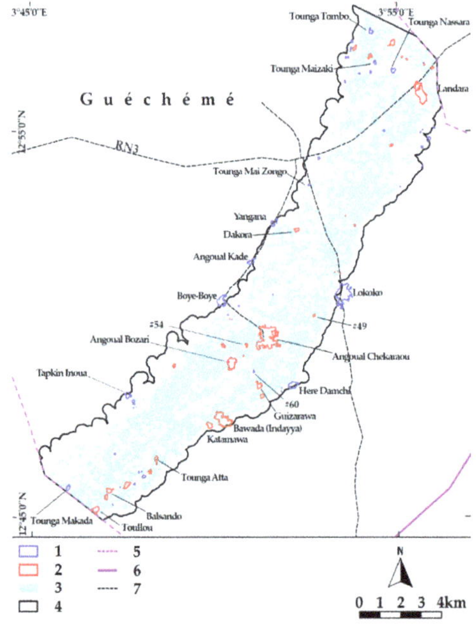

Figure 7. Settlement localization in the Maouri Dallol of Guéchémé: in non-flood-prone areas (1), in flood-prone areas (2), flood scenario RP200 (3), study area (4), municipal border (5), national border (6), unpaved tracks (7), and indication of analyzed settlements.

Of the 87 settlements present in 2019 in the study area, 24—split equally between the FPA and the NFPA—recorded an expansion in surface area greater than 25%. This sample, analyzed in-depth, revealed a 52% increase, particularly in the FPA (+71%) compared to the NFPA (+30%) of the built-up area. The number of buildings increased by 12% (0% in the NFPA and 21% in the FPA) (Table 7). The wide differences of increment between built-up area and buildings led to a decrease in building density (from 40.6 to 29.4 build/ha) as a result of a consolidation process through the installation of buildings with corrugated iron sheet roofs. Between 2009 and 2019, these buildings increased by 3% to 21%, without substantial differences between the FPA and the NFPA. The roofs made from corrugated iron sheets thus increased by 592% against a 10% increase in buildings. In some locations, corrugated iron sheets are now used in almost half of the buildings: an increase of 1400% in the decade considered (Table 8).

Table 7. Building expansion of 24 settlements (12 in the flood-prone area (FPA) and 12 in the non-flood-prone area (NFPA)) in the Guéchémé Dallol between 2009 and 2019.

Settlements	In the Non-Flood-Prone Area			In the Flood-Prone Area		
	2009	2019	$\Delta_{2009\text{–}2019}$	2009	2019	$\Delta_{2009\text{–}2019}$
Area (ha)	76.16	99.36	+30%	88.09	150.58	+71%
Buildings (n.)	3566	3578	+0%	3096	3760	+21%
Density (build./ha)	46.8	36.0	−23%	35.1	24.8	−29%

Table 8. Roof dynamics of 24 settlements (12 in FPA and 12 in NFPA) in the Guéchémé Dallol between 2009 and 2019.

Roof	2009		2019		$\Delta_{2009\text{–}2019}$
	no.	%	no.	%	%
Straw	2369	36	886	12	−63
Mud	4067	61	4867	67	+20
Corrugated iron sheet	226	3	1564	21	+592
Total	6662	100	7317	100	+10

The exposure of buildings to flooding saw an increase of 29% (40% more than the total increase) in the area with a low probability of flooding RP200. At the same time, the exposure of buildings with corrugated iron sheet roofs increased by 695% in the same RP200 area, which was slightly more than the total increase (+609%). The consolidation of buildings was generally higher in the FPA than in the NFPA: the maximum consolidation occurred in the area with a high probability of flooding RP2 (+27%), followed by that with a low probability of flooding RP200 (+12%), and only lastly in the area with a medium probability of flooding RP20 (+9%) (Table 9).

Table 9. Flood exposure of buildings in the Guéchémé Dallol between 2009 and 2019, comparison between building expansion and building consolidation in the FPA.

Flood Exposure		2009	2019	$\Delta_{2009\text{–}2019}$	Consolidation Factor [1]
		no.	no.	%	-
Buildings	Total	3096	3740	+21	-
	RP2	474	540	+14	0.94
	RP20	755	932	+23	1.02
	RP200	1030	1327	+29	1.07
% in the flood-prone area [2]		33%	35%	-	-
Corr. iron sheet	Total	115	815	+609	-
	RP2	11	99	+800	1.27
	RP20	25	194	+676	1.09
	RP200	37	294	+695	1.12
% in the flood-prone area [2]		32%	36%	-	-

[1] RPx $\Delta_{2009\text{–}2019}$/Total $\Delta_{2009\text{–}2019}$; [2] RP200 scenario referenced data.

3.4. Main Dynamics

In the Guéchémé Dallol, in the decade considered, three main settlement dynamics emerged (Table 10, Figures 8–10):

1. Building expansion (a), expressed by a strong increase in the built-up area and/or the number of buildings. The expansion was low density and the buildings used corrugated iron sheets. Angoual Chekaraou was the most significant example, with an increase of 147% in surface area, 108% in buildings, and 1300% in corrugated iron sheet roofs. The expansion occurred above all in the FPA RP20.
2. Building consolidation (b), expressed by the replacement of precarious buildings with semi-permanent buildings (roofs made from corrugated iron sheets). This process reduced building density and sometimes the number of buildings. Lokoko testified to this process, with an increase of 400% in buildings with corrugated iron sheet roofs and a 23% reduction in building density.
3. Expansion and building consolidation (c), expressed by an increase in the built-up area (sometimes with already semi-permanent constructions) and a building replacement of the existing fabric. Toullou, for example, increased its surface area (+35%) and replaced straw roofs (−60%) with those made from corrugated iron sheets (+350%). By virtue of building consolidation, the percentage of buildings in the FPA was unchanged.

Table 10. Settlement dynamics of representative settlements for each main dynamic between 2009 and 2019.

Dynamics	a Expansion		b Consolidation		c Expansion + Consolidation	
Significant Settlements	Angoual Chekaraou		Lokoko		Toullou	
Data	2009	$\Delta_{2009-2019}$	2009	$\Delta_{2009-2019}$	2009	$\Delta_{2009-2019}$
Area (ha)	28.6	+147%	38.9	+22%	5.7	+35%
Buildings (no.)	416	+108%	1599	−6%	285	+9%
Corrugated iron sheets (no.)	28	+1318%	68	+410%	8	+350%
Straw (no.)	75	−91%	771	−85%	125	−60%
% in the flood-prone area	18	+7pp [1]	-	-	54	+0pp

[1] Percentage points.

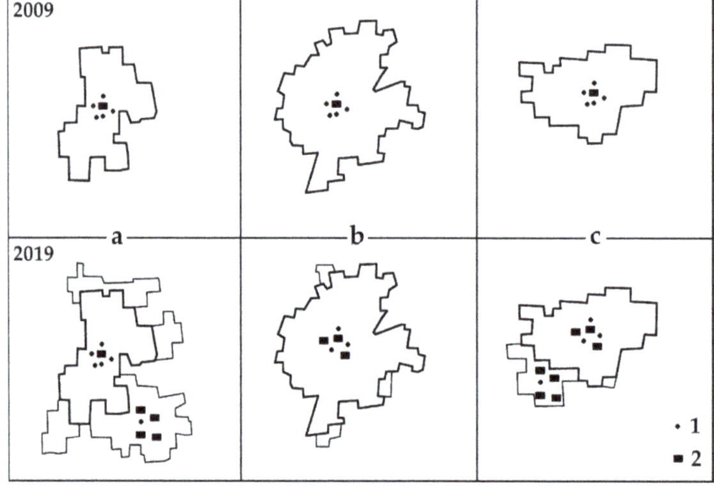

Figure 8. *Cont.*

Angoual Chekaraou	Lokoko	Toullou
Tounga Tombo	Angoual Bozari	Here Damchi
Dakora	Bawada (Indayya)	Guizarawa
Tounga Atta	Boye-Boye	Balsando
Tounga Maizaki	Katamawa	Angoual Kade
	Tapkin Inoua	Tounga Nassara
	Yangana	Tounga Makada
	Tounga Mai Zongo	

Figure 8. Schematization of (**a**) less dense building expansion with dense pre-existing built-up area; (**b**) building consolidation with reduction in density of the pre-existing fabric due to building replacement; (**c**) expansion and building consolidation with reduction of the pre-existing fabric due to building replacement and new less dense expansions; precarious building (1), semi-permanent building (2), list of representative settlements for each main dynamic.

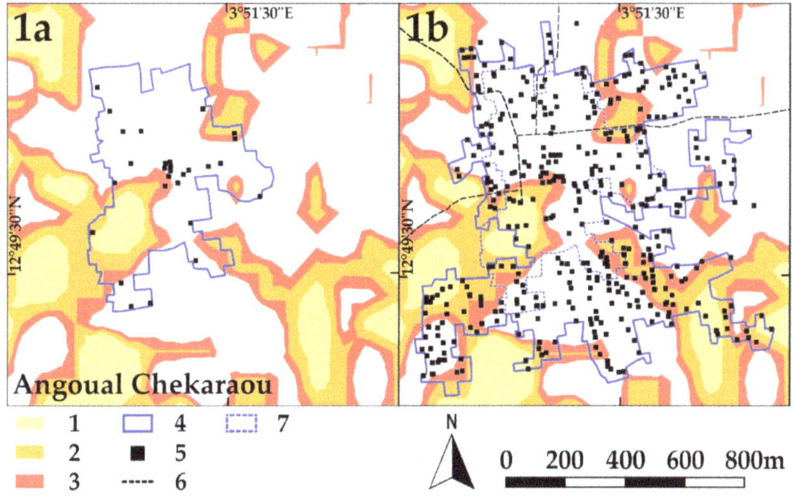

Figure 9. Building expansion in the FPA in Angoual Chekaraou between 2009 (**1a**) and 2019 (**1b**): high (1), medium (2), and low (3) probability of flooding, built-up area (4), buildings with iron sheet roofs (5), unpaved tracks (6), and 2009 built-up area (7).

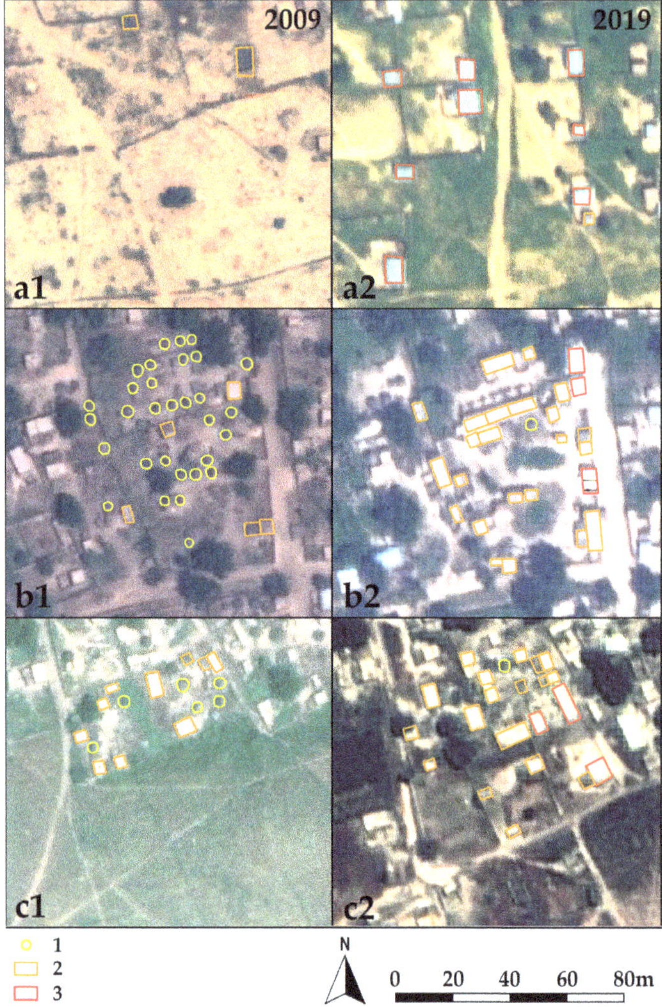

Figure 10. Details of the three main settlement dynamics of straw (1), mud (2), and corrugated iron sheet (3) roofs in Angoual Chekaraou (**a1**,**a2**), Lokoko (**b1**,**b2**), and Toullou (**c1**,**c2**) between 2009 and 2019.

A summary of the settlements data is reported in Table 11 (further information is available in the Supplementary Material).

Table 11. Surface areas, buildings, and roofs in FPA RP200 and in the NFPA.

Settlement	Area 2019		Buildings 2019		Iron Sheet Roofs 2019	
	Total	In the Flood-Prone Area	Total	In the Flood-Prone Area	Total	In the Flood-Prone Area
	ha	%	no.	%	no.	%
Dakora	2.69	49	150	49	10	50
Angoual Chekaraou	70.71	31	866	25	397	31
Illela Tounga Alou [1]	-	-	-	-	-	-
Angoual Bozari	16.92	21	517	18	58	21
Bawada (Indayya)	32.95	45	1047	38	205	40
Katamawa	5.41	38	196	41	31	23
Guizarawa	5.07	28	227	19	19	47
Toullou	7.69	59	310	54	36	78
Balsando	4.44	20	226	21	40	25
Tounga Atta	2.5	100	145	100	7	100
#49 [2]	0.85	100	11	100	4	100
#54	1.35	94	45	91	8	75
Lokoko	47.51	-	1510	-	347	-
Tounga Tombo	3.22	-	129	-	65	-
Tounga Nassara	3.44	-	146	-	15	-
Tounga Maizaki	1.95	-	77	-	10	-
Tounga Mai Zongo	0.41	-	19	-	7	-
Yangana	3.44	-	158	-	27	-
Angoual Kade	4.86	-	212	-	47	-
Boye-Boye	15.19	-	547	-	107	-
Tapkin Inoua	5.25	-	273	-	30	-
Tounga Makada	3.01	-	111	-	13	-
Here Damchi	10.53	-	385	-	80	-
#60	0.55	-	11	-	1	-

[1] Disappeared between 2009 and 2019. [2] Settlements marked with a '#' are not reported in the National Repertoire of Localities (RENALOC).

4. Discussion

In three-quarters of the countries South of the Sahara, the rural population today is still in the majority and strongly increasing [9,18,82]. In these countries, the primary sector remains strategic for development and for reducing poverty [20]. This suggests the refocusing of attention from the urban sector [15,26,36] to the rural sector. Sustainable rural development involves protecting settlements from increasingly frequent flooding. Ascertaining if and how far rural settlements are occupying the FPA [3,4,7] should be a preliminary step for identifying appropriate protection and prevention measures: an aspect still investigated little by peer-reviewed literature [83].

The objective of this study was to verify the recent changes occurring in the settlements and housing stock in one of these areas: the Maouri Dallol in the municipality of Guéchémé, Niger. The quantitative analysis of the settlement dynamics between 2009 and 2019 made it possible to achieve this aim thanks to the availability of multi-temporal satellite images. The municipality's valley receives, on average, 595 mm of rainfall per year, slightly more than the regional average of the Dosso region [43,44]: a favorable condition for rain crops in a semi-arid zone, such as the Niger Sahel [79]. However, in the last twenty years, extreme rainfall (95th percentile) has increased by about 20 mm of intensity [6]. The effect of these rains on degraded soil increases the runoff and leads to the greater flooding of the Maouri Dallol. Over the last ten years in the Dallol, settlements have increased in number (+7%) and size.

The first question to be answered in this study was if, in a rural context in demographic growth, the exposure of settlements to flooding had increased: an aspect recognized by literature [2,42,64,84] but rarely quantified [34,35]. It has been ascertained that in the last decade, in the Guéchémé Maouri Dallol, the expansion of the 24 observed settlements occurred for 71% in the FPA and 30% in the NPFA. Therefore, the exposure strongly increased due to human activity.

The second question to be answered concerned the flood resistance of buildings constructed in the last decade in terms of building consolidation. In the Maouri Dallol between 2009 and 2019, there was a marked consolidation of buildings expressed by the transition from mud roofs to those made from corrugated iron sheets (from 3% to 21% in just 10 years), without distinction between the FPA and the NFPA. This process is decreasing the building density of all the analyzed settlements. The consolidation in the FPA sees an average of +720%, with peaks of up to +3500% in the area with a medium probability of flooding (RP20) in Angoual Chekaraou. This consolidation protects homes from the impact of heavy rainfall on the roof, but not from the runoff and the water table rise. Even though the low flood velocities observed do not represent a decisive factor in terms of the potential damage caused by floods, the considerable flood depths (up to nearly 2 m) pose a real threat to buildings: if water enters homes with mud walls, they are guaranteed to collapse. This demonstrates that the collapse of buildings occurs only because they are built from precarious materials and that concrete materials should be used for the full height of buildings to ensure flood protection.

Many similar studies on floodplain settlement dynamics agree on the process of settlement expansion in flood-prone areas, both in African contexts [85] and in other regions of the world [12], but a peculiar characteristic of the Maouri Dallol is a decreased building density of the settlements compared to other contexts [86]. The consolidation of buildings and its implications on flood exposure is still poorly explored in the Sahel. However, in a broader context, the water effects on buildings after a flood event is fairly recognized [39,41].

One of the interpretative hypotheses of the two observed phenomena is that in the context of a growing population in a rural area and the consequent increased need to put more and more land into cultivation, the new nuclei need to settle close to the fields (and water) during the agricultural season, despite the occurrence of flooding [8]. The consolidation of residences with corrugated iron sheet roofs is undoubtedly an improvement to which everyone aspires as soon as their economic conditions allow it. However, in the event of flooding, the invested capital is also at risk of being lost. The meetings with the local authorities and with the community of Guéchémé highlighted the clear local perception of rainfall changes and the degradation of the soil, but not the increase in the extent of the FPA due to the ever more frequent extreme events. These areas, which are today estimated to be low in flooding probability, may become likely to be flooded if the processes of erosion and soil degradation continue in the future. Precisely for this reason, knowing the FPA according to different rainfall intensities and the settlement dynamics in place within them becomes important for deciding upon the measures to be adopted in the municipal development plan, particularly if extreme events grow in intensity and the population exposed to them increases [23]. The seasonality and high variability of surface flows in the Maouri Dallol may be one of the factors that determines a lack of adaptive responses by the local population compared to watercourses with a more regular regime, as observed in the Sirba River watershed in Niger [65].

The urgency to act is also dictated by the speed (just a decade for the discussed case study) and extent to which the expansion and consolidation of inhabited areas in the FPA have occurred. The municipal development plan could include flood risk prevention (information on at-risk areas, resettlement of exposed inhabitants, or protection of inhabited areas). The three main dynamics among the settlements of the Guéchémé Dallol facilitate the identification of specific measures to reduce the flooding risk. Delocalization or retrofitting of the individual buildings with entry barriers and raised basements is recommended for settlements that spread in low density in the FPA, while the construction of embankments and dikes to protect against runoff water is more suitable for settlements that are still dense [87].

Among the results of this study, the role of satellite remote sensing, fundamental for multi-temporal analysis (in the absence of other reliable reference data: e.g., cadastral data), is highlighted, thanks to the now well-established availability of historical archives (starting from around the year 2000) of high-resolution satellite images. On the other hand, low thematic accuracy has been obtained in the application of semi-automatic standard classification algorithms for extracting building footprints from high-resolution satellite images. These accuracies are undoubtedly affected by the geographical context, characterized by the presence of numerous buildings with mud-covered roofs (or temporarily used for forage conservation) and unpaved roads, both having similar radiometric responses and, therefore, difficult to isolate. Conversely, more promising results are observed for corrugated iron sheet roofs, which can easily be isolated. These results led to an approach based on photointerpretation and manual digitization in a GIS environment. In order to increase the level of the automation of this phase of the methodology, the use of image segmentation techniques based on deep learning algorithms is considered extremely promising [88], also by exploiting the better spectral resolution of recent satellite sensors.

The main limit of the analysis is the accuracy of the perimeter of the FPA. This is affected by the lack of detailed information on both soil morphology and hydrology. In particular: (1) the low planimetric resolution and vertical accuracy of the DEM; (2) the lack of surface flow measurements; and (3) the lack of surface flow measurements makes the determination of the FPA inaccurate. For point (1), the use of high-resolution satellite stereoscopic pairs, which facilitate the extraction of DEM with planimetric resolutions and vertical accuracy in the order of a few meters, is considered promising. For points (2) and (3), it is hoped that this initial characterization of the FPA of the Maouri Dallol floodplain will increase the authorities' awareness of the importance of investing time and resources both in field observations and in flood protection measures in this area.

5. Conclusions

In recent years, the rural Sahel has been struck by catastrophic floods, which have slowed down the already stunted development process. Until now, the literature has explained these catastrophes by studying climate changes and variations as well as alterations of land cover. Knowledge of floodplain settlement dynamics has not advanced in parallel, especially in rural contexts. This study focuses on a 135 km^2 rural area in Niger, with a high population growth and flood risk, during the decade 2009–2019. The settlement dynamics were ascertained through the photointerpretation of multi-temporal high-resolution satellite images and superimposed on the flood boundaries according to three return period scenarios (2, 20, and 200 years).

A significant increase in the intensity of heavy rainfall in the 95th percentile over the last 20 years (+20 mm/day) and four catastrophic floods during the last decade was observed. However, the settlements increased by extension (+52%) and 70% of this expansion occurred in areas with a medium and high probability of flooding. The number of buildings also increased by 21% in the same areas. Many buildings (+592%) were consolidated by installing corrugated iron sheet roofs in place of traditional mud roofs, which could be a proxy indicator of improved economic prosperity. This consolidation process was more intense in areas with a high probability of flooding (up to +800%). However, this improvement is not sufficient to cope with rain-induced floods, as the runoff water can make the walls of the buildings, which are still built from mud, collapse. Therefore, the investment to improve the roof increases the exposed capital but reduces the exposure to damage caused by floods to a limited extent in a context where the hazard has increased.

The significance of these results should draw the attention of the local authorities, who need to identify and implement prevention and preparation measures in the local development plans. The better knowledge and awareness of flood-prone areas and of how settlements develop around them will help the authorities to protect people and assets from natural hazards and facilitate economic development.

Historical satellite images enabled the extraction of detailed building information at the beginning and end of the study period, but the traditional methods of semi-automatic extraction of individual buildings from high resolution satellite images did not produce adequate results in this geographical context. On the other hand, the hydraulic model developed in this study was able to delimit the floodplain settlement dynamics with sufficient accuracy for local planning with respect to the methods currently applied in the Sahel, such as the creation of a buffer around the rivers and the overlay of images of historical floods over low resolution satellite images on the ground. This type of analysis can be replicated in other rural contexts with high population density, allowing local administrations to identify and localize flood prevention and preparation actions.

Finally, the relevance of a multi-disciplinary research group, which allowed for the synergistic integration of spatial planning, hydraulics, and satellite remote sensing skills to extract value-added information in support of decision-making processes, should be underlined.

Supplementary Materials: The following are available online at http://www.mdpi.com/2071-1050/12/14/5632/s1, Table S1: Twentyfour settlement dynamics in the Maouri dallol at Guéchémé, Niger (2009–2019).

Author Contributions: Conceptualization, A.G.; methodology, A.G., G.M., and F.G.T.; software, G.M., F.G.T.; validation, A.G.; writing—original draft preparation, A.G.; writing—review and editing, A.G., G.M., and F.G.T.; All authors have read and agreed to the published version of the manuscript.

Funding: This research was funded by DIST-Politecnico and University of Turin, Italy within the ANADIA 2.0 project.

Acknowledgments: The authors would like to thank the Italian Agency for Development Cooperation for cofounding the ANADIA 2.0 project that allowed for the development of this study, Katiellou Gaptia Lawan (Directorate National for Meteorology of Niger-DMN), Aliou Tankari (Ministry of Agriculture and Livestock of Niger-MAE), and Maurizio Tiepolo (DIST-Politecnico and University of Turin) for sharing the results of meetings with the Guéchémé municipality with the authors, the Guéchémé community, and for providing new satellite imagery of the area, and Maurizio Tiepolo for his help during the preparation of this article.

Conflicts of Interest: The authors declare no conflict of interest. The funders had no role in the design of the study; in the collection, analyses, or interpretation of data; in the writing of the manuscript, or in the decision to publish the results.

References

1. Amogu, O.; Descroix, L.; Yéro, K.S.; Le Breton, E.; Mamadou, I.; Ali, A.; Vischel, T.; Bader, J.C.; Moussa, I.B.; Gautier, E.; et al. Increasing River Flows in the Sahel? *Water* **2010**, *2*, 170–199. [CrossRef]
2. Tschakert, P.; Sagoe, R.; Ofori-Darko, G.; Codjoe, S.N.A. Floods in the Sahel: An analysis of anomalies, memory, and anticipatory learning. *Climatic Change* **2010**, *103*, 471–502. [CrossRef]
3. The Emergency Events Database, EM-DAT. Available online: http://www.emdat.be/emdat_db/ (accessed on 17 April 2020).
4. Fiorillo, E.; Crisci, A.; Issa, H.; Maracchi, G.; Morabito, M.; Tarchiani, V. Recent Changes of Floods and Related Impacts in Niger Based on the ANADIA Niger Flood Database. *Climate* **2018**, *6*, 59. [CrossRef]
5. Tamagnone, P.; Massazza, G.; Pezzoli, A.; Rosso, M. Hydrology of the Sirba River: Updating and Analysis of Discharge Time Series. *Water* **2019**, *11*, 156. [CrossRef]
6. Taylor, C.; Belušić, D.; Guichard, F.; Parker, D.J.; Vischel, T.; Bock, O.; Harris, P.P.; Janicot, S.; Klein, C.; Panthou, G. Frequency of extreme Sahelian storms tripled since 1982 in satellite observations. *Nature* **2017**, *544*, 475–478. [CrossRef] [PubMed]
7. Aich, V.; Liersch, S.; Vetter, T.; Andersson, J.C.M.; Müller, E.N.; Hattermann, F.F. Climate or Land Use?—Attribution of Changes in River Flooding in the Sahel Zone. *Water* **2015**, *7*, 2796–2820. [CrossRef]
8. Wilcox, C. Evaluating Hydrological Changes in Semi-Arid West Africa: Detection of Past Trends in Extremes and Framework for Modeling the Future. Ph.D. Thesis, Docteur de la Communauté Université Grenoble Alpes, Grenoble, France, 1 July 2019.
9. Onywere, S.M.; Getenga, Z.M.; Mwakalila, S.S.; Twesigye, C.K. Assessing the Challenge of Settlement in Budalangi and Yala Swamp Area in Western Kenya Using Landsat Satellite Imagery. *Open Environ. Eng. J.* **2011**, *4*, 97–104. [CrossRef]
10. Descroix, L.; Guichard, F.; Manuela, G.; Lambert, L.; Panthou, G.; Mahe, G.; Gal, L.; Dardel, C.; Quantin, G.; Kergoat, L.; et al. Evolution of Surface Hydrology in the Sahelo-Sudanian Strip: An Updated Review. *Water* **2018**, *10*, 748. [CrossRef]
11. Praskievicz, S.; Chang, H. A review of hydrological modelling of basin-scale climate change and urban development impacts. *Prog. Phys. Geogr. Earth and Environ.* **2009**, *33*, 650–671. [CrossRef]
12. Früh-Müller, A.; Wegmann, M.; Koellner, T. Erratum to: Flood exposure and settlement expansion since pre-industrial times in 1850 until 2011 in north Bavaria, Germany. *Reg. Environ. Chang.* **2014**, *15*, 183–193. [CrossRef]
13. Huong, H.T.L.; Pathirana, A. Urbanization and climate change impacts on future urban flooding in Can Tho City, Vietnam. *Hydrol. Earth Syst. Sci. Discussions* **2011**, *8*, 10781–10824. [CrossRef]

14. Mbow, C.; Diop, A.; Diaw, A.T.; Niang, C.I. Urban sprawl development and flooding at Yeumbeul suburb (Dakar-Senegal). *Afr. J. Environ. Sci. Technol.* **2008**, *2*, 75–88.
15. IPCC. *Global Warming of 1.5 °C. An IPCC Special Report on the Impacts of Global Warming of 1.5 °C above Pre-Industrial Levels and Related Global Greenhouse Gas Emission Pathways, in the Context of Strengthening the Global Response to the Threat of Climate Change, Sustainable Development, and Efforts to Eradicate Poverty*; Masson-Delmotte, V., Zhai, P., Pörtner, H.-O., Roberts, D., Skea, J., Shukla, P.R., Pirani, A., Moufouma-Okia, W., Péan, C., Pidcock, R., et al., Eds.; IPCC: Geneva, Switzerland, 2018. (in press)
16. Adelekan, I.O. Vulnerability of poor urban coastal communities to flooding in Lagos, Nigeria. *Environ. Urban.* **2010**, *22*, 433–450. [CrossRef]
17. Tamagnone, P.; Comino, E.; Rosso, M. Rainwater harvesting techniques as an adaptation strategy for flood mitigation. *J. Hydrol.* **2020**, *586*, 124880. [CrossRef]
18. UNDESA-United Nations Department of Economic and Social Affairs. *World Urbanization Prospects. The 2018 Revision*; United Nations: New York, NY, USA, 2018; p. 128.
19. Jidauna, G.G.; Dabi, D.D.; Dia, R.Z. The effect of climate change on agricultural activities in selected settlements in the Sudano-Sahelian Region of Nigeria. *Archi. Appl. Sci. Res.* **2011**, *3*, 154–165.
20. Tiepolo, M.; Braccio, S. Mainstreaming disaster risk reduction into local development plans for rural tropical Africa: A systematic assessment. *Sustainability* **2020**, *12*, 2196. [CrossRef]
21. Thompson, H.E.; Berrang-Ford, L.; Ford, J.D. Climate change and food security in Sub-Saharan Africa: A systematic literature review. *Sustainability* **2010**, *2*, 2719–2733. [CrossRef]
22. Connolly-Boutin, L.; Smit, B. Climate change, food security, and livelihoods in sub-Saharan Africa. *Reg. Environ. Chang.* **2016**, *16*, 385–399. [CrossRef]
23. Tiepolo, M.; Rosso, M.; Massazza, G.; Belcore, E.; Issa, S.; Braccio, S. Flood Assessment for Risk-Informed Planning along the Sirba River, Niger. *Sustainability* **2019**, *11*, 4003. [CrossRef]
24. Taubenböck, H.; Wurm, M.; Netzband, M.; Zwenzner, H.; Roth, A.; Rahman, A.; Dech, S. Flood risks in urbanized areas—Multi-sensoral approaches using remotely sensed data for risk assessment. *Nat. Hazards Earth Syst. Sci.* **2011**, *11*, 431–444. [CrossRef]
25. Kamusoko, C.; Gamba, J. Simulating Urban Growth Using a Random Forest-Cellular Automata (RF-CA) Model. *ISPRS Int. J. Geo-Inf.* **2015**, *4*, 447–470. [CrossRef]
26. Sliuzas, R.; Flacke, J.; Jetten, V. Modelling urbanization and flooding in Kampala, Uganda. In Proceedings of the 14th N-AERUS/GISDECO Conference, Enschede, The Netherlands, 12–14 September 2013; pp. 1–16.
27. Popa, M.C.; Peptenatu, D.; Drăghici, C.C.; Diaconu, D.C. Flood Hazard Mapping Using the Flood and Flash-Flood Potential Index in the Buzău River Catchment, Romania. *Water* **2019**, *11*, 2116. [CrossRef]
28. Annis, A.; Nardi, F.; Petroselli, A.; Apollonio, C.; Arcangeletti, E.; Tauro, F.; Belli, C.; Bianconi, R.; Grimaldi, S. UAV-DEMs for Small-Scale Flood Hazard Mapping. *Water* **2020**, *12*, 1717. [CrossRef]
29. Peruzzi, C.; Castaldi, M.; Francalanci, S.; Solari, L. Three dimensional hydraulic characterisation of the Arno River in Florence. *J. Flood Risk Manag.* **2019**, *12*, 1–13. [CrossRef]
30. Mohammed, T.A.; Al-Hassoun, S.; Ghazali, A.H. Prediction of Flood Levels Along a Stretch of the Langat River with Insufficient Hydrological Data. *Pertanika J. Sci. Technol.* **2011**, *19*, 237–248.
31. Tarchiani, V.; Massazza, G.; Rosso, M.; Tiepolo, M.; Pezzoli, A.; Housseini Ibrahim, M.; Katiellou, G.L.; Tamagnone, P.; De Filippis, T.; Rocchi, L.; et al. Community and Impact Based Early Warning System for Flood Risk Preparedness: The Experience of the Sirba River in Niger. *Sustainability* **2020**, *12*, 1802. [CrossRef]
32. Comino, E.; Dominici, L.; Ambrogio, F.; Rosso, M. Mini-hydro power plant for the improvement of urban water-energy nexus toward sustainability—A case study. *J. Clean. Prod.* **2020**, *249*, 119416. [CrossRef]
33. Tamagnone, P.; Comino, E.; Rosso, M. Landscape Metrics Integrated in Hydraulic Modeling for River Restoration Planning. *Environ. Model. Assess.* **2020**, *25*, 173–185. [CrossRef]
34. Güneralp, B.; Güneralp, I.; Liu, Y. Changing global patterns of urban exposure to flood and drought hazards. *Glob. Environ. Chang.* **2015**, *31*, 217–225. [CrossRef]
35. Röthlisberger, V.; Zischg, A.P.; Keiler, M. Spatiotemporal aspects of flood exposure in Switzerland. In Proceedings of the E3S Web of Conferences, Lyon, France, 17–21 October 2016; pp. 1–5.
36. IPCC. *Climate Change 2014: Synthesis Report. Contribution of Working Groups I, II and III to the Fifth Assessment Report of the Intergovernmental Panel on Climate Change*; Pachauri, R.K., Meyer, L.A., Eds.; IPCC: Geneva, Switzerland, 2014; p. 151.

37. Amoako, C.; Inkoom, D.K.B. The production of flood vulnerability in Accra, Ghana: Re-thinking flooding and informal urbanisation. *Urban Stud.* **2018**, *55*, 2903–2922. [CrossRef]
38. Castro Correa, C.P.; Ibarra, I.; Lukas, M.; Véliz, J.O.; Sarmiento, J.P. Disaster Risk Construction in the Progressive Consolidation of Informal Settlements: Iquique and Puerto Montt (Chile) Case Studies. *Int. J. Disaster Risk Reduct.* **2015**, *13*, 109–127. [CrossRef]
39. Balasbaneh, A.T.; Abidin, A.R.Z.; Ramli, M.Z.; Khaleghi, S.J.; Marsono, A.K. Vulnerability assessment of building material against river flood water: Case study in Malaysia. In *IOP Conference Series: Earth and Environmental Science, Proceeding of the 2nd International Conference on Civil & Environmental Engineering, Langkawi, Malaysia, 20–21 November 2019*; IOP Publishing Ltd.: Bristol, UK, 2019; pp. 1–8.
40. Dangol, N.; Day, J. Flood adaptation by informal settlers in kathmandu and their fear of eviction. *Int. J. Saf. Secur. Eng.* **2017**, *7*, 147–156. [CrossRef]
41. López, Ó.L.; Torres, I. Ventilation system for drying out buildings after a flood: Influence of the building material. *Dry. Technol.* **2017**, *35*, 867–876. [CrossRef]
42. Eakin, H.; Lerner, A.M.; Murtinho, F. Adaptive capacity in evolving peri-urban spaces: Responses to flood risk in the Upper Lerma River Valley, Mexico. *Glob. Environ. Chang.* **2010**, *20*, 14–22. [CrossRef]
43. Direction de la Météorologie Nationale du Niger (DMN). Guéchémé Pluviométrie 1981–2018. 2019.
44. Climate Data, Dosso Climate. Available online: https://en.climate-data.org/africa/niger/dosso/dosso-26232/#temperature-graph (accessed on 8 December 2019).
45. Institut National de la Statistique. *Niger: Répertoire National des Localités (RENALOC)*, 1st ed.; Ministère des Finances: Niamey, Niger, 2014; p. 718.
46. WMO Guidelines on the Definition and Monitoring of Extreme Weather and Climate Events (Final Draft). 2018. Available online: http://www.wmo.int/pages/prog/wcp/ccl/references.php (accessed on 5 May 2020).
47. BDINA. Base de Données sur les Inondations Niger ANADIA. Available online: https://www.inondations-niger.org/content.php?page=32 (accessed on 5 May 2020).
48. Lehner, B.; Verdin, K.; Jarvis, A. New global hydrography derived from spaceborne elevation data. *Eos Trans. Am. Geophys. Union J.* **2008**, *89*, 93–94. [CrossRef]
49. USGS; WWF. Hydrological Data and Maps based on Shuttle Elevation Derivates at Multiple Scales (HydroSHEDS). Available online: https://hydrosheds.cr.usgs.gov/ (accessed on 22 March 2020).
50. NASA; DRL; ASI. Shuttle Radar Topography Mission (SRTM). Available online: https://www2.jpl.nasa.gov/srtm/ (accessed on 22 March 2020).
51. Woodward, D.E.; Hoeft, C.C.; Humpal, A.; Cerrelli, G. Chapter 15—Time of Concentration. In *National Engineering Handbook Part 630 Hydrology*; USDA Natural Resources Conservation Service, Ed.; NRCS: Fort Worth, TX, USA, 2010; pp. 1–29.
52. Salimi, E.T.; Nohegar, A.; Malekian, A.; Hoseini, M.; Holisaz, A. Estimating time of concentration in large watersheds. *Paddy Water Environ.* **2017**, *15*, 123–132. [CrossRef]
53. Ravazzani, G.; Boscarello, L.; Cislaghi, A.; Mancini, M. Review of Time-of-Concentration Equations and a New Proposal in Italy. *J. Hydrol. Eng.* **2019**, *24*, 04019039. [CrossRef]
54. Shiau, J. Return period of bivariate distributed extreme hydrological events. *Stoch. Environ. Res. Risk Assess.* **2003**, *17*, 42–57. [CrossRef]
55. Subyani, A.M. Hydrologic behavior and flood probability for selected arid basins in Makkah area, western Saudi Arabia. *Arab. J. Geosci.* **2011**, *4*, 817–824. [CrossRef]
56. Thas, O.; Ottoy, J.P. Some generalizations of the Anderson–Darling statistic. *Stat. Probab. Lett.* **2003**, *64*, 255–261. [CrossRef]
57. Ožanić, N.; Rubinić, J.; Karleuša, B.; Holjević, D. Problems of High Water Appearances in Urban Areas. In Proceedings of the IX International Symposium on Water Management and Hydraulic Engineering, Ottenstein, Austria, 4–7 September 2005; pp. 395–402.
58. US Army Corps of Engineers—Hydrologic Engineering Center, HEC-HMS. Available online: https://www.hec.usace.army.mil/software/hec-hms/ (accessed on 18 June 2020).
59. Boughton, W.C. A review of the USDA SCS curve number method. *Soil Res.* **1989**, *27*, 511–523. [CrossRef]
60. Woodward, D.E.; Hawkins, R.H.; Jiang, R.; Hjelmfelt, A.T.; Van Mullem, J.A.; Quan, Q.D. Runoff Curve Number Method: Examination of the Initial Abstraction Ratio. In Proceedings of the World Water and Environmental Resources Congress 2003 and Related Symposia, Philadelphia, PA, USA, 23–26 June 2003; pp. 1–16.

61. Copernicus Global Land Service: Africa Land Cover. Available online: https://africa.lcviewer.vito.be/2015 (accessed on 2 July 2020).
62. Soulis, K.X.; Valiantzas, J.D. SCS-CN parameter determination using rainfall-runoff data in heterogeneous watersheds—The two-CN system approach. *Hydrol. Earth Syst. Sci.* **2012**, *16*, 1001–1015. [CrossRef]
63. Ling, L.; Yusop, Z.; Yap, W.-S.; Tan, W.L.; Chow, M.F.; Ling, J.L. A Calibrated, Watershed-Specific *SCS-CN* Method: Application to Wangjiaqiao Watershed in the Three Gorges Area, China. *Water* **2020**, *12*, 60. [CrossRef]
64. Mamadou, I.; Gautier, E.; Descroix, L.; Noma, I.; Moussa, I.B.; Maiga, O.F.; Genthon, P.; Amogu, O.; Malam Abdou, M.; Vandervaere, J. Exorheism growth as an explanation of increasing flooding in the Sahel. *Catena* **2015**, *131*, 130–139. [CrossRef]
65. Massazza, G.; Tamagnone, P.; Wilcox, C.; Belcore, E.; Pezzoli, A.; Vischel, T.; Panthou, G.; Housseini Ibrahim, M.; Tiepolo, M.; Tarchiani, V.; et al. Flood Hazard Scenarios of the Sirba River (Niger): Evaluation of the Hazard Thresholds and Flooding Areas. *Water* **2019**, *11*, 1018. [CrossRef]
66. US Army Corps of Engineers—Hydrologic Engineering Center, HEC-RAS. Available online: https://www.hec.usace.army.mil/software/hec-ras/ (accessed on 22 March 2020).
67. Ardiçlioglu, M.; Kuriqi, A. Calibration of channel roughness in intermittent rivers using HEC-RAS model: Case of Sarimsakli creek, Turkey. *SN Appl. Sci.* **2019**, *1*, 1080. [CrossRef]
68. GeoEye, GeoEye-1 Scene 010474140010, Corrected, Bundle, Longmont, Colorado, CO, USA: GeoEye, 6–14 October 2009. Available online: https://brocku.ca/library/wp-content/uploads/sites/51/MDG-How-to-Reference.pdf. (accessed on 10 July 2020).
69. WorldView, WorldView-2 Scene 1030010097333C00, Corrected, Bundle, Munich, Germany: EUSI GmbH, 1–6 September 2019. Available online: https://brocku.ca/library/wp-content/uploads/sites/51/MDG-How-to-Reference.pdf. (accessed on 10 July 2020).
70. ESRI. Pansharpening Function. Available online: https://pro.arcgis.com/en/pro-app/help/data/imagery/pansharpening-function.htm (accessed on 3 June 2020).
71. SDSTATE. Radiometric Calibration. Available online: https://www.sdstate.edu/jerome-j-lohr-engineering/radiometric-calibration (accessed on 10 February 2020).
72. ESRI. *Dictionary of GIS Terminology*; Karman, M., Amdahl, G., Eds.; ESRI Press: Redlands, CA, USA, 2000; p. 119.
73. Yan, G.; Mas, J.-F.; Maathuis, B.H.P.; Xiangmin, Z.; Van Dijk, P.M. Comparison of pixel-based and object-oriented image classification approaches—A case study in a coal fire area, Wuda, Inner Mongolia, China. *Int. J. Remote Sens.* **2006**, *27*, 4039–4055. [CrossRef]
74. Ying, T. Chapter 11—Applications. In *Gpu-Based Parallel Implementation of Swarm Intelligence Algorithms*, 1st ed.; Ying, T., Ed.; Morgan Kaufmann: Burlington, MA, USA, 2016; pp. 167–177.
75. Harris Geospatial Solutions, Spectral Indices. Available online: https://www.harrisgeospatial.com/docs/SpectralIndices.html (accessed on 26 November 2019).
76. Fernández, D.S.; Lutz, M.A. Urban flood hazard zoning in Tucumán Province, Argentina, using GIS and multicriteria decision analysis. *Eng. Geol.* **2010**, *111*, 90–98. [CrossRef]
77. Casas, A.; Benito, G.; Thorndycraft, V.R.; Rico, M. The topographic data source of digital terrain models as a key element in the accuracy of hydraulic flood modelling. *Earth Surf. Process. Landf.* **2006**, *31*, 444–456. [CrossRef]
78. Zouari, K.; Moulla, A.S.; Smati, A.; Adjomayi, P.A.; Boukari, M.; Thiam, A.; Kone, S.; Rabe, S.; Bobadji, I.; Maduabuchi, C.M.; et al. *Integrated and Sustainable Management of Shared Aquifer Systems and Basins of the Sahel Region, Iullemeden Aquifer System*; International Atomic Energy Agency (IAEA): Vienna, Austria, 2017; p. 114.
79. FAO. Mission to the Sahel. Available online: http://www.fao.org/3/f7795e/f7795e05.htm (accessed on 24 June 2020).
80. Tiepolo, M.; Tarchiani, V. *Risque et Adaptation Climatique Dans la Région Tillabéri, Niger*; L'Harmattan: Paris, France, 2016; p. 282.
81. Tiepolo, M.; Bacci, M.; Braccio, S. Multihazard Risk Assessment for Planning with Climate in the Dosso Region, Niger. *Climate* **2018**, *6*, 67. [CrossRef]
82. Tiepolo, M.; Braccio, S.; Tarchiani, V. *Lo Sviluppo delle aree Rurali Remote: Petrolio, Uranio e Governance Locale in Niger*; FrancoAngeli: Milan, Italy, 2009; p. 224.

83. Okoro, B.C.; Ibe, O.P.; Ekeleme, A.C. Development of a Modified Rational Model for Flood Risk Assessment of Imo State, Nigeria Using Gis and Rs. *Int. J. Eng. Sci.* **2014**, *3*, 1–8.
84. Jayne, T.S.; Chamberlin, J.; Headey, D.D. Land pressures, the evolution of farming systems, and development strategies in Africa: A synthesis. *Food Policy* **2014**, *48*, 1–17. [CrossRef]
85. Tiepolo, M.; Braccio, S. 12. Flood Risk Preliminary Mapping in Niamey, Niger. In *Planning to Cope with Tropical and Subtropical Climate Change*, 1st ed.; Tiepolo, M., Ponte, E., Cristofori, E., Eds.; De Gruyter: Berlin, Germany, 2016; pp. 201–220.
86. Mbanga, L.A. Human Settlement Dynamics in the Bamenda III Municipality, North West Region, Cameroon. *J. Settl. Spat. Plan.* **2018**, *9*, 47–58.
87. Kuriqi, A.; Ardiçlioglu, M.; Muceku, Y. Investigation of seepage effect on river dike's stability under steady state and transient conditions. *Pollack Period.* **2016**, *11*, 87–104. [CrossRef]
88. Ghassemi, S.; Sandu, C.; Fiandrotti, A.; Giulio Tonolo, F.; Boccardo, P.; Francini, G.; Magli, E. Satellite Image Segmentation with Deep Residual Architectures for Time-Critical Applications. In Proceedings of the 2018 26th European Signal Processing Conference (EUSIPCO), Rome, Italy, 3–7 September 2018; pp. 2235–2239.

Article

An Interdisciplinary Approach to the Sustainable Management of Territorial Resources in Hodh el Chargui, Mauritania

Chiara Caselle [1,*], Sabrina Maria Rita Bonetto [1], Domenico Antonio De Luca [1], Manuela Lasagna [1], Luigi Perotti [1], Arianna Bucci [2] and Stefano Bechis [3]

1 Department of Earth Science, University of Torino, 10125 Torino, Italy; sabrina.bonetto@unito.it (S.M.R.B.); domenico.deluca@unito.it (D.A.D.L.); manuela.lasagna@unito.it (M.L.); luigi.perotti@unito.it (L.P.)
2 Geodata Engineering S.p.A., Corso Bolzano 14, 10121 Torino, Italy; aria.bucci@gmail.com
3 Interuniversity Department of Regional and Urban Studies and Planning (DIST), Politecnico and University of Torino, 10125 Torino, Italy; stefano.bechis@unito.it
* Correspondence: chiara.caselle@unito.it

Received: 28 May 2020; Accepted: 17 June 2020; Published: 23 June 2020

Abstract: The present study proposes an analytical investigation of the natural resources and social framework of the Hodh el Chargui region (Mauritania), aiming to offer a useful instrument for planning and management to the local authorities. The situation of the region was evaluated by means of a participatory survey carried out among the local inhabitants. The obtained results include a collection of data about population, territorial organization, access to basic education and health services, infrastructure, main economic activities, and natural resources (in terms of water, both surface and groundwater, duration and intensity of rainfalls, soil types, and vegetal resources). The survey outcomes were completed with an integrated approach based on Earth Observation (EO) data supports, such as digital elevation models (DEMs) and Landsat8 imagery. The interdependence among the different data was evaluated and discussed, with regard to the influence of the availability of natural resources on the development of agricultural activities and on the general social welfare. The results are organized in the form of digital maps and a user-friendly webmap platform to facilitate access for all the technical and nontechnical actors involved in the project.

Keywords: natural resources; Mauritania; resource management; regional planning; participatory approach; EO data

---

1. Introduction

The regional planning and management of natural resources require us to consider the interactions among human needs, ecosystem dynamics and resource sustainability, keeping a balance between the different elements. In particular, in tropical rural or semirural areas of sub-Saharan Africa, the shortage of natural resources requires specific attention to correct planning and a proper direction of the interventions [1–4]. In these areas, the frequent drought periods, the ephemeral nature of surface water and the poorness of natural vegetation often generate severe risk scenarios. In addition, the vulnerability of rural and agro-pastoral communities living in these environments is often high due to the low adequacy of infrastructure that should provide access to the basic needs of the population (e.g., water, health care, education, transport).

The assessment of natural resource availability and sustainable use is, however, insufficient to carry out a correct decision-making and intervention process that should also include the evaluation of social, economic and cultural factors [5,6].

At a regional scale, the identification of more exposed communities, with suggestions on how to deal with risks, could help the proper direction of aid, helping the local authorities in risk-informed decision-making that favours the sustainable development of the territory. An effective planning process requires the combined consideration of environmental, technological, economic and socio-political factors.

With these purposes, the European Project "Renforcement Institutionnel en Mauritanie vers la Résilience Agricole & Pastorale" (RIMRAP) developed an interdisciplinary approach for the assessment of the vulnerability to risk of the population in Mauritania, with the aim of facilitating risk-informed intervention planning. The present study proposes a review of the results obtained in the analysis of southeastern Wilaya (Region), denominated as Hodh el Chargui (Figure 1).

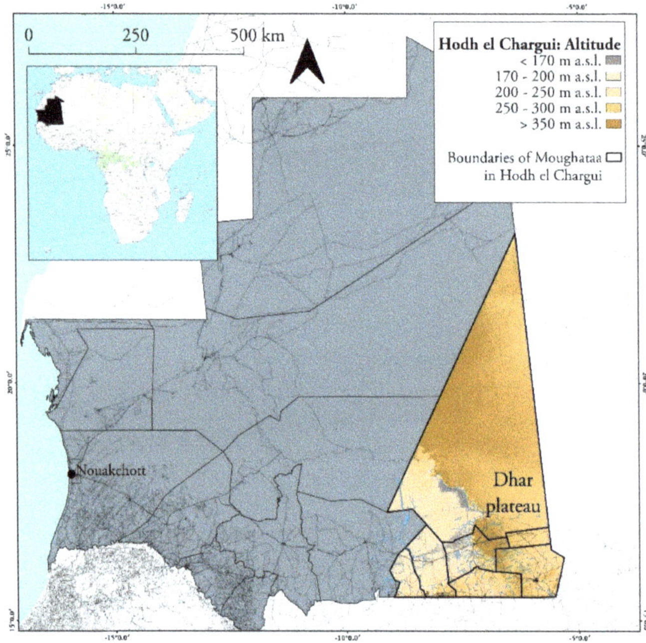

Figure 1. Location of the Hodh el Chargui region in southeastern Mauritania.

The region, one of the poorest and most remote of Mauritania, is located about 1000 km east of the capital city, Nouakchott. As a consequence, the area suffers from weaknesses in public investments and, hence, of basic social services (e.g., schools and health centres), transport and communications. The administrative organization sees a division in 7 Moughataa (Provinces), subdivided, in turn, into 31 minor territorial units (municipalities). The area is mainly rural, being occupied by small villages (communities) inhabited by both nomadic and settled people, whose principal activities are pastoralism and, only secondarily, agriculture [7].

From a morphological point of view, the territory is characterised, in the northeastern part, by a high plateau (Dhar Plateau), with maximum heights of 420 m above sea level (a.s.l. in all the maps). The western cliff of the plateau delimitates a large flat area, with altitudes ranging between 150 and 250 m a.s.l., which corresponds to the most lively and actively populated part of the region.

The area is mainly covered by sand dunes, especially in the north of the region, where the aeolian deposits may reach thicknesses of more than 100 m. In the south of the region, the thicknesses are lower, and the deposits are mainly fixed dune fields extended in an ENE–WSW direction that progressively leave their place to the emerging rocky basement. The surface water resources of the region mainly consist of wadis, which, during the dry season, dry-up, leaving only sanded riverbeds. Very often, they are subject to an enlargement of the riverbed and to the erosion of the banks caused by occasional intensive precipitation and by degraded vegetation cover [8,9].

The climate is controlled by a seasonal alternation, with a long dry season (from October to June) and a short humid season (from July to September), with rainy precipitations that may sometimes be violent and catastrophic.

The hard climatic conditions and the poor development of agricultural activities imply a low coverage of cereal needs (on average, 30% of needs) and considerable imports of rice, oil, wheat flour and other commodities. This affects the country's balance of payments and exposes it to risks of external crisis. In addition, the rural exodus deprives the countryside of manpower and prevents the execution of the traditional mechanisms of development, maintenance and conservation of these fragile ecosystems, threatening the livelihoods of rural populations in these areas, undermining their resilience and exacerbating conflicts for natural resources [7,10].

The present study aims to approach this reality and create a reliable and accurate cartographical representation of the demographical, administrative, socio-economic features and availability of natural resources. The objective is to create a realistic framework of the human and natural resources in the Hodh el Chargui, in order to provide a reliable reference point to the local authorities for risk-informed planning of the interventions. The approach involved onsite data collection through a participatory survey proposed to the local inhabitants. The results of the survey were integrated and completed with the help of data retrieved with Earth Observation techniques.

2. Materials and Methods

2.1. Onsite Survey

The onsite survey was conducted with a participative approach, involving the local inhabitants, in order to obtain diffuse information on the local conditions. A specific questionnaire was developed, including inquiries on the population (total number and repartition between women and men), the percentage of literate people, the type of habitations, the infrastructures for water, energy, health, education, transportation and telecommunications, and the principal economic activities, with specific attention to pastoralism and agriculture. Besides this information, a more specific set of questions was directed to the analysis of natural resources, considering the duration of the rainy season, the perception of the climatic changes in the last 20 years and the features of soil, vegetal resources and water (surface and groundwater). This general questionnaire was recalibrated for the survey of the 31 municipalities and of all agro-pastoral communities with at least 300 inhabitants. In the former case, the survey was conducted in the chief towns, collecting information about the entire area of the municipalities. In the latter case, the survey is more specifically referred to as the single community, allowing for a capillary reconstruction of the features of the region. The complete question forms may be found in Supplementary Materials S1 and S2.

In addition, a specific questionnaire was prepared for the investigation of the features of water wells in the area (Supplementary Material S3); this survey was specifically aimed at the reconstruction of groundwater characteristics and their accessibility and governance. The collected information was completed and integrated with the available literature data that included similar surveys on the wells of the region [8,11].

2.2. Integration of Collected Information with EO Data

Despite the large distribution of the communities on the territory of the Hodh el Chargui, the data collected with the onsite survey only provided specific information. To complete the framework about the distribution on natural resources in the region, data were integrated through the analysis and interpretation of USGS (United States Geological Survey) Landsat imagery and with digital elevation models (DEMs) of the region.

More in detail, the Landsat images were standard Landsat-8 operational land imager (OLI) data products, radiometrically and geometrically corrected and referred to as Level-1TP (L1TP). The L1TP products are digital numbers with coefficients provided to convert data to either radiance or reflectance. The images, obtained from USGS Landsat collection service, are considered suitable for time-series analysis. The georegistration is indeed consistent and within prescribed tolerances (<12 m root mean square error — RMSE). The downloaded images are radiometrically calibrated and orthorectified, using ground control points (GCPs) and digital elevation model (DEM) data to correct relief displacement. GCPs were derived from the Global Land Survey 2000 (GLS2000) dataset. The onboard sensor reflective operational land imager (OLI) bands are 30-m resolution, and DNs are stored as 16-bit signed integers that can be linearly scaled to the top of atmosphere (TOA) reflectance. The selected images were acquired on 11 September 2015 for the rainy season and 15 April 2019 for the dry season, and their positions correspond to 199-049 and 200-049 of the Landsat repository. The Landsat-8 L1TP image digital numbers for each band were converted to top of atmosphere (TOA) reflectance, using the scaling factors stored in the metadata using QGIS SCP (semiclassification plugin, free version). Finally, dark subtraction with dark object minimum atmospheric correction was applied in order to obtain final surface reflectance Landsat 8 band stacks [12].

The DEM was the USGS EROS Archive Digital Elevation Model—Shuttle Radar Topography Mission (SRTM)—1 Arc-Second Global version, obtained from USGS EarthExplorer. The Shuttle Radar Topography Mission (SRTM) was flown aboard the space shuttle Endeavour 11 on 22 February 2000. The SRTM elevation data offer worldwide coverage of void-filled data at a resolution of 1 arc-second (30 m) and provide open distribution of this high-resolution global dataset [13,14].

These data were analysed and interpreted to obtain:

- a cartographic base for specific data collection and interpretation;
- hydrological analysis of the surface water finalised to the creation of a watershed basins map;
- a reconstruction of the land use of the project area, focusing on vegetation cover in different seasons.

The automatic or semiautomatic definition of watershed basins starting from a DEM is a common practice in the GIS applicative world. However, particular cases as large flat areas, with complex river networks, flat terrains, crossed and looped channels and a large number of polders may create complex hydrographic conditions that make it impossible to accomplish automatic river network extraction and watershed delineation by using DEMs with automatic processes [15–17]. To face this problem, a specific procedure was developed, including the first phase of manual delineation of a river network, using as input the available literature—e.g., historical sources, reports, maps—and the USGS Landsat8 images. The second phase of the procedure consisted of a semiautomatic watershed definition with the ArcGis Watershed Delineation Tool [18,19] based on the DEM and the manually delineated river network. Results were then attentively checked to verify the correspondence between the automatic result and the real morphology suggested by the Landsat images.

For the analysis of the vegetation and the creation of the use of land maps, the Semiautomatic Classification QGIS plugin [20] was employed. The method requires the definition of reference areas with known land cover features that are used by the software to train the classification algorithm. In this application, we took as reference the data about natural vegetation from the surveys in the communities, using the Spectral Angle Map algorithm, with a threshold angle of 20° for the classification. The final Landsat 8 image classification includes water, bare soil/sand dunes and vegetated areas. The classification was repeated for Landsat images corresponding to the end of the rainy season and the end of the dry season to observe the respective distribution of water and vegetation covers in these two climatic end-members.

2.3. Publication of Results on a Webmap for Dissemination and Evaluation

Project results of an area necessarily pass through a cartographic representation. In recent years, dynamic and web mapping have constantly grown as a direct consequence of the development of digital technologies and the wide diffusion of the Internet. This has created new methods of map production, instead of pure desktop GIS, making them more accessible, both technologically and economically. This new cartography has become one of the best tools for disseminating information and making it accessible to all [21]. Dynamic webmaps allow for a definition of the different dimensions of the project, involving wide, intermediate and detail scales, public and private corporate players, and different kinds of actions [22]. In this kind of project, one of the main goals is to make data accessible and easy to use by the general public and for NGO (non-governmental organizations) staff. Hence, it is very important to angle towards an application usable by the general public but, at the same time, preserve scientific rigour.

For these reasons, the main results of this project were uploaded on a webmap platform, accessible at the link www.geositlab.unito.it/rimrap, to facilitate their easy access by the local authorities and technical operators. The platform was obtained using the open-source QGIS plugin "QGIS2web" to create an html page that contained the cartographical data in an interactive and user-friendly map. This final product is accessible online but was also delivered to the local actors in an offline version in order to facilitate accessibility, even in the absence of a good internet connection.

3. Results

3.1. Onsite Survey

The map in Figure 2 shows the 31 municipalities' chief towns, the administrative boundaries of the municipalities, the 265 agro-pastoral communities and the water wells surveyed in this research and in previous studies [8,11].

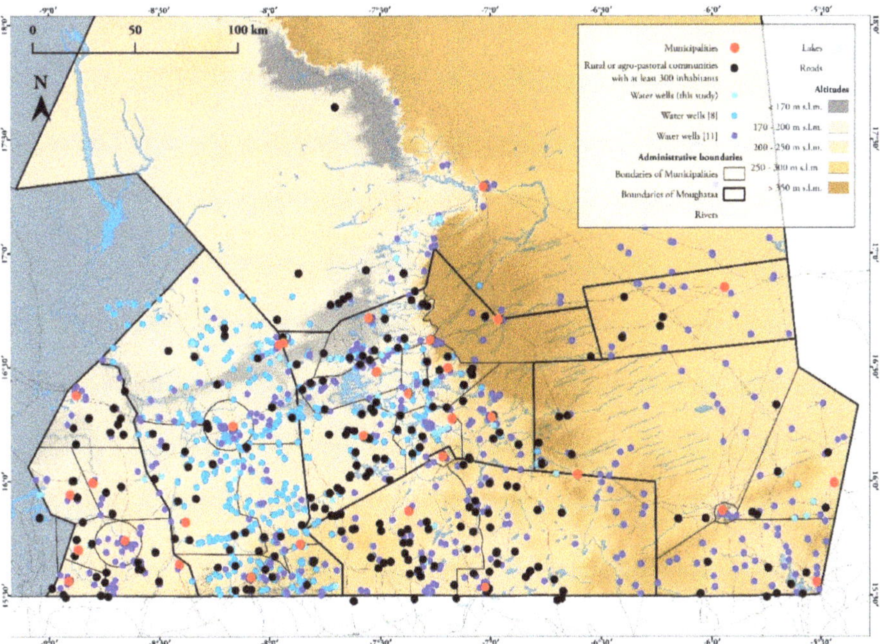

Figure 2. Administrative maps of the Hodh el Chargui, with the surveyed municipalities, communities and water wells.

3.1.1. Demography and General Information

The survey in the municipalities of Hodh el Chargui recorded 469,477 people, with 47% males and 53% females. The resulting repartition of this population on the 31 municipalities of the region is shown in Figure 3a, while Figure 3b shows the human density of each of the municipalities.

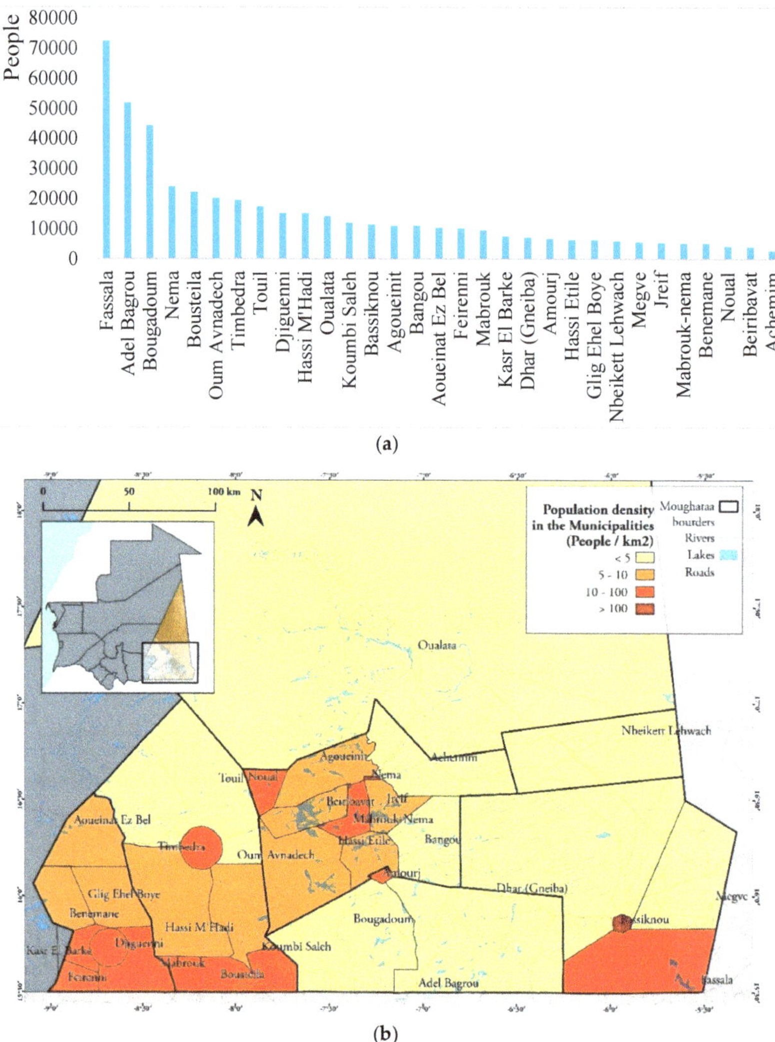

(a)

(b)

Figure 3. (**a**) Number of people living in each of the municipalities; (**b**) population density in the 31 municipalities of the Hodh el Chargui region.

The map in Figure 4 shows the percentages of literate people (i.e., people who have received basic school education) for men and women, respectively. The overall literate population is around 25%, with values around 30% for the male population and around 20% for the female population. The highest percentages of literate people (42% and 38%) may be found in the Moughataas of Nema and Djiguenni, respectively, in the municipalities of Mabrouk and Djiguenni itself.

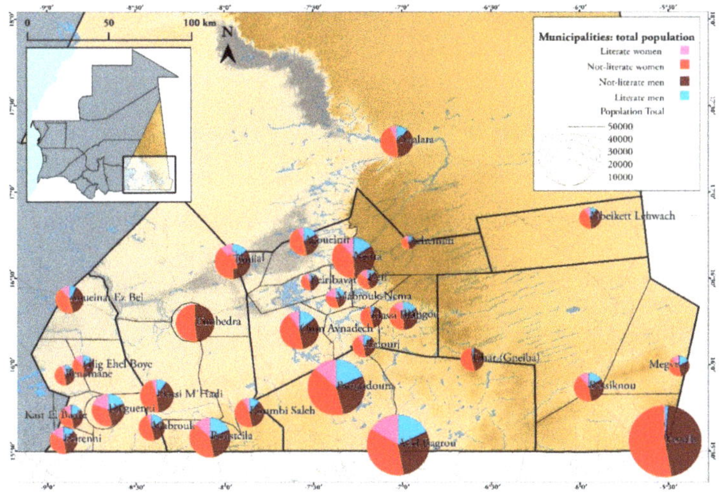

Figure 4. Percentage of males and females and literate and not-literate people in the 31 municipalities of the region.

The features of the habitations in the agro-pastoral communities were surveyed, distinguishing between "En banco" habitations, hard dwellings, semihard dwellings, stone buildings, tents and sheds. The map in Figure 5 shows, for each community, the principal type. As can be seen in the pie chart, the majority of the communities are characterized by habitations "En banco", followed by sheds and, in a lower proportion, stone buildings. A small percentage of the communities see the prevalence of hard buildings, while none of them described tents or semihard buildings as the principal type of habitation.

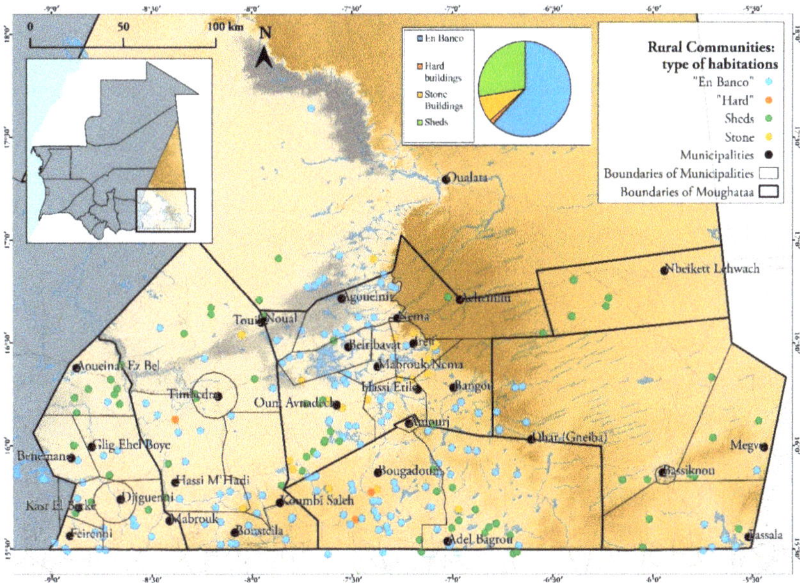

Figure 5. Principal type of habitation in all the rural communities with at least 300 inhabitants of the region.

The survey also enquired about the availability of toilets, attesting that they are available in 60% of the communities.

3.1.2. Infrastructure

The availability of infrastructure in the rural communities is an important parameter, well-representing the resilience of a society, since a capillary distribution of services in the territory means easier access for all the inhabitants of the region. The surveyed infrastructures involve water management, energy, health, education, transportation and telecommunications.

In the Hodh el Chargui, the main water resource is groundwater, since surface water usually shows an ephemeral character. The most important infrastructure for water management is, therefore, the water wells for access to drinking water for both human and animal use. In the survey, three different types of water wells were considered:

- well: hand-excavated wells and, therefore, generally shallow, with manual systems for the water extraction;
- drainage well: large diameter wells (1–2 m), shallow, hand-dug, where the groundwater flows by gravity;
- borehole: small-diameter hole advanced below the ground surface by various drilling rigs, mostly equipped with a pump.

The total number of these types of wells and their activity status are reported in Figure 6.

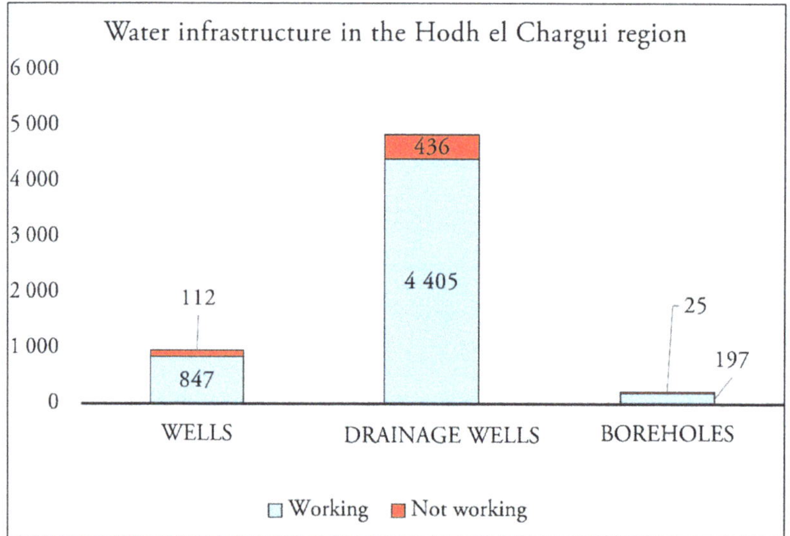

Figure 6. Total number of wells, cistern wells and boreholes in the region and their activity status.

In addition, the survey collected data about springs (i.e., places where water emerges naturally from the rock or soil).

As shown in the map in Figure 7a, many of the communities have more than one water well in the neighbouring area, with a mean number of 17. However, the absence of controlled and scientific-based management and organization of the number of wells often affects their activity status.

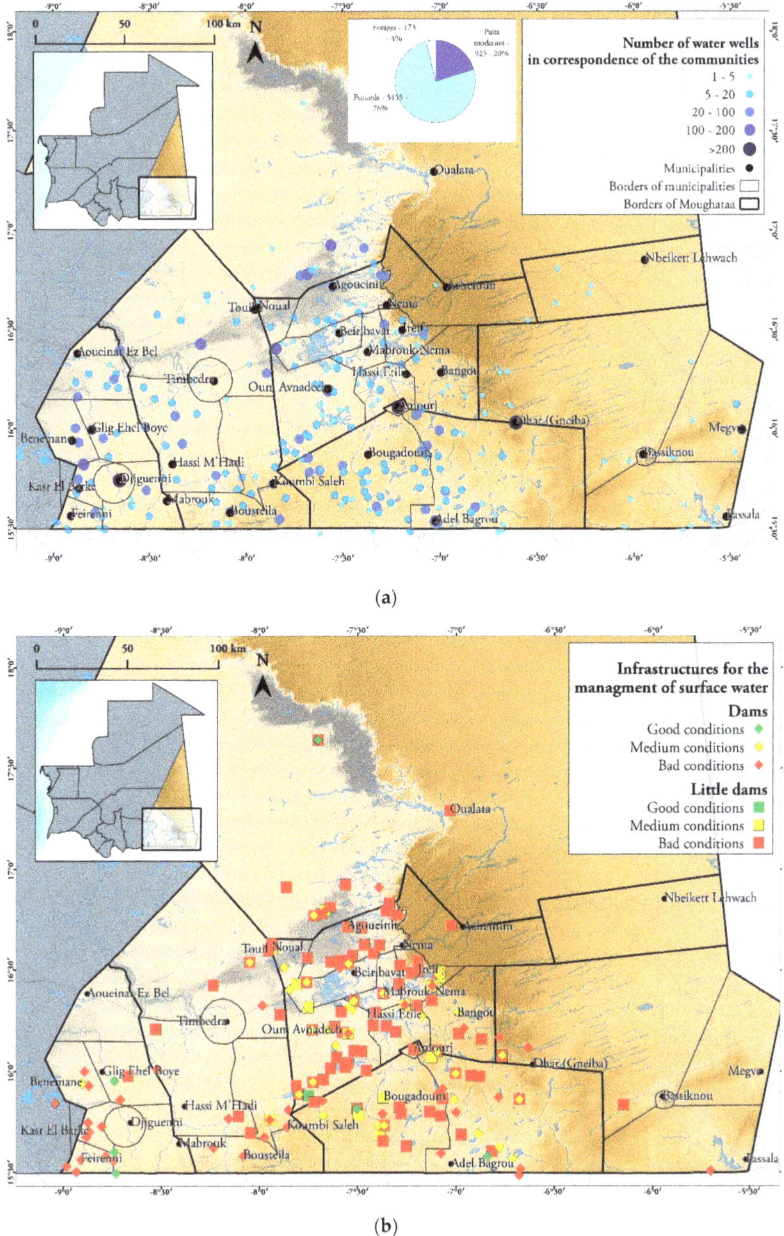

Figure 7. (**a**) Maps of infrastructure for the access to groundwater; (**b**) maps of infrastructure for the management of surface water.

Despite the predominant role covered by the groundwater, a specific relevance is connected to the infrastructure for the management of surface water (e.g., dams, barriers) for agriculture purposes (e.g., irrigation). The distribution and the conditions of this kind of infrastructure are reported in Figure 7b. As shown in the map, despite a large number of structures, the conditions are often bad, affecting the real state of operation.

The infrastructure for energy mainly includes photovoltaic systems. Other energy sources were surveyed (i.e., wind systems and generators), but their presence was registered only in the municipalities of Adel Bagrou and Aoueinat Ez Bel, respectively. The photovoltaic systems, on the other hand, are present in 48% of the municipalities, while only 35% have a connection to the electricity network. The photovoltaic systems are also well developed in the rural communities, especially in the region of Nema (Figure 8).

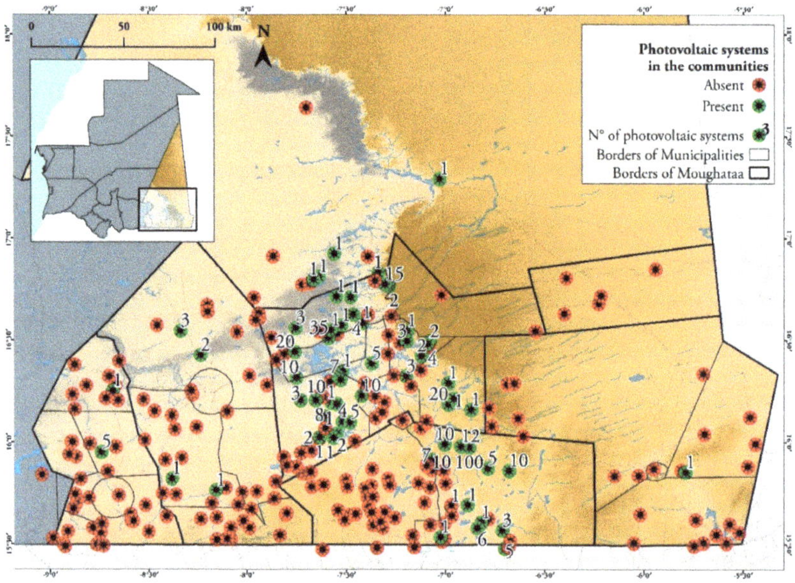

Figure 8. Availability of photovoltaic systems in the communities.

The infrastructure dedicated to healthcare (i.e., hospitals, dispensaries and health units) is present in 43% of the surveyed communities, following the distribution reported in Figure 9a. Among the remaining 57% (i.e., 151 communities), 96 communities declared the availability of a means of transport to the nearest hospital in case of emergencies (i.e., handcart, car, or on the back of an animal), following the chart in Figure 9b.

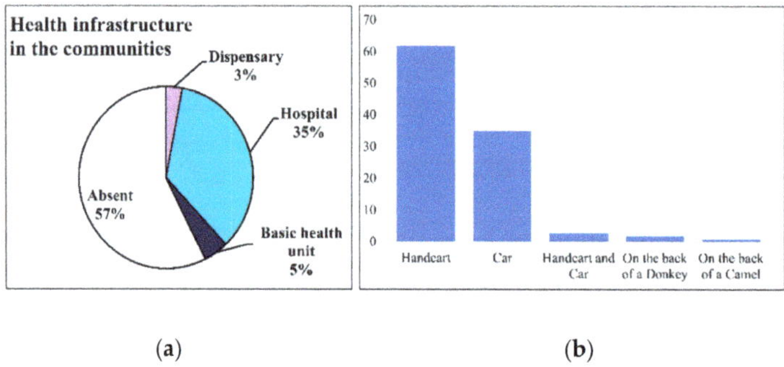

(**a**) (**b**)

Figure 9. (**a**) Percentages of communities with dispensaries, hospitals and basic health units; (**b**) means of transport available in the communities without infrastructure for healthcare onsite.

The education infrastructure (i.e., schools) represents one of the most important elements for the improvement of the local communities. The percentage of communities provided with a school is high (i.e., 94%). However, the conditions of the school buildings are bad in 48% of the cases (Figure 10).

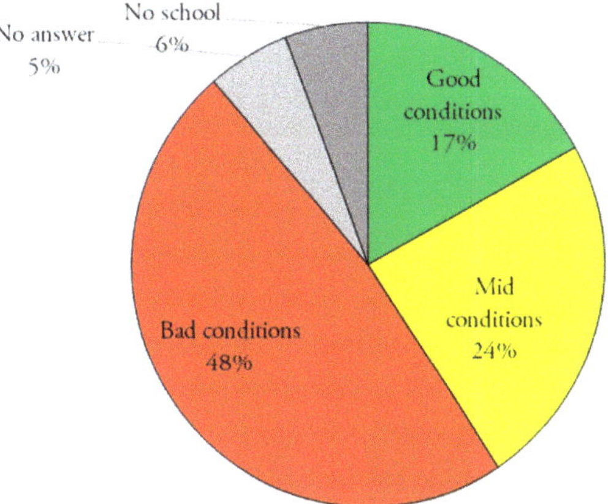

Figure 10. Conditions of the dwellings of the schools in the communities.

Regarding transport infrastructure, the presence of asphalt roads was attested in only 23 of the 265 surveyed rural communities (Figure 11). The remaining communities are usually crossed by a natural road. These kinds of roads are often interrupted by the effect of the rainfall, for mean time periods that range between 1 day and 6 months, as reported on the map in Figure 11.

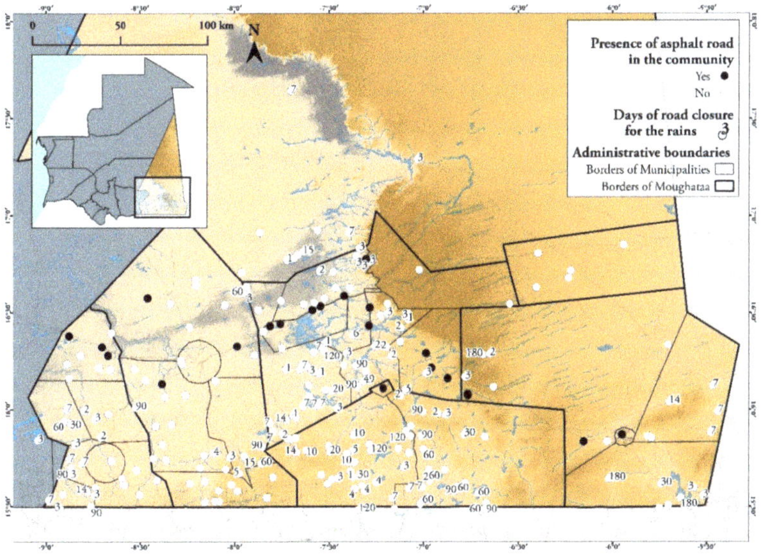

Figure 11. Map of availability of asphalt roads in the communities. For the communities without the presence of asphalt roads, the mean number of days of road closure caused by seasonal rainfall is reported.

Eventually, a good distribution of communication infrastructure was surveyed on the territory, at least in terms of mobile phone networks, which are present in 84.5% of the communities.

Internet network, radio and television are less diffused, with percentages of 17.0%, 37.4% and 18.9%, respectively.

3.1.3. Principal Economic Activities

As shown in Figure 12, the first economic activity in the majority of the communities is pastoralism, followed by agriculture. Other types of activities, such as commerce and artisanship, are, however, present, but to a lower extent. The mean percentages of men and women performing the different professions are reported in Table 1.

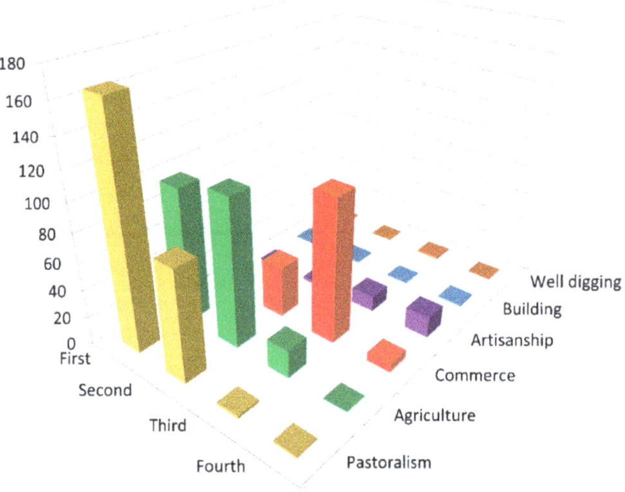

Figure 12. Number of communities that consider each economic activity, respectively, the first, the second, the third, or the fourth as important for their local economy.

Table 1. Repartition of the principal professions between men and women.

Profession	Men (%)	Women (%)
Baker	96	4
Butcher	55	45
Trader	61	39
Carpenter	100	0
Tailor	82	18
Weaver	8	92
Ironworker	100	0
Tanner	0	100
Builder	100	0
Digger	100	0

Hence, the main economic activity of the region is pastoralism, a sector that is well established and largely produces for export to other regions of the country and abroad. The activity is, however, strongly affected by seasonality. As shown in Figure 13a, indeed, the distance of the pastures from the communities is one order of magnitude higher during the dry season. In these periods, most of the shepherds migrate to the southern pastures, and only a small part of the livestock remains in the communities for requirements of milk, cheese, and meat (Figure 13b).

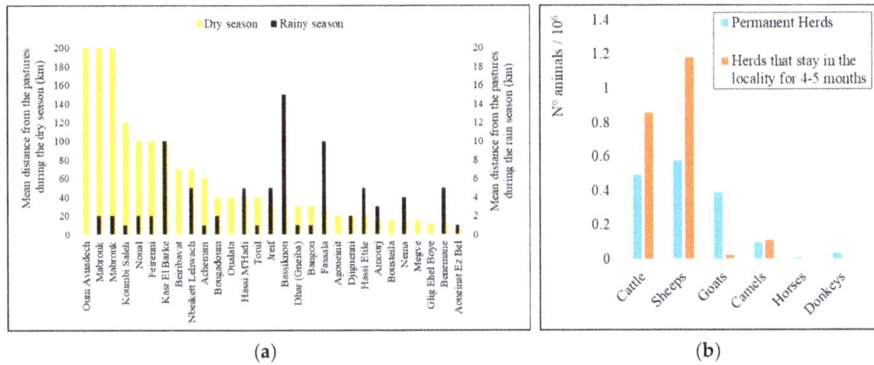

(**a**) (**b**)

Figure 13. (**a**) Mean distance of the communities from the pastures during the dry and rainy seasons. (**b**) Features of the herds in the communities.

Agriculture, on the other hand, is strongly affected by the dry climate and the want of surface water resources. Hence, it is less developed, not being able to supply the needs of the region entirely. Recently, however, some experiences of local agricultural production have started. They mainly involve the production of vegetable crops, especially short cycle plants, which give their product in a few weeks and are therefore less sensitive to the possibility of lack of water.

In other frameworks with similar environmental conditions (e.g., [23]), the success of agricultural activities was mainly related to the introduction of innovative technologies, for instance, the use of specific seeds for semiarid areas or the application of innovative irrigation techniques (e.g., "drip solar" irrigation systems). Similar techniques may also allow the success of agricultural production in the Hodh el Chargui region, despite the arid conditions.

3.1.4. Natural Resources

The availability of natural resources (i.e., surface water, groundwater and vegetation) mainly depends on seasonal precipitations. The rainy season starts between July and August and ends between September and October, as attested by the survey in the communities (Figure 14).

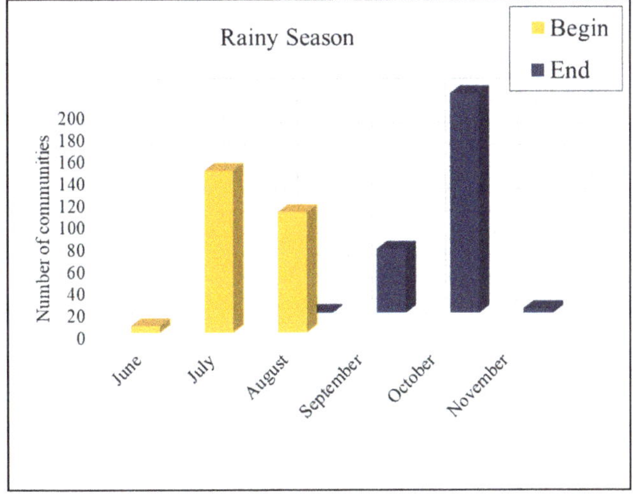

Figure 14. Answers of the communities about the beginning and end of the rainy season.

The survey also proposes an evaluation of the vegetal resources in the area: each community is invited to identify the principal type of natural or seminatural vegetation present in the neighbouring area (Figure 15). As can be seen, the majority of the communities (51%) identify the presence of shrub savannah, with low shrubs and grazing grass. These kinds of spaces are good for used as animal pastures during the rainy season. After long periods of drought, however, this low vegetation usually disappears, and in the late dry season, these pastures are usually completely dry.

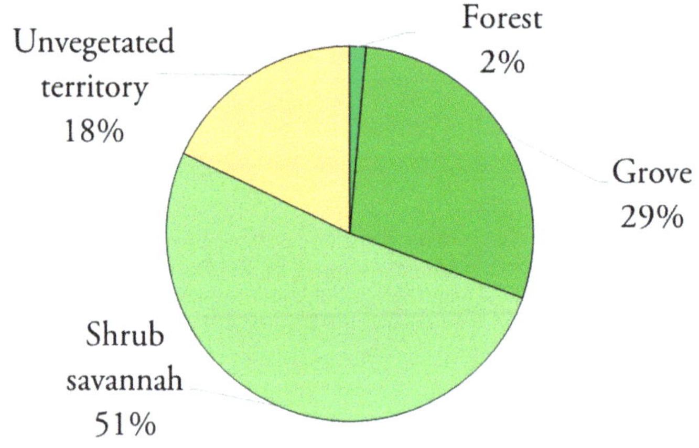

Figure 15. Types of seminatural vegetation present in correspondence of the communities.

A high percentage of the communities (29%) attested the presence of a grove, similar to the shrub savannah, but with the presence of a few tall trees. Additionally, in these conditions, the long dry seasons usually dry the grass used as pastures during the rains. Only 2% of the communities describe the presence of "forest", usually in correspondence to a river of a water stream that, during the rainy season, assures a large availability of water and a more abundant growth of vegetation, with tall trees.

The absence of any kind of vegetation, even during the rainy season, or the presence of sand dunes is attested in 18% of communities.

The availability of water resources mainly depends on the groundwater that should also guarantee the hydric supply during the long months of the dry season.

Nevertheless, for the majority of the considered wells, the survey identified the existence of a season when wells run dry. As can be seen in Figure 16, this period is between April and May/June, i.e., in correspondence with the end of the long dry season. This phenomenon may be due to several reasons, e.g., excessive water extraction, long drought periods or the short distance between the wells.

The discharge of the pumps (reported for 22% of the wells) is mainly lower than 1 m^3/h (23%) or in the range of 1 and 10 (57%) m^3/h. Values higher than 20 m^3/h are mainly related to boreholes or modern wells, emphasising the importance of modern infrastructure for water supply needs.

From a geographical point of view, clear differences in water depth can be recognised between the eastern and western parts of the region. In Figure 17, the depth of the static level for the surveyed wells is reported, highlighting in the eastern plateau a majority of wells with depth values higher than 40 m (i.e., less favourable conditions for the extraction of groundwater). In the western sector, on the other hand, values are generally lower than 30 m, and depth values are widely inhomogeneous.

This situation is coherent with the conceptual hydrogeological model proposed by [8], which includes three main hydrogeological units: the Upper Neoproterozoic hydrogeological unit, within the marine sediments that make up the ancient basement; the Mesozoic hydrogeological unit, within the Jurassic dolerites and Cretaceous sandstones that characterise the Dhar of Nema;

the Quaternary hydrogeological unit, mainly consisting of aeolian sandy deposits and ancient and recent alluvial deposits, emerging in the wadis.

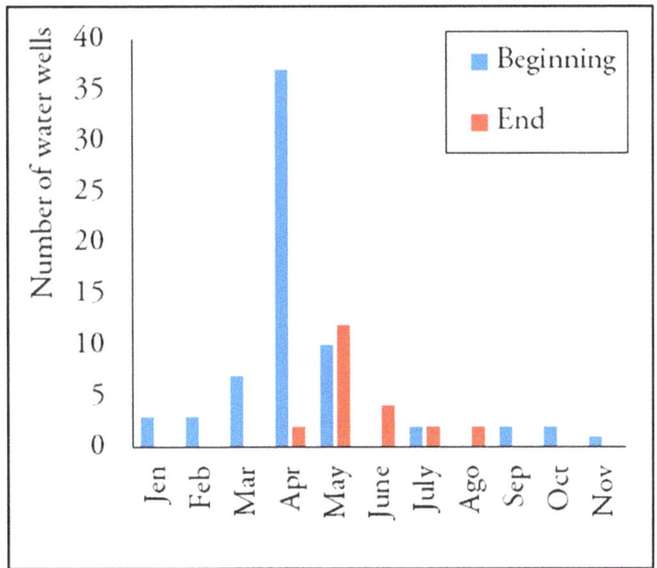

Figure 16. Beginning and end of the season when wells run dry.

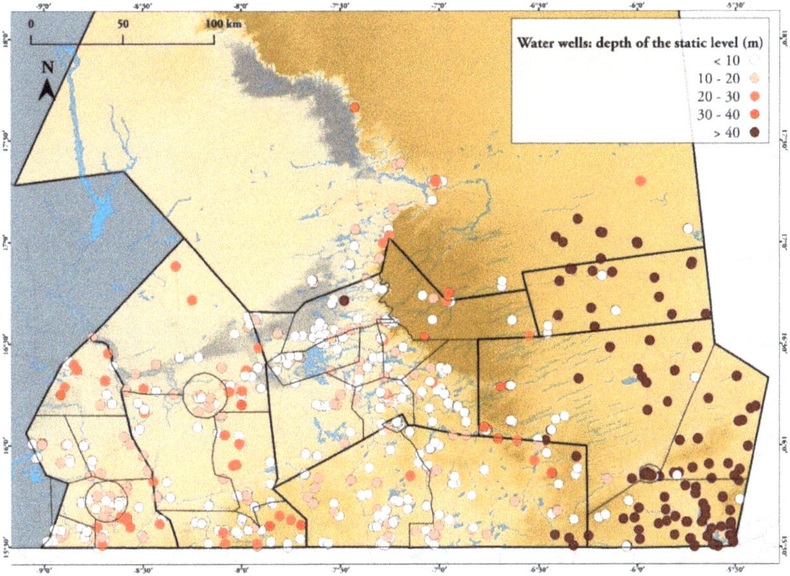

Figure 17. Map of the depth of the static level in the water wells surveyed in this study and in [11].

Hence, the higher depths registered in the eastern area should correspond to aquifers in correspondence to the Mesozoic sediments of the Dhar plateau, while the shallow features and the inhomogeneous distribution of the measurements in the western area confirm the existence of multiple aquifers within the recent aeolian and alluvial deposits, with higher geological variability.

In addition to the quantitative features of groundwater resources, the survey also provided information about the water qualitative parameters, collecting both objective (e.g., electrical conductivity) and subjective (e.g., taste, smell, colour) ones.

The electrical conductivity provides important information about the salinity and, consequently, the drinkability of groundwater. For human use, no WHO health-based guideline value was established; however, a value of 2500 μS cm^{-1} at 20 °C is indicated in a WHO directive [24], reflecting what is both achievable and acceptable to consumers.

For animals, generally, higher salt content is accepted, with maximum values between 5000 and 10,000 microS/cm, depending on the type of animal [25]. Even though the majority of the surveyed wells returned values lower than 2500 microS/cm, the maximum value registered in the area is very high, up to 15,000 microS/cm (i.e., not acceptable values, not even for animals). As shown in Figure 18, these anomalously high values are randomly present in the whole region. The absence of a clear distribution is probably due to the low continuity of the water bodies and the different depths of the wells.

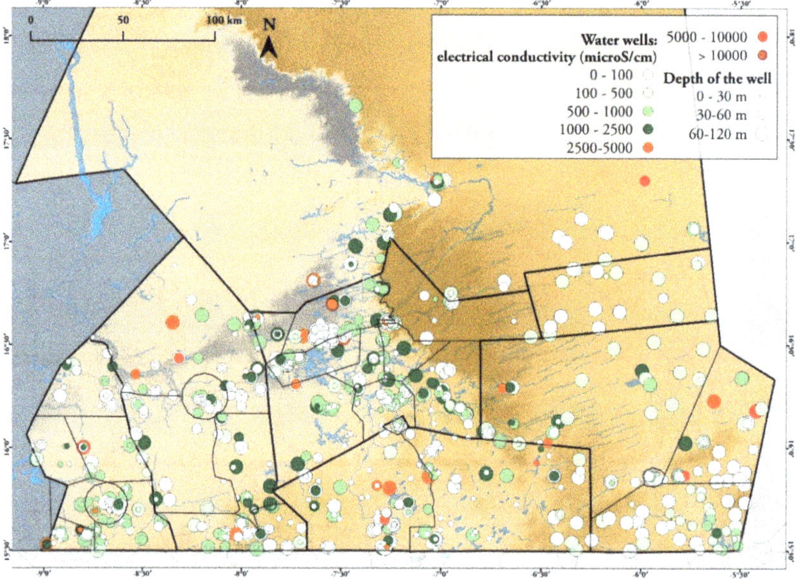

Figure 18. Map of the conductivity and the total depth of the wells surveyed in this study and in [11].

The qualitative evaluations of the water taste return values are in line with the measures of electrical conductivity, again highlighting the existence of the problem of high salinity water, as reported in Table 2.

Table 2. Qualitative classification of water salinity in wells, based on the subjective evaluation of taste.

Taste	Number of Water Wells
Very salty	1
Salty	2
Slightly salty	39
Heavy	12
Sweet	44

3.1.5. Influence of Climatic Change

The rainfall and meteorological characteristics of the two seasons of the year influence the hydrological risks. The survey on the communities investigated how these characteristics have changed in the last 10 years as a consequence of worldwide climate change. The results, summarised in Tables 3 and 4, testify to a general increase in temperatures. In addition, the rainy season was described, in the majority of the cases, as shorter and with less intense and less frequent rainfall events. As a consequence, floods (and related damages) were described as constant or in diminution, while an increase in drought events (and related damages) was registered in the majority of the communities.

Table 3. Answers of the communities in the survey about climate change during the last 20 years.

	First Answer	Second Answer	Third Answer
Rainy season	Shorter (92%)	As usual (3%)	Longer (2%)
Dry season	Longer (96%)	As usual (3.5%)	Shorter (0.5%)
Temperatures	Higher (96%)	As usual (4%)	Longer (2%)

Table 4. Data about the increase or decrease of damage due to atmospheric events (wind storms, floods and droughts).

	More Frequent (% of Answers)	As Usual (% of Answers)	Less Frequent (% of Answers)
Storms (strong wind)	83	10	6
Floods	5	58	37
Flood damages	3	58	39
Drought	88	11	2
Drought damages	83	16	2

According to the results of the survey in the 31 municipalities (Figure 19), the increasing drought events mostly caused damage to vegetation and pastures. Hence, a high impact is also registered for animals in need of pastures for their feeding. Agricultural products are in the third position, presumably because of the lower diffusion of this economic activity when compared to pastoralism. Water resources (i.e., mainly groundwater) having longer charging time and, therefore, a longer response to rain were less affected. Eventually, some records of people affected by drought events were reported, even though the survey does not investigate the specific reasons.

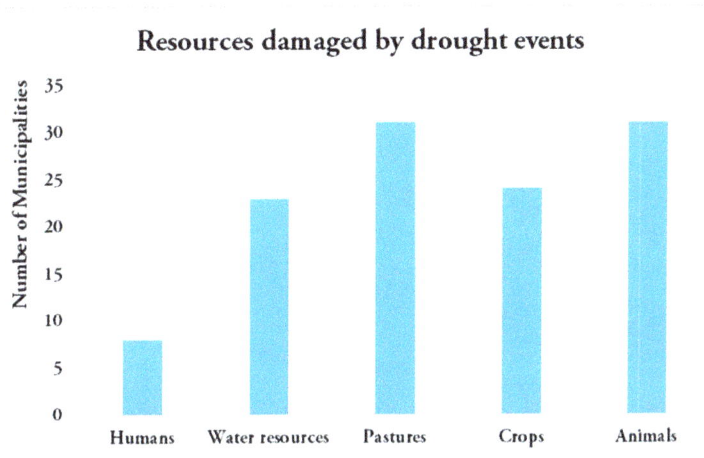

Figure 19. Most affected resources by the effects of drought.

3.2. Additional Analyses with EO Satellite Data

3.2.1. Watershed Basins

The watershed basins extracted with the semiautomatic procedure for the area of Nema are shown in Figure 20. The river network was manually delineated following the Landsat8 images, using different false colour composites. The ephemeral rivers are complex and ramified in the northeastern part of the map, in correspondence with the Nema plateau. In this area, the presence of altitude variations favours the water flow along definite paths and increases the energy of water, enhancing its erosive power. In the flat area on the south-west, on the other hand, the water flows slowly, creating large swampy areas and lakes that, during the rainy period, are rich in water and vegetation.

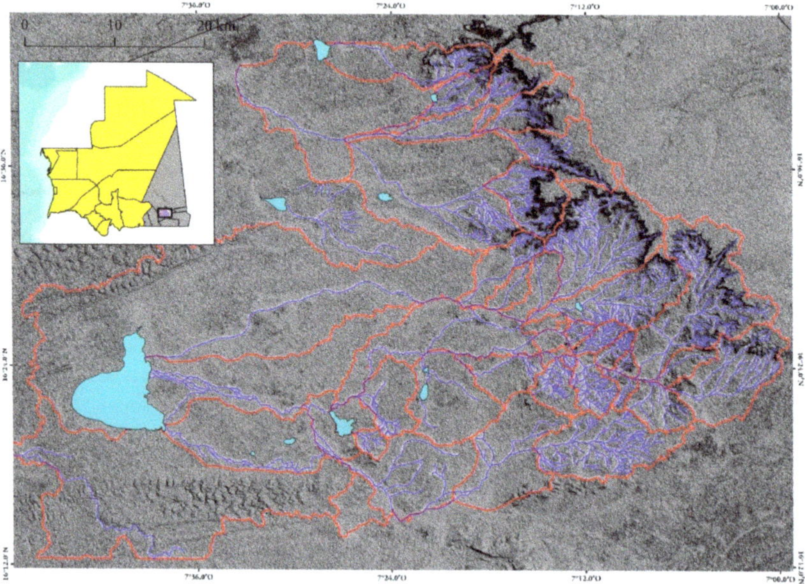

Figure 20. River network (in blue) and watershed basins (in red) defined for the area of Nema.

3.2.2. Use of Land Cover Classification Maps

Figure 21 shows the land cover classification for the area of the Hodh el Chargui. The map distinguishes between areas with vegetation, which may be used as pastures, and unvegetated areas, consisting of bare soil or sand dunes. Tables 5 and 6 report the size of the training samples (i.e., reference data) and the accuracy of the classification for the rainy season and the dry season, respectively. The drastically different results obtained for the rainy season (Figure 21a) and the dry season (Figure 21b) confirm the strong effect of seasonality on the availability of surface water and vegetation for the pastures.

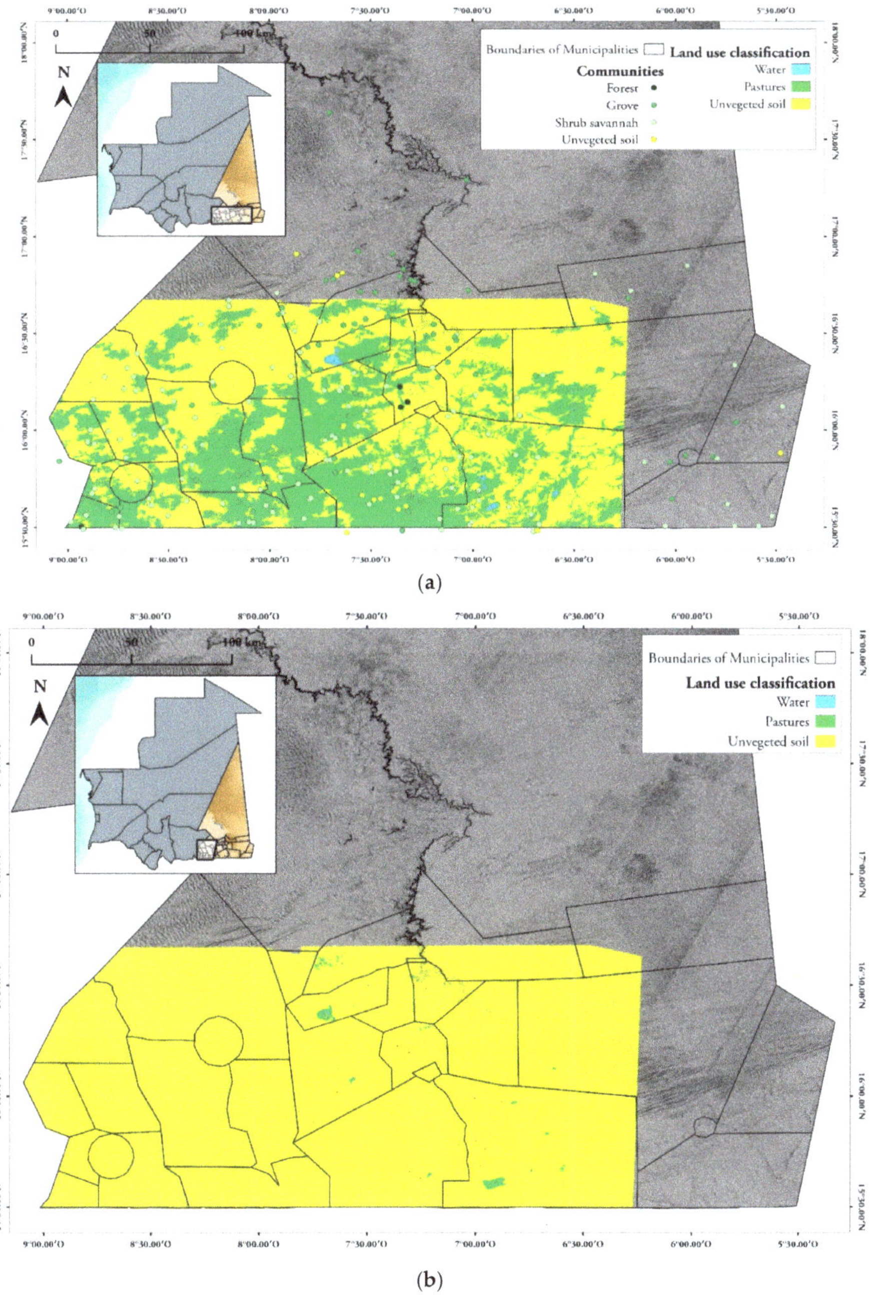

Figure 21. (**a**) Map of use of land classification at the end of the rainy season. (**b**) Map of use of land cover classification at the end of the dry season.

Table 5. Error matrix of the classification procedure for the rainy season.

	Reference Data				
Classified Data	**Soil**	**Pastures**	**Water**	**Row Total**	**User's Accuracy**
Unclassified	0	0	1112	1112	
Soil	2854	0	693	3547	80.46%
Pastures	0	3511	418	3929	89.36%
Water	0	0	1120	1120	100.00%
Column Total	2854	3511	3343	9708	
Producer's Accuracy	100.00%	100.00%	33.50%	**Overall Accuracy** = 77.10%	
				Overall Kappa Index = 0.68	

Table 6. Error matrix of the classification procedure for the dry season.

	Reference Data				
Classified Data	**Soil**	**Pastures**	**Water**	**Row Total**	**User's Accuracy**
Unclassified	0	0	0	0	
Soil	2854	338	180	3372	84.64%
Pastures	0	3017	636	3653	82.59%
Water	0	156	2527	2683	94.91%
Column Total	2854	3511	3343	9708	
Producer's Accuracy	100.00%	85.93%	75.59%	**Overall Accuracy** = 86.51%	
				Overall Kappa Index = 0.80	

3.3. Publication of Results on a Webmap for Dissemination and Evaluation

The online web–GIS map is accessible at the link www.geositlab.unito.it/rimrap. The data reported on the maps include:

- information on the schools and hospitals available in the area;
- the surveyed wells, classified in terms of the state of activity, with an interactive form reporting the main information about the well features (depth, perennity, distance from the village, modality of water extraction) and the water features (temperature, electrical conductivity, taste);
- the location of the 265 surveyed rural communities, classified in terms of population, with an interactive form, including some basic information (e.g., name of the village, principal economic activity);
- a record of the improvement interventions performed in the framework of the RIMRAP project.

4. Conclusions

This study proposes an analysis of the vulnerability of the local communities in the Hodh el Chargui region (Mauritania) through the collection of heterogeneous data concerning environmental and socio-economic information. Data were collected with the combined use of an onsite survey, which involved the local population through a participatory approach, and techniques of Earth Observation.

The results provide a reliable framework of the local reality in terms of demographical, administrative and socio-economic features and availability of natural resources. The analysis includes an evaluation of the repartition between women and men, the percentage of the literate population, the infrastructure for the basic needs of the population and the principal economic activities, with specific attention to pastoralism and agriculture. In addition, a more specific insight on natural resources is obtained by considering the duration of the rainy season, the climatic changes in the last 20 years, the features of the vegetation cover in the different seasons and the water resources, both surface and groundwater.

This set of data aims at the identification of the main vulnerabilities of the local population, allowing for proper territorial planning and sustainable management of the resources by means of the development of correct interventions for the improvement of the territory. These data represent, indeed, the first step to the creation of specific plans of territorial development and improvement of local communities, both at the regional and local scales. Based on the described results, a "Plan of Municipal Development" will be prepared for each of the 31 municipalities of the region, emphasizing the main vulnerabilities of the municipal territory and assessing the priority of the interventions to be realized. In the framework of the RIMRAP project, part of the funds have already been dedicated to the execution of interventions considered as urgent or priority on the basis of the collected information. These interventions mainly involve the management of water (e.g., construction or repair of dams and other surface water infrastructure, water well fixing) and food security (with the construction or improvement of cereal banks).

The general overview of the region provided by this study creates the conditions for a more thoughtful and risk-informed decision-making process, bringing an improvement of the local economy, population welfare and accessibility to natural resources and human infrastructure.

Supplementary Materials: The following are available online at http://www.mdpi.com/2071-1050/12/12/5114/s1. File S1: Survey on the municipalities; File S2: Survey on the communities; File S3: File on the water wells.

Author Contributions: Conceptualization: S.B., S.M.R.B., and D.A.D.L.; methodology: S.B., M.L., S.M.R.B., and L.P.; investigation: S.B., S.M.R.B., and M.L.; formal analysis: A.B. and C.C.; writing—original draft preparation: C.C.; writing—review and editing: S.B., M.L., S.M.R.B., A.B., and L.P.; visualization: C.C., A.B., and L.P. All authors have read and agreed to the published version of the manuscript.

Funding: This research was funded by the European Project "Renforcement Institutionnel en Mauritanie vers la Résilience Agricole et Pastorale" (RIMRAP)—Financement de la Commission Européenne: 11e Fonds européen de développement. Référence: FED/137269/ACT/MR—Contract n° FED/2016/373-942: "Reduction de la vulnérabilité Agropastorale et amélioration de la résilience dans le Hodh de Chargui".

Acknowledgments: The authors desire to thank Terre Solidali Onlus for the useful contribution in the onsite activities.

Conflicts of Interest: The authors declare no conflict of interest.

References

1. Bonetto, S.M.R.; De Luca, D.A.; Lasagna, M.; Lodi, R. Groundwater distribution and fluoride content in the west arsi zone of the oromia region (Ethiopia). In *Engineering Geology for Society and Territory*; Springer Science and Business Media LLC: Cham, Switzerland, 2014; Volume 3, pp. 579–582.
2. Lasagna, M.; Dino, G.A.; Perotti, L.; Spadafora, F.; De Luca, D.A.; Yadji, G.; Dan-Badjo, A.T.; Moussa, I.; Harouna, M.; Konaté, M.; et al. Georesources and environmental problems in niamey city (Niger): A preliminary sketch. *Energy Procedia* **2015**, *76*, 67–76. [CrossRef]
3. Perotti, L.; Dino, G.A.; Lasagna, M.; Moussa, K.; Spadafora, F.; Yadji, G.; Dan-Badjo, A.T.; De Luca, D.A. Monitoring of urban growth and its related environmental impacts: Niamey case study (Niger). *Energy Procedia* **2016**, *97*, 37–43. [CrossRef]
4. Bonetto, S.M.R.; Facello, A.; Cristofori, E.I.; Camaro, W.; Demarchi, A. An approach to use earth observation data as support to water management issues in the ethiopian rift. In *Climate Change Management*; Springer Science and Business Media LLC: Cham, Switzerland, 2017; Volume 316, pp. 357–374.
5. Fonjong, L.N.; Emmanuel, N.N.; Fonchingong, C.C. Rethinking the contribution of indigenous management in small-scale water provision among selected rural communities in Cameroon. *Environ. Dev. Sustain.* **2005**, *6*, 429–451. [CrossRef]
6. Del Amo-Rodríguez, S.; Tenorio, M.D.C.V.; Ramos-Prado, J.M.; Porter-Bolland, L. Community landscape planning for rural areas: A model for biocultural resource management. *Soc. Nat. Resour.* **2010**, *23*, 436–450. [CrossRef]

7. Bechis, S.; Bonetto, S.M.R.; Bucci, A.; Canone, D.; Isotta Cristofori, E.; De Luca, D.A.; Demarchi, A.; Garnero, G.; Guerreschi, P.; Lasagna, M.; et al. Improving governance of access to water resources and their sustainable use in Hodh el Chargui communities (South East Mauritania). In Proceedings of the EGU General Assembly Conference Abstracts, Vienna, Austria, 4–13 April 2018.
8. Ghiglieri, G.; Carletti, A. Integrated approach to choosing suitable areas for the realization of productive wells in rural areas of sub-Saharan Africa. *Hydrol. Sci. J.* **2010**, *55*, 1357–1370. [CrossRef]
9. Tiepolo, M.; Bacci, M.; Braccio, S.; Bechis, S. Multi-hazard risk assessment at community level integrating local and scientific knowledge in the Hodh Chargui, Mauritania. *Sustainability* **2019**, *11*, 5063. [CrossRef]
10. Demarchi, A.; Bechis, S.; Perotti, L.; Garnero, G.; Isotta Cristofori, E.; Alunno, L.; Facello, A.; Semita, C.; Bonetto, S.M.R.; Guerreschi, P.; et al. An interdisciplinary approach to the analysis of agro pastoral resilience in the Hodh el Chargui region (Mauritania). In Proceedings of the EGU General Assembly Conference Abstracts, Vienna, Austria, 4–13 April 2018.
11. Friedel, M.J. *Inventory and review of existing PRISM hydrogeologic data for the Islamic Republic of Mauritania, Africa*; Open File Report 2008-1138; U. S. Geological Survey: Reston, VA, USA, 2008.
12. Zhang, H.K.; Roy, D.P.; Yan, L.; Li, Z.; Huang, H.; Vermote, E.; Skakun, S.; Roger, J.-C. Characterization of Sentinel-2A and Landsat-8 top of atmosphere, surface, and nadir BRDF adjusted reflectance and NDVI differences. *Remote Sens. Environ.* **2018**, *215*, 482–494. [CrossRef]
13. Yang, L.; Meng, X.; Zhang, X. SRTM DEM and its application advances. *Int. J. Remote Sens.* **2011**, *32*, 3875–3896. [CrossRef]
14. USGS the Shuttle Radar Topography Mission (SRTM) Collection User Guide. 2015. Available online: https://lpdaac.usgs.gov/documents/179/SRTM_User_Guide_V3.pdf (accessed on 25 May 2020).
15. Radwan, F.; Alazba, A.; Mossad, A. Watershed morphometric analysis of Wadi Baish Dam catchment area using integrated GIS-based approach. *Arab. J. Geosci.* **2017**, *10*, 256. [CrossRef]
16. Lai, Z.; Li, S.; Lv, G.; Pan, Z.; Fei, G. Watershed delineation using hydrographic features and a DEM in plain river network region. *Hydrol. Process.* **2015**, *30*, 276–288. [CrossRef]
17. Liu, X.; Wang, N.; Shao, J.; Chu, X. An automated processing algorithm for flat areas resulting from DEM Filling and interpolation. *ISPRS Int. J. Geo Inf.* **2017**, *6*, 376. [CrossRef]
18. Djokic, D.; Ye, Z.; Dartiguenave, C. *Arc Hydro Tools Overview, Version 2.0*; ESRI Press: Redlands, CA, USA, 2011.
19. *Arc Hydro Geoprocessing Tools-Tutorial-Version 2.0*; ESRI Press: Redlands, CA, USA, 2011.
20. Congedo, L. Semi-automatic classification plugin documentation. *Release* **2016**, *4*, 29.
21. Ghiraldi, L.; Giordano, E.; Perotti, L.; Giardino, M. Digital tools for collection, promotion and visualisation of geoscientific data: Case study of seguret valley (Piemonte, NW Italy). *Geoheritage* **2014**, *6*, 103–112. [CrossRef]
22. Giuffrida, S.; Gagliano, F.; Giannitrapani, E.; Marisca, C.; Napoli, G.; Trovato, M.R. Promoting research and landscape experience in the management of the archaeological networks. A project-valuation experiment in Italy. *Sustainability* **2020**, *12*, 4022. [CrossRef]
23. Dan-Badjo, A.T.; Diadie, H.O.; Bonetto, S.M.R.; Semita, C.; Cristofori, E.I.; Facello, A. Using improved varieties of pearl millet in rainfed agriculture in response to climate change: A case study in the tillabéri region in Niger. In *Climate Change Research at Universities*; Springer International Publishing: Cham, Switzerland, 2017; pp. 345–358.
24. World Health Organization (WHO). *Drinking Water Parameter Cooperation Project. Support to the Revision of Annex I Council Directive 98/83/EC on the Quality of Water Intended for Human Consumption (Drinking Water Directive)*; WHO: Geneva, Switzerland, 2017.
25. Agriculture and Food. Water Quality for Livestock. Available online: https://www.agric.wa.gov.au/livestock-biosecurity/water-quality-livestock (accessed on 15 May 2020).

Article

Flood Assessment for Risk-Informed Planning along the Sirba River, Niger

Maurizio Tiepolo [1,*], Maurizio Rosso [2], Giovanni Massazza [1], Elena Belcore [1], Suradji Issa [3] and Sarah Braccio [1]

1 DIST, Politecnico and University of Turin, viale Mattioli 39, 10125 Turin, Italy
2 DIATI, Politecnico of Turin, Corso Duca degli Abruzzi 24, 10129 Turin, Italy
3 Direction Departementale de l'Agriculture, Gothèye, Niger
* Correspondence: maurizio.tiepolo@polito.it; Tel.: +39-0110907491

Received: 29 June 2019; Accepted: 22 July 2019; Published: 24 July 2019

Abstract: South of the Sahara, flood vulnerability and risk assessments at local level rarely identify the exposed areas according to the probability of flooding or the actions in place, or localize the exposed items. They are, therefore, of little use for local development, risk prevention, and contingency planning. The aim of this article is to assess the flood risk, providing useful information for local planning and an assessment methodology useful for other case studies. As a result, the first step involves identifying the information required by the local plans most used south of the Sahara. Four rural communities in Niger, frequently flooded by the Sirba River, are then considered. The risk is the product of the probability of a flood multiplied by the potential damage. Local knowledge and knowledge derived from a hydraulic numerical model, digital terrain model, very high resolution multispectral orthoimages, and daily precipitation are used. The assessment identifies the probability of fluvial and pluvial flooding, the exposed areas, the position, quantity, type, replacement value of exposed items, and the risk level according to three flooding scenarios. Fifteen actions are suggested to reduce the risk and to turn adversity into opportunity.

Keywords: climate change; contingency plan; flood risk; local development plan; risk management; sustainable rural development

1. Introduction

In the first decade of this century, floods have struck 11.5 million people south of the Sahara [1]. It is not surprising that community preparedness appears in the Agenda of the African Union [2]. Nevertheless, between 2013 and 2017, official development aid spent just 0.1 million Euros on this activity against 12.4 billion Euros used for disaster risk reduction in the Subcontinent [3]. Various multilateral organizations have urged or are supporting the preparation of local disaster risk reduction and contingency plans [4,5]. At present, there is little information on the state of local planning [6]. So far, the scientific community has been engaged above all in flood vulnerability and risk assessments, important activities which are, however, difficult to coordinate with local planning. For some time, attempts have been made to improve communication between climatologists and planners, also providing to the latter simplified analysis methods [7], aiming for greater collaboration [8,9], and urging the analysts to recommend specific actions [10]. The community-based approach used in many assessments has not produced the quantitative information expected by planning [11] and seems inadequate to appreciate the risks caused by climatic changes that communities have not yet experienced [12]. Ultimately, the use of vulnerability and risk assessments in planning remains a critical point [13–16]. The systematic review of flood assessments in the sub-Saharan context produced in recent years (Table 1) reveals that just one in four identifies the exposed areas according to the probability of flooding. The threshold (flood level or amount of precipitation) above which flooding occurs is not identified. Therefore, the assessments cannot be used for early warning. One assessment out of four identifies the exposed items, but none considers among these the infrastructures and crops. Climate and land use/land cover changes are not considered. Just one assessment out of three identifies the actions in place to reduce risk. No assessment considers the opportunities offered by floods for sustainable rural development.

Table 1. Sub-Saharan Africa, 2010–2019. Consistency of 23 vulnerability and risk assessments with the needs of local plans.

Information Required by Planners and Decision Makers		**Information Provided by Assessments (Reference)**
Community resources	Capacities	[14,17–22]
	Assets	[17,19]
Hazard	Probability of occurrence	[21,23–25]
Flood-prone area map		[18,23,25–29]
	According probability of occurrence	[21,23,29–32]
Flood vulnerability map		[21,30,31,33–39]
Exposed items	Location	[18,24,32]
	Type	[18,22,24,25,27]
	Quantity	[18,22,24,25,32]
	Replacement value	[24,25]
Risk index		[21,25]
Vulnerability index		[19]
Opportunities		-
Actions	Ongoing	[18,22,25,27,38,40,41]
	Future	[18,20,21,25,27,29,30,33,38,40–42]

In summary, the problem is that the findings of the assessments are not aligned with the information requirements of planners and decision makers. One solution may be to abandon the practice of conducting assessments in an isolated manner and to develop them by coordinating the interests of several organizations. The pool assessments thus implemented would be more likely to contribute to disaster risk reduction (DRR) strategies [6]. A quicker solution would be to identify preliminarily the information required by the local plans and to establish the types of findings expected from the assessments accordingly.

The aim of this article is therefore to propose a flood risk assessment oriented at local development, risk prevention, and contingency plans. A detailed and focused flood risk assessment assists the evaluation (risk level, identification of actions) and decision-making process of risk preparation (giving priority to actions, localizing them, and planning them over time).

From here onwards, using the term 'local level', we will refer to the minimum administrative jurisdiction (the municipality) and the term 'contingency plan' will be considered synonymous with emergency plan.

The flood risk assessment is developed with four rural communities distributed along 30 km terminals of the Sirba River: Tallé (population 2603 in 2012), Garbey Kourou (4634), Larba Birno (4713), Touré (4065). All communities are on the left bank and belong to the municipality of Gothèye, Niger. The Sirba has a transboundary watershed of 39,138 km^2, whose upstream part is in Burkina Faso (93%) and terminal part is in Niger (7%) (Figure 1).

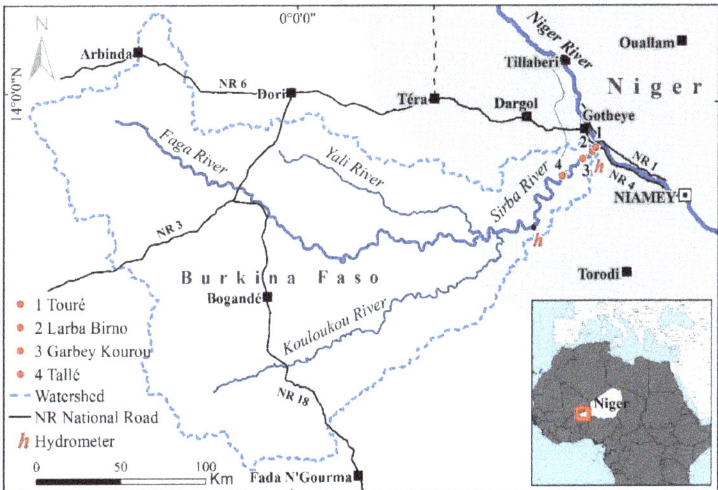

Figure 1. The transboundary watershed of the Sirba River and the four communities.

Although these communities are primarily exposed to drought [22], they were flooded by the Sirba in August 2012, by heavy rainfall in July 2018, and by the backwater of the Niger River in January 2019 [43].

The risk assessment considers the risk (R) as 'the probability of occurrence of hazardous events or trends multiplied by the consequences if these events occur' [44]. The risk is therefore the product of the hazard (H), or 'the potential occurrence of a natural or human-induced physical event or trend, or physical impact, that may cause loss of life, injury, or other health impacts, as well as damage and loss to property, infrastructure, livelihoods, service provision, and environmental resources' and the potential damages (D): $R = H \times D$. The damages from flooding have already been considered as a determinant of the risk in Niger [45], in the Global South [46–51], and in the OECD member states [52,53]. The equation used is an alternative to the one that includes exposure, vulnerability, and adaptation. Potential damages constitute the ultimate effect of exposure, vulnerability, and adaptation [54]. The damage calculation is done on the reconstruction or/and replacement cost of each item exposed to flood. Thus, the assessment considers only the tangible, direct costs that could be generated in buildings, infrastructures, and crops.

The risk assessment is a process of definition of the scope, criteria, comprehension of the context, identification, analysis, and assessment of the risk [55]. As a result, the assessment is organized into four phases.

The first phase identifies the information that should be used in planning according to national guidelines for the preparation of local plans and according to literature [56]. This makes it possible to establish the findings expected from the risk assessment. The flood risk levels used and the local plans required by law in Niger are therefore ascertained.

The second phase identifies the flood risks (sources, causes, and events) and the resources held by each community (capacity and assets) to address them.

The third phase is dedicated to analysis. The probabilities of fluvial and pluvial flooding are calculated and the consequent flood zones are mapped, the individual exposed items are identified, their replacement value is estimated, and the risk level is determined.

The fourth and final phase identifies the possible risk reduction actions, considering the ongoing adaptation efforts developed by each community.

Risk treatment (selection of actions, distribution over time, and implementation) is not the aim of the assessment as this belongs to the planning phase.

The findings of the assessment are obtained integrating local knowledge with methodologies and information still little used in sub-Saharan Africa but which are promising for their quality and appropriateness to local development, risk preparation, and contingency plans. In the next sections, the materials and methods, results achieved, discussion, and conclusions are presented.

2. Materials and Methods

The assessment of the fluvial and pluvial flooding risk in Tallé, Garbey Kourou, Larba Birno, and Touré uses local and technical knowledge according to four phases.

The first phase compares the results of the systematic review of the published vulnerability and risk assessments with the information required by the national guidelines for the preparation of development plans in Benin, Burkina Faso, Cameroon, Madagascar, Mauritania, Niger, Senegal, South Africa, Uganda, and that used in a selection of local plans. This first step highlights the outputs required from a planning-oriented risk assessment. As regards the risk criteria, the three flooding scenarios used by the Niger Basin Authority were considered: frequent (yellow), severe (orange), and catastrophic (red). The categories of local plans required by law in Niger were identified by questioning the Sustainable Development and Environment National Council (CNEDD, according to the French acronym) and the Directorate General for Civil Protection (DGPC, in French).

The second phase identifies, through meetings with each community, the hydro-climatic threats, the past catastrophic events, the rainfall threshold, and the flood level over which damages are produced and local resources mobilized to address them (Figure 2).

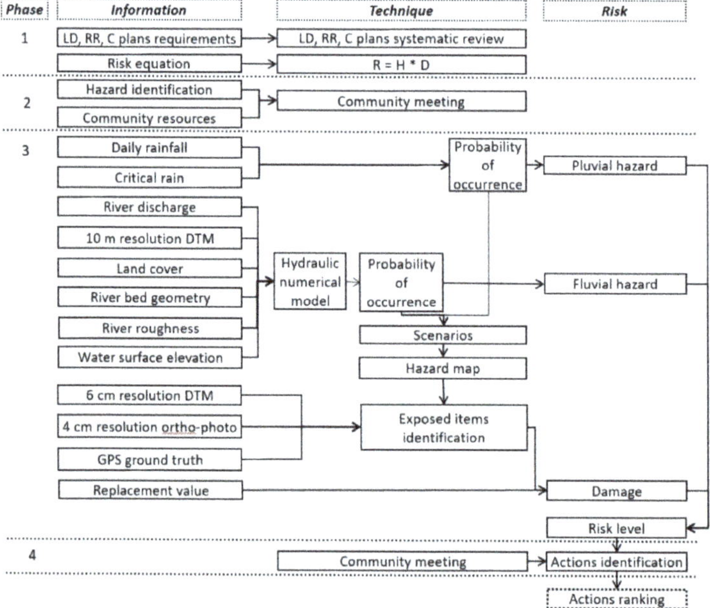

Figure 2. Flood risk assessment flowchart.

The third phase firstly calculates the probability of occurrence (1/return period) of a flood according to three scenarios of fluvial and pluvial flooding (frequent, severe, catastrophic). The probability of flooding of the Sirba and the backwater of the Niger Rivers is calculated based upon the discharge recorded respectively at the stations of Garbey Kourou in the period 1956–2018 and of Niamey in the period 1929–2019 with a Generalized Extreme Values approach [23].

The probability of pluvial flooding is calculated starting from the daily rainfall and related damages recorded by each community in a register provided by the National Directorate for Meteorology. The probability of occurrence of frequent, severe, and catastrophic rainfall is then calculated with respect to the daily precipitations recorded at the meteorological station of Gothèye (30 km away) between 1961 and 2015. Severe rainfall forms ephemeral water bodies in Larba Birno and Touré that may cause the collapse of buildings and the leakage of contents of the latrine pit.

The area exposed to fluvial flooding according to the three hazard scenarios was identified through the hydraulic numerical model developed on the HEC-RAS software in a 1D configuration [23]. The model calculates the water surface elevations for the different discharges and extends them on the riverbed geometry. The geometry was constructed on a Digital Terrain Model—DTM with 10 m of horizontal resolution, detailed by way of river cross-sections detected each kilometer and hydraulic structures. The roughness was based on riverbed granulometry and the downstream conditions considering the hydraulic levels of the Niger River. This model was calibrated with the hydraulic levels measured at the Bossey Bangou and Garbey Kourou hydrometric stations and in the communities during the 2018 rainy season. This tool allows, towards unsteady simulations conducted with measured hydrographs, the estimation of the propagation and submergence times during the fluvial flooding.

The area exposed to pluvial flooding was identified firstly with an inspection aimed at localizing the biggest ephemeral water bodies in late August 2018, when three-quarters of the annual precipitation had already been accumulated. The identification continued by taking red, green, blue (RGB) images with a 24.3 megapixel camera and RGB near-infrared (NIR) images with a specially created low-cost 5 MP optical sensor from an unmanned aerial vehicle (UAV) [55]. Two flights were made on each community on 14 and 15 September 2018 at 270 m above ground level with the RGB camera and at 120 m with the NIR camera. The collected data were processed with SfM software and orthophotos to 6 cm in RGBN and 4 cm in RGB were obtained for each community. From the RGB data, a DTM of 6 cm of resolution was generated [57]. The water was identified using the normalized differential water index (NDWI = (Green − NIR)/(Green + NIR)) [58] on the multispectral images obtained. Finally, the area surrounding the ephemeral water bodies was estimated by way of the DTM (Supplementary File 1).

The exposed items were identified by way of photo interpretation of the orthoimages at 4 cm resolution recorded by the UAV (Supplementary File 2). Finally, the areas of potential water stagnation (i.e., maximum expected extension of stagnation) were estimated on the basis of the DTM. Those images identify the latrines, granaries, boreholes, wells, fountains, photovoltaic miniplants, and rain-fed and irrigated crops that could not be identified on high-resolution satellite images (Figure 3).

The estimate of damages to the buildings is not based upon the stage-damage function, nor does it consider the flow velocity since the dwellings collapse once they are flooded, as the majority are built from crude earth masonry. The flooding duration and the depth of the water are considered in order to estimate the damages to the crops.

The replacement value of the buildings is estimated through a focus group in each community, considering the construction cost of a standard 24 m^2 crude earth house, a latrine, a shower, and a granary. The crop yields are estimated by the Departmental Directorate for Agriculture of Gothèye, which also monitors the prices of agricultural products on the markets. Yields and prices of millet and paddy originate from statistics of the Ministry of Agriculture [59].

The risk level is obtained by multiplying the hazard by the damages in the case of fluvial and pluvial flooding according to the frequent, severe, and catastrophic scenarios.

The fourth phase identifies actions to reduce the risk associated with the type of flooding which threatens each community and with the exposed items as proposed by the individual communities during a participative meeting held on 25–26 June 2019 and integrated, after discussion, with some recurring actions in the examined plans.

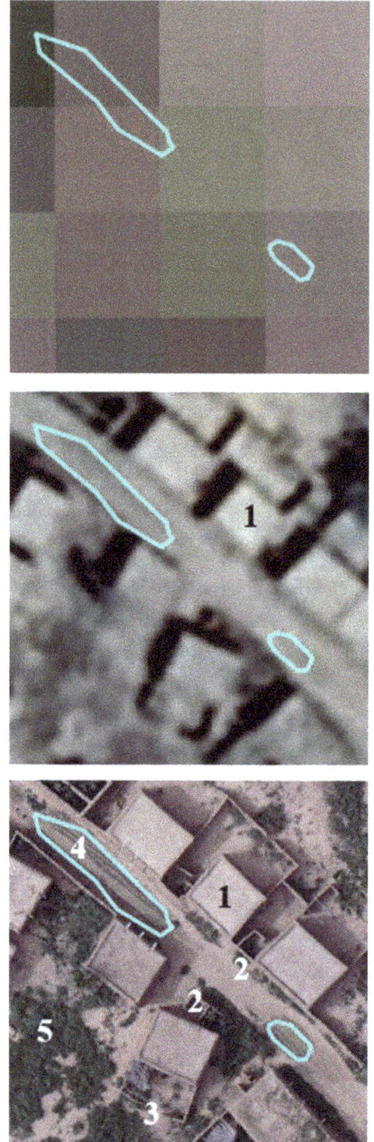

Figure 3. Touré, 2018. Images for exposed items' identification: Sentinel satellite image with 10 m resolution (top); satellite high-resolution image freely available from Google Earth showing (1) earth construction (center); very high resolution image taken from a UAV showing (1) hearth house, (2) latrines, (3) house entrance, (4) ephemeral water body, (5) millet (bottom).

3. Results

3.1. Aligning Flood Risk Assessments with Local Plans Requirements

The reduction of the hydro-climatic risk in Niger is essentially entrusted to the municipal development plan (MDP). From 2015, MDPs must identify the impact of climate change and actions to reduce it [60]. The local administrations are struggling to perform this task effectively. It must be considered that in Niger the municipalities are vast jurisdictions that include many rural settlements in strong expansion and multiplication. Gothèye, for example, as an administrative jurisdiction of 3600 km^2, includes 146 rural settlements and over 93,000 inhabitants (2012). The rural municipalities and the consultants used by them to draw up development plans have access to scant geographical information and often do not have the tools, resources, and time to collect additional information and to process it. As a result, the risk reduction actions are identified based upon the needs expressed by the communities during rapid consultations. In the best cases, the planned actions concern the reduction of the runoff (half-moons, stone lines, trapezoidal bunds) and of riverbank erosion (gabions, tree planting), and the increase in tree cover, measures that have no resolving effects on the fluvial flooding risk.

Community adaptation plans as introduced firstly by the ANADIA project (Italian development aid) and then by the BRACED project (English development aid) are experimental, voluntary tools. Development plans for individual settlements, as tested in Togo, and risk reduction plans, quite common in Latin America, are uncommon in Niger. The latter are usually made up of a presentation report, a map of the flood-prone areas, a map of the exposed items, a zoning plan which subdivides the territory depending on the flood probability, and a regulation, which specifies what is permitted, prohibited, and worth doing in each zone. The risk reduction plans usually recommend the preparation of a contingency plan.

In future, the contingency plan, as tested on a municipal scale in Burundi, South Africa and on a scale of individual settlements in Mali, may be required by law in Niger to address the hydro-climatic hazards in the most densely populated flood risk areas. This plan should be coordinated with the early warning system (EWS) and constructed taking account of local capacities and assets. An emergency committee, a map of the flood areas according to flooding probability, a map of exposed items to guide information, the drills, and making accessible the refuge sites to households settled in flood-prone areas are recurrent components of this plan.

These components of the plans, as taken from the national guidelines of Niger and eight other African countries and from the literature, need specific information from vulnerability and risk assessments (Figure 4).

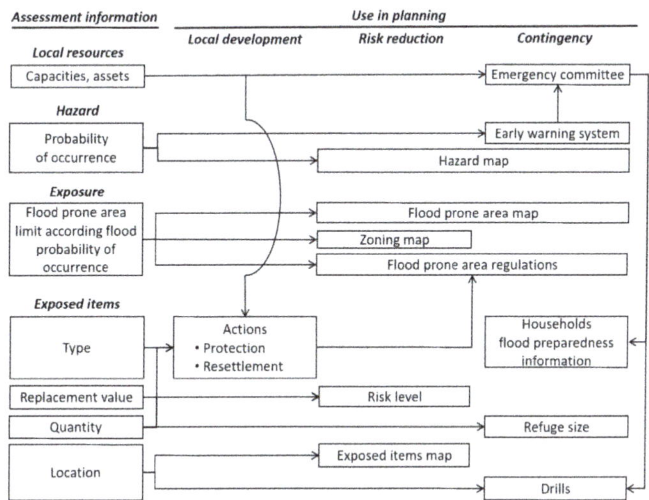

Figure 4. Use of vulnerability and risk assessments findings according to national guidelines for local development plan preparation in Niger and in eight other African countries.

In the nine considered countries, municipal development plans made 98% of local plans, and are a consolidated planning tool, now in its third generation. Contingency, risk prevention, and adaptation plans at the local level remain still occasional tools.

The assessment thus provides useful information firstly for the MDPs and, prospectively, for the risk prevention and contingency plans.

3.2. Flood Hazards

The four communities along the Sirba are exposed to fluvial flooding during the rainy season. Tallé is exposed to fluvial flooding also in the dry season, due to the backwater of the Niger River. Larba Birno and Touré are also exposed to pluvial flooding (Table 2, Figures 5–8).

Table 2. Four communities along the Sirba River. Hazard.

Hazard	Probability		Return Period	Discharge	Rainfall
	Level	%	Years	m^3/s	mm
Fluvial Sirba	Yellow	10	10	800	
	Orange	3	30	1500	
	Red	1	100	2400	
Backwater	Red	5	20	2238	
Pluvial	Yellow	10	10	-	90
	Orange	6	17	-	100
	Red	2	50	-	200

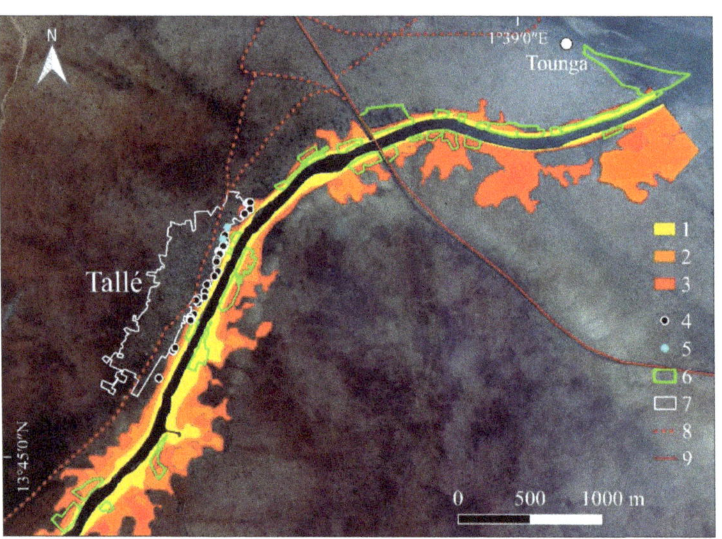

Figure 5. Tallé, 2018. Hazard and exposed items map: frequent (1), severe (2), catastrophic (3), flood-prone area, exposed house (4), well or borehole (5), field (6), built-up area (7), unpaved roads (8), and paved roads (9).

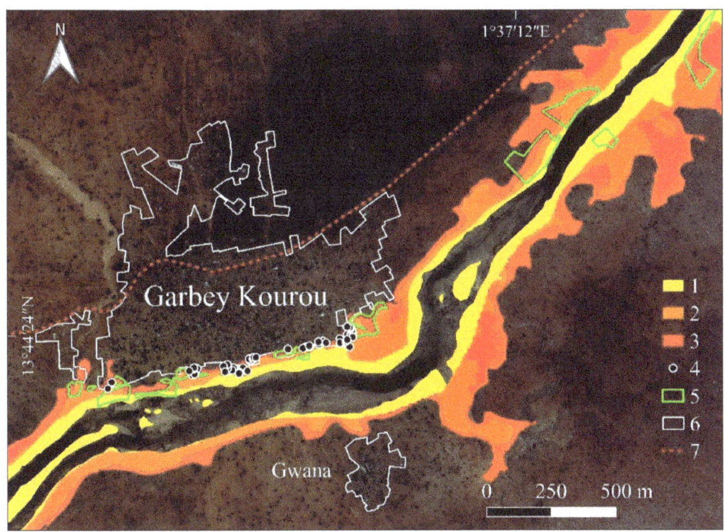

Figure 6. Garbey Kourou, 2018. Hazard and exposed items map: frequent (1), severe (2), catastrophic (3), flood-prone area, exposed house (4), field (5), built-up area (6), unpaved road (7).

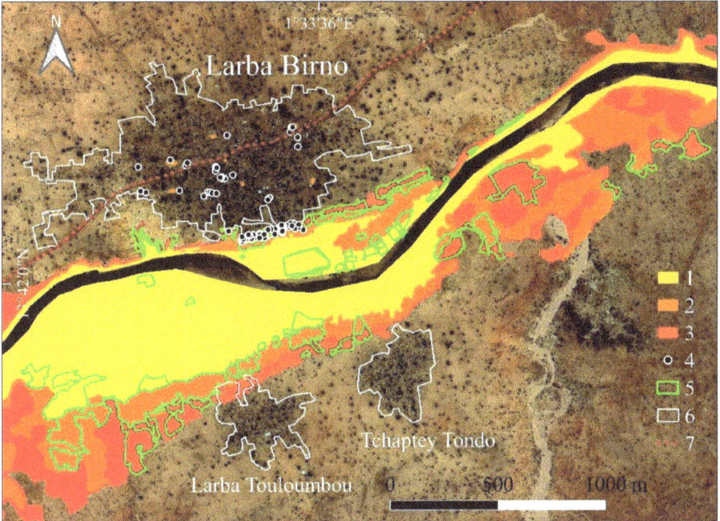

Figure 7. Larba Birno, 2018. Hazard and exposed items map: frequent (1), severe (2), catastrophic (3), flood-prone area, exposed house (4), field (5), built-up area (6), unpaved road (7).

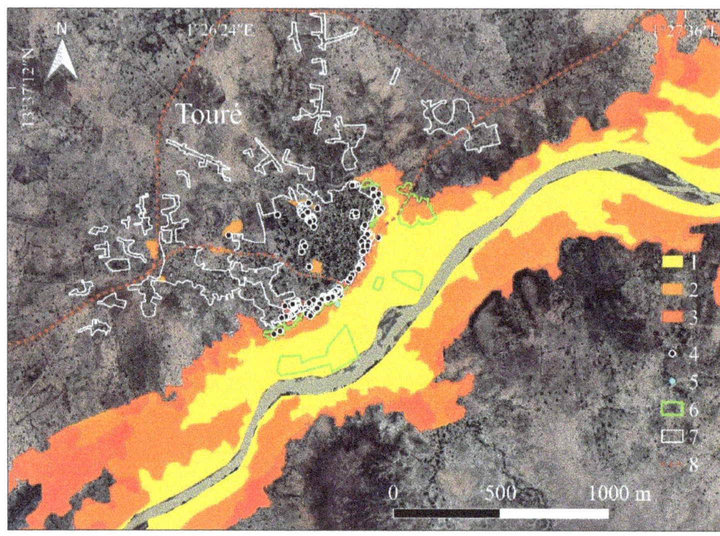

Figure 8. Touré, 2018. Hazard and exposed items map: frequent (1), severe (2), catastrophic (3), flood-prone area, exposed house (4), well or borehole (5), field (6), built-up area (7), unpaved roads (8).

The analysis of the historical series of the Sirba discharge identifies that a flood of 800 m^3/s has a return period (RP) of 10 years and can be considered frequent, a flood of 1500 m^3/s has an RP of 30 years and is severe, and a flood of 2400 m^3/s has an RP of 100 years and is catastrophic. The analysis of local precipitations highlights that a daily precipitation with 90 mm of rain has an RP of 10 years; it is frequent and can cause some damage. A daily precipitation with 100 mm of rain has an RP of 17 years, and should be considered a severe event as it generates great damage. A day with 200 mm of rain has an RP of 50 years and generates such damage as to be considered a catastrophic event (Table 2).

The hydraulic numerical model identifies areas exposed to frequent (RP of 10 years), severe (RP of 30 years), and catastrophic (RP of 100 years) flooding in circumstances of homogeneity of the historical series of discharges (stationary approach). The RP reduces respectively to 2, 5, and 10 years [43] if the nonstationary approach is adopted, necessary to consider strong changes in hydrology which reflect both changes in the climate [61] and in the land use/land cover [62–64].

The Sirba flood reaches its peak in 3–9 days and returns to normal level in 4–11 days. The propagation time from Bossey Bangou (108 km upstream of the Niger River, where a hydrometric station has operated since 2018) to Touré is 20 h, to Larba Birno is 26 h, and to Garbey Kourou and Tallé is 28 h.

3.3. Flood Damage

The very high resolution orthoimages facilitate the identification of exposed items that are not visible on satellite images (such as those available for free on Google Earth) and the identification of up to one-third more exposed items. The dwellings appear to have been withdrawn from areas exposed to frequent and severe flooding, but the crops, particularly the commercial crops in Tallé, wells, and boreholes at Larba Birno are plentiful (Table 3).

The area exposed to catastrophic flooding contains the bulk of the exposed items in two out of four communities. This stock of dwellings has the floor at ground level, the door has no threshold, and the walls and roof are made from crude earth. These constructions collapse as soon as the water enters inside or rains too intensely. Boreholes, wells, and fountains without a foundation above water level cannot be used for weeks. The lengthy duration of the flood destroys the crops, once submerged.

Table 3. Four communities along the Sirba River, 2018. Houses in flood-prone areas.

Flood Hazard		Houses in Flood Prone Zones			
Type	Level	Garbey Kourou n.	Larba Birno n.	Tallé n.	Touré n.
Fluvial	Yellow	0	0	6	0
Fluvial	Orange	24	3	20	3
Fluvial	Red	33	41	48	79
Backwater	Red	0	0	-	-
Pluvial	Orange	-	23	-	28

The total amount of damage is the highest in Tallé in the event of backwater of the Niger River, followed by Touré in the case of fluvial flooding, and by Larba Birno in the case of pluvial flooding (Table 4). Buildings account for the bulk of the damages in all scenarios and communities (from 84% to 92%) with the sole exception of the flood due to the backwater of the Niger River (Table 5).

Table 4. Four communities along the Sirba River. Replacement value of the exposed items K €.

Flood Hazard	Replacement Value of the Exposed Items			
	Garbey Kourou K €	Larba Birno K €	Tallé K €	Touré K €
Fluvial yellow	0.2	1.3	13	0.3
Fluvial orange	8.2	4.1	6.9	1.3
Fluvial red	13.5	16.9	20.6	34.4
Backwater red	0	0	40.7	0
Pluvial orange	0	13.6	0	11.6

Table 5. Four communities along the Sirba River. Exposure of houses over all exposed items (% of total replacement value, K €).

Flood Hazard	Exposure of Houses over All Exposed Items			
	Garbey Kourou %	Larba Birno %	Tallé %	Touré %
Fluvial yellow	0	0	14	0
Fluvial orange	88	28	86	84
Fluvial red	92	91	69	87
Backwater red	-	0	0	-
Pluvial orange	-	100	-	100

3.4. Flood Risk Level

Larba Birno presents the highest risk due to a high value of exposed items in an area with medium probability of pluvial flooding (Table 6). Touré follows, with numerous dwellings in an area with medium probability of pluvial flooding. Then comes Tallé, with a high value of exposed items in an area with low probability of river flooding, and finally Garbey Kourou, with a low value of exposed items in an area with low probability of flooding.

Table 6. Four communities along the Sirba River. Risk level.

Flood Hazard	Risk Level				
	Probability	Garbey Kourou M €	Larba Birno M €	Tallé M €	Touré M €
Fluvial yellow	10	0.02	0.13	1.3	0.03
Fluvial orange	3	0.27	0.12	0.23	0.04
Fluvial red	1	0.14	0.17	0.21	0.34
Backwater red	6	-	-	0.65	-
Pluvial orange	6	-	0.82	-	0.69

3.5. Identification of Actions for Risk Reduction

Community meetings have shown that the pluvial flood opens up opportunities for recession farming (okra) in Garbey Kourou, in the ephemeral pond upstream of the road, and the fluvial flood allows crops of cowpeas, corn, and squash in Touré, especially on the right bank of the Sirba River.

The proposed actions consider these results and the actions in progress, limited to an emergency committee, flood drills, and a hydrometric station at Garbey Kourou, a manual rain gauge and a river gauge in every community, along with the withdrawal of dwellings from areas exposed to frequent and severe flooding (Table 7), a flood and drought multihazard community adaptation plan (2014) in Garbey Kourou and Tallé.

Table 7. Four communities along the Sirba River. Ongoing adaptation actions.

Flood Hazard	Action	On-Going Adaptation Actions			
		Garbey Kourou	Larba Birno	Tallé	Touré
Fluvial flood	Committee	●			
	Setback	●			
	Hydrometric station	●			
Pluvial flood	River gauge	●	●	●	●
	Manual pluviometer	●	●	●	●
	Adaptation plan	●	●	●	
	Drills	●			

Other aspects to consider are that the inhabitants settled in the flood-prone area do not want to leave the river because they go there every day to do their laundry and to wash themselves, and there is the desire to remain in the large initial lots of the settlement, in which the ancestors lived. Finally, even in the case of destruction of the house by the river or heavy rains, the tendency is to reconstruct a little further upstream to reuse the clay of the old bricks that elsewhere would not be available or too tiring to transport. In many cases, therefore, it is not access to land that pushes some inhabitants into areas at risk but rather reasons linked to daily life, the cost of transfer, and the link with the place.

The limit of catastrophic flooding (which may occur with a probability 10 times higher according to a nonstationary approach) should be reported and the platform of the boreholes, wells, fountains, and photovoltaic miniplants should be raised as a result. The ephemeral water bodies within the built-up areas should be treated by a storm water drainage. Rain-fed crops should be moved away from the riverbanks. Paddy fields should be protected, remaking the river levees.

When farmers use their own land for commercial gardens, having them set back from the riverbanks should be possible. Cassava should be preferred to other crops since it demands less water.

Any flood risk prevention plan should contain a map of the flood area, the exposed actions, a zoning plan according to flooding probability, and a regulation.

The plan should prohibit further construction and the replacement of dwellings made from crude earth with durable dwellings in the area exposed to catastrophic flooding, unless the floor is above the level of the water.

As regards the contingency plan, an early warning system should be prepared and implemented using local rural radio and by spreading door-to-door information on how to prepare for flooding for all households settled in the area exposed to catastrophic flooding.

A contingency plan always includes a refuge site dimensioned on the number of inhabitants present in the area at risk. However, the communities along the Sirba do not need this structure because during the season in which a flood may occur, the settlements are emptied and the inhabitants move near the fields in temporary settlements to follow the crops. Drills should be organized every year and the emergency committee reactivated or established (Figure 9).

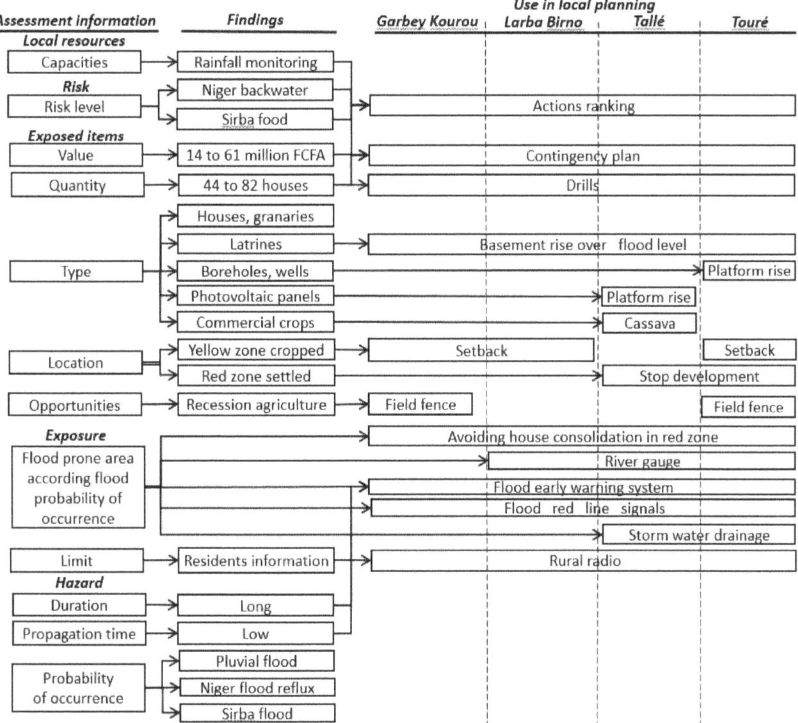

Figure 9. Use of risk assessment for local planning.

4. Discussion

Exceptions aside, the flood vulnerability and risk assessments considered in the systematic review are deficient in identifying the flood-prone area according to the frequency of the flood and critical rain, the exposed items, and the ongoing actions and they are, therefore, of little use for MDPs and, prospectively, for risk prevention and contingency plans at local level [14].

The initial problem was firstly to align the expected findings of the assessment with the necessary information to reduce flood risk. As a result, the requirements of information for each category of local plans in use in sub-Saharan Africa and in Niger, in particular, were identified. The assessment for Garbey Kourou, Larba Birno, Tallé, and Touré was developed using methods and information based upon those needs: a hydraulic numerical model, the river discharge time series, the long-term daily pluviometric series, DTM, and very high resolution orthoimage, completed by inspections and meetings with the communities. The adopted methodology is totally replicable in order to support the sustainable development of each rural community both in Sahelian areas and in similar contexts.

In this way, it was ascertained that the waters of the Sirba and the Niger Rivers slowly reach the flood peak. This involves the loss of crops once they are submerged. The flood subsides slowly. This means that recession agriculture after Niger River backwater is not possible and that the flood does not bring benefits to Tallé, but only damage. On the contrary, the fluvial flood in Larba Birno and the pluvial flood in Garbey Kourou allow recession agriculture. In these cases, climate change and the consequent flooding constitute an opportunity—an issue still little investigated by the literature. The slow river regime does, however, have an advantage: a flood propagation time long enough to allow for the mayor and the communities downstream to be alerted with sufficient warning to react, even if the flood peak occurs at night.

The multihazard assessment (fluvial flooding with two seasonal peaks and pluvial flooding) highlights that within a river reach of just 30 km, each community is exposed to specific hazards, has different exposed items in the flood prone area, and sometime offers rural development opportunities, and thus requires equally specific measures, a result opposite to that found in other studies [14].

Commercial crops are the main exposed items in the zone with high probability of fluvial flooding for all considered communities and, in the case of Tallé, also for the backwater of the Niger River.

Commercial agriculture is mainly carried out by women and is of crucial importance for rural households. Revenues obtained from the sale of cassava, potatoes, cabbage, squash, and tomatoes make it possible to compensate for the losses of rain-fed cereal crops in the event of agricultural drought and to sustain health, education, and clothing costs. The importance of damage to commercial crops that we have highlighted constitutes a paradox on which the literature has not sufficiently dwelt: To guarantee sustainable rural development in a semiarid context, it is essential to consider the risk of flooding as well as the risk of drought. However, the actions to protect commercial crops from flooding are expensive: the further away you go from the flood-prone area, the more need for wells whose realization, in the specific context of the Sirba, characterized by frequent rocky outcrops, can be very expensive.

The current assessments do not quantify the damages in such a detailed way due to the low-resolution images used, which prevent the individual exposed items from being localized.

The presence of many dwellings in the area exposed to catastrophic flooding, thus of many households, is scantly evidenced in the literature [24,32].

There are few risk reduction actions in progress. As to the actions to be developed, the list is longer and more specific (information, awareness, contingency plan, drills, basement/platform rise, storm water drainage, flood signals, prohibition of house consolidation, field metal fences) than that proposed by the assessments considered which mostly recommend early warning systems [20,21,29,38], resettlement [18,27,30], dam and levee construction [29,31,38], and house improvements [21,40]. These actions are not sufficient in the case study since they do not refer to specific hazards, exposed items, and plan categories.

The assessment identified the discharges at the upstream hydrometer located at Bossey Bangou, close to the Niger–Burkina Faso border, beyond which damages may occur downstream and their probability of occurrence linked the three hazard levels (frequent, severe, catastrophic) to the early warning.

The assessment demonstrates the potential of very high resolution DTM and orthoimages still little used in vulnerability and risk assessments at the local level south of the Sahara. This technology reduces the Information Technology divide if used by local firms using low-cost sensors. In addition, it offers information on the flood-prone areas and on the exposed items which high-resolution satellite images are unable to provide or which require lengthy and costly inspections.

The method that combines local knowledge and technical knowledge presents at least three advantages: it provides a complete photograph of each settlement, which inspections are unable to produce, it is feasible even in the sweltering months (April–May) during which it is not possible to perform lengthy inspections, and it provides a quantitative basis for risk monitoring. Nevertheless, the assessment does have two limits. First of all, the replacement value of the exposed items does not consider domestic vegetable gardens, plentiful above all on the right bank at Larba Birno, the service interruption of the boreholes, the wells, the fountains, the photovoltaic miniplants, or the impacts of ephemeral water bodies on health. Second, the value of production in recession agriculture was not calculated because the assessment took place too far after the last major flood (2012) and therefore the fields cultivated in this mode were no longer observable. These aspects should be addressed by future assessments. The estimate of damages is thus conservative. In addition, in the space of a few years, more high-value exposed items could be found in the area at low risk of river flooding in the absence of actions to limit their proliferation.

The findings of the assessment are above all of interest for local development, risk prevention, and contingency planning, since they provide precise quantities in defined locations and for a localized population in order to decide on a set of proposed actions. Secondly, the results of the assessment are used to validate the hydraulic numerical model: the detailed identification of the exposed items and the flooded areas allows for the optimization of their calibration and the observation of changes following future floods [65].

5. Conclusions

In Africa, south of the Sahara, flood risk assessments at the local level do not provide information aligned with those required by local development plans, risk prevention plans, and contingency plans. A recurrent case concerns the detail of the information provided by the assessments, which is often more appropriate to the regional scale than to the local one. Furthermore, the areas exposed to flooding, the items they contain, and the proposal of specific actions to reduce the risk of flooding appropriate to the particular characteristics of the investigated contexts are too often absent from the assessments. Assessments are therefore not very useful to support an informed decision-making process in risk reduction. To overcome this problem, the simplest solution is to firstly identify the information required by local plans and consequently establish the type of results that the assessment should achieve. Furthermore, this involves reviewing the assessment approach, which must consequently integrate local knowledge with scientific knowledge, for example, by supporting the local ITC (UAV images) which provide more detailed information than anything obtained through commercial high-resolution images.

The flood risk assessment developed along the Sirba River considers and meets the needs of local planning as it establishes the probability of occurrence according to the flooding scenarios established by the Niger River Basin Authority; it consequently identifies the flood-prone areas and exposed items, assesses the potential damages and risk level, and identifies 15 risk-reduction actions to be proposed to the decision makers. Contrary to conventional assessments, this work identifies the opportunities for rural development offered by an event generally considered adverse, such as flooding. The use of very high-resolution images, thereby limits the inspections to a few, targeted receptors, saving time.

The multihazard approach, the integration of local and scientific knowledge, and the techniques used to measure the level of risk and its determinants illustrated in this case study can inspire all those flood assessments on a community scale that must be realized in contexts where information is scarce.

Author Contributions: Conceptualization, M.T.; Methodology, M.T., M.R., and E.B.; Validation, G.M. and S.I.; Investigation, M.T., G.M., E.B., S.I., and S.B.; Writing—original draft preparation, M.T. and G.M.; Writing—review and editing, G.M.; Visualization, S.B.; Funding acquisition, M.T.

Funding: This research was funded by the Italian Agency for Development Cooperation, by Institute of Bioeconomy (IBE)-National Research Council of Italy (CNR) at Florence (leader), by the National Directorate for Meteorology of Niger (DMN), and by DIST-Politecnico and University of Turin within the project ANADIA2.

Acknowledgments: We would like to thank Vieri Tarchiani (IBE-CNR) and Katiellou Gaptia Lawan (DMN) for facilitating the field activities, Moussa Mouhimouni (DMN) and Mohamed Housseini Ibrahim (Directorate National of Hydraulics) for the information, Aziz Kountché (Africa Drone Service) for the UAV activity, Abdou Hidjo (chief of Garbey Kourou), Salou Mounkaila (chief of Larba Birno), Harouna Mounkeila (chief of Tallé), Issa Sirfi (chief of Touré), and Oumarou Issaka (Departmental directorate for environment, Gothèye) for participation at meeting discussion.

Conflicts of Interest: The authors declare no conflict of interest. The funders had no role in the design of the study; in the collection, analyses, or interpretation of data; in the writing of the manuscript, or in the decision to publish the results.

References

1. EM DAT. The Emergency Events Database. 2019. Available online: www.emdat.be (accessed on 18 May 2019).
2. African Union. Agenda 2063: The Africa We Want. 2013. Available online: https://au.int/agenda2063/goals (accessed on 18 May 2019).
3. OECD 2019. Stats. Available online: https://stats.oecd.org/qwids (accessed on 18 May 2019).

4. UNISDR, AAPDRR. African Union 2018. In Proceedings of the Declaration of the Sixth High Level Meeting on Disaster Risk Reduction, Tunis, Tunisia, 13 October 2018. Available online: www.unisdr.org/conference/2018/apr-acdrr (accessed on 18 May 2019).
5. R-UNDG-Regional United Nations Development Group 2018. Strategic Framework to Support Resilient Development in Africa. 2018. Available online: https://www.preventionweb.net/files/57759_undgframeworkforresilientdevelopmen.pdf (accessed on 23 July 2019).
6. UNDRR. *Global Assessment Report on Disaster Risk Reduction*; UNDRR: Geneva, Switzerland, 2019; ISBN 978-92-1-004180-5.
7. Ferrier, N.; Emdad Haque, C. Hazard risks assessment methodology for emergency managers: A standardized framework for application. *Natl. Hazards* **2003**, *28*, 271–290. [CrossRef]
8. Næss, L.O.; Norland, I.T.; Lafferty, W.M.; Aall, C. Data and process linking vulnerability assessment to adaptation decision-making on climate change in Norway. *Glob. Environ. Chang.* **2006**, *16*, 221–233. [CrossRef]
9. Ciurean, R.L.; Schröter, D.; Glade, T. Conceptual frameworks of vulnerability assessments for natural disaster reduction. In *Approaches to Disaster Management—Examinig the Implication of Hazards, Emergencies and Disasters*; Tiefenbacher, J., Ed.; Intech Open: London, UK, 2013.
10. Fussel, H.-M.; Klein, R.J.T. Climate change vulnerability assessments: An evolution of conceptual thinking. *Clim. Chang.* **2006**, *75*, 301–329. [CrossRef]
11. Van Aalst, M.; Cannon, T.; Burton, I. Community level adaptation to climate change: The potential role of participatory community risk assessment. *Glob. Environ. Chang.* **2008**, *18*, 165–179. [CrossRef]
12. Forsyth, T. Community-based adaptation: A review of past and future challenges. *WIREs Clim. Chang.* **2013**, *4*, 439–446. [CrossRef]
13. Hay, J.E.; Mimura, N. Vulnerability, risk and adaptation assessment methods in the Pacific islands region: Past approaches, and considerations for the future. *Sustain. Sci.* **2013**, *8*, 391–405. [CrossRef]
14. Mwale, F.D.; Adeloye, A.J.; Beevers, L. Quantifying vulnerability of rural communities to flooding in SSA: A contemporary disaster management perspective applied to the Lower Shire valley, Malawi. *Int. J. Disaster Risk Reduct.* **2015**, *12*, 172–187. [CrossRef]
15. McDowell, G.; Ford, J.; Jones, J. Community-level climate change vulnerability research: Trends, progress, and future directions. *Environ. Res. Lett.* **2016**, *11*, 33001. [CrossRef]
16. Adger, W.N.; Brown, I.; Surminski, S. Advances in risk assessment for climate change adaptation policy. *Philos. Trans. R. Soc. A* **2016**, *376*, 20180106. [CrossRef] [PubMed]
17. Kienberger, S. Spatial modelling of social and economic vulnerability to floods at the district level in Búzi, Mozambique. *Nat. Hazards* **2012**, *64*, 2001–2019. [CrossRef]
18. Osman, A.; Nyarko, B.K.; Mariwah, S. Vulnerability and risk levels of communities within Ankobra estuary of Ghana. *Int. J. Disaster Risk Reduct.* **2016**, *19*, 133–144. [CrossRef]
19. Nabegu, A.B. Analysis of vulnerability to flood disaster in Kano State, Nigeria. *Greener J. Phys. Sci.* **2012**, *4*, 22–29.
20. Ntajal, J.; Lamptey, B.L.; Sogbedji, J.M. Flood vulnerability mapping in the lower Mono river basin in Togo, West Africa. *Int. J. Sci. Eng. Res.* **2016**, *7*, 1553–1562.
21. Komi, K.; Amisigo, B.A.; Diekkrüger, B. Integrated flood risk assessment of rural communities in the Oti river basin, West Africa. *Hydrology* **2016**, *3*, 42. [CrossRef]
22. Hahn, M.B.; Riederer, A.M.; Foster, S.O. The livelihood vulnerability index: A pragmatic approach to assessing risks from climate variability and change—A case study in Mozambique. *Glob. Environ. Chang.* **2009**, *19*, 74–88. [CrossRef]
23. Massazza, G.; Tamagnone, P.; Wilcox, C.; Belcore, E.; Pezzoli, A.; Vischel, T.; Panthou, G.; Ibrahim, M.H.; Tiepolo, M.; Tarchiani, V.; et al. Flood scenarios of the Sirba river (Niger) evaluation of the hazard tresholds and flooding areas. *Water* **2019**, *11*, 1018. [CrossRef]
24. Rudari, R.; Beckers, J.; De Angeli, S.; Rossi, L.; Trasforini, E. Impact of modelling scale on probabilistic flood risk assessment: The Malawi case. *E3S Web Conf.* **2016**, *7*, 04015. [CrossRef]
25. Tiepolo, M.; Braccio, S. Analyse-évaluation du risqué d'inondation et de sécheresse à micro-échelle: Garbey Kourou et Tallé au Niger. In *Risque et adaptation climatique dans la region Tillabéri, Niger*; Tarchiani, V., Tiepolo, M., Eds.; L'Harmattan: Paris, France, 2016; pp. 205–232. ISBN 978-2-343-08-493-0.
26. Olatona, O.O.; Obiora-Okeke, O.A.; Adewumi, J.R. Mapping of flood risk zones in Ala river basin Akure, Nigeria. *Am. J. Eng. Appl. Sci.* **2017**, *11*, 210–217. [CrossRef]

27. Olabode, A.D.; Ajibade, L.T.; Yunisa, O. Analysis of flood risk zones (Frzs) around Asa river in Ilorin using geographic information systems (GIS). *Int. J. Innov. Sci. Eng. Technol.* **2014**, *1*, 621–628.
28. Ibrahim, B.A.; Tiki, D.; Mamdem, L.; Leumbe, O.L.; Bitom, D.; Lazar, G. Multicriteria analysis (MCA) approach and GIS for flood risk assessment and mapping in Mayo Kani division, Far North region of Cameroon. *Int. J. Adv. Remote Sens. GIS* **2018**, *7*, 2793–2808. [CrossRef]
29. Akpejiori, J.I.; Ehiorobo, J.O.; Izinyon, O.C. Flood monitoring and flood risk assessment in Agenebode, Edo State, Nigeria. *Int. J. Eng. Appl.* **2017**, *7*, 53–59. [CrossRef]
30. Oriola, E.; Chibuike, C. Flood risk analysis of Edu local government area (Kwara State, Nigeria). *Sustainability* **2017**, *3*, 106–116. [CrossRef]
31. Koumassi, D.H.; Tchibozo, A.E.; Vissin, E.W.; Houssou, C. SIG et télédétéction pour l'optimisation de la cartographie des risques d'inondation dans le basin de la Sota au Benin. *Rev. Ivoir. Sci. Technol.* **2014**, *23*, 137–152.
32. Rakotoarisoa, M.M.; Fleurant, C.; Taibi, A.N.; Rouan, M.; Caillaut, S.; Razakamanana, T.; Ballouche, A. Un modèle multi-agents pour évaluer la vulnérabilité aux inondations: Le cas des villages aux alentours du fleuve Fiherenana (Madagascar). *Cybergeo* **2018**. [CrossRef]
33. Ugoyibo, O.D.V.; Enyinnaya, O.C.; Souleman, L. Spatial assessment of flood vulnerability in Anambra local government area, Nigeria using GIS and remote sensing. *Br. J. Appl. Sci. Technol.* **2017**, *19*, 1–11. [CrossRef]
34. Gebeyehu, A.; Teshome, G.; Birke, H. Flood hazard and risk assessment in Robe watershed using GIS and remote sensing, north Shewa zone, Amhara region in Ethiopia. *J. Humanit. Soc. Sci.* **2018**, *23*, 34–44. [CrossRef]
35. Oyatayo, K.T.; Iguisi, E.O.; Sawa, B.A.; Ndabula, C.; Jidauna, G.; Irokua, S.A. Assessment of parametric flood vulnerability pattern of Makurundi town, Benue State, Nigeria. *Conflu. J. Environ. Stud.* **2019**, *12*, 11–28.
36. Ejenma, E.; Sunday, V.N.; Okeke, O.; Eluwah, A.N.; Onwuchekwa, I.S. Mapping flood vulnerability arising from land use/land covers change along river Kaduna, Kaduna State, Nigeria. *J. Humanit. Soc. Sci.* **2014**, *19*, 155–160.
37. Orewole, M.O.; Alaigba, D.B.; Oviasu, O.U. Riparian corridors encroachment and flood risk assessment in Ile-Ife: A GIS perspective. *Open Trans. Geosci.* **2015**, *2*, 19–20.
38. Mayomi, I.; Dami, A.; Maryah, U.M. GIS based assessment of flood risk and vulnerability of communities in the Benue floodplains, Adamawa State, Nigeria. *J. Geogr. Geol.* **2013**, 5. [CrossRef]
39. Yahya, S.; Ahmad, N.; Abdalla, R.F. Multicriteria analysis for flood vulnerable areas in Hadejia-Jama'are river basin, Nigeria. *Eur. Sci. Res.* **2010**, *42*, 71–83.
40. Afriyie, K.; Ganle, J.K.; Santos, E. The floods came and we lost everything': Weather extremes and households' asset vulnerability and adaptation in rural Ghana. *Clim. Dev.* **2018**, *10*, 259–274. [CrossRef]
41. Dumenu, W.K.; Obeng, E.A. Climate change and rural communities in Ghana: Social vulnerability, impacts, adaptations and policy implications. *Environ. Sci. Policy* **2016**, *44*, 208–217. [CrossRef]
42. Yawson, D.O.; Adu, M.O.; Armah, F.A.; Kusi, J.; Ansah, I.G.; Chiroro, C. A needs-based approach for exploring vulnerability and response to disaster risk in rural communities in low income countries. *Australas. J. Disaster Trauma Stud.* **2015**, *19*, 27–36.
43. Tamagnone, P.; Massazza, G.; Pezzoli, A.; Rosso, M. Hydrology of the Sirba river: Updating and analysis of discharge time series. *Water* **2019**, *11*, 156. [CrossRef]
44. IPCC. Annex II: Glossary. In *Climate Change 2014: Synthesis Report. Contribution of Working Groups I, II and III to the Fifth Assessment Report of the Intergovernmental Panel on Climate Change*; IPCC: Geneva, Switzerland, 2014; pp. 117–130.
45. Tiepolo, M.; Braccio, S. Local and scientific knowledge integration for multi-risk assessment in rural Niger. In *Renewing Local Planning to Face Climate Change in the Tropics*; Tiepolo, M., Pezzoli, A., Tarchiani, V., Eds.; Springer Open: Cham, Switzerland, 2017; pp. 227–245.
46. Acosta, L.A.; Eugenio, E.A.; Macandog, P.B.M.; Magcale-Macandog, D.B.; Hui Lin, E.K. Loss and damage from typhoon-induced floods and landslides in the Philippines: Community perceptions on climate impacts and adaptation options. *Int. J. Glob. Warm.* **2016**, *9*, 33–65. [CrossRef]
47. Amirebrahimi, S.; Rajabifard, A.; Mendis, P.; Ngo, T. A framework for a microscale flood damage assessment and visualization for a building using BIM-GIS integration. *Int. J. Digit. Earth* **2016**, *9*, 364–386. [CrossRef]
48. Brower, R.; Akter, S.; Brander, L.; Haque, E. Socioeconomic vulnerability and adaptation to environmental risk: A case study of climate change and flooding in Bangladesh. *Risk Anal.* **2007**, *27*, 313–326. [CrossRef]

49. Mahmood, S.; Khan, A.; Mayo, S.M. Exploring underlying causes and assessing damages of 2010 flash flood in the upper zone of Panjkora river. *Nat. Hazards* **2016**, *83*, 1213–1227. [CrossRef]
50. Perera, E.D.P.; Hiroe, A.; Srestha, D.; Fukami, K.; Basnyat, D.B.; Gautam, S.; Hasegawa, A.; Uenoyama, T.; Tanaka, S. Community-based flood damage assessment approach for lower west Rapti river basin in Nepal under the impact of climate change. *Nat. Hazards* **2015**, *75*, 669–699. [CrossRef]
51. Rahman, A.; Khan, A. Analysis of flood causes and associated socio-economic damages in the Hindukush region. *Nat. Hazards* **2011**, *59*, 1239–1260. [CrossRef]
52. Garrote, J.; Alvarenga, F.M.; Diez-Herrero, A. Quantification of flash flood economic risk using ultra-detailed stage-damage functions and 2-D hydraulic models. *J. Hydrol.* **2016**, *541*, 611–625. [CrossRef]
53. Ernst, J.; Dewals, B.J.; Detrembleur, S.; Archenbeau, P.; Erpicum, S.; Pirotton, M. Micro-scale flood risk analysis based on detailed 2D hydraulic modelling and high resolution geographic data. *Nat. Hazards* **2010**, *55*, 181–209. [CrossRef]
54. Zimmermann, M.; Loat, R. Flood Mapping. *Global Water Partnership, World Meteorological Organization: Geneva*, 2013. Available online: https://library.wmo.int/index.php?lvl=notice_display&id=16352#.XS7Wb3vOOUk (accessed on 18 May 2019).
55. ISO 2018. *ISO 31000: Risk Management-Guidelines*, 2nd ed.; ISO: Geneva, Switzerland, 2018.
56. Alexander, D. Towards the development of a standard in emergency planning. *Disaster Prev. Manag.* **2005**, *14*, 158–175. [CrossRef]
57. Piras, M.; Belcore, E.; Pezzoli, A.; Massazza, G.; Rosso, M. Raspberry PI 3multispectral low-cost sensor for UAV based remote sensing. Case study in South-West Niger. *Int. Arch. Photogramm. Remote. Sens. Spat. Inf. Sci.* **2019**, *4213*, 207–214. [CrossRef]
58. McFeeters, S.K. The use of the Normalized Difference Watrer Index (NDWI) in the delineation of open water features. *Int. J. Remote Sens.* **1996**, *17*, 1425–1432. [CrossRef]
59. Republic of Niger; Ministry of Agriculture and Livestock. *Rapport D'évaluation Préliminaire des Récoltes et Résultats Provisoires de la Campagne Agricole d'Hivernage*; Ministère de l'Agriculture et de l'Élevage: Niamey, Niger, 2018.
60. Republic of Niger; MDCAT; DGDRL. *Directives Pour la Replanification D'un Plan de Développement Communal-PDC*; Ministère du Développement Communautaire et de l'Aménagement du Territoire: Niamey, Niger, 2015.
61. Bacci, M.; Mouhaïmouni, M. Hazard events characterization in Tillaberi region, Niger: Present and future projections. In *Renewing Local Planning to Face Climate Change in the Tropics*; Tiepolo, M., Tarchiani, V., Pezzoli, A., Eds.; Springer: Cham, Switzerland, 2017; pp. 41–56.
62. Aich, V.; Liersch, S.; Vetter, T.; Andersson, J.C.M.; Müeller, E.N.; Hattermann, F.F. Climate or land use? Attribution of changes in river flooding in the Sahel zone. *Water* **2015**, *7*, 2796–2820. [CrossRef]
63. Descroix, L.; Guichard, F.; Grippa, M.; Lambert, L.A.; Panthou, G.; Mahe, G.; Gal, L.; Dardel, C.; Quantin, G.; Kergoat, L.; et al. Evolution of surface hydrology in the Sahelo-Sudanian strip: An updated review. *Water* **2018**, *10*, 748. [CrossRef]
64. Wilcox, C.; Panthou, G.; Bodian, A.; Bodian, A.; Blanchet, J.; Descroix, L.; Quantin, G.; Cassé, C.; Tanimoun, B.; Kone, S. Trends in hydrological extremes in the Senegal and Niger rivers. *J. Hydrol.* **2018**, *566*, 531–545. [CrossRef]
65. De Moel, H.; Jongman, B.; Kreibich, H.; Merz, B. Flood risk assessment at different spatial scales. *Mitig. Adapt. Strat. Glob. Chang.* **2015**, *20*, 865–890. [CrossRef]

Article

Multi-Hazard Risk Assessment at Community Level Integrating Local and Scientific Knowledge in the Hodh Chargui, Mauritania

Maurizio Tiepolo [1,*], Maurizio Bacci [1,2], Sarah Braccio [1] and Stefano Bechis [1]

1 Interuniversity Department of Regional and Urban Studies and Planning (DIST)-Politecnico and University of Turin, 10125 Turin, Italy; maurizio.bacci@cnr.it (M.B.); sarah.braccio@polito.it (S.B.); stefano.bechis@unito.it (S.B.)

2 Institute of Bio Economy, National Research Council, 50145 Florence, Italy

* Correspondence: maurizio.tiepolo@polito.it

Received: 26 July 2019; Accepted: 9 September 2019; Published: 16 September 2019

Abstract: Hydro-climatic risk assessments at the regional scale are of little use in the risk treatment decision-making process when they are only based on local or scientific knowledge and when they deal with a single risk at a time. Local and scientific knowledge can be combined in a multi-hazard risk assessment to contribute to sustainable rural development. The aim of this article was to develop a multi-hazard risk assessment at the regional scale which classifies communities according to the risk level, proposes risk treatment actions, and can be replicated in the agropastoral, semi-arid Tropics. The level of multi-hazard risk of 13 communities of Hodh Chargui (Mauritania) exposed to meteorological, hydrological, and agricultural drought, as well as heavy precipitations, was ascertained with an index composed of 48 indicators representing hazard, exposure, vulnerability, and adaptive capacity. Community meetings and visits to exposed items enabled specific indicators to be identified. Scientific knowledge was used to determine the hazard with Climate Hazards Group Infra-Red Precipitation with Station (CHIRPS) and Tropical Rainfall Measuring Mission (TRMM) datasets, Landsat images, and the method used to rank the communities. The northern communities are at greater risk of agricultural drought and those at the foot of the uplands are more at risk of heavy rains and consequent flash floods. The assessment proposes 12 types of actions to treat the risk in the six communities with severe and high multi-hazard risk.

Keywords: agricultural drought; climate change; heavy rains; hydrological drought; meteorological drought; risk assessment; Sahel; sustainable rural development

1. Introduction

In the semi-arid rural areas of tropical Africa, drought reduces access to water for human consumption and affects both livestock and rain-fed crops. In the same areas, more and more frequent flash floods destroy irrigated crops, damage hydraulic works, and consequently prevent recession agriculture with which smallholder farmers compensate the deficit of rain-fed crops [1]. The coexistence of different hazards is so frequent that the Sendai framework for Disaster Risk Reduction (2015) recommends a sustainable livelihood development based upon multi-hazard risk assessments [2]. However, the risk assessments published on tropical Africa (Table 1) deal with a single hazard at a time, which in 90% of cases, is the flood hazard and, for the remainder, the drought hazard.

Table 1. Twenty-six risk assessments at regional-scale in Tropical Africa, 2005–2018.

Risk Assessment Phase	Determinant	Hydro-Climatic Threat	Reference
Analysis	Hazard	Flood, drought	[3]
		Fluvial, pluvial flood	[1]
		Flood	[4–17]
		Fluvial flood	[7,8,11,18–25]
		Drought	[26,27]
		Meteorological drought	-
		Hydrological drought	[27]
		Probability	[1,3,11,18,19,27]
	Adaptive capacity		[1,9,17,18]
Evaluation	-		[1,4,7,11,13,19,21]

The published assessments usually follow two approaches. The first approach performs distance assessments through satellite images, digital elevation models, soil maps, precipitations, and stream-flow datasets. When the approach involves using indicators, it obtains them from literature rather than from ground truth. It neglects the risk reduction actions in progress and proposes few in case of drought. The second approach listens to the community. This is not sufficient to appreciate the risk of events that the community has not yet experienced [28], to assess the probability of disaster (which is a constitutive element of the concept of risk), or to identify the exposed areas.

Although the importance of local knowledge in adapting to climate change [29] is unquestionable and there is a broad consensus regarding the need to combine it with scientific knowledge in risk reduction [30–33], assessments in tropical Africa which integrate both types of knowledge are still rare [1,3,18].

The Sendai framework recommends improving the understanding of risks at all spatial scales [10]. Indeed, after Sendai, the subnational assessments have increased. The regional scale is important for action, as it is that in which official development aid and the local authorities themselves operate most often with medium-term programs and projects. However, regional scale risk assessments struggle to estimate the hazard when climate information is scarce [34], to identify truly expressive indicators of the risk determinants, to take account of the risk reduction measures in place, and to suggest pertinent ones for the future (Table 1).

At the regional scale, assessments which identify the most exposed communities and suggest how to deal with the risk could help official development aid and the local authorities in risk-informed decision-making, which would certainly aid sustainable livelihood development.

The aim of this article was to develop a multi-hazard risk assessment ranking the communities according to the risk level and identify risk reduction actions using local and scientific knowledge and techniques adapted to unskilled operators and scant information.

To achieve this objective, the multi-hazard risk index (MHRI) was proposed. The risk equation used combines hazard (H), meaning the "potential occurrence of a natural physical event that may cause loss of life, injury or damage to property" [35]; exposure (E), or "the presence of people, livelihoods, species or ecosystems, environmental functions, services, and resources, infrastructure, or economic, social, or cultural assets in places and settings that could be adversely affected" [35]; vulnerability (V), "the propensity or predisposition to be adversely affected" [35]; and adaptive capacity (AC), namely "the ability of systems, institutions, humans and other organisms to adjust to potential damage, to take advantage of opportunities, or to respond to consequences" [35]: $R = H \times (E + V - AC)$. The equation is an adaptation of that proposed by Crichton [36]. Each risk determinant is expressed by indicators, which were identified after discussion with the communities and visits to the exposed items.

The multi-hazard risk assessment was carried out with 13 rural communities of the municipalities of Adel Bagrou, Agoueinit, Bougadoum, Oum Avnadech, in Hodh Chargui, Mauritania, a landlocked region located 1100 km from the Atlantic Coast (Figure 1).

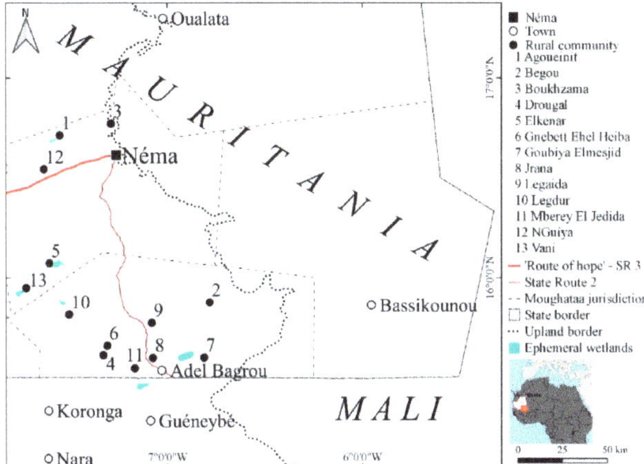

Figure 1. The 13 rural communities of Hodh Chargui where the multi-hazard risk assessment was developed.

Location names in local documents change. This article used the names adopted by the general census of the population of the Islamic Republic of Mauritania in 2013 [37]. These communities were among the most in need of reducing the hydro-climatic risk, according a preliminary survey on 271 communities (2017) of the Hodh Chargui. The 13 communities can be aggregated into three clusters: One around Néma, the capital of Hodh Chargui; one to the south-west of Néma; and one to the south, around Adel Bagrou, close to border with Mali. The Hodh Chargui is experiencing strong demographic growth: The region increased from 212,000 inhabitants in 1988 to 431,000 inhabitants in 2013 [37].

The 13 agropastoral communities have between 400 and 2600 inhabitants. These are settlements that have been established mostly in the last 40 years, constituted by the agglomeration of stone dwellings (sometimes made from crude earth bricks). Each stone dwelling is flanked by a construction with a two-pitched roof, under which life takes place during the warm months, and an enclosure for the animals around a tree in the branches of which the fodder is stored.

The key elements of each community are the wells which are, in general, traditional (uncovered, without water pumps), and are, in some cases, flooded during the wet season. In the dry season, they may be found some kilometers away. It is rare for a community to have a borehole with a respective water reservoir. The mosque, the school, and the community leader's residence are the other significant locations of each community. During the dry season (November–May), the shepherds go to the southern pastures. Only a small part of the livestock remains in the communities for requirements of milk, cheese, and meat. To sell the animals, the shepherds go as far as Senegal, traveling between 100 and 300 km [38].

Between June and August, rain-fed agriculture is practiced. The herds return to the pastures around the villages of origin. Half of the communities use an ephemeral wetland, at the edges of which they dig wells for pastoral and sometimes human use. In this season, some communities remain isolated because the conditions of the tracks do not allow vehicular access. Poor rainfall leads communities to pick up the runoff with earth embankments to practice recession agriculture from October onward. The embankments are, however, exposed to trampling by the herds which cross them or by the heavy rains. When the embankments are damaged, they no longer retain water and recession agriculture is reduced to small surfaces. The households that can afford it enclose the fields with metal fences to protect the crops from stray livestock. Those who do not have such resources resort to the use of thorny branches, which is prohibited, to preserve the scarce arboreal and shrubby coverage of the region. Later, when the dry season takes hold, irrigated commercial agriculture commences.

These activities are constantly exposed to drought, which manifests primarily with "an abnormal precipitation deficit" [35] (meteorological drought). This has effects that continue well beyond the wet season: The availability of fodder (trees, shrubs, grass) drops, the ephemeral wetlands where livestock are watered do not fill up, the surface aquifer does not refill, the pastoral wells and traditional wells used by many communities dry up or have poor quality water, and the possibility of recession agriculture and irrigated gardening fades.

Drought can also manifest with a "shortage of precipitation during the runoff and percolation season primarily affecting water surfaces" (hydrological drought) and with a "shortage of precipitation during the growing season" (agricultural drought) [35].

The three forms of drought co-exist with heavy precipitations, which damage the earth embankments, as well as make the wells in the flood areas inaccessible and, if flooded, unusable. All these events threaten the livelihood of the Hodh Chargui communities (Figure 2).

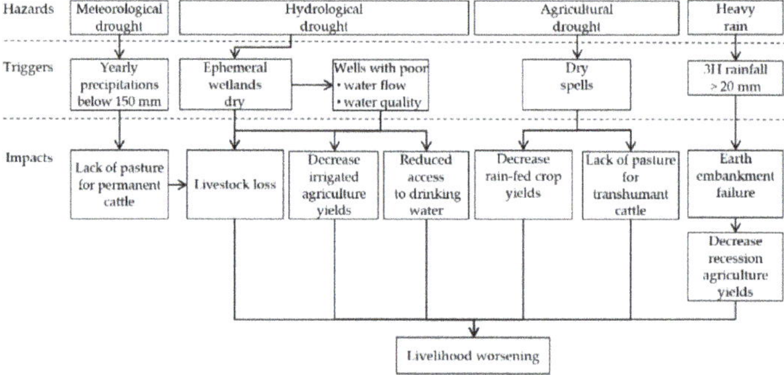

Figure 2. Main hazards and their impacts on livelihoods in the Hodh Chargui, Mauritania.

This assessment thus considered meteorological, hydrological, and agricultural drought and heavy precipitations. The four hazards are combined in the multi-hazard risk index (MHRI), respecting the same importance of each risk determinant and the quantitative measurement of the indicators [39].

The assessment is organized into four phases (Figure 3).

The first phase identifies the context: The trend of daily precipitation, the variation in the extension of surface waters, the risk criteria (equation, probability of occurrence, level), and the technique to be used. The second phase identifies the risk: Which datasets to use to determine the hazard and how to ascertain exposure, vulnerability, and adaptive capacity. The third phase identifies the indicators for each risk determinant, collects information from the 13 communities, processes the data, produces the MHRI, and represents it on the map. The final phase identifies the risk reduction actions.

The most significant results obtained by the assessment are the reproducibility of the methodology and the existence of very different risk levels in an apparently homogeneous territory, influenced, above all, by meteorological drought and heavy rains, and the proposal of 12 risk reduction actions.

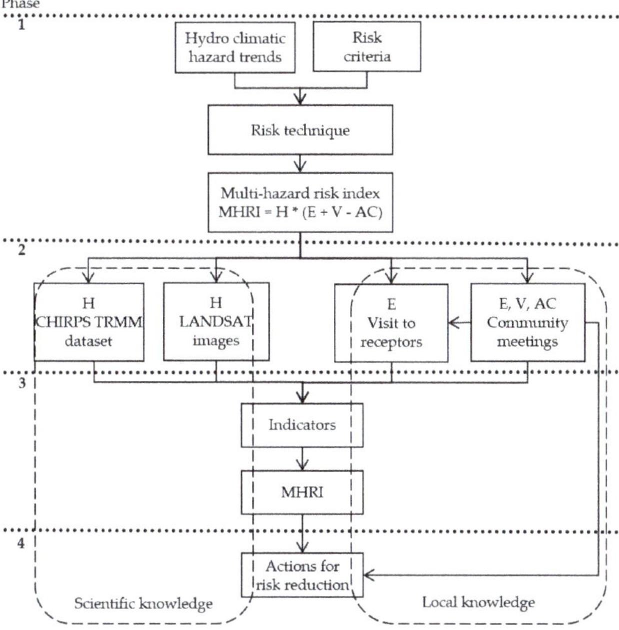

Figure 3. Multi-hazard risk assessment flowchart.

2. Materials and Methods

The multi-hazard risk was ascertained using the index technique [40]. This technique fits unskilled operators and can be used in other regions of Mauritania.

The MHRI adds up the risk indices of the three types of drought and heavy precipitations. The index is made up of 48 indicators that quantitatively scored hazard, exposure, vulnerability, and adaptive capacity. The actual values found for each indicator were normalized within a 0–1 scale. The indicators were then added and normalized in a 0–1 scale for exposure, vulnerability, and adaptive capacity (Supplementary file 1). Twenty-nine vulnerability and exposure indicators were acquired through a survey in each community (April 2017), fifteen exposure and adaptive capacity indicators were measured during a visit at the end of the dry season (May 2018), and four hazard indicators were acquired from datasets on the daily and three-hourly rainfall and from satellite images (Figure 4). The survey (April 2017) and the subsequent visit (May 2018) encountered between 12 and 20 inhabitants in each community gathered in two separate groups: Herders (men), and horticulturalists, goat and sheep breeders, and sun-dried tomatoes, milk, yogurt, cheese, dried meat, leather, livestock feed sellers (mainly women). Each group was asked a series of questions (Supplementary file 2).

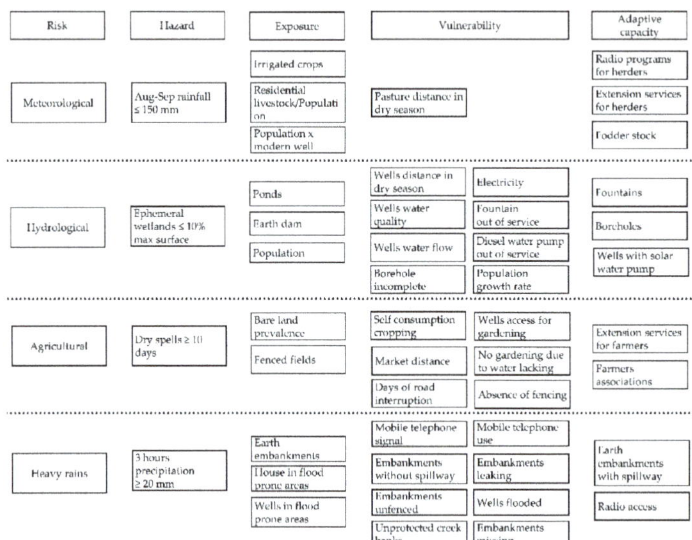

Figure 4. Indicators used in the multi-hazard risk index for 13 communities in the Hodh Chargui, Mauritania.

The meteorological drought hazard is expressed by the probability of occurrence of rainfall accumulation during the months of July, August, and September of less than 150 mm. That limit was identified based upon the quantity of rain considered the minimum amount necessary to produce plant biomass in an arid Sahelian environment [41–44]. Rainfall distribution is derived from the Climate Hazards Group Infra-Red Precipitation with Station (CHIRPS) dataset in the 1981–2018 period. Using these values, we calculated the annual rainfall accumulation and evaluated it on the 38-year historical series (1981–2018) to determine how many years there was a rainfall of less than 150 mm for each community.

The hydrological drought hazard was calculated on six ephemeral wetlands of reference for most of the communities considered. Its determination, due to the absence of localized information, did not use the current indices of hydrological drought [45]. The analysis was based upon the extension of the ephemeral wetlands as identified by calculating the Normalized Difference Water Index (NDWI) [46] on the Landsat satellite images. The availability of a limited series of images led us to identify the rainfall accumulation, which determined the years in which the surfaces of the ephemeral wetlands were less extensive. Then, we searched for the frequency of that value in the 1981–2018 rainfall series.

First, the annual surface profile of each ephemeral wetland in a dry year (2014) and in a wet year (2015) was determined. The satellite images were taken on different dates (t) each year, requiring the construction of the profile by interpolating the surface data with the formula:

$$Surface_t = Surface_{t-1} + \left(\frac{Surface_{t+1} - Surface_{t-1}}{N.Days_{[(t+1)-(t-1)]}} \right) \times N.Days_{[(t)-(t-1)]}$$

The results allowed us to construct the average surface area growth values (Figure 5).

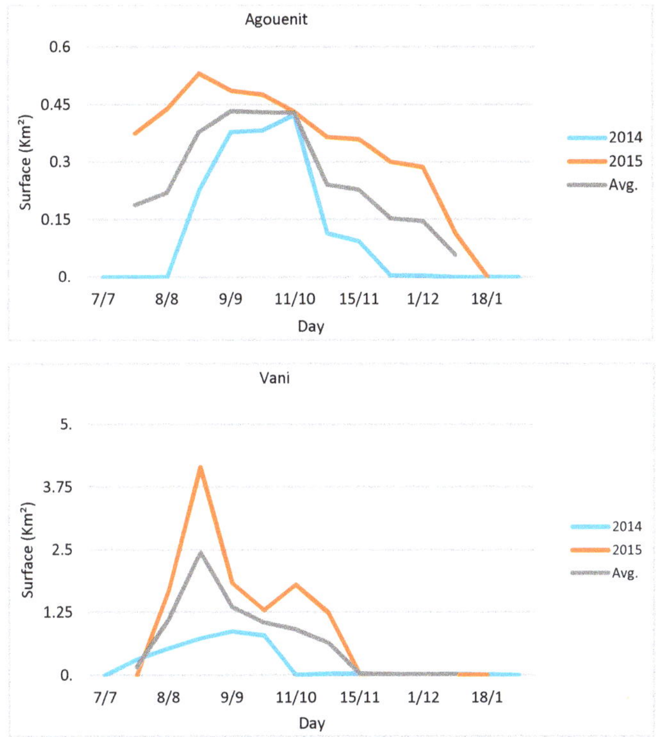

Figure 5. Interpolated profile of Agoueinit and Vani ephemeral wetlands area in 2014 and 2015 and the average growth curve.

Using those average values, it was possible to calculate a standardized profile of the daily step curves for the July–November period. The standardization was calculated daily, using the maximum value recorded for each ephemeral wetland as a reference (Figure 6).

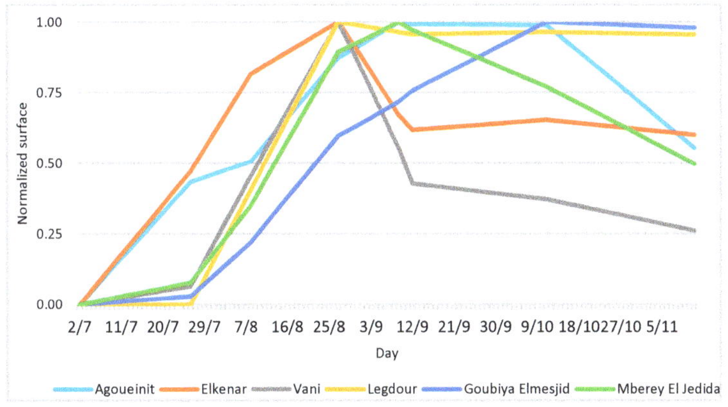

Figure 6. Normalized growth curves of the six ephemeral wetlands.

With only the 2014 and 2015 years available, we found a rough curve of variation of the surfaces of the ephemeral wetlands. However, with this method, it was possible to discriminate between the various ephemeral wetlands.

Using the CHIRPS estimated rainfall dataset, the rain signal was decomposed at various time intervals from the accumulation of 7, 14, 30, and 60 days before the measurement of the surface of the ephemeral wetland in question, the accumulation of the two central months of the rainy season (August–September), and the entire season. With those values available, we sought to understand which rainfall time interval most influenced the filling of each individual ephemeral wetland, and then analyzed the correlation. The results show a different sensitivity of each individual ephemeral wetland to the August–September precipitations (Table 2).

Table 2. Correlation between the precipitation-ephemeral wetland surface.

Ephemeral Wetland	7 Days	14 Days	30 Days	60 Days	August–September	Season
Agoueinit	0.61	−0.01	0.16	−0.06	0.39	0.40
Elkenar	−0.44	0.02	0.48	0.32	0.54	0.35
Goubiye	−0.54	−0.20	0.10	0.41	0.44	0.46
MBoreye	−0.33	0.15	0.44	0.45	0.58	0.48
Vani	−0.14	0.16	0.44	0.28	0.42	0.13
Average	−0.33	0.15	0.44	0.45	0.58	0.48

The period of rainfall that, on average, most influenced the filling of the ephemeral wetland was the accumulation of rain in the months of August and September. That time interval was used to identify the rainfall that characterized the three years with less surface of the ephemeral wetland. The average value of these three years was used to identify the critical rainfall and, consequently, the probability of occurrence on the entire series (1981–2018).

The adopted method remains more suitable to measure changes in water bodies over time than that proposed by the European Commission Joint Research Centre, which reports the status of the individual pixels that make up the water bodies without, however, reporting the precise date of which they are observed [47].

The agricultural drought hazard was measured with the probability of occurrence of dry spells of at least 10 consecutive days during the months of July, August, and September, ascertained using the CHIRPS dataset for the period 1981–2018 for each community. The choice of this spell length reflected the need to consider a threshold that could generate a negative impact on rain-fed crops in this region, which is strictly dependent on the specific crops' resistance to drought stress [48].

The heavy precipitation hazard is expressed by the probability of occurrence of three-hourly rainfalls higher than 20 mm ascertained using the Tropical Rainfall Measuring Mission (TRMM) dataset for the 1991–2014 period at each ephemeral wetland.

The exposure to meteorological drought is given by the presence of irrigated crops, the number of inhabitants per well, and tropical livestock units [49] which remain in each community in the dry season: The higher the number, the more the community is exposed to drought. The exposure to hydrological drought is represented by the number of ponds, earth dams, and inhabitants of each community. As for exposure to agricultural drought, this is given by the presence of horticultural activities protected by barbed wire fencing against the intrusion of stray cattle and the presence of pasture and arable surfaces. A proxy indicator is the share of bare land in the territory of each community: The lower it is, the greater the exposure. The exposure to heavy precipitations is given by the number of earth embankments and by the wells and houses in flood prone areas, which could become inaccessible or be flooded.

The vulnerability to meteorological drought is given by the distance of the pastures from the village in the dry season. The vulnerability to hydrological drought is expressed by distant wells with poor water flow and quality and by the lack of boreholes, functioning fountains, or by broken diesel water pumps. It is also expressed by the population growth rate of the community in question, which increases the demand for water. The vulnerability to agricultural drought is still linked to the availability and accessibility of wells for irrigation, the practice of cropping for self-consumption only, the rate of unfenced lots, the distance to the market, and the number of days of road interruption. The vulnerability to heavy precipitations is expressed by the possibility of receiving an early warning by telephone (therefore, the coverage of the area with a mobile phone signal), by the lack of protection of the earth embankments from the crossing of livestock, by the absence of spillways and locks, which reduce the pressure of flash floods on the hydraulic works, and by the presence of creeks without bank protection.

The adaptive capacity is of three types [50]: Capacity to anticipate risk, to respond to risk, and to recover and to change. For meteorological drought, the existence of radio programs aimed at farmers who report where vaccines and vaccination parks for livestock (anticipate), pastures, water, and fodder banks (recover) are available is paramount. For hydrological drought, the existence of boreholes, fountains or mini aqueducts (respond) which cover the demand for water by drawing from deep aquifers, especially if powered by solar water pumps, which have lower operating costs than diesel water pumps, is important.

Agricultural drought can be overcome if there are extension services (anticipate) and farmers' associations (recover). For heavy precipitations, radio access counts to receive early warning (anticipate). Spillways and locks in the earth embankments allow the pressure of flash floods on the earth embankments to be regulated, preserving them from collapse (respond).

In order to compare the different hazards, we took a series of precautions [39]. Each indicator and each determinant has the same significance. The probability of occurrence of each hydro-climatic hazard was calculated observing the same timeframe (1981–2018), except for heavy precipitations (1998–2014).

For each individual risk, the value of every individual determinant ranges between 0 and 1, irrespective of the number of indicators that describe it. Each indicator has the same significance. The MHRI was obtained by adding the values of the four single risks. Its value can theoretically vary from 0 to 8. The absolute interval between the maximum and minimum value was divided into four equal parts to indicate low (0–2), moderate (2–4), high (4–6), and severe (6–8) risk. In reality, the highest value can never be reached because the hazard is always less than 1 and all the communities have risk reduction actions in place that bring the value of the adaptive capacity higher than 0.

The calculation allowed to identify which risk, determinant, and indicators have the greatest effect on the MHRI. The indicators that present the highest value (exposure, vulnerability) or lowest value (adaptive capacity) oriented the identification of risk treatment actions among the best practices developed in the Hodh Chargui region. Each of these were then assessed with the communities, considering the expected impact on sustainable rural development, the successful use in the region, the community participation in construction works, the maintenance requirements, the maintenance local capacities, and the community acceptance. One point was attributed to each criteria. The resulting ranking identified priority actions.

3. Results

3.1. Hazard

3.1.1. Meteorological Drought

It is somewhat challenging to define meteorological drought in Hodh Chargui. The scarcity of observed data does not allow us to understand the quantity of precipitation that can generate a negative influence on the production system. Furthermore, the production systems are naturally resistant to extreme drought conditions. Literature identified 100–150 mm of the annual rainfall as the threshold within which plant species that are most resistant to drought can produce biomass even in extreme conditions [41–44]. For this work, we decided to use the 150 mm threshold for meteorological drought.

Using the CHIRPS series, the rain profile of the region was extracted. In the last 38 years, the Hodh Chargui had its driest period during the 1980s, with a minimum of 102 mm in 1983. In the last 13 years, there has been a trend toward the recovery of rainfall, with values that never dropped below 150 mm per year commencing from 2006 (Figure 7).

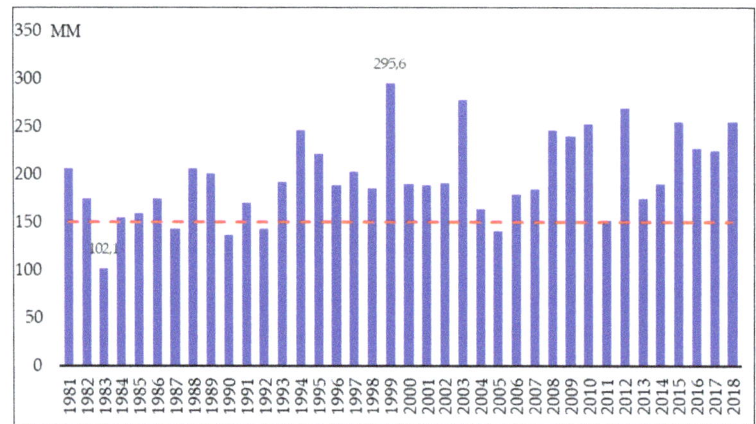

Figure 7. Yearly precipitation 1981–2018 in the southern Hodh Chargui region by Climate Hazards Infra-Red Precipitation with Station (CHIRPS) dataset and 150 mm limit.

The rainfall distribution follows a south-north gradient starting from 300 to 150 mm/year. Nevertheless, during the 2010–2018 period, there was a recover in rainfall of 30–50 mm/year throughout the region (Figure 8).

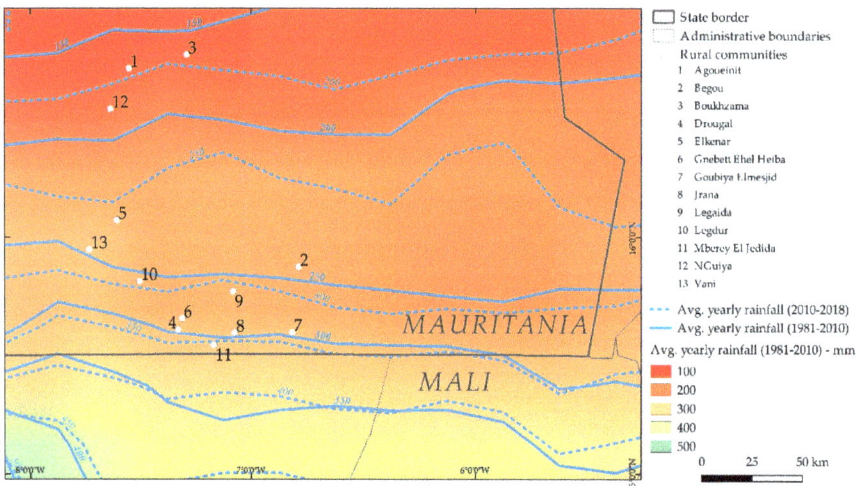

Figure 8. The difference in average accumulated rainfall during 2010–2018 and 1981–2010.

Boukahzama 1, Agoueinit, and NGuiya are the northernmost communities and are most likely to have rainfall of less than 150 mm. Conversely, the five communities on the border with Mali (Drougal, Gnebett Ehel Heiba, Jrana, Mberey El Jedida, and Goubya Elmesjid) have a very low probability of meteorological drought (Table 3).

Table 3. Meteorological drought probability for 13 communities of the Hodh Chargui.

Community	n. Years ≤ 150 mm	Probability
Boukhzama 1	26	0.68
Agoueinit	14	0.37
NGuiya	14	0.37
Begou	4	0.11
Elkenar	4	0.11
Legaida	3	0.08
Legdur	2	0.05
Vani	2	0.05
Mberey El Jedida	1	0.03
Goubya Elmesjid	1	0.03
Jrana	1	0.03
Drougal	0	0.00
Gnebett Ehel Heiba	0	0.00

3.1.2. Hydrological Drought

Six ephemeral wetlands constitute the surface water resources of reference for the investigated communities. These are semi-permanent and shallow water bodies of maximum extension between 6 and 30 km^2. All are characterized by weak depth. Despite this, they are a fundamental resource for human and pastoral water supplies, for fishery resources, and in the case of Agoueinit, for recession agriculture. The flood regime is not the same. It is rare for a dry or wet year to affect all six ephemeral wetlands. In 2003, the ephemeral wetlands reached, in total, 78% of the maximum surface, followed by 2011 (70%), 2009 (57%), and 2012 (55%). In 1987, five ephemeral wetlands out of six had a surface reduced to less than 10% of the maximum observed extension. In 2005, it was so dry that the average surface of the ephemeral wetlands dropped to 1% of the maximum average. In 2014, a hydrological drought was experienced by half of the ephemeral wetlands, while two out of six experienced a hydrological drought in 2016. The two southernmost ephemeral wetlands have not suffered drought in the last eight years, while the northernmost has been dry for six years out of eight, and the center-south has been dry for one or two years out of eight (Figure 9).

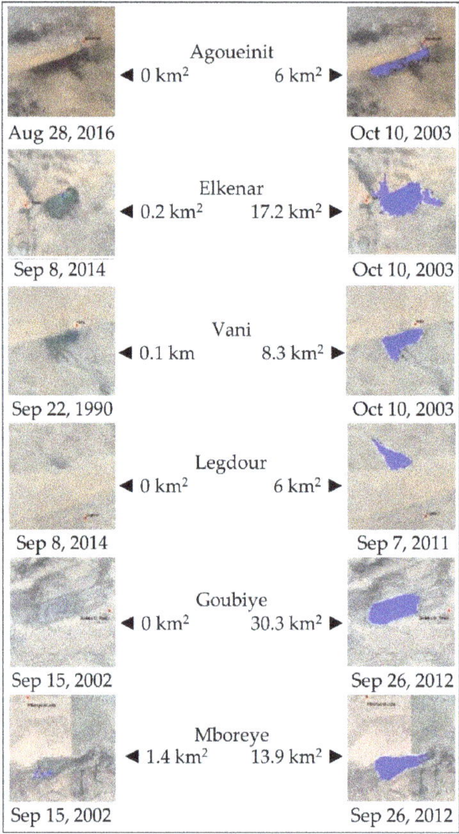

Figure 9. Minimal (**left**) and maximal (**right**) surface of six ephemeral wetlands in the Hodh Chargui, Mauritania, 2001–2018.

According to the described methodology, which attributes the extension of six ephemeral wetlands to rainfall, it is possible to establish the drought probability on the same period used to estimate the meteorological drought probability. It follows that the ephemeral wetland of Agoueinit is particularly prone to drying out compared to all the others (Table 4).

Table 4. Hydrological drought probability for the six ephemeral wetlands in the Hodh Chargui.

Ephemeral Wetland	Probability
Agoueinit	0.47
Elkenar	0.13
Vani	0.09
Goubye Elmesjid	0.09
Jrana	0.09
Legdour	0.06
Mberey El Jedida	0.06

3.1.3. Agricultural Drought

Agricultural drought occurs when drought affects agricultural and pastoral production. The 13 communities in question cultivate in rain-fed conditions, in the form of recession agriculture, and of irrigated gardens. The first type of agriculture is influenced by the rainfall distribution. Pastoral production, which plays a major role in the region's economy, is influenced by the presence of pastures for transhumant herds during the wet season. Biomass production is therefore relegated to spontaneous herbaceous and shrub species, which are naturally very resistant to water stress conditions. However, lengthy dry spells during the wet season can reduce the availability of fodder. Using the CHIRPS dataset, the daily series of the different communities was extracted and the maximum length of the dry spell during the season was assessed (Figure 10).

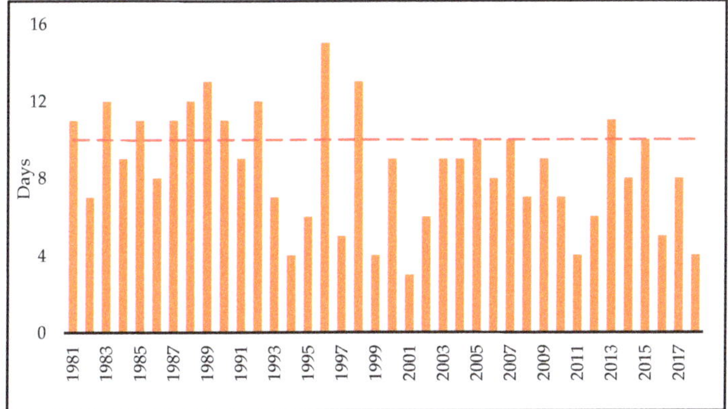

Figure 10. Maximum dry spell length in July, August, and September in Hodh Chargui using the CHIRPS dataset.

From 2006 onward, rainfall is favorable, although it is always below 300 mm/year. However, there are three years with dry spells equal to or greater than ten days. The frequency of dry spells follows a north-south distribution. Dry spells are more frequent in the northern communities of Boukhzama 1, Agoueinit, and NGuiya, and are less frequent in the southern communities of Drougal, Gnebett Ehel Heiba, Goubya Elmesjid, Jrana, and Mberey El Jedida (Table 5).

Table 5. Agricultural drought probability expressed by dry spells frequency in 13 communities of Hodh Chargui, 1981–2018.

Community	Dry Spell of 10 Consecutive Dry Days n. Years	Probability
Boukhzama 1	34	0.89
Agoueinit	28	0.74
NGuiya	28	0.74
Legaida	22	0.58
Begou	21	0.55
Elkenar	17	0.45
Legdur	16	0.42
Vani	16	0.42
Jrana	14	0.37
Mberey El Jedida	14	0.37
Goubya Elmesjid	13	0.34
Drougal	12	0.32
Gnebett Ehel Heiba	12	0.32

3.1.4. Heavy Precipitations

It is difficult to define heavy precipitation in Hodh Chargui. The data are scarce and only one study has allowed for the definition of a threshold of extreme rainfall (37 mm/day) [51]. That value does not necessarily involve the generation of conditions favorable to flash floods. Those phenomena are linked to high rainfall intensity in periods of less than one day. As a consequence, the frequency of three-hourly rainfall higher than 20 mm was verified by analyzing the extractions of the three-hourly values from the Tropical Rainfall Measuring Mission (TRMM) dataset in the 1991–2014 period for each of the 13 communities. It follows that Boukhzama 1, Drougal, and Gnebett Ehel Heiba exceed this threshold more frequently than Elkenar, Jrana, and Mberey El Jedida. In this case, there is no decreasing distribution of the frequency of the hazard as it proceeds from north to south (Table 6).

Table 6. Heavy precipitations (>20 mm in 3 h) probability according the Tropical Rainfall Measuring Mission (TRMM) dataset, 1998–2014.

Community	> 20 mm in 3 h n. of Years	Probability
Boukhzama 1	7	0.41
Drougal	6	0.35
Gnebett Ehel Heiba	6	0.35
Begou	5	0.29
Legaida	5	0.29
Legdur	5	0.29
Vani	5	0.29
Agoueinit	4	0.24
Goubya Elmesjid	4	0.24
NGuiya	4	0.24
Elkenar	3	0.18
Jrana	3	0.18
Mberey El Jedida	3	0.18

3.2. *Exposure*

The exposure to different hazards is represented by 11 indicators: Three for meteorological drought (irrigated crops, residential livestock, and number of inhabitants per well), three for hydrological drought (ponds, earth dams, and population), two for agricultural drought (bare land rate, fenced fields) and three for heavy rains (earth embankments, houses, and wells in flood prone area). The highest values of exposure to meteorological, hydrological, and agricultural drought and to heavy rains are reached respectively in Agoueinit, Vani, Agoueinit, and Boukhzama 1. The exposure to all hazards shows that Agoueinit and Legdur have the highest values and Mberey El Jedida has the lowest value.

3.3. *Vulnerability*

The vulnerability to the different hazards is represented by 23 indicators: One for meteorological drought (distance to pasture in dry season), eight for hydrological drought (electricity access, distance to wells, borehole not working, diesel water pump broken down, irregularly functioning fountain, wells water flow and quality, population growth rate), six for agricultural drought (wells access for gardening, absence of gardening due to lack of water, gardens fencing, distance to market, cropping for self-consumption, road interruptions), eight for heavy rains (mobile telephone signal and use, earth embankment absence, leaking, lacking spillway, fence, wells flooded, unprotected creek banks). The highest values of vulnerability to meteorological drought are reached by Legaida, while those to hydrological drought are reached by Vani, those to agricultural drought by Goubya Elmejid, and those to heavy rains by Legdur. The vulnerability to all hazards shows that Legaida and Legdur have the highest values, while Agoueinit and Vani have the lowest values.

3.4. *Adaptive Capacity*

The adaptive capacity to the different hazards is represented by 10 indicators: Three for the meteorological drought (herders/farmers radio programs, extension services for herders, fodder stock), two for the hydrological drought (fountain and boreholes), two for the agricultural drought (solar water pumps and small household farmer's associations) and two for heavy rains (radio access, and earth embankments provided with spillway). The highest values of adaptation to meteorological drought are reached by Elkenar and NGuiya, those to hydrological drought by Agouenit, those of adaptive capacity to agricultural drought by Boukhzama 1, and those of adaptive capacity to heavy rains by Drougal, Elkenar, Gnebett Ehel Heiba, Goubya Elmesjid, Legaida, Legdur, Nguiya, and Vani. The adaptive capacity to all hazards is highest in Goubya Elmesjid and lowest in Agoueinit, Begou and Legaida.

3.5. Multi-Hazard Risk Level

The interval between the maximum and minimum value of the multi-hazard risk index (MHRI) was split into four parts of the same breadth to represent the severe (0.83–1.09), high (0.56–0.82), moderate (0.28–0.55), and low risk (0–0.27). It follows that NGuiya, Agueinit, and Begou are at severe risk, Legdur, Boukhzama 1 and Legaida are at high risk, Gnebett Ehel Heiba, Jrana, and Elkenar are at moderate risk, and all other communities are at low risk (Figure 11, Table 7). Therefore, it can be said that the most northern communities tend to have the highest risk levels and the five southernmost communities tend to have a low to moderate risk level. The value of the risk index was substantially determined by that of agricultural drought and heavy rains (Table 7).

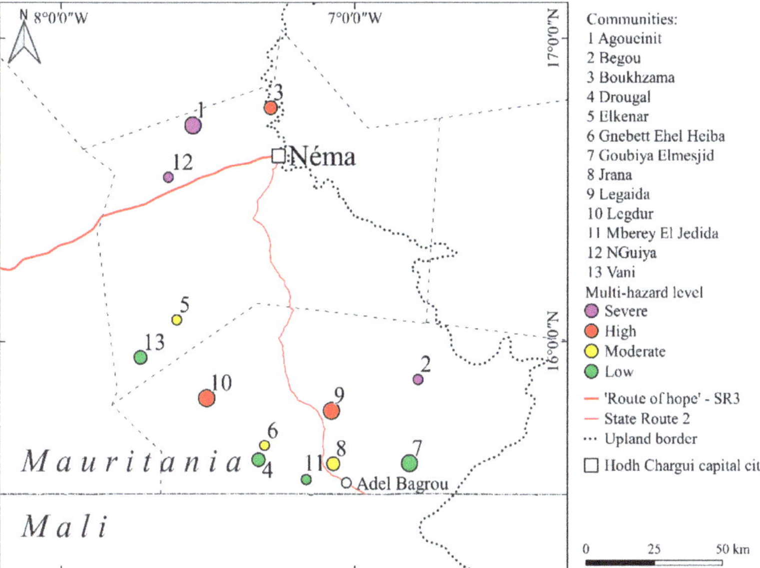

Figure 11. Thirteen rural communities at multi-hazard risk in the Hodh Chargui, Mauritania.

Table 7. Multi-hazard risk index for 13 communities of Hodh Chargui, Mauritania.

Community	Meteorological Drought	Hydrological Drought	Agricultural Drought	Heavy Rain	MHRI
					Score
12 NGuiya	0	0.06	1.13	0	1.09
1 Agoueinit	0.04	0.08	0.64	0.24	1
2 Begou	0	0.04	0.6	0.35	0.97
10 Legdur	0.03	0.04	0.49	0.19	0.78
3 Boukhzama 1	0	0.00	0	0.74	0.74
9 Legaida	0.1	0.09	0.35	0.16	0.7
6 Gnebett Ehel Heiba	0	0.06	0.3	0	0.36
8 Jrana	0.01	0.08	0.32	−0.12	0.29
5 Elkénar	−0.05	0.12	0.29	−0.08	0.28
13 Vani	−0.02	0.13	0.2	−0.13	0.17
11 Mberey El Jedida	0	0.02	0.15	0	0.16
7 Goubya Elmesjid	0	0.08	0.09	−0.12	0.14
4 Drougal	0	0.04	0.1	−0.16	0.02

3.6. The Use of the Multi-Hazard Risk Index for the Identification of Risk Treatment Actions

The risk assessment was aimed at the official development aid active in the region and at the regional administration. However, the method may also be of interest for other contexts exposed to similar hazards. In addition to the ranking of the communities according to risk level, the assessment process proposes thirteen risk treatment actions for the six communities at severe and high risk. These actions were identified after a visit to the exposed items. This involves acting on the water supply (well deepening, building or raising the apron of the pastoral wells, covering them, equipping them with pedals or solar-powered water pumps, water troughs for cattle watering, guaranteeing access even in the wet season), protect crops from stray cattle (fencing in metal barbed weir), and reduce the impact of heavy rains (repairing the earth embankments and equipping them with spillways or repairing the existing spillway, locks and fencing in metal barbed wire, gabion wall to protect the riverbank) (Figure 12, Table 8).

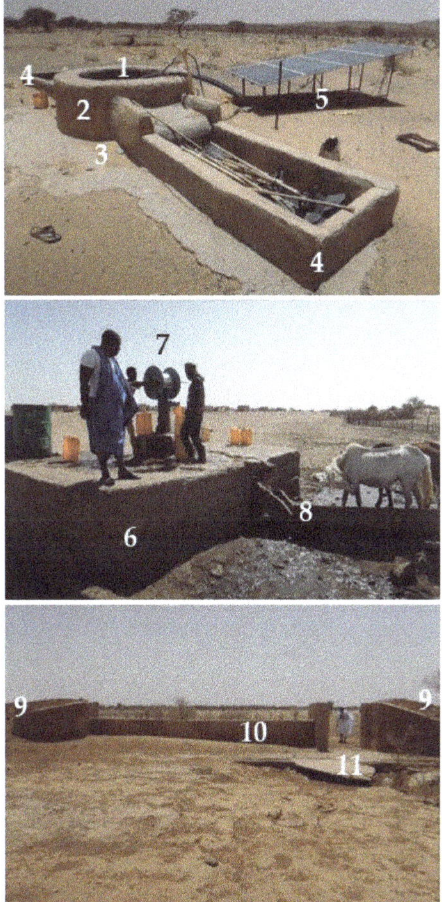

Figure 12. Open well (**1**) showing low apron (**2**) and basement (**3**), water trough for cattle watering (**4**), solar panels (**5**) in Boukhzama 1; a well provided with raised basement (**6**), a manual water pump (**7**), and a broken water trough (**8**); an earth embankment (**9**) provided with spillway (**10**) and lock (**11**) in Agoueinit.

Table 8. Risk treatment for the five communities at severe and high risk of the Hodh Chargui, Mauritania.

Community	Exposure	Vulnerability	Risk Reduction
NGuiya	Borehole	Diesel pump out of service	Solar powered water pump
	Wells	Poor water flow No basement No pump No water trough	Well deepening Basement Pedal powered pump Water trough construction
	Earth embankments	Deteriorated No fence in barbed wire	Reshaping the earth embankment Fence in barbed wire
Agoueinit	1st earth embankment	Lack of spillway and lock	Spillway and lock construction Fence in barbed wire
	2nd earth embankment	Spillway deteriorated	Spillway reparation
Begou	Wells	Wells flooded	Well deepening Elevating the apron Solar powered water pump Access to well in wet season
Legdur	Wells	Wells flooded	Covering the well Elevating the apron Solar-powered water pump
	Earth embankments	Lack of spillway, lock	Spillway and lock
Boukhzama 1	House, wells	Riverbank erosion	Gabion walls
	Wells	No water trough	Water trough for cattle
Legaida	Wells	Poor water flow	Well deepening Apron construction Water trough for cattle watering Solar-powered water pump

These measures, compared with those proposed by the literature, are more specific and directly implementable (Table 9).

Table 9. Comparison of risk reduction actions identified for Hodh Chargui with those suggested by other assessments.

Hazard	Hodh Chargui Assessment	Other Assessments [Reference]
Meteorological drought		-
Hydrological drought	Wells deepening Apron construction Water trough for cattle watering	-
Agricultural drought	Wells raised basement Wells cover Pedal powered water pump Solar powered water pump Wells access in wet season	-
Pluvial flood	Spillway Lock Fence in barbed wire	Ground pad with pill way [10] Dry stone cords [10] Dry stone thresholds [10] Tabias [10] Creek banks planting [13] Increasing tree vegetation [4] Avoid building in flood prone areas [14]
Fluvial flood	Gabion wall to protect riverbank	Dykes, embankments, ditches [6,13,21] Resettlement [6,21] Early warning [18,19,21,25] Environmental education [6,24] Land use planning [24] Watershed management plan [19] Flood monitoring [24] Emergency preparedness [19]

4. Discussion

This study reviewed the problems common to the majority of risk assessments published thus far in tropical Africa. The study then proposed a multi-hazard vision, the integration of local and scientific knowledge, a drought and flooding probability estimate, indicators representing the risk determinants, and 12 actions to deal with the risk.

The objective was to define a replicable methodology that would allow us to produce a ranking of the communities and to suggest actions for those that would have been at greatest risk. Therefore, it was necessary to refer to specific communities with their own hazards, exposures, vulnerabilities, and adaptive capacities, rather than representing communities as points of a risk map constructed by superposition of low-resolution information layers, as is often done in assessments at regional scale.

This holistic approach found little evidence in the literature consulted at the regional scale. Local knowledge and visits to the exposed items facilitated the determination of the exposure, vulnerability, and adaptive capacity.

Scientific knowledge was used to determine which of the four hazards pose the highest threat to the sustainable development of livelihoods in the rural Hodh Chargui and considered the method used to rank the communities. Contrary to the indications of the Sendai framework (2015) [2], the integration of knowledge is still unusual on a regional scale. The systematic review highlighted that only one assessment out of four published on tropical African regions estimated the probability of flooding or drought [1,3,11,18,19,27]. This is likely to result from poor access to local data. In the case of the Hodh Chargui, for example, the only surviving weather station with a continuous series of more than 30 years of daily precipitation data is that of the airport of Néma, which is too little to represent a vast territory such as that in which the 13 communities are distributed. The use of estimated precipitation from satellite (CHIRPS dataset) and the surface area of the ephemeral wetlands (Landsat images) allowed for the estimation of the probability of meteorological, hydrological, and agricultural drought, and the TRMM dataset allowed for the estimation of the probability of heavy precipitation. The low spatial resolution of the Landsat images used (30 m) was adequate for the size of water bodies observed (6 to 30 km^2).

The literature review ascertained that in risk assessments at the regional scale, the indicators are chosen according to the information most easily accessible, rather than according to the information that best represents the risk determinants. In particular, it is rare to find specific indicators for meteorological, hydrological, and agricultural drought. Meeting with the communities and the visits to the exposed items allowed for indicators specific to the context to be identified.

The considerable differences in the level of agricultural drought and heavy rains risks among the 13 communities have generated a differentiated multi-hazard index. The northernmost communities were found to have greater probability of agricultural drought risk, compared to the five southern communities on the border with Mali. However, they are also closer to the large market of Néma (22,000 inhabitants in 2013,) which demands many horticultural products in a region in which they are scarce. Therefore, they would have greater opportunities to diversify their livelihood with commercial gardening if they could improve their access to water. The communities at the foot of the uplands (Boukhzama 1 and Begou) are more exposed to the risk of heavy rains and, therefore, to flash floods.

The discussions with the communities and the visit to the receptors enabled the identification of twelve actions for the six communities at severe and high risk, which are rarely found in literature [11]. These first concern the improvement of access to water: Well deepening, apron elevation, covering, providing a pedal or a solar water pump, a water trough for cattle watering, and facilitating access in flood prone area during the wet season. Second, they concern earth embankments (creation of spillways and locks, protection with metal barbed wire) and the protection of the riverbanks (gabion wall).

The assessment methodology may be of interest for other semi-arid, agropastoral regions of the Tropics.

The assessment presents four main limits. First, it was based on a review that has only considered the published literature. Grey literature contains other assessments, but its dissemination is ephemeral and it has a temporary inspiring effect on assessment practices. For these reasons, it was not taken into consideration.

Second, the hydrological hazard was calculated in a very simplified manner, seeking to achieve a result despite the scarcity of available information at local scale. The correlation between the dynamics of the surface area of ephemeral wetlands and precipitations in a Sahelian context has been questioned by literature, particularly due to the alteration of the vegetation cover and related erosive processes that have increased the runoff over time [52–54]. However, in the specific context of southern Hodh Chargui, the almost flat orography may have limited the erosive processes compared to other areas at the foot of the uplands as Boukhzama 1 and Begrou [54].

Third, the occurrence probability of heavy precipitations was calculated on a shorter series than the conventional one (24 years instead of 38) due to the time limitations of the dataset used.

Fourth, water flow and the quality of the wells were appreciated qualitatively. Measuring these parameters is possible and would give further solidity to the assessment. However, as the number of investigations increases, the duration of the survey is extended and may risk limiting the comparison between the first and the last investigated communities, which was instead the objective of the assessment.

5. Conclusions

The recommendation of the Sendai framework (2015) to develop risk knowledge on a subnational scale is having effects in tropical Africa. In the last four years, hydro-climatic risk assessments on a regional scale have been increasingly practiced. However, the holistic approach, the integration of local and scientific knowledge, the assessment of the hazard, and the recommendation of specific actions are rarely practiced. The objective of this article was to propose a multi-hazard risk assessment method replicable in other semi-arid contexts of the Tropics that adopted a holistic approach by integrating local and scientific knowledge, ranking the communities according to the risk level, and listing actions to reduce the risk.

Of the thirteen communities where the assessment was developed, six were found to be severe and high risk according the relative intervals (low risk according the absolute intervals). The level of multi-hazard risk varies significantly, as it is mainly influenced by the risk of heavy rains and agricultural drought. The proposed actions are consequently detailed and largely differ from the generic remedies proposed by the literature. They concern the improvement of water access for agropastoral and human use, and the protection of hydraulic works and riverbanks from flash floods.

The proposed method can be extended to other Hodh Chargui communities and can be replicated in semi-arid agropastoral regions of the Tropics. The methodology requires less time and lower costs compared to assessments conducted exclusively through community surveys and provides more precise and articulated results.

The assessment helps to prioritize the actions depending on the value of the MHRI and to identify them in accordance with the prevailing risk. The assessment can also allow decisionmakers to monitor and assess the changes in the risk level occurring over time.

Supplementary Materials: The following are available online at http://www.mdpi.com/2071-1050/11/18/5063/s1, Table S1: MHRI-Multi-Hazard Risk Index; Table S2: Survey questions.

Author Contributions: Conceptualization, M.T.; Methodology, M.T. and M.B.; Investigation, M.T., M.B., S.B. (Stefano Bechis) and S.B. (Sarah Braccio); Writing—original draft preparation, M.T. and M.B.; Writing—review and editing, M.T., M.B., M.B. and S.B. (Sarah Braccio); Visualization, S.B. (Sarah Braccio); Funding acquisition, M.T.

Funding: This research was funded by DIST-Politecnico and University of Turin.

Acknowledgments: We are grateful to the two anonymous reviewers who offered enormously helpful comments to an earlier draft. We are, of course, responsible for any errors. We would like to thank Laura Alunno (Terre Solidali), Yeslem Hamadi, Lebatt Moubark (RIMRAP project) for facilitating the field activities, Souleymanou Cheikh Shed Bouh (Mayor of Agoueinit municipality) and Mohamed Lemine ould Abeid (Office National de la Météorologie) for information.

Conflicts of Interest: The authors declare no conflict of interest. The funders had no role in the design of the study; in the collection, analyses, or interpretation of data; in the writing of the manuscript, or in the decision to publish the results.

References

1. Tiepolo, M.; Rosso, M.; Massazza, G.; Belcore, E.; Issa, S.; Braccio, S. Flood assessment for risk-informed planning along the Sirba river, Niger. *Sustainability* **2019**, *11*, 4003. [CrossRef]
2. United Nations. *Sendai Framework for Disaster Risk Reduction 2015–2030*; UNISDR: Geneva, Switzerland, 2015.
3. Tiepolo, M.; Bacci, M.; Braccio, S. Multihazard risk assessment for planning with climate in the Dosso region, Niger. *Climate* **2018**, *6*, 67. [CrossRef]
4. Gebeyehu, A.; Teshome, G.; Birke, H. Flood hazard and risk assessment in Robe watershed using GIS and remote sensing, North Shewa zone, Amhara region, in Ethiopia. *J. Humanit. Soc. Sci.* **2010**, *23*, 1441–1460. [CrossRef]
5. Ibrahim, B.A.; Tiki, D.; Mamdem, L.; Leumbe Leumbe, O.; Bitom, D.; Lazar, G. Multicriteria analysis (MCA) approach and GIS for flood risk assessment and mapping in Mayo Kani division, Far North region of Cameroon. *Int. J. Adv. Remote Sens. GIS* **2018**, *7*, 2793–2808. [CrossRef]
6. Oriola, E.; Chibuike, C. Flood risk analysis of Edu local government area (Kwara State, Nigeria). *Geogr. Environ. Sustain.* **2016**, *9*, 106–116. [CrossRef]
7. Sakakun, S.; Kussul, N.; Shelestov, A.; Kissul, O. Flood hazard and flood risk assessment using a time series of satellite images: A case study in Namibia. *Risk Anal.* **2014**, *34*, 1521–1537. [CrossRef]
8. Wondim, Y.K. Flood hazard and risk assessment using GIS and remote sensing in lower Awash sub-basin, Ethiopia. *J. Environ. Earth Sci.* **2016**, *6*, 69–85, ISSN 2225-0948.
9. Adeloye, A.J.; Mwale, F.D.; Dulanya, Z. A metric-based assessment of flood risk and vulnerability of rural communities in the Lower Shire Valley, Malawi. *Proc-IAHS* **2015**, *370*, 139–145. [CrossRef]
10. Sami, K.; Mohsen, B.A.; Afef, K.; Fouad, Z. Hydrological modeling using GIS for mapping flood zones and degree flood risk in Zeuss-Koutine basin (South of Tunisia). *J. Environ. Prot.* **2013**, *4*, 1409–1422. [CrossRef]
11. Tiepolo, M.T.; Braccio, S.B. Flood risk assessment at municipal level in the Tillabéri region, Niger. In *Planning to Cope with Tropical and Subtropical Climate Change*; Tiepolo, M., Ponte, E., Cristofori, E., Eds.; De Gruyter Open: Berlin, Germany, 2016; pp. 221–245.
12. Koumassi, D.H.; Tchibozo, A.E.; Vissin, E.W.; Houssou, C.S. SIG et télédétection pour l'optimisation de la cartographie des risques d'inondation dans le bassin de la Sota au Bénin. *Rev. Ivoir. Sci. Technol.* **2014**, *23*, 137–152, ISSN 1813-3290.
13. Leumbe Leumbe, O.; Bitom, D.; Mamdem, L.; Tiki, D.; Ibrahim, A. Cartographie des zones à risque d'inondation en zone soudano-sahelienne: Cas de Maga et ses environs dans la region de l'extrême-nord Cameroun. *Afr. Sci.* **2015**, *11*, 45–61, ISSN 1813-548X.
14. Bachir Saley, M.; Kouamé, F.K.; Penven, M.J.; Biémi, J.; Kouadio, H.B. Cartographie des zones à risque d'inondation dans la région semi-montagneuse à l'Ouest de la Côte d'Ivoire: Apports des MNA et de l'imagerie satellitaire. *Télédétection* **2005**, *5*, 53–67.
15. Bachir Saley, V.H.N.B.; Wade, S.; Valere, D.E.; Kouame, F.; Affian, K. Cartographie du risque d'inondation par une approche couplée de la télédétection et des systèmes d'informations géographiques (SIG) dans le département de Sinfra (centre-ouest de la Cote d'Ivoire). *Eur. Sci. J.* **2014**, *10*, 170–191, ISSN 1857-7431.
16. Okoro, B.C.; Ibe, O.P.; Ekeleme, A.C. Development of a modified rational model for flood risk assessment of Imo State, Nigeria using GIS and Rs. *Int. J. Eng. Sci.* **2014**, *3*, 1–8, ISSN 2319-1813.
17. Bunting, E.; Steele, J.; Keys, E.; Muyengwa, S.; Child, B.; Southworth, J. Local perception of risk to livelihoods in the semi-arid landscape of southern Africa. *Land* **2013**, *2*, 225–251. [CrossRef]
18. Ntajal, J.; Lamptey, B.L.; Mahamadou, I.B.; Nyarko, B.K. Flood disaster risk mapping in the lower Mono river basin in Togo, West Africa. *Int. J. Disaster Risk Reduct.* **2017**, *23*, 93–103. [CrossRef]
19. Alfa, M.I.; Ajibike, M.A.; Daffi, R.E. Application of analytic hierarchy process and geographic information system techniques in flood risk assessment: a case of Ofu river catchment in Nigeria. *J. Degrad. Min. Land Manag.* **2018**, *5*, 1363–1372. [CrossRef]
20. Behanzin, I.D.; Thiel, M.; Sarzynski, J.; Boko, M. GIS-based mapping of flood vulnerability and risk in the Benin Niger river valley. *Int. J. Geomat. Geosci.* **2015**, *6*, 1653–1669, ISSN 0976-4380.

21. Mayomi, I.; Dami, A.; Maryah, U.M. GIS based assessment of flood risk and vulnerability of communities in the Benué floodplains, Adamawa State, Nigeria. *J. Geogr. Geol.* **2013**, *5*, 148–160, ISSN 1916-9787. [CrossRef]
22. Nkeki, F.N.; Henah, P.J.; Ojeh, V.N. Geospatial techniques for the assessment and analysis of flood risk along the Niger-Benue basin in Nigeria. *J. Geogr. Inf. Syst.* **2013**, *5*, 123–135. [CrossRef]
23. Olatona, O.O.; Obiora-Okeke, O.A.; Adewumi, J.R. Mapping of flood risk zones in Ala river basin Akure, Nigeria. *Am. J. Eng. Appl. Sci.* **2018**, *11*, 210–217. [CrossRef]
24. Udo, E.A.; Eyoh, A. Spatial analysis of river inundation and flood risk potential along Kogi state river Niger-Benoue basin using Geospatial techniques. *J. Environ. Sci. Comput. Sci. Eng. Technol.* **2017**, *6*, 351–364. [CrossRef]
25. Komi, K.; Neal, J.; Trigg, M.A.; Diekkrüger, B. Modeling of flood hazard extent in data sparse areas: a case study of the Oti river basin, West Africa. *J. Hydrol. Reg. Stud.* **2017**, *10*, 122–132. [CrossRef]
26. Gedif, B.; Hadish, L.; Addisu, S.; Suryabhagavan, K.V. Drought risk assessment using remote sensing and GIS: the case of southern zone, Tigray region, Ethiopia. *J. Nat. Sci. Res.* **2014**, *4*, 87–94, ISSN 2225-0921.
27. Edossa, D.C.; Babel, M.S.; Gupta, A.D. Drought analysis in the Awash river basin, Ethiopia. *Water Resour. Manag.* **2010**, *24*, 1441–1460. [CrossRef]
28. Forsyth, T. Community-based adaptation: A review of past and future challenges. *WIREs Clim. Change* **2013**. [CrossRef]
29. Nyong, A.; Adesina, F.; Elasha, B.O. The value of indigenous knowledge in climate change mitigation and adaptation strategies in the African Sahel. *Mitig. Adapt. Strat. Glob. Change* **2007**, *12*, 787–797. [CrossRef]
30. Mercer, J.; Kelman, I.; Taranis, L.; Suchet-Parson, S. Framework for integrating indigenous and scientific knowledge for disaster risk reduction. *Disasters* **2010**, *34*, 214–239. [CrossRef]
31. Walshe, R.A.; Nunn, P.D. Integration of indigenous knowledge and disaster risk reduction: a case study from baie Martelli, Pentecost island, Vanauatu. *Int. J. Disaster Risk Sci* **2012**, *3*, 185–194. [CrossRef]
32. Hiwasaki, L.; Luna, E.; Syamsidik; Shaw, R. Process for integrating local and indigenous knowledge with science for hydro-meteorological disaster risk reduction and climate change adaptation in coastal and small island communities. *Int. J. Disaster Risk Reduct.* **2014**, *10*, 15–27. [CrossRef]
33. United Nations Office for Disaster Risk Reduction. *Global Assessment Report on Disaster Risk Reduction 2019*; UNDRR: Geneva, Switzerland, 2019.
34. Dinku, T.; Hellmuth, M. *Informing Climate-Resilient Development in Data Sparse Regions*; Working Paper; USAID: Washington, DC, USA, 2013.
35. IPCC. Glossary. In *Climate Change 2014: Synthesis Report. Contribution of Working Groups I, II and III to the Fifth Assessment Report of the Intergovernmental Panel on Climate Change*; Mach, K.J., Planton, S., Von Stechow, C., Eds.; IPCC: Geneva, Switzerland, 2014; pp. 117–130.
36. Crichton, D. The risk triangle. In *Natural disaster management*; Ingleton, J., Ed.; Tudor Rose: London, UK, 1999; pp. 102–103.
37. Office National de la Statistique-ONS. *Recensement Général de la Population et de l'Habitat (RGPH 2013), Localités Habitées en Mauritanie*; ONS: Nouakchott, République Islamique de Mauritanie, 2015. Available online: www.ons.mr (accessed on 10 July 2019).
38. Apolloni, A.; Nicolas, G.; Coste, C.; El Mamy, A.B.; Yahya, B.; El Arbi, A.S.; Gueya, M.B.; Gilbert, M.; Lancelot, R. Towards the description of livestock mobility in Sahelian Africa: Some results from a survey in Mauritania. *PLoS ONE* **2016**, *13*, e0191565. [CrossRef] [PubMed]
39. Kappes, M.S.; Keller, M.; von Elverfeldt, K.; Glade, T. Challenges of analyzing multi-hazard risk: a review. *Nat. Hazards* **2010**, *64*, 1925–1958. [CrossRef]
40. International Electrotechnical Commission/International Organization for Standardization. *31010 Risk management—Risk assessment techniques*, 1st ed.; IEC: Geneva, Switzerland, 2009.
41. Sivakumar, M.V.K. Agroclimatic Aspects of Rainfed Agriculture in the Sudano-Sahelian zone. In Proceedings of the International workshop on soil, crop, and water management systems for rainfed agriculture in the Sudano-Sahelian zone, Niamey, Niger, 7–11 January 1987.
42. Nicholson, S.E.; Tucker, C.J.; Ba, M.B. Desertification, drought, and surface vegetation: An example from the West African Sahel. *Bull. Ame. Meteorol. Soc.* **1998**, *79*, 815–830. [CrossRef]
43. Le Houerou, H.N. *The Grazing Land Ecosystems of the African Sahel*; Springer–Verlag: Berlin/Heidelberg, Germany, 2012.

44. UNESCO. *Map of the World Distribution of Arid Regions. Map at Scale 1:25,000,000 with Explanatory Note*; UNESCO: Paris, France, 1979; p. 54. ISBN 92-3-101484-6.
45. Svoboda, M.; Fuchs, B.A. *Handbook of Drought Indicators and Indices*; World Meteorological Organization: Geneva, Switzerland, 2016; ISBN 978-91-87823-24-4.
46. McFeeters, S.K. The use of the Normalized Difference Water Index (NDWI) in the delineation of open water features. *Int. J. Remote Sens.* **1996**, *17*, 1425–1432. [CrossRef]
47. European Commission, Joint Research Centre. Global Surface Water–Data Access 1984–2018. Available online: https://global-surface-water.appspot.com/download (accessed on 10 July 2019).
48. Sivakumar, M.V.K. Empirical analysis of dry spells for agricultural applications in West Africa. *J. Clim.* **1992**, *5*, 532–539. [CrossRef]
49. Janke, H.E. Livestock production system. In *Livestock Development in Tropical Africa*; Kieler Wissenschaftsverlag Vauk: Kiel, Germany, 1982.
50. Cardona, O.D.; Van Aalst, M.K.; Birkmann, J.; Fordham, M.; McGregor, G.; Perez, R.; Pulwarty, R.S.; Schipper, E.L.F.; Sinh, B.T. *Determinants of Risk: Exposure and Vulnerability in Managing the Risks of Extreme Events and Disasters to Advance Climate Change Adaptation*; Field, C.B., Ed.; A special report of working groups I and II on the IPCC; Cambridge University Press: Cambridge, UK, 2014; pp. 66–108.
51. Ozer, P.; Hountondji, Y.C.; Gassani, J.; Djaby, B.; De Longueville, F. Evolution récente des extrêmes pluviométriques en Mauritanie (1933–2010). In Proceedings of the XXVIIe colloque de l'Association internationale de climatologie, Dijon, France, 2–5 July 2014.
52. Gardelle, J.; Hiernaux, P.; Kergoat, L.; Grippa, M. Less rain, more water in ponds: a remote sensing study of the dynamics of surface waters from 1950 to present in pastoral Sahel (Gourma region, Mali). *Hydrol. Earth Syst. Sci.* **2010**, *14*, 309–324. [CrossRef]
53. Gal, L.; Grippa, M.; Hiernaux, P.; Peugeot, C.; Mougin, E.; Kergoat, L. Changes in lakes water volume and runoff over ungauged Sahelian watersheds. *J. Hydrol.* **2016**, *540*, 1176–1188. [CrossRef]
54. Gal, L.; Grippa, M.; Hiernaux, P.; Pons, L.; Kergoat, L. Modeling the paradoxical evolution of runoff in pastoral Sahel. The case of the Agoufou watershed, Mali. *Hydrol. Earth Syst. Sci. Discuss* **2016**, *623*. [CrossRef]

Article

Risk Aversion, Inequality and Economic Evaluation of Flood Damages: A Case Study in Ecuador

Vito Frontuto [1,*], Silvana Dalmazzone [1], Francesco Salcuni [1] and Alessandro Pezzoli [2]

[1] Dipartimento di Economia e Statistica, Università di Torino, 10153 Torino, Italy; silvana.dalmazzone@unito.it (S.D.); salcuni.francesco@gmail.com (F.S.)

[2] Dipartimento Interateneo di Scienze, Progetto e Politiche del Territorio, Università di Torino e Politecnico di Torino, 10125 Torino, Italy; alessandro.pezzoli@polito.it

* Correspondence: vito.frontuto@unito.it

Received: 14 October 2020; Accepted: 27 November 2020; Published: 2 December 2020

Abstract: While floods and other natural disasters affect hundreds of millions of people globally every year, a shared methodological approach on which to ground impact valuations is still missing. Standard Cost-Benefit Analyses typically evaluate damages by summing individuals' monetary equivalents, without taking into account income distribution and risk aversion. We propose an empirical application of alternative valuation approaches developed in recent literature, including equity weights and risk premium multipliers, to a case study in Ecuador. The results show that accounting for inequality may substantially alter the conclusions of a standard vulnerability approach, with important consequences for policy choices pertaining damage compensation and prioritization of intervention areas.

Keywords: natural disasters; flooding; flood vulnerability; inequality; risk premium; expected annual damages; certainty equivalent annual damages; equity weight expected annual damages; equity weight certainty equivalent annual damage

---

1. Introduction

Flooding, defined by the Intergovernmental Panel on Climate Change (IPCC) [1] as 'the overflowing of the normal confines of a stream or other body of water or the accumulation of water over areas that are not normally submerged', is one of the most common and destructive natural disasters. Estimates of both affected people and economic losses vary widely. According to the Organization for Economic Co-operation and Development (OECD) [2], floods affect up to 250 million people in the world every year. In 2019, floods caused over 5000 casualties worldwide [3].

Population growth is driving an increase in the number of people living in areas susceptible to flooding, with a consequent surge in impacts on lives, properties and productive assets. Urbanization and development reduce the water retention capacity of soils and increase runoff [4]. Climate change is increasing the frequency and intensity of flood disasters throughout the world, which nearly doubled in 2000–2009 compared to the previous decade [5]. This combination of demographic, development and climatic drivers challenges societal resilience to catastrophic flood events. New data released by the World Resource Institute in April 2020 forecast the number of people harmed by floods to double globally by 2030. According to the projections obtained in 2019 by the Aqueduct Floods modeling tool of the World Resource Institute [6], damages to urban property are expected to rise from USD 174 to USD 712 billion per year.

The structure of impacts is not uniform across the world: low-income countries suffer higher fatalities, whereas high-income countries register higher values of damage to properties and infrastructures. Low or lower-middle-income countries accounted for 49 percent of flood events recorded in the International Disaster Database EM-DAT between 1971 and 2015 and for more than 60 percent of all deaths. High and upper-middle-income countries accounted for just under 80 percent of the monetary value of all reported material damages from flood events [2].

The socio-economic significance of the issue and the expectation of an escalating trend stimulated a vast and fast-growing literature on economic impacts of flooding, particularly in urban contexts. McClymont et al. [7] provide a thorough account of the literature on flood risk management and resilience. Hennighausen and Suter [8] explore the impact of flood risk perception in the housing market in the US. Shatkin [9] develops a conceptual framework for assessing the implications of flood risk for urban development, considering issues of property rights, informality, neoliberalization and financialization and the role of the state, with a particular focus on Asian megacities. Goh [10] explores the interrelationships between biophysical factors (ecological scales of the watershed) and socio-political factors (infrastructural scales associated with flood protection, social and spatial marginalization) behind urban flood risk, based on field research in Indonesia. Chen et al. [11] study flooding-migration relationships by combining nationally representative survey data with inundation measures derived from weather stations and satellites. Oosterhaven and Tobben [12] propose a method to estimate the indirect impacts of flood disasters and apply it to the major 2013 flooding event of southern and eastern Germany. Kashyap and Mahanta [13] provide an in-depth review of previous literature.

As both latitude and poverty play a major role in explaining exposure to natural disasters, a number of case studies have focused on developing regions: Ogie et al. [14] on coastal megacities of developing nations, Cobian Alvarez and Resosudarmo [15] on Indonesia, Reynaud et al. [16] on Vietnam, De Silva and Kawasaki [17] on Sri Lanka, Erman et al. [18] on Tanzania, Kurosaki [19] on Pakistan, to cite a few.

A number of studies have also examined the vulnerability and response of different socio-economic groups to natural disasters (e.g., Rasch [20]; Rodriguez-Oreggia et al. [21]; Glave et al. [22]; Lopez-Calva and Ortiz-Juarez [23]; Carter et al. [24]; Brouwer et al. [25]; Masozera et al. [26]) as well as the relationship between poverty and disasters (Tahira and Kawasaki [27]; Borgomeo et al. [28]; Henry et al. [29]; Patnaik and Narayanan [30]; Hallegatte et al. [31]).

There is however, in our view, a yet understudied area of enquiry—the one concerning the methodological aspects of the valuation of economic impacts. Monetary estimates of economic losses from flooding play a crucial role in informing decisions and setting priorities on risk mitigation investments as well as in determining post-disaster compensations. Yet, there are no generally agreed principles on which to ground impact valuations, which partly explains the very large variance across estimates provided even by the most authoritative sources. Particularly lacking, in our view, is a shared methodological approach to account for income inequality in determining the real welfare impact of natural disasters. Simply summing individuals' monetary equivalents is likely to provide a misleading picture of relative impacts and inappropriate policy implications when flooding disproportionately affects the poor, for whom even the loss of everything may amount to small absolute monetary values.

In fact, in standard Cost-Benefit Analyses (CBA), as commonly implemented by governments and international agencies, policies are typically evaluated by summing individuals' monetary equivalents without any distributional concern (e.g. The guidelines for CBA issued by the OECD [32], the European Commission [33], the U.S. Environmental Protection Agency [34]) The same considerations hold generally also for guidelines specific to flood damage assessments (e.g. [35,36]).)

The issue of using distributional weights in CBA dates back to the 1950s [37], but recent literature shows that this discussion has been largely ignored in real world practice (inter alia Drupp et al. [38] and Adler [39]). Kind et al. [40] have suitably tackled the issue and proposed a social welfare approach to CBA for flood and other disaster risk management, showing with a simulation how considering income distribution can lead to different conclusions 'on who to target, what to do, how much to invest and how to share risks' (p. 1). If confirmed, their results would enable decision makers to improve the effectiveness and equitability of flood management policies. However, their methodological approach has not yet been tested in real world studies.

The objective of our work is to contribute to fill this gap. After presenting the methodological options through which we can consider income distribution in the evaluation of flood damages, we offer an illustration based on empirical data from a region of high flood vulnerability and significant income inequality, the Duràn Canton in the Guayas province of Ecuador. The analysis confirms that accounting for inequality substantially alters the ranking of different areas in terms of vulnerability to flood damages and thus provides important insights for policy choices pertaining damage compensation and prioritization of intervention areas.

The paper is organized as follows. In Section 2, we formally describe the four alternative evaluation methodologies proposed in previous studies to estimate flood damages. In Section 3, we present the context of the case study and the data on which the analysis is based. Then we develop the empirical analysis, by calculating (in Section 4) the equity weights and the risk premium multipliers required for the inequality-adjusted evaluation of damages, the results of which are illustrated and discussed in Section 5. Section 6 concludes the paper.

2. Evaluation Methodologies

Following Kind et al. [40], we consider four different methodologies to estimate costs and benefits of flood risk reduction.

The first is the standard estimation of the Expected Annual Damage (EAD). Damages are derived from the stage-damage (or depth-damage) function, which provides estimates of the total damages due to a flood given its depth. Total damages are then divided by the probability of flooding (inverse of the return period). EAD focuses on damages to buildings and it does not take into account diminishing marginal utility of income or risk premia. It is the procedure generally used to evaluate damages in a standard CBA (for applications to flood risk assessment, see for example Skovgård et al. [41], Dupuits et al. [42], Alian et al. [43]). Even though it does not accurately reflect welfare economics theory, it may represent a satisfying proxy in situations where the institutional setting provides compensations for flood damages and the latter do not represent a major share of disposable incomes.

A first factor neglected in standard valuations of expected damages, as already discussed in Schulze and Kneese [44], is risk aversion. Risk-averse people, in order to protect themselves from adverse events, are willing to pay an amount larger than the expected damage (ED)—which is what makes insurance markets feasible. Additional Willingness to Pay (*WTP*) above the reduction of ED is the risk premium. We assume a typical [45] risk-averse utility function—a concave curve that becomes flatter as income increases—with constant elasticity:

$$U(Y) = \frac{Y^{1-\gamma}}{1-\gamma} \tag{1}$$

where Y is income and γ is the elasticity of marginal utility of income—the variation of utility in response to changes in income. For this utility function we can express the risk premium multiplier (*RM*), following the European Commission's guidelines to CBA [33], as:

$$RM = \frac{WTP}{EAD} = \frac{1 - \left\{1 + P\left[(1-Z)^{(1-\gamma)} - 1\right]\right\}^{\frac{1}{(1-\gamma)}}}{PZ} \tag{2}$$

where the numerator is the *WTP* for flood risk reduction, the denominator is the expected damage, P is the probability of flood occurrence (inverse of the return period) and Z is the share of income eroded by the flood—the commonly adopted measure of vulnerability. The multiplier increases more than proportionally with vulnerability.

One possible monetary evaluation approach accounting for risk aversion consists in evaluating costs and benefits of disaster prevention or remediation policies on the ground of a certainty equivalent, calculated by multiplying the expected damage by the risk premium multiplier defined above. The resulting measure, called by Kind et al. [40] Certainty Equivalent Annual Damages (CEAD), weighs *WTP* by a factor that increases more than proportionally with the fraction of household income lost, so as to account for the fact that economic theory and empirical evidence make us expect more socio-economically vulnerable individuals to be more risk averse. When compensation programs are insufficient to cover actual damages and these damages may erode a significant portion of incomes, adopting CEAD in CBA is a useful improvement over EAD.

The two approaches above do not take into account that marginal disutility of losses may vary substantially with the income of affected households, as predicted by welfare economics (and estimated in over 50 countries by Layard et al. [46]). The limits of CBAs weighing all benefits and costs equally regardless to whom they accrue—an issue thoroughly discussed in theory, besides Adler [39], also by Fleurbaey and Abi-Rafeh [47], Anthoff et al. [48] and the UK Greenbook [49]—become increasingly relevant in contexts where compensation is negligible, socio-economic vulnerability is high and income distribution is strongly unequal.

Given a standard utilitarian welfare function $W = f(U_1, U_2, \ldots, U_N)$, a change in social welfare can be written as the sum of the marginal contribution to social welfare of the variation in utility of each individual:

$$\partial W = \left(\frac{\partial W}{\partial U_1} \partial U_1 + \frac{\partial W}{\partial U_2} \partial U_2 + \cdots + \frac{\partial W}{\partial U_N} \partial U_N \right) \tag{3}$$

If we consider a change in income:

$$\partial W = \left(\frac{\partial W}{\partial U_1} \frac{\partial U_1}{\partial Y_1} \partial U_1 + \frac{\partial W}{\partial U_2} \frac{\partial U_2}{\partial Y_2} \partial U_2 + \cdots + \frac{\partial W}{\partial U_N} \frac{\partial U_N}{\partial Y_N} \partial U_N \right) \tag{4}$$

Equity weights can be derived, as done, for example, in Fleurbaey and Abi-Rafeh [47] and the European Commission [33], by summing one monetary unit to a person's annual income and calculating the variation in utility:

$$\partial W = \left(\omega_{U_1} \cdot \omega_{Y_1} \cdot \partial Y_1 + \omega_{U_2} \cdot \omega_{Y_2} \cdot \partial Y_2 + \cdots + \omega_{U_N} \cdot \omega_{Y_N} \cdot \partial Y_N \right) \tag{5}$$

where $\omega_{U_i} = \frac{\partial W}{\partial U_i}$ and $\omega_{Y_i} = \frac{\partial U}{\partial Y_i}$. According to the approximation suggested by OECD [50], the equity weight ω for a marginal increase in income for a person with income Y_i can be computed as:

$$\omega_{Y_i} = \left({}^{Y_i}/{}_{Y_{avg}} \right)^{-\gamma} \tag{6}$$

By introducing this equity weight in the calculation of EADs, one obtains an alternative measure, named by Kind et al. [40] Equity Weight Expected Annual Damages (EWEAD). EWEADs are obtained as the product of EAD and the equity weight, and they represent the weight assigned to a dollar loss by the affected individual.

A further alternative measure can be obtained by combining the three approaches above, so as to include both considerations of varying marginal disutility of losses, which may be important when damages are a significant share of incomes and these incomes are unfairly distributed, and of risk aversion, relevant when available compensations are insufficient and, again, distribution of income is significantly unequal. The resulting measure, called Equity Weight Certainty Equivalent Annual Damage (EWCEAD) [40], can be calculated by multiplying the EAD by the equity weight and the risk premium multiplier.

To sum up, the four alternative evaluation methodologies can be expressed as:

(i) Expected Annual Damage (EAD) = TD/Pr(e)
(ii) Certainty Equivalent Annual Damage (CEAD) = EAD × Risk Premium Multipliers
(iii) Equity Weights Expected Annual Damage (EWEAD) = EAD × Equity weights
(iv) Equity Weights Certainty Equivalent Annual Damage (EWCEAD) = EAD × Equity weights × Risk Premium Multipliers.

In the following sections, we implement them in an empirical valuation of flood damages in our case study, we analyze and compare the results obtained and we highlight the implications of alternative methodological choices.

3. Data

3.1. The Research Context

This research was developed in connection with the project "Climatic Resilience of Duran" (RESCLIMA DURAN), to which the University of Turin contributed with a study on the economic valuation of damages complementing the hydrological, geotechnical and community perception analyses developed by local experts (e.g., Tauzer et al. [51]) and by several other European and North American universities and research institutes (a project description is available at: https://www.researchgate.net/project/CLIMATE-RESILIENCE-FOR-CITIES-IN-ECUADOR-Case-of-Duran-RESCLIMA). The Duràn Canton, our study area, is part of the Guayas province in Ecuador, in the estuarine region of the Guayas River (Figure 1). The total area is 331.22 km^2, of which 58.14 km^2 of urban area and 273.08 km^2 of rural area. 97.91 percent of the about 272,000 inhabitants are urbanized. It represents a growing municipality within the largest urban center in Ecuador, Guayaquil, characterized by demographic and socio-economic dynamics—in terms of urbanization trends, segregation between modernized sectors and marginal areas, insecurity, high inequality [52]—typical of large cities in tropical areas.

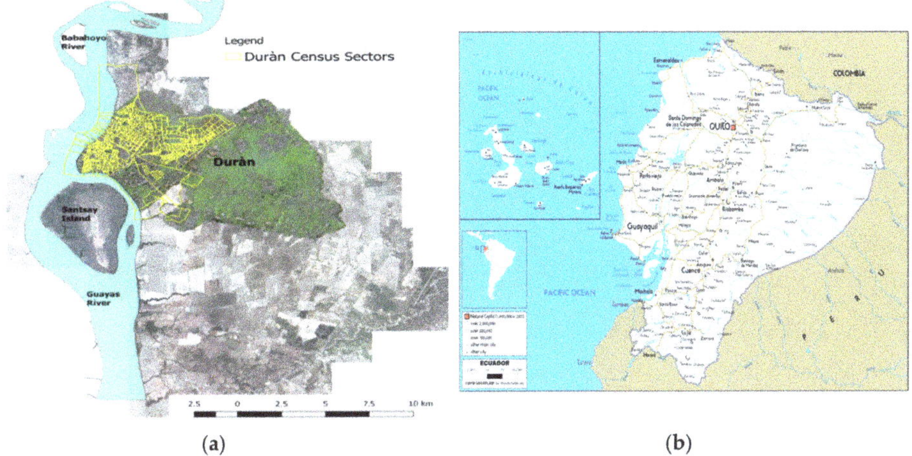

Figure 1. Duràn Canton, Ecuador. (**a**) Map of Duràn urban area; (**b**) Map of Ecuador.

The Canton is composed of 531 census sectors, but the latest Ecuador census (Encuesta Naciònal de Ingresos y Gastos de los Hogares Urbanos y Rurales; Instituto National de Estadistica y Censos (INEC) 2011 [53]) covers only 18 of them. In these sectors, between 10 and 13 families per sector were surveyed, for a total of 213 household observations, which constitute our sample. The survey contains data on population, education level, persons employed, monthly income, monthly expenditure on food and house typology. Houses are classified into four main typologies: villas, independent houses (smaller than villas), apartments in buildings, and houses made of wood or canes. Considering the predominant construction material, houses are further divided in concrete houses, brick-only houses, wooden houses, and cane houses (Table 1).

The average households' annual income is around USD 8000. The sampled houses measure, on average, 68 m^2 and are mostly built with concrete (81 percent), although 16 percent of the houses is still made of wood or canes. Out of the 213 household observations, 153 are house owners (72 percent) and the remaining 60 (28 percent) are tenants.

Latitude and the combination of the cold Humboldt current with the hot currents in Gulf of Panama and the El Niño Southern Oscillation (ENSO) phenomenon give Ecuador, with the exception of the Andean regions, a tropical climate, with heavy precipitations between January and May leading to frequent overflows of the Guayas river and the region's inner waterways. Coastal Ecuador is one of the highest hydraulic risk locations in Latin America, and cities along the mouth of the Guayas river rank among the most vulnerable areas to flooding worldwide [54]. The urban area of Duràn Canton is at an altitude varying between 0 and 88 meters above sea level. Unstructured urbanization has pushed the poor into the risk prone lowest-lying areas [51,55].

Table 1. Descriptive Statistics (Source: our elaboration on [53]).

		Mean	St.Dev
Average Annual Income ($/2011)		8153	3490
Gender			
	Female	0.506	
	Male	0.494	
Age Group			
	0–14	0.309	
	15–64	0.647	
	65+	0.043	
House Dimension (sqm)		68.13	48.68
House typology			
	Villas	0.633	
	Independent houses	0.061	
	Apartments in buildings	0.140	
	Wood and cane houses	0.164	
Construction material			
	Concrete	0.817	
	Brick-only	0.014	
	Wood	0.014	
	Cane	0.156	
House ownership			
	Owner	0.718	
	Tenant	0.282	

3.2. *Return Period, Stage-Damage Function and Flood Inundation Map*

According to hydrological models developed by the local government [56], the largest part of the Duràn Canton territory experiences extremely frequent flooding, with estimated return periods of five years (blue area in Figure 2). The most urbanized census sectors are mainly subject to return periods of up to 25 years. Arnell et al. [57] report that the frequency of river flooding in the period 1961–1990 will likely double by 2050 in Central and Eastern Europe, Central America, Brazil and some parts of Western and Central Africa. According to data reported in the EM-DAT database, the average annual number of flood events worldwide has increased from under 30 between 1971–1980 to almost 50 between 1981–1990 to over 140 between 2011 and 2015.

The stage-damage (or depth-damage) function, as mentioned above, is a function that connects damages to the depth of flood water. The database of the Joint Research Center of the European Commission (JRC) created by Huizinga et al. [58] contains damage factors of the function for all Latin American countries. The maximum damage value is estimated for Ecuador in USD 436 per square meter. This value—the highest in Latin America—represents the sum of structural and house or other building content damages, with structural damages estimated at USD 291/sqm and content damages at USD 145/sqm. We have adjusted damage values, as suggested by the JRC guidelines [58], considering rural versus urban context and the predominant material of buildings. The stage-damage function for Latin America is reported in Figure 3.

The values of flood depth in Duràn Canton for a return period of five years, described by the color gradient in Figure 4, were obtained from maps developed by Tapia [59].

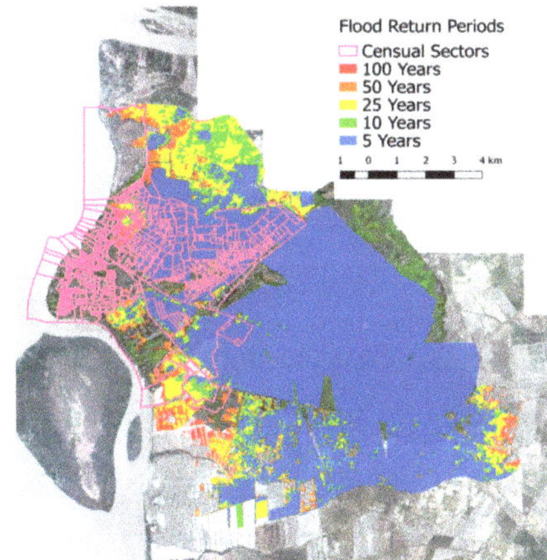

Figure 2. Return period map for Duràn Canton. (Source: our re-elaboration on [56]).

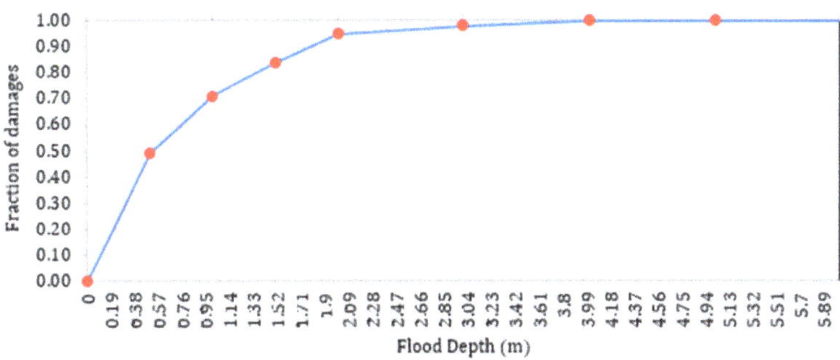

Figure 3. Stage-damage function for Latin America (Source: our adaptation on [58]).

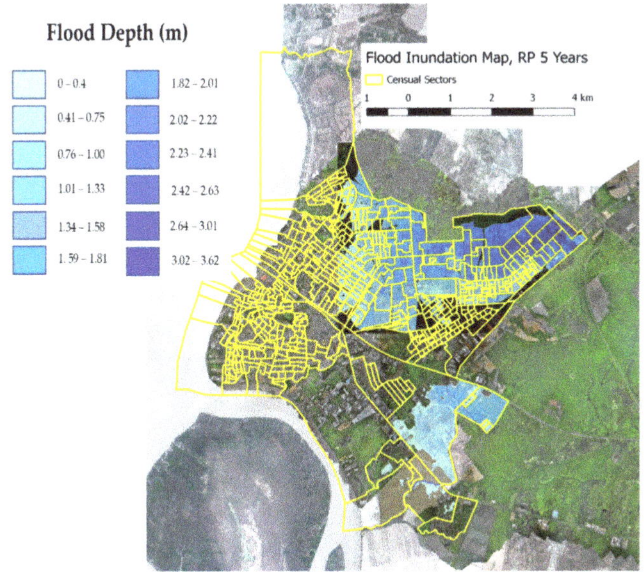

Figure 4. Flood inundation map for five years return period (Source: Our elaboration on [43]).

Total damages were derived from the stage-damage function and the inundation maps, for each censual sector and for each return period. Total damages were calculated dividing the house dimensions (square meters) by the return period of floods. The result is the Expected Annual Damage.

4. Empirical Equity Weights and Risk Premium Multipliers

In order to compute the risk premium multipliers of Equation (2) and the equity weights of Equation (6), we need empirical values for the elasticity of marginal utility (γ) and for the standard vulnerability (Z). We compute the equity weights, starting from the annual income per census sector, considering also risk aversion and income distribution. The elasticity of marginal utility, which must be $\gamma > 0$ and $\gamma \neq 1$, varies across countries and with the level of development. An estimated value for Ecuador is not available in the literature. Existing empirical estimates include Evans [60], who provides an average value of 1.4 in 20 OECD countries; Kula [61], who estimates a value of 1.64 for India; and Lopez [62], who computes the elasticity of marginal utility for nine Latin American countries with values between 1.1 and 1.9, as shown in Table 2.

Table 2. Elasticity of marginal utility in Latin American countries (Source: [62], p. 12).

Countries	γ
Argentina	1.3
Bolivia	1.5
Brazil	1.8
Chile	1.3
Colombia	1.9
Honduras	1.1
Mexico	1.3
Nicaragua	1.4
Peru	1.9

In order to select a value of γ appropriate for Ecuador, we conduct a sensitivity analysis by varying γ in the range 1.1–1.9, the interval of values estimated for Latin American countries by Lopez [62]. The results of the sensitivity analysis are available on request from the corresponding author. The results, in terms of expected damages, remain almost unchanged as the value of γ increases. Then we assume a value of γ = 1.5, considering that the income distribution and the Gini Index in Ecuador are comparable to the ones reported for other countries in South America (e.g., Bolivia, Nicaragua, Mexico) that show elasticities of marginal utility in the range 1.3–1.5 [62]. The resulting equity weights for each census sector are reported in Table 3.

From the latest Ecuador National Survey of Income and Expenditure of Urban and Rural Homes (2011) [53], we retrieved information also on each household status of house owner or tenant, whose descriptive statistics were reported in Table 1.

An important methodological issue highlighted by our Duràn Canton case study, but of high general significance particularly for natural disasters in developing countries, is that standard vulnerability, computed as share of income eroded by annual flood damages (however computed), $Z = \mathit{Flood\ damages}/Y_i$, is unable to account for damages higher than the annual income. Indeed, in our empirical analysis we find that, in poor neighborhoods, the case of households hit by flood damages to their properties (houses or their contents) higher than the family's annual income is all but infrequent. This implies a term $Z > 1$ and hence a negative risk multiplier: in this way, standard analytical tools truncate the accounting of fractional losses suffered by the poorest.

In order to overcome this limitation, we substitute the share of income lost due to the flood with the fractional value of flood damages over total wealth (TW), $Z = \mathit{Flood\ damages}/TW_i$. If the house is owned, the total wealth includes both income and the damageable value of the house, and potential flood damages are relative both to the structure and the contents. If the house is not owned, potential flood damages can only reach the maximum damage value for the contents, and total wealth is given by the sum of income and the damageable part of the contents. As a proxy of total wealth, therefore, we use the sum of annual income and the maximum value of potential flood damage obtained from the stage-damage function. In the case of households owning their house, the maximum value includes both structural and contents damage (USD 436/sqm); tenant households can only suffer contents damage (the maximum value of which is estimated in USD 145/sqm).

Table 3. Empirical equity weights.

Census Sector	Equity Weight
2002	1.388
4002	1.188
6011	0.798
9003	1.647
11006	1.111
11007	1.566
14004	1.359
17007	1.865
18002	1.785
20007	1.503
22005	1.997
28008	1.568
35004	1.103
39002	0.405
41001	1.146
42010	0.731
49009	0.345
55012	0.408

The substitution of income lost to flood damages with the share of total wealth lost is an innovation with respect to standard approaches, which allows us to have a value of vulnerability Z always between 0 and 1, obtaining valid values for the risk multiplier also for the poorest population quantiles.

The average risk premium multipliers present a slightly rising trend as the return time increases (Figure 5) due to more intense flooding and greater damages to buildings. However, given the peculiarities of our case study, the variability of average risk premium is limited. Figure 5 also shows the census sectors not impacted at low return times (sectors 39002 and 09003), in which the average risk premium is zero.

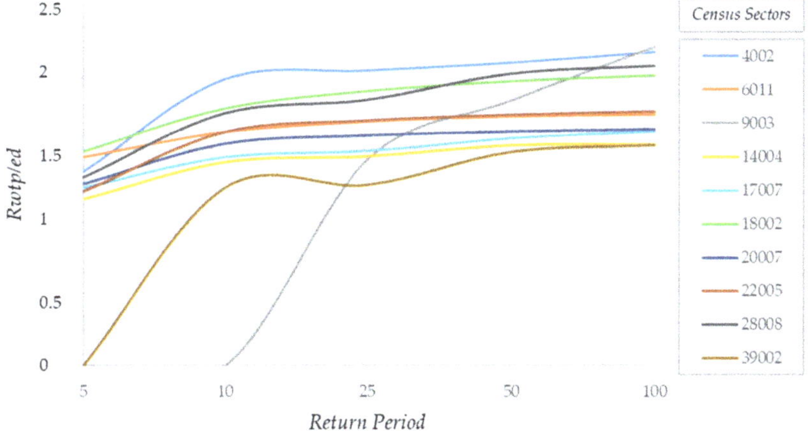

Figure 5. Average risk premium multipliers for census sectors presenting damages. Return periods between 5 and 100 years and $\gamma = 1.5$.

5. Results

To summarize, our empirical analysis combines information on (i) income and house owner or tenant status for the 213 household observations in the Duràn Canton covered by the INEC 2011 census; (ii) damage factors from Arnell and Lloyd-Hughes [57]'s Latin America stage-damage function; and (iii) values of flood depth in Duràn Canton for a return period of five years, from the inundation maps [59]. We compare the resulting evaluation of flood damages obtained with the four alternative methodologies discussed in Section 2, for return periods of 10, 25, 50 and 100 years and under the assumption of a constant elasticity of marginal utility of income of 1.2.

Figures 6–9 display the damage profiles for Expected Annual Damages, Certainty Equivalent Annual Damages, Equity Weights Expected Annual Damages and Equity Weights Certainty Expected Annual Damages, respectively.

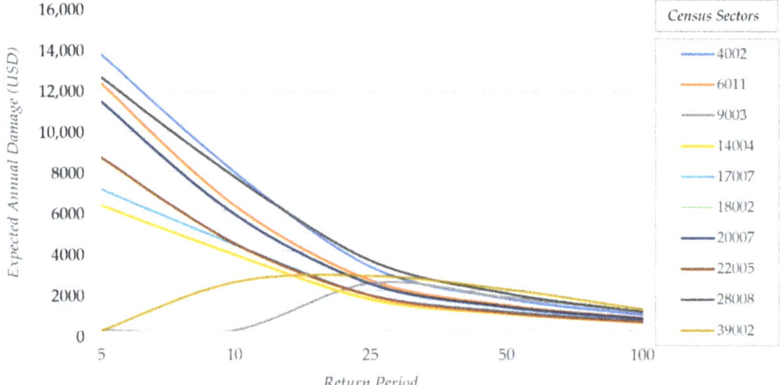

Figure 6. Damage profile evaluated with Expected Annual Damage (EAD), by census sector and return period.

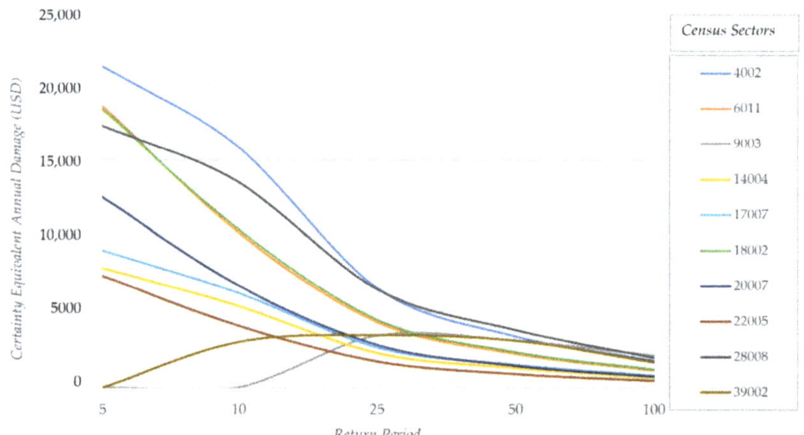

Figure 7. Damage profile evaluated with Certainty Equivalent Annual Damage (CEAD), by return period.

Figures 6–9 show a rapid reduction in the estimated damages as return times lengthen, regardless of the calculation method used. This happens because, in the specific context of the Duràn Canton, flood events are already particularly severe with low return times and they decrease with longer times. In particular, if we look at the case of EAD, which is the ratio between total damages and the probability of occurrence (Figure 6), it becomes clear that if damages do not increase as the return time increases, the ratio of these two measures will tend to decrease. This result is definitely site-specific and it depends on both the orographic characteristics of the case study and the simulated inundation maps. We also observe that some census sectors are not affected by inundations for return periods of 5 and 10 years but they are with longer periods (i.e., 39002 and 09003).

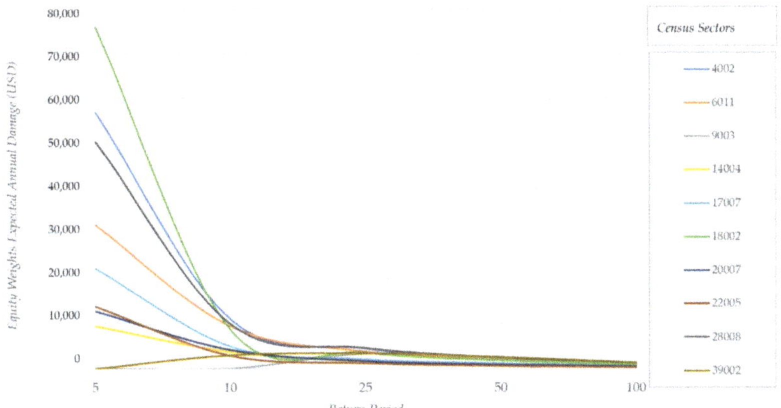

Figure 8. Damage profile evaluated with Equity Weights Expected Annual Damage (EWEAD), by return period.

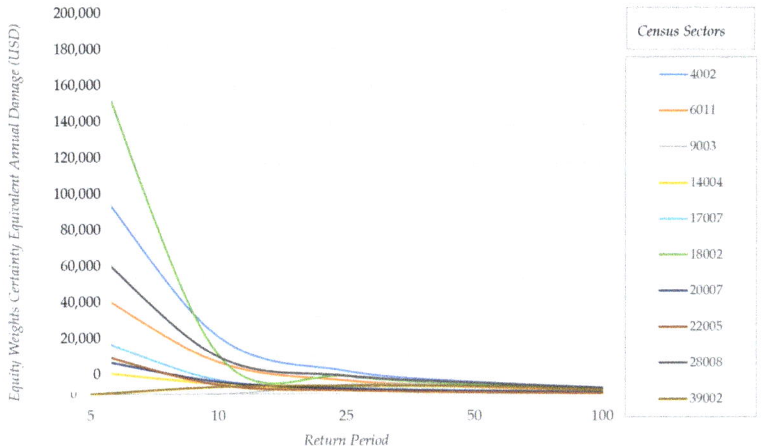

Figure 9. Damage profile evaluated with Equity Weights Certainty Equivalent Annual Damage (EWCEAD), by return period.

Finally, we can notice two main differences among the methods used to compute expected damages. When we take into account income distribution and risk premium, the ranking of sectors by intensity of damage is significantly altered by the choice of evaluation methodology. Moreover, the shape of the curves tends to be more complex when only risk premium multipliers are considered (CEAD in Figure 7) because risk premium multipliers are more heterogeneous among return times and they tend to be more clearly traced when we introduce the distribution of income through equity.

In order to allow an explicit comparison of damage evaluations conducted with the four alternative methodologies, in Table 4 we report the results for all sectors for a return period of five years. Out of the 18 sectors of Duràn Canton, eight are inundated with a return period of five years. The other sectors are never inundated or are inundated for longer return periods: a return period of five years maximizes the area interested by floods (Figure 2).

Table 4. Ranking of sectors by flood damages, for each estimation methodology.

Sector	Average Household Income (*USD*)	EAD (*USD*)	Average EAD (*USD*)	Median EAD (*USD*)	Sector	Average Household Income (*USD*)	EWEAD (*USD*)	Average EWEAD (*USD*)	Median EWEAD (*USD*)
4002	7153	13,901	1263	447	18002	5452	76,014	6334	1421
28008	5943	12,723	1060	980	4002	7153	57,056	5186	841
6011	9324	12,433	1130	1177	28008	5943	50,473	4206	1328
18002	5452	11,502	958	1166	6011	9324	31,884	2898	1751
20007	6113	8695	724	556	17007	5295	22,235	1853	1164
17007	5295	7106	592	560	22005	5058	13,775	1147	397
14004	6538	6297	524	210	20007	6113	12,618	1051	543
22005	5058	4953	413	177	14004	6538	9456	788	467
Sector	**Average Household Income (*USD*)**	**CEAD (*USD*)**	**Average CEAD (*USD*)**	**Median CEAD (*USD*)**	**Sector**	**Average Household Income (*USD*)**	**EWCEAD (*USD*)**	**Average EWCEAD (*USD*)**	**Median EWCEAD (*USD*)**
4002	7153	21,223	1929	535	18002	5452	180,015	15,001	1944
6011	9324	18,592	1690	1728	4002	7153	111,872	10,170	1065
18002	5452	18,361	1530	1571	28008	5943	76,053	6337	1731
28008	5943	17,262	1438	1390	6011	9324	54,220	4929	2592
20007	6113	12,541	1045	662	17007	5295	29,487	2457	1458
17007	5295	9043	753	720	22005	5058	21,903	1825	432
14004	6538	7889	657	226	20007	6113	18,374	1531	664
22005	5058	7359	613	189	14004	6538	11,816	984	526

The area suffering the highest damages is Sector 4002, with total EAD of USD 13,901 and average EAD of USD 1263. Sector 4002 is not the most frequently and severely inundated sector, but it is the sector, along with 6011, with the highest average annual per household income, larger houses and where a bigger share of families are house owners. The predominant construction material is concrete, which makes for houses of higher value with respect to brick-only, wooden or cane constructions more frequent in lower income sectors. Due to the very high value of Expected Annual Damages, Sector 4002 ranks as the most damaged sector also under the CEAD methodology, even though it does not have the highest risk premium multiplier.

However, when equity weights are considered, Sector 4002 is no longer the most impacted sector.

Sectors 28008 and 18002, areas with high equity weights and risk premium multipliers, which rank second and fourth respectively under the Expected Annual Damage framework, become the first and third most severely affected areas if equity weights and risk premium multipliers are accounted for in the evaluation of damages (EWEAD and EWCEAD). Conversely, Sector 6011 (the sector with the highest average income per household), which would be considered the second most damaged area under a standard EAD approach, slides down to fourth position in the ranking if damages are evaluated with equity weights. The adoption of methodologies that incorporate information on income distribution does alter significantly the outcome of evaluations and the ranking of target areas for compensation and reconstruction.

In Table 4, we report also the median value for each of the alternative methodologies used to compute expected damages. This measure of central tendency helps us to identify the census sectors presenting low-income households suffering severe damages and, in general, more unequal income distributions. This is the case for Sector 4002, which presents the highest average EAD but is among the sectors with the lowest median EAD; in this sector, the presence of few households with very low annual income exerts a strong effect on the mean which is instead mitigated by the median.

6. Conclusions

The EAD framework represents the procedure to evaluate damages from natural disasters in a typical CBA. Indeed, standard CBA is a satisfying procedure when adequate schemes are in place for the compensation of damages, income distribution is fair and damages are moderate. However, this is not the case in many instances—particularly in urban areas with low average income and marked inequality. By testing EAD and three alternative evaluation methodologies on data from a particularly significant case study—a coastal tropical urban area among the most vulnerable to flooding worldwide—we provide evidence of general value and a framework replicable in any other relevant context. Our results show that alternative measures of monetary damages from natural disasters, more coherent with economic theory of individual preferences and a social welfare perspective, can substantially modify both compensations and the ranking of priority areas of intervention. Our empirical implementation of the theoretical framework proposed in Adler [39] and Kind et al. [40] shows that the observation of income distribution, specifically via its reflection on marginal utility of income and on risk aversion, may provide a different view from the commonly adopted approach and it allows decision makers to pursue mitigation, adaptation and compensation policies more closely, reflecting a social welfare objective.

Obviously, this study also leaves room for further improvements. We have used a general stage-damage function fitted to Latin American countries, whereas more sophisticated, ad hoc studies could develop specific stage-damage functions fitted to the specific evaluation area—Ecuador or Duràn Canton data, in this case. We have used the latest available census, published in 2011 [53]; the study could be validated and updated by using the new census data which will become available in 2021-22. The sensitivity analysis could be enriched: particularly (i) a specific value of the γ parameter for the area of interest could be calculated from original data; and (ii) the analysis could be repeated with different utility functions. Further studies, replicating the analysis in other contexts and perhaps refined along these lines, would contribute to strengthening the case for revisiting the way CBA is performed in the presence of high-income inequality. We hope this first empirical investigation will spur further research interest on alternative approaches for the monetary valuation of the impacts of floods and other natural disasters on people's livelihoods.

Author Contributions: Individual contributions to the research are as follows: Conceptualization, V.F.; Data curation, F.S.; Methodology, V.F. and S.D.; Supervision, S.D. and A.P.; Writing—original draft, F.S.; Writing—review & editing, V.F., S.D. and A.P. All authors have read and agreed to the published version of the manuscript.

Funding: This research was developed in connection with the project RESCLIMA, funded by the Municipality of Duràn and the Escuela Superior Politecnica del Litoral (ESPOL) and executed by the Pacific International Center for Disaster Risk Reduction (CIP-DRR). However, it was self-financed through a co-funding of the University of Turin and the Polytechnic of Turin.

Acknowledgments: The authors would like to thank Prof. Mercy Borbor-Cordova, coordinator of the project RESCLIMA DURÀN, colleagues from the ESPOL (Escuela Superior Politécnica del Litoral, Guayaquil, Ecuador) and Angel Valdiviezo from the General Direction of Risk Management of Duràn Canton for sharing local experience, information and data necessary to implement the case study.

Conflicts of Interest: The authors declare no conflict of interest.

References

1. IPCC. *Managing the Risks of Extreme Events and Disasters to Advance Climate Change Adaptation. A Special Report of Working Groups I and II of the Intergovernmental Panel on Climate Change*; Cambridge University Press: Cambridge, UK; New York, NY, USA, 2012.
2. OECD. *Financial Management of Flood Risk*; OECD Publishing: Paris, France, 2016. [CrossRef]
3. CRED—Centre for Research on the Epidemiology of Disasters. *Disaster Year in Review 2019*; CRED: Brussels, Belgium, 2020; Available online: https://www.cred.be/publications (accessed on 2 September 2020).
4. Pereira, P.; Barcelò, D.; Panagos, P. Soil and water threats in a changing environment. *Environ. Res.* **2020**, *186*, 109501. [CrossRef]
5. Keating, A.; Campbell, K.; Mechler, R.; Michel-Kerjan, E.; Mochizuki, J.; Kunreuther, H.; Bayer, J.; Hanger, S.; McCallum, I.; See, L.; et al. *Operationalizing Resilience against Natural Disaster Risk: Opportunities, Barriers, and a Way forward*; Zurich Flood Resilience Alliance; IIASA: Zurich, Switzerland, 2014.
6. World Resource Institute. *Aqueduct Water Risk Atlas (Aqueduct 3.0)*; WRI: Washington, DC, USA, 2019; Available online: https://wriorg.s3.amazonaws.com/s3fs-public/uploads/aqueduct-whats-new.pdf (accessed on 18 November 2020).
7. McClymont, K.; Morrison, D.; Beevers, L.; Esther, C. Flood Resilience: A Systematic Review. *J. Environ. Plan. Manag.* **2020**, *63*, 1151–1176. [CrossRef]
8. Hennighausen, H.; Suter, J.F. Flood Risk Perception in the Housing Market and the Impact of a Major Flood Event. *Land Econ.* **2020**, *96*, 366–383. [CrossRef]
9. Shatkin, G. Futures of Crisis, Futures of Urban Political Theory: Flooding in Asian Coastal Megacities. *Int. J. Urban Reg. Res.* **2019**, *43*, 207–226. [CrossRef]
10. Goh, K. Urban Waterscapes: The Hydro-Politics of Flooding in a Sinking City. *Int. J. Urban Reg. Res.* **2019**, *43*, 250–272. [CrossRef]
11. Chen, J.J.; Mueller, V.; Jia, Y.; Tseng, S.K.-H. Validating Migration Responses to Flooding Using Satellite and Vital Registration Data. *Am. Econ. Rev.* **2017**, *107*, 441–445. [CrossRef]
12. Oosterhaven, J.; Tobben, J. Wider Economic Impacts of Heavy Flooding in Germany: A Non-linear Programming Approach. *Spat. Econ. Anal.* **2017**, *12*, 404–428. [CrossRef]
13. Kashyap, S.; Mahanta, R. Vulnerability Aspects of Urban Flooding: A Review. *Indian J. Econ. Dev.* **2018**, *14*, 578–586. [CrossRef]
14. Ogie, R.I.; Adam, C.; Perez, P. A Review of Structural Approach to Flood Management in Coastal Megacities of Developing Nations: Current Research and Future Directions. *J. Environ. Plan. Manag.* **2000**, *63*, 127–147. [CrossRef]
15. Cobian Alvarez, J.A.; Resosudarmo, B.P. The Cost of Floods in Developing Countries' Megacities: A Hedonic Price Analysis of the Jakarta Housing Market, Indonesia. *Environ. Econ. Policy Stud.* **2019**, *21*, 555–577. [CrossRef]

16. Reynaud, A.; Nguyen, M.-H.; Aubert, C. Is There a Demand for Flood Insurance in Vietnam? Results from a Choice Experiment. *Environ. Econ. Policy Stud.* **2018**, *20*, 593–617. [CrossRef]
17. De Silva, M.M.G.T.; Kawasaki, A. Socioeconomic Vulnerability to Disaster Risk: A Case Study of Flood and Drought Impact in a Rural Sri Lankan Community. *Ecol. Econ.* **2018**, *152*, 131–140. [CrossRef]
18. Erman, A.E.; Tariverdi, M.; Obolensky, M.A.B.; Chen, X.; Vincent, R.C.; Malgioglio, S.; Maruyama Rentschler, J.E.; Hallegatte, S.; Yoshida, N. *Wading Out the Storm: The Role of Poverty in Exposure, Vulnerability and Resilience to Floods in Dar Es Salaam*; Policy Research Working Paper Series 8976; World Bank: Washington, DC, USA, 2017.
19. Kurosaki, T. Vulnerability of Household Consumption to Floods and Droughts in Developing Countries: Evidence from Pakistan. *Environ. Dev. Econ.* **2015**, *20*, 209–235. [CrossRef]
20. Rasch, R. Income Inequality and Urban Vulnerability to Flood Hazard in Brazil. *Soc. Sci. Q.* **2017**, *98*, 299–325. [CrossRef]
21. Rodriguez-Oreggia, E.; De La Fuente, A.; De La Torre, R.; Moreno, H.A. Natural Disasters, Human Development and Poverty at the Municipal Level in Mexico. *J. Dev. Stud.* **2013**, *49*, 442–455. [CrossRef]
22. Glave, M.; Fort, R.; Rosemberg, C. *Disaster Risk and Poverty in Latin America: The Peruvian Case Study*; Group for the Analysis of Development (GRADE): Lima, Peru, 2009.
23. Lopez-Calva, L.F.; Ortiz-Juarez, E. *Evidence and Policy Lessons on the Links between Disaster Risk and Poverty in Latin America*; MDG-01-2009; UNDP Regional Bureau for Latin America and the Caribbean: New York, NY, USA, 2009.
24. Carter, M.R.; Little, P.D.; Mogues, T.; Negatu, W. Poverty traps and natural disasters in Ethiopia and Honduras. *World Dev.* **2007**, *35*, 835–856. [CrossRef]
25. Brouwer, R.; Akter, S.; Brander, L.; Haque, E. Socioeconomic vulnerability and adaptation to environmental risk: A case study of climate change and flooding in Bangladesh. *Risk Anal.* **2007**, *27*, 313–326. [CrossRef]
26. Masozera, M.; Bailey, M.; Kerchner, C. Distribution of impacts of natural disasters across income groups: A case study of New Orleans. *Ecol. Econ.* **2007**, *63*, 299–306. [CrossRef]
27. Tahira, Y.; Kawasaki, A. The impact of the Thai flood of 2011 on the rural poor population living on the flood plain. *J. Disaster Res.* **2017**, *12*, 147–157. [CrossRef]
28. Borgomeo, E.; Hall, J.W.; Salehin, M. Avoiding the water-poverty trap: Insights from a conceptual human-water dynamical model for coastal Bangladesh. *Int. J. Water Resour.* **2017**, *34*, 1–23. [CrossRef]
29. Henry, M.; Kawasaki, A.; Takigawa, I.; Meguro, K. The impact of income disparity on vulnerability and information collection: An analysis of the 2011 Thai flood. *Flood Risk Manag.* **2015**, *10*, 339–348. [CrossRef]
30. Patnaik, U.; Narayanan, K. Vulnerability and Coping to Disasters: A Study of Household Behaviour in Flood Prone Region of India. *Munich Pers. RePEc Arch.* **2010**, *21992*, 1–22.
31. Hallegatte, S.; Henriet, F.; Patwardhan, A.; Narayanan, K.; Ghosh, S.; Karmakar, S.; Patnaik, U.; Abhayankar, A.; Pohit, S.; Corfee-Morlot, J. *Flood Risks, Climate Change Impacts and Adaptation Benefits in Mumbai: An Initial Assessment of Socio-Economic Consequences of Present and Climate Change Induced Flood Risks and of Possible Adaptation Options*; OECD Publishing Office: Paris, France, 2010.
32. OECD. *Cost-Benefit Analysis and the Environment: Further Developments and Policy Use*; OECD Publishing: Paris, France, 2018. [CrossRef]
33. European Commission. *Guide to Cost-Benefit Analysis of Investment Projects*; European Commission: Brussels, Belgium, 2015. Available online: https://ec.europa.eu/regional_policy/sources/docgener/studies/pdf/cba_guide.pdf (accessed on 15 February 2020).
34. U.S. Environmental Protection Agency, National Center for Environmental Economics. *Guidelines for Preparing Economic Analyses*; EPA 240-R-00-003; U.S. Environmental Protection Agency, National Center for Environmental Economics: Washington, DC, USA, 2000. Available online: https://nepis.epa.gov/Exe/ZyPDF.cgi/P1004DN9.PDF?Dockey=P1004DN9.PDF (accessed on 18 November 2020).
35. European Commission. *Floods and Economics: Appraising, Prioritising and Financing Flood Risk Management Measures and Instruments*; Working Group F on Floods, Thematic Workshop 25-26/10/2010; European Commission: Ghent, Belgium, 2011. Available online: https://ec.europa.eu/environment/water/water-framework/economics/pdf/WGF11-3-BE-Floods_and_economics_workshop.pdf (accessed on 15 February 2020).
36. European Commission. *Directorate General Humanitarian Aid and Civil Protection (DG-ECHO). Integrating CBA in the Development of Standards for Flood Protection & Safety (FLOOD-CBA2)*; Final Report; European Commission: Ghent, Belgium, 2017. Available online: http://www.floodcba2.eu/site/ (accessed on 18 November 2020).

37. Meade, J.E. *Trade and Welfare: Mathematical Supplement*; Oxford University Press: Oxford, UK, 1955.
38. Drupp, M.A.; Meyac, J.N.; Baumgärtner, S.; Quaas, M.F. Economic Inequality and the Value of Nature. *Ecol. Econ.* **2018**, *150*, 340–345. [CrossRef]
39. Adler, M.D. Benefit–Cost Analysis and Distributional Weights: An Overview. *Rev. Environ. Econ. Policy* **2016**, *10*, 264–285. [CrossRef]
40. Kind, J.; Botzen, W.J.W.; Aerts, J.C.J.H. Accounting for risk aversion, income distribution and social welfare in cost-benefit analysis for flood risk management. *WIREs Clim. Chang.* **2017**, *8*, e446. [CrossRef]
41. Skovgård, O.A.; Zhou, Q.; Linde, J.J.; Arnbjerg-Nielsen, K. Comparing Methods of Calculating Expected Annual Damage in Urban Pluvial Flood Risk Assessments. *Water* **2015**, *7*, 255. [CrossRef]
42. Dupuits, E.J.C.; Diermanse, F.L.M.; Kok, M. Economically optimal safety targets for interdependent flood defences in a graph-based approach with an efficient evaluation of expected annual damage estimates. *Nat. Hazards Earth Syst. Sci.* **2017**, *17*, 1893–1906. [CrossRef]
43. Alian, N.; Ahmadi, M.M.; Bakhtiari, B. Uncertainty Analysis of Expected Annual Flood Damage for Flood Risk Assessment (A Case Study: Zayande Roud Basin). *J. Water Soil Sci.* **2019**, *23*, 141–152.
44. Schulze, W.D.; Kneese, A.V. Risk in Benefit-Cost Analysis. *Risk Anal.* **1981**, *1*, 81–88. [CrossRef]
45. Arrow, K.J. *The Theory of Risk Aversion. Aspects of the Theory of Risk Bearing*; Yrjo Jahnssonin Saatio: Helsinki, Finland, 1965; Reprinted in: *Essays in the Theory of Risk Bearing*; Markham: Chicago, IL, USA, 1971; pp. 90–109.
46. Layard, R.; Mayraz, G.; Nickell, S. The marginal utility of income. *J. Public Econ.* **2008**, *92*, 1846–1857. [CrossRef]
47. Fleurbaey, M.; Abi-Rafeh, R. The use of distributional weights in benefit–cost analysis: Insights from welfare economics. *Rev. Environ. Econ. Policy* **2016**, *10*, 286–307. [CrossRef]
48. Anthoff, D.; Hepburn, C.; Tol, R.S. Equity weighting and the marginal damage costs of climate change. *Ecol. Econ.* **2009**, *68*, 836–849.
49. Her Majesty's Treasury (HMT). *The Green Book. Appraisal and Evaluation in Central Government*; TSO: London, UK, 2011.
50. OECD. *Cost-Benefit Analysis and the Environment: Recent Developments*; OECD Publishing: Paris, France, 2006. [CrossRef]
51. Tauzer, E.; Borbor-Cordova, M.J.; Mendoza, J.; De La Cuadra, T.; Cunalata, J.; Stewart-Ibarra, A.M. A participatory community case study of periurban coastal flood vulnerability in southern Ecuador. *PLoS ONE* **2019**, *14*, e0224171. [CrossRef]
52. Moser, C. *Ordinary Families, Extraordinary Lives: Assets of Poverty Reduction in Guayaquil, 1978–2004*; Brookings Institution Press: Washington, DC, USA, 2009.
53. INEC. *Encuesta Naciònal de Ingresos y Gastos de los Hogares Urbanos y Rurales*; Instituto Nacional de Estadística y Censos, Gobierno de la Republica del Ecuador: Quito, Ecuador, 2011. Available online: https://www.ecuadorencifras.gob.ec/encuesta-nacional-de-ingresos-y-gastos-de-los-hogares-urbanos-y-rurales/ (accessed on 15 February 2020).
54. Calil, J.; Reguero, B.G.; Zamora, A.R.; Losada, I.J.; Méndez, F.J. Comparative Coastal Risk 719 Index (CCRI): A multidisciplinary risk index for Latin America and the 720 Caribbean. *PLoS ONE* **2017**, *12*, e0187011. [CrossRef]
55. Hardoy, J.E.; Mitlin, D.; Satterthwaite, D. *Environmental Problems in an Urbanizing World: Finding Solutions in Cities in Africa, Asia and Latin America*; Routledge: London, UK, 2013. [CrossRef]
56. Gobierno Autonomo Descentralizado Durán (GAD-Durán). Mapa de Amenazas por Inundaciones: Durán, Ecuador; 2014. Available online: http://preventionweb.net/go/40953 (accessed on 15 February 2020).
57. Arnell, N.W.; Lloyd-Hughes, B. The global-scale impacts of climate change on water resources and flooding under new climate and socio-economic scenarios. *Clim. Chang.* **2014**, *122*, 127–140. [CrossRef]
58. Huizinga, J.; de Moel, H.; Szewczyk, W. *Global Flood Depth-Damage Functions: Methodology and the Database with Guidelines*; JRC Working Papers JRC105688; Joint Research Centre of the European Commission (JRC): Seville, Spain, 2017.
59. Tapia, A.J.C. *Hydrologic Modelling of an Experimental Area in the Guayas River Basin to Quantify Liquid and Solid Flow Production*; Universidad Nacional de La Plata: Quito, Ecuador, 2012.
60. Evans, D.J. The Elasticity of Marginal Utility of Consumption: Estimates for 20 OECD Countries. *Fisc. Stud.* **2005**, *26*, 197–224. [CrossRef]
61. Kula, E. Estimation of a Social Rate of Interest for India. *J. Agric. Econ.* **2004**, *55*, 91–99. [CrossRef]

62. Lopez, H. *The Social Discount Rate: Estimates for Nine Latin American Countries*; Policy Research working paper WPS 4639; World Bank: Washington, DC, USA, 2008. Available online: http://documents.worldbank.org/curated/en/135541468266716605/The-social-discount-rate-estimates-for-nine-Latin-American-countries (accessed on 2 September 2020).

Publisher's Note: MDPI stays neutral with regard to jurisdictional claims in published maps and institutional affiliations.

Article

Community Perception and Communication of Volcanic Risk from the Cotopaxi Volcano in Latacunga, Ecuador

Juan Camilo Gomez-Zapata [1,2,*], Cristhian Parrado [3], Theresa Frimberger [4], Fernando Barragán-Ochoa [5,6], Fabio Brill [7,8], Kerstin Büche [9], Michael Krautblatter [4], Michael Langbein [10], Massimiliano Pittore [1,11], Hugo Rosero-Velásquez [12], Elisabeth Schoepfer [10], Harald Spahn [13] and Camilo Zapata-Tapia [14,15]

1 Seismic Hazard and Risk Dynamics, Helmholtz Centre Potsdam GFZ German Research Centre for Geosciences, 14473 Potsdam, Germany
2 Institute for Geosciences, University of Potsdam, 14469 Potsdam, Germany
3 Grupo FARO, Quito 170518, Ecuador; cparrado@grupofaro.org
4 Landslide Research Group, Technical University of Munich, 80333 Munich, Germany; theresa.frimberger@tum.de (T.F.); m.krautblatter@tum.de (M.K.)
5 Asociación de Profesionales de Gestión de Riesgos de Ecuador, Quito 170519, Ecuador; fbarraganochoa@gmail.com
6 Instituto de Altos Estudios Nacionales, Quito 170135, Ecuador
7 Hydrology, Helmholtz Centre Potsdam GFZ German Research Centre for Geosciences, 14473 Potsdam, Germany; fabio.brill@gfz-potsdam.de
8 Institute for Environmental Science and Geography, University of Potsdam, 14476 Potsdam-Golm, Germany
9 Geomer GmbH, 69126 Heidelberg, Germany; kerstin.bueche@geomer.de
10 German Remote Sensing Data Center (DFD), German Aerospace Center (DLR), 82234 Oberpfaffenhofen, Germany; Michael.Langbein@dlr.de (M.L.); Elisabeth.Schoepfer@dlr.de (E.S.)
11 EURAC Research, 39100 Bolzano, Italy; massimiliano.pittore@eurac.edu
12 Engineering Risk Analysis Group, Technical University of Munich, 80333 Munich, Germany; hugo.rosero@tum.de
13 Independent Consultant, 26689 Apen, Germany; harald.spahn@web.de
14 College of Social Sciences and Humanities, Universidad San Francisco de Quito, Quito 170901, Ecuador; camilozapatatapia@gmail.com
15 GADPC, Gobierno Autónomo Descentralizado Provincial de Cotopaxi, Latacunga 050101, Ecuador
* Correspondence: jcgomez@gfz-potsdam.de

Citation: Gomez-Zapata, J.C.; Parrado, C.; Frimberger, T.; Barragán-Ochoa, F.; Brill, F.; Büche, K.; Krautblatter, M.; Langbein, M.; Pittore, M.; Rosero-Velásquez, H.; et al. Community Perception and Communication of Volcanic Risk from the Cotopaxi Volcano in Latacunga, Ecuador. *Sustainability* **2021**, *13*, 1714. https://doi.org/10.3390/su13041714

Academic Editor: Maurizio Tiepolo
Received: 31 December 2020
Accepted: 1 February 2021
Published: 5 February 2021

Abstract: The inhabitants of Latacunga living in the surrounding of the Cotopaxi volcano (Ecuador) are exposed to several hazards and related disasters. After the last 2015 volcanic eruption, it became evident once again how important it is for the exposed population to understand their own social, physical, and systemic vulnerability. Effective risk communication is essential before the occurrence of a volcanic crisis. This study integrates quantitative risk and semi-quantitative social risk perceptions, aiming for risk-informed communities. We present the use of the RIESGOS demonstrator for interactive exploration and visualisation of risk scenarios. The development of this demonstrator through an iterative process with the local experts and potential end-users increases both the quality of the technical tool as well as its practical applicability. Moreover, the community risk perception in a focused area was investigated through online and field surveys. Geo-located interviews are used to map the social perception of volcanic risk factors. Scenario-based outcomes from quantitative risk assessment obtained by the RIESGOS demonstrator are compared with the semi-quantitative risk perceptions. We have found that further efforts are required to provide the exposed communities with a better understanding of the concepts of hazard scenario and intensity.

Keywords: risk communication; volcanic hazards; social risk perception; resilience; demonstrator; scenario; multi-risk analysis

1. Introduction

An active volcanic environment is prone to produce cascading and compound natural hazards. Cascading hazards comprise a primary hazard triggering a secondary one [1], whilst compound hazards refer to events (not necessary interdependent) events whose spatiotemporal footprints overlap (i.e., they occur almost simultaneously and affect the same or neighbouring locations) [2]. For instance, increasing volcanic activity can occur in company with seismic activity and continuous gas emissions and lightning and ultimately trigger lava flow, pyroclastic density currents, tephra (including volcanic ash and ballistics), debris avalanches (sector collapse), tsunami (for submarine volcanoes or at the seaside), and lahars [3]. Syneruptive lahars (also called primary lahars) can happen due to glacier melting during a volcanic eruption, whilst secondary lahars are commonly triggered by heavy rainfalls [4,5]. The forecast of cascading and/or compound volcanic hazards is very diverse and with time dependencies. Models are heavily tailored towards the specific volcanic system [6]. To explore the possible consequences before the actual occurrence of the events, risk scenarios are instrumental for risk communication practises. A risk scenario, as stated in [7], is considered as a situation picture in which a hazardous event with a certain probability would occur and cause some damage. The appropriate communication of risk scenarios can ultimately contribute to territory planning, response planning, design of evacuation routes, and enhancing overall preparedness.

Consequences of volcanic events can be severe, especially when the affected community is not well prepared. One of the most widely known examples of physical damage on assets and human losses due to a lack of effective risk communication occurred during the 1985 eruption of Nevado del Ruiz volcano in Colombia, during which 25,000 people died due to primary lahars [8–10]. However, volcanoes do not only affect the communities in their proximities, but have also generated systemic infrastructure failures and cascading effects (cascading effects can be defined as the disruptions consequent upon the preceding event that can have an acting large-scale across sectoral boundaries [11]) on a large-scale. A clear example of this type of effect occurred during the eruption of the Eyjafjallajökull volcano in Iceland. During two months in 2010, about 100,000 flights between Europe and North America were cancelled due to the sustained ash emission, causing more than $1.7 billion losses in lost revenues for airlines [12]. A further example of volcanic multi-hazard risk is the 2018 eruption of the Anak Krakatau volcano in Indonesia, which induced its own flank collapse, triggering a tsunami that resulted in the death of 430 people mostly in the western area of Java Island [13]. A tsunami threat from the Krakatau volcano was not unknown, since a similar historical event happened in 1883. However, it was not taken up in a broader discussion on how to deal with such a risk scenario [14].

Monitoring of volcanic activity has been significantly improved in recent years through denser and more widespread networks [15,16]. Moreover, there have been increasing research activities on the interaction between volcanic hazards (e.g., [17,18]). However, the impacts caused by volcanic hazards are rarely assessed in a comprehensive manner due to the lack of worldwide unified exposure models [19] and the scarce damage data collection on the exposed assets needed to constrain vulnerability models [6]. These difficulties are even more pronounced in a multi-risk context, where there is still a gap in the investigation of the interactions at the vulnerability level [20]. Hence, only a few examples of quantitative damage assessment have been reported in the scientific literature (e.g., [21,22]). Furthermore, there is a lack of tools for simulating representative volcanic scenarios in order to analyse the extent and spatial distribution of the expected consequences, needed for decision making and planning. Therefore, scenario-based approaches for a volcanic multi-hazard risk environment are not always available or might not be effectively communicated to the exposed communities before the occurrence of a volcanic crisis [23]. On the one hand, setting up these methods in a consistent scenario-based approach is a challenging task on its own. On the other hand, effectively communicating the potential direct damages and losses and the associated likely disruptions of critical infrastructure is also a daunting task, which in turn depends on the availability of scenario-based risk outcomes.

Rural communities of economically developing countries are particularly prone to encounter more difficulties throughout every single step of the multi-hazard risk chain (e.g., [24,25]). The social vulnerability perception of rural inhabitants might not be always taken into consideration by the local planners, partially due to their remoteness, i.e., typical large distances from the main urban centres [26], or socio-economic factors such as their alphabetization level [27], poor access to information systems, or even the basic lack of knowledge of what potential hazards may impact their communities [28]. These characteristics are common in areas exposed to volcanic hazards. In 2015, roughly 415 million people, most of them located in rural areas, lived within a 100 km radius from the 220 active volcanoes listed in the NOAA Significant Volcanic Eruption Database [29]. Hence, rural communities worldwide are more prone to suffer damaging effects from volcanic eruptions [30]. These consequences are not only expected to impact individual components such as buildings [31] and agricultural fields [32], but also critical infrastructure (e.g., power networks, roads, and water supply systems) for which the evaluation of systemic vulnerability is also required (e.g., [33,34]).

Cascading effects may further drastically change the health quality, as well as economic and social activities of the exposed communities [35]. For example, the continuous emissions of volcanic ashes can interrupt agro-industrial activities, which are the most typical source of income of rural communities [36]. These communities may also experience low serviceability of lifeline networks [37] and/or suffer from physical isolation from neighbouring communities, e.g., due to damaged bridges. Only in a few cases, the cascading effects due to volcanic eruptions have been analysed in a systematic manner [38]. Therefore, there is an urgent need to effectively communicate the scientific results of a volcanic risk assessment while simultaneously addressing the social perception and understandings, by the exposed communities, of different risk factors [39]. As stated in [40], clear risk communication in all the components of a multi-risk chain (i.e., hazards, exposure, and physical and systemic vulnerabilities) with the directly exposed communities, local decision-makers, and planners is fundamental to construct more resilient communities.

Although community participation is considered an essential component of effective resilience planning to natural hazard-risks [41,42], only in recent years some studies have integrated scientific approaches with the active participation of the community, local planners, decision-makers, and actors of the civil society (e.g., [43–47]). The specific community perceptions of vulnerability and risk related to volcanic hazards have been investigated in former works (e.g., [48–51]) through "top-down" approaches. In [52], it has been suggested that scientists should have a transversal role and a stronger presence in the communication of volcanic hazards and risks from "bottom-up" approaches. To the best of the authors' knowledge, these practices have been documented in a few works for rural communities (i.e., [53,54]). Hence, we realise that there is still significant work to be carried out to strengthen the risk-informed communities exposed to volcanic hazards. With this background, we present throughout this work an integrative framework between scientific approaches that study the possible damaging effects from volcanic scenarios with the local knowledge and social risk perceptions. The study area of Latacunga, capital of the Cotopaxi province in Ecuador, with mainly rurally composed communities, and exposed to the Cotopaxi volcano has been investigated in order to enhance a risk-informed community and awareness and contribute to increasing their resilience.

2. Framework and Objectives

Volcanic eruptions pose an enormous risk to Ecuador, because most of the exposed human settlements in the central and northern highlands are situated less than 25 km from an active volcano. Cities previously affected by volcanic eruptions include Quito, Latacunga, Salcedo, Cayambe, Ibarra-Otavalo, Ambato, Riobamba, and Baños [55]. Lahars have been among the deadliest volcanic hazards, but the emission of volcanic ash has been more frequent in the Ecuadorian Andes [56]. Ash falls do not only have direct consequences on the inhabitants' health and on the exposed infrastructure, but also on agriculture and animal husbandry, which is particularly important for the rural communities in Ecuador. Ash falls have hit the rural communities settled in the vicinity of the most active Ecuadorian volcanoes (i.e., Tungurahua, Reventador, Sangay, and Cotopaxi). Moreover, poverty, marginality, and high inequality of the exposed communities coexist with their physical and systemic vulnerabilities [57].

2.1. Description of the Study Area

The Cotopaxi volcano is an active stratovolcano (5897 m.a.s.l) located in the Cordillera Real of the Ecuadorian Andes (Figure 1) and is covered by an extensive, but diminishing glacier cap. Cotopaxi is one of the most dangerous volcanoes worldwide [58] with average recurrence intervals for eruptions between 117–147 years [59]. It can produce syneruptive lahars triggered by explosive eruptions, which can travel hundreds of kilometres [60]. Three drainage systems originate on Cotopaxi (Figure 2), which have all been inundated by lahars in prehistoric times [61]. However, only the northern and southern drainage systems are densely populated: The largest urban agglomeration encountered by the northern system is "El Valle de Los Chillos" (with about 400,000 inhabitants) in the vicinity of southern Quito, whilst the southern drainage system encounters the Latacunga canton (with about 300,000 inhabitants). The last major eruption of the Cotopaxi volcano in the historical records occurred in 1877. It induced syneruptive lahars that severely affected the proximal rural communities [62], with more than 1000 deaths registered, and caused a severe economic crisis [63]. If a similar scenario occurred nowadays, the social and economic consequences would be far more catastrophic due to the high population density and the central importance of Latacunga and the Cotopaxi region for the economic development of the country [58].

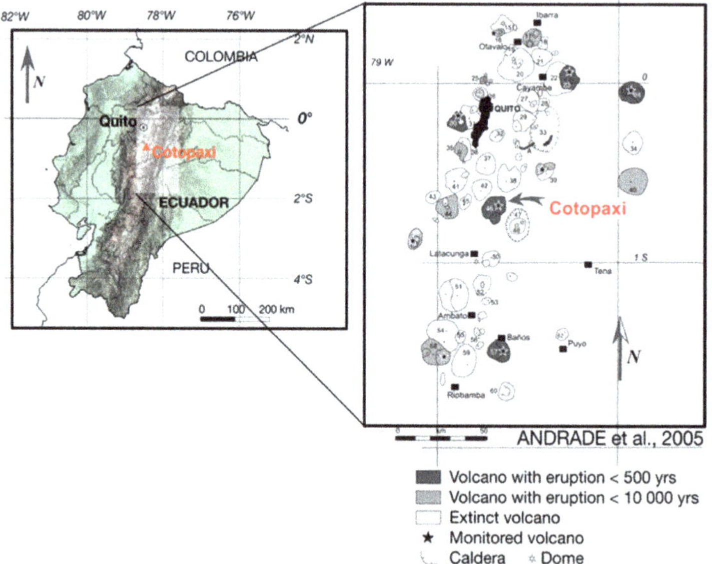

Figure 1. Location of the main volcanic systems in the Ecuadorian Andes highlighting the location of the Cotopaxi volcano and Latacunga. Modified after [64].

Latacunga is the largest city of the Latacunga canton (second Ecuadorian administrative division) and it is the capital of the Cotopaxi province (first division). It is located at a 14 km distance from the Cotopaxi volcano. For the year 2020, and based on the population projections of the National Institute of Statistics and Censuses [65], the city has an inferred population of approximately 205,600 inhabitants, with a major rural composition (59.8%). The last peak of volcanic activity of the Cotopaxi volcano occurred in mid-April 2015 and lead to a crisis in risk management in Latacunga and neighbouring municipalities [66]. Firstly, an increase in the seismic activity of the volcano was accompanied by the emission of sulphur dioxide and ash fall for some weeks [56]. Subsequently, authorities and local press communicated to the inhabitants of the communities in the vicinity of the Cotopaxi volcano that it was necessary to evacuate their homes promptly due to the imminent occurrence of lahars [66]. This generated social chaos due to the ignorance of the evacuation routes, the uncontrolled behaviour of the citizens (due to generalised fear of looting), as well as a very low level of trust in government representatives [67]. Eventually, the 2015 activity never surpassed a magnitude VEI 2 and no large syneruptive lahar flows occurred [68]. The lesson learned from this experience was the need for adequate evacuation protocols and local authorities with an understanding of the complexity of the risk in the area. Moreover, it was realized how important it is for citizens to understand their own social, physical, and systemic vulnerability [68].

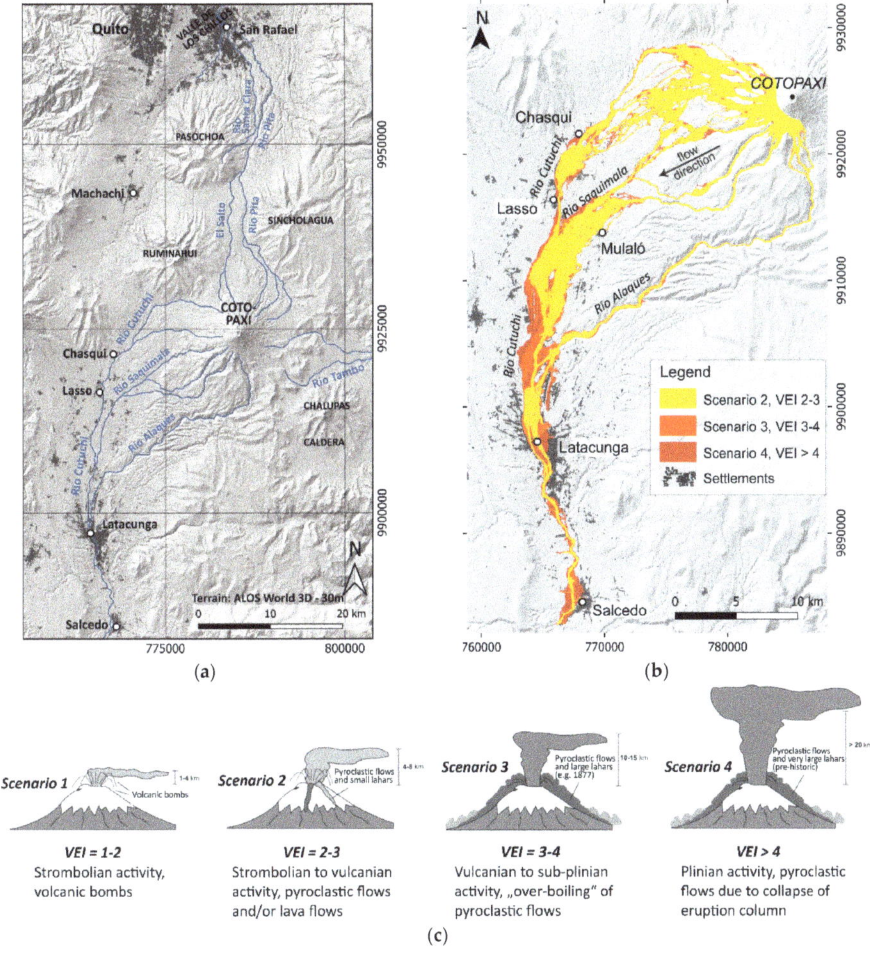

Figure 2. (**a**) Location of the Cotopaxi volcano and the main drainages and populated centres. (**b**) Estimated lahar footprints in the southern drainage system from three scenarios as function of the VEI (Volcanic Explosivity Index). (**c**) Brief description of the eruption scenarios expected at Cotopaxi in terms of the VEI. Modified after [64,69].

Latacunga is settled on ancient and recent geological materials formed by volcanic material. Some of the most representative and better-exposed stratigraphic formations of ancient ashes and lahar deposits originated from the previous volcanic activity of the Cotopaxi volcano were visited (Figure 3) with the guidance of experts from the Geophysical Institute of the National Polytechnic School (IG-EPN (Instituto Geofísico de la Escuela Politécnica Nacional (Quito, Ecuador))) and the Decentralized Autonomous Government of the Province of Cotopaxi (GADPC (GADPC, Gobierno Autónomo Descentralizado Provincial de Cotopaxi, Latacunga, Ecuador)). Some of these deposits are from pre-historical times, whilst the shallower ones date from the 1877 event that destroyed Latacunga [70]. Official maps of the Geological and Energy Research Institute (IIGE) and IG-EPN [71] were used during the field reconnaissance. This field trip was relevant to visualize the geological characteristics of the study area, as well as to strengthen the cooperation and idea exchanges with the local experts.

Figure 3. (**Left**): Channel of the Cutuchi River in the city centre of Latacunga. An old textile factory is visible, which has been buried up to the forth story by the 1877 lahar. (**Right**): Thick sequence of lahar deposits, scoria flow deposits, and tephra beds exposed in a quarry along the Rio Saquimala close to Mulalo. (Photos: Theresa Frimberger, 2018).

Latacunga is not only exposed to the natural hazards imposed by the Cotopaxi volcano, but also to other geodynamic (e.g., landslides and earthquakes) and hydro-climatologic hazards (e.g., frosts and droughts). As reported in [72], there has been an intensification in the variability of precipitations, droughts, and frosts in Latacunga. This has been evidenced in the period between the years 1981–2014, during which the average air temperature has increased about 0.8 °C. These ongoing phenomena related to climate change have generated negative consequences mainly in the rural area and in agriculture areas [72].

2.2. Objectives

The understanding of disaster risk based on the independent investigation of their dimensions, hazards, exposure, and vulnerability, guided by a multi-hazard risk approach with risk-informed decision-makers, is the key advice of the Sendai Framework for Disaster Risk reduction (2015–2030) [73]. Having in mind the aforementioned limitations on volcanic risk assessment as well as the lack of exploration tools for risk communication, two initiatives, namely the programme "Sustainable Intermediate Cities—CIS" and the research project "Multi-Risk Analysis and Information System Components for the Andes Region—RIESGOS" have been working in Latacunga, Ecuador, with the aim of increasing awareness and preparedness and enhancing the coping capacities of the communities exposed to the Cotopaxi volcano. The particular objectives of this integrative study are:

1. Presenting a comprehensive risk communication process, from scenario-based volcanic risk analysis along with active participation of the exposed communities, while also investigating the risk perception of the exposed communities.
2. Providing a recent measurement of spatially distributed risk perception in the Cotopaxi area (results from the CIS questionnaire)

3. Testing the applicability of the RIESGOS demonstrator, a decentralized web-service architecture that allows for integrating local expert knowledge and locally designed models in a scenario-based multi-risk analysis, for the purpose of interactive communication
4. Merging the results of 2 and 3 to investigate how well the simulated quantitative risk matches the subjectively perceived risk in a common area.

3. Materials and Methods

An integrative framework between scientific approaches and risk communication practices with the exposed society has been set up in Latacunga (Ecuador) by two different initiatives, (1) the CIS (Sustainable Intermediate Cities) programme and (2) the RIESGOS project (Multi-risk analysis and information system components for the Andes region).

3.1. The CIS Programme: The Creation of a Local Laboratory to Evaluate the Social Perception of Risk and Resilience

"The Latacunga Laboratory: Risk management, resilience, and adaptation to climate change (Laboratorio Urbano de Latacunga: Gestión de riesgos, resiliencia y adaptación al cambio climático)" has been created within the CIS programme, as part of the joint initiatives of GIZ ("Deutsche Gesellschaft für Internationale Zusammenarbeit") and Grupo FAR (Ecuadorian NGO (https://grupofaro.org/ (accessed on 1 January 2021)). The creation of so-called resilience observatories for exposed communities to natural hazards is a relatively new trend [45,74]. Similarly, the Latacunga Laboratory seeks to contribute to the risk management to natural hazards that are likely to occur in the territory, while aiming to contribute in the long-term to the development of the city embracing its urban–rural ties. With that goal, initial contributions related to social risk perceptions have been documented in [75] as a joint effort between the Latacunga Laboratory, the local government, academic institutions, and local actors.

3.1.1. Comparative Analysis of the Social Risk Perception Factors to Natural Hazards and the Spatial Distribution of Volcanic-Related Risk Factors

We conducted a survey by means of a custom-designed questionnaire, a fundamental tool for acquiring information on public knowledge of the community [76]. It is composed of a series of multiple-choice questions in Spanish. The survey was carried out in the field and online to collect data about the individual knowledge, attitudes, and risk perceptions of the inhabitants of Latacunga. The online survey was promoted on social media and was available on the official website of the CIS Latacunga Laboratory (https://latacungaresiliente.com/ (accessed on 1 January 2021)) for a month. In the meantime, the field survey was carried out only in the urban agglomeration of Latacunga. The collected data is used for two main objectives: (1) as input to perform the semi-quantitative method proposed in [77] that ranks the social perception of volcanic risk factors (i.e., hazard recurrence, exposure, vulnerability, and resilience) among other natural hazards likely to occur in the study area (i.e., earthquakes, drought, frost, floods, and landslides), and (2) to map the spatial distribution of volcano-related risk perception into comprehensive categories (i.e., easily understandable by the exposed communities).

A design of the field surveying site was carried out. According to the last official census available [65] and population projections by the survey elaboration date (September 2019), 50,442 inhabitants over the age of 18 years were considered as qualified informants. In order to get a statistically representative sample, a confidence level of 95% and a margin of error of 5% were selected. On this basis, we estimated that a sample for the field surveys not smaller than 380 inhabitants had to be selected. Considering 10% additional surveys, a final sample size of 420 people was chosen. The population density (Figure 4a) was used to constrain the spatial distribution of the field surveys within the urban blocks with a residential occupancy (Figure 4b). The surveys were carried out by 55 students of the ISTC (Instituto Superior Tecnológico Cotopaxi) in September 2019.

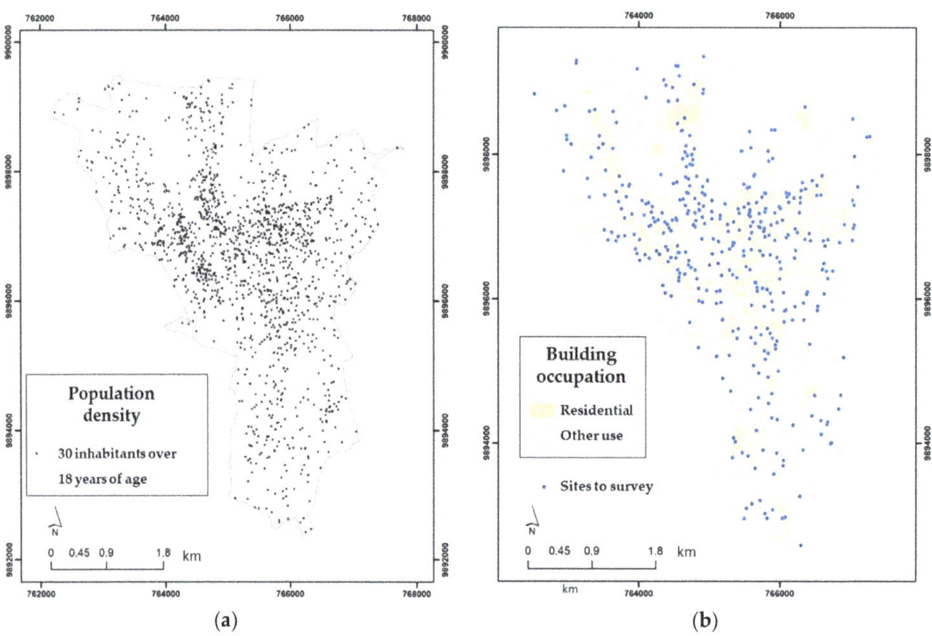

Figure 4. (**a**) Population qualified for the survey to evaluate the social risk perception in the urban centre of Latacunga. (**b**) Sites to survey within the residential buildings. Modified after [75].

It is worth mentioning that, on the one hand, some drawbacks have been found when the community perception of exposure, vulnerability, and resilience are independently addressed for large-scale studies [78,79]. On the other hand, there have been also reported benefits of this separation for mapping the social risk perception to natural hazards (e.g., [80,81]) when bottom-up approaches are carried out. Therefore, we have decided to independently investigate the social perceptions towards these components through separated questions. The Likert scale is used in this context to obtain a quantifiable level of perception of each risk factor. An integer numerical score (1, 2, or 3) is assigned to every possible answer. Although the passage from a qualitative perception to an index can be questioned, several recent studies have shown the usefulness of the Likert scale [82–86]. Notably, in [87], it was found to provide a good compromise between the quality of the information collected and the accessibility to respondents, while the bias in responses decreases, and there is consistency across different measurements and research domains of disaster risk reduction.

Subsequently, the average is computed for every question to obtain the perception of every component. These values are inputs to the computation of the risk perception pre-index through the use of Equation (1), where P stands for "perception". An example subset of the questions is presented in Table A1 (Appendix A). The questions and answers were validated by local risk management experts from the Association of Risk Management Professionals of Ecuador (Asociación de Profesionales de Gestión de Riesgos de Ecuador, APGR).

$$P(Risk) = \left(\frac{P(Hazard) \times P(Exposure) \times P(Vulnerability)}{P(Resilience)} \right) \tag{1}$$

The numerator of Equation (1) can have a maximum possible value of 27, whilst the minimum for the resilience term in the denominator is 1. Therefore the maximum risk perception value that this method admits is 27. The values in the range from 1–27 form a "pre-index". To obtain a more comprehensive numerical value, a "reduced index" in the 0–3 range is obtained through the application of Equation (2).

$$Reduced\ index = log_3(preindex\ value) \quad (2)$$

The relations between the "pre-index" and the "reduced index" is shown in Figure 5. For mapping purposes an "equal interval" classification for the reduced index scale is introduced with five classes of length 0.6 for finally presenting the spatialized perception of every risk factor in a compressive manner to the community. The calculated results for every answered question at each survey location (Figure 4b) are used to map the spatial distribution of the perception of hazard recurrence, exposure, vulnerability and resilience, and the risk index (computed with Equation (1). Subsequently, they were interpolated through the use of the ordinary kriging geostatistical algorithm [88].

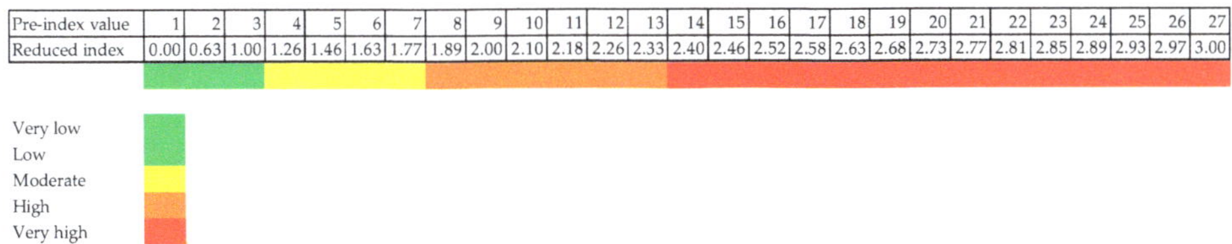

Pre-index value	1	2	3	4	5	6	7	8	9	10	11	12	13	14	15	16	17	18	19	20	21	22	23	24	25	26	27
Reduced index	0.00	0.63	1.00	1.26	1.46	1.63	1.77	1.89	2.00	2.10	2.18	2.26	2.33	2.40	2.46	2.52	2.58	2.63	2.68	2.73	2.77	2.81	2.85	2.89	2.93	2.97	3.00

Figure 5. Graphical scale and correspondence between the pre-index value and the reduced index.

3.2. The RIESGOS Project: Iterative Simulation Improvement and Enhanced Communication

The idea of constructing a web-tool, the RIESGOS demonstrator, as a decentralised and intraoperative environment for the exploration of the consequences in Latacunga from different volcanic hazard scenarios was proposed to the local stakeholders who participated in four participative workshops. Two of them were held in Latacunga (7 December 2018; 25 November 2019) in the headquarters of GADPC (Decentralized Autonomous Government of the Cotopaxi Province) and two workshops took place in Quito on 11 December 2018, and on 27 November 2019, respectively. The participants ranged from research partners, representatives of the rural municipalities (parishes) of the Cotopaxi province, public authorities, environment secretaries, actors of the civil society such as local representatives of agriculture associations, and urban and rural leaders. Similarly, as recently presented in [46], the workshops were used as a means to implement a user-centred iterative approach, seeking a continuous redesign of the RIESGOS demonstrator that has been guided by the needs of potential users and practical applicability. This has been ensured by a comprehensive analysis of user requirements (e.g., open-source, user-friendly graphical user interface and transferability).

3.2.1. The RIESGOS Demonstrator Tool for Quantitative Multi-Risk Analysis

The iteratively constructed RIESGOS demonstrator for a multi-risk information system is based on a modular and scalable concept in which the different hazards, the related exposure models, and vulnerability schemas are each represented by one individual web service. These independent and distributed web-services (managed and maintained by individual research institutions) are based on the quantitative methodologies developed within the RIESGOS framework for multi-risk analysis (i.e., [69,89–94]). Therefore, their integration into the RIESGOS demonstrator simulates the multi-risk environment of Latacunga. This modular approach offers the possibility to integrate different web services into already existing system environments.

Currently, the graphical user interface of the demonstrator can be accessed from a web browser only by users with special rights. The main screen of the graphical user interface is divided into three main display areas: the central map window, the configuration wizard for the control of each web service to the left, and the results panel to the right (e.g., see Figure 6). The code of the graphical user interface (RIESGOS frontend) is openly published on GitHub (https://github.com/riesgos/dlr-riesgos-frontend (accessed on 1 January 2021)). The use of standardized web services such as geospatial web services defined by the Open Geospatial Consortium (OGC) allows users accessing open and flexible multi-risk information and data products. Web-services and exposed data resources can be accessed using a variety of means from a simple command-line tool, over a web browser, to existing graphical user interfaces of public authorities and companies, which are equipped with a map user. OGC web services allow all kinds of geospatial functionality out-of-the-box including data access, data display, styling, and processing. Web services can easily be integrated into existing clients. The providers of web services define their products, display options, and configuration items. More details of this integrative process are reported in [94]. Through the clear separation in competencies between web services and user-interface, modularity and scalability are increased.

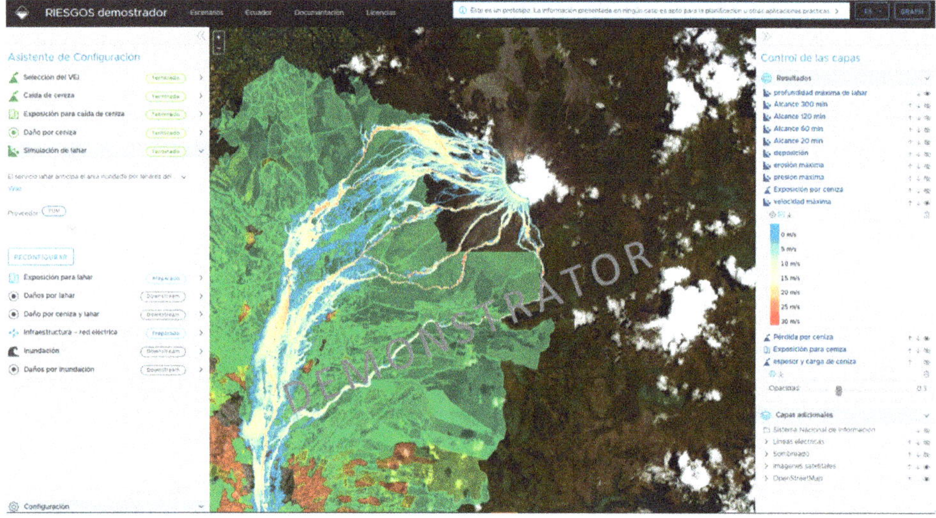

Figure 6. Example of the graphical representation of loss distribution due to ash fall scenario in the RIESGOS demonstrator (as of December 2020) from a previously selected VEI. Reddish and greenish aggregation areas representing higher and lower values, respectively. On top of these results, the lahar model (with the same VEI) is displayed as input to calculate the cumulative damage over the same geo-cells exposed to both perils.

Precomputed hazard models of ash-falls and lahars are displayed by the RIESGOS demonstrator after the selection of a scenario in terms of the expected for an eruption of the Cotopaxi volcano. Local probabilistic ashfall models for the Cotopaxi volcano generated by the IG-EPN (following the method of [95] with 20-year observation of wind flow directions) are currently integrated as twelve explorative scenarios. They are represented by isolines (Figure A1). The lahar models described in [69] are incorporated, showing the maximum possible values of five physical properties (i.e., flow velocity, flow depth, pressure, erosion, and deposition (see Figure A2).

The exposure model provides the input to calculate the direct losses over residential building portfolios classified in specific building classes for every hazard. An example for lahar-building classes is depicted in Figure A3. These models were constrained through the use of taxonomic characteristics available in the official cadastral dataset of the GADPC (Gobierno Autónomo Descentralizado Provincial de Cotopaxi, Latacunga, Ecuador), such as roof and wall materials, and the proportions of the predominant building materials suggested for Latacunga in [96]. No further details are provided on the manner the building exposure models were constructed, since this is out of the scope of this paper.

The vulnerability analysis of the typical residential buildings is performed using representative building exposure models with their respective fragility functions and suitable economical consequence models. Specifically, this approach is an extension of the Performance-Based Earthquake-Engineering (PBEE) method developed by [97], which has more recently been adapted to other kinds of natural hazards. The fragility model proposed in [98] is used in lahar fragility, whilst the one in [99] is used in ash fall fragility for typical residential buildings that can be encountered in the study area. The demonstrator ultimately obtains the spatial distribution of damage and losses per individual hazard, plus the option of obtaining the cumulative damage and losses due to the action of both hazardous events using the novel method outlined in [90]. Some examples are depicted in Figures 6 and A4. No further technical details are provided, because it is out of the scope of this work. Furthermore, the demonstrator enables the visualization of the areas that might potentially get disconnected from different networks and thus the identification of cascading effects on the economic activity. The method of implementation in the systemic vulnerability analysis applied in this case is similar to the one proposed in [100]. This information can be related with census data for estimating the population that might be affected by a blackout [101]. One example of this process is depicted in Figure A5 for the interruption probabilities of the electrical power network due to the impact of a lahar.

4. Results

4.1. The Recognition of the Latacunga Local Laboratory by the Local Actors of the Community

"The Latacunga Laboratory: Risk management, resilience, and adaptation to climate change" has strengthened its presence in the territory through several continuous participative activities that are aligned with the objectives mentioned in Section 3.1. For instance, the Laboratory has been recently working in materialising initiatives that were proposed by local entrepreneurs. One of them is currently working on the recovery of "Relatos de una erupcion" (Tales of an eruption), which works on rescuing the historical memory of what happened in the eruption of the Cotopaxi volcano in 1877. This has been carried out through audio-visual stories that are told by direct descendants who survived this event. This initiative enhances co-responsibility and respect for historical memory. The oral transmission of this information is an important input to generate awareness. Details about these initiatives can be found in the Latacunga Laboratory website (https://latacungaresiliente.com/rescate-de-la-memoria-historica-de-la-erupcion-del-volcan-cotopaxi/ (accessed on 1 January 2021)).

Comparative Analysis of the Social Risk Perception Factors to Natural Hazards and the Spatial Distribution of Volcanic-Related Risk Factors

The method described in Section 3.1.1 was applied to rank the volcanic risk perception for the most densely populated area in Latacunga conurbation. Making use of the 420 processed surveys as input data, the social perception to the recurrence of hazards, exposure, vulnerability, and resilience for six natural hazards likely to occur in Latacunga (i.e., earthquakes, volcanic eruptions, droughts, frosts, landslides, and floods) has been investigated. This is presented in the form of the comparative matrix shown in Table 1, which reports the mean values (for all the surveys) related to the perception of every component, as well as the computed risk index for every considered hazard. The higher the value, the greater the perception of risk. In the case of resilience, the interpretation is the opposite: the higher the value, the higher perception of resilience after a hazardous event.

The greatest concern among the inhabitants of Latacunga is their own perceived vulnerability to volcanic hazards. Remarkably, their resilience after a volcanic event scores the lowest value. This implies the community is aware that they would have great difficulty (or impossibility) to recover from the related damages. It is worth noticing that despite the fact that in the questionnaires there was no distinction made in terms of the type of volcanic hazards (lahar, ash fall/tephra fall, or ballistics) or in terms of their intensity, the collective imaginary always tended to associate the occurrence of a destructive lahar as "the volcanic hazard". Most likely, the oral transmission of the experiences of the survivors from the 1877 event has permeated the mental construction of their descendants.

Table 1. Hazard matrix and perception of risk factors towards natural hazards in the urban area of Latacunga.

	Risk Factors	Perception of Hazard Recurrence	Perception of Exposure	Perception of Vulnerability	Perception of Resilience	Perception of Risk
			Scale: 0–3			Scale: 1–27
Natural Hazard	Volcanoes	2.61	2.73	2.83	1.93	10.45
	Earthquakes	2.58	2.75	2.77	1.98	9.93
	Frost	2.62	2.27	2.14	2.38	5.35
	Drought	2.33	2.27	2.23	2.32	5.08
	Floods	2.00	2.05	2.04	2.27	3.68
	Landslides	1.99	2.05	2.08	2.29	3.71

The mean results in terms of the percentage for the answered questionnaire that makes up the field and online surveys are depicted in Table A2. Contrary to the field survey, the online surveys score large values in the basic knowledge and reconnaissance of their exposed environment (i.e., evacuation routes, emergency committee, the existence of initiatives for risk reduction). The field surveys express that 64% of the inhabitants consider the volcanic related hazards as events that are likely to happen in the city within their lifetimes. Furthermore, 86% answered that they believe an eventual eruption of the Cotopaxi volcano would cause very serious damaging effects to the city. Likewise, 75% considered they will have very serious impacts directly on their families and themselves. 25% of the population considers that recovery from a serious volcanic event would be impossible, whilst 57% think it would be difficult to overcome. Regarding knowledge, 67% of the population know safe places in the event of a possible disaster, while 61% know evacuation routes. However, only 34% ensure there are emergency plans in their neighbourhood. Half of the respondents do not even know if they live in a volcanic hazard zone. Additionally, 56% of the field-surveyed inhabitants and 69% of the online-respondents consider they would have rapid reaction capacities. Finally, ~42% of the population talks about how to act in case of emergency with their families.

Every answer of the 420 field surveys was spatially distributed onto the survey locations (Figure 4). Their associated numerical values of the Likert scale were interpolated through the use of the ordinary kriging geostatistical algorithm [88]. Subsequently, every numerical value is converted to the equivalent categories presented in Figure 5. The spatially explicit categories represent the social perception of volcanic hazard recurrence, exposure, vulnerability, and resilience in the study area. They are respectively depicted in Figure 7a–d. The former factors are integrated through Equation (1) to generate Figure 7e, which represents the semi-quantitative volcanic risk perception index proposed in [77]. In general terms, the perceptions of hazard recurrence, exposure, and vulnerability are quite similar. However, in the central–easternmost and northernmost zones, there is a high perception of hazard recurrence, a very low perception of resilience, and a moderate perception of vulnerability, whilst in the southernmost part (where the Cutuchi River flows), the four assessed factors show high and very high values that ultimately lead to a generalized "very high" category in the volcanic risk index. This is contrary to what is observed in the central–western and northern parts. Due to the increasing distances from the main drainages, there is a strong anti-correlation between the higher resilience levels (reddish areas) and the other risk factors. Hence, despite the last volcanic crisis in 2015, there is still a generalised very low perception among the inhabitants that they cannot suffer any direct impact or damaging effect after increasing volcanic activity, because they consider the occurrence of lahars (within their lifetimes) to be impossible. Clearly, the inhabitants of that sector are not aware of the large intensities the Cotopaxi volcano can achieve (e.g., a Plinian activity (VEI > 4) in Figure 2).

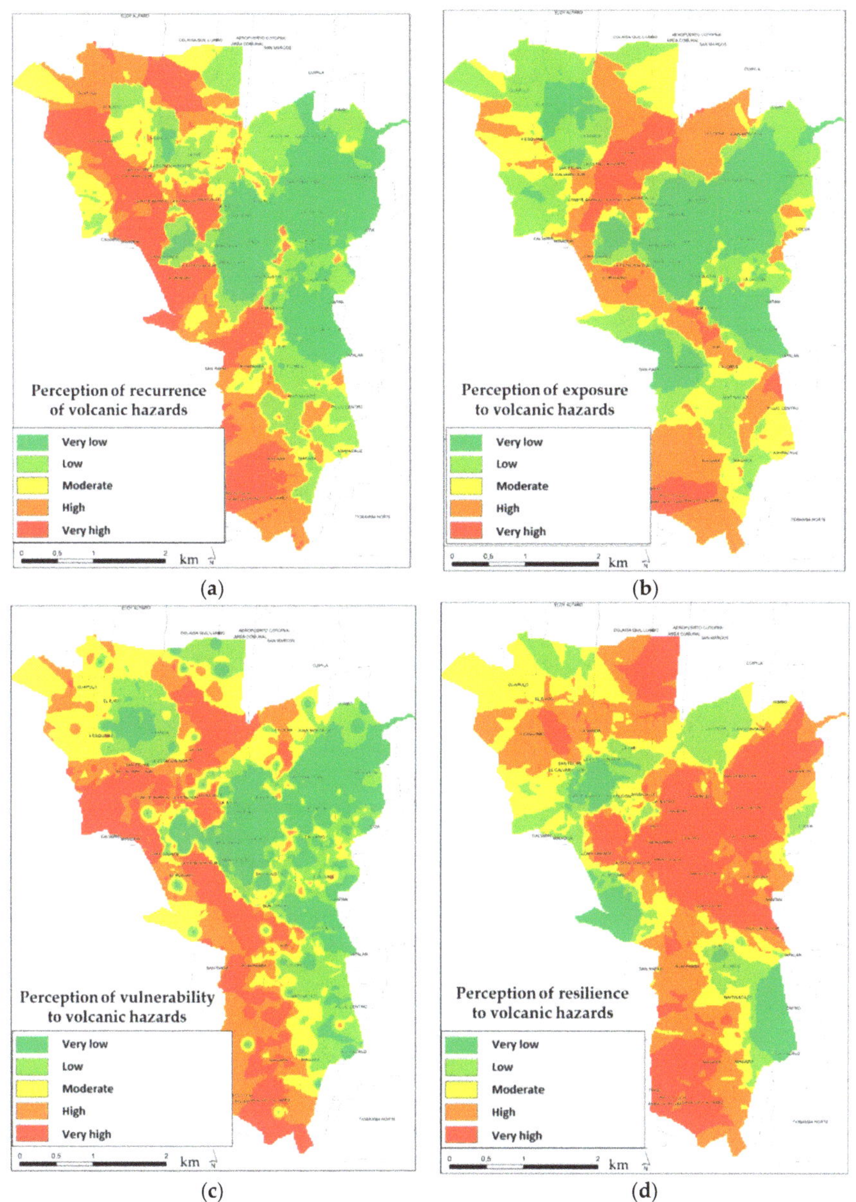

(a) (b) (c) (d)

Figure 7. *Cont.*

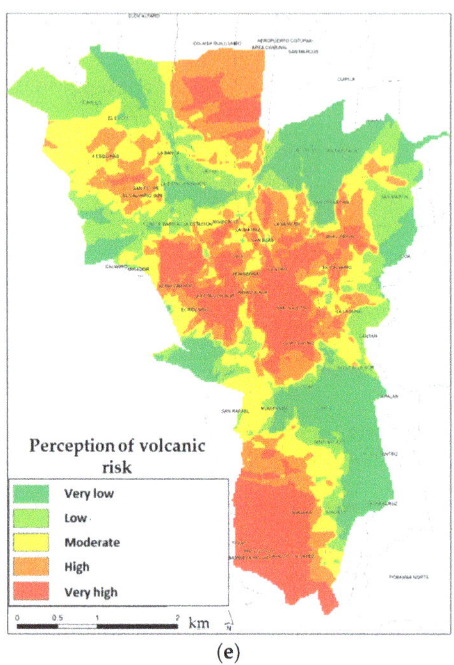

(e)

Figure 7. Spatial representation of the perception of volcanic hazards in the urban centre of Latacunga in terms of (**a**) recurrence; (**b**) exposure level; (**c**) vulnerability; (**d**) resilience; (**e**) risk index calculated using the former components as inputs. Modified after [75].

4.2. The Commnuication of the Scenario-Based Risk Assessment Concept with Local Stakeholders

During the four RIESGOS participative workshops, the invited stakeholders expressed interest in understanding the impacts of an extreme volcanic eruptions on the exposed elements such as buildings and critical infrastructure. Brainstorming exercises were carried out during the two first workshops. The participants were invited to imagine a future potential eruption with the emission of ash fall and occurrence of lahars. Thereafter, based on their perspectives and local knowledge, it was asked which physical, systemic, and cascading damaging effects they would expect on their built environment, infrastructure systems, and socioeconomic activities.

Some basic concepts of the probabilistic method, as an open-source web-service assesses the vulnerability of the exposed residential buildings (see Section 3.2.1), were presented to the local stakeholders. Due to the iterative approach used in constructing the demonstrator, some of the details that have been presented in this work as methods are actually the initial outputs of the first participative workshops. In this regard, the adaptability of "foreign" lahar vulnerability models (i.e., not developed for Ecuador) in the study area (e.g., [21,31,102]) was initially discussed with the representatives of the scientific local institutions. Due to the absence of locally developed ash fall vulnerability models for the residential buildings in the surroundings of the Cotopaxi volcano, the use of vulnerability models for the southern Colombian Galeras volcano [99] was perceived suitable to be implemented in the risk calculations rather than the fragility functions frequently developed for other areas (e.g., Italy [103]). With this feedback, the web-tool was redesigned. This is an example of how the engagement of local participants can improve both the technical development of quantitative methods (by agreeing on a proper model) as well as the understanding of such methods by the community.

Possible cascading effects that would occur in the case of critical infrastructure failure were debated. For instance, the participants realised that assessing the vulnerability of electric networks to ash falls and lahars is fundamental because of the further consequences on daily social and economic activities. However, the most debated topic was the reliability of the road system that, in the case of failure, may induce physical disruption and affect evacuation and emergency response during a volcanic crisis. Other public infrastructures that would be affected by Cotopaxi's lahars include the Army headquarters "Brigada Patria", Latacunga hospital, and the new penitentiary [58]. The interest in relocating some of the exposed assets was discussed.

"Hands-on" sessions took place during the two last workshops. The participants could experience on their own the use of the RIESGOS demonstrator. They selected different scenarios to visually compare every hazard footprint and intensity (i.e., for ash fall and lahars) as well as their associated risk outcomes on residential buildings and electric power networks. This was done through the selection of individual and successive hazard scenarios addressing cumulative damage. During the "hands-on" session, the participants recognized the potential of the demonstrator as an exploration tool for risk communication.

5. Discussion

The CIS and RIESGOS projects have independently addressed the domain of risk communication in Latacunga (Ecuador) at different geographical scales. The investigation and mapping of the perception of volcanic risk factors led by the Latacunga Laboratory (created by CIS) was carried out in a focused area (urban area) due to the necessity of having control points (where field-surveys were carried out) for a further geostatistical interpolation process, whilst, in the framework of the RIESGOS project, the construction of the hazard, exposure, and vulnerability approaches for scenario-based multi-risk calculations have been carried out at the canton level. Despite that, the community perceptions of the entire canton and province can be assessed in the future through field surveys for other urban centres (e.g., Pujili, Saquisili, and Salcedo), a meaningful spatially explicit perception of volcanic risk factors could only be mapped for the urban centres. This is because, due to the scattered location of the residential buildings in the rural areas, conventional geostatistical interpolation algorithms would carry significant bias in the results. For the commonly investigated area by RIESGOS and CIS, we can see that the exposed community recognise that they are under a variable level of risk regarding volcanic events depending on their location. These perceptions match the lahar footprints from the scenarios with higher probabilities of occurrence (VEI < 3). However, for larger intensities, (e.g., lahar footprints from a VEI > 4 scenario, see Figure 2b), we observe a mismatch with the spatially explicit community perceptions of volcanic risk factors (Figure 7). For instance, the easternmost areas of the urban centre of Latacunga show low and very low reconnaissance of volcanic risk factors due to their increasing distance respect to the main drainages. The inhabitants of that particular sector have perceived as impossible the occurrence of and suffering from consequences of lahars. The ignorance of the lahar footprints expected from these large intensity scenarios means that the concepts of "safe place" and evacuation routes are not applicable for either. These results should not be interpreted as fixed or permanent, but they rather constitute a temporal reading of the collective mental construction of the inhabitants at the time the surveys were carried out. Nevertheless, considering that the community is placed in ancient lahar deposits, as well as the relatively short time since the last 2015 volcanic crisis, one can realise that from the comparison of the respective outcomes arises the need to prioritize some zones where further divulgation activities should be made in the future regarding the possible scenarios and intensities that the Cotopaxi volcano can actually produce.

The formulated questions comprised in the survey forms are locally revised by experts from the APGR while paying attention to the use of collectively known terminology and the cultural characteristics of the community. In this work, we have implemented a simple numerical expression (Equation (1)) that equally ranks the risks factors of the different volcanic risk factors. This selection carries epistemic uncertainties. For instance, a customisation weighting schema to each factor, the selection of the median or mode instead of the mean value (herein adopted), together with a broader range in the Likert scale (e.g., 1 to 7 as explored in [104]) could be alternative approaches to be compared or even integrating each other into condition trees as proposed in [105]. The selection of the Likert scale to rank the answers and ultimately map the community perceptions implied an ordinal scale that was further converted into a nominal one based on the "equal-scale" (Equation (2)). This decision was made because, since the methods and results are aimed to be divulged, the categories have been found to be comprehensive, easily understandable, and culturally accepted by the community. Although the Likert scale has been extensively and recently used to successfully assess the community perception (e.g., [82–86]), there are several limitations in its adoption. For instance, as stated in [87], this kind of scale, despite maximizing the reliability of answers, also sacrifices the level of detail. However, it should be noted that through the simple possible answers related to the vulnerability perception and the nominal categories, we are only proposing a very simple categorization. More robust approaches that have addressed spatial multi-criteria analysis (as presented in [106]) have shown the impact of addressing diverse socioeconomic variables that we have not addressed in in our approach. A similar situation occurs with the resilience perception, which as discussed in [24], can be decomposed into very heterogeneous variables in economically developed countries.

Therefore, we are not claiming that our results related to the community perception of risk factors are exhaustive, but instead, they should be used as a basis for developing more complex analyses in future stages. For instance, even though we have already observed clear behaviour differences between the responses from online and field surveys, with explicitly designed survey and accounting variables such as work location, alphabetisation level, and economic activity, we could in the future classify the population into different social groups and find similarities and differences in their behaviour within a social environment to carry out more sophisticated methods, as proposed in [107]. Thereby, for each group, we could expect different reactions to a future volcanic crisis and then propose particular resilience practices. However, these kinds of approaches will largely depend on the data availability, which is particularly difficult in the rural tropics [25,57].

As described in recent participative experiences to assess the community perception to natural hazards (e.g., [45,47]) we have also experienced that the workshops carried out were allowed to go beyond a simple exchange of information. They paved the way for a better divulgation of concepts such as triggering and cascading hazards, dynamic vulnerability, cumulative damage, and cascading effects. These understandings in turn facilitated the knowledge flows and feedback acquisition to continuously design the RIESGOS demonstrator guided by increasingly risk-informed decision-makers. With this bottom-up iterative approach in the web-tool design, we are following the suggestions of the Sendai Framework for Disaster Risk Reduction (2015–2030) [73]. The outcomes of the demonstrator are not static hazard maps that are delivered to the exposed population from top-down approaches (e.g., [48–50]), but rather scenario-based online computations that can dynamically change based upon the continuous integration of local datasets and models.

During the "hands-on" sessions, the potential users perceived the RIESGOS demonstrator to intended prompt risk communication processes. For the study area, only hazard models have been typically available, and the few risk outcomes obtained in the past have been reported in tables and not in a spatially explicit manner [58]. Therefore, this work is providing the community with the availability of scenario-based risk models based on the vulnerability of the exposed elements in graphical and user-friendly interphase, which is an added value for the local community. The integrated scenario-based lahar footprints per VEI [69] and the locally developed probabilistic ash falls models [95] are themselves useful outcomes for civil protection and local-planners. They can be used to identify which human settlements and agricultural plantations might be affected or even discuss the relocation of some of the exposed components of critical infrastructure. Although we have not accounted for the conditional probabilities between triggering and cascading hazards as proposed in [17], we have instead presented fixed risk scenarios. For such a purpose, the demonstrator is served by a novel method that calculates and disaggregates the cumulative damage when there are interactions at the vulnerability level. In the specific volcanic context, although the concept of dynamic vulnerability had been already theoretically sketched in the work of [108], to the best of the authors' knowledge, we have first presented an example case of cumulative damage for risk-informed communities exposed to compound and cascading volcanic hazards. This is an innovative approach that not only contributes to reducing the generalized gap in the interactions at the vulnerability level [35], but also to communicating the results to the local stakeholders. With these contributions, the potential users could identify the most vulnerable areas for further mitigation strategies. It is worth mentioning that, since the RIESGOS demonstrator is currently not an operational tool, but rather shows the scientific and technological capabilities, the economic loss estimations for every exposure geo-cell (where residential buildings are aggregated) should not be used as definitive results. Therefore, due to the underlying uncertainties in these results, there is still the permanent necessity pointed out in [39,52] of having expert local users and scientists who can analyse and effectively communicate this information.

The technology transfer of the activities included in the CIS and RIESGOS programmes is highly relevant. The modular software architecture is particularly relevant for this aspect, for which the databases and methodologies of local Ecuadorian institutions may be ultimately integrated. However, the applicability of the demonstrator in the long-term will depend on how the local authorities will "give life" to the initiative, considering the local legal aspects. For future communication initiatives, due to the intrinsic interoperative sequence of inputs and outputs, the demonstrator can be a pedagogic tool to divulge multi-risk situations as similarly carried out by audio-visual approaches (e.g., [53,109]). Nevertheless, these kinds of local actors should be the first ones to understand the aforementioned concepts of "scenario" and "intensity" within the multi-risk chain, and most importantly, that they can be further contrasted with future and continuous spatially explicit social risk perceptions monitoring initiatives.

6. Conclusions

We have presented an integrative framework of qualitative community risk perceptions (carried out by the CIS Latacunga Laboratory) and scenario-based quantitative multi-hazard risk assessment (developed by the RIESGOS project). These initiatives have jointly worked on comprehensive volcanic risk communication processes in Latacunga, a city with a mainly rurally composed population, exposed to volcanic hazards from the Cotopaxi volcano.

Online and field surveys were carried out to rank the volcanic risk factors to investigate the individual knowledge and attitudes in Latacunga. Only the geo-located interviews in the field were used to map the community risk perceptions and to calculate a spatially explicit risk perception index through a semi-quantitative approach.

The participative workshops allowed the potentially affected communities to identify how their exposed assets, depending on their physical and systemic vulnerabilities, would be differently affected by several volcanic hazard scenarios. The iteratively customised RIESGOS demonstrator proved to be a useful tool for the communication of quantitative risk scenarios, raising the awareness of potentially affected population for the concept of scenarios and intensity. Its outcomes facilitate discussions among the participants on topics such as relocation of critical infrastructure elements. The demonstrator is not only enhancing the awareness of the communities, but also the user involvement in its development, improving the quality of the software. Although the development of the CIS and RIESGOS methodologies started independently, the respective outcomes of this collaborative work has allowed identifying areas where risk perception and scenario-based risk models are in disagreement. Thus, the need to continue assessing the social risk perception along with future risk communication efforts in the Cotopaxi region is highlighted.

Author Contributions: Conceptualization: J.C.G.-Z., M.P., and C.P.; methodology: M.P., C.P., E.S., F.B.-O., and H.S.; software: M.L.; validation: T.F., C.Z.-T., and C.P.; formal analysis: J.C.G.-Z., F.B.-O., and C.P.; investigation: T.F., F.B., K.B., and M.K.; resources: C.Z.-T.; data curation: J.C.G.-Z., M.L., and H.R.-V.; writing—original draft preparation: J.C.G.-Z., M.L., and C.P.; writing—review and editing: T.F., F.B., and J.C.G.-Z.; visualization: T.F., F.B.-O., and M.L.; supervision: H.S.; project management: E.S. All authors have read and agreed to the published version of the manuscript.

Funding: The research and development project RIESGOS (Grant No. 03G0876) is funded by the German Federal Ministry of Education and Research (BMBF) as part of the funding programme "CLIENT II—International Partnerships for Sustainable Innovations". The development of the Latacunga Laboratory: Risk management, resilience and adaptation to climate change (within the program "Sustainable Intermediate Cities—CIS) is funded by the Deutsche Gesellschaft für Internationale Zusammenarbeit (GIZ) programme of the GIZ Federal Ministry for Economic Cooperation and Development of Germany (BMZ).

Institutional Review Board Statement: Not applicable.

Informed Consent Statement: Informed consent was obtained from all subjects involved in the study.

Data Availability Statement: The data presented in this study are available on request from the corresponding author. The data are not publicly available due to the continuous development of the RIESGOS demonstrator and community risk perception carried out by GIZ.

Acknowledgments: The authors want to express their gratitude to the 55 students from *Instituto Superior Tecnológico Cotopaxi* (ISTC) who carried out the field surveys, as well as to all the participants of the workshops and the inhabitants in Latacunga who replied the surveys. Thanks to Daniel Straub (TUM), Jörn Lauterjung, Heidi Kreibich, and Fabrice Cotton (GFZ) for their advice during the elaboration of this work. Special thanks to Benjamin Bernard and Sebastian Averdunk (TUM) for the ash fall simulations inputs, as well as to Daniel Andrade and Patricia Mothes (IG-EPN) for the valuable feedback throughout the development of this work. Thanks to Daniela de Gregorio (UNINA), Roberto Torres-Corredor (SGC), Susanna Jenkins (EOS), and Robin Spence (Cambridge A.R) for having kindly provided sets of ash fall fragility functions. Thanks to Karl Heinz Gaudry (GIZ/CIM), Martin Cordovez Dammer, Marta Correa, and Edwin León (IIGE) for the discussions about critical infrastructure and cascading effects during the former volcanic crisis in the study area during the author's visits to Ecuador. Thanks to Hugo Yepes, Pablo Palacios, Jose Marrero, and Juan Carlos Singaucho (IG-EPN) for the feedback about exposure modelling and loss visualization. Thanks to Dorothea Kallenberger (GIZ), Ana Patricia (Grupo FARO), and Cristopher Velasco (president of the APGR) for having supported the research of the community risk perception study in Latacunga. Thanks to Luis Chasi for leading the initiative "Relatos de una erupcion". Thanks to Diego Molina (GADPC) for promoting discussions about the social risk perception of the local actors of the Cotopaxi province during the workshops held in Latacunga. Thanks to all SENESCYT and SNGER for the co-organization of the workshops held in Quito. Special thanks to Nils Brinkmann and Matthias Rüster (GFZ) for having participated in the construction of some of the computer codes implemented in this work.

Conflicts of Interest: The authors declare no conflict of interest. The funders had no role in the design of the study; in the collection, analyses, or interpretation of data; in the writing of the manuscript; or in the decision to publish the results.

Appendix A

Table A1. Example of the procedure for calculating the risk perception pre-index.

Questions	Answer Options	Quantitative Score	Answer						Value Transformation						Perception of Risk Factor		Pre-Index of Risk Perception (in the Scale 1–27)
			Earthquake	Volcanoes	Drought	Frost	Landslides	Floods	Earthquake	Volcanoes	Drought	Frost	Landslides	Floods			
Perception of hazard recurrence	Certainly yes	3	x	x	x	x		x	3	3	3	3		3	P(Hazard)	2.83	12.60
	It might occur	2					x						2				
	Impossible	1															
Perception of their impacts	Very serious	3	x	x				x	3	3				3	P(Vulnerability)	2.5	
	Moderate	2			x	x	x				2	2	2				
	No effects	1															
Perception of their effects to your family	Direct	3	x	x		x		x	3	3		3		3	P(Exposure)	2.67	
	Indirect	2			x		x				2		2				
	No effects	1															
Perception of recovery from them	Impossible	1	x	x				x	1	1				1	P(esilience)	1.5	
	Difficult	2			x	x	x				2	2	2				
	Likely	3															

Table A2. Questionnaire within the survey to assess the social risk perception to volcanic risk in the urban area of Latacunga. The mean values of the entire survey are reported. Adapted after [75].

Questions	Possible Answer	Type of Survey (%) Online	Field	Aggregated
% **P(Hazard)**. Perception of volcanic hazards recurrence. Do you think a volcanic eruption (from the Cotopaxi) can occur?	Certainly yes	54.43	63.61	62.04
	It might occur	44.30	34.29	36.01
	Impossible	1.27	2.09	1.95
% **P(Vulnerability)**. How do you consider the effects after a volcanic eruption would be?	Very serious	91.36	85.56	86.56
	Moderate	8.64	12.86	12.12
	No effects	0.00	1.57	1.30
% **P(Exposure)**. How do you consider the effects after a volcanic eruption would impact your family and yourself?	Very serious	81.48	75.39	76.46
	Moderate	17.28	22.51	21.60
	No effects	1.23	2.09	1.94
% **P(resilience)**. How do you consider the recovery process from the effects after a volcanic eruption?	Impossible	11.11	24.87	22.46
	Difficult	58.02	57.33	57.45
	Likely	30.86	17.80	20.09
Do you know if your home is in a volcanic hazard zone?	Yes	69.70	45.80	49.70
	No	32.1	54.2	50.3
Are there emergency plans in your neighbourhood?	Yes	19.75	36.65	33.69
	No	38.27	39.27	39.09
	Do not know	41.98	24.08	27.21
Are there safe places in the vicinity where you live? (in case of a volcanic eruption)	Yes	69.14	66.49	66.95
	No	16.05	18.85	18.36
	Do not know	14.81	14.66	14.69
Are there evacuation routes to safe sites?	Yes	70.37	60.47	62.20
	No	8.64	21.47	19.22
	Do not know	20.99	18.06	18.57

Table A2. *Cont.*

Questions	Possible Answer	Type of Survey (%)		
		Online	Field	Aggregated
Is there an emergency committee in your neighbourhood?	Yes	9.88	32.98	28.94
	No	46.91	35.34	37.37
	Do not know	43.21	31.68	33.69
Do you know if there are initiatives, actions, or works to reduce the risks from volcanic eruptions in Latacunga?	Yes	66.70	56.30	58.10
	No	33.30	43.70	41.90
Do you think you are capable of having a fast react during a volcanic eruptions?	Yes	69.10	53.10	55.90
	No	30.90	46.90	44.10
How often do you talk to your family about how to behave in the event of an emergency?	Never	0.00	17.63	14.66
	Rarely	23.38	41.05	38.07
	Sometimes	50.65	22.89	27.57
	Usually	25.97	18.42	19.69

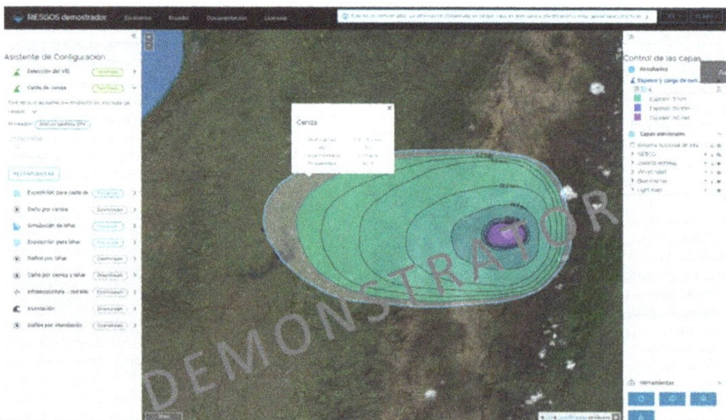

Figure A1. Example of the graphical representation of the spatially distributed ash fall intensities as isolines in the RIESGOS demonstrator (as of December 2020) from a previously selected VEI. The thickness values are displayed. Once a point within the isolines is clicked, the expected load (kPa) value is also shown.

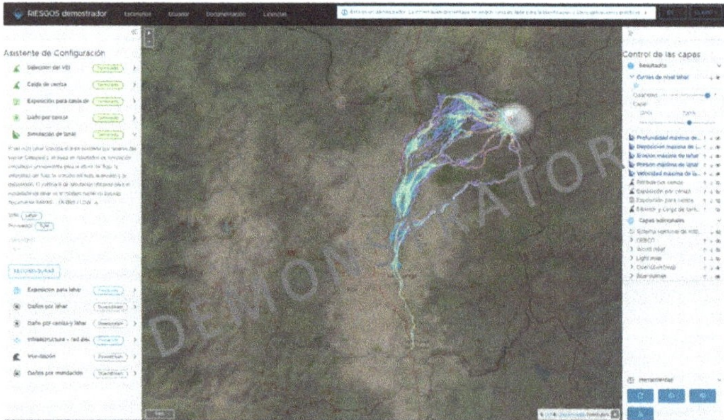

Figure A2. Example of the graphical representation of the footprint and intensities of the lahars in the RIESGOS demonstrator (as of December 2020) from a previously selected VEI. On the top-right side of the window, the outputs of the lahar simulation are listed (i.e., lahar flow velocity, flow depth, pressure, erosion, and deposition).

Figure A3. Example of the graphical representation of the residential building exposure model in the RIESGOS demonstrator (as of December 2020). It is represented into aggregation areas based on the official rural and urban administrative divisions of Latacunga. There are displayed the quantities of every ash fall risk oriented building class proposed in [99] within a selected area.

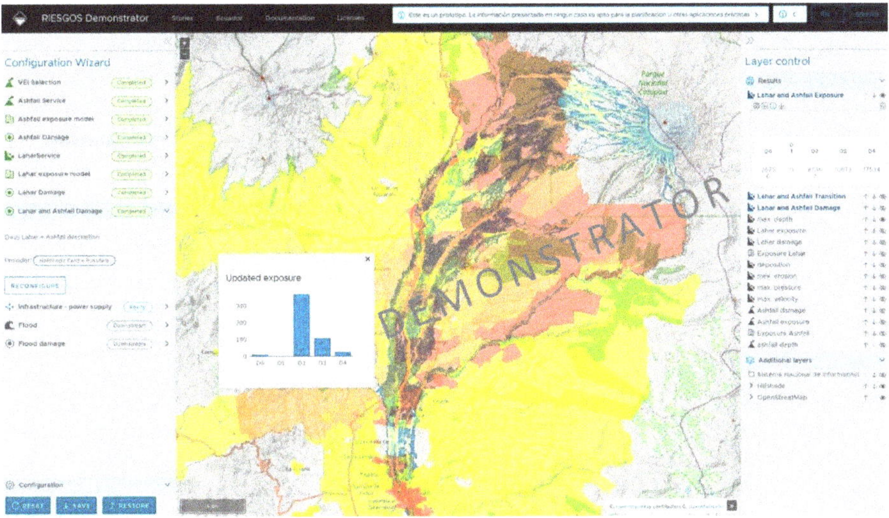

Figure A4. Example of the graphical representation of damage state distribution due to the combined effect of ash falls and lahar scenarios in the RIESGOS demonstrator (as of December 2020) calculated using the method proposed in [90], [92].

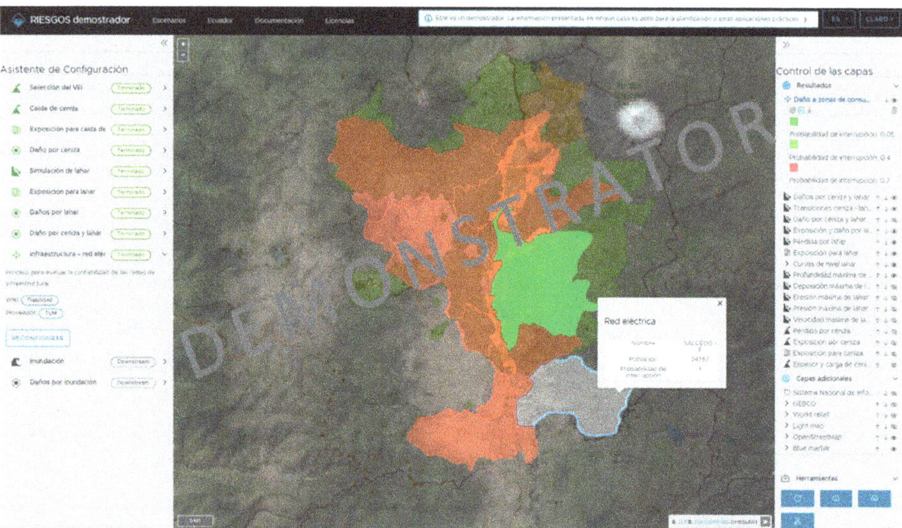

Figure A5. Example of the visualization of the expected interruption probabilities in the RIESGOS demonstrator (as of December 2020) of the electrical power network due to the action of a lahar scenario.

References

1. Gill, J.C.; Malamud, B.D. Hazard interactions and interaction networks (cascades) within multi-hazard methodologies. *Earth Syst. Dyn.* **2016**, *7*, 659–679. [CrossRef]
2. Pescaroli, G.; Alexander, D. A definition of cascading disasters and cascading effects: Going beyond the "toppling dominos" metaphor. *Planet Risk* **2015**, *3*, 58–67.
3. Ward, P.J.; Blauhut, V.; Bloemendaal, N.; Daniell, J.E.; de Ruiter, M.C.; Duncan, M.J.; Emberson, R.; Jenkins, S.F.; Kirschbaum, D.; Kunz, M.; et al. Review article: Natural hazard risk assessments at the global scale. *Nat. Hazards Earth Syst. Sci.* **2020**, *20*, 1069–1096. [CrossRef]
4. Cando-Jácome, M.; Martínez-Graña, A. Determination of primary and secondary lahar flow paths of the Fuego volcano (Guatemala) using morphometric parameters. *Remote Sens.* **2019**, *11*, 727. [CrossRef]
5. Mothes, P.A.; Vallance, J.W. Chapter 6—Lahars at Cotopaxi and Tungurahua Volcanoes, Ecuador: Highlights from stratigraphy and observational records and related downstream hazards. In *Volcanic Hazards, Risks and Disasters*; Shroder, J.F., Papale, P., Eds.; Elsevier: Boston, MA, USA, 2015; pp. 141–168, ISBN 978-0-12-396453-3.
6. Merz, B.; Kuhlicke, C.; Kunz, M.; Pittore, M.; Babeyko, A.; Bresch, D.N.; Domeisen, D.I.V.; Feser, F.; Koszalka, I.; Kreibich, H.; et al. Impact forecasting to support emergency management of natural hazards. *Rev. Geophys.* **2020**, *58*, e2020RG000704. [CrossRef]
7. Li, M.; Wang, J.; Sun, X. Scenario-based risk framework selection and assessment model development for natural disasters: A case study of typhoon storm surges. *Nat. Hazards* **2016**, *80*, 2037–2054. [CrossRef]
8. Lowe, D.R.; Williams, S.N.; Leigh, H.; Connort, C.B.; Gemmell, J.B.; Stoiber, R.E. Lahars initiated by the 13 November 1985 eruption of Nevado del Ruiz, Colombia. *Nature* **1986**, *324*, 51–53. [CrossRef]
9. García, C.; Mendez-Fajury, R. If I understand, i am understood: Experiences of volcanic risk communication in Colombia. In *Observing the Volcano World: Volcano Crisis Communication*; Fearnley, C.J., Bird, D.K., Haynes, K., McGuire, W.J., Jolly, G., Eds.; Springer International Publishing: Cham, Switzerland, 2018; pp. 335–351. ISBN 978-3-319-44097-2.
10. Pierson, T.C.; Janda, R.J.; Thouret, J.-C.; Borrero, C.A. Perturbation and melting of snow and ice by the 13 November 1985 eruption of Nevado del Ruiz, Colombia, and consequent mobilization, flow and deposition of lahars. *J. Volcanol. Geotherm. Res.* **1990**, *41*, 17–66. [CrossRef]
11. Pescaroli, G.; Alexander, D. Critical infrastructure, panarchies and the vulnerability paths of cascading disasters. *Nat. Hazards* **2016**, *82*, 175–192. [CrossRef]
12. Bolić, T.; Sivčev, Ź. Eruption of Eyjafjallajökull in Iceland: Experience of European Air Traffic Management. *Transp. Res. Rec.* **2011**, *2214*, 136–143. [CrossRef]
13. Walter, T.R.; Haghighi, M.H.; Schneider, F.M.; Coppola, D.; Motagh, M.; Saul, J.; Babeyko, A.; Dahm, T.; Troll, V.R.; Tilmann, F.; et al. Complex hazard cascade culminating in the Anak Krakatau sector collapse. *Nat. Commun.* **2019**, *10*, 4339. [CrossRef] [PubMed]
14. Lauterjung, J.; Spahn, H. Tsunami Hazard and Its Challenges for Preparedness. Available online: https://www.thejakartapost.com/academia/2019/01/08/tsunami-hazard-and-its-challenges-for-preparedness.html (accessed on 16 December 2020).

15. Poland, M.P.; Anderson, K.R. Partly cloudy with a chance of lava flows: Forecasting volcanic eruptions in the twenty-first century. *J. Geophys. Res. Solid Earth* **2020**, *125*, e2018JB016974. [CrossRef]
16. Biass, S.; Scaini, C.; Bonadonna, C.; Folch, A.; Smith, K.; Höskuldsson, A. A multi-scale risk assessment for tephra fallout and airborne concentration from multiple Icelandic volcanoes—Part 1: Hazard assessment. *Nat. Hazards Earth Syst. Sci.* **2014**, *14*, 2265–2287. [CrossRef]
17. Zuccaro, G.; Cacace, F.; Spence, R.J.S.; Baxter, P.J. Impact of explosive eruption scenarios at Vesuvius. *J. Volcanol. Geotherm. Res.* **2008**, *178*, 416–453. [CrossRef]
18. Marzocchi, W.; Garcia-Aristizabal, A.; Gasparini, P.; Mastellone, M.L.; Di Ruocco, A. Basic principles of multi-risk assessment: A case study in Italy. *Nat. Hazards* **2012**, *62*, 551–573. [CrossRef]
19. Pittore, M.; Wieland, M.; Fleming, K. Perspectives on global dynamic exposure modelling for geo-risk assessment. *Nat Hazards* **2017**, *86*, 7–30. [CrossRef]
20. Gallina, V.; Torresan, S.; Critto, A.; Sperotto, A.; Glade, T.; Marcomini, A. A review of multi-risk methodologies for natural hazards: Consequences and challenges for a climate change impact assessment. *J. Environ. Manag.* **2016**, *168*, 123–132. [CrossRef]
21. Zuccaro, G.; De Gregorio, D. Time and space dependency in impact damage evaluation of a sub-Plinian eruption at Mount Vesuvius. *Nat. Hazards* **2013**, *68*, 1399–1423. [CrossRef]
22. Gehl, P.; Quinet, C.; Le Cozannet, G.; Kouokam, E.; Thierry, P. Potential and limitations of risk scenario tools in volcanic areas through an example at Mount Cameroon. *Nat. Hazards Earth Syst. Sci.* **2013**, *13*, 2409–2424. [CrossRef]
23. Doyle, E.E.H.; McClure, J.; Paton, D.; Johnston, D.M. Uncertainty and decision making: Volcanic crisis scenarios. *Int. J. Disaster Risk Reduct.* **2014**, *10*, 75–101. [CrossRef]
24. Ran, J.; MacGillivray, B.H.; Gong, Y.; Hales, T.C. The application of frameworks for measuring social vulnerability and resilience to geophysical hazards within developing countries: A systematic review and narrative synthesis. *Sci. Total Environ.* **2020**, *711*, 134486. [CrossRef] [PubMed]
25. Li, X.; Li, Z.; Yang, J.; Li, H.; Liu, Y.; Fu, B.; Yang, F. Seismic vulnerability comparison between rural Weinan and other rural areas in Western China. *Int. J. Disaster Risk Reduct.* **2020**, *48*, 101576. [CrossRef]
26. Papathoma-Koehle, M.; Maris, F.; Fuchs, S. Remoteness and austerity: A major driver of vulnerabilities to natural hazards. In Proceedings of the EGU General Assembly Conference Abstracts, Online, 4–8 May 2020; p. 1577.
27. Parham, M.; Teeuw, R.; Solana, C.; Day, S. Quantifying the impact of educational methods for disaster risk reduction: A longitudinal study assessing the impact of teaching methods on student hazard perceptions. *Int. J. Disaster Risk Reduct.* **2020**. [CrossRef]
28. Papathoma-Köhle, M.; Schlögl, M.; Fuchs, S. Vulnerability indicators for natural hazards: An innovative selection and weighting approach. *Sci. Rep.* **2019**, *9*, 15026. [CrossRef]
29. National Geophysical Data Center; World Data Service (NGDC/WDS) NCEI/WDS Global Significant Volcanic Eruptions Database. NOAA National Centers for Environmental Information. Available online: https://www.ngdc.noaa.gov/hazard/volcano.shtml (accessed on 8 May 2020).
30. Pesaresi, M.; Ehrilich, D.; Kemper, T.; Siragusa, A.; Florcyk, A.; Freire, S.; Corbane, C. *Atlas of the Human Planet 2017. Global Exposure to Natural Hazards*; EUR 28556 EN; European Comission: Luxembourg, 2017.
31. Jenkins, S.F.; Spence, R.J.S.; Fonseca, J.F.B.D.; Solidum, R.U.; Wilson, T.M. Volcanic risk assessment: Quantifying physical vulnerability in the built environment. *J. Volcanol. Geotherm. Res.* **2014**, *276*, 105–120. [CrossRef]
32. Craig, H.; Wilson, T.; Stewart, C.; Villarosa, G.; Outes, V.; Cronin, S.; Jenkins, S. Agricultural impact assessment and management after three widespread tephra falls in Patagonia, South America. *Nat. Hazards* **2016**, *82*, 1167–1229. [CrossRef]
33. Wilson, G.; Wilson, T.M.; Deligne, N.I.; Blake, D.M.; Cole, J.W. Framework for developing volcanic fragility and vulnerability functions for critical infrastructure. *J. Appl. Volcanol.* **2017**, *6*, 14. [CrossRef]
34. Wilson, T.M.; Stewart, C.; Wardman, J.B.; Wilson, G.; Johnston, D.M.; Hill, D.; Hampton, S.J.; Villemure, M.; McBride, S.; Leonard, G.; et al. Volcanic ashfall preparedness poster series: A collaborative process for reducing the vulnerability of critical infrastructure. *J. Appl. Volcanol.* **2014**, *3*, 10. [CrossRef]
35. Terzi, S.; Torresan, S.; Schneiderbauer, S.; Critto, A.; Zebisch, M.; Marcomini, A. Multi-risk assessment in mountain regions: A review of modelling approaches for climate change adaptation. *J. Environ. Manag.* **2019**, *232*, 759–771. [CrossRef] [PubMed]
36. Thompson, M.A.; Lindsay, J.M.; Wilson, T.M.; Biass, S.; Sandri, L. Quantifying risk to agriculture from volcanic ashfall: A case study from the Bay of Plenty, New Zealand. *Nat. Hazards* **2017**, *86*, 31–56. [CrossRef]
37. Deligne, N.I.; Horspool, N.; Canessa, S.; Matcham, I.; Williams, G.T.; Wilson, G.; Wilson, T.M. Evaluating the impacts of volcanic eruptions using RiskScape. *J. Appl. Volcanol.* **2017**, *6*, 18. [CrossRef]
38. Scaini, C.; Biass, S.; Galderisi, A.; Bonadonna, C.; Folch, A.; Smith, K.; Höskuldsson, A. A multi-scale risk assessment for tephra fallout and airborne concentration from multiple Icelandic volcanoes—Part 2: Vulnerability and impact. *Nat. Hazards Earth Syst. Sci.* **2014**, *14*, 2289–2312. [CrossRef]
39. Doyle, E.E.H.; McClure, J.; Johnston, D.M.; Paton, D. Communicating likelihoods and probabilities in forecasts of volcanic eruptions. *J. Volcanol. Geotherm. Res.* **2014**, *272*, 1–15. [CrossRef]
40. Thomalla, F.; Boyland, M.; Johnson, K.; Ensor, J.; Tuhkanen, H.; Gerger Swartling, Å.; Han, G.; Forrester, J.; Wahl, D. Transforming development and disaster risk. *Sustainability* **2018**, *10*, 1458. [CrossRef]

41. Horney, J.; Nguyen, M.; Salvesen, D.; Tomasco, O.; Berke, P. Engaging the public in planning for disaster recovery. *Int. J. Disaster Risk Reduct.* **2016**, *17*, 33–37. [CrossRef]
42. Kwok, A.H.; Paton, D.; Becker, J.; Hudson-Doyle, E.E.; Johnston, D. A bottom-up approach to developing a neighbourhood-based resilience measurement framework. *Disaster Prev. Manag. Int. J.* **2018**, *27*, 255–270. [CrossRef]
43. Lévy, J. Science + Space + Society: Urbanity and the risk of methodological communalism in social sciences of space. *Geogr. Helv.* **2014**, *69*, 99–114. [CrossRef]
44. Pescaroli, G. Perceptions of cascading risk and interconnected failures in emergency planning: Implications for operational resilience and policy making. *Int. J. Disaster Risk Reduct.* **2018**, *30*, 269–280. [CrossRef]
45. Heinzlef, C.; Robert, B.; Hémond, Y.; Serre, D. Operating urban resilience strategies to face climate change and associated risks: Some advances from theory to application in Canada and France. *Cities* **2020**, *104*, 102762. [CrossRef]
46. Gill, J.C.; Malamud, B.D.; Barillas, E.M.; Guerra Noriega, A. Construction of regional multi-hazard interaction frameworks, with an application to Guatemala. *Nat. Hazards Earth Syst. Sci.* **2020**, *20*, 149–180. [CrossRef]
47. Fleming, K.; Abad, J.; Booth, L.; Schueller, L.; Baills, A.; Scolobig, A.; Petrovic, B.; Zuccaro, G.; Leone, M.F. The use of serious games in engaging stakeholders for disaster risk reduction, management and climate change adaption information elicitation. *Int. J. Disaster Risk Reduct.* **2020**, *49*, 101669. [CrossRef]
48. Jóhannesdóttir, G.; Gísladóttir, G. People living under threat of volcanic hazard in southern Iceland: Vulnerability and risk perception. *Nat. Hazards Earth Syst. Sci.* **2010**, *10*, 407–420. [CrossRef]
49. Paton, D.; Smith, L.; Daly, M.; Johnston, D. Risk perception and volcanic hazard mitigation: Individual and social perspectives. *J. Volcanol. Geotherm. Res.* **2008**, *172*, 179–188. [CrossRef]
50. Bronfman, N.C.; Cisternas, P.C.; López-Vázquez, E.; Cifuentes, L.A. Trust and risk perception of natural hazards: Implications for risk preparedness in Chile. *Nat. Hazards* **2016**, *81*, 307–327. [CrossRef]
51. Leonard, G.S.; Stewart, C.; Wilson, T.M.; Procter, J.N.; Scott, B.J.; Keys, H.J.; Jolly, G.E.; Wardman, J.B.; Cronin, S.J.; McBride, S.K. Integrating multidisciplinary science, modelling and impact data into evolving, syn-event volcanic hazard mapping and communication: A case study from the 2012 Tongariro eruption crisis, New Zealand. *J. Volcanol. Geotherm. Res.* **2014**, *286*, 208–232. [CrossRef]
52. Pierson, T.C.; Wood, N.J.; Driedger, C.L. Reducing risk from lahar hazards: Concepts, case studies, and roles for scientists. *J. Appl. Volcanol.* **2014**, *3*, 16. [CrossRef]
53. Hicks, A.; Armijos, M.T.; Barclay, J.; Stone, J.; Robertson, R.; Cortés, G.P. Risk communication films: Process, product and potential for improving preparedness and behaviour change. *Int. J. Disaster Risk Reduct.* **2017**, *23*, 138–151. [CrossRef]
54. Armijos, M.T.; Phillips, J.; Wilkinson, E.; Barclay, J.; Hicks, A.; Palacios, P.; Mothes, P.; Stone, J. Adapting to changes in volcanic behaviour: Formal and informal interactions for enhanced risk management at Tungurahua Volcano, Ecuador. *Glob. Environ. Chang.* **2017**, *45*, 217–226. [CrossRef]
55. Lupiano, V.; Chidichimo, F.; Machado, G.; Catelan, P.; Molina, L.; Calidonna, C.R.; Straface, S.; Crisci, G.M.; Di Gregorio, S. From examination of natural events to a proposal for risk mitigation of lahars by a cellular-automata methodology: A case study for Vascún valley, Ecuador. *Nat. Hazards Earth Syst. Sci.* **2020**, *20*, 1–20. [CrossRef]
56. Bernard, B.; Battaglia, J.; Proaño, A.; Hidalgo, S.; Vásconez, F.; Hernandez, S.; Ruiz, M. Relationship between volcanic ash fallouts and seismic tremor: Quantitative assessment of the 2015 eruptive period at Cotopaxi volcano, Ecuador. *Bull. Volcanol.* **2016**, *78*, 80. [CrossRef]
57. Frontuto, V.; Dalmazzone, S.; Salcuni, F.; Pezzoli, A. Risk Aversion, Inequality and Economic Evaluation of Flood Damages: A Case Study in Ecuador. *Sustainability* **2020**, *12*, 68. [CrossRef]
58. Rodriguez, F.; Toulkeridis, T.; Sandoval, W.; Padilla, O.; Mato, F. Economic risk assessment of Cotopaxi volcano, Ecuador, in case of a future lahar emplacement. *Nat. Hazards* **2017**, *85*, 605–618. [CrossRef]
59. Barberi, F.; Coltelli, M.; Frullani, A.; Rosi, M.; Almeida, E. Chronology and dispersal characteristics of recently (last 5000 years) erupted tephra of Cotopaxi (Ecuador): Implications for long-term eruptive forecasting. *J. Volcanol. Geotherm. Res.* **1995**, *69*, 217–239. [CrossRef]
60. Hall, M.; Mothes, P. The rhyolitic–andesitic eruptive history of Cotopaxi volcano, Ecuador. *Bull. Volcanol.* **2008**, *70*, 675–702. [CrossRef]
61. Sierra, D.; Vasconez, F.; Andrade, S.D.; Almeida, M.; Mothes, P. Historical distal lahar deposits on the remote eastern-drainage of cotopaxi volcano, Ecuador. *J. South Am. Earth Sci.* **2019**, *95*, 102251. [CrossRef]
62. Aguilera, E.; Pareschi, M.T.; Rosi, M.; Zanchetta, G. Risk from Lahars in the Northern Valleys of Cotopaxi Volcano (Ecuador). *Nat. Hazards* **2004**, *33*, 161–189. [CrossRef]
63. Doocy, S.; Daniels, A.; Dooling, S.; Gorokhovich, Y. The human impact of volcanoes: A historical review of events 1900–2009 and systematic literature review. *PLoS Curr.* **2013**, *5*. [CrossRef] [PubMed]
64. Andrade, D.; Hall, M.; Mothes, P.; Troncosco, L.; Eissen, J.-P.; Samaniego, P.; Efred, J.; Ramon, P.; Rivero, D.; Yepes, H. *Los Peligros Volcánicos Asociados con el Cotopaxi*; Los peligros volcánicos en el Ecuador; IG-EPN: Quito, Ecuador, 2005.
65. INEC. *Censo de Poblacion y Vivienda. Estadisticas de Vivienda y Hogares*; Estadisticas de Vivienda y Hogares: Quito, Ecuador, 2010.
66. Czerny, M.; Czerny, A. Urbanisation processes in zones threatened by volcanic activity: The case of Latacunga at the foot of Cotopaxi in Ecuador. *Misc. Geogr.* **2020**, *24*, 183–192.

67. Christie, R.; Cooke, O.; Gottsmann, J. Fearing the knock on the door: Critical security studies insights into limited cooperation with disaster management regimes. *J. Appl. Volcanol.* **2015**, *4*, 19. [CrossRef]
68. Mothes, P.A.; Ruiz, M.C.; Viracucha, E.G.; Ramón, P.A.; Hernández, S.; Hidalgo, S.; Bernard, B.; Gaunt, E.H.; Jarrín, P.; Yépez, M.A.; et al. Geophysical footprints of Cotopaxi's unrest and minor eruptions in 2015: An opportunity to test scientific and community preparedness. In *Volcanic Unrest: From Science to Society*; Gottsmann, J., Neuberg, J., Scheu, B., Eds.; Springer International Publishing: Cham, Switzerland, 2019; pp. 241–270. ISBN 978-3-319-58412-6.
69. Frimberger, T.; Andrade, D.; Weber, D.; Krautblatter, M. Modelling future lahars controlled by different volcanic eruption scenarios at Cotopaxi (Ecuador) calibrated with the massively destructive 1877 lahar. *Print Earth Surf. Process. Landf.* **2020**. [CrossRef]
70. Pistolesi, M.; Cioni, R.; Rosi, M.; Cashman, K.V.; Rossotti, A.; Aguilera, E. Evidence for lahar-triggering mechanisms in complex stratigraphic sequences: The post-twelfth century eruptive activity of Cotopaxi Volcano, Ecuador. *Bull. Volcanol.* **2013**, *75*, 698. [CrossRef]
71. Mothes, P.; Espin, P.; Hall, M.; Vasconez, F.; Sierra, D.; Andrade, D. *Mapa Regional De Amenazas Volcánicas Potenciales del Volcán Cotopaxi, Zona Sur. Mapa de Peligros*; Instituto Geofisico de la Escuela Politenica Nacional y el Instituto Geografico Militar: Quito, Ecuador, 2016.
72. Heifer Fundation. *Páramos de Cotopaxi y Cambio Climático. Experiencias Campesinas de Adaptación al Cambio Climático*; Fundación Heifer: Quito, Ecuador, 2018.
73. UNISDR. *Sendai Frameworkfor Disaster Risk Reduction2015–2030*; United Nations International Strategy for Disaster Reduction: Sendai, Japan, 2015.
74. Heinzlef, C.; Barocca, B.; Leone, M.; Glade, T.; Serre, D. Resilience issues and challenges into built environments: A review. *Nat. Hazards Earth Syst. Sci. Discuss.* **2020**, *2020*, 1–35.
75. Grupo FARO; Asociación de Profesionales de Gestión de Riesgos de Ecuador, and Instituto Superior Tecnológico Cotopaxi. *Percepción social del riesgo en la ciudad de Latacunga*; Grupo FARO: Quito, Ecuador, 2020; Available online: https://grupofaro.org/portfolio/estudio-de-percepcion-social-del-riesgo-en-la-ciudad-de-latacunga/ (accessed on 26 November 2020).
76. Bird, D.K. The use of questionnaires for acquiring information on public perception of natural hazards and risk mitigation—A review of current knowledge and practice. *Nat. Hazards Earth Syst. Sci.* **2009**, *9*, 1307–1325. [CrossRef]
77. Carreño, M.L.; Cardona, O.D.; Barbat, A.H. *Sistema de Indicadores Para la Evaluación de Riesgos*; Centre Internacional de Mètodes Numèrics en Enginyeria (CIMNE): Barcelona, Spain, 2005; ISBN 84-95999-70-6.
78. Kelman, I. Lost for words amongst disaster risk science vocabulary? *Int. J. Disaster Risk Sci.* **2018**, *9*, 281–291. [CrossRef]
79. Shaw, K. "Reframing" Resilience: Challenges for planning theory and practice. *Plan. Theory Pract.* **2012**, *13*, 299–333.
80. Weichselgartner, J.; Kelman, I. Geographies of resilience: Challenges and opportunities of a descriptive concept. *Prog. Hum. Geogr.* **2014**, *39*, 249–267. [CrossRef]
81. Sajjad, M.; Chan, J.C.L.; Kanwal, S. Integrating spatial statistics tools for coastal risk management: A case-study of typhoon risk in mainland China. *Ocean Coast. Manag.* **2020**, *184*, 105018. [CrossRef]
82. Oláh, J.; Virglerova, Z.; Popp, J.; Kliestikova, J.; Kovács, S. The assessment of non-financial risk sources of SMES in the V4 countries and Serbia. *Sustainability* **2019**, *11*, 4806. [CrossRef]
83. Frazier, T.G.; Wood, E.X.; Peterson, A.G. Residual risk in public health and disaster management. *Appl. Geogr.* **2020**, *125*, 102365. [CrossRef]
84. Wuni, I.Y.; Shen, G.Q.; Osei-Kyei, R.; Agyeman-Yeboah, S. Modelling the critical risk factors for modular integrated construction projects. *Int. J. Constr. Manag.* **2020**, 1–14. [CrossRef]
85. Marín-Monroy, E.A.; Trejo, V.H.; de la Pena, M.A.O.R.; Polanco, G.A.; Barbara, N.L. Assessment of socio-environmental vulnerability due to tropical cyclones in La Paz, Baja California Sur, Mexico. *Sustainability* **2020**, *12*, 1575. [CrossRef]
86. Cano, A.M.; Sagredo, R.R.; García-Carrión, R.; Garcia-Zapirain, B. Social impact assessment of HealthyAIR tool for real-time detection of pollution risk. *Sustainability* **2020**, *12*, 9856. [CrossRef]
87. Pescaroli, G.; Velazquez, O.; Alcántara-Ayala, I.; Galasso, C.; Kostkova, P.; Alexander, D. A likert scale-based model for benchmarking operational capacity, organizational resilience, and disaster risk reduction. *Int. J. Disaster Risk Sci.* **2020**, *11*, 404–409. [CrossRef]
88. Issaks, E.H.; Srivastava, R.M. *An Introduction to Applied Geostatistics*; University Press: Oxford, NY, USA, 1989.
89. Rosero-Velásquez, H.; Straub, D. Representative Natural Hazard Scenarios for Risk Assessment of Spatially Distributed Infrastructure Systems. In Proceedings of the 29th European Safety and Reliability Conference (ESREL 2019), Hannover, Germany, 22–26 September 2019.
90. Langbein, M.; Gomez-Zapata, J.C.; Frimberger, T.; Brinckmann, N.; Torres-Corredor, R.; Andrade, D.; Zapata-Tapia, C.; Pittore, M.; Schoepfer, E. Scenario- based multi- risk assessment on exposed buildings to volcanic cascading hazards. In Proceedings of the EGU General Assembly Conference Abstracts, Online, 4–8 May 2020; p. 19861.
91. Brill, F.; Pineda, S.P.; Cuya, B.E.; Kreibich, H. A data-mining approach towards damage modelling for El Niño events in Peru. *Geomat. Natural Hazards Risk* **2020**, *11*, 1966–1990. [CrossRef]
92. Gomez-Zapata, J.C.; Pittore, M.; Brinckmann, N.; Shinde, S. Dynamic physical vulnerability: A Multi-risk Scenario approach from building- single- hazard fragility- models. In Proceedings of the EGU General Assembly Conference Abstracts, Online, 4–8 May 2020; p. 18379.

93. Pittore, M.; Zapata, J.C.G.; Brinckmann, N.; Weatherill, G.; Babeyko, A.; Harig, S.; Mahdavi, A.; Proß, B.; Velasquez, H.F.R.; Straub, D.; et al. Towards an integrated framework for distributed, modular multi-risk scenario assessment. In Proceedings of the EGU General Assembly Conference Abstracts, Online, 4–8 May 2020; p. 19097.
94. Brinckmann, N.; Pittore, M.; Rüster, M.; Proß, B.; Gomez-Zapata, J.C. Put your models in the web—less painful. In Proceedings of the EGU General Assembly Conference Abstracts, Online, 4–8 May 2020; p. 8671.
95. Tadini, A.; Roche, O.; Samaniego, P.; Guillin, A.; Azzaoui, N.; Gouhier, M.; de' Michieli Vitturi, M.; Pardini, F.; Eychenne, J.; Bernard, B.; et al. Quantifying the Uncertainty of a coupled plume and tephra dispersal model: PLUME-MOM/HYSPLIT simulations applied to andean volcanoes. *J. Geophys. Res. Solid Earth* **2020**, *125*, e2019JB018390. [CrossRef]
96. Yepes-Estrada, C.; Silva, V.; Valcárcel, J.; Acevedo, A.B.; Tarque, N.; Hube, M.A.; Coronel, G.; María, H.S. Modeling the residential building inventory in South America for seismic risk assessment. *Earthq. Spectra* **2017**, *33*, 299–322. [CrossRef]
97. Cornell, C.A.; Krawinkler, H. Progress and challenges in seismic performance assessment. *Peer Cent. News* **2000**, *3*, 1–3.
98. Mavrouli, O.; Fotopoulou, S.; Pitilakis, K.; Zuccaro, G.; Corominas, J.; Santo, A.; Cacace, F.; De Gregorio, D.; Di Crescenzo, G.; Foerster, E.; et al. Vulnerability assessment for reinforced concrete buildings exposed to landslides. *Bull. Eng. Geol. Environ.* **2014**. [CrossRef]
99. Torres-Corredor, R.A.; Ponde-Villarreal, P.; Gomez-Martinez, D.M. Vulnerabilidad física de cubiertas de edificaciones de uso de ocupación normal ante caídas de ceniza en la zona de influencia del Volcán Galeras. *Bol. Geol.* **2017**, *39*, 2.
100. Crucitti, P.; Latora, V.; Marchiori, M. Model for cascading failures in complex networks. *Phys. Rev. E* **2004**, *69*, 045104. [CrossRef]
101. Poljanšek, K.; Bono, F.; Gutiérrez, E. Seismic risk assessment of interdependent critical infrastructure systems: The case of European gas and electricity networks. *Earthq. Eng. Struct. Dyn.* **2012**, *41*, 61–79. [CrossRef]
102. Thouret, J.-C.; Antoine, S.; Magill, C.; Ollier, C. Lahars and debris flows: Characteristics and impacts. *Earth-Sci. Rev.* **2020**, *201*, 103003. [CrossRef]
103. Jenkins, S.F.; Phillips, J.C.; Price, R.; Feloy, K.; Baxter, P.J.; Hadmoko, D.S.; de Bélizal, E. Developing building-damage scales for lahars: Application to Merapi volcano, Indonesia. *Bull. Volcanol.* **2015**, *77*, 75. [CrossRef]
104. Croasmun, J.; Ostrom, L. Using likert-type scales in the social sciences. *J. Adult Educ.* **2011**, *40*, 19–22.
105. Beven, K.J.; Aspinall, W.P.; Bates, P.D.; Borgomeo, E.; Goda, K.; Hall, J.W.; Page, T.; Phillips, J.C.; Simpson, M.; Smith, P.J.; et al. Epistemic uncertainties and natural hazard risk assessment—Part 2: What should constitute good practice? *Nat. Hazards Earth Syst. Sci.* **2018**, *18*, 2769–2783. [CrossRef]
106. Armaș, I.; Gavriș, A. Social vulnerability assessment using spatial multi-criteria analysis (SEVI model) and the Social Vulnerability Index (SoVI model)—A case study for Bucharest, Romania. *Nat. Hazards Earth Syst. Sci.* **2013**, *13*, 1481–1499. [CrossRef]
107. Moscato, V.; Picariello, A.; Sperli, G. An emotional recommender system for music. *IEEE Intell. Syst.* **2020**. [CrossRef]
108. Zuccaro, G.; De Gregorio, D.; Leone, M.F. Theoretical model for cascading effects analyses. *Int. J. Disaster Risk Reduct.* **2018**, *30*, 199–215. [CrossRef]
109. Mercorio, F.; Mezzanzanica, M.; Moscato, V.; Picariello, A.; Sperli, G. DICO: A Graph-DB framework for community detection on big scholarly data. *IEEE Trans. Emerg. Top. Comput.* **2019**. [CrossRef]

Article

Community and Impact Based Early Warning System for Flood Risk Preparedness: The Experience of the Sirba River in Niger

Vieri Tarchiani [1,*], Giovanni Massazza [2], Maurizio Rosso [3], Maurizio Tiepolo [2], Alessandro Pezzoli [2], Mohamed Housseini Ibrahim [4], Gaptia Lawan Katiellou [5], Paolo Tamagnone [3], Tiziana De Filippis [1], Leandro Rocchi [1], Valentina Marchi [1] and Elena Rapisardi [1]

1. Institute of BioEconomy–National Research Council (IBE-CNR), 50019 Sesto Fiorentino (FI), Italy; tiziana.defilippis@ibe.cnr.it (T.D.F.); leandro.rocchi@ibe.cnr.it (L.R.); valentina.marchi@ibe.cnr.it (V.M.); elena.rapisardi@ibe.cnr.it (E.R.)
2. Interuniversity Department of Regional and Urban Studies and Planning (DIST), Politecnico and University of Turin, 10125 Turin, Italy; giovanni.massazza@polito.it (G.M.); maurizio.tiepolo@polito.it (M.T.); alessandro.pezzoli@polito.it (A.P.)
3. Department of Environment, Land and Infrastructure Engineering (DIATI), Politecnico of Turin, 10129 Turin, Italy; maurizio.rosso@polito.it (M.R.); paolo.tamagnone@polito.it (P.T.)
4. Hydrology Directorate (DH), Ministry of Hydraulics and Sanitation, Niamey BP 257, Niger; housseiniibrahimmohamed@yahoo.fr
5. National Directorate of Meteorology (DMN), Ministry of Transports, Niamey BP 218, Niger; katielloulaw@gmail.com

* Correspondence: vieri.tarchiani@ibe.cnr.it; Tel.: +39-0553033711

Received: 7 February 2020; Accepted: 26 February 2020; Published: 28 February 2020

Abstract: Floods have recently become a major hazard in West Africa (WA) in terms of both their magnitude and frequency. They affect livelihoods, infrastructure and production systems, hence impacting on Sustainable Development (SD). Early Warning Systems (EWS) for floods that properly address all four EWS components, while also being community and impact-based, do not yet exist in WA. Existing systems address only the main rivers, are conceived in a top-down manner and are hazard-centered. This study on the Sirba river in Niger aims to demonstrate that an operational community and impact-based EWS for floods can be set up by leveraging the existing tools, local stakeholders and knowledge. The main finding of the study is that bridging the gap between top-down and bottom-up approaches is possible by directly connecting the available technical capabilities with the local level through a participatory approach. This allows the beneficiaries to define the rules that will develop the whole system, strengthening their ability to understand the information and take action. Moreover, the integration of hydrological forecasts and observations with the community monitoring and preparedness system provides a lead time suitable for operational decision-making at national and local levels. The study points out the need for the commitment of governments to the transboundary sharing of flood information for EWS and SD.

Keywords: early warning; flood risk; hydrology; local communities; Niger river basin; rural development; Sahel

1. Introduction

As clearly stated by Ban Ki-moon, United Nations Secretary-General, on 1st September 2015: "The Sendai framework has important implications. It shifts the emphasis from disaster management to disaster risk management." This paradigm shift puts an emphasis on understanding the risks as the underpinning drivers for investing in resilience and preparedness, rather than in response and recovery.

Early Warning Systems (EWS) are a pillar of Disaster Risk Reduction (DRR) being "an integrated system of hazard monitoring, forecasting and prediction, disaster risk assessment, communication and preparedness activities systems and processes that enables individuals, communities, governments, businesses and others to take timely action to reduce disaster risks in advance of hazardous events" [1]. EWS contribute in reducing vulnerability to floods in urban and rural areas, which can also affect Sustainable Development in the latter. The Sendai framework for Disaster Risk Reduction 2015–2030 recognized seven strategic targets including EWS. The framework also identified four priorities, the fourth of which embeds the concept of Climate Services (CS) as a powerful tool for more effective disaster preparedness and the 'build back better' principle. In this respect, the European research and innovation roadmap for Climate Services expands the contribution of CS, particularly "hydrometeorological services", to the Sendai framework and EWS, a building block of preparedness [2].

The scientific literature presents two main approaches for EWS: top-down and hazard-centered and bottom-up people-centered [3,4]. Kelman and Glantz [5] also categorized EWS by "First Mile" and "Last Mile" approaches, where the former involves communities from the beginning and the latter concentrates on technical solutions and include communities only at the end of the process. As is currently widely acknowledged, top-down or last-mile approaches concentrating on developing forecasts, methodologies and models show many limits in effectively reaching and engaging communities [6,7].

Experiences of bottom-up flood EWS can be retrieved from South Asia, particularly Nepal, India and Bangladesh, all of which practice flood EWS on different scales and use different approaches. Some trans-border flood EWS experience also exists between those countries [8]. In Asia, a strong emphasis on people-centered EWS was observed after the Third International Conference on EWS in Bonn in 2006 and is described as "a complete set of components that connects those who need to receive messages from others who compile and track the hazard information of which messages are composed" [9].

A new paradigm shift was achieved with the concept of Community-Based Early Warning Systems (CBEWS) supported by international non-governmental organizations. A CBEWS is "an early warning system (EWS) where communities are active participants in the design, monitoring and management of the EWS, not just passive recipients of warnings" [10]. The implementation of CBEWS is documented by many authors in developing countries, such as Malawi [11], Nepal [10,12], Indonesia [13], India [14] and Cambodia [15]. CBEWS have proved successful in saving lives, but are often limited in their forecasting time, reducing their suitability for saving assets or livelihoods [16]. Top-down and bottom-up approaches, therefore, need to be integrated: people-centered approaches have to be coupled with robust flood monitoring and forecasting systems.

Since 2015, and according to the recommendation of the World Meteorological Organization (WMO) [17], the need has arisen for EWS to be built on impact-based forecasting and warning services. The aim is to bridge the gap between hydrometeorological forecasts and the potential consequences of the forecasted hazard on specific sectors. This approach, linking forecast information to decision-relevant impact thresholds for users, improves uptake and effectiveness [18]. Although only recently introduced, best practices can be found in the national meteorological services of developed countries, such as the United Kingdom Met Office and National Weather Service of the United States, and are being tested in South Asia [19].

In West Africa, EWS have been conceived and implemented since the 1980s in the sector of food security, mainly addressing agricultural drought [20]. However, efforts surrounding flood management in West Africa have, for the most part, focused on rescue and relief during and after events, while scientific and technical attempts to simulate runoff and forecast flood behavior are limited due to the poor gauging of rainfall and discharge. In recent years, some international initiatives have addressed flood forecasting at global [21] continental [22] or river basin [23] levels to respond to the growing need for flood risk early warning. Despite this effort, none of the web-based systems that are used for ongoing transnational flood forecasting are connected to local EWS, even if they can provide valuable inputs for them. In West Africa, and probably across the whole continent, CBEWS for floods conceived through impact-based forecasts and warnings are not yet documented by the scientific or gray literature.

The objective of our research was to demonstrate that it is possible to set up a comprehensive Community and Impact Based EWS (CIBEWS), responding to the key points and indicators described in the literature, by enhancing existing tools, experience and knowledge in a remote rural area of a poor developing country, such as Niger. This paper describes the approach adopted by the Niger government, with the support of other technical partners, in the setting-up of a CIBEWS in Niger on a Sahelian tributary of the Niger river: the Sirba river. The Local Floods Early Warning System for Sirba, called SLAPIS (Système Locale d'Alerte Precoce contre les Inondations de la Sirba), has been set up within the ANADIA2 (Adaptation to Climate Change and Disaster Risk Reduction for Food Security—Phase 2) project by the National Directorate for Hydrology (DH, Niger) in collaboration with the National Directorate for Meteorology (DMN, Niger), the Interuniversity Department of Regional and Urban Studies and Planning (DIST) Politecnico and University of Turin (Italy) and the Institute for the BioEconomy of the National Research Council (IBE-CNR, Italy). The project was funded by the Italian Agency for Development Cooperation (AICS). The advantages of attaining the SLAPIS objectives are reducing the impacts of floods in both rural (contributing to Sustainable Development) and urban areas, namely that of Niamey, which is a few kilometers downstream of the Sirba–Niger confluence.

2. Materials and Methods

2.1. Study Area and the Hydrological Context

The area of interest is located in the Sirba river basin, the main tributary of the Niger river in the Middle Niger River Basin (MNRB). The Sirba river basin covers an area of approximately 39,000 km^2 across Burkina Faso (93% of the basin) and Niger in the central Sahel. The territory has a granitic substrate and a slight height variation between the upper level of 444 m a.s.l. and the lower of 181 m a.s.l.. The climate is semi-arid with a long dry season and a rainy season concentrated in 3 to 4 months, between June and September, and an annual rainfall between 400 and 700 mm [24]. The Sahelian climate is characterized by strong rainfall variability with persistent dry spells and extreme rainfall events [25]. Therefore, the hydrology of the Sirba river is determined by the monsoon season and its spatio-temporal variability. The flood magnitude is more influenced by surface runoff than by groundwater flow [26].

The Nigerien sector of the Sirba river was chosen as the study area. The reach covers 108 km, from the state border with Burkina Faso, a few kilometers downstream of the confluence of the three main tributaries (Yali, Faga and Koulouko rivers), to the confluence with the Niger river (Figure 1). According to the last census (2012), the Nigerien part of the Sirba basin has 171 villages with a total population of 88,863. The majority (97 settlements) are distributed in riverine areas, meaning that 61,703 people, belonging to 7732 households, live in potentially flood-prone zones.

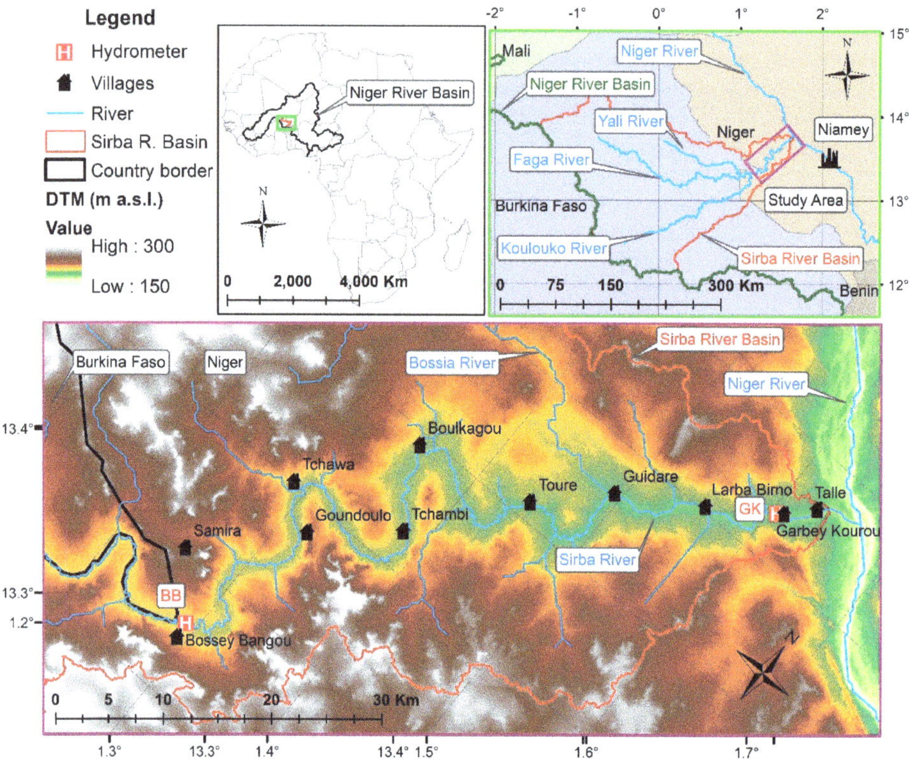

Figure 1. The geographical framework of the study area [27]. Bossey Bangou hydrometer (BB); Garbey Kourou hydrometer (GK).

Since the beginning of the century, as first highlighted by Tharule [28], extreme floods have been a crucial issue in the development of Sahelian countries. Indeed, an increasing number of flood events and flood-related impacts has been reported by many authors [29] and the frequency of flood events is particularly alarming in the MNRB [30–32]. These events often have disastrous consequences for the population, infrastructure, environment and economic sectors. During the last decade, the floods that struck Ouagadougou and Bobo Dioulasso (Burkina Faso) in 2009, the series of floods that hit Niamey (Niger) in 2010, 2012, 2017 and 2019 and those affecting Mopti and Bamako (Mali) in 2019 were particularly significant.

The high occurrence of these catastrophic events, despite the limited recovery of the climatic trends from the long drought that affected the region in the 1970s and 1980s, is referred to as the "Sahel Paradox" [33,34]. Hydrological studies conducted over the last 25 years clearly show two opposing phenomena: a runoff reduction in Sudano–Guinean catchments and an increase in Sahelian catchments [35–37]. Many researchers claim that, in the Sahel, besides the recent recovery of rainfall, which is still below the pre-1970 levels, and the increasing occurrence of extreme rainfall events, the main driver of floods is the strong land/vegetation degradation that has progressively reduced the water-holding capacity of the soil, leading to greater and faster runoff [33,38,39]. Indeed, even in the context of a so-called "regreening" of the Sahel, the recent increase in seasonal greenness at the Sahelian regional scale [40], investigations have highlighted that this vegetation evolution is not spatially uniform, and large areas remain affected by degradation, such as the northeast of Burkina Faso and the southwest of Niger.

The joint impact of land degradation and extreme rainfall increase produced an extension of the drainage network and the rupture of endorheic basins that caused a further discharge increase [39]. The right bank tributaries of the Niger river, and among these the Sirba river, show discharges 150% higher and runoff coefficients three times higher than those observed up to 50 years ago [31].

In Niger, the increase in flood events has been demonstrated to be country-wide by Fiorillo et al. [41], analyzing official data collected by the government on damages from 1998 to 2017. Regarding the regional and sub-regional impacts of floods, the southwestern areas of the country were confirmed to be the most exposed to flood risk. Over the last 20 years, the scientific literature has focused mainly on changes in Niger river flood magnitudes, trying to understand both the changes underway in regional hydrological characteristics and the main factors triggering the increase in floods in the area. However, Tiepolo et al. [42,43] demonstrated that the Niger river is just one of the causes of the flood risk, with other mechanisms and triggers being present.

2.2. Methods for System Set-Up

According to the United Nations International Strategy for Disasters Reduction (UNISDR), the four pillars of EWS, SLAPIS has been set up through a progressive process (Figure 2), addressing (1) risk knowledge, (2) risk monitoring and warning, (3) risk information dissemination and communication, and (4) the response capacity of communities and the authorities to respond to the risk information. Approaches, methods, data collected and analysis are described herein, according to each of the four pillars.

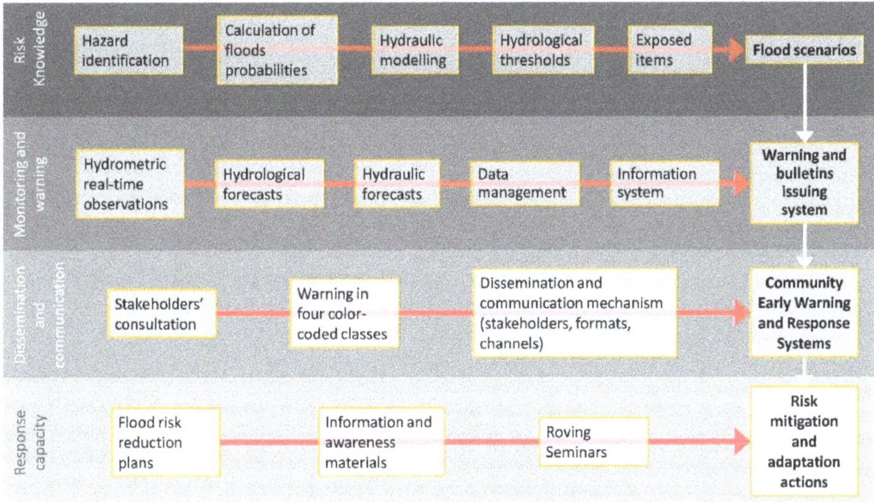

Figure 2. Conceptual framework of Système Locale d'Alerte Precoce contre les Inondations de la Sirba (SLAPIS) organized on the four pillars of EWS (United Nations International Strategy for Disasters Reduction).

2.2.1. Risk Knowledge: Risk Assessment at Local Level and Flood Scenarios

The risk assessment activities have been performed in the four main rural communities distributed along the last 40 km of the Sirba river: Tallé (population 2603 in 2012), Garbey Kourou (4634), Larba Birno (4713) and Touré (4065). The methodology adopted, as described by Tiepolo et al. [44], considers the risk (R) as "the probability of occurrence of hazardous events or trends multiplied by the consequences if these events occur" [45]. The potential damages have already been used as a component of the risk function in both Niger [46] and the developing countries of South Asia.

Risk assessment is conceived as a process starting from the understanding of the local risk governance framework including local planning processes, national guidelines and the literature. The second step is the identification, through meetings with each community, of the hydro-climatic threats, past catastrophic events, rainfall threshold, and flood level above which damages are registered and local resources mobilized by each community (capacity and assets) to address them. The last phase is the calculation of flood probabilities and the establishment of flood scenarios. Flood scenarios were calculated through the development of an ad-hoc hydraulic numerical model simulating the river behavior for each discharge threshold in the Hydrologic Engineering Center's River Analysis System (HEC-RAS) [47] environment. The hydraulic model was implemented on a topography based on a 10 m Digital Terrain Model (DTM) detailed with Geographical Positioning System (GPS) topographical surveys and calibrated with discharge and level observations, as described by Massazza et al. [27]. In order to take into account changes in the hydrological behavior of the Sirba river over time, as described by Tamagnone et al. [26], non-stationary analyses were conducted to identify the probability of occurrence of the assigned hazard thresholds [27,48,49]. The hazard thresholds, as already described by Massazza et al. [27], are shown in Table 1.

Table 1. Hazard thresholds for different flow conditions of the Sirba river [27]. Discharge (Q); Flow Duration Curves (FDC); Stationary Generalized Extreme Value (SGEV); Stationary Return Period (S-RT); Non-Stationary Generalized Extreme Value (NSGEV); Non-Stationary Return Period (NS-RT).

Color	Q max (m^3/s)	Index			Magnitude	Expected Damages
		FDC (Q_{XX})	S_{GEV} (S-RT_{XX})	NS_{GEV} (NS-RT_{XX})		
green	600	15	5	/	normal condition	/
yellow	800	5	10	2	frequent flood	fish nets, water pumps, livestock
orange	1500	/	30	5	severe flood	wells, boreholes, low-altitude houses, barns, and crops
red	2400	/	100	10	catastrophic flood	extended area at medium–low altitude (houses, barns, and crops)

Each threshold was simulated using the hydraulic model in order to define the area and relative hazard level to which riverine populations are subjected. Hydrological thresholds of flood scenarios were linked to field impacts, according to the national flood hazard classification [50] and international guidelines, such as the WMO Guidelines on Multi-Hazard Impact-Based Forecast and Warning Services [17]. Lastly, the risk level characterizing each scenario was obtained by matching together the information regarding the extent of flood-prone zones, the identification of exposed assets and their value [44].

2.2.2. Monitoring and Warning Service: Hydrological Observations and Forecasts, Data and Information Management

The monitoring component of the system relies on two automatic gauging stations at Bossey Bangou (upstream, at the Burkina Faso border) and Garbey Kourou (downstream, near the confluence with the Niger river). The Garbey Kourou hydrometer was installed in 1956 and is equipped with two water pressure measuring devices, one controlled by the Niger Basin Authority (NBA) and the other by the DH of the Republic of Niger, while the Bossey Bangou gauging station was installed in June 2018, in the framework of SLAPIS, and is managed by DH. The Bossey Bangou (2018–2019) and Garbey Kourou (1956–2019) updated discharge series and a set of 14 discharge measurements were used for hydrological and hydraulic modelling [26]. The headwater of the Sirba river in Burkina Faso is equipped only with hydrometric stations that are non-operational for the real-time monitoring of discharge and, therefore, are useless for the EWS.

Further information on water depth, maximum water levels and flooding extent were collected from local observations and field surveys made at the main localities along the Sirba river. Colored hydrometric staffs (ladders) were installed in May 2019 in five villages along the Sirba river: Touré, Larba Birno, Garbey Kourou and Tallé in the municipality of Gotheye and Larba Toulombo in the municipality of Namaro. The staffs are marked with the four different colored flood scenarios (green, yellow, orange and red). The levels of the colored staffs were defined thanks to fixed topographical points identified during the land surveys. A volunteer observer was appointed within the Community Early Warning and Response System (SCAP-RU) of the village and was trained.

Concerning forecasts, the system relies on two types: hydraulic model forecasts (related to observations of upstream hydrometric stations) and hydrological model predictions (derived from hydrological models acting on the Sirba basin). The first consists of the warning that should be conveyed to villages in the case of the river passing the hazard threshold at the upstream hydrometer. The hydraulic model allowed the flood propagation time to be calculated and, thus, the warning time for each village [44]. This type of forecast has a higher level of certainty but may give only a few hours or up to one day of notice to the riparian villages downstream.

Hydrological model forecasts have a major uncertainty but can give indications towards the evolution of the hydrology up to 10 days in advance. At present, the early warning system bases its forecasts on the global hydrological model GloFAS 2 [21]. Preliminary analysis shows that the gap between observed and forecasted discharge is quite significant. This suggested the post-processing of forecasts in order to decrease the bias and improve the EWS reliability. GloFAS forecasts are adjusted with corrective factors, improving their reliability according to historical series and real-time measured data. The optimization process was conducted through the linear regression method over homogeneous periods of the rainy season and was based on 10 years of simulations (2008–2018). The optimization allows quality improvement with an increase in Root Mean Square Error (RMSE) and the Probability of Detection (POD) of extreme events and, at the same time, reduces the False Alarm Rate (FAR), as described by Passerotti et al. [51]. A further improvement in the forecasting system is foreseen with the integration of a second model, Niger-HYPE [23].

Data management and services are ensured by a Spatial Data Infrastructure (SDI) based on interoperable and open source solutions and Open Geospatial Consortium (OGC) web services [52] for the management of observed and forecasted data and the establishment of a hydrological warning communication service. Methodologically, the implementation steps were the conceptual and formal data model design, the development of the SDI, the setting up of some Open Web Services (OWS) standards through the development of services and procedures for data flow management, the forecast data optimization and geoprocessing functions.

The SLAPIS client–server architecture (Figure 3) is based on open source technologies and software components which allow it to interact between data providers and end-users, including three main layers: data retrieval and storage, web services and user interface. All data are managed by a central open source geodatabase, which is the core of the SLAPIS server. Geoprocessing routines and data optimization procedures have been implemented on the data layer in order to ensure that the observed and forecasted data are uploaded into the system data model. Furthermore, Application Programming Interfaces (APIs) have been developed both to transfer forecast data from Niger-HYPE and GloFAS platforms and to foster the communication among the SDI components. For retrieving data from providers not equipped with standard and interoperable web services, we used the File Transfer Protocol (FTP).

For the front-end of the system, a customized Graphical User Interface (GUI) was designed and implemented for monitoring, in quasi-real-time, the observed and forecasted data and their visualization in graphic and tabular formats. The customized functions allow the users to retrieve (from the GUI) the entire data set for further analysis or applications. SLAPIS also has an open data portal which, using the Comprehensive Knowledge Archive Network (CKAN) [53] open source data catalogue, allows access to the available data, including raw and intermediate research data, as well as complementary studies on the area. Each dataset recorded in CKAN contains a description of the data and other useful information, such as available formats, the producer (if they are freely available) and the topic. Finally, a simple information box is available on the main page and it is automatically updated by the system with the current state of vigilance.

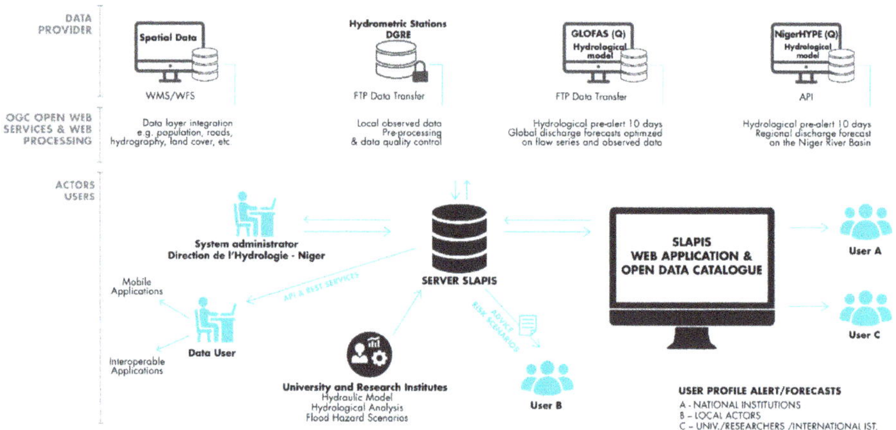

Figure 3. The SLAPIS information system architecture.

2.2.3. Dissemination and Communication: Stakeholders' Consultation

The dissemination and communication mechanism of SLAPIS was defined by a stakeholders' consultation and an analysis of the national alert mechanism. As indicated by the National Alert Code [54], the warning measures are disseminated by the decision of the Minister of Civil Protection at a national level, by Governors in the regions, Prefects in departments and Mayors in the municipalities (Article 5). Alert messages are prepared from information provided by technical institutions at different administrative levels.

An analysis of the needs of the actors in terms of information on the flood risk was performed through semi-structured interviews with national stakeholders, technical workshops with local administrations and focus groups with the communities involved. The result was the definition of the SLAPIS communication and information plan, stating that information produced by SLAPIS should be accessible to all stakeholders through specific tools and channels. Subsequently, information on the state of vigilance is transformed into alerts by the competent institutions according to the magnitude and amplitude of the forecasted flood risk.

Concerning the last-mile of communication, challenges include access to the information, the ability to understand the warning, and the ability to take action [55,56]. All these issues were dealt with via focus groups with local governments and community representatives. A set of actions were defined in order to create awareness at a community level about the flood scenarios and the actions to be taken in the case of a warning. Among different approaches, visualization is the one that has been preferred to aid the interpretation of flood scenarios [57]. In the context of SLAPIS, visualization includes the adoption of the four-color classification for scenarios. They are associated with warning content (the core of the message is the color), with the colored hydrometric staff gauges (qualitative gauging staffs) installed along the river, as well as information panels in the villages indicating priority actions to be taken. We adopted the four color-coded classes currently used by different countries (i.e., United Kingdom—the Met Office; EU—Meteoalarm, Philippine—Pegasa, Italy—Protezione Civile, India Meteorological Department) [19] and responding to the international standards (International Organization for Standardization - ISO 22324:2015). The classes are related to discharge, return periods and impacts on the main riverine settlements according to the classification, as described by Massazza et al. [27]—essentially, green stands for the normal condition, meaning a no-impact scenario, yellow (Stationary Return Period 10 years) stands for minor impacts, orange (Stationary Return Period 30 years) stands for significant impacts and red (Stationary Return Period 100 years) stands for severe impacts. Moreover, the installation of colored hydrometric staffs aims to increase awareness of the flood risk among communities by showing the levels of the hazard thresholds—the height that the flood can reach. Local hydrometric staffs also aim to establish a local communication system, building on the approach described by many authors in Asia [15,58], between upstream and downstream villages.

2.2.4. Response Capability: Communities Preparedness and Action

According to Girons Lopez [59], response capability was based on social preparedness for flood loss mitigation. Community flood risk reduction plans were prepared for the four main villages of Touré, Larba Birno, Garbey Kourou and Tallé in the municipality of Gotheye. The plans have the objective of associating the flood scenarios and the stakes to underline the specific criticalities of each village and propose measures to reduce potential damage. As described by Tiepolo et al. [44], the plans were drawn up with a multi-step methodology: participatory hazard identification, probability of flood occurrence, flood-prone areas, asset (mostly housing and crops) identification and risk reduction actions. The assets are identified in the flood zone by municipal technicians integrated using very high photointerpretation, as described by Belcore et al. [60]. Actions include both risk prevention and the preparedness actions known by the target communities, as well as best practices from the reference literature.

According to Fakhruddin et al. [61], the participatory development of flood risk reduction plans has the objective of enabling people to act, empowering communities with basic knowledge of the flood risks and of more urgent actions to be taken according to each scenario. Actions to reduce risk are associated with the flood scenario and the four color-coded classes of warnings. Therefore, the warnings embed both physical information (water depth and flood zones) and social information (such as community assets likely to be affected and community actions to be taken).

Community preparedness has also been strengthened by adopting and adapting the approach developed by Stitger et al. [62,63] for drought risk management through Roving Seminars on agrometeorology and agroclimatology. A new concept of Roving Seminars for flood risk management has been developed. The seminars take the form of a one-day meeting in a village, which the whole community is invited to attend. The objective is to make communities become more self-reliant in dealing with hydrometeorological issues related to floods that affect human life, habitats, assets, livestock and crops, and to increase the interaction between the community and the National Meteorological and Hydrological Services.

3. Results

The results are reported in the following sections, relating to each of the four pillars of EWS.

3.1. Risk Knowledge

The first main result in the definition of risk level was the assessment of flood scenarios. They were defined on the hazard thresholds, fixed on both the statistical analysis of discharge and impacts on the main riverine settlements according to the four color-coded classes. Flood hazard scenarios were mapped, showing the extension of flood-prone areas (Figure 4). The bulk of the assets are located on the left bank of the Sirba river (houses, community services, infrastructure, fields and vegetable gardens) while the assets on the right bank are essentially fields and orchards with few settlements.

The hydraulic numerical model was also used to calculate the conveyance time of the flood wave: the upstream hydrometer of Bossey Bangou provides notice of between 20 (Touré), 26 (Larba Birno) and 28 hours (Garbey Kourou and Tallé) [27]. Scenarios include the identification of exposed assets (houses, orchards, crops, pits, barns and wells) and their value, as described by Tiepolo et al [46]. Table 2 reports the value of the assets that could be damaged in the four main riverine villages by a flood event with a magnitude equal to the hazard threshold. Reported amounts should be considered, keeping in mind that the average annual GDP per capita in Niger is 430€ (2018) and the minimum wage is 46€ per month.

Figure 4. Atlas of flood-prone areas of the Sirba river. Tallé village (Gotheye municipality, Niger). This image reproduces the Table C1 of the Atlas which is in the official language of Niger (French). The Atlas is annexed as a Supplementary Material.

Table 2. Cumulative value (k€) of exposed assets in the four main villages along the Sirba river (adapted from Tiepolo et al. [44]).

Zone	Garbey Kourou	Larba Birno	Tallé	Touré	All
Yellow	0.2	1.3	13.0	0.3	14.8
Orange	8.4	5.4	19.9	1.6	35.3
Red	21.9	22.3	41.7	36.0	120.7

3.2. Monitoring and Warning Service

The observed and forecasted data are accessible by stakeholders using the SLAPIS web platform (www.slapis-niger.org), with specific characteristics that make it unique in the panorama of web tools developed for alerting at local level, because it integrates the following five levels (Figure 5):

1. Forecasts: the system downloads the daily forecasts from hydrological models (GloFAS and Niger-HYPE) and stores them in the central Postgres geodatabase. Data are postprocessed according to the optimization procedures and finally shown as the discharge in a graphic format (currently only GloFAS);
2. Observations: the system downloads real-time observations from hydrometric stations each hour and stores them in the Postgres geodatabase. Data are shown as the water height and discharge in both graphs and tables;
3. Levels of vigilance: levels of vigilance are set by the system once the thresholds are exceeded on observed data at the upstream station;

4. Flood scenarios: the system automatically displays on the map the flood scenario related to the exceeded vigilance threshold;
5. Vigilance bulletin: Exceedances in hazard thresholds activate a vigilance bulletin in the system and an automatic email message to the Directorate of Hydrology. The vigilance bulletins can be used as they are or can be edited to activate the alerting chain of national and local authorities, as defined by the National Alert Code.

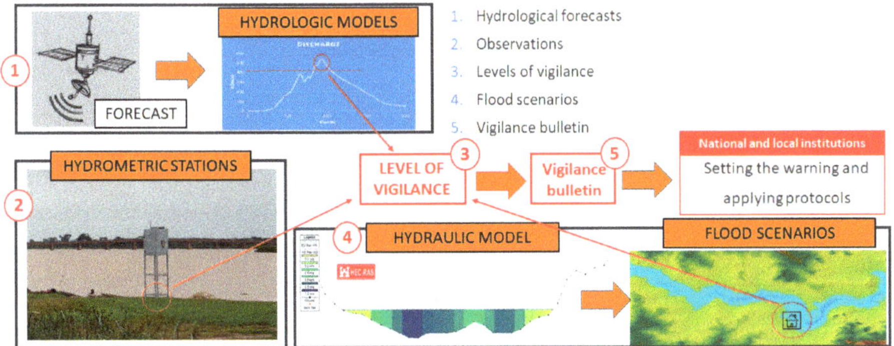

Figure 5. Structure of the SLAPIS monitoring and warning services.

The observed data are also accessible through the CKAN catalogue (http://sdicatalog.fi.ibimet.cnr.it:5003/fr/dataset?groups=slapis_prj), as well as other geographical data used by the system in different formats (SHP, GeoJSON, JSON and CSV).

The network of local observers is composed of two at the gauging stations and five village observers at the colored hydrometric staffs (Figure 6).

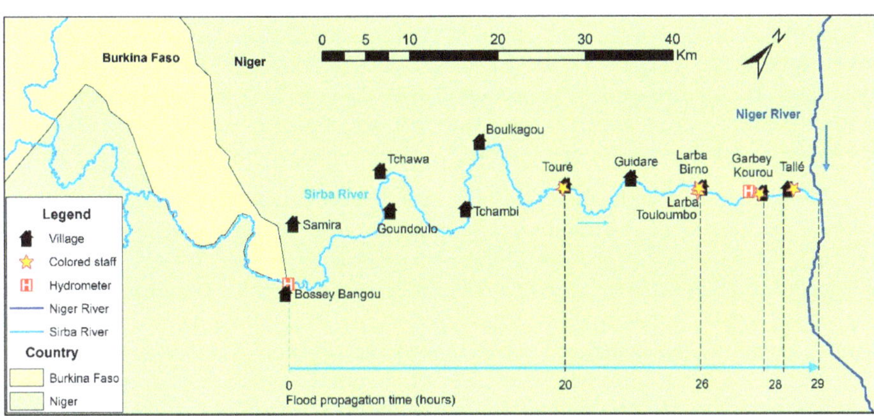

Figure 6. Map of the observation network and the flood propagation time.

In order to increase awareness of the flood risk among communities and integrate the observation network with additional local measures, colored hydrometric staffs (Figure 7) have been installed in five villages along the Sirba river with the objective of showing the levels of the hazard thresholds, the height that the flood can reach, to increase awareness of the danger of flooding and communicate the level of risk to villages downstream. The observers have been trained to read the scales and communicate any rise in the waters to the DH, the Vulnerability Monitoring Observatory (OSV) located in the municipality and the SCAP-RUs of the villages downstream, according to a specific observing protocol. In 2019, they used a WhatsApp group to send the information as a photo; however, in 2020, a system of visualization of the observed water levels at the colored staffs will be integrated into the SLAPIS platform.

Figure 7. Colored hydrometric staff at Garbey Kourou, Niger.

3.3. Dissemination and Communication

SLAPIS is integrated in the national alert system; indeed, its information mechanism was defined thanks to an analysis of the national alert mechanism and an analysis of the needs of the actors in terms of information on the flood risk. The communication and dissemination plan of SLAPIS defined that the flood scenarios produced in the framework of SLAPIS and the related warnings on the level of vigilance are accessible to all stakeholders with specific tools. Subsequently, the state of vigilance is transformed into an alert by the competent institutions. Figure 8 summarizes the SLAPIS information mechanism. In particular, it indicates the actors to whom the information is communicated directly, through which channel and in what format.

The information on the state of vigilance is then transformed into an alert by the competent institutions according to their protocol and communicated through institutional channels, as recommended by Rahman et al. [64]. According to Oktari et al. [65], the communication system is multi-channeled in order to ensure maximum outreach. Specific communication channels are established with different types of stakeholders. There are four main official information flows:

- To Civil Protection Directorate (DGPC) at a central level, that sends the alert to the Regional Directorate, then to the Departmental Directorate, which then communicates it to the Mayor;
- To the National Food Crisis Prevention and Management System (DNPGCA) at central level, that sends the alert to the Regional Committee (CRPGCA), then to the Departmental Committee (CSRPGCA), who communicates it to the Mayor;
- To the Ministry of Humanitarian Action and Disaster Management (MAH) which, not having any de-centered structure, transmits the alert via all the means of communication provided by the National Alert Code to all levels including the population.
- To the Mayor of affected municipalities who, once he has received the information concerning the level of vigilance, alerts:
    - The OSV through an emergency meeting. The OSV alerts sectoral infrastructure (schools, water supply, etc.);
    - Community Radio by telephone to dictate an alert release;
    - Chiefs of concerned villages by telephone. The village chief mobilizes all available means to alert the population (loudspeaker, town crier, etc.);

- o The Prefecture by telephone and/or SMS;
- o SCAP-RUs (through the OSV).

In addition to the information from the information system, SLAPIS integrates local observations made at the colored staffs installed in the main villages bordering the Sirba. The observer appointed within the SCAP-RU is responsible for communicating the possible rise in the water level to the OSV located in the municipality and to downstream villages using the color code of the flood scenarios (green, yellow, orange and red).

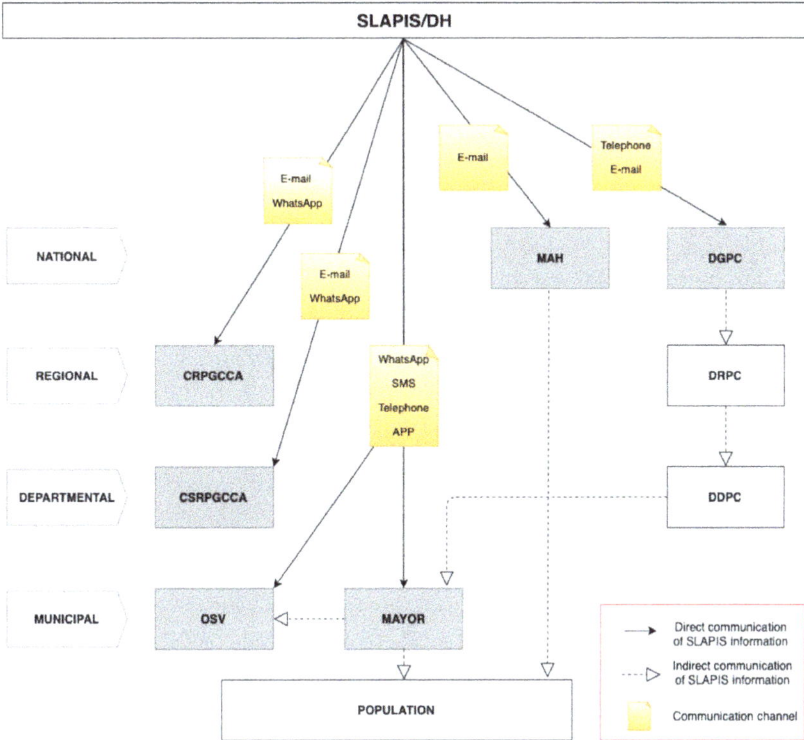

Figure 8. Information flow (Hydrology Directorate (DH); Ministry of Humanitarian Action and Disaster Management (MAH); Regional Committee of Food Crisis Prevention and Management System (CRPGCCA); Departmental Committee of Food Crisis Prevention and Management System (CSRPGCA); Vulnerability Monitoring Observatory (OSV); General Directorate for Civil Protection (DGPC); Regional Directorate for Civil Protection (DRPC); Department Directorate for Civil Protection (DDPC)).

3.4. Response Capability: Communities Preparedness and Action

Since 2006, the WMO has been promoting the organization of one-day Roving Seminars on weather and climate for farmers. We adapted the concept of Roving Seminars to the hydrological risks, with the main objective of strengthening communities' self-reliance in dealing with floods and other extreme hydrometeorological risks. Moreover, the seminars increase the interaction between the local communities and local staff of the National Meteorological and Hydrological Services, building trust and confidence. The seminars are organized in two parts. The first focuses on the weather, climate and hydrology of the region, particularly on the relationship between rainfall and discharge on the whole river basin and on local sub-basins. The aspects of change in land-use/land-cover and changes in hydrological processes are also well developed in order to promote a better comprehension of the dynamics, in which farmers are a key actor. The second part of the day focuses on the communities' perception of weather and hydrological information/alerts provided by SLAPIS. The main objective is to obtain feedback from the communities by a free and frank exchange of ideas and information. This part of the seminar is designed to engage all the participants in discussions and obtain suggestions on additional information and ways and means of improving future communication to facilitate effective operational decision making.

SCAP-RU are key actors in monitoring and coordinating actions in an emergency and in peacetime. SLAPIS invested in the empowerment of SCAP-RU, both technically and socially. From a technical point of view, SCAP-RU have been trained in risk assessment, monitoring and impact observation. Each SCAP-RU has been equipped with a smartphone and access to the internet to collect and transmit observations and receive information and alerts. SCAP-RU members have been charged with specific roles before, during and after emergencies, reinforcing their status and becoming a reference within the community.

In Niger, local plans to address hydro-climatic hazards are not required by law. Within SLAPIS, we developed local flood risk reduction plans for the main villages exposed to flood risk. The plans were coordinated with the communities and organized considering local capacities and assets. Even if the plans are very simple, they include a contingency part involving the setting up of an emergency committee, a map of the areas according to flooding probability and a map of exposed assets and basic emergency instructions. The plans also include structural prevention measures, such as the protection or displacement of boreholes, wells, fountains and photovoltaic plants [44]. The visualization of contingency actions in relation to risk levels was realized through local information panels, which report the main actions to be taken by communities in text and graphics.

Another result was the production of a Cartographic Atlas of flood-prone areas (published as a Supplementary Material to this paper) over the 108 km of the Nigerien reach of the Sirba river. Maps are compiled at large (1: 100,000) and detailed (1: 20,000) scales. The most exposed riverine villages of Tallé, Garbey Kourou, Larba Birno, Larba Toulombo, Guidare, Toure, Boulkagou and Bossey Bangou are also represented at greater detail (1: 5000).

4. Discussion

The results of the study are discussed herein according to the indicators identified by Sai et al. [19] for steering each of the four components of effective EWS (See Table A1 in Appendix A).

Concerning risk knowledge, the indicators identified in the literature are:

- Local risk assessment—interaction of vulnerability and hazard scenarios for determining the risk to the exposed elements with a detailed resolution [66]: this is a key element to connect top-down and bottom-up approaches. It was addressed through participatory local risk assessment for main localities along the river [44] and flood hazard scenarios, including hazard thresholds and flooding areas [27,44];
- Hazard mapping—hazard maps for different scenarios need to be developed to identify exposure to different hazard magnitudes [61,67]: flood hazard scenarios were mapped (www.slapis-niger.org) and an Atlas of flood-prone areas of the Sirba river produced (see Supplementary Materials);
- Vulnerability mapping—vulnerable elements and critical infrastructure need to be mapped, documented and periodically updated [68]: this point was addressed through the involvement of municipalities in listing, georeferencing and describing assets for each flood scenario in the main localities [44,60]; for future sustainability, the challenge is to keep the lists updated, which mostly depends on the willingness of the municipalities involved.

Concerning the monitoring and warning components, the indicators proposed by the literature are:

- Timely and accurate forecast—good quality data have to be collected and processed in real or quasi-real-time to produce meaningful, timely and accurate forecasts [3]: hydrological forecasts can ensure a lead time of 10 days, but their accuracy is lower than hydraulic forecasts based on the upstream hydrometer [27], which have a lead time of 28 hours. The accuracy of hydrological forecasts has been improved, assimilating real-time data observed at the gauging stations at Bossey Bangou and Garbey Kourou;

- Impact-based thresholds—warnings must be prepared and issued based on expected impacts severity, enabling end-users to take appropriate risk mitigation actions [17]: SLAPIS Hydrological warnings are based on four color-coded classes related to flood scenarios and assets at flood risk [27,44]. For citizens already used to that approach, the meaning is intuitive, but in a Sahelian rural area, the understanding is not evident. Indeed, many awareness-raising activities have been organized with local communities;
- Geographic-specific warnings—a dense monitoring network ensures a good coverage of the forecast, being more specific and issuing warnings to localized targets [65,67,69]: addressed by two automatic hydrometers on the Sirba river (108 km distance = ~ 28 hours. propagation time) at Bossey Bangou (13.73° N, 1.60° E) and Garbey Kourou (13.35° N, 1.29° E), real-time data collection in the online geo-database, real-time display of hydrological data (www.slapis-niger.org), real-time data available on CKAN at http://sdicatalog.fi.ibimet.cnr.it:5003/fr/dataset/bosse-bangou and some colored hydrological staffs installed between the stations for local observations (av. 35 km distance = ~ 8 hours. propagation time). The use of colored staffs for empowering communities in local warning is well documented in the literature and it also raises awareness about flood risk;
- Sector-specific warning—warning contents have been prepared and issued to different end-user groups with the same needs [3]: hydrological warnings are issued when a threshold is exceeded. The same information is sent to different groups of users with different format and communication channels according to users' needs and the preference expressed during the stakeholders' consultation.

Concerning dissemination and communication, the indicators collected by Sai et al. [19] are:

- Robust standing operating procedure—government policy establishes the warning dissemination pathway and the rules for defining specific impact-based warnings [64–66]: the main challenge is to involve all stakeholders, avoiding conflicts of mandate and responsibility. The National Alert Code is, theoretically, clear but practically different institutions claim the same mandate on flood risk management. Dissemination and communication plans have been developed according to the National Alert Code and the shared solution we found was to provide the same information to all stakeholders, leaving scope for the further dissemination to their own internal procedures. Therefore, all stakeholders in the alerting chain are involved through specific communication channels. The following alerting process reflects the chain of each stakeholder;
- Complete and timely dissemination—warnings must reach all exposed communities, including those in remote areas, in good time before the hazardous event occurs [64,65]: the pre-alert system based on forecasts has a lead time of 10 days, while the alert based on hydrological observation has a lead time between 6 to 28 hours. according to distance from the Bossey Bangou gauging station. The communication system for the last mile is mainly by mobile phone, now common even in more remote rural areas;
- Multiple dissemination channels—exposed communities must be warned via different media according to their ability/possibility of using them [65]: a communication system is in place to reach municipalities in multiple ways. At municipal level, Mayors can deploy multiple communication channels (Radio FM, Smartphone text and voice calls to village chiefs and SCAP-RU, public criers). At local level SCAP-RUs are equipped with colored hydrological staffs and smartphones (WhatsApp) to alert SCAP-RUs of downstream villages and OSV;
- End-user's dissemination and communication needs—message content, communication and dissemination means are tailored to end-user needs, ensuring higher warning understanding [61,64,67,69]: alerts rely on color-coded classes, associated with scenarios, impacts

and actions. Each SCAP-RU has knowledge of the scenarios associated with alert colors, flooding area maps, assets at risk and priority actions to be taken. Participatory meetings proved to be really useful in building trust.

Concerning response capability, the indicators addressed are:

- Information on impacts and advice—messages must contain information on the expected impacts and advice on how to implement risk mitigation actions [3,70]: Each SCAP-RU has knowledge of the color-coded classes scenarios. In each village information panels with infographics are displayed to report priority actions to be taken. Roving Seminars on flood risk proved to be very useful in raising awareness and building understanding of actions to be taken according to different scenarios.
- Community and volunteers education—volunteers must ensure coordination at local level, helping communities to effectively respond to alerts [66,69]: the SCAP-RU (composed of volunteers) are in charge of activity coordination at a local level (villages), local observations and exchanges with other villages, alert dissemination within the village, and the taking of urgent actions by the community.
- Preparedness and contingency plans—hazard and vulnerability maps are used as a management tool for improving the response and coordinating emergencies [66,71]: municipalities are empowered with flooding area maps at different scales, the list of assets at risk for each village and flood scenarios and the flood risk reduction plans for the main villages in their territory. The challenge for municipalities is to keep the lists of assets for each flood scenario updated.

Finally, some cross-cutting indicators are also addressed:

- Local community participation—end users can actively contribute to all four components of EWS [3,61,64,71]: communities are key actors in Participatory Local Risk assessment [44], discharge monitoring at a local level (SCAP-RU), communicating the observed levels of vigilance (SCAP-RU) and defining the contingency actions of risk reduction [46]. Furthermore, the technical architecture and GUI of the SLAPIS are designed to integrate (i) top-down and bottom-up approaches, (ii) hydrological forecasts and observations with local perception of populations and a lead time suitable for operational decision-making processes at national and local levels. According to many authors, involving communities and local stakeholders is the main challenge for achieving the purpose of the EWS and SLAPIS proved that it can be effectively achieved by appointing local observers, organizing meetings with communities, involving them in the risk knowledge phase and jointly defining communication and risk reduction plans.

5. Conclusions

According to Cools et al. [71], the key point to enhance the effectiveness of a flood EWS is: "a better match between the available risk information, the forecasting system and the response capability of authorities and the at-risk population". Cools et al. suggested, and the present study demonstrates, that "Engaging local communities and authorities in the EWS design can improve the effectiveness of the whole early warning process and hence results in a higher response to an alert warning".

The study proves that such key points can be operationally addressed, leveraging existing resources, local stakeholders and knowledge using simple but effective approaches and integrating state-of-the-art hydrologic-hydraulic scientific results in a decisional scheme for Sahelian rural areas. This mechanism will be replicable in each context, even if characterized by knowledge and structural deficits, creating a better capacity to exchange data and information and by directly connecting available technical capabilities with the local level. Beneficiaries are, therefore, able to define the rules that will develop the whole system, which, in any case, needs to be consistent with the legislation in force in the country and with internationally recognized best practices.

This study suggests that, instead of developing new forecasting tools, it can be preferable to enhance those already operating on the basin and calibrate them on the local scale by adding real-time observation control points and to connect discharge thresholds, field observations and hydrological forecasts with potential impacts through flood scenarios. This multidisciplinary approach ensures a greater level of suitability and sustainability. Indeed, it allows us to enhance the resonance of hydrological models already developed by the scientific community, which are not usually exploited by local technical structures and to concentrate efforts on: (1) the downscaling of forecasts, (2) topographical surveys and discharge measurements on-site, (3) the quality improvement of observations and (4) the implementation of hydraulic models to guide the planning of mitigation and adaptation strategies. The strength of simplicity also lies in not having to spread complex messages, but simply the reference risk scenario and, finally, its color-code (according to the international standards of ISO 22324:2015), which already embeds all of the other information.

The main limit of the study is that it focuses only on the Sirba reach in Niger. The absence of trans-border flood risk assessment and the lack of real-time hydrological monitoring upstream in Burkina Faso prevent the flood risk information being spatially extended upstream. Moreover, the limited lead time provided by the observation at the Bossey Bangou gauging station to the main downstream villages is a limitation of the EWS on the Sirba river.

The ongoing improvements of SLAPIS, already planned before the 2020 rainy season, include implementation of: (1) a second optimization based on real-time observed discharge and (2) the operational integration of the forecasting system of the regional hydrological model Niger-HYPE, ensuring more resilience and more accurate discharge forecast in the system. Further future developments of this study include the improvement and extension of the existing flood EWS to the whole Sirba basin. Naturally, this implies an urgent need for systematic flood monitoring and communication between the two countries (Burkina Faso and Niger), as well as a coherent flood risk assessment performed upstream of the boundary. In this respect, the respective governments' commitment to sharing of information and effectively disseminating it to flood-prone communities is necessary. At broader level, the study found that there is a need for institutionalizing and strengthening the existing practices of sharing of flood information between different countries for EWS purposes. The simple and integrated approach illustrated in this case study, bridging the gap between top-down and bottom-up approaches described by the literature, can inspire governments, local administrations and development partners to invest in the improvement of existing tools and knowledge in order to strengthen cooperation, collaboration and coordination, reduce hazard impacts and sustain the development of rural and urban areas.

Supplementary Materials: The following are available online at http://www.mdpi.com/2071-1050/12/5/1802/s1, Atlas of flood-prone areas of the Sirba river (Atlas cartographique des zones inondables de la Rivière Sirba.

Author Contributions: Conceptualization: V.T.; Methodology: V.T., M.T., M.R., A.P., T.D.F. G.L.K.; Developing: L.R., T.D.F., E.R., G.M., P.T.; Validation: L.R., E.R., G.M. P.T.; Investigation: G.M., P.T., M.H.I., V.M.; Writing—original draft preparation: V.T.; Writing—review and editing, M.T., T.D.F., L.R., V.M., G.M., P.T.; Funding acquisition: V.T. All authors have read and agreed to the published version of the manuscript.

Funding: This research was co-funded by the Italian Agency for Development Cooperation, by the Institute of Bioeconomy (IBE)-National Research Council of Italy (CNR), by the National Directorate for Meteorology of Niger (DMN) and by DIST Politecnico and the University of Turin within the project ANADIA2.

Acknowledgments: We would like to thank Moussa Labo (former Director of DMN) for trusting in the project. We would like to express our deepest gratitude to the Joint Research Center of European Commission and European Centre for Medium-Range Weather Forecasts (JRC-ECMWF) and the operational flood forecasting and alerts in West Africa (FANFAR) Project for belief in a collaborative approach and providing access to GloFAS and Niger-HYPE forecasts. We thank Bruno Guerzoni, Hamidou Issa Mossi (Mayor of Gotheye), Abdoul Wahab Oumaru (Mayor of Namaro), Ibrahima Djibo (Mayor of Torodi), Souradji Issa (DDA/Gotheye), for support and participation. Finally, the authors want to thank, with affection, Steffen Muller (GIZ), discussions with whom led to the idea of SLAPIS being born in the hot Nigerien evenings.

Conflicts of Interest: The authors declare no conflict of interest. The funders had no role in the design of the study; in the collection, analyses, or interpretation of data; in the writing of the manuscript, or in the decision to publish the results.

Appendix A

Table A1. Indicators of effective EWS identified by Sai et al. [19] and the approach adopted to address them within SLAPIS.

EWS Component	Indicator	Description	References	SLAPIS
Risk knowledge	Local risk assessment	Interaction of vulnerability and hazard scenarios for determining risk to the exposed elements with a detailed resolution.	Scolobig et al. [66]	Participatory Local Risk Assessment for main localities along the river [46] Flood hazard scenarios, including hazard thresholds and flooding areas [27,46]
	Hazard mapping	Hazard maps for different scenarios need to be developed to identify exposure for different hazard magnitudes.	Fakhruddin et al. [61] Koks et al. [67]	Flood hazard scenarios mapping (www.slapis-niger.org) Flooding areas atlas of the Sirba (See Supplementary Materials)
	Vulnerability mapping	Vulnerable elements and critical infrastructures need to be mapped, documented and periodically updated.	Bhuiyan and Al Baky [68]	Listing, georeferencing and describing assets for each flood scenario [46,60]
Monitoring and warning	Timely and accurate forecast	Good quality data have to be collected and processed in a real or quasi-real-time to produce meaningful, timely and accurate forecasts.	Basher [3]	Use of two hydrological models (GLOFAS and Niger-HYPE) assimilating real-time data observed at the gauging stations of Bossey Bangou and at Garbey Kourou [27] and hydraulic forecasts
	Impact-based thresholds	Warnings must be prepared and issued based on expected impacts severity, enabling end user to take appropriate risk mitigation actions.	WMO [17]	Hydrological warnings based on four color-coded classes related to flood scenarios and assets at flood risk [27,46]
	Geographic-specific warnings	A dense monitoring network ensures a good coverage of the forecast being more specific, issuing warnings to localized targets.	Cumiskey et al. [69] Oktari et al. [65] Shah et al. [70]	2 automatic gauging stations on the Sirba river (100 km distance = ~ 28 hours. propagation time) at Bossey Bangou (lat° 13.73, lon° 1.60) and Garbey Kourou (lat° 13.35, 1.29 lon°) Real-time data collected in online geo-database SLAPIS Real-time display of hydrological data (www.slapis-niger.org) Real-time data available on CKAN at http://sdicatalog.fi.ibimet.cnr.it:5003/fr/dataset/bosse-bangou 5 colored hydrological staffs between the stations for local observations (av. 35 km distance = ~ 8 hours propagation time)
	Sector specific warning	Warning contents have been prepared and issued to different end users clusters with same needs.	Basher [3]	Hydrological warning issued when a threshold is exceeded. The same information is sent to different groups of users with different format and communication channel according to the preference expressed during the stakeholders' consultation.

Table A1. *Cont.*

EWS Component	Indicator	Description	References	SLAPIS
Dissemination and communication	Robust standing operating procedure	Government policy establishes the warning dissemination pathway and the roles for defining specific impact-based warnings.	Oktari et al. [65] Rahman et al. [64] Scolobig et al. [66]	Dissemination and communication plan developed according to the National Alert Code. All stakeholders in the alerting chain are involved. Specific communication channels established with different types of stakeholders. Alerting process reflects the chain of each stakeholder
	Complete and timely dissemination	Warnings must reach all exposed communities, including those in remote areas, in time before the hazardous event occurs.	Oktari et al. [65] Rahman et al. [64]	Pre-alert system based on forecasts has a lead time of 10 days, while the alert based on hydrological observation and hydraulic forecasts has a lead time between 6 to 28 hours. according to the distance from the upstream gauging station.
	Multiple dissemination channels	Exposed communities must be warned via different media according to their ability/possibility of using them.	Oktari et al. [65]	A communication system in place to reach municipalities in multiple ways. At municipal level, Mayors can deploy multiple communication channels (FM Radio, smartphone texts and voice call to village chiefs and SCAP-RU, public criers). At a local level, SCAP-RUs are empowered with colored hydrological staffs and smartphones (WhatsApp) to alert SCAP-RUs of downstream villages and OSV
	End user's dissemination and communication needs	Message content, communication and dissemination means are tailored on end user needs, ensuring higher warning understanding.	Cumiskey et al. [69] Koks et al. [67] Fakhruddin et al. [61] Rahman et al. [64]	Alerts rely on four color-coded classes, associated with scenarios, assets and actions. Each SCAP-RU has knowledge of scenarios associated with alert colors, flooding areas maps, assets at risk and priority actions to be taken.
Response capability	Information on impactsand advices	Messages must contain information on the expected impacts and the advices on how to implement risk mitigation actions.	Shah et al. [70] Basher [3]	Each SCAP-RU has knowledge of color-coded class scenarios. In each village, information panels with infographics are displayed to imply priority actions to be taken. Roving Seminars on flood risk to raise awareness and build trust and understanding
	Community and volunteers education	Volunteers must ensure the coordination at local level, helping communities to effectively respond to alerts.	Cumiskey et al. [69] Scolobig et al. [66]	SCAP-RU are charged with activity coordination at local level (villages): local observations and exchange with other villages, alert dissemination within the village, urgent actions by community
	Preparedness and contingency plans	Hazard and vulnerability maps are used as a management tool for improving response and coordinating emergencies.	Cools et al. [71] Scolobig et al. [66]	Municipalities are empowered with flooding area maps at different scales, the list of assets at risk for each village and flood scenarios and flood risk reduction plans for the main villages in the territory [46].

Table A1. *Cont.*

EWS Component	Indicator	Description	References	SLAPIS
Cross-cuttingcomponent	Local community participation	End users can actively contribute to all four EWS components	Cools et al. [71] Fakhruddin et al. [61] Rahman et al. [64] Basher [3]	Communities are key actors in: • Participatory local risk assessment [46] • Discharge monitoring at local level (SCAP-RU) • Communicating the observed levels of vigilance (SCAP-RU) • Defining the contingency actions of risk reduction [46] GUI of the SLAPIS are designed to integrate i) top-down and bottom-up approaches

References

1. UNISDR. *Terminology on Disaster Risk Reduction*; UNISDR: Geneva, Switzerland, 2009.
2. Street, R.B.; Buontempo, C.; Mysiak, J.; Karali, E.; Pulquério, M.; Murray, V.; Swart, R. How could climate services support disaster risk reduction in the 21st century. *Int. J. Disaster Risk Reduct.* **2019**, *34*, 28–33. [CrossRef]
3. Basher, R. Global early warning systems for natural hazards: Systematic and people-centred. *Philos. Trans. A Math. Phys. Eng. Sci.* **2006**, *364*, 2167–2182. [CrossRef] [PubMed]
4. Garcia, C.; Fearnley, C.J. Evaluating critical links in early warning systems for natural hazards. *Environ. Hazards* **2012**, *11*, 123–137. [CrossRef]
5. Kelman, I.; Glantz, M.H. Early Warning Systems Defined. In *Reducing Disaster: Early Warning Systems for Climate Change*; Zommers, Z., Singh, A., Eds.; Springer: Berlin/Heidelberg, Germany, 2014; pp. 89–108.
6. Marchezini, V.; Horita, F.E.A.; Matsuo, P.M.; Trajber, R.; Trejo-Rangel, M.A.; Olivato, D. A review of studies on Participatory Early Warning Systems (P-EWS): Pathways to support citizen science initiatives. *Front. Earth Sci.* **2018**, *6*, 184. [CrossRef]
7. Baudoin, M.A.; Henly-Shepard, S.; Fernando, N.; Sitati, A.; Zommerset, Z. From top-down to "community-centric" approaches to early warning systems: Exploring pathways to improve disaster risk reduction through community participation. *Int. J. Disaster Risk Sci.* **2016**, *7*, 163–174. [CrossRef]
8. Rahman, M.; Gurung, G.B.; Ghimire, G.P. *Trans-border Flood Early Warning System in South Asia: Practices, Challenges and Prospect*; Practical Action: Kathmandu, Nepal, 2018.
9. NOAA. *Flash Flood Early Warning System Reference Guide*; University Corporation for Atmospheric Research: Silver Spring, MD, USA, 2010.
10. Smith, P.J.; Brown, S.; Dugar, S. Community-based early warning systems for flood risk mitigation in Nepal. *Nat. Hazards Earth Syst. Sci.* **2017**, *17*, 423–437. [CrossRef]
11. Brown, S.; Cornforth, R.; Boyd, E.; Standley, S.; Allen, M.; Clement, K.; Gonzalo, A.; Erwin, G.; Michelle, S.; Haseeb, I. *Science for Humanitarian Emergencies and Resilience (SHEAR) Scoping Study: Annex 3-Early Warning System and Risk Assessment Case Studies*; Evidence on Demand: UK, 2014. [CrossRef]
12. Gautam, D.K.; Phaiju, A.G. Community based approach to flood early warning in West Rapti River Basin of Nepal. *J. Integr. Disaster Risk Manag.* **2013**, *3*, 155–169. [CrossRef]
13. Practical Action and Mercy Corps. *Community Based Early Warning Systems in South and South East Asia*; Practical Action and Mercy Corps: Rugby, UK, 2012. Available online: http://flagship4.nrrc.org.np/sites/default/files/documents/best-practice-learning-in-community-based-EWS.pdf (accessed on 18 November 2019).
14. Shukla, Y.; Mall, B. Enhancing Frontline Resilience: Transborder Community-Based Flood Early Warning System in India and Nepal. In *Technologies for Development, Proceedings of the UNESCO Chair Conference on Technologies for Development, Lausanne, Switzerland, 4–6 May 2016*; Hostettler, S., Najih Besson, S., Bolay, J.C., Eds.; Springer: Berlin/Heidelberg, Germany, 2018; pp. 201–212.
15. IFRC. *Community Early Warning Systems: Guiding Principles*; International Federation of Red Cross and Red Crescent Societies: Geneva, Switzerland, 2012. Available online: http://www.ifrc.org/PageFiles/103323/1227800-IFRC-CEWS-Guiding-Principles-EN.pdf (accessed on 18 November 2019).
16. ISET International; ISET-Nepal; Practical Action Nepal. *Zurich Risk Nexus: Urgent Case for Recovery: What We Can Learn from the August 2014 Karnali River Floods in Nepal*; Zurich Insurance Company Ltd.: Zurich, Switzerland, 2015. Available online: http://reliefweb.int/sites/reliefweb.int/files/resources/risk-nexus-karnali-river-floods-nepal-july-2015.pdf (accessed on 10 January 2020).
17. WMO. *WMO Guidelines on Multi-Hazard Impact-Based Forecast and Warning Services*; World Meteorological Organization: Geneva, Switzerland, 2015. Available online: http://library.wmo.int/pmb_ged/wmo_1150_en.pdf (accessed on 20 November 2019).
18. Nkiaka, E.; Taylor, A.; Dougill, A.J.; Antwi-Agyei, P.; Fournier, N.; Bosire, E.N.; Konte, O.; Lawal, K.A.; Mutai, B.; Mwangi, E.; et al. Identifying user needs for weather and climate services to enhance resilience to climate shocks in sub-Saharan Africa. *Environ. Res. Lett.* **2019**, *14*, 123003. [CrossRef]
19. Sai, F.; Cumiskey, L.; Weerts, A.; Bhattacharya, B.; Haque Khan, R. Towards impact-based flood forecasting and warning in Bangladesh: A case study at the local level in Sirajganj district. *Nat. Hazards Earth Syst. Sci. Discuss.* **2018**, *2018*, 1–20. [CrossRef]

20. Genesio, L.; Bacci, M.; Baron, C.; Diarra, B.; Di Vecchia, A.; Alhassane, A.; Hassane, I.; Ndiaye, M.; Philippon, N.; Tarchiani, V.; et al. Early warning systems for food security in West Africa: Evolution, achievements and challenges. *Atmo. Sci. Lett.* **2011**, *12*, 142–148. [CrossRef]
21. Alfieri, L.; Burek, P.; Dutra, E.; Krzeminski, B.; Muraro, D.; Thielen, J.; Pappenberger, F. GloFAS—Global ensemble streamflow forecasting and flood forecasting. *Hydrol. Earth Syst. Sci.* **2013**, *17*, 1161–1175. [CrossRef]
22. Bartholmes, J.C.; Thielen, J.; Ramos, M.H.; Gentilini, S. The European flood alert system EFAS—Part 2: Statistical skill assessment of probabilistic and deterministic operational forecasts. *Hydrol. Earth Syst. Sci.* **2009**, *13*, 141–153. [CrossRef]
23. Andersson, J.C.M.; Arheimer, B.; Traoré, F.; Gustafsson, D.; Ali, A. Process refinements improve a hydrological model concept applied to the Niger River basin. *Hydrol. Process.* **2017**, *31*, 1–15. [CrossRef]
24. Tiepolo, M.; Tarchiani, V. *Risque Et Adaptation Climatique Dans La Région Tillabéri, Niger*; L'Harmattan: Paris, France, 2016; ISBN 978-2-343-08493-0.
25. Bigi, V.; Pezzoli, A.; Rosso, M. Past and future precipitation trend analysis for the city of Niamey (Niger): An overview. *Climate* **2018**, *6*, 73. [CrossRef]
26. Tamagnone, P.; Massazza, G.; Pezzoli, A.; Rosso, M. Hydrology of the Sirba river: Updating and analysis of discharge time series. *Water* **2019**, *11*, 156. [CrossRef]
27. Massazza, G.; Tamagnone, P.; Wilcox, C.; Belcore, E.; Pezzoli, A.; Vischel, T.; Panthou, G.; Ibrahim, M.H.; Tiepolo, M.; Tarchiani, V.; et al. Flood hazard scenarios of the Sirba River (Niger): Evaluation of the hazard thresholds and flooding areas. *Water* **2019**, *11*, 1018. [CrossRef]
28. Tarhule, A. Damaging rainfall and flooding: The other Sahel hazards. *Clim. Chang.* **2005**, *72*, 355–377. [CrossRef]
29. Nka, B.N.; Oudin, L.; Karambiri, H.; Paturel, J.E.; Ribstein, P. Trends in floods in West Africa: Analysis based on 11 catchments in the region. *Hydrol. Earth Syst. Sci.* **2015**, *19*, 4707–4719. [CrossRef]
30. Amogu, O.; Descroix, L.; Yéro, K.S.; Le Breton, E.; Mamadou, I.; Ali, A.; Vischel, T.; Bader, J.-C.; Moussa, I.B.; Gautier, E.; et al. Increasing river flows in the Sahel? *Water* **2010**, *2*, 170–199. [CrossRef]
31. Descroix, L.; Genthon, P.; Amogu, O.; Rajot, J.L.; Sighomnou, D.; Vauclin, M. Change in Sahelian rivers hydrograph: The case of recent red floods of the Niger River in the Niamey region. *Glob. Planet. Chang.* **2012**, *98*, 18–30. [CrossRef]
32. Aich, V.; Liersch, S.; Vetter, T.; Andersson, J.C.M.; Müller, E.N.; Hattermann, F.F. Climate or land use?—Attribution of changes in river flooding in the Sahel zone. *Water Switz.* **2015**, *7*, 2796–2820. [CrossRef]
33. Descroix, L.; Guichard, F.; Grippa, M.; Lambert, L.; Panthou, G.; Mahé, G.; Gal, L.; Dardel, C.; Quantin, G.; Kergoat, L.; et al. Evolution of surface hydrology in the Sahelo-Sudanian Strip: An updated review. *Water* **2018**, *10*, 748. [CrossRef]
34. Descroix, L.; Moussa, I.B.; Genthon, P.; Sighomnou, D.; Mahé, G.; Mamadou, I.; Vandervaere, J.P.; Gautier, E.; Maiga, O.F.; Rajot, J.L.; et al. Impact of Drought and Land—Use Changes on Surface—Water Quality and Quantity: The Sahelian Paradox. In *Current Perspectives in Contaminant Hydrology and Water Resources Sustainability*; Bradley, P.M., Ed.; Intech: Rijeka, Croatia, 2013; pp. 243–271.
35. Mahe, G.; Lienou, G.; Descroix, L.; Bamba, F.; Paturel, J.E.; Laraque, A.; Meddi, M.; Habaieb, H.; Adeaga, O.; Dieulin, C.; et al. The rivers of Africa: Witness of climate change and human impact on the environment. *Hydrol. Process.* **2013**, *27*, 2105–2114. [CrossRef]
36. Descroix, L.; Mahé, G.; Lebel, T.; Favreau, G.; Galle, S.; Gautier, E.; Olivry, J.C.; Albergel, J.; Amogu, O.; Cappelaere, B.; et al. Spatio-temporal variability of hydrological regimes around the boundaries between Sahelian and Sudanian areas of West Africa: A synthesis. *J. Hydrol.* **2009**, *375*, 90–102. [CrossRef]
37. Descroix, L. *Processus Et Enjeux D'eau En Afrique De l'Ouest Soudano-Sahélienne*; Editions des archives contemporaines: Paris, France, 2018. [CrossRef]
38. Panthou, G.; Vischel, T.; Lebel, T. Recent trends in the regime of extreme rainfall in the Central Sahel. *Int. J. Climatol.* **2014**, *34*, 3998–4006. [CrossRef]
39. Mahe, G.; Paturel, J.E.; Servat, E.; Conway, D.; Dezetter, A. The impact of land use change on soil water holding capacity and river flow modelling in the Nakambe River, Burkina-Faso. *J. Hydrol.* **2005**, *300*, 33–43. [CrossRef]
40. Herrmann, S.M.; Anyamba, A.; Tucker, C.J. Recent trends in vegetation dynamics in the African Sahel and their relationship to climate. *Glob. Environ. Chang.* **2005**, *15*, 394–404. [CrossRef]

41. Fiorillo, E.; Crisci, A.; Issa, H.; Maracchi, G.; Morabito, M.; Tarchiani, V. Recent changes of floods and related impacts in Niger based on the ANADIA Niger Flood Database. *Climate* **2018**, *6*, 59. [CrossRef]
42. Tiepolo, M.; Bacci, M.; Braccio, S. Multihazard Risk assessment for planning with climate in the Dosso Region, Niger. *Climate* **2018**, *6*, 67. [CrossRef]
43. Tiepolo, M.; Braccio, S. Flood Risk Assesment at Municipal Level in the Tillabéri Region, Niger. In *Planning to Cope with Tropical and Subtropical Climate Change*; Tiepolo, M., Ponte, E., Cristofori, E., Eds.; De Gruyter: Berlin, Germany, 2016; pp. 221–242.
44. Tiepolo, M.; Rosso, M.; Massazza, G.; Belcore, E.; Issa, S.; Braccio, S. Flood Assessment for Risk-Informed Planning along the Sirba River, Niger. *Sustainability* **2019**, *11*, 4003. [CrossRef]
45. IPCC. Annex II: Glossary. In *Contribution of Working Groups I, II and III to the Fifth Assessment Report of the Intergovernmental Panel on Climate Change*; Climate Change 2014: Synthesis Report; IPCC: Geneva, Switzerland, 2014; pp. 117–130.
46. Tiepolo, M.; Braccio, S. Local and Scientific Knowledge Integration for Multi-Risk Assessment in Rural Niger. In *Renewing Local Planning to Face Climate Change in the Tropics*; Tiepolo, M., Pezzoli, A., Tarchiani, V., Eds.; Springer Open: Cham, Switzerland, 2017; pp. 227–245.
47. HEC-RAS Home Page. Available online: http://www.hec.usace.army.mil/software/hec-ras (accessed on 14 February 2020).
48. Wilcox, C.; Vischel, T.; Panthou, G.; Bodian, A.; Blanchet, J.; Descroix, L.; Quantin, G.; Cassé, C.; Tanimoun, B.; Kone, S. Trends in hydrological extremes in the Senegal and Niger Rivers. *J. Hydrol.* **2018**, *566*, 531–545. [CrossRef]
49. Yang, J.; Zhang, H.; Ren, C.; Nan, Z.; Wei, X.; Li, C. A Cross-Reconstruction Method for Step Changed Runoff Series to Implement Frequency Analysis under Changing Environment. *Int. J. Environ. Res. Public Health* **2019**, *16*, 4345. [CrossRef] [PubMed]
50. Niger Basin Authority. Available online: http://www.abn.ne/ (accessed on 4 March 2019).
51. Passerotti, G.; Massazza, G.; Pezzoli, A.; Bigi, V.; Zsoter, E.; Rosso, M. Hydrological model application in the Sirba River: Early Warning System and GloFAS improvements. *Water* **2020**, *12*, 620. [CrossRef]
52. Open Geospatial Consortium. Available online: https://www.opengeospatial.org/standards (accessed on 10 January 2020).
53. CKAN Association. Available online: https://ckan.org/about/association/ (accessed on 10 January 2020).
54. *République Du Niger Décret Du Ministère De l'Intérieur, De La Sécurité Publique, De La Décentralisation Et Des Affaires Coutumières Et Religieuses, Définissant Le Code D'alerte National*; République du Niger: Niamey, Niger, 2019.
55. Budimir, M.; Brown, S.; Dugar, S. Communicating Risk Information and Early Warnings: Bridging the Gap between Science and Practice. In *Disaster Risk Reduction: A road of opportunities*; United Nations Major Group for Children and Youth: New York, NY, USA, 2017; pp. 13–18. Available online: https://www.preventionweb.net/go/53923 (accessed on 17 November 2019).
56. Moser, S.C. Communicating climate change: History, challenges, processes and future directions. *WIREs Clim. Chang.* **2010**, *1*, 31–53. [CrossRef]
57. Budimir, M.; Donovan, A.; Brown, S.; Shakya, P.; Gautam, D.; Uprety, M.; Cranston, M.; Sneddon, A.; Smith, P.; Dugar, S. Communicating complex forecasts for enhanced early warning in Nepal. *Geosci. Commun. Discuss.* **2019**, *2019*, 1–32.
58. Mercy Corps and Practical Action. *Establishing Community Based Early Warning System: Practitioner's Handbook*; Mercy Corps and Practical Action: Rugby, UK, 2010. Available online: http://www.preventionweb.net/educational/view/19893 (accessed on 18 November 2019).
59. Girons Lopez, M.; Di Baldassarre, G.; Seibert, J. Impact of social preparedness on flood early warning systems. *Water Resour. Res.* **2017**, *53*, 522–534. [CrossRef]
60. Belcore, E.; Piras, M.; Pezzoli, A.; Massazza, G.; Rosso, M. Raspberry PI 3 multispectral low-cost sensor for UAV based remote sensing. Case study in south-west Niger. *Int. Arch. Photogramm. Remote Sens. Spat. Inf. Sci.* **2019**, *42*, 207–214. [CrossRef]
61. Fakhruddin, S.; Kawasaki, A.; Babel, M.S. Community responses to flood early warning system: Case study in Kaijuri Union, Bangladesh. *Int. J. Disaster Risk Reduct.* **2015**, *14*, 323–331. [CrossRef]

62. Stigter, C.J. A Decade of Capacity Building Through Roving Seminars on Agro-Meteorology/-Climatology in Africa, Asia and Latin America: From Agrometeorological Services via Climate Change to Agroforestry and Other Climate-Smart Agricultural Practices. In *Implementing Climate Change Adaptation in Cities and Communities*; Leal Filho, W., Adamson, K., Dunk, R., Azeiteiro, U., Illingworth, S., Alves, F., Eds.; Springer: Berlin/Heidelberg, Germany, 2016. [CrossRef]
63. Stigter, C.J.; Tan, Y.; Das, H.P.; Zheng, D.; Rivero Vega, R.E.; Van Viet, N.; Bakheit, N.I.; Abdullahi, Y.M. Complying with Farmers' Conditions and Needs Using New Weather and Climate Information Approaches and Technologies. In *Managing Weather and Climate Risks in Agriculture*; Sivakumar, M.V.K., Motha, R., Eds.; Springer: Berlin/Heidelberg, Germany, 2007. [CrossRef]
64. Rahman, M.M.; Goel, N.K.; Arya, D.S. Study of early flood warning dissemination system in Bangladesh. *J. Flood Risk Manag.* **2012**, *6*, 290–301. [CrossRef]
65. Oktari, R.S.; Munadi, K.; Ridha, M. Effectiveness of dissemination and communication element of tsunami early warning system in Aceh. *Proc. Econ. Finan.* **2014**, *18*, 136–142. [CrossRef]
66. Scolobig, A.; De Marchi, B.; Borga, M. The missing link between flood risk awareness and preparedness. Findings from case studies in an Italian Alpine Region. *Nat. Hazards* **2012**, *63*, 499–520. [CrossRef]
67. Koks, A.A.; Jongman, B.; Husnt, T.G.; Botzen, W.J.W. Combining hazard, exposure and social vulnerability to provide lessons for flood risk management. *Environ. Sci. Policy* **2015**, *47*, 42–52. [CrossRef]
68. Bhuiyan, S.R.; Al Baky, A. Digital elevation based flood hazard and vulnerability study at various return periods in Sirajganj Sadar Upazila, Bangladesh. *Int. J. Disaster Risk. Reduct.* **2014**, *10*, 48–58. [CrossRef]
69. Cumiskey, L.; Werner, M.; Meijer, K.; Fakhruddin, S.H.M.; Hassan, A. Improving the social performance of flash flood early warnings using mobile services. *Int. J. Disaster Res. Built Environ.* **2015**, *6*, 57–72. [CrossRef]
70. Shah, M.A.R.; Douven, W.J.A.M.; Werner, M. Flood warning responses of farmer households: A case study in Uria Union in the Brahmaputra flood plain, Bangladesh. *J. Flood Risk Manag.* **2012**, *5*, 258–269. [CrossRef]
71. Cools, J.; Innocenti, D.; O'Brien, S. Lessons from flood early warning systems. *Environ. Sci. Policy* **2016**, *58*, 117–122. [CrossRef]

 *sustainability*

Article

Agrometeorological Forecast for Smallholder Farmers: A Powerful Tool for Weather-Informed Crops Management in the Sahel

Maurizio Bacci [1,2,*], Youchaou Ousman Baoua [3] and Vieri Tarchiani [1]

1 Institute of Bio Economy, National Research Council, 50145 Florence, Italy; vieri.tarchiani@ibe.cnr.it
2 Interuniversity Department of Regional and Urban Studies and Planning (DIST)-Politecnico and University of Turin, 10125 Turin, Italy
3 Direction de la Météorologie Nationale du Niger, Niamey BP 218, Niger; ousmanebaoua@yahoo.fr
* Correspondence: maurizio.bacci@cnr.it; Tel.: +39-055-303-3711

Received: 26 February 2020; Accepted: 12 April 2020; Published: 16 April 2020

Abstract: Agriculture production in Nigerien rural areas mainly depends on weather variability. Weather forecasts produced by national or international bodies have very limited dissemination in rural areas and even if broadcast by local radio, they remain generic and limited to short-term information. According to several experiences in West Africa, weather and climate services (WCSs) have great potential to support farmers' decision making. The challenge is to reach local communities with tailored information about the future weather to support strategic and tactical crop management decisions. WCSs, in West Africa, are mainly based on short-range weather forecasts and seasonal climate forecasts, while medium-range weather forecasts, even if potentially very useful for crop management, are rarely produced. This paper presents the results of a pilot initiative in Niger to reach farming communities with 10-day forecasts from the National Oceanic and Atmospheric Administration—Global Forecast System (NOAA-GFS) produced by the National Centers for Environmental Prediction (NCEP). After the implementation of the download and treatment chain, the Niger National Meteorological Directorate can provide 10-day agrometeorological forecasts to the agricultural extension services in eight rural municipalities. Exploiting the users' evaluation of the forecasts, an analysis of usability and overall performance of the service is described. The results demonstrate that, even in rural and remote areas, agrometeorological forecasts are valued as powerful and useful information for decision-making processes. The service can be implemented at low cost with effective technologies making it affordable and sustainable even in developing countries. Nonetheless, the service's effectiveness depends on several aspects mainly related to the way information is communicated to the public.

Keywords: climate services; local drought risk reduction; smallholder farmers; agrometeorological forecast; Niger

1. Introduction

In West Africa, climate change and variability are a major concern for sustainable development affecting agriculture and other key sectors with direct implications on the health and food security of rural populations. In the Sahel, major drought events caused famine in 1973–1974 and 1984 [1] and, since then, rainfall variability has been a key stressor in agricultural activities and is projected to intensify with future climate change [2–4]. Sahelian countries, such as Niger, where most of the population lives in rural areas and their main livelihood is rainfed agriculture, are particularly vulnerable to climate risks. In Niger, 84% of the population live in rural areas [5], and 89% of them are considered multidimensionally poor [6], experiencing "multiple deprivations at the household and individual level in health, education and standard of living" following the definition adopted by the UNDP for the multidimensional index of poverty [7]. In Niger, the combination of rainfall variability and poverty dooms rural communities to a chronic vulnerability to food insecurity [8,9].

For two decades, climate-smart agriculture (CSA) has been included in the development agenda with the aim of sustainably increasing agricultural productivity and building resilience of agricultural systems to climate change [10,11]. Among CSA best practices, weather and climate services (WCSs) have been acknowledged as a key opportunity to reduce farmers' vulnerability in Africa [10]. Many authors [12–17] provide evidence that agrometeorological information tailored to farmers can improve agricultural productivity and increase their income thereby reducing the impacts of climate change and minimizing the risk of food insecurity. However, at the farm level, agrometeorological services are often not really relevant, adapted, and usable for decision making in crop management. Indeed, national institutions are not well-equipped to respond to users' needs in terms of both forecasting expertise and dissemination channels.

National meteorological services in West Africa usually concentrate their forecasting efforts on short term and seasonal lead times. Short-term weather forecasts, one up to three days, as usually provided by met services, are not sufficient to take some strategic decisions on crop management planning and risk reduction strategies, which require 10 days to seasonal lead times [10]. Ten-day forecasts are not usually produced, even if they are particularly relevant for farmers and the benefit of their application in crop management, in terms of impact, is demonstrated to be even greater than seasonal forecasts [18]. Moreover, 10-day forecasts can be used for integrating the information from seasonal forecasts with a major increase in their effective application for WCSs [19].

Ten-day forecasts are generally considered a difficult time range, since the lead time is long enough for uncertainty from the initial conditions to arise and bias the numerical simulations [20]. Meanwhile, the models do not have enough time to "warm up" and correctly reproduce the global atmospheric dynamics [21]. The skill of such forecasts is often considered too low by met services and they do not often produce them. The main concern is to deal with the intrinsic uncertainty of medium-range weather forecasts. To properly communicate uncertainty, it is recommended that quantitative information be encoded in a way that fosters accurate decoding and prevents deterministic misinterpretations of uncertainty and the level of detail is chosen according to the forecast's skill [22]. Therefore, recognized good practices recommend translating quantitative values into categories through a participatory approach after an initial stage of testing and performance evaluation in the specific context.

A review on perceptions and use of WCSs in agricultural decision making [23] highlights that farmers seek contextual and location-specific forecasts to aid decision making and "would like access to forecasts at the correct time to facilitate decisions (such as predicted harvest conditions) and predict impacts, so they can be mitigated before they become severe".

The time-gap between issuing of the forecast and its reception by farmers has long been a critical challenge for agrometeorological warning systems, particularly in the least-developed countries, such as Niger, where information dissemination networks are weak, distances are large, and the extension service has suffered a chronic lack of financial resources for many years. The wide diffusion of mobile phone technology throughout sub-Saharan Africa offers the potential to reduce the time-gap and costs associated with the dissemination of forecasts to farmers [24,25], changing the extension service paradigm. Indeed, the agriculture extension service has for a long time been characterized by, and often criticized because of, its linear knowledge transfer approach and its incapacity to use systematic approaches for service delivery [26]. According to this approach, the proximity of agriculture extension officers to farmers was considered the main factor influencing the adoption of climate-informed cropping practices [18]. Currently, the concept of proximity is changing from a geographical perception to a social one, mediated by social networks and the media. New Information and Communication Technologies (ICTs) are enhancing the interaction and information exchange within the agriculture innovation system and changing the relations between information providers, extension officers, and farmers [27]. Moreover, digital technology could allow communities to provide timely feedback on information received and its performance, thereby improving their engagement in the whole agrometeorological monitoring system.

During the last decade, several pilot initiatives, by national and international bodies, have been set up to provide WCSs to farmers in West Africa, mainly based on short-term weather forecasts or climate seasonal forecasts addressing the main hazards that affect agricultural activities such as a delay in the onset of the rainy season, dry spells, and an early end of the rainy season. Nevertheless, none of these initiatives addresses the issue of producing and disseminating medium-range weather forecasts and related agrometeorological information for smallholder farmers.

Given the hypothesis that 10-day agrometeorological forecasts can effectively enhance a WCS's capability to improve agricultural productivity and increase farmers' incomes, this study aimed to answer the following research questions:

1. How can a national met service produce and disseminate 10-day agrometeorological forecasts in remote rural areas in a timely and sustainable way?
2. Do 10-day agrometeorological forecasts meet the needs of farmers in supporting their decision-making process?
3. How can the encoding of agrometeorological information disseminated to farmers be improved?

In order to answer these questions, this paper describes the service developed within the ANADIA (adaptation to climate change and disaster risk reduction in agriculture for food security) project in Niger. The project, funded by the Italian Agency for Development Cooperation, provided technical support to the Niger National Meteorological Directorate (DMN) to improve its capacity to reduce climate risk at the local level. The project included eight municipalities in Tillaberi and Dosso regions. Within this initiative, a case study was set up aiming to (i) fill the gap in agrometeorological medium-range forecasting at DMN, (ii) improve the usability of such information by the local communities and (iii) tailor information dissemination at the municipality level. The results coming from this experience demonstrated that, even if some improvements are needed to make this process fully operational, the forecasts issued are globally appreciated by the local communities, who judged them useful for their decision-making processes.

The paper is organized in five sections. After this introduction, Section 2 describes the materials and methods and Section 3 presents the main results. In Section 4, we discuss the findings according to the three research questions and, lastly, in Section 5, some conclusions and indications on the way forward are provided.

2. Materials and Methods

2.1. The Approach

According to Vaughan and Dessai [28], WCSs are more effective if co-produced through a collaboration between forecasters and users. The approach we used was inspired by the theory of co-production described by Vincent et al. [29], which implies an adaptation of the workflow "moving from supply-driven to demand-driven models" and also adopting adaptive programming "to incorporate learning into activities in real time".

The aim was to produce a few comprehensible agrometeorological indexes useful for farmers' decision-making processes. The set of indexes was defined through participatory meetings with local extension services covering crop farming, livestock rearing, and the environment. The indexes identified through the participatory process are described in Table 1.

Table 1. Agrometeorological indexes issued from Global Forecast System (GFS) 10-day weather forecasts.

Index	Description	Values
Cumulative 10 days rainfall (mm);	Total amount of precipitation forecasted for the next 10 days expressed in mm	0–999
Number of rainy days;	The number of days, during the next 10 days, with a forecasted daily precipitation ≥ 1 mm	0–10
Number of rainy days above 20 mm;	The number of days, during the next 10 days, with a forecasted precipitation ≥ 20 mm	0–10
Maximum number of consecutive dry days;	The maximum number of consecutive days, during the next 10 days, with a forecasted precipitation < 1 mm	0–10
Number of dry periods of at least 5 consecutive dry days.	Number of periods, during the next 10 days, of at least 5 consecutive days with forecasted precipitation < 0.1 mm	0–2

Identified indexes were implemented and tested during a first cropping season and then discussed again with users adopting a design-based research method, iteratively improving the design through trials [30], repeated in the second year. The users selected for the test were the municipal extension officers, who can identify the advantages and gaps of the service.

The dissemination mechanism was jointly defined with users: every 10 days during the cropping season (from 1 May to 21 October), the information was sent by DMN as a plain text message by mobile phone (SMS or instant messaging application, i.e., WhatsApp, according to users' preferences). Dissemination at local level was then performed by local media and social networks.

In order to ensure the sustainability of the service, the entire process of issuing forecasts was transferred to the DMN. The procedure was conceived as simple as possible in order to be run with low computing power and also open and flexible to be customized and extended to other municipalities in Niger. The operating system, programming environments, and data used for implementing the procedure were all open and free of charge. Particular attention was given to building the capacities of local technicians in programming and meteorological data processing. Experts from DMN were trained in Bash programming language (syntax and semantic); machine, operating system, and software configuration; and gridded data processing. After the training, a manual describing all the steps was written as course material. In a second phase of the training in Niger, some days were devoted to assisting the DMN in the configuration of the machines in their labs and testing the procedure with the local internet connection. This training represents a cornerstone for the sustainability of the service.

2.2. Methods

Ten-day forecasts are available from different models and institutions worldwide, but, in this initiative, we chose to test the Global Forecast System (GFS) by the National Centers for Environmental Prediction (NCEP) of the National Oceanic and Atmospheric Administration (NOAA) forecasts because they are provided free of charge and in real time. The GFS is a weather forecast model made available at three resolutions—0.25, 0.5 and 1 degree—with 4 runs per day, at 00, 06, 12, and 18 (GMT), by internet and it is possible to configure a machine to automatically download just the parameters of interest, in this case, the accumulated precipitation (APCP) output. The GFS forecasts contain 325 fields per forecast time and each time step is about 350 MB, so the complete 10-day forecast bundle reaches several gigabytes of data to download. Given the limitation of the internet band in Niger, it is impossible to complete the download in a useful timeframe. We optimized the download to about 30 MB of data by adapting the script "get_gfs.pl" produced by Wesley Ebisuzaki and available on the Climate Prediction Center (CPC) web page [31].

The download system worked on Linux operating system with Perl programming language [32]; CURL [33], a command line tool and library for transferring data with URLs; and Wgrib [34], which is a program to process, inventory, and decode GRIB files. The machine configuration did not require high performance so it is also possible to use obsolete configurations to launch the procedure (we successfully tested the procedure with a Pentium machine). Once the rainfall forecast images at 0.25′ resolution from 0 to 240 h were downloaded, we used a script, developed in Climate Data Operator (CDO) [35] language, to process and calculate the different indexes required and, finally, to produce a text message for each pilot municipality. As the script was executed, it automatically extracted the 10-day forecast of the grid that contains the centroid of the municipality. The forecast products elaborated for each municipality were:

1. Cumulative 10 days rainfall (mm);
2. Number of rainy days;
3. Number of rainy days above 20 mm;
4. Maximum number of consecutive dry days;
5. Number of dry periods of at least 5 consecutive dry days.

As pilot initiative, there was no available information about the performance of the GFS 10-day forecasts in Niger. In the first stage, it was not possible to apply any bias correction to the forecast because the debiasing process demands an initial training period. For the same reason, we decided to disseminate the forecasts in uncategorized numeric values (numbers) because no historical series of forecasts was available for statistical categorization and also to reduce misleading communication. Indeed, users did not have any experience in the interpretation of categorized forecast.

Nevertheless, the perspective was to debias and categorize the forecasts and this research contributes with an ex-post analysis, using the comparison between predicted and observed values for each dekad and municipality. Two different approaches were used. The first was based on the definition of 5 classes, valid both for forecast and observation time series, then we assigned the class for each record in the time series. The values were defined based on the quantiles of the historical observations time series only, because historical forecasts time series were not available. The classes could be eventually defined with different values following the farmers' needs, however, the limit is that this approach did not ensure any bias correction. The classes are presented in Table 2.

Table 2. Categorized classification and label.

Class	Range	Label
1	0–2 mm	Dry
2	2–18 mm	Mild rainfall
3	18–39 mm	Humid
4	39–66 mm	Very humid
5	Above 66 mm	Extremely humid

The second classification used the values of observation and forecast for the 2019 campaign categorized in 5 classes (using the 20th, 40th, 60th, and 80th quantiles) each (using forecast quantiles for forecasts and observations quantiles for the observed values). This approach also allowed a very simple quantile—quantile bias correction of the forecast. The classes for 2019 are reported in Table 3.

Table 3. Categorized classification for forecasts and observations and labels.

Class	Quantile Range Forecasts	Quantile Range Observations	Label
1	0–3.1 mm	0–2 mm	Dry
2	3.1–9 mm	2–18 mm	Mild rainfall
3	9–21.7 mm	18–39 mm	Humid
4	21.7–50.6 mm	39–66 mm	Very humid
5	Above 50.6 mm	Above 66 mm	Extremely humid

Short text messages were chosen as output of the script because they can easily be spread via SMS or instant messaging applications (e.g., WhatsApp) according to user preference. The whole procedure lasted around 30 min with the available internet band in Niger. It was run by the DMN at the beginning of each 10 days (1st, 11th, and 21st days of the month) and, at the same time, users gave their feedback on the previous 10-day forecast by filling in a form. The form provided qualitative and quantitative evaluation of the previous 10-day forecast. Quantitative feedback consisted of a comparison of forecasted and observed rainfall (collected with a rain gauge of the national observation network that they manage) filling the form in Appendix A (Figure A1), qualitative feedback consisted of users' appreciation of each product, rating the performance from 1 (lowest) to 5 (highest), and other comments and suggestions, filling the form in Appendix B (Figure A2).

The test of the 10-day forecast started in May 2018 with an initial 5 municipalities in the Dosso region. During the first year, DMN regularly sent the forecast to local focal points, receiving their feedback during the whole cropping season. Then, in November 2018 during a participatory meeting with all local stakeholders, the result of the first test season was discussed. The participants greatly appreciated the service and based on the feedback and requests from other municipalities, the service was extended to another 3 in the Tillaberi region (Figure 1).

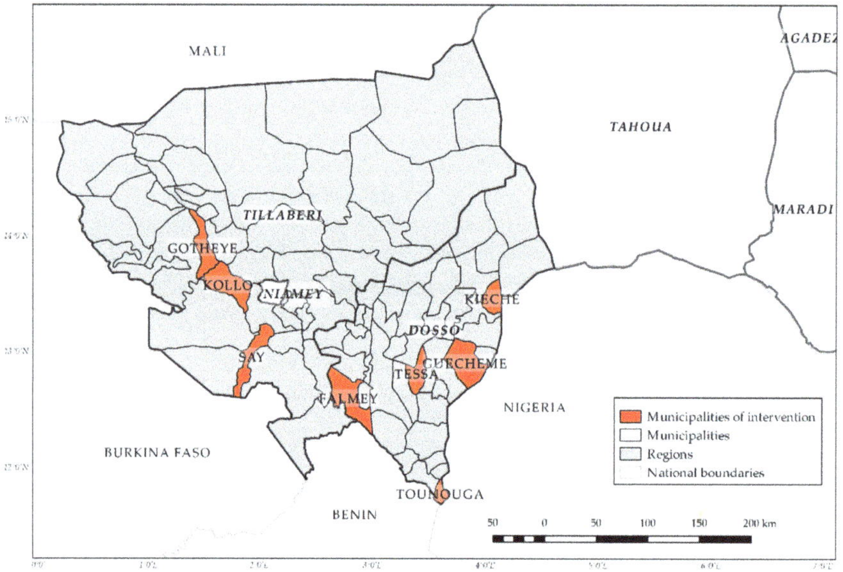

Figure 1. Niger and the eight municipalities served.

3. Results

3.1. Skill of Forecasted Indexes

During the two years, 210 10-day agrometeorological forecasts were issued, distributed as follows: 90 for the first year in five municipalities of the Dosso Region (18 dekads from 1 May to 21 October 2018) and 120 in the second year in eight municipalities of the Dosso and Tillaberi regions from 1 May to 21 October 2019. At the beginning of 2019, we had some technical problems in accessing input forecasts due to changes in the GFS data format and also some communication difficulties in the municipalities of Ouro Guéladio, Gotheye, Namaro, and Tounouga.

Analyzing the forecasts distributed during the 2019 rainy season in comparison with the observed data, an overall underestimation of precipitation was observed in almost all municipalities except for Ouro Guéladio (Table 4).

Table 4. Comparison of forecasted and observed annual cumulated rainfall per municipality in 2019.

Municipality	Cumulative Rainfall Forecast (mm)	Rainfall Recorded (mm)	Difference mm	Difference %	Number of Dekads
Falmey	552.4	662.8	−110.4	−17%	16
Gotheye	378.8	384.3	−5.5	−1%	15
Guéchémé	556.6	586.0	−29.4	−5%	16
Ouro Guéladio	402.6	392.4	10.2	3%	11
Kiéché	433.3	510.6	−77.2	−15%	16
Namaro	329.7	388.0	−58.3	−15%	15
Tessa	557.4	793.0	−235.6	−30%	16
Tounouga	498.9	688.5	−189.6	−28%	15

Even if in some municipalities (Ouro Guéladio, Gotheye) the difference in the total amount of precipitation is not significant, different skills of the forecasts were observed during the rainy season with an irregular distribution among municipalities (Figure 2).

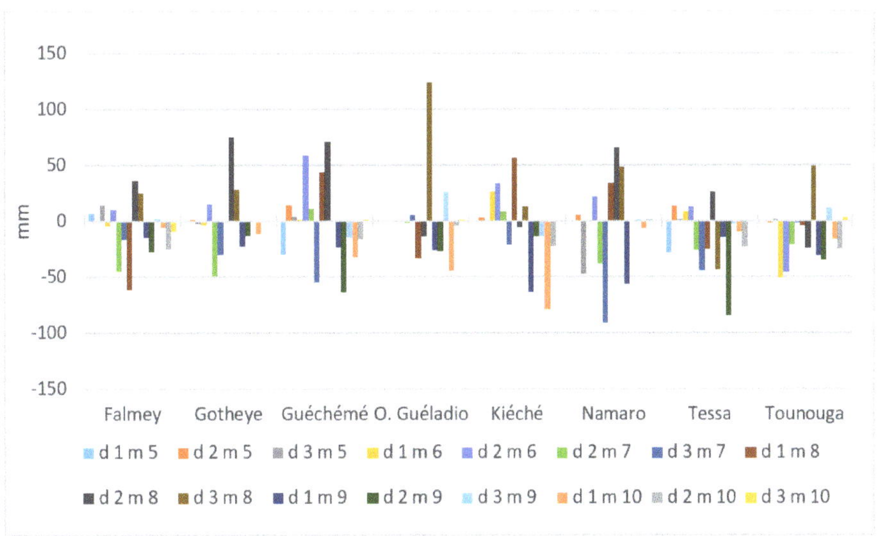

Figure 2. Difference between dekadal forecast and observed rainfall amount per municipality during 2019. season; negative values mean that forecasts underestimated the observed data. Abbreviations: d—dekad, m—month, O. Guéladio—Ouro Guéladio.

Despite this irregularity in the error distribution of rainfall amount, the forecasts were satisfactory on the most important aspect having a direct impact on farmer's choices, such as the indication if the next dekad will be wet (>1 mm) or dry (≤1 mm) (Table 5).

Table 5. Contingency table of wet and dry dekads (all municipalities in 2019).

	Forecast Dry	Forecast Wet
Observation dry	4	14
Observation wet	1	101

In general, forecasts were able to correctly predict the weather for the next dekad with a percentage of 87.5. This information is crucial for farmers because they can plan their agricultural activities or take measures against dry spells in the coming dekad (delay of sowing, use of fertilizers, etc.). In general, GFS forecasting skills were irregular among municipalities during the rainy season. The skill of other indexes is reported in Table 6.

Table 6. Index skill in 2019 (120 observations).

Index	Forecast	Observed	Difference	Diff. Avg. per Dekad
Number of rainy days	439	293	146	1.2
Number of rainy days above 20 mm	31	54	−23	−0.2
Number of the maximum consecutive dry days	512	665	−153	−1.3

Concerning dry days, there was a systematic underestimation of less than 2 dry days per dekad, which was probably due to an overestimation of the number of drizzly days. In fact, the forecast models in West Africa seemed to overestimate the number of days with just a few millimeters of rainfall [36].

Regarding intense phenomena, the capacity of the model to predict very heavy precipitation days, in which the rainfall amount was more than 20 mm, in the 10-day period is shown in Table 7.

Table 7. Contingency table of dekads with at least one rainy day above 20 mm (all municipalities in 2019).

	Forecast None	Forecast Event
Observation none	57	9
Observation event	32	22

In 2019, the model correctly predicted 65.8% of the occurrence of heavy precipitation in the next dekad.

3.2. Perception of the Performance of the Different Indexes

Although the accuracy of forecasts might not seem very satisfactory, the feedback collected from the 120 surveys distributed during the cropping season, about the perception of the different indexes' performance, showed a different and more positive picture. The answers to the following questions, rated from 1 = very weak to 5 = very high, are displayed in Table 8:

(1a) Rate the performance of the 10-day cumulative rainfall forecast;
(1b) Rate the performance of the number of rainy days forecast;
(1c) Rate the performance of the number of rainy days with more than 20 mm forecast;
(1d) Rate the performance of the maximum duration of dry spell forecast.

Table 8. Average rate of answers about index performance in 2019.

Municipality	Average 1a	Average 1b	Average 1c	Average 1d
Falmey	Medium	Medium	Medium	High
Gotheye	Medium	Medium	High	High
Guecheme	Weak	High	High	Medium
Gueladio	Weak	High	High	High
Kieche	Medium	High	High	High
Namaro	Weak	Medium	High	High
Tessa	Medium	High	Medium	High
Tounouga	Medium	Medium	High	Medium
Total average	Medium	Medium	High	High

The rating of each index was the result of a judgment process that is related to the user's expertise in the field, which depends on the agronomic characteristics of the municipality and the specific vulnerability to climate variability of each cropping stage. The overall appreciation of the performance of the indexes was medium or high. In particular, users rated best the capacity of the model to forecast dry spells and very heavy precipitation days. Despite the results shown in Tables 3 and 4, the end users appreciated the ability of the forecasts to predict extreme phenomena. This information, even if not accurate in terms of numerical values, was sufficient to take some actions (i.e., waiting until the next dekad to sow, harvesting in advance, distributing fertilizers, etc.). In general, the decision that farmers make with one or more episodes of heavy precipitation forecast in a dekad is the same, so it became more essential to know if a phenomenon was expected in the next 10 days instead of knowing exactly the number of episodes.

An important aspect is to understand how the end user rates the performance of each index. Comparing, for each dekad, the rating of the user and the difference recorded with the rain gauges, we could identify the threshold implicitly adopted by users to rate the performance. Table 9 shows the range of precipitation difference, in absolute values, that were recorded during the survey for the eight municipalities.

Table 9. Users' rate of the precipitation amount forecast (1a) in terms of range of differences (mm) between forecast and observed data in 2019 (8 municipalities).

Rate	Number of Ratings	Minimum Difference (Absolute Value)	Maximum Difference (Absolute Value)	Average Difference (Absolute Value)
Very weak	29	6	124.5	42.4
Weak	19	1.1	70.9	25.7
Medium	19	1.3	48.5	16.1
High	23	0.5	43.1	12.0
Very high	12	0	5.4	1.6

It might be astonishing to observe that 1.1 mm of rainfall amount difference in 10 days is set as a weak forecast performance or 43.1 mm of difference set as a high forecast performance. However, these thresholds have to be contextualized in the framework of crops and season development and the relative importance of the concomitance of a dry or very wet dekad, where such a difference can produce impacts on crops (or not). In general, users rate as a very high forecast performance up to 5 mm of difference and very weak when the difference is over 40 mm but the rating is highly dependent on the period of the rainy season.

Consequently, the further question we asked ourselves is whether the users had sufficient skill to correctly evaluate the performance of the indexes. Grouping the users' ratings by their score, we analyzed the dispersion plot of each pair, forecast and observation, for the five rating classes (Figure 3). The combined dispersion plot shows a general coherence by users in rating the forecast performance with just a few outliers. The best performance of the dispersion plot of the "very high" score (blue dots) and the diminishing quality of the following classes until the "very weak" score (red dots) that record the worst performance, demonstrate the overall capacity of users in their qualitative evaluation of the forecast performance through their own expertise.

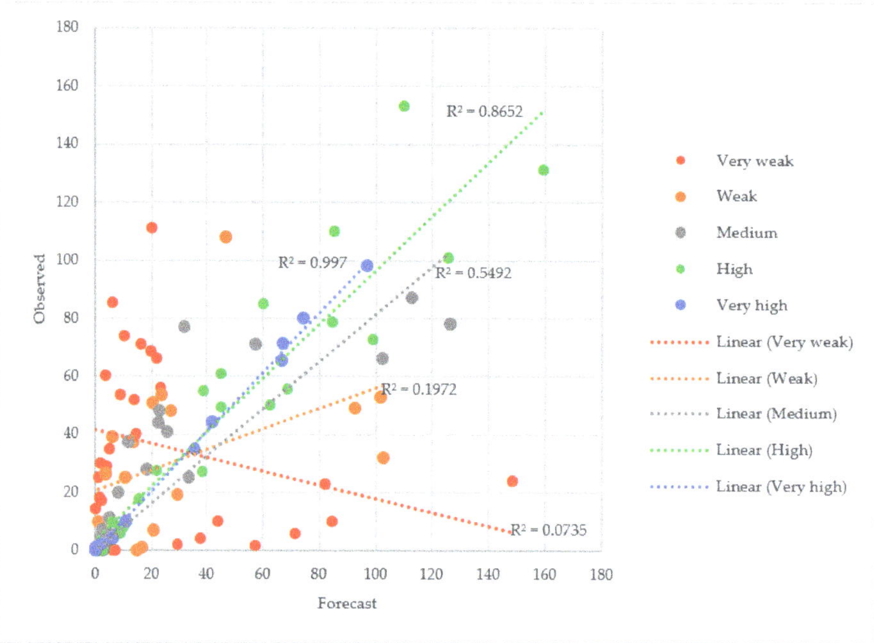

Figure 3. Dispersion plot of forecast vs. observed dekadal values grouped by user rating (very weak, weak, medium, high, and very high) with trends and correlation.

For the second point of the survey, "Do you find this information about the forecast indexes pertinent/useful?", in all (120) forms received the answer was a positive "yes". This is the most encouraging aspect of the initiative, which demonstrated that reaching the farmers with tailored information was very important despite the uncertainties of the global models. Farmers already use their traditional knowledge to predict the evolution of the season [37,38] and these forecasts were mainly integrated with their own knowledge to make their choices.

3.3. *Test of Forecasted Rainfall Amount Categorization*

At the end of the rainy season, an ex-post analysis was made using the comparison between the predicted and observed values of dekadal rainfall amounts. Two different approaches were tested: (i) five homogeneous classes for observations and forecast values and (ii) the definition of the five classes through the application of the quantile–quantile correction for the two series, introducing a very simple bias correction. The skill evaluation between the two classification methods was done through the difference (distance) in classes from the class of forecast vs. the class of observation. Figure 4 shows the results using the 120 dekads available for the eight municipalities in 2019. The application of the quantile—quantile classification method allowed a very good probability (83.3%) to be obtained that the forecasts were in the same class as the observation. Only in 2.5% of the dekads did the forecast differ by more than one class with respect to the observation. While with the fixed classification only 38.3% of the forecasts were in the same class as the observation.

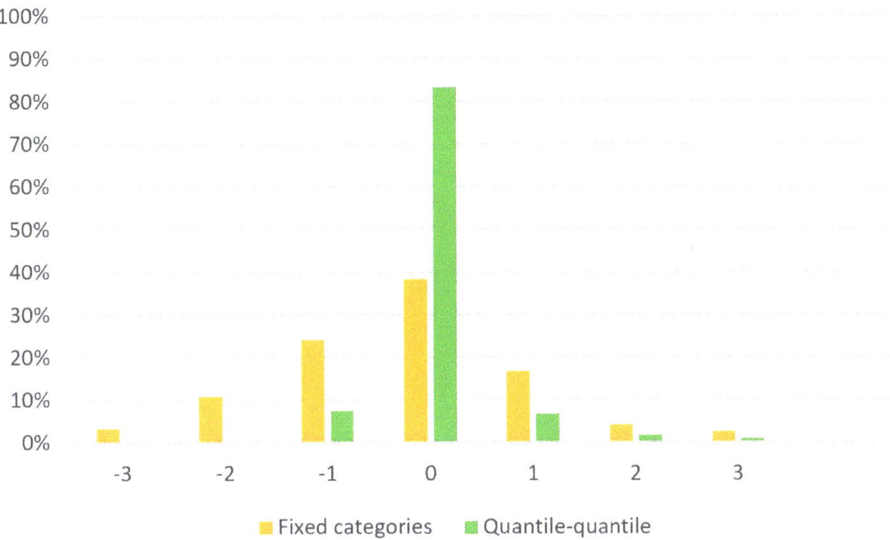

Figure 4. Histogram of the difference in performance between using the fixed categories or quantile–quantile classification. The x-axis values are the number of classes of difference (0 = same class; n = n categories of difference).

4. Discussion

The quantitative results of the performance analysis of the 10-day forecast indexes were not very different from the results already published in the literature on the skills of medium-range weather forecasts [39–42]. The hypothesis we made at the beginning was that in a context where any weather forecast with a lead time longer than one day exists, 10-day agrometeorological forecasts, even with low accuracy, can sustain local decision making if appropriately applied. Therefore, assuming that the skill of the forecasts is good enough to guide cropping practices, some answers to our research questions are discussed hereinafter.

4.1. How Can a National Met Service Produce and Disseminate 10-Day Agrometeorological Forecasts in Remote Rural Areas in a Timely and Sustainable Way?

The question contains two different aspects concerning the availability of dissemination technology and technical expertise. Both challenges were tackled adopting simple solutions, contributing to the sustainability of the process. Very simple dissemination technology could be used effectively. Even in remote rural areas, mobile phones are widely used and even smartphones are common. The dissemination mechanism worked quite well during the two years. We had some issues at the beginning of 2019 related to internet connectivity in a few municipalities. This showed that the mechanism is more sustainable if users work with their own smartphone rather than providing new ones for the purpose of dissemination. Prepaid internet credit can simplify and encourage users to be connected but it is not strictly necessary.

The choice of plain text messages simplified the process of communication from the central to the local level. Dissemination was then ensured by local media and social networks. A key role in agrometeorological information dissemination to the public was therefore played by rural radios, which are the main information media in Niger rural areas and identified by farmers as the preferred channel to access WCSs [43,44]. For this reason, radio stations were fully engaged in the dissemination process since the beginning of the initiative.

The agrometeorological indexes were actually disseminated at the beginning of each 10 days (1st, 11th, and 21st days of the month) while the 10-day forecasts were produced each day by GFS. It would be possible to produce the agrometeorological indexes daily and send them to farmers. However, sending indexes with a daily frequency to farmers would not be sustainable, and probably also not useful. The communication mechanism was designed according to the existing DMN chain of information dissemination, so that during the two years of experimentation, the forecasts were produced and disseminated with the decadal bulletins sent by DMN each 1st, 11th, and 21st days of the month. Nonetheless, the dissemination of updates with a shorter delay, i.e., 5 days, should be considered as a possible improvement in the whole process.

The training program and further knowledge transfer were demonstrated to be critical to successfully set up and maintain a semi-automatic procedure and replicate it in other areas. We found that new technical problems could arise daily, such as changes in the data distribution and format by the NOAA, which failed to undermine the sustainability of the system. However, strong collaboration and on-the-job training allowed the procedure to be modified and fixed in a reasonable time.

In the context of weak national institutions, another important aspect of sustainability is the cost to maintain the system. The use of open and freely available programming languages and simple procedures not demanding high calculating power ensured low cost, as well as the use of GFS products that were freely available. A final aspect is the replicability and sustainability of extension to other areas, even to the whole country. The entire procedure took 30 min to run even on obsolete Pentium hardware, but the time was not proportional to the number of localities because the bottleneck is internet access (data download from GFS), not the calculating time for processing the forecasts. The system is therefore easily extendable to the whole country without a significant increase in running time.

4.2. Do 10-Day Agrometeorological Forecasts Meet the Needs of Farmers in Supporting Their Decision-Making Process?

Users perceived the agrometeorological 10-day forecasts as important for their decision-making processes, even if they were not quantitatively accurate. Indeed, the feedback received from users underlined that they are able to manage forecast uncertainties in their daily activities. Although this should not be an excuse for not improving the forecast (the forecasting skills could be improved by using a regional model and intensifying the network of observation stations to allow a better initialization of the model), it seemed to us more profitable and pertinent to tailor existing information to specific end-users' needs.

The service was generally and unanimously perceived as positive and useful for local communities, even if some problems related to the forecast accuracy were real. The bidirectional exchange of information, strengthening collaboration between the national and local levels, was a positive outcome of the initiative. The participatory approach made local communities feel that they were engaged in a process and therefore contributed to building trust and credibility. Indeed, according to Wall [45], the process is often as important as the outcome.

According to the feedback from users, further development will concern new forecast information, such as strong winds and temperatures, which farmers perceive as useful for crop management.

4.3. How Can the Encoding of Agrometeorological Information Disseminated to Farmers Be Improved?

Forecast skill can be improved by using quantile–quantile categorized classification of the forecast dekadal rainfall amount. This approach allows misunderstanding due to the provision of quantitative values to be reduced and also to introduce a simple bias correction. It was not possible to adopt such an approach from the beginning because of the lack of forecasts time series. In the coming years, it will be possible to train the system to recalculate the centile every year in order to obtain more robust statistics for translating quantitative into qualitative information. However, the opportunity to translate quantitative information into qualitative (i.e., very wet, wet, almost dry, dry, etc.) must be verified with users because the language gap could become a real barrier. In many cases, the translation of the categories concept into local languages (Hausa and Djerma) could be misleading. Moreover, it is important to underline that the classified information does not have an absolute meaning for farmers because different crops and cropping systems have different needs in different periods that drive specific strategic choices. For this reason, further studies should explore the most effective way to communicate such forecasts maintaining a high level of usability.

5. Conclusions

The overall evaluation by users of the service provided was positive because, for the first time, they could access agrometeorological forecasts tailored to their own municipality. Users were really interested in accessing 10-day forecasts and were aware of their utility for crop management, particularly for extreme meteorological phenomena that could have a strong impact on their crops. Despite the uneven accuracy of the forecast, which is a common issue of 10-day forecasts, the main meteorological trends were detected and the information was perceived as useful for local decision-making processes. Indeed, farmers are used to dealing with weather uncertainties and they do not feel uncomfortable about the difference between forecast and observation. They particularly appreciated the forecast of very intense phenomena and dry spells during the next 10 days. Moreover, the local communities liked to be engaged in such a participatory process and they actively contributed to the progressive improvement of the service. This is particularly encouraging given that they were in charge of spreading the information to the wider farming community. The involvement of rural radio stations in the dissemination was shown to be really effective, thanks also to the strong collaboration between the DMN, extension officers, and journalists for the translation of information into local languages.

According to the feedback received from users, a further development of this WCS for smallholders will be to test the performance of temperature and strong wind forecasts during the 2020 rainy season. Moreover, the results showed that the adoption of categorized indexes could be very profitable for correcting biases. Nevertheless, before the setting up of an operational service, further investigation is needed to understand the most efficient way to communicate this information to smallholders in order to avoid misunderstandings.

Author Contributions: Conceptualization: M.B.; Methodology: M.B., Y.O.B., and V.T.; Developing: M.B. and Y.O.B.; Validation: M.B. and Y.O.B.; Investigation: M.B. and Y.O.B.; Writing—original draft preparation: M.B.; Writing—review and editing Y.O.B. and V.T.; Funding acquisition, V.T. All authors have read and agreed to the published version of the manuscript.

Funding: This research was co-funded by the Italian Agency for Development Cooperation, by the Institute of Bioeconomy (IBE)-National Research Council of Italy (CNR) at Florence, by the National Directorate for Meteorology of Niger (DMN), and by DIST-Politecnico and University of Turin within the project ANADIA2 [AID 010848].

Acknowledgments: The authors would like to thank the Italian Agency for Development Cooperation for supporting the ANADIA project and the actions that allowed the development of this activity. We are thankful to Gaptia Lawan Katiellou (National Directorate for Meteorology) for his support. We would like to express our deepest gratitude to Adamou Aïssatou Sitta, Binta Adamou, and Aissa Liman Diallo (National Directorate for Meteorology) for their help in the development of the activity and the personnel of the extension services of Dosso and Tillaberi for having shared data and information with us.

Conflicts of Interest: The authors declare no conflict of interest.

Appendix A

Fiche de comparaison pluviométrique des cumuls prévus et observés

Commune : ____________

Mois : _________________

Jour	Quantité de Pluie à la commune	Nombre de villages de la commune ayant enregistré de la Pluie
DECADE		
Cumul total décadaire de la commune		
Cumul décadaire prévu au niveau de la commune		

Figure A1. The form distributed in every municipality to collect the daily amount of precipitation and the number of villages receiving the rainfall.

Appendix B

FICHE D'EVALUATION DES INDICATEURS

COMMUNE : ______________________ Date de transmission de la fiche : __________

Mois : ______________________ Décade : __

AGENT : ______________________

1. **Comment a été la performance des indicateurs ci-dessous :**
   a. Quelle est la performance de la prévision du cumul de pluie dans les 10 jours ?

   ☐ Très élevé ☐ élevé ☐ moyen ☐ faible ☐ très faible

   b. Quelle est la performance de la prévision du nombre de jours pluvieux ?

   ☐ Très élevé ☐ élevé ☐ moyen ☐ faible ☐ très faible

   c. Quelle est la performance de la prévision du nombre de jours avec plus de 20mm de pluie ?

   ☐ Très élevé ☐ élevé ☐ moyen ☐ faible ☐ très faible

   d. Quelle est la performance de la prévision du maxi de jours consécutifs sans pluie ?

   ☐ Très élevé ☐ élevé ☐ moyen ☐ faible ☐ très faible

2. **Trouvez-vous utile/pertinente cette information sur les indicateurs ?**

   ☐ Oui ☐ Non

3. **Dans cette période de la saison agricole auriez-vous préféré recevoir d'autres indicateurs ?**

   ☐ Oui ☐ Non

   Si Oui, quel sont les indicateurs que vous désirez recevoir ?

   - ____

Figure A2. The form distributed about the perception of the performance of the different indexes of the forecast.

References

1. Grolle, J. Historical case studies of famines and migrations in the West African Sahel and their possible relevance now and in the future. *Popul. Environ.* **2015**, *37*, 181–206. [CrossRef]
2. Hulme, M. Climate perspectives in Sahelian desiccation. *Glob. Environ. Chang.* **2011**, *11*, 19–29. [CrossRef]
3. Shanahan, T.M. Atlantic forcing of persistent drought in West Africa. *Science* **2009**, *324*, 377–380. [CrossRef] [PubMed]
4. Brooks, N. *Drought in the African Sahel: Long Term Perspectives and Future Prospects*; Tyndall Center for Climate Change Research: Norwich, UK, 2004; Volume 61.
5. Institut National de la Statistique du Niger. *Recensement Général de la Population et de L'habitat 2012*; INS: Niamey, Niger, 2015. Available online: http://www.stat-niger.org/statistique/file/RGPH2012/ETAT_STRUCTURE_POPULATION.pdf (accessed on 22 January 2020).
6. Institut National de la Statistique du Niger. *Rapport National sur le Développement Humain (RNDH) Niger 2016, Développement Humain et Résilience des Ménages à L'insécurité Alimentaire au Niger*; INS and UNDP: Niamey, Niger, 2016. Available online: http://www.stat-niger.org/statistique/file/RNDH/RNDH_2016.pdf (accessed on 22 January 2020).

7. Alkire, S.; Santos, M.E. Measuring Acute Poverty in the Developing World: Robustness and Scope of the Multidimensional Poverty Index. *World Dev.* **2014**, *59*, 251–274. [CrossRef]
8. Zakari, S.; Ying, L.; Song, B. Factors influencing household food security in West Africa: The case of Southern Niger. *Sustainability* **2014**, *6*, 1191–1202. [CrossRef]
9. Gambo Boukary, A.; Diaw, A.; Wünscher, T. Factors Affecting Rural Households' Resilience to Food Insecurity in Niger. *Sustainability* **2016**, *8*, 181. [CrossRef]
10. FAO. *Climate Smart Agriculture: Policies, Practices and Financing for Food Security, Adaptation and Mitigation*; FAO: Rome, Italy, 2010. Available online: http://www.fao.org/3/i1881e/i1881e00.pdf (accessed on 24 March 2020).
11. FAO. *Regional Overview of Food Insecurity: African Food Insecurity Prospects Brighter Than Ever*; FAO: Accra, Ghana, 2015. Available online: http://www.fao.org/3/a-i4635e.pdf (accessed on 24 March 2020).
12. Vaughan, C.; Hansen, J.; Roudier, P.; Watkiss, P.; Carr, E. Evaluating agricultural weather and climate services in Africa: Evidence, methods, and a learning agenda. *WIREs Clim. Chang.* **2019**, *10*, 586. [CrossRef]
13. Lo, M.; Dieng, M. *Impact Assessment of Communicating Seasonal Climate Forecasts in Kaffrine, Diourbel, Thies and Fatick (Niakar) Regions in Senegal*; CCAFS: Copenhagen, Denmark, 2015; p. 70.
14. Anuga, S.W.; Gordon, C. Adoption of climate-smart weather practices among smallholder food crop farmers in the Techiman municipal: Implication for crop yield. *Res. J. Agric. Environ. Manag.* **2016**, *5*, 279–286.
15. Ouedraogo, I.; Diouf, N.S.; Ouédraogo, M.; Ndiaye, O.; Zougmoré, R. Closing the gap between climate information producers and users: Assessment of needs and uptake in Senegal. *Climate* **2018**, *6*, 13. [CrossRef]
16. Tarchiani, V.; Rossi, F.; Camacho, J.; Stefanski, R.; Mian, K.; Pokperlaar, D.; Coulibaly, H.; Sitta Adamou, A. Smallholder Farmers Facing Climate Change in West Africa: Decision-Making between Innovation and Tradition. *J. Innov. Econ. Manag.* **2017**, *24*, 151–176. [CrossRef]
17. Tarchiani, V.; Camacho, J.; Coulibaly, H.; Rossi, F.; Stefanski, R. Agrometeorological services for smallholder farmers in West Africa. *Adv. Sci. Res.* **2018**, *15*, 15. [CrossRef]
18. Roudier, P.; Alhassane, A.; Baron, C.; Louvet, S.; Sultan, B. Assessing the benefits of weather and seasonal forecasts to millet growers in Niger. *Agric. For. Meteorol.* **2016**, *223*, 168–180. [CrossRef]
19. Genesio, L.; Bacci, M.; Baron, C.; Diarra, B.; Di Vecchia, A.; Alhassane, A.; Hassane, I.; Ndiaye, M.; Philippon, N.; Tarchiani, V.; et al. Early warning systems for food security in West Africa: Evolution, achievements and challenges. *Atmos. Sci. Lett.* **2011**, *12*, 142–148. [CrossRef]
20. Lorenz, E.N. Deterministic nonperiodic flow. *Atmos. Sci.* **1963**, *20*, 131–140. [CrossRef]
21. Meehl, G.A.; Goddard, L.; Murphy, J.; Stouffer, R.J.; Boer, G.; Danabasoglu, G.; Dixon, K.; Giorgetta, M.A.; Greene, A.M.; Hawkins, E.D.; et al. Decadal prediction: Can it be skillful? *Bull. Am. Meteorol. Soc.* **2009**, *90*, 1467–1485. [CrossRef]
22. Fundel, V.J.; Fleischhut, N.; Herzog, S.M.; Göber, M.; Hagedorn, R. Promoting the use of probabilistic weather forecasts through a dialogue between scientists, developers and end-users. *Q. J. R. Meteorol. Soc.* **2019**, *145*, 210–231. [CrossRef]
23. Mase, A.S.; Prokopy, L.S. Unrealized Potential: A Review of Perceptions and Use of Weather and Climate Information in Agricultural Decision Making. *Weather Clim. Soc.* **2014**, *6*, 47–61. [CrossRef]
24. Baumüller, H. Agricultural Service Delivery through Mobile Phones: Local Innovation and Technological Opportunities in Kenya. In *Technological and Institutional Innovations for Marginalized Smallholders in Agricultural Development*; Gatzweiler, F., von Braun, J., Eds.; Springer: Cham, Switzerlad, 2016; pp. 143–162.
25. Etwire, P.M.; Buah, S.; Ouédraogo, M.; Zougmoré, R.; Partey, S.T.; Martey, E.; Dayamba, S.D.; Bayala, J. An assessment of mobile phone-based dissemination of weather and market information in the Upper West Region of Ghana. *Agric. Food Secur.* **2017**, *6*, 8. [CrossRef]
26. Leeuwis, C.; Aarts, N. Rethinking communication in innovation processes: Creating space for change in complex systems. *J. Agric. Educ. Ext.* **2011**, *17*, 21–36. [CrossRef]
27. Munthali, N.; Leeuwis, C.; van Paassen, A.; Lie Asare, R.R.; van Lammeren, R.; Schut, M. Innovation intermediation in a digital age: Comparing public and private new-ICT platforms for agricultural extension in Ghana. *NJAS Wagening. J. Life Sci.* **2018**, *86–87*, 64–76. [CrossRef]
28. Vaughan, C.; Dessai, S. Climate Services for Society: Origins, Institutional Arrangements, and Design Elements for an Evaluation Framework. *WIREs Clim. Chang.* **2014**, *5*, 587–603. [CrossRef]
29. Vincent, K.; Daly, M.; Scannell, C.; Leathes, B. What can Climate Services learn from theory and practice of co-production? *Clim. Serv.* **2018**, *12*, 48–58. [CrossRef]

30. Barab, S. Design-based research. In *The Cambridge Handbook of the Learning Sciences*; Sawyer, R.K., Ed.; Cambridge University Press: Cambridge, UK, 2006; pp. 153–170. [CrossRef]
31. Climate Prediction Center Fast Downloading of Grib. Available online: https://www.cpc.ncep.noaa.gov/products/wesley/get_gfs.html (accessed on 10 February 2020).
32. Perl. Available online: https://www.perl.org/ (accessed on 10 February 2020).
33. Curl. Available online: http://www.curl.com/ (accessed on 10 February 2020).
34. Wgrib. Available online: https://www.cpc.ncep.noaa.gov/products/wesley/wgrib.html (accessed on 10 February 2020).
35. Climate Data Operators (CDO). Available online: https://code.mpimet.mpg.de/projects/cdo (accessed on 10 February 2020).
36. Druyan, L.M.; Fulakeza, M.; Lonergan, P.; Worrell, R. Regional model nesting within GFS daily forecasts over West Africa. *Open Atmos. Sci. J.* **2010**, *4*, 1–11. [CrossRef]
37. Nyong, A.; Adesina, F.; Osman Elasha, B. The value of indigenous knowledge in climate change mitigation and adaptation strategies in the African Sahel. *Mitig. Adapt. Strateg. Glob. Chang.* **2007**, *12*, 787–797. [CrossRef]
38. Mertz, O.; Mbow, C.; Reenberg, A.; Diouf, A. Farmers' Perceptions of Climate Change and Agricultural Adaptation Strategies in Rural Sahel. *Environ. Manag.* **2009**, *43*, 804–816. [CrossRef]
39. Fan, Y.; van den Dool, H. Bias Correction and Forecast Skill of NCEP GFS Ensemble Week-1 and Week-2 Precipitation, 2-m Surface Air Temperature, and Soil Moisture Forecasts. *Weather Forecast.* **2011**, *26*, 355–370. [CrossRef]
40. Fanglin, Y. *Review of GFS Forecast Skills in 2017*; Global Modeling and Data Assimilation Branch, Environmental Modeling Center, National Centers for Environmental Prediction: Court College Park, MD, USA, 2017.
41. Novak, D.R.; Bailey, C.; Brill, K.F.; Burke, P.; Hogsett, W.A.; Rausch, R.; Schichtel, M. Precipitation and Temperature Forecast Performance at the Weather Prediction Center. *Weather Forecast.* **2014**, *29*, 489–504. [CrossRef]
42. Historical Performances of Global NWP Models. Available online: https://www.emc.ncep.noaa.gov/gmb/STATS_vsdb/longterm/ (accessed on 10 February 2020).
43. Oyekale, A.S. Access to risk mitigating weather forecasts and changes in farming operations in East and West Africa: Evidence from a baseline survey. *Sustainability* **2015**, *7*, 14599–14617. [CrossRef]
44. Tarhule, A.; Lamb, P.J. Climate research and seasonal forecasting for West Africans: Perceptions, dissemination, and use? *Bull. Am. Meteorol. Soc.* **2003**, *84*, 1741–1759. [CrossRef]
45. Wall, T.U.; Meadow, A.M.; Horganic, A. Developing evaluation indicators to improve the process of coproducing usable climate science. *Weather Clim. Soc.* **2017**, *9*, 95–107. [CrossRef]

Article

Theorising Indigenous Farmers' Utilisation of Climate Services: Lessons from the Oil-Rich Niger Delta

Eromose Ehije Ebhuoma [1,*], Mulala Danny Simatele [2], Llewellyn Leonard [1], Osadolor Obiahon Ebhuoma [3], Felix Kwabena Donkor [1] and Henry Bikwibili Tantoh [1]

1 Department of Environmental Sciences, College of Agriculture and Environmental Sciences, University of South Africa (UNISA), Johannesburg 1709, South Africa; Llewel@unisa.ac.za (L.L.); Felixdonkor2002@yahoo.co.uk (F.K.D.); bikwibilith@gmail.com (H.B.T.)

2 Global Change Institute (GCI), University of the Witwatersrand, Johannesburg 2050, South Africa; Mulala.Simatele@wits.ac.za

3 School of Agricultural, Earth and Environmental Sciences, University of KwaZulu-Natal, Westville Campus, Westville, Durban 4000, South Africa; osadolorebhuoma@gmail.com

* Correspondence: eromose2012@gmail.com; Tel.: +27-11-471-2346

Received: 22 July 2020; Accepted: 4 September 2020; Published: 8 September 2020

Abstract: In the wake of a rapidly changing climate, climate services have enabled farmers in developing countries to make informed decisions, necessary for efficient food production. Climate services denote the timely production, translation, delivery and use of climate information to enhance decision-making. However, studies have failed to analyse the extent to which Indigenous farmers residing and producing their food in an environment degraded by multinational corporations (MNCs) utilise climate services. This study addresses this gap by analysing Indigenous farmers' utilisation of climate services in Igbide, Olomoro and Uzere communities, in the oil-rich Niger Delta region of Nigeria. Focus group discussions and semi-structured interviews were used to obtain primary data. Findings suggest that although the activities of Shell British petroleum, a MNC, have compromised food production, other factors have fuelled farmers' unwillingness to utilise climate services. These include their inability to access assets that can significantly scale up food production and lack of weather stations close to their communities needed to generate downscaled forecasts, amongst others. This paper argues that failure to address these issues may stifle the chances of actualising the first and second sustainable development goals (no poverty and zero hunger) by 2030 in the aforementioned communities.

Keywords: climate services; indigenous farmers; multinational corporations; systems thinking; Nigeria; sub-Saharan Africa

1. Introduction

Evidence in the literature suggests that climate variability and change have significantly compromised effective food production in sub-Saharan Africa (SSA) [1–3]. The radical transformation of a region once able to ensure that farmers could produce their food efficiently, to one where food productivity and output can no longer be guaranteed, due to climate variability and change, is partly responsible for the increased food and nutrition insecurity in SSA [4–6]. This has set in motion numerous unwanted responses in farming communities. These include climate-induced migration [3,7], engaging in more off-farm activities [8] and heavy reliance on government's food aid stamps [1], among others. These unwanted responses will likely be worsened, in part, by the aggravated difficulties expected to be associated with future food production due to increased occurrences of climate variability and change. Climate projections for SSA indicate that increased and unprecedented occurrences of extreme weathers will become the new normal by 2050 [9].

The growing concerns around SSA farmers' increased susceptibility to current and future climate variability and change have catalysed investments in climate services [2]. Climate services refer to the 'timely production, translation, delivery and use of climate data and information to enhance decision-making' [10] (p. 1). A huge wave of optimism exists within the scientific community about the capability of climate services to facilitate informed decisions, necessary to enhance food production among farmers [11,12]. Several studies underline the significance of climate services in scaling-up food production in the face of increased climate variability [2,13–15]. Yet, to our knowledge, no study has analysed the extent to which Indigenous farmers residing in communities with a history of fierce contestations with multinational corporations (MNCs) utilise climate services. By Indigenous, we mean 'those communities that claim a historical continuity with their traditional territories' [16] (p. 290). Indigenous people possess a peculiar culture and knowledge distinct to their community that have been tried and tested with real-life scenarios [17–19]. It is usually passed on from one generation to the next through oral communication and repetitive engagement [18].

MNCs, in their quest to exploit rich natural resources, have aggressively degraded the immediate environment that Indigenous farmers rely on to produce their food [20,21]. As highlighted in the Niger Delta region of Nigeria [22,23], Katanga in the Democratic Republic of Congo [24], Western Ghana [25] and Limpopo, South Africa [26], the activities of MNCs have adversely compromised farm yields and productivity. It is, therefore, crucial to ascertain if these cohorts of farmers trust and utilise climate services disseminated by government institutions. As Dutfield [19] (p. 24) argues, the exploitation of natural resources occurs through an 'unholy alliance of corporations with governments.'

Failure to deeply analyse the extent to which farmers, residing in regions where the activities of MNCs have degraded their natural environment, utilise climate services may undermine the actualisation of both the first and second sustainable development goals (SDGs) (no poverty and zero hunger) by 2030. This is especially if such farmers fail to utilise climate services to make informed farming decisions. Arguably, providers of climate services may be unaware of several underlying issues in communities affected by natural resources exploitation that may have been lingering for decades, which limits farmers' ability to produce food efficiently e.g., see [27–29]. This, in turn, may fuel scepticism among farmers regarding their respective government's agenda to invest in climate services over addressing the underlying issues that have negatively impacted food production, and consequently, undermine the use of such services.

Against this background, this paper investigates Indigenous farmers' perception of climate services. Also, it unpacks how the negative impacts of MNC activities influence Indigenous farmers' use of climate services. Further, it demystifies the attributes of those receptive to climate services. This paper analyses primary data obtained from the oil-rich Niger Delta region of Nigeria. The Niger Delta serves as an important case study especially since the region has been besieged by land degradation, for decades, due to crude oil exploration and exploitation by MNCs [19,22,30], which, in turn, has severely compromised food production [4,31]. Also, 'since 1970, oil revenues from Niger Delta have contributed over $350 billion to Nigeria while the region remained one of the most impoverished parts of the country' [32] (p. 221). It is hoped that this paper will trigger discussions regarding factors that deter Indigenous farmers, whose livelihood activities and way of life have been adversely affected by the activities of MNCs, from utilising climate services. Such discussions are desperately needed to unravel effective interventions necessary to build the trust and confidence of Indigenous farmers in climate services.

This paper proceeds in five parts. The first provides a snapshot of existing literature regarding the role of climate services in upscaling food production. The second part succinctly illustrates how oil exploration activities by Shell British petroleum, a MNC, which results in oil spillages and gas flaring, have compromised food production and quality of crops harvested in the study areas, situated in the Niger Delta. It also highlights the frustrations of the people as they have not been beneficiaries of resources that can enable them produce food effectively as well as improve their quality of life. The third part provides a synopsis of the study areas and how primary data were obtained. The fourth part brings to the fore the results. It aptly underlines the various factors responsible for the non-utilisation of climate services by Indigenous farmers. It also showcases the attributes of those willing to rely on climate services, if the factors undermining its use are decisively addressed. The final part discusses the implication of the findings.

2. The Role of Climate Services in Scaling-Up Food Production

A climate service is a decision aide that supports ex-ante climate risk management in agriculture, health, water, transport and other vulnerable sectors of the global economy [33]. The endorsement of the global framework for climate services, whose mission is 'to strengthen the production, availability, delivery and application of science-based climate prediction and services,' by delegates of 155 countries at the 2009 World Climate Conference III buttresses this point [34]. In SSA, the erratic and unprecedented occurrences of extreme weather conditions, especially in the last two decades, have crystallised the need for farmers to rely on climate services to make informed farming decisions [2,5,14,15]. Also, the drive to utilise climate services has been fuelled by the increased unreliability of Indigenous meteorological forecasts in some locations. e.g., see [35,36]. As Ouédraogo et al. [15] argue, climate services can assist farmers in reducing uncertainty and taking advantage of favourable weather information by planting intensively, and thus, minimise loss during unfavourable weather forecasts. In the case of extreme weather predictions, several pieces of literature have shown that climate services empower local and Indigenous farmers by allowing them to take measures aimed at protecting lives, livelihoods and household properties [37–39]. Climate services, therefore, support climate-resilient development.

In a simulation exercise conducted in Senegal, West Africa, Roudier et al. [2] found that farmers who used climate forecasts maximised benefits from predicted favourable conditions. In Burkina Faso, 'farmers exposed to climate information changed their farm practices based on the information they received, and that translated into better management of inputs to increase their farm productivity and improve their resilience to climate variability' [40] (p. 4). Somewhat similar findings have emerged from studies conducted in the Republic of Benin [41], India [42] and Uganda [14] respectively. Arguably, just when the future of food production seemed bleak for smallholder farmers due to the rapidly changing climate coupled with the increasing unreliability of Indigenous meteorological forecasts in some locations, climate services have rekindled farmers' hope to obtaining a meaningful livelihood. To substantiate this point, farmers in some SSA countries are now willing to pay for climate services [15,36]. This reinforces the World Meteorological Organization's [43] standpoint that the benefits of investments in climate services for agriculture and food security greatly outweigh the cost.

It is, however, noteworthy that the studies cited above were conducted in areas without substantial abundance of natural resources to trigger the influx of MNCs into such locations. If the theorisation of message interpretation by Sellnow et al. [44] is anything to go by, the chances of Indigenous farmers—who have and continue to be victims of the negative effects of the activities of MNCs—relying on climate services to make an informed decision will be slim. The reason for this, according to Sellnow et al. [44], is that for every message received, the audience draws conclusion by interpreting the conveyed message using availability heuristic, among others. Availability heuristic refers to 'a rule of thumb that allows people to solve problems based on what they remember and how easily their memory is retrieved, and how readily available that memory is' [45] (p. 54). The ease of retrievability is closely linked to significant landmarks in a person's life and could create signposts in one's memory, thereby making an experience easy to retrieve [46]. Thus, availability heuristic, which influences trust and credibility [47,48], may negatively impact the adoption and utilisation of climate services by Indigenous farmers whose livelihoods have been compromised by the activities of MNCs.

3. The Delta State: Background Information

The Delta state comprises one of the nine states in Nigeria that make up the oil-rich Niger Delta region. Crude oil—the mainstay of the nation's foreign reserves and GDP—is mainly obtained from the Niger-Delta [23,27], a region which constitutes approximately 8% of the nation's total landmass

(Figure 1). Anecdotal evidence suggests that the second crude oil discovery happened in Uzere in 1958, after the first discovery was made in Oloibiri, Bayelsa state, in 1956 [49]. Igbide, Uzere and Olomoro communities have approximately 62 oil wells between them. The aforementioned communities, as well as the entire Niger Delta region, have contributed immensely to Nigeria's gross domestic product (GDP) due to no less than four decades of consistent oil exploration and exploitation activities, amounting to 90, 000 barrels of oil daily [50,51]. However, these communities have gained a global reputation for the negative impacts that crude oil exploration has had on the livelihoods and general welfare of its inhabitants [23]. Specifically, oil spillages and gas flaring, which have adversely compromised the health of local people, their ability to produce food effectively and the nutritional quality of crops harvested, have consistently made headlines. Consequently, this has metamorphosed a relatively tranquil region, in the 1960s, to one prone to conflicts with MNCs, since the 1990s.

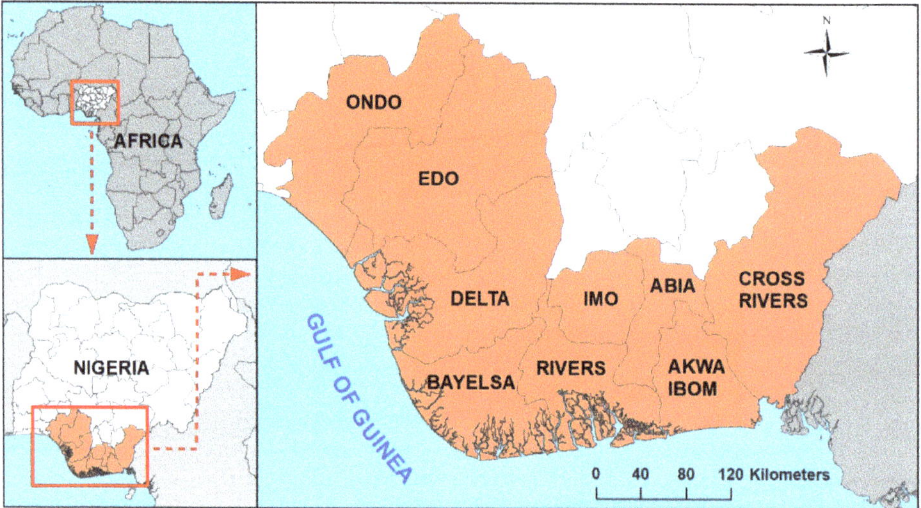

Figure 1. Map of the Niger-Delta region in Southern Nigeria.

No fewer than 30 protests and demonstrations against MNCs (some relatively peaceful and others extremely violent) as well as intermittent disruption of oil exploration in all three communities have been recorded since the 1990s [32,51,52]. According to the president-general of Isoko Development Monitoring Group (IDMG):

'Oil exploration has literally killed the fruitfulness of Isokoland. Our farmers are crying, lamenting their ordeal in the hands of the oil companies in their lands that have refused to pay them meaningful compensations' [53] (p. 1).

Compounding the woes of farmers is their inability to easily access fundamental assets that can significantly improve their welfare. This is a factor at play due to corrupt practices among government officials. Fundamental assets refer to the financial, natural, social, human and natural assets or capital (Table 1). As a result, these communities, including their community members, continue to live in abject poverty without basic infrastructures such as good roads, potable water and constant power supply [23]. Hence, these communities and the entire Niger Delta are classified as a paradox; oil-rich but impoverished [30].

Table 1. Definition of the fundamental bundle of assets or asset portfolio.

Asset or Capital	Definition
Physical	This includes equipment and infrastructures such as road networks and other productive resources owned by individuals, households, communities or the country itself.
Financial	This refers to financial resources available and easily accessible to individuals that include loans, access to credit and savings in a bank or any other financial institutions.
Human	This refers to the level of education, skills, health status and nutrition of individuals. Labour is closely associated with human capital investments. The health status of individuals impact either positively or negatively on their ability to work, while skills, including their local and indigenous knowledge systems (LIKS), and level of education are crucial because it influences the return individuals derive from labour.
Social	This refers to the norms, rules, obligations, mutuality and trust embedded in social relations, social structures and societies' institutional disposition.
Natural	This refers to the atmosphere, land, minerals, forests, water and wetlands. For the rural poor, land is an essential asset.

Sources: Bebbington [54]; Thornton et al. [55]; Moser and Satterthwaite [56]; Moser [57].

4. Materials and Methods

4.1. Study Area

Igbide, Uzere and Olomoro communities are located in Isoko south local government area (ISLGA) in the Delta state of Nigeria (Figure 2). *Isoko* is the local dialect spoken in these areas. The region has a mean annual precipitation of 2500–3000 mm [4]. The weather patterns can be categorised into rainy and dry seasons [58]. Normally, the rainy season commences fully in June and lasts until October. This is closely followed by the dry season which commences in November until early March. In the dry season, dry and dusty conditions known as *harmattan* occur between mid-December and late January.

Small-scale farming is the primary economic driver in these three communities. Cassava (*manihot esculenta*) and groundnuts (*arachis hypogea*) are the major crops produced annually together with okra, pepper, sweet potatoes, yam and plantain. Cassava constitutes 60% of calories consumed. With the exception of cassava, the crops produced depend predominantly on the early rains. Groundnuts, however, are highly sensitive to rising temperatures. Some elderly farmers (above 50 years old) lamented about how the scorching sun, between February and April, has significantly reduced the crop's output by nearly 50% in comparison to the bountiful harvests they were used to in the 1960s. Presently, when they harvest, some of the groundnuts are empty pods. The farmers' explained that reduced groundnut output, which has become the norm, started from the 1990s.

The farmlands in Igbide and Uzere are low-lying, while Olomoro comprises both low and high-lying farmlands. The low-lying farmlands in each of the study areas experience seasonal flooding from mid-June, at the earliest, to the last week in October every year. In extreme conditions, the low-lying farmlands remain inundated until the third week in November. Thus, cassava, which requires a minimum of six months to attain maturity, is usually planted in December and harvested between June and August each year on the low-lying farmlands.

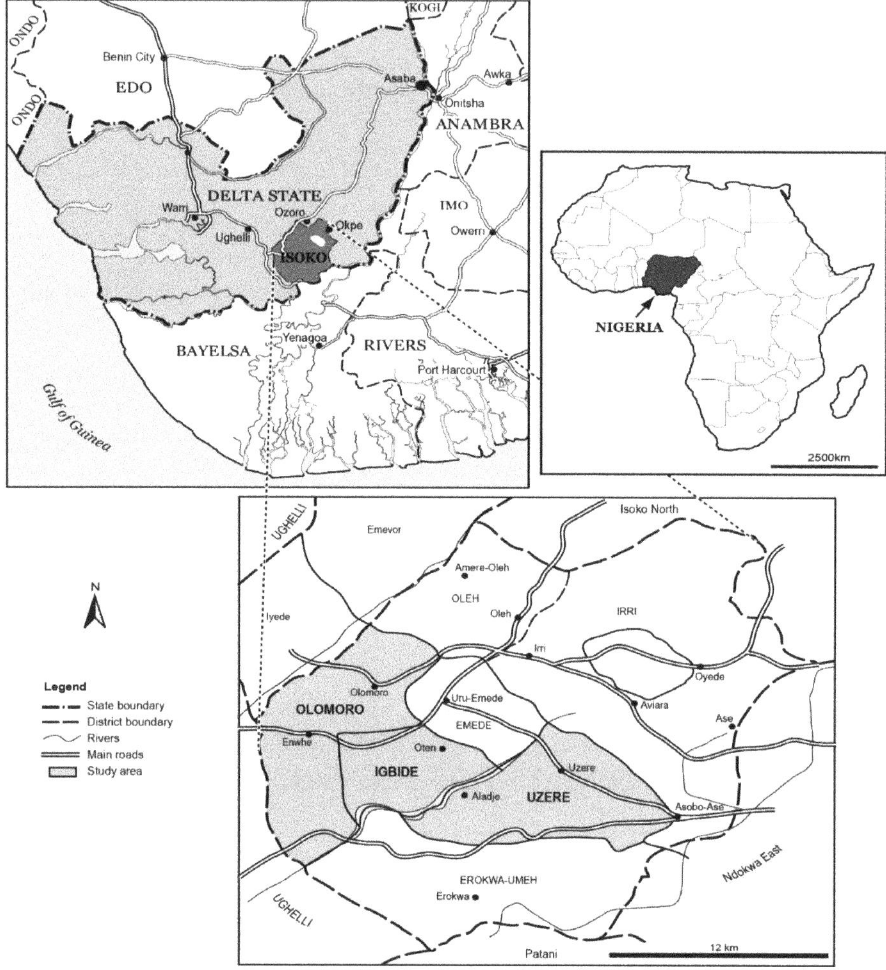

Figure 2. Map of the study areas.

4.2. Methodology

Qualitative methods were adopted to offer deep insights usually hidden from statistical analysis. The study on which this article is based, formed part of a larger undertaking comprising 35 focus group discussions (FGDs) and 14 one-on-one semi-structured interviews, conducted between June and October 2015. Follow up interviews were conducted in July 2016. These discussions explored the participants' perceptions of climate-related threats to food production, how they adapt to such threats, and the extent to which they use government-issued seasonal forecasts, among other issues. Each focus group comprised between three and twelve participants. Two-thirds of the participants were between the ages of 42 and 85 (median = 64) and had no formal education. This article draws on 22 FGDs and five one-on-one, semi-structured interviews where participants provided valuable insights regarding their perception of climate-related threats to food production, how they adapt to such threats, and the extent to which climate services influence their farming decisions, among other issues. Key questions asked include *do you have any local sign(s) that you rely on to know when the early rainfall or rainy season is about to start so that you can start preparations for production or the best time to start planting? How accurate has the local sign(s) been in predicting the weather? Have you ever relied on weather information from NIMET when preparing and planning for a planting season? Did you receive the 2012 flood warning by NIMET? Did you receive the 2013 flood warning predicted by NIMET?* If yes, *how did it influence your farming decisions?* And *how willing are you to rely on scientific weather predictions in the future?* Some FGDs and semi-structured interviews were conducted in *Isoko* and translated to *Pidgin* English by field assistants to enable the first author to understand their responses. Other interviews were conducted in *Pidgin* English. Both the FGDs and interviews were edited to formal English during transcription of the audio recordings and have been presented as edited.

Eligible participants were selected based on the following criteria: the individuals had to have been farming in one of the study areas for a minimum of ten years, gender (both had to be represented), their household assets and livelihoods had to have been adversely affected by the 2012 flood disaster, those who produced most of their food on low-lying farmlands, and their willingness to participate in the study. The low-lying farmlands are usually inundated around June to October annually. The qualitative data obtained were analysed using thematic analysis.

5. Results

5.1. Perception of Climate Services

When participants were asked if they relied on climate services to make farming decisions, they explicitly highlighted that they do not utilise the information. When probed as to why they do not utilise such information, most revealed that they rely on their local and indigenous knowledge systems (LIKS). These include in-depth historical knowledge of the weather patterns, croaking of frogs and appearance of red-like millipedes. Others include the flowering of rubber trees (*ficus elastic decora*), greening of cassava leaves, lunar observation in December, and the height at which the weaver bird constructs its nest on a tree [59,60]. These Indigenous meteorological indicators help to forecast the commencement of the rainy season, as well as the anticipated total rainfall in a farming season. While the participants acknowledged that their LIKS have not always been entirely accurate, the majority, however, revealed that they are not keen to switch to utilising climate services for the following reasons:

5.1.1. 'Suffering in the Midst of Plenty' Syndrome

For the majority, it defied human comprehension for the Nigerian government to invest in climate services when households continue to live in abject poverty in the midst of plenty. Despite consistent exploitation of crude oil for over four decades, their communities have remained inconceivably underdeveloped due to a dearth of communal physical assets. Because they practice annual cropping, harvesting of cassava coincides with the rainy season. Thus, it is exasperating and expensive for households to transport their produce to neighbouring communities to sell as transport drivers are forced to take alternative routes that take much longer. This is due to the numerous potholes on the roads that are tantamount to death traps during the rainy season when they are filled with rainwater. Also, constant power supply is non-existent in ISLGA. This has resulted in the few privileged farmers—who own refrigerators—not being able to cultivate perishable crops like okra and pepper in large quantities, as they cannot prevent most of the produce from rotting away. Furthermore, farmers' inability to access financial assets is a major deterring factor in their struggle to live above the global poverty line of USD 2 daily.

While participants' acknowledged that the government occasionally made provisions to furnish farmers with farm loans and other incentives, such as fertilizers and farm machinery, the majority stated that they have never been beneficiaries. Some argued that they only became aware of such opportunities after the disbursement process was over. Others revealed that the loan distribution process is marred by corruption and that the key beneficiaries were 'ghost' farmers and relatives of those in charge of disbursing the loans. In venting his frustration, one participant between the ages of 40 and 50 lamented:

'Government is not helping anybody from this area despite being an oil producing community ... We only hear it on the radio of the various incentives given to assist farmers, but it is not getting to those on the grassroots. Instead, the beneficiaries are those in the hierarchy such as the executives of small-scale farmers' group association who use it for their private farms.'

It is important to note that some of the participants skilled in generating Indigenous weather forecasts attributed the increased anomalies of their LIKS to gas flaring in Oleh, a community approximately 20 km away from the study areas. An elderly participant, aged between 60 and 70 years, revealed:

'Gas exploration and flaring have seriously affected the trees we use to predict the weather because they do not grow well as they are supposed to. It becomes difficult for our predictions to be correct.'

Also, another elderly participant aged between 70 and 80 years asserted that gas flaring impairs her vision, which is critical to predicting if it would rain or not, through cloud observation. However, an elderly participant also aged between 70 and 80 years argued that gas flaring could not be the reason for the inconsistencies of their LIKS, but admitted that it kills their crops and damages the land. She further pointed out that the quality of *garri* produced has declined because the soil nutrients have been compromised due to oil exploitation activities, a viewpoint shared all the study participants. A few participants, however, added that farmers' inability to practice bush fallowing, due to sporadic increase in population, is also a contributing factor. In this regard, an elderly participant between the ages of 60 and 70 years argued:

'Before Shell commenced oil exploration activities, we were told that at a particular stage, they would fertilise the whole land ... They said they would inject the land for them so that it will replenish the soil. The aim was to ensure that the soil will always be fertile for agricultural production ... But up till date, this has not been carried out.'

Also, focus group participants in Igbide revealed that gas flaring expedites the corrosion and decay of zinc roofing sheets—the most affordable materials—used by households to roof their houses. A participant between the ages of 50 and 60 years categorically stated:

'If you roof your house with iron zinc, within 4 to 6 years, it will start decaying. When it starts raining, the roofs start leaking, and then you have no option but to change the roof. This is causing people economic setback because the money they are supposed to use for other things are now used to repair or replace their damaged roofs. If you don't have money to buy the aluminium Cameroon zinc, before you leave this world you will probably have to roof your house for about seven times.'

The adverse effects of oil exploration activities on food production (Figure 3) coupled with the history of neglect by various government regimes inter alia made most of the participants question the rationale behind investing in climate services. They cannot reconcile how their region can be endowed with huge reservoirs of crude oil and yet the indigenes continue to live in abject poverty. From the participants' perspective, providing and ensuring easy accessibility to assets should be the first step to enhancing food production for farmers because they understand the weather patterns in their community. In this regard, a participant between the ages of 40 and 50 years stated:

'It is only a foolish person that does not understand the terrain. When you plant cassava on 10 plots, like me, as from next month [referring to July] I will start to harvest my cassava. I will not wait till August when the flood usually inundates the farmlands, that is how we cope. A wise person will remove his/her cassava from the farmland before that time.'

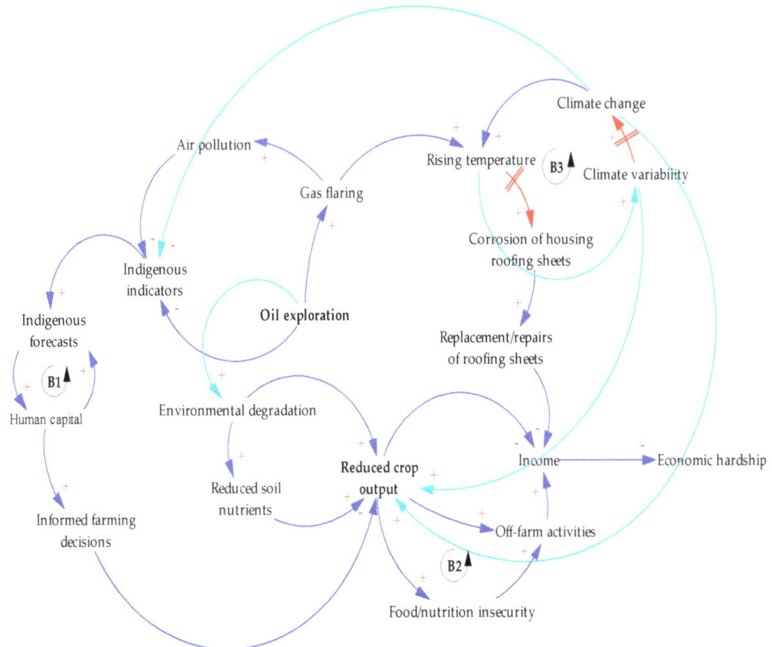

Figure 3. Causal loop diagram illustrating how oil exploration compromises food production (B = balancing feedback loop). N.B. If an increase in one variable leads to an increase in another variable, it is denoted by '+' close to the arrow head, while if it results in a decrease in another variable, it is denoted by '−' close to the arrow head. The double parallel red lines indicate a delay between cause and effect.

5.1.2. Doubts Surrounding the Accuracy of Meteorological Forecasts

The credibility of meteorologists in accurately predicting useful weather information was severely questioned in the aftermath of the 2013 floods that never made landfall in the Delta state as forecasted by the Nigerian meteorological agency (NIMET). In the seasonal climate outlook for 2013, NIMET stated that 31 states were likely to be hit with the same magnitude of flood that occurred in 2012. The 2012 floods wreaked devastating havoc in 31 states of the federation, that was accurately predicted [61,62]. The 2013 flood forecast pinpointed the Delta state as one of the hotspots where the worst impact of the looming disaster would occur. The forecast spread like wildfire in Igbide, Uzere and Olomoro communities, where the 2012 flood wreaked devastating havoc partly because most participants did not receive any warning due to the government's overwhelming reliance on radio and television broadcasts, which was a gross mismatch for the farmers.

This time around, participants became aware of the 2013 forecast through word of mouth from relatives living in urban areas as well as fellow villagers. Consequently, the majority embarked on the inconvenient but inevitable task of adopting proactive measures as they were determined not to be caught unawares again, as they were in 2012. Thus, most participants completely abandoned their farmlands in the low-lying regions for fear of their produce being destroyed by floodwaters. They also sought high-lying farmlands in locations both within and outside their communities that were unaffected by the 2012 floods. The quest to hire or lease viable farmlands in high-lying areas resulted in the sporadic hike in rental price. A few participants, however, produced food on their low-lying farmlands, but in negligible quantities and focused more on crops that matured early. As one participant between the ages of 50 and 60 years explained:

'Farmers who used to cultivate in farmlands close to Urie Lake [farmlands in this region are inundated to about 2–2.5 m between August and October annually] were afraid of planting when they heard that another flood was coming. They left Igbide to neighbouring communities that were unaffected by the 2012 floods looking for farmlands to grow their food. Some were already contemplating relocating from the community. Others planted on their low-lying farmlands but in small quantities. The sudden panic amongst farmers made it extremely costly to plant during that period.'

However, the participants' decision to adopt several pro-active measures was met with unpleasant consequences (Table 2), which had severe implications for their livelihoods as the flood did not occur, at least in the Delta state. To compound their woes, farmers were neither provided with an explanation why the forecast was imprecise nor given relief measures to minimise the hardship brought about by 'inaccurate forecast-induced food insecurity.' In narrating the impact that the wrong forecast had on household food security, a participant between the ages of 50 and 60 years stated:

Table 2. Proactive measures adopted by farmers in response to the seasonal climate forecast in 2013.

Assets	Planned Adaptation Measures Adopted	Implications for the Farmers' Asset Portfolio
Natural	Most farmers completely abandoned the farmlands in the low-lying region. Nonetheless, a few cultivated crops like maize and groundnut that mature within four months on the low-lying farmlands but, in negligible quantities.	Inability to produce adequate food meant increased risk of malnutrition and undernourishment, which in turn can increase susceptibility to illnesses because of reduced immunity especially among children. In order words, the meagre food production has huge implications for the human capital of farmers and their households.
Financial	Constructed 'sophisticated' wooden platforms in their houses in order to preserve household assets when the flood arrives. Some farmers hired high-lying farmlands in spaces unaffected by the flood disaster in 2012 in order to continue in food production. Even after securing high-lying farmlands, the majority refused to cultivate cassava in large quantities because the impacts of the 2012 flood disaster were still clearly visible in the respective communities.	Inability to meet other livelihood objectives such as paying children's tuition fees and paying dues for other social obligations, which can implicitly and explicitly erode social capital. Due to inability to secure government loans because of inadequate assets, among others, they are constrained to borrow funds from money lenders and social clubs to engage effectively in food production and address other pertinent issues such as paying children's tuition fees. However, the loans are usually at astronomical interest rates (about 40% interest over six months). Inability of some farmers to continue with their monthly monetary contributions (*Osusu*). This is an agreed upon sum of money which is given to a trust worthy person every week. At the end of each month, the lump sum is given to a member of the scheme based on prior arrangement. *Osusu* tends to alleviate their financial woes by ensuring that they are able to continue to produce food annually.
Social	Searched for high-lying farmlands in neighbouring communities and farmlands in areas unaffected by the 2012 floods in their communities.	Increased outward migration can undermine social cohesion among community members. Weakened social capital is also likely to arise when privileged owners of farmlands in high-lying areas exploit vulnerable farmers by abruptly increasing the price to hire a plot of farmland.
Human	Increased migration to other communities in order to produce food. Engaging more intensively in off-farm activities such as brick-laying and bush clearing.	Engaging intensively in off-farm activities can compromise the health of an individual.

'As a result of the news, people refused to go and plant. This resulted in severe hunger for a lot of people in the community. Me, myself, I did not plant because of the news.'

In the same vein, another participant between the ages of 40 and 50 years explained:

'We use to supply *garri* (processed cassava) to Bayelsa and Edo states prior to that 2013 forecast. Because we adhered to the rumour, we were now the ones going to those states to buy garri. The painful part was that the cost to purchase the produce was extremely expensive. So this time around, if the government distributes any forecast that says there will be extreme flooding, I will ignore the forecast and plant normally.'

The experience has instilled in the minds of the people that their welfare is of little or no concern to the government, and consequently, has eroded their trust in climate services.

5.1.3. Mediums of Disseminating Weather Information

The channel of communication used to disseminate weather and climate information to households in the Delta state may be a factor undermining its use to make informed farming decisions. Radio and television broadcasts are the mediums used to convey weather information to households. This may arguably be the reason why nearly all the participants, with the exception of one person, did not receive the 2012 flood warning, although it is not entirely clear if they would have adopted pro-active behaviours as their LIKS did not predict that such event would occur. With the exception of Sundays when they attend church services and social gatherings as well as market days—once in four days—when they sell their crops, the farmers' typical daily routine entails leaving their homes before 7 a.m. for their respective farmlands. This is to ensure that as much work as possible is done before the scorching noonday and afternoon sun emerges. Thereafter, they seek shade to get a reprieve from the heat before continuing their work. Consequently, the farmers usually return home late in the afternoons or evenings extremely fatigued, and thus, barely have any iota of strength left in them to listen to both radio and television broadcasts.

Also, the lack of constant power supply is another factor that compromises the effectiveness of radio and television broadcasts in the region. In lamenting on the despicable state of power supply in the region, a participant aged between 60 and 70 years argued:

'We are just paying for ordinary power cables that do not provide us with electricity. For more than four days now, there has been no electricity.'

In terms of the most effective channels for communicating important information to households, the participants unequivocally argued that the use of community town-criers and town-hall meetings remain the most effective mediums of communication in their respective communities. In addition, a participant from Uzere, in his early 80s, recommended that local people should be trained on how to interpret weather information. This is because:

'NIMET and agricultural extension workers will not have the time to go round the nooks and cranny of the community to inform people of what the weather outlook for the forthcoming farming season will be like, as against their people that reside in the community.'

5.1.4. Absence of Weather Stations

At the time of data collection, the closest weather station to the study areas was located in Warri, a city approximately 60 km away from the study sites. The other station was in Asaba, about 140 km away from the study sites. The implication is that the forecasts that were generated were not tailored to ISLGA. It is, therefore, not inconceivable to argue that the lack of downscaled forecasts may have deterred some farmers from listening to both radio and television broadcasts for weather-related information. Also, this may partly explain why the participant (a woman from Igbide, in her 50s) who heard the flood warning over the radio failed to act proactively. She remarked:

'Before the flood occurred, it was announced on the radio that there was going to be a flood incident, which would affect most parts of the nation. However, due to a lack of scientific understanding of the message's content and not being certain that Igbide will be affected, I ignored the warning.'

It is, however, noteworthy that a small minority was open to the idea of using climate services to inform farming decisions.

5.2. Characteristics of Participants Willing to Utilise Climate Services

Two groups of participants were open to using climate services to make farming decisions. The first group open to the idea of utilising climate services were young male participants (under 40 years) who had obtained at least a high school education. The young male participants assisted their parents in their respective fields while being actively involved in off-farm activities, such as electronics repairs, carpentry and welding. With regards to the perceived role climate services can play in facilitating effective food production, a participant, in his early 30s, commented:

'If farmers in this community can be getting the seasonal rainfall predictions, it will be extremely helpful in protecting our crops and other livelihood assets ... serious farmers will need this sort of information.'

The next group of farmers that were willing to utilise climate services to make informed farming decisions were the elderly people (above 50 years old) that were educated and had a viable alternative source of livelihood. Nonetheless, an individual that fell within this bracket sounded disinterested. The elderly participants interested in climate services were retired civil servants who received a regular monthly pension. They had properties such as vast hectares of farmlands, houses and fishing boats they leased out to earn regular income. In this regard, a pensioner from Uzere aged between 60 and 70 years, who had obtained a university degree, expressed his willingness to utilise climate information by stating that:

'If farmers can get the meteorological information before the planting season (November), it will play a significant role in protecting the assets of the small-scale farmers. If I can get the weather information for this community, I will take full advantage of it.'

6. Discussion

In the wake of a rapidly changing climate, local farmers' utilisation of climate services in SSA is crucial to facilitating informed farming decisions necessary to scale-up food production [2,14,41,42]. In contrast to their counterparts in various SSA countries that have embraced this technology, local and indigenous farmers in the Delta state are yet to adopt climate services, with the majority not keen to do so. The indigenous farmers are disillusioned that despite their communities' significant contribution to Nigeria's GDP, basic amenities and infrastructures—consistent power supply, potable water and good roads—are virtually non-existent. Also, despite the farmers' overwhelming dependence on food production to obtain a livelihood, the majority have never been beneficiaries of vital assets such as farm loans and machinery. This is against the backdrop of the Delta state government and federal government making such schemes readily available [23,63,64]. Thus, the notion that climate services—an 'abstract entity'—can scale-up food production seems utopic. From the farmers' viewpoint, their community symbolises an express tunnel for capital accumulation, with MNCs and government being the sole beneficiaries. Also, the fact that their LIKS used in predicting weather conditions have been negatively affected by oil exploration and exploitation activities, as well as farmers being the victims of a wrong forecast cast further doubt on the efficacy of climate services to upscale food production.

Studies conducted in various developing countries that acknowledge indigenous farmers' willingness to rely on [33,65] and actual utilisation [14,15,36] of climate services, the environment they depend upon to generate indigenous meteorological forecasts have not been exploited by MNCs. As studies in Thailand [66], Ethiopia [67] and Nigeria [68] show, a catalyst to a decline and or loss of LIKS is anthropogenically-induced modification of the environment due to natural resources exploitation. Nonetheless, the indigenous farmers in the Delta state may change their current standpoint of not being willing to utilise climate services if access to fundamental assets that can significantly improve their welfare are made readily available. This hinges on the fact that, despite lack of trust in the government coupled with inability to access capital assets, they still relied on the 2013 forecast to make farming decisions.

As documented by Singh et al. [46] (p. 2428), people's memories give 'weightage to recent events because the consequences are perceived more strongly than those in the distant past'. Thus, easy access to assets, coupled with situating weather stations in ISLGA that are 30 km^2 apart (recommended guideline to generate accurate local downscaled forecasts [69]), may cause farmers to recalibrate the incorrect 2013 forecast experience signpost in their memory as a one-time event with a low probability of recurrence. To advance the adoption of climate services, it might be necessary to work closely with those willing to utilise climate services, perhaps through pilot studies as a useful starting point. Showcasing the effectiveness of climate services in upscaling food production versus relying solely on LIKS might be the propelling force that will get other farmers to change their standpoint. Such pilot studies will address Serra and McKune's [70] (p. 4) concern that, 'while much attention has focused on improving coverage of climate services, we should equally invest resources and efforts in tackling the complex domain of individual and collective perception of information, and consequent behavioural change.'

If and after succeeding in convincing farmers on the effectiveness of climate services, and in the eventuality of another wrong future forecast, although rare if downscaled forecasts are produced but not entirely inevitable, see [71], government must realise that the adverse impacts of such occurrence on farmers' livelihood almost mirrors the consequences of an extreme weather disaster. Therefore, such occurrences should be handled in similar magnitude as an extreme weather event, which often triggers government's immediate response in mobilising resources to assist victims. If farmers are convinced that there will be a safety net in the aftermath of an erroneous forecast, the likelihood that they will utilise climate services may be high.

However, caution should be exercised when interpreting the findings of this study because the farmers' advocacy for assets over climate services may be attributed to the fact that only groundnut production has been adversely affected by climate variability and change. Thus, they may have become accustomed to the shock, especially since the reduction in groundnut production has occurred for over three decades. The regular occurrences of low groundnut outputs, which have reduced the monetary income received from its sale, may have desensitized farmers to the shock. This experience is in stark contrast to cassava—the main staple consumed by farmers—whose production has not been aggressively impacted by climate variability and change. The major deterrent to maximising cassava production is the annual seasonal flooding (mid-June to last week in October) of their low-lying farmlands, which is not perceived as a challenge because it has been the norm in the respective communities [4]. This may provide a valuable justification why they are not keen to utilise climate services, unlike farmers in India and Bangladesh [39] where the major staples consumed are highly sensitive to climate variability and change. As a result, these farmers rely on the technology to improve farming decisions. Nonetheless, this study offers valuable insights into factors that may undermine the use of climate services in areas that have been exploited by MNCs.

7. Conclusions

In the face of increasingly palpable impacts of climate variability and change amongst farmers in developing nations, climate services are regarded as essential to such farmers making informed decision for efficient food production. Moreover, climate services have recorded tremendous success to the point that farmers are now willing to pay for such services. Despite this development, studies have failed to understand the extent to which Indigenous farmers residing in communities with a history of fierce contestations with MNCs utilise climate services to increase agricultural productivity. It has been observed that MNCs, in their quest to exploit rich natural resources, aggressively degrade the immediate environment that Indigenous farmers rely on for food production. Since the exploitation of natural resources by MNCs occur through an alliance with a country's government, it is necessary to ascertain how the negative impacts of MNC activities influence Indigenous farmers' perception and utilisation of climate services.

The findings of this study suggest that indigenous farmers in the research sites where data for this study was collected do not rely on climate services but on their LIKS such as the croaking of frogs, and lunar observation in December amongst others. Although these farmers do acknowledge that their LIKS have not always been entirely accurate, the majority suggested that they were not keen on utilising climate services. A reason for this standpoint is because the participants continue to be victims of what we refer to as 'suffering in the midst of plenty' syndrome. They cannot reconcile how people continue to live in abject poverty, and their communities have remained underdeveloped despite the significant contributions crude oil explored from their locality make to Nigeria's GDP. While the farmers acknowledge that the government occasionally made provisions to provide farmers with farm loans and other assets like farm machinery, most have never been beneficiaries. Also, participants asserted that the quality of *garri* produced has declined because the soil nutrients have been compromised due to oil exploitation activities. Thus, the adverse effects of oil exploration activities on food production and the influence it has on the accuracy of their LIKS (Figure 3) coupled with the history of neglect by various government regimes inter alia made most participants question the rationale behind investing in climate services.

Furthermore, other factors beyond the activities of MNCs have fuelled farmers' unwillingness to utilise climate services. These include non-usage of their local channels of communication—town-hall meetings and town-criers—to disseminate weather information, lack of weather station close to the communities to generate accurate downscaled forecasts and being victims of an erroneous scientific forecast in 2013. Notwithstanding, the study finds that young males (under 40 years) and a few older adults (over 50 years old) are willing to utilise climate services if the factors undermining the production of accurate forecasts are addressed. The commonalities among these cohorts of farmers are that they have obtained at least a high school education and have viable alternative sources of livelihood outside of farming. Given this scenario, addressing the challenges that hamper farmers' utilisation of climate services in Igbide, Olomoro and Uzere communities is now a matter of urgency especially if the Nigerian government is keen on actualising the first and second SDG goals by 2030. Failure to address these issues will be a massive blow for the incumbent government that has been investing heavily in agriculture, deemed a viable alternative to successfully diversify Nigeria's economy from its extensive reliance on crude oil for most of its foreign exchange. Besides erecting weather stations in ISLGA that are 30 km^2 apart, a useful starting point will be to conduct pilot studies with those willing to utilise climate services. Highlighting the effectiveness of climate services in upscaling food production versus relying solely on LIKS might be a catalyst that will convince other farmers to use climate services in making informed farming decisions.

Author Contributions: Conceptualization, visualization, design of research instruments, E.E.E.; methodology, E.E.E. and O.O.E.; data collection, writing—original draft preparation, E.E.E.; writing—review and editing, F.K.D. and H.B.T.; supervision, M.D.S. and L.L. All authors have read and agreed to the published version of the manuscript.

Funding: The research funding provided by the National Institute for the Humanities and Social Sciences and the Social Sciences Council for the Development of Social Science Research in Africa (NIHSS-CODESRIA) is gratefully acknowledged. The first author acknowledges the South African System Analysis Centre (SASAC) and the funding provided by the National Research Foundation (NRF) and the Department of Science and Technology (DST) of South Africa.

Acknowledgments: The generous comments of all three reviewers are deeply appreciated.

Conflicts of Interest: The authors declare no conflict of interest.

References

1. Bryan, E.; Deressa, T.; Gbetibouo, G.; Ringler, C. Adaptation to climate change in Ethiopia and South Africa: Options and constraints. *Environ. Sci. Policy* **2009**, *12*, 413–426. [CrossRef]
2. Roudier, P.; Muller, B.; d'Aquino, P.; Roncoli, C.; Soumaré, M.A.; Batté, L.; Sultan, B. The role of climate forecasts in smallholder agriculture: Lessons from participatory research in two communities in Senegal. *Clim. Risk Manag.* **2014**, *2*, 42–55. [CrossRef]
3. Simatele, D.; Simatele, M. Migration as an adaptive strategy to climate variability: A study of the Tonga-speaking people of Southern Zambia. *Disasters* **2015**, *39*, 762–781. [CrossRef] [PubMed]
4. Ebhuoma, E.; Simatele, D. Defying the odds: Climate variability, asset adaptation and food security nexus in the Delta State of Nigeria. *Int. J. Disaster Risk Reduct.* **2017**, *21*, 231–242. [CrossRef]
5. Elum, Z.A.; Modise, D.M.; Marr, A. Farmer's perception of climate change and responsive strategies in three selected provinces of South Africa. *Clim. Risk Manag.* **2017**, *16*, 246–257. [CrossRef]
6. Mavhura, E. Systems analysis of vulnerability to hydrometeorological threats: An exploratory study of vulnerability drivers in northern Zimbabwe. *Int. J. Disaster Risk Reduct.* **2019**, *10*, 204–219. [CrossRef]
7. Saeed, F.; Salik, K.M.; Ishfaq, S. Climate Induced Rural-to-Urban Migration in Pakistan. Sustainable Development Policy Institute (SDPI) Working Paper, Pakistan. Available online: http://www.prise.odi.org/wp-content/uploads/2016/01/Low_Res-Climate-induced-rural-to-urban-migration-in-Pakistan.pdf (accessed on 13 June 2020).
8. Antwi-Agyei, P.; Stringer, L.C.; Dougill, A.J. Livelihoods adaptation to climate variability: Insights from households in Ghana. *Reg. Environ. Chang.* **2014**, *14*, 1615–1626. [CrossRef]
9. Intergovernmental Panel on Climate Change (IPCC). Africa. Intergovernmental Panel on Climate Change. Available online: https://www.ipcc.ch/pdf/assessment-report/ar5/wg2/WGIIAR5-Chap22_FINAL.pdf (accessed on 19 October 2014).
10. Vaughan, C.; Buja, L.; Kruczkiewicz, A.; Goddard, L. Identifying research priorities to advance climate services. *Clim. Serv.* **2016**, *4*, 65–74. [CrossRef]
11. Ziervogel, G.; Downing, T.E. Stakeholder networks: Improving seasonal climate forecasts. *Clim. Chang.* **2004**, *65*, 73–101. [CrossRef]
12. Choi, H.S.; Schneider, U.A.; Rasche, L.; Cui, J.; Schmid, E.; Held, H. Potential effects of perfect seasonal climate forecasting on agricultural markets, welfare and land use: A case study of Spain. *Agric. Syst.* **2015**, *133*, 177–189. [CrossRef]
13. Ziervogel, G.; Calder, R. Climate variability and rural livelihoods: Assessing the impact of seasonal climate forecasts in Lesotho. *Area* **2003**, *35*, 403–417. [CrossRef]
14. Tall, A.; Hansen, J.; Jay, A.; Campbell, B.; Kinyangi, J.; Aggarwal, P.K.; Zougmore, R. *Scaling up Climate Services for Farmers: Mission Possible. Learning from Good Practice in Africa and South Asia*; CCAFS Report No. 13; CGIAR Research Program on Climate Change, Agriculture and Food Security (CCAFS): Copenhagen, Denmark, 2014. Available online: https://cgspace.cgiar.org/bitstream/handle/10568/42445/CCAFS%20Report%2013%20web.pdf?sequence=7&isAllowed=y (accessed on 21 July 2020).
15. Ouédraogo, M.; Barry, S.; Zougmoré, R.; Partey, S.; Somé, L.; Baki, G. Farmers' willingness to pay for climate information services: Evidence from cowpea and sesame producers in Northern Burkina Faso. *Sustainability* **2018**, *10*, 611. [CrossRef]
16. Wilson, N.J.; Mutter, E.; Inkster, J.; Satterfield, T.A. Community-Based Monitoring as the practice of Indigenous water governance: A case study of Indigenous water quality monitoring in the Yukon River Basin. *J. Environ. Manag.* **2018**, *210*, 290–298. [CrossRef] [PubMed]
17. Hiwasaki, L.; Luna, E.; Syamsidik; Shaw, R. Process for integrating local and indigenous knowledge with science for hydro-meteorological disaster risk reduction and climate change adaptation in coastal and small island communities. *Int. J. Disaster Risk Reduct.* **2014**, *10*, 15–27. [CrossRef]
18. Warren, D.M. Using Indigenous Knowledge in Agricultural Development. Available online: http://documents1.worldbank.org/curated/en/408731468740976906/pdf/multi-page.pdf (accessed on 25 August 2020).
19. Dutfield, G. *"Between a Rock and Hard Place: Indigenous Peoples, Nation States and the Multinationals" in Food and Agriculture Organization of the United Nations (FAO) Medicinal Plants for Conservation and Health Care*; FAO: Rome, Italy, 1995. Available online: http://www.fao.org/3/a-w7261e.pdf (accessed on 29 May 2020).

20. Ikelegbe, A. The economy of conflicts in the oil rich Niger Delta region of Nigeria. *Afr. Asian Stud.* **2006**, *5*, 87–122. [CrossRef]
21. Okwechime, I. Environmental Conflict and Internal Migration in the Niger Delta Region of Nigeria. Bielefeld: COMCAD. Working Papers—Centre on Migration, Citizenship and Development. Available online: https://www.uni-bielefeld.de/soz/ab6/ag_faist/downloads/WP_119.pdf (accessed on 15 May 2020).
22. Omotola, S. Liberation movements and rising violence in the Niger Delta: The new contentious site of oil and environmental politics. *Stud. Confl. Terror.* **2010**, *33*, 36–54. [CrossRef]
23. Ebhuoma, E.; Simatele, M.; Tantoh, H.; Donkor, F. Asset vulnerability analytical framework and systems thinking as a twin methodology for highlighting factors that undermine efficient food production. *Jàmbá* **2019**, *11*, a597. [CrossRef]
24. Scheele, F.; de Haan, E.; Kiezebrink, V. *Cobalt Blues Environmental Pollution and Human Rights Violations in Katanga's Copper and Cobalt Mines*; Centre for Research on Multinational Corporations (SOMO): Ansterdam, The Netherlands, 2016. Available online: https://www.somo.nl/wp-content/uploads/2016/04/Cobalt-blues.pdf (accessed on 13 May 2020).
25. Schueler, V.; Kuemmerle, T.; Schroder, H. Impacts of surface gold mining on land use systems in Western Ghana. *AMBIO* **2011**, *40*, 528–539. [CrossRef]
26. Mtero, F. Rural livelihoods, large-scale mining and agrarian change in Mapela, Limpopo, South Africa. *Resour. Policy* **2017**, *53*, 190–200. [CrossRef]
27. Omeje, K. Oil conflict in Nigeria: Contending issues and perspectives of the local Niger Delta people. *New Political Econ.* **2005**, *10*, 321–334. [CrossRef]
28. Fulmer, A.M.; Godoy, A.S.; Neff, P. Indigenous rights, resistance, and the law: Lessons from a Guatemalan mine. *Lat. Am. Politics Soc.* **2008**, *50*, 91–121. [CrossRef]
29. Aha, B.; Ayitey, J.Z. Biofuels and the hazards of land grabbing: Tenure (in)security and indigenous farmers' investment decisions in Ghana. *Land Use Policy* **2017**, *60*, 48–59. [CrossRef]
30. Obi, C. Nigeria's Niger Delta: Understanding the complex drivers of violent oil-related conflict. *Afr. Dev.* **2009**, *2*, 103–128. [CrossRef]
31. Duru, C.U. Environmental Degradation: Key Challenge to Sustainable Economic Development in the Niger Delta. Ph.D. Thesis, Walden University, Minneapolis, MN, USA, 2014. Available online: https://scholarworks.waldenu.edu/cgi/viewcontent.cgi?referer=https://www.google.com/&httpsredir=1&article=1113&context=dissertations (accessed on 15 June 2020).
32. Obi, C.I. Oil extraction, dispossession, resistance, and conflict in Nigeria's oil-rich Niger Delta. *Can. J. Dev. Stud.* **2010**, *30*, 219–236. [CrossRef]
33. Tall, A.; Coulibaly, J.Y.; Diop, M. Do climate services make a difference? A review of evaluation methodologies and practices to assess the value of climate information services for farmers: Implications for Africa. *Clim. Serv.* **2018**, *11*, 1–12. [CrossRef]
34. World Meteorological Organization (WMO). What do We Mean by Climate Services? Available online: https://public.wmo.int/en/bulletin/what-do-we-mean-climate-services (accessed on 17 August 2020).
35. Roncoli, C.; Ingram, K.; Kirshen, P. Reading the rains: Local knowledge and rainfall forecasting in Burkina Faso. *Soc. Nat. Resour.* **2002**, *15*, 409–427. [CrossRef]
36. Mahoo, H.; Mbungu, W.; Yonah, I.; Recha, J.; Radeny, M.; Kimeli, P.; Kinyangi, J. *Integrating Indigenous Knowledge with Scientific Seasonal Forecasts for Climate Risk Management in Lushoto District in Tanzania*; CCAFS Working Paper 103; CGIAR Research Program on Climate Change, Agriculture and Food Security (CCAFS): Copenhagen, Denmark, 2015. Available online: https://cgspace.cgiar.org/bitstream/handle/10568/56996/Working%20Paper%20103.pdf (accessed on 21 July 2020).
37. Carr, E.R.; Abrahams, D.; De la Poterie, A.T.; Suarez, P.; Koelle, B. Vulnerability assessments, identity and spatial scale challenges in disaster-risk reduction. *Jàmbá J. Disaster Risk Stud.* **2015**, *7*, 201. [CrossRef]
38. Carr, E.R.; Onzere, S.; Kalala, T.; Owusu-Daaku, K.N.; Rosko, H. *Assessing Mali's l'Agence Nationale de la Météorologie's (Mali Meteo) Agrometeorological Advisory Program: Final Report in the Farmer Use of Advisories and the Implications for Climate Service Design*; Development Experience Clearinghouse, USAID (DEC): Washington, DC, USA, 2015.
39. Hansen, J.W. Making Climate Information Work for Agricultural Development. *World Politics Rev.* **2012**. Available online: https://www.adaptation-undp.org/sites/default/files/downloads/making_climate_information_work_for_agricultural_development_-_hansen_2012.pdf (accessed on 11 July 2020).

40. Ouédraogo, M.; Zougmoré, R.; Barry, S.; Somé, L.; Baki, G. *The Value and Benefits of Using Seasonal Climate Forecasts in Agriculture: Evidence from Cowpea and Sesame Sectors in Climate-Smart Villages of Burkina Faso*; CCAFS Info Note; CGIAR Research Program on Climate Change, Agriculture and Food Security (CCAFS): Copenhagen, Denmark, 2015. Available online: https://cgspace.cgiar.org/bitstream/handle/10568/68537/CCAFS%20info%20note.pdf (accessed on 21 July 2020).
41. Amegnaglo, C.; Anaman, K.; Mensah-Bonsu, A.; Onumah, E.E.; Gero, F. Contingent valuation study of the benefits of seasonal climate forecasts for maize farmers in the Republic of Benin, West Africa. *Clim. Serv.* **2017**, *6*, 1–11. [CrossRef]
42. Sivakumar, M.V.K.; Collins, C.; Jay, A.; Hansen, J. *Regional Priorities for Strengthening Climate Services for Farmers in Africa and South Asia*; CCAFS Working Paper No. 71; CGIAR Research Program on Climate Change, Agriculture and Food Security (CCAFS): Copenhagen, Denmark, 2014. Available online: https://cgspace.cgiar.org/bitstream/handle/10568/41599/CCAFS%20WP%2071.pdf (accessed on 21 July 2020).
43. World Meteorological Organization. Benefits of Investments in Climate Services for Agriculture and Food Security Outweigh Costs. Available online: https://public.wmo.int/en/media/press-release/benefits-of-investments-climate-services-agriculture-and-food-security-outweigh (accessed on 17 August 2020).
44. Sellnow, T.L.; Ulmer, R.R.; Seeger, M.W.; Littlefield, R.S. (Eds.) *Effective Risk Communication. A Message-Centered Approach*; Springer: New York, NY, USA, 2009.
45. Marx, S.M.; Weber, E.U.; Orlove, B.S.; Leiserowitz, A.; Krantz, D.H.; Roncoli, C.; Philips, J. Communication and mental processes: Experiential and analytic processing of uncertain climate information. *Glob. Environ. Chang.* **2007**, *17*, 47–58. [CrossRef]
46. Singh, C.; Osbahr, H.; Dorward, P. The implications of rural perceptions of water scarcity on differential adaptation behaviour in Rajasthan, India. *Reg. Environ. Chang.* **2018**, *18*, 2417–2432. [CrossRef]
47. Slovic, P.; Finucane, M.L.; Peters, E.; MacGregor, D.G. Risk as analysis and risk as feelings: Some thoughts about affect, reason, risk, and rationality. *Risk Anal.* **2004**, *24*, 311–322. [CrossRef] [PubMed]
48. Ebhuoma, E.; Simatele, D.M. 'We know our terrain': Indigenous knowledge systems preferred to scientific systems of weather forecasting in the Delta State of Nigeria. *Clim. Dev.* **2019**, *11*, 112–123. [CrossRef]
49. Akpabio, E.M.; Akpan, N.S. Governance and Oil Politics in Nigeria's Niger Delta: The Question of Distributive Equity. *J. Hum. Ecol.* **2010**, *30*, 111–121. [CrossRef]
50. Ejobowah, B.J. Who owns the oil? The politics of ethnicity in the Niger-Delta of Nigeria. *Afr. Today* **2000**, *47*, 28–47. [CrossRef]
51. Ojo, G.U. Uzere Community Protest Against Shell, Nigeria. Available online: https://ejatlas.org/conflict/uzere-community-protest-against-shell-nigeria (accessed on 23 May 2020).
52. Adeniran, A.I. Niger Delta, protest movements. *Int. Encycl. Rev. Protest* **2009**, *1*, 1–3.
53. Ererobe, E. Federal Government, Multinationals and Isoko Nation. Available online: https://www.vanguardngr.com/2009/08/fg-multinationals-and-isoko-nation/ (accessed on 3 April 2020).
54. Bebbington, A. Capitals and capabilities: A framework for analysing peasant viability, rural livelihoods and poverty. *World Dev.* **1999**, *27*, 2021–2044. [CrossRef]
55. Thornton, P.K.; Jones, P.G.; Owiyo, T.; Kruska, R.L.; Herrero, M.; Orindi, V.; Bhadwal, S.; Kristjanson, P.; Notenbaert, A.; Bekele, N.; et al. Climate change and poverty in Africa: Mapping hotspots of vulnerability. *Afr. J. Agric. Resour. Econ.* **2008**, *2*, 24–44.
56. Moser, C.; Satterthwaite, D. Towards Pro-Poor Adaptation to Climate Change in the Urban Centers of Low and middle-Income Countries. Available online: http://pubs.iied.org/pdfs/10564IIED.pdf (accessed on 16 September 2014).
57. Moser, C. A Conceptual and Operational Framework for Pro-Poor Asset Adaptation to Urban Climate Change. Available online: http://siteresources.worldbank.org/INTURBANDEVELOPMENT/Resources/336387-1256566800920/6505269-1268260567624/Moser.pdf (accessed on 29 September 2014).
58. Adejuwon, J. Rainfall seasonality in the Niger Delta Belt, Nigeria. *J. Geogr. Reg. Plan.* **2012**, *5*, 51–60.
59. Fitchett, J.; Ebhuoma, E. Phenological cues intrinsic in indigenous knowledge systems for forecasting seasonal climate in the Delta State of Nigeria. *Int. J. Biometeorol.* **2018**, *62*, 1115–1119. [CrossRef]
60. Ebhuoma, E. A framework for integrating scientific forecasts with indigenous systems of weather forecasting in southern Nigeria. *Dev. Pract.* **2020**, *30*, 472–484. [CrossRef]
61. Ebhuoma, E.; Leonard, L. An operational framework for communicating flood warnings to indigenous farmers in southern Nigeria: A systems thinking analysis. *GeoJournal* **2020**. [CrossRef]

62. Nigerian Meteorological Agency. Seasonal Rainfall Prediction. Available online: http://www.nimet.gov.ng/sites/default/files/publications/2016%20SRP.pdf (accessed on 26 June 2017).
63. Olaitan, M.A. Finance for small and medium enterprises: Nigeria's agricultural credit guarantee scheme fund. *J. Int. Farm Manag.* **2006**, *3*, 30–38.
64. Philip, D.; Nkonya, E.; Pender, J.; Oni, A. *Constraints to Increasing Agricultural Productivity in Nigeria: A Review*; International Food Policy Research Institute.: Washington, DC, USA, 2009. Available online: http://dspace.africaportal.org/jspui/bitstream/123456789/31819/1/NSSP%20Background%20Paper%206.pdf?1 (accessed on 7 April 2017).
65. Kolawole, O.D.; Motsholapheko, M.K.; Ngwenya, B.N.; Thakadu, O.; Kgathi, D.L. Climate variability and rural livelihoods: How households perceive and adapt to climate shocks in the Okavango Delta, Botswana. *Weather Clim. Soc.* **2016**, *18*, 131–145. [CrossRef]
66. Wongbusarakum, S. Loss of traditional practices, loss of knowledge, and the sustainability of cultural and natural resources: A case of Urak Lawoi people in the Adang Archipelago, Southwest Thailand. In *Learning and Knowing in Indigenous Societies Today*; Bates, P., Chiba, M., Kube, S., Nakashima, D., Eds.; UNESCO: Paris, France, 2009; pp. 1–128.
67. Doda, Z. Indigenous Knowledge, the Environment and Natural Resource Management: The Adverse Impacts of Development Interventions. In Proceedings of the Sensitization Workshop (Center for the Study of Environment and Society), Awassa, Ethiopia, May 2005. Available online: https://www.researchgate.net/profile/Zerihun_Doffana/publication/267337336_Indigenous_Knowledge_the_Environment_and_Natural_Resource_Management_the_Adverse_Impacts_of_Development_Interventions_A_Review_of_Literature_Paper_Presented_at_the_Sensitization_Workshop_Center_for_th/links/544d4a600cf2bcc9b1d8e971.pdf (accessed on 21 July 2020).
68. Mmon, P.C.; Arokoyu, S.B. Mangrove forest depletion, biodiversity loss and traditional resources management practices in the Niger Delta, Nigeria. *Res. J. Appl. Sci. Eng. Technol.* **2010**, *2*, 28–34.
69. Van De Giesen, N.; Hut, R.; Andreini, M.; Selker, J.S. Trans-African Hydro-Meteorological Observatory (TAHMO): A Network to Monitor Weather, Water, and Climate in Africa. Available online: https://ui.adsabs.harvard.edu/abs/2013AGUFM.H52C..04V/abstract (accessed on 10 December 2019).
70. Serra, R.; McKune, S. Climate Information Services and Behavioural Change: The Case of Senegal. Sahel Research Group Working Paper No. 010. Working Paper Series for Development, Security and Climate Change in the Sahel. Available online: https://sahelresearch.africa.ufl.edu/files/SRG_WorkingPaper_Serra_McKune_Final.pdf (accessed on 21 July 2020).
71. Barzin, R.; Chen, J.J.J.; Young, B.R.; Farid, M.M. Application of weather forecast in conjunction with price-based method for PCM solar passive buildings—An experimental study. *Appl. Energy* **2016**, *163*, 9–18. [CrossRef]

Article

Groundwater Resources in the Main Ethiopian Rift Valley: An Overview for a Sustainable Development

Sabrina Maria Rita Bonetto, Chiara Caselle *, Domenico Antonio de Luca and Manuela Lasagna

Department of Earth Science, University of Turin, 10125 Turin, Italy; sabrina.bonetto@unito.it (S.M.R.B.); domenico.deluca@unito.it (D.A.d.L.); manuela.lasagna@unito.it (M.L.)
* Correspondence: chiara.caselle@unito.it

Abstract: In arid and semi-arid areas, human health and economic development depend on water availability, which can be greatly compromised by droughts. In some cases, the presence of natural contaminants may additionally reduce the availability of good quality water. This research analyzed the water resources and hydrochemical characteristics in a rural area of the central Main Ethiopian Rift Valley, particularly in the districts of Shashemene, Arsi Negelle, and Siraro. The study was developed using a census of the main water points (springs and wells) in the area and the sampling and physico-chemical analysis of the water, with particular regard to the fluoride concentration. In many cases, fluoride content exceeded the drinking water limits set by the World Health Organization, even in the absence of anthropogenic contamination. Two different aquifers were recognized: A shallow aquifer related to the eastern escarpment and highlands, and a deep aquifer in the lowland areas of the rift valley on the basis of compositional changes from Ca–Mg/HCO_3 to Na–HCO_3. The distribution of fluoride, as well as pH and EC values, showed a decrease from the center of the lowlands to the eastern highlands, with similar values closely aligned along an NNE/SSW trend. All these data contribute to creating awareness among and sharing information on the risks with rural communities and local governments to support the adequate use of the available water resources and to plan appropriate interventions to increase access to fresh water, aimed at the sustainable human and rural local development of the region.

Keywords: groundwater resources; fluoride; main Ethiopian Rift Valley

Citation: Bonetto, S.M.R.; Caselle, C.; de Luca, D.A.; Lasagna, M. Groundwater Resources in the Main Ethiopian Rift Valley: An Overview for a Sustainable Development. *Sustainability* **2021**, *13*, 1347. https://doi.org/10.3390/su13031347

Academic Editor: Maurizio Tiepolo
Received: 20 December 2020
Accepted: 25 January 2021
Published: 28 January 2021

Publisher's Note: MDPI stays neutral with regard to jurisdictional claims in published maps and institutional affiliations.

1. Introduction

The regional planning and the management of natural resources require considering the interactions among human needs, ecosystem dynamics, and resource sustainability. Human needs and economic activities, such as industry, agriculture, and animal husbandry, require a continuous water supply. Over the past decades, water use has more than doubled [1], and the water demand will further increase due to a growing global human population. However, both the availability and the quality of water resources will be affected by socio-economic and technological developments, climate change and increasing climate extremes, such as droughts and floods, particularly in developing countries [2,3]. Demographic growth and unsustainable economic practices are affecting the water quantity and quality, making water an increasingly scarce and expensive resource, especially for the poor, the marginalized, and the vulnerable [4]. These factors will make it difficult to achieve the Sustainable Development Goal SDG 6 (one of the United Nations' Sustainable Development Goals (SDGs) for the year 2030), which requires sustainable management of clean accessible water for all. To achieve Goal 6, broad and in-depth knowledge of the global dynamics of water use and availability is necessary. Sustainable management of water resources for different uses will not only need to account for demand in water quantity, but also for water temperature, nutrient levels, and other pollutants [5].

The availability and quality of fresh water are particularly important in arid and hyper-arid environments, where groundwater plays an important role in supporting the economy, being the main source of water [6,7]. In these environments, the geological setting commonly contains bedrock of crystalline rocks characterized by secondary porosity due to the presence of interconnected fractures, faults, and shear zones, which represent a favorable setting for hosting and channeling groundwater [8,9]. Therefore, groundwater flow and yield in these aquifers are strictly related to the presence of lineaments and structural features in the basement [10–12]. An additional supply of water is represented by loose deposits, such as fluvial or lacustrine deposits, that are randomly distributed on the surface [13].

The success of expensive drilling campaigns is influenced by the complexity of the geological, structural, and hydrogeological setting of many arid environments and by the lack of subsoil investigations [14]. More specifically, geological features greatly affect the yield and quality of groundwater resources. As a consequence, high concentrations of many chemical elements can occur naturally in groundwater due to the local geology and to water interactions with rocks. Several studies [15–19] have shown that water in areas with particular geological features did not meet established drinking water limits for the presence of natural geogenic contaminants without influence from anthropogenic causes.

This study focuses on the characteristics of groundwater quality in the Oromia Region of the south-center portion of the Main Ethiopian Rift (MER), which represents the northern sector of the East African Rift System. In this region, rocks are mostly volcanic, and volcanic activity is still ongoing. Only a few perennial streams are present at the surface and, despite the presence of many lakes, the water supply in the area mainly depends on rainfall because of the poor physical and chemical quality of surficial water. More specifically, due to the scarce data about water availability, the paper is mainly focused on groundwater quality in the study area.

Natural sources of elevated fluoride concentrations in groundwater and lakes in the MER have been reported in the scientific literature [20–23] and, therefore, water is not always suitable for human consumption. Indeed, prolonged ingestion of drinking water containing F^- ions that exceed the tolerance limits can cause dental and skeletal fluorosis, with negative effects on children and young people [24].

The concentration and mobilization of F^- ions and other elements in groundwater are dependent on the chemical and physical processes that occur between the water and the geological environment. The occurrence of F^- in water is generally associated with volcanic rocks, particularly in the high-temperature geothermal settings that are common at convergent plate boundaries, as well as in intraplate areas characterized by extensional tectonic activity, such as the African Rift [22].

The World Health Organization recommends a maximum concentration limit of 1.5 mg/L, in the case of hot water (above 25 °C), or in tropical countries where the daily intake of drinking water is high, the value decreases to 0.7 mg/L [25]. High concentrations of fluoride affect human health. Fluoride may give rise to mild dental fluorosis at concentrations between 0.9 and 1.2 mg/L in drinking water [26]. This is particularly true in warmer areas, where dental fluorosis occurs with lower concentrations in drinking water because of the greater amount of water consumed [27,28]. Fluoride can also have more serious effects on skeletal tissues (bones and teeth) with long-term ingestion, particularly if the drinking water contains 3–6 mg of fluoride per liter [28].

In reason of the reported presence of fluoride contaminations in surface and groundwater in neighboring areas, the present study aims to propose a census of wells and springs in the Oromia Region and qualitative analysis of the water resource available for the local communities, with particular attention to the presence and distribution of fluorides. The outcomes of the study may provide important elements for a thoughtful decision-making process aimed at the sustainable human and rural local development of the region. We focused, more specifically, on a countryside area of the West Arsi Zone (Oromia Region) in the rural district of Arsi Negelle, Shashamane, and Siraro. The area is characterized by flat land, and the economy is based upon the integration of smallholding agriculture and livestock. As a consequence, life in the community and rural activities are tightly connected to the water supply [29–32]. This rural area is very populated, but only 30% of the people have access to water, and rural communities are at greater risk from meteorological, hydrological, and agricultural drought [33,34]. The results of the geographical distribution and physico-chemical features of the water resource were interpreted and discussed on the basis of the geological setting of the area, creating a global picture of the water availability and quality.

2. Geological Setting

The study area is located in the central sector of the Main Ethiopian Rift Valley (MER). The MER is a NNE-SSW trending segment of the Rift Valley, and it is bordered by the Ethiopian Plateau to the west and the Somali Plateau to the east by means of evident fault slopes belonging to the main Rift Valley System Fault [35].

Lowlands (approximately 1600 m a.s.l.), transitional escarpments and highlands (approximately 2500 m a.s.l.) are the main geomorphologic environments of the MER. Many volcanic lakes are present in the lowlands. The lakes have had a recent evolution marked by transgressive and regressive phases, each phase followed by deposition of fluvio-lacustrine deposits.

The bedrock in the highlands consists predominantly of basic volcanic rocks (lava and ash, mainly of Tertiary age), whereas the bedrock in the lowlands consists mainly of acidic volcanic rocks (peralkaline silica-rich rhyolitic ignimbrites, including ash and pumice). Weathered and reworked volcanic rocks, silicic tephra, and small alluvial fans occur along the escarpments and the border of the highlands (Figure 1). Eluvial lateritic crusts are also present and consist of clay, silt, and fine sand [20,21].

The MER is geographically divided into three sectors: (i) NMER (northern MER), from the Afar depression to near Lake Koka; (ii) CMER (central MER), from Lake Koka to Lake Awasa through the lake region; and (iii) SMER (southern MER), from Lake Awasa to the broadly rifted zone of southern Ethiopia [36] (Figure 1).

The part of the Oromia Region is included in the CMER and is located in the Zway-Shala basin area. This sector is characterized by bedrock mainly consisting of volcanic products (pyroclastic rocks, rhyolites, tuffs, and basaltic lava flows), and volcano-lacustrine and fluvio-lacustrine deposits (clay, silt, sand, and gravel interbedded layers) [35,37–39]. In particular, the study area contains the Langano, Abijata, Awasa, and Shala Lakes, which are the remnants of a single wide ancient lake that extended to the area where upper Quaternary fluvio-lacustrine deposits are now observed [35]. Some lakes are connected by an ephemeral drainage system consisting of a few perennial streams and are closely linked with the groundwater system [40].

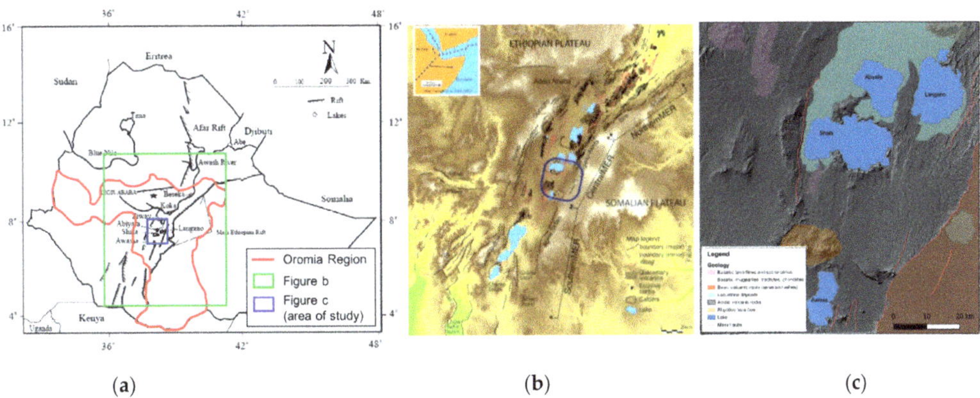

(a) (b) (c)

Figure 1. (a) Location of the Oromia Region, (b) sketch of the three geographic sectors of the Main Ethiopian Rift, in the blue square, the studied area (c) geological map of the study area (modified from [41]).

The extensional tectonics are responsible for the presence of lineaments that are subparallel to the NE–SW-trending rift axis. In particular, at least three sets of faults have been recognized in the area: NNE–SSW, N–S, and NNW–SSE trending faults. The first two sets are dominant and extend for long distances, following the axis of a tectonically active fault system called the Wonji Fault Belt (WFB). In addition, E-W- to NW-SE-trending, cross-rift oblique-slip fault systems have been locally observed [38,39]. The marginal escarpments have well-defined, steep normal-fault scarps [40].

Older structural trends have been identified in the highlands, where faults with different orientations are reported [34,41].

Active faulting within the rift (i.e., Wonji Fault Belt) causes geothermal and fumarolic activity, and high-temperature thermal springs occur along the border of some lakes and in connection with scattered volcanic centers [21].

3. Hydrological Setting

Two major aquifer classes can be identified in the MER: (i) Extensive aquifers with intergranular permeability (unconsolidated sediment: alluvium, eluvium, colluvium, and lacustrine sediment), and (ii) extensively fractured and weathered volcanic rocks (basalts, rhyolites, trachytes, and ignimbrites) [42]. The latter shows variable transmissivity and different hydraulic conductivities in relation to both the degree of fracturing and weathering, and the presence of interbedded palaeosols and alluvial or pyroclastic deposits within the volcanic series [43].

In fractured volcanic rocks, the flow is predominantly fault-controlled and laterally discontinuous: The groundwater flows parallel and subparallel to the main trending faults, influencing the subsurface hydraulic connection among the rift lakes and the relationship between the river and groundwater. Local palaeochannels drive the groundwater flow (i.e., the palaeochannel along the Bulbula River, which connects the Ziway and Abiyata Lakes) [40,44].

Based on geological characteristics and structural settings, distinct hydrogeological systems are recognized in the highlands and lowlands [45]. Lowlands are characterized both by fractured basaltic and ignimbritic aquifers, which are mainly unconfined or semi-confined, and permeable alluvial and colluvial deposits or lacustrine deposits forming the main shallow aquifers. Conversely, in the highlands, alluvial deposits (in alluvial plains or strips along river courses) and weathered volcanic rocks (with limited interbedded alluvial gravels and sands) are present, forming multilayer confined, semi-confined, and unconfined aquifers.

In general, the main sources of recharge are rainfall and river channel losses [42]. The rate of recharge is strongly influenced by the distribution and amount of rainfall, the permeability of the rocks, the geomorphological setting, and the presence of surface water close to major unconfined and semi-confined aquifers. The main groundwater recharge occurs in the highlands and preferentially moves along the rift within the large regional faults parallel and subparallel to the axis of the rift. In most cases, the escarpments act as discharge areas, which is manifested by the occurrence of high-discharge and fault-controlled springs along them. The rift floor represents a regional discharge area, and indirect recharge from rivers and lakes occurs in the highly faulted area. Recharge and groundwater flow in the weathered upper zone of the highlands are the driving force for much of the hydrology of the Zway-Shala basin, rather than deep upwelling flow systems with long residence times. The sharp decrease in elevation favors fast drainage of the groundwater in the form of springs. The springs located along the steep boundary between the rift and escarpment have very high seasonal variations in discharge (Figure 2). The same situation exists in highland areas close to deeply incised and large river valleys [40].

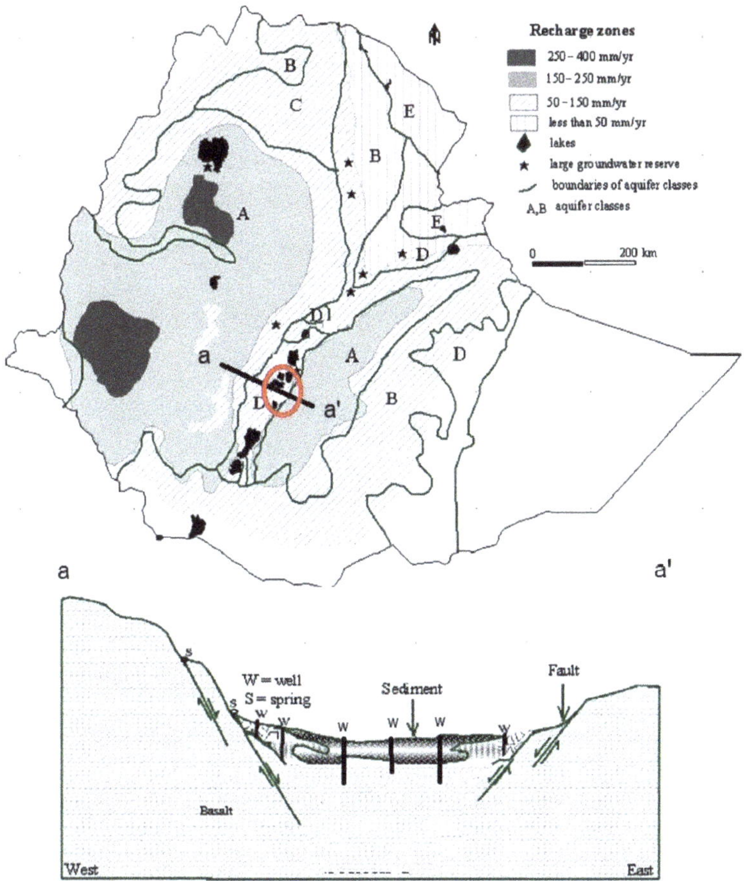

Figure 2. Map of groundwater recharge. A = widespread good quality groundwater at a relatively shallow depth (dominantly highland volcanic aquifers recharged by high rainfall). B = large groundwater reserve with fair to bad quality often localized in lower elevation areas (rift valley and volcanics in pediment covered with thick sediments and intermountain grabens), C = low to moderate groundwater reserve with fair quality (highland trap series volcanic aquifer with less sediment cover and recharge), D = medium to high groundwater reserve in the volcanics and sediments recharged by rainfall and rivers in places with serious salinity problem, E = low groundwater reserve with moderate quality recharged by seasonal floods and streams. aa': Profile sketching the hydrogeological setting of the area (modified from [40]). The red circle shows the investigated area.

4. Materials and Methods

A field survey and groundwater sampling were performed in the districts of Shashemene, Arsi Negelle, and Siraro. The field work consisted of a census of the existing water points (wells and springs). The wells were classified into deeply drilled wells (ranging from 20 to 400 m deep) and shallow manually dug wells (up to 20 m deep).

The location of water points is reported in Figure 3. Corresponding to each water point, information about the location (geographic coordinates) and the water point features (depth, diameter, discharge) was collected. Moreover, a picture showing the conditions of the wells and springs was included. The data collection was realized with the support of the administrative staff of the Woredas and the cooperation of social promoters (LVIA).

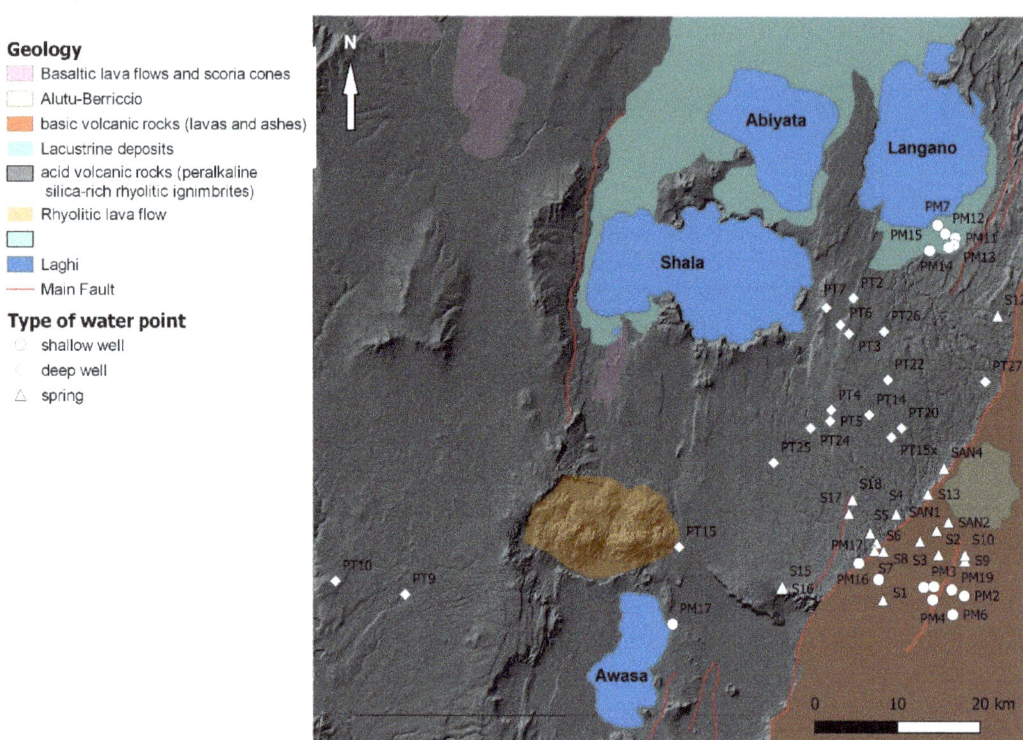

Figure 3. Types and Distribution of the Water Sampling Points in the Study Area.

To verify the groundwater quality, the following physical-chemical parameters (pH, electrical conductivity [EC], temperature, and fluoride concentration) were measured in situ with field instruments. The equipment for field measurements consisted of:

- handheld GPS instruments to locate points on a Landsat map through ArcPad,
- a portable fluoride meter (HI98402 by Hanna Instruments, USA) to measure the fluoride content in the water (ppm),
- a conductivity meter (HI99300 by Hanna Instruments, USA) to measure the water EC in the field (µS/cm),
- a freatimeter to measure the depth to groundwater in the wells;
- a litmus test for pH.

The collected field data were reported in a specific field card for each censused water point (Annex 1).

A total of 52 water points was sampled: 16 groundwater samples from shallow wells, 17 from deep wells, and 19 from springs. Groundwater from shallow wells was sampled using a polyvinyl chloride (PVC) bailer, 91 cm long and with a diameter of 4 cm. Bailer sampling techniques required a gentle lowering of the instrument into the water column of the well to reduce potential problems due to fluid turbulence and a proper transfer of water from the bottom of the bailer to sample containers. Bailers, when properly used, are an acceptable sampling tool [46,47]. Water sampling in deep wells was realized by means of pumps after at least five minutes of pumping, or until temperature and electrical conductivity EC remained stable, according to [48]. Sampling from springs was performed directly in correspondence with the water emergence from the soil. Groundwater was stored in 100-mL polyethylene bottles.

Water analyses were carried out at the Hydrochemical Laboratory of the Department of Earth Sciences at the University of Turin. Major anions (Br, Cl, F, NO_2, NO_3, SO_4) were analyzed using ion chromatography. The concentrations of Ca, Mg, CO_3, and HCO_3 were measured by titration. Other ions were analyzed by spectrometry. Fluoride concentration was measured both in situ and in the laboratory.

The chemical analyses were followed by the ionic balance calculation. The level of error was calculated by using the following formula:

$$\text{Error of ion balance} = ((\Sigma\text{cations} - \Sigma\text{ anions})/(\Sigma\text{cations} + \Sigma\text{ anions})) \times 100$$

An error of up to ±5% was considered as tolerable [49].

5. Results

The census of the water points in the Arsi Negele, Shashemene, and Siraro districts showed the presence of springs, shallow handmade wells, and deeply drilled wells, with a defined geographical distribution. Specifically, in this study we censused 16 shallow wells, primarily located along the eastern escarpment or along the highlands (PM in Annex 2), 30 deeply drilled wells, mainly located in the lowlands, (PT in Annex 2) and 21 springs, mainly located in the easternmost sector of the study area, particularly close to the eastern escarpment of the old caldera rift of Awasa Lake [50] (S or SAN in Annex 2). One of the censused springs (S15/c1) showed thermal physical-chemical features. Water in shallow wells may be contained in altered, highly fractured, or reworked volcanic rocks, residual soils or eluvial/colluvial deposits, while deep wells are mainly drilled in fractured volcanic rocks, intercepting water from the fractures.

The physical features of water (EC, pH, and temperature) showed an appreciable range of variability (Annex 2). The EC ranged from 57 to 1969 μS/cm, with the lowest values measured in the springs (from 57 to 287 μS/cm) and the highest values in the shallow and deep wells (from 109 to 1453 μS/cm). The pH values ranged from 5.5 to 9 in wells and from 5.5 to 6 in springs, with the only exception of the thermal spring, which was strongly alkaline (pH = 8). The distribution of pH and EC showed a decrease from the center of the lowlands (western-central sector of the study area) to the eastern highlands, with similar values closely aligned along an NNE/SSW trend. The temperature ranged from 17.6 to 30 °C, with higher values in the deep wells (from 21 to 30 °C, 25.4 °C on average), and lower values in the shallow wells (from 17.6 to 24 °C, 20 °C on average) and in the springs (from 18.2 to 25 °C). The thermal spring reaches a temperature of 90 °C.

The results of the chemical analyses, reported in Annex 2 and summarized in Figures 4 and 5, identified different hydrochemical facies for the different kinds of well. In the springs and in the shallow wells, water samples can be described as both sodium bicarbonate and calcium bicarbonate, even if a higher concentration of HCO_3 was registered in the wells. In the deep wells, on the other hand, a prevalence of sodium bicarbonate facies was observed (with the only exception of p27/C1 well), suggesting the presence of a deep groundwater circulation influenced by ion exchange (i.e., Ca-Na ion exchange).

The Schoeller diagram (Figure 5) shows a similar trend for all of the samples, with the exception of the S15c/1 (i.e., thermal spring), which had high Na, Cl, and HCO_3 concentrations and very low Mg content.

In all the performed tests, the error of ion balance was inferior to ±5%, and consequently, it was considered tolerable.

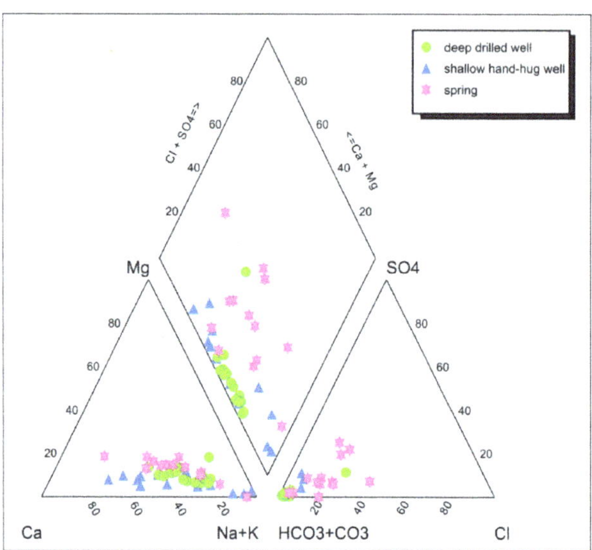

Figure 4. Piper classification diagram for different water types in the investigated area.

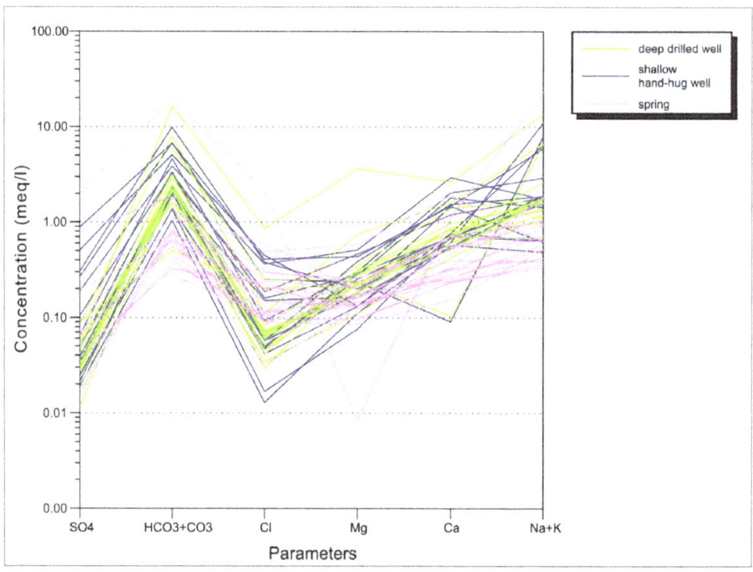

Figure 5. Schoeller classification diagram for different water types in the investigated area.

In the deep wells, despite the low concentrations of Ca and Mg, the concentration of F was usually high. Fluoride exceeding the WHO maximum acceptable concentration (1.5 mg/L) was highlighted in 25% of the samples (69% were from deeply drilled wells, 23% from shallow wells, and 8% from springs). More than 21% of the samples ranged between 0.7 and 1.5 mg/L in fluoride concentration (45% were from shallow wells and 55% from deeply drilled wells). A fluoride concentration between 1.5 and 4 mg/L was observed in the area south of Langano Lake (Arsi Negele District), whereas concentrations higher than 4 mg/L were mainly observed in the Awasa volcanic district, in the westernmost study area

(Siraro District), and southwest of Shala Lake, close to the old caldera rift [50]. The lowest values (less than 0.7 mg/L) were observed along the eastern escarpment and highlands, while the maximum concentrations were in the thermal spring of Awasa (S15/c1) and in the deep well PT15x/c1, located northeast of Awasa Lake, with values of 28.6 mg/L and 13.1 mg/L, respectively (Figure 6). However, with the exception of the thermal one, the censused springs generally showed low salinity and low fluoride concentration. The measured concentrations of fluoride describe a distribution similar to the one observed for pH and EC values, with a decrease from the center of the lowlands (western-central sector of the study area) to the eastern highlands, with similar values closely aligned along an NNE/SSW trend.

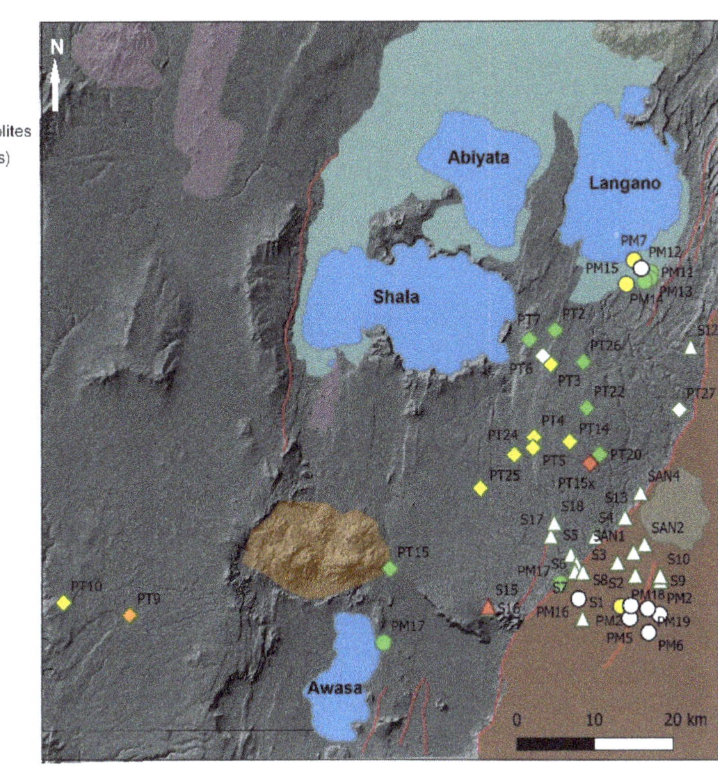

Figure 6. Geographical distribution of the fluoride values recorded in the investigated area.

The concentration of nitrates was usually below the drinking water standards (50 mg/L, [51]). Both hand-dug shallow wells and deep wells showed low nitrate concentrations, inferior to 15 mg/l (with the only exceptions of PM17/c4, having a nitrate level of 36.5 mg/L and PT27/c1, with a nitrate concentration of 45.6 mg/L). In the springs, on the other hand, the nitrate level was generally higher, often showing a concentration greater than 10 mg/L. However, only one of the springs (S16/c1) presented a nitrate concentration higher than the limits, equal to 70.1 mg/L.

6. Discussion

The geographical distribution of shallow and deep wells and springs in the Oromia region (Figures 3–6) suggests the presence of two different aquifers, respectively exploited with the different kinds of wells. The maps show, indeed, a preferential location of shallow manually-dug wells and cold springs in correspondence with the eastern escarpment and highlands, while the deep wells and the thermal springs are mainly located in the lowland areas of the rift. The physical and chemical features of water support the hypothesis of two different water circuits, as also observed by Ayenew [40,45] and Rango et al. [21]. Water samples from the shallow wells and from the cold springs showed, indeed, lower temperatures (17 to 25 °C) and Ca–HCO_3 or Na–HCO_3 hydrochemical facies. In contrast, the hot spring and most of the groundwater samples in the rift displayed a Na–HCO_3 fingerprint, with Na and HCO_3 concentrations representing more than 80% of the ionic species in the solution. Consequently, it is possible to suggest a compositional change from Ca–Mg/HCO_3 to Na–HCO_3 along the path of groundwater flow from the highlands to the rift floor.

In accordance with the hydrogeological model proposed by [40], the shallow aquifer is hosted in altered rocks, residual soils, and fluvio-lacustrine deposits and is usually present in the uppermost tens of meters, where water flows in porous media. This aquifer is mainly located in the highland end-rift escarpments and appears more productive than those in the deep escarpments, with a decrease in permeability with the degree of alteration. In contrast, the deep aquifer of the lowlands is hosted in fractured volcanic bedrock where water is intercepted at different depths, generally hundreds of metres below the surface (up to 250–300 m). In the northern part of lowland areas, near to the lakes, however, additional shallow porous aquifers may be hosted in the fluvio-lacustrine deposits, as confirmed by the shallow wells PM7, PM11, PM13, PM15.

Data obtained from chemical analyses showed that the increase in Na observed in the deep aquifer affected the F concentration (Figure 7). Similarly, water with relatively low contents of Ca and Mg had a higher concentration of fluoride. High fluoride concentration values reflect a strong removal of Ca from the solution and a concurrent enrichment in Na. An increase in EC and temperature were also associated with the F concentration: The water temperature generally increased with the depth of the wells, as well as with the fluoride concentration (Figure 8).

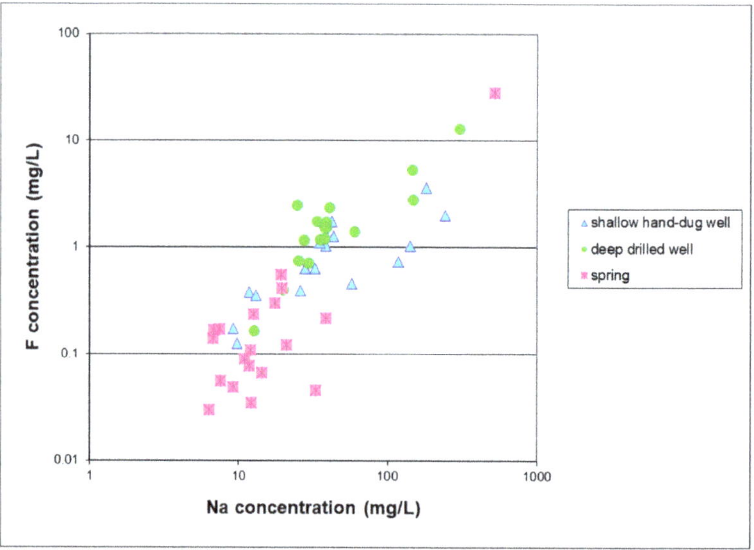

Figure 7. Scatter plot between sodium and fluoride (linear trend with equation $y = 0.04x - 0.7$ and $R^2 = 0.8$).

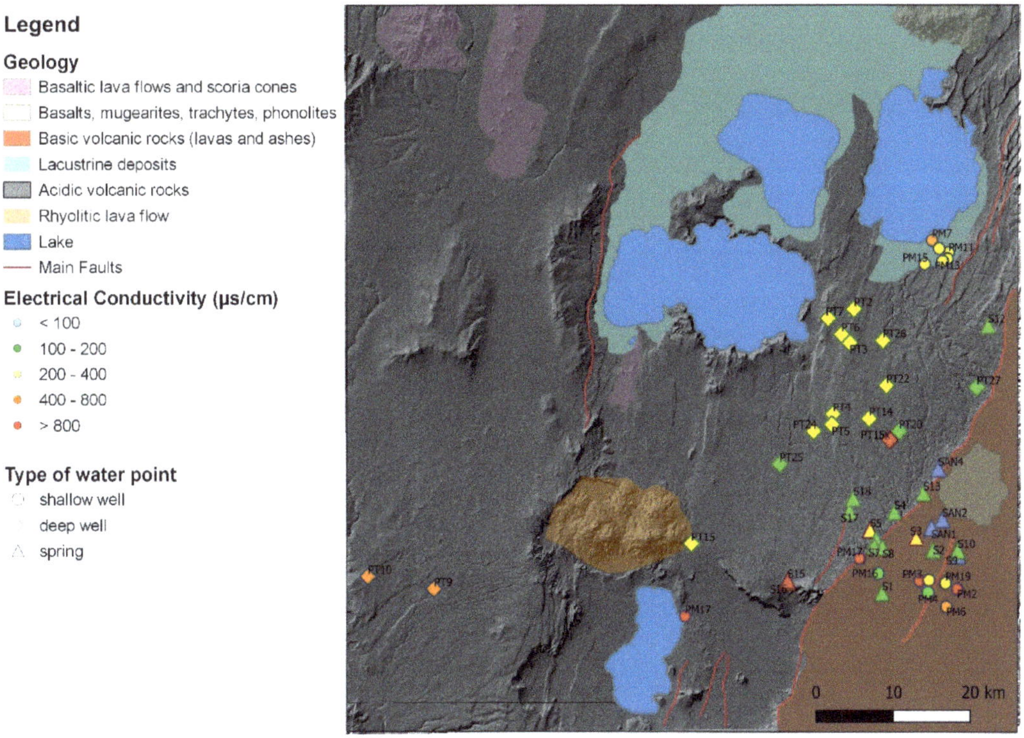

Figure 8. Geographical distribution of ce values recorded in the investigated area.

The decreasing trend of fluoride and pH and EC values from the center of the lowlands to the eastern highlands is consistent with the geological and structural framework of the area, in accordance with the type of bedrock, the distribution of surface deposits, and the orientation of the main faults (NNE/SSW to NE/SW trend).

For the springs, various hydrochemical features were recorded, suggesting two different origins, which are also reported in the literature. Springs can have a superficial flow circuit with low electrical conductivity and fluoride concentration, and relatively high nitrate levels, or a deep flow circuit of thermal water with high electrical conductivity, fluoride concentration, and temperature, and low nitrate levels. These data support an anthropic origin of nitrate concentration, probably due to the oxidation of nitrogenous waste products in human and animal excreta [52–54]. The discrimination between animal manure and domestic sewage was not possible using the current data. However, a groundwater isotopic analysis (i.e., using nitrate and boron isotopes) could be a valuable help to distinguish between these two nitrate sources and to implement management actions for groundwater protection [55,56].

The fluoride content has a geogenic origin, which is supported by the geological setting and the low anthropogenic pressure in the region, as well as by the low contents of nitrate and sulfate. During magmatic processes, fluoride does not easily fit into the crystal structure of minerals and is preferentially partitioned into melt, remaining in the residual magma until its crystallization. Consequently, fluoride is more concentrated in evolved sialic rocks, such as the bedrock of the lowland Oromia Region, rather than in primitive basic rocks (e.g., basalts) [22]. The increase in pH and temperature favors the interaction between sialic rock and water because they influence the reactions of dissolution and adsorption-desorption, which contribute to the mobility of the fluoride contained in the primary mineral phases (such as fluoride, apatite, amphibole, biotite, and volcanic glass) or secondary minerals (such as Fe^-, Al^- and Mn^- oxides and hydroxides, and clay minerals), particularly in arid or semi-arid climates [21,57,58].

Locally, the leaching of fluoride in groundwater is connected to the presence of active faults where the circulation of CO_2 enhances the dissolution process of silicates, oxides, and hydroxides [59]. In the study area, which is located in an active rift zone, similar conditions are common and can contribute to fluoride mobilization from rock and sediment into groundwater. The alignment of water points (springs and wells) with similar F^- ion concentrations along preferential NNE-SSW trends (consistent with the orientation of the main faults) highlights the possible tectonic contribution to the high content of fluorides in the groundwater.

7. Conclusions

In arid and semi-arid areas, human health and economic development depend on water availability, which can be strongly compromised by droughts [60]. Many people have no access to fresh water. Surface water and precipitation are insufficient, and in some cases, the quality of the water is compromised by natural contaminants.

The present research analyzed the physical and chemical characteristics of groundwater in a rural area of the central MER, particularly in the districts (Woredas) of Shashemene, Arsi Negelle, and Siraro, to create awareness of the availability, quality, and risks connected to water resource management. The study area is located in a symmetrical rift basin where groundwater flows from the highlands to the rift valley, across well-defined marginal faults that form escarpments, and into the fractured rift floor, which is dotted by volcanic cones. The study was developed using a census of the main water points (springs and wells) in the area with sampling and physical-chemical analysis of the water samples, with particular regard to the fluoride concentration. The hydrochemical analyses of the groundwater showed large variability in both geographical location and depth of the water points. Regarding the geographical distribution, most springs and shallow wells were located in relation to the eastern highland and at the bottom of the eastern plateau slopes. The deep-drilled wells were, instead, concentrated in the center lowlands of the rift valley. The depth of the groundwater surface varied within a wide range. The depth of the groundwater surface ranged from 250 to 300 m in the lowlands to a few meters in local shallow aquifers with scarce productivity, whereas, in most of the highland plains, the water depth did not exceed 50–60 m. Fluoride represents the main pollutant of groundwater. The highest fluoride contents were measured in the deep-drilled wells of the lowlands, which were related to the fractured aquifer, or in the thermal springs. Moreover, similar fluoride concentrations were primarily aligned in an NNE-SSW direction, similar to the orientation of the main fault systems of the MER. Low fluoride contents (<0.7 mg/L) were normally measured in connection with the shallow and porous aquifer of the highlands and eastern escarpment. Commonly, high fluoride concentrations were associated with Na–HCO3-type water. Two different origins were observed for the springs: A superficial flow circuit with low electrical conductivity and fluoride concentration and relatively high nitrate levels, or a deep flow circuit of thermal water with high electrical conductivity, fluoride concentration and temperature, and low nitrate levels.

The study describes the hydrogeological setting of the study area, providing a framework of the water-point distribution and water quality. The collected data provide information on the risks with rural communities and local governments. Moreover, these results can be useful to support the adequate use of the available water resources and to plan proper actions to increase access to fresh water (new water points, locations, expected depths of the water level, associated costs, etc.), aimed at the sustainable human and rural local development of the region.

Future studies in this region may include the measures of the static level of the water table in different periods of the year in order to evaluate the influence of the seasonality (i.e., dry season and rain season) on the water resource. In addition, quantitative analysis through water flow measurements would be important to assess the real amount of available water in correspondence with the different types of wells.

Supplementary Materials: The following are available online at https://www.mdpi.com/2071-1050/13/3/1347/s1.

Author Contributions: Conceptualization, S.M.R.B. and M.L.; methodology, S.M.R.B. and M.L.; software, M.L., C.C.; validation, S.M.R.B., M.L., D.A.d.L.; formal analysis, C.C.; data curation, C.C.; writing—original draft preparation, C.C., S.M.R.B., M.L.; writing—review and editing, C.C.; supervision, D.A.d.L.; project administration, S.M.R.B. All authors have read and agreed to the published version of the manuscript.

Funding: This research received no external funding.

Institutional Review Board Statement: Not applicable.

Informed Consent Statement: Not applicable.

Data Availability Statement: Data is contained within the article or supplementary material.

Acknowledgments: The authors desire to thank LVIA Onlus for the useful contribution in the on-site activities.

Conflicts of Interest: The authors declare no conflict of interest.

References

1. Gleick, P.H. Water Use. *Annu. Rev. Environ. Resour.* **2003**, *28*, 275–314. [CrossRef]
2. Kundzewicz, Z.W.; Krysanova, V. Climate change and stream water quality in the multi-factor context. *Clim. Chang.* **2010**, *103*, 353–362. [CrossRef]
3. Lasagna, M.; Ducci, D.; Sellerino, M.; Mancini, S.; De Luca, D.A. Meteorological Variability and Groundwater Quality: Examples in Different Hydrogeological Settings. *Water* **2020**, *12*, 1297. [CrossRef]
4. Unesco. Available online: https://sustainabledevelopment.un.org/topics/waterandsanitation (accessed on 15 November 2020).
5. van Vliet, M.T.H.; Flörke, M.; Wada, Y. Quality matters for water scarcity. *Nat. Geosci.* **2017**, *10*, 800–802. [CrossRef]
6. Bechis, S.; Bonetto, S.; Bucci, A.; Canone, D.; Cristofori, E.; De Luca, D.A.; Demarchi, A.; Garnero, G.; Guerreschi, P.; Lasagna, M. Improving governance of access to water resources and their sustainable use in Hodh el Chargui communities (South East Mauritania). In Proceedings of the 20th EGU General Assembly 2018, Vienna, Austria, 4–13 April 2018; Volume 20. EGU2018-15888-1.
7. Datta, S.; Ajaz, A. Geospatial Data Assimilation and Mapping Groundwater Vulnerability in High Plains Aquifer Using DRASTIC Model. *Fundam. Appl. Agric.* **2019**, *4*, 933–942. [CrossRef]
8. Sultan, M.; Wagdy, A.; Manocha, N.; Sauck, W.; Gelil, K.A.; Youssef, A.; Becker, R.; Milewski, A.; El Alfy, Z.; Jones, C. An integrated approach for identifying aquifers in transcurrent fault systems: The Najd shear system of the Arabian Nubian shield. *J. Hydrol.* **2008**, *349*, 475–488. [CrossRef]
9. Tessema, A.; Mengistu, H.; Chirenje, E.; Abiye, T.A.; Demlie, M. The relationship between lineaments and borehole yield in North West Province, South Africa: Results from geophysical studies. *Hydrogeol. J.* **2011**, *20*, 351–368. [CrossRef]
10. Fernandes, A.J.; Rudolph, D.L. The influence of Cenozoic tectonics on the groundwater-production capacity of fractured zones: A case study in Sao Paulo, Brazil. *Hydrogeol. J.* **2001**, *9*, 151–167. [CrossRef]
11. Prabu, P.; Rajagopalan, B. Mapping of lineaments for groundwater targeting and sustainable water resource management in hard rock hydrogeological environment using RS-GIS. In *Climate Change and Regional/Local Responses*; Pallav, R., Ed.; InTech: London, UK, 2013; pp. 235–247.
12. Bonetto, S.M.R.; Facello, A.; Umili, G. The contribution of CurvaTool semi-automatic approach in structural and groundwater investigations. A case study in the Main Ethiopian Rift Valley. *Egypt. J. Remote Sens. Space Sci.* **2020**, *23*, 97–111. [CrossRef]
13. Bonetto, S.; Facello, A.; Cristofori, E.I.; Camaro, W.; Demarchi, A. An Approach to Use Earth Observation Data as Support to Water Management Issues in the Ethiopian Rift. In *Climate Change Adaptation in Africa: Fostering Resilience and Capacity to Adapt. Climate Change Management*; Leal Filho, W., Belay, S., Kalangu, J., Means, W., Munishi, P., Musiyiwa, K., Eds.; Springer: Cham, Switzerland, 2016; pp. 357–374. [CrossRef]
14. Lasagna, M.; Bonetto, S.M.R.; Debernardi, L.; De Luca, D.A.; Semita, C.; Caselle, C. Groundwater Resources Assessment for Sustainable Development in South Sudan. *Sustainability* **2020**, *12*, 5580. [CrossRef]
15. Bretzler, A.; Johnson, C.A. The Geogenic Contamination Handbook: Addressing arsenic and fluoride in drinking water. *Appl. Geochem.* **2015**, *63*, 642–646. [CrossRef]
16. Nickson, R.; McArthur, J.M.; Ravenscroft, P.; Burgess, W.G.; Ahmed, K. Mechanism of arsenic release to groundwater, Bangladesh and West Bengal. *Appl. Geochem.* **2000**, *15*, 403–413. [CrossRef]

17. Hussain, I.; Arif, M.; Hussain, J. Fluoride contamination in drinking water in rural habitations of Central Rajasthan, India. *Environ. Monit. Assess* **2011**, *184*, 5151–5158. [CrossRef] [PubMed]
18. Missimer, T.M.; Teaf, C.; Beeson, W.T.; Maliva, R.G.; Woolschlager, J.; Covert, D. Natural Background and Anthropogenic Arsenic Enrichment in Florida Soils, Surface Water, and Groundwater: A Review with a Discussion on Public Health Risk. *Int. J. Environ. Res. Public Health* **2018**, *15*, 2278. [CrossRef] [PubMed]
19. Alarcón-Herrera, M.T.; Martin-Alarcon, D.A.; Gutiérrez, M.; Reynoso-Cuevas, L.; Martín-Domínguez, A.; Olmos-Márquez, M.A.; Bundschuh, J. Co-occurrence, possible origin, and health-risk assessment of arsenic and fluoride in drinking water sources in Mexico: Geographical data visualization. *Sci. Total Environ.* **2020**, *698*, 134168. [CrossRef]
20. Ayenew, T. The distribution and hydrogeological controls of fluoride in the groundwater of central Ethiopian rift and adjacent highlands. *Environ. Earth Sci.* **2007**, *54*, 1313–1324. [CrossRef]
21. Rango, T.; Bianchini, G.; Beccaluva, L.; Tassinari, R. Geochemistry and water quality assessment of central Main Ethiopian Rift natural waters with emphasis on source and occurrence of fluoride and arsenic. *J. Afr. Earth Sci.* **2010**, *57*, 479–491. [CrossRef]
22. Bianchini, G.; Brombin, V.; Marchina, C.; Natali, C.; Godebo, T.R.; Rasini, A.; Salani, G.M. Origin of Fluoride and Arsenic in the Main Ethiopian Rift Waters. *Minerals* **2020**, *10*, 453. [CrossRef]
23. Bonetto, S.M.R.; De Luca, D.A.; Lasagna, M.; Lodi, R. Groundwater Distribution and Fluoride Content in the West Arsi Zone of the Oromia Region (Ethiopia). In *Engineering Geology for Society and Territory*; Springer: Cham, Switzerland, 2014; Volume 3, pp. 579–582.
24. Rango, T.; Bianchini, G.; Beccaluva, L.; Ayenew, T.; Colombani, N. Hydrogeochemical study in the Main Ethiopian Rift: New insights to the source and enrichment mechanism of fluoride. *Environ. Earth Sci.* **2009**, *58*, 109–118. [CrossRef]
25. WHO. *Volume 1—Recommendations. Guidelines for Drinking Water Quality*, 3rd ed.; WHO: Geneva, Switzerland, 2004.
26. Moulton, F.R. *Fluorine and Dental Health*; American Association for the Advancement of Science: Washington, DC, USA, 1942.
27. Environmental Protection Agency (EPA). *Office of Drinking Water. Drinking Water Criteria Document on Fluoride*; TR-823-5; US Environmental Protection Agency: Washington, DC, USA, 1985.
28. WHO. *Fluorine and Fluorides*; Environmental Health Criteria: Geneva, Switzerland, 1984; No. 36.
29. Bonetto, S.; Facello, A.; Camaro, W.; Cristofori, E.I.; Demarchi, A. An approach to integrate spatial and climatological data as support to drought monitoring and agricultural management problems in South Sudan. In Proceedings of the EGU General Assembly, Vienna, Austria, 17–22 April 2016; Volume 18. EGU2016-16952-2.
30. Dan-Badjo, A.T.; Diadie, H.O.; Bonetto, S.M.R.; Semita, C.; Cristofori, E.I.; Facello, A. Using Improved Varieties of Pearl Millet in Rainfed Agriculture in Response to Climate Change: A Case Study in the Tillabéri Region in Niger. In *Climate Change Research at Universities: Addressing the Mitigation and Adaptation Challenges*; Leal Filho, W., Ed.; Springer International Publishing: Cham, Switzerland, 2017; pp. 345–358.
31. Lasagna, M.; Dino, G.A.; Perotti, L.; Spadafora, F.; De Luca, D.A.; Yadji, G.; Dan-Badjo, A.T.; Moussa, I.; Harouna, M.; Konaté, M.; et al. Georesources and Environmental Problems in Niamey City (Niger): A Preliminary Sketch. *Energy Procedia* **2015**, *76*, 67–76. [CrossRef]
32. Perotti, L.; Dino, G.A.; Lasagna, M.; Moussa, K.; Spadafora, F.; Yadji, G.; Dan-Badjo, A.T.; De Luca, D.A. Monitoring of Urban Growth and its Related Environmental Impacts: Niamey Case Study (Niger). *Energy Procedia* **2016**, *97*, 37–43. [CrossRef]
33. Demarchi, A.; Bechis, S.; Perott, L.; Garnero, G.; Isotta Cristofori, E.; Alunno, L.; Facello, A.; Semita, C.; Bonetto, S.; Guerreschi, P. An interdisciplinary approach to the analysis of agro pastoral resilience in the Hodh el Chargui region (Mauritania). In Proceedings of the 20th EGU General Assembly 2018, Vienna, Austria, 4–13 April 2018; Volume 20. EGU2018-15808-1.
34. Caselle, C.; Bonetto, S.M.R.; De Luca, D.A.; Lasagna, M.; Perotti, L.; Bucci, A.; Bechis, S. An Interdisciplinary Approach to the Sustainable Management of Territorial Resources in Hodh el Chargui, Mauritania. *Sustainability* **2020**, *12*, 5114. [CrossRef]
35. Benvenuti, M.; Carnicelli, S.; Belluomini, G.; Dainelli, N.; Di Grazia, S.; Ferrari, G.; Iasio, C.; Sagri, M.; Ventra, D.; Atnafu, B.; et al. The Ziway–Shala lake basin (main Ethiopian rift, Ethiopia): A revision of basin evolution with special reference to the Late Quaternary. *J. Afr. Earth Sci.* **2002**, *35*, 247–269. [CrossRef]
36. Keranen, K.; Klemperer, S.L. Discontinuous and diachronous evolution of the Main Ethiopian Rift: Implications for develop-ment of continental rifts. *Earth Planet Sci. Lett.* **2008**, *265*, 96–111. [CrossRef]
37. Abebe, B.; Boccaletti, M.; Mazzuoli, R.; Bonini, M.; Tortorici, L.; Trua, T. *Geological Map of the Lake Ziway—Asela Region (Main Ethiopian Rift)*; 1:50000 scale; C.N.R., ARCA: Firenze, Italy, 1998.
38. Boccaletti, M.; Bonini, M.; Mazzuoli, R.; Trua, T. Pliocene-Quaternary volcanism and faulting in the northern Main Ethiopian Rift (with two geological maps at scale 1:50,000). *Acta Vulcanol.* **1999**, *11*, 83–97.
39. Abbate, E.; Bruni, P.; Sagri, M. Geology of Ethiopia: A Review and Geomorphological Perspectives. In *Landscapes and Landforms of Norway*; Billi, P., Ed.; Springer: Dordrecht, The Netherlands, 2015; pp. 33–64.
40. Ayenew, T.; Demlie, M.; Wohnlich, S. Hydrogeological framework and occurrence of groundwater in the Ethiopian aquifers. *J. Afr. Earth Sci.* **2008**, *52*, 97–113. [CrossRef]
41. Dainelli, N.; Benvenuti, M.; Sagri, M. *Geological Map of the Ziway-Shala Lakes Basin*; 1:250,000. European Commission (EC), STD3 Project—Contract TS3—CT92-0076, 2001; Italian Ministry for University and Scientific and Technological Research (MURST) DB Map; Department of Earth Science, University of Florence: Firenze, Italy, 2001.
42. EIGS. *Hydrogeological Map of Ethiopia*; 1:2,000,000 scale; Ethiopian Institute of Geological Surveys: Addis Ababa, Ethiopia, 1993.

43. Kebede, S.; Travi, Y.; Alemayehu, T.; Ayenew, T. Groundwater recharge, circulation and geochemical evolution in the source region of the Blue Nile River, Ethiopia. *Appl. Geochem.* **2005**, *20*, 1658–1676. [CrossRef]
44. Chernet, T.; Travi, Y. Preliminary observations concerning the genesis of high flouride contents in the Ethiopian Rift. In *Geoscientific Research in Northeast Africa*; Thorweiche, U., Schandlmeier, H., Eds.; Balkema: Rotterdam, The Netherlands, 1993; Volume 8, pp. 651–654.
45. Ayenew, T. Major ions composition of the groundwater and surface water systems and their geological and geochemical controls in the Ethiopian volcanic terrain. *SINET Ethiop. J. Sci.* **2006**, *28*, 171–188. [CrossRef]
46. Lee, W.E., III. A tale of two samplers-part I: A comparison of bailer and peristaltic pump groundwater sampling protocols. *Pollut. Eng.* **2002**, *34*, 4.
47. Lasagna, M.; De Luca, D.A. The use of multilevel sampling techniques for determining shallow aquifer nitrate profiles. *Environ. Sci. Pollut. Res.* **2016**, *23*, 20431–20448. [CrossRef]
48. Yeskis, D.; Zavala, B. Ground-Water Sampling Guidelines for Superfund and RCRA Project Managers. In *Ground Water Forum Issue Paper, EPA*; US EPA: Washington, DC, USA, 2002; EPA 542-S-02-001, 53p.
49. Appelo, C.; Postma, D. *Geochemistry, Groundwater and Pollution*; Balkema: Rotterdam, The Netherlands, 2004.
50. Boccaletti, M.; Getaneh, A.; Mazzuoli, R.; Tortorici, L.; Trua, T. Chemical variations in a bimodal magma system: The Plio-Quaternary volcanism in the Dera Nazret area (Main Ethiopian Rift, Ethiopia). *Afr. Geosci. Rev.* **1995**, *2*, 37–60.
51. WHO. *Guidelines for Drinking-Water Quality*, 4th ed.; WHO Library Cataloguing-in-Publication Data: Geneva, Switzerland, 2011.
52. WHO. Nitrate and Nitrite in Drinking-Water. Background Document for Development of WHO Guidelines for Drinking Water Quality. Available online: https://www.who.int/water_sanitation_health/dwq/chemicals/nitratenitrite2ndadd.pdf (accessed on 20 January 2021).
53. Martinelli, G.; Dadomo, A.; De Luca, D.; Mazzola, M.; Lasagna, M.; Pennisi, M.; Pilla, G.; Sacchi, E.; Saccon, P. Nitrate sources, accumulation and reduction in groundwater from Northern Italy: Insights provided by a nitrate and boron isotopic database. *Appl. Geochem.* **2018**, *91*, 23–35. [CrossRef]
54. Lasagna, M.; De Luca, D.A.; Franchino, E. Intrinsic groundwater vulnerability assessment: Issues, comparison of different methodologies and correlation with nitrate concentrations in NW Italy. *Environ. Earth Sci.* **2018**, *77*, 277. [CrossRef]
55. Lasagna, M.; De Luca, D.A.; Franchino, E. The role of physical and biological processes in aquifers and their importance on groundwater vulnerability to nitrate pollution. *Environ. Earth Sci.* **2016**, *75*, 961. [CrossRef]
56. Lasagna, M.; De Luca, D.A. Evaluation of sources and fate of nitrates in the western Po plain groundwater (Italy) using nitrogen and boron isotopes. *Environ. Sci. Pollut. Res.* **2019**, *26*, 2089–2104. [CrossRef] [PubMed]
57. Amini, M.; Mueller, K.; Abbaspour, K.C.; Rosenberg, T.; Afyuni, M.; Møller, K.N.; Sarr, M.; Johnson, C.A. Statistical Modeling of Global Geogenic Fluoride Contamination in Groundwaters. *Environ. Sci. Technol.* **2008**, *42*, 3662–3668. [CrossRef]
58. Belete, A.; Beccaluva, L.; Bianchini, G.; Colombani, N.; Fazzini, M.; Marchina, C.; Natali, C.; Rango, T. Water-rock intercation and lake hydrochemistry in the Main Ethiopian Rift. In *Landscape and Landforms of Ethiopia*; Paolo, B., Ed.; Springer: Berlin/Heidelberg, Germany, 2015; pp. 307–321.
59. Sracek, O.; Wanke, H.; Ndakunda, N.; Mihaljevic, M.; Buzek, F. Geochemistry and fluoride levels of geothermal springs in Namibia. *J. Geochem. Explor.* **2015**, *148*, 96–104. [CrossRef]
60. Caselle, C.; Lasagna, M.; Bonetto, S.M.R.; De Luca, D.A.; Bechis, S. Groundwater features in Hoedh el Chargui, Mauritania. *Acque Sotter. Ital. J. Groundw.* **2020**, *9*, 43–51. [CrossRef]

Article

Development of a Real-Time, Mobile Nitrate Monitoring Station for High-Frequency Data Collection

Martin Jason Luna Juncal [1], **Timothy Skinner** [1], **Edoardo Bertone** [1,2,3,*] and **Rodney A. Stewart** [1,2]

1 School of Engineering and Built Environment, Griffith University, Gold Coast Campus, Southport 4222, Australia; martin.lunajuncal@griffithuni.edu.au (M.J.L.J.); timothy.skinner@griffithuni.edu.au (T.S.); r.stewart@griffith.edu.au (R.A.S.)
2 Cities Research Institute, Griffith University, Gold Coast Campus, Southport 4222, Australia
3 Australian Rivers Institute, Griffith University, Gold Coast Campus, Southport 4222, Australia
* Correspondence: e.bertone@griffith.edu.au; Tel.: +61-7555-28574

Received: 29 May 2020; Accepted: 16 July 2020; Published: 17 July 2020

Abstract: A mobile monitoring station was developed to measure nitrate and physicochemical water quality parameters remotely, in real-time, and at very high frequencies (thirty minutes). Several calibration experiments were performed to validate the outputs of a real-time nutrient sensor, which can be affected by optical interferences such as turbidity, pH, temperature and salinity. Whilst most of these proved to play a minor role, a data-driven compensation model was developed to account for turbidity interferences. The reliability of real-time optical sensors has been questioned previously; however, this study has shown that following compensation, the readings can be more accurate than traditional laboratory-based equipment. In addition, significant benefits are offered by monitoring waterways at high frequencies, due to rapid changes in analyte concentrations over short time periods. This, combined with the versatility of the mobile station, provides opportunities for several beneficial monitoring applications, such as of fertiliser runoff in agricultural areas in rural regions, aquaculture runoff, and waterways in environmentally sensitive areas such as the Great Barrier Reef.

Keywords: agriculture; Nitrate runoff; real-time monitoring; water quality

---

1. Introduction

Nitrate fluxes in the water system due to fertiliser mismanagement create substantial environmental and health issues where, particularly in rural communities, efficient and reliable water treatment processes are not always available [1]. Reliable monitoring of nitrates is therefore critical to understanding and predicting nutrient runoff and, in turn, poor fertiliser application practices.

Traditional methods rely on time-intensive sampling and analysis processes, usually through surface water samples which are collected and analysed in a laboratory (Table A1, Appendix A) [2,3]. Real-time nutrient sensors have the potential to generate widespread benefits for agriculture and aquaculture industries, while also providing a rapid source of high-frequency data that can assist in mitigating environmental damage in sensitive areas.

One such method of automation involves the use of high-frequency optical sensors. However, the reliability and potential successes of optical nitrate sensors is understudied and has been questioned previously [4]. One project relied on real-time sensors for pH, conductance, temperature, turbidity, chlorophyll and dissolved oxygen monitoring in water streams; however, to complete nutrient analyses, manual sampling was required to develop a regression model that could estimate bacterial concentrations [5]. As a result, the study may have benefited from the use of real-time nutrient sensors due to the potential ability to correlate temporal data with the water quality probe results. Another study implemented real-time monitoring buoys to obtain nutrient data [6]. While innovative, these data are limited due to the small measurement range of the instrument (0–5 mg L^{-1} of NO_3), as well as a maximum autonomous monitoring period of six months [7].

Currently, there is not a complete acceptance of real-time nutrient sensors for reliable mainstream water quality monitoring [8]. As a result, this perceived unreliability leads to skepticism, which in turn hinders opportunities for the widespread implementation of real-time nutrient sensors. However, previous studies [9] have shown that readings from optical sensors can be compensated for optical interferences while simultaneously deployed remotely. As a consequence, there is the potential, through a combination of targeted experiments and data-driven modelling, to improve the reliability of in-situ optical sensors, such as those targeting nitrates. In turn, this would assist in allowing responses to environmental issues to occur in real-time.

Consequently, the overarching goal of this study was to design, construct, calibrate and field test a real-time nitrate measurement instrument to enable high-frequency and remote nitrate monitoring applications to occur in the future. This goal was achieved through a number of research tasks, including comparing the accuracy of a NiCaVis 705 IQ sensor (from YSI, Yellow Springs, OH, USA) to a spectrophotometer, under laboratory conditions. The laboratory experiments also contributed to the second aim, which was to design a compensation model for nitrate optical sensors. Finally, this testing and the subsequent compensation models contributed to the third aim of the study, which was to develop and deploy a mobile monitoring station to obtain and transmit water quality data, especially nitrate data, in real-time and at very high-frequencies (30-minute intervals). The achievement of these three aims contributed to our goal to foster the deployment of reliable real-time nitrate sensors for a range of important applications.

2. Materials and Methods

2.1. Phase I. Laboratory Calibration

To start the development of a mobile monitoring system, a comprehensive laboratory calibration of the NiCaVis 705 IQ sensor was conducted. This sensor specialises in identifying the concentration of several water quality parameters in real-time, at high-frequency intervals. The focus of this particular study was on calibrating its nitrate detection capability [10]. As a result, the sensor was specifically calibrated to account for the expected waterway conditions of the region, which are likely to differ significantly depending on the location of deployment. However, most rivers tend to share similar water quality ranges; hence, the goal was to develop a compensation range that could make the nitrate readings from NiCaVis reliable and consistent with what is typically observed in these waterbodies [11,12].

To achieve this, potassium nitrate solutions with concentrations ranging between 0.05 mg L^{-1} and 500 mg L^{-1} were created, thus allowing a set of samples capable of simulating eutrophication conditions (concentrations > 2 mg L^{-1} [13]) to be developed, while also enabling the assessment of the efficacy of the NiCaVis 705 IQ sensor through a broad enough concentration spectrum [14–18].

The preliminary calibration and analysis was conducted by measuring potassium nitrate solutions with the NiCaVis 705 IQ sensor as well as a laboratory spectrophotometer and comparing the results as a standardisation measurement. Upon developing this baseline for both devices, interference sources consisting of alterations to the pH, salinity, turbidity, humic acid content, bromide content and temperature were made to calibrate the sensor under different conditions, particularly since studies have shown that these parameters may reduce the accuracy of optical sensors [9,19–22]. Thereafter, a repetitive measurement method was followed to obtain the sought-after results. Firstly, 50 mL of potassium nitrate was placed in tubes, where an interference source was applied and thoroughly mixed to create a homogenous solution. Each sample was then analysed by a laboratory spectrophotometer and the NiCaVis 705 IQ, and the results were compared afterwards. Finally, the remaining liquid was discarded, and the process was repeated for the remaining interferences.

Upon completing the laboratory calibration, the data were analysed to validate the sensor's readings and to develop compensation models as required. In doing so, a multiple regression model was developed to predict real nitrate concentrations based on two input variables (turbidity and raw nitrate readings).

2.2. Phase II. Mobile Trailer Development

The development of the mobile trailer was conceptualised to address some of the issues regarding the accessibility and implementation of traditional water quality monitoring methods. Sampling from rural locations at regular intervals is costly, as is the implementation of methods to maintain the health of the waterways in these isolated areas. Thus, a mobile trailer was considered useful due to its movability to different locations, as well as its capability to remotely transmit nitrate and other water quality data via a cloud-based system, thus eliminating the need for continual trips to rural areas.

However, to develop the trailer, two aspects had to be considered: the exterior casing and the interior space. For the exterior, the aim was to obtain a relatively inexpensive trailer that had the capability to be adapted and modified to suit the project's needs, while the interior required sufficient spacing to hold two sensors, an analysis container, communications/pumping systems, as well as additional auxiliary electronics. Figure 1 displays the exterior modifications, where solar panels were added to a trailer and connected to batteries to provide sufficient power supply to the system. Figure 1 also depicts the overall housing for the sensors, where a steel-cased trailer was selected to ensure that a robust, durable system could be setup.

Figure 1. Trailer exterior modified with solar panels.

Figure 2 then displays the interior design that was adopted to house the electronics, communication system and sensors.

Figure 2. Interior spacing of the trailer, showing the wet area, communications box and sensor equipment.

Overall, Figures 1 and 2 display relatively simple housing for the required systems, where an additional compartment was used to store a pumping system. The design of the mobile trailer was created with the intention that the entirety of the water quality monitoring instruments could be held within the one compartment. Therefore, only a small pump had to be placed outside of the trailer to extract water into the wet area, where the sensors could identify the nitrate and water quality parameter concentrations at half-hourly intervals. The data were then streamed through the communications datalogger and automatically uploaded online to a cloud-based server, known as Eagle.IO (https://eagle.io/). Further details regarding the piping and instrumentation layout are displayed in Figure A1 of Appendix B.

Overall, the mobile trailer was designed and constructed to be robust and reliable for field research applications. While the design could be refined further and optimised, the focus of the current study was on its capability and accuracy to remotely collect high frequency water quality parameter data. Future design optimisation work could be conducted on the trailer shelter, water sample pumping system or the renewable energy system, where reduced sizes and capacities could result in cost-savings without compromising operational performance.

2.3. Phase III. Pilot Field Study

The pilot field study took place at Hilliards Creek in Queensland, Australia. Located in the small town of Ormiston (population < 6000), the site provided an ideal opportunity to test the monitoring trailer in a secure and secluded area. Figure A2 in Appendix C displays a map of the pilot location, which was chosen due to the site's proximity (3.4 km) to an upstream wastewater treatment plant (WWTP). The WWTP is known to discharge nitrate-containing effluent into Hilliards Creek, and hence the location provided the opportunity to achieve this study's goals by providing a location to collect accurate, high-frequency nitrate data.

In order to collect the water quality data, the sampling pump was placed in the water at approximately 30 cm above the sediment bed. Agrometeorological data were collected at half-hourly intervals, which included nitrate concentrations (from the NiCaVis 705 IQ), as well as the pH, turbidity, salinity and temperature (obtained by Xylem Analytics' EXO2 Sonde, from YSI, Yellow Springs, OH, USA) of the waterway. Furthermore, rainfall data were obtained at the same frequency from a gauge located adjacent to the WWTP. Prior to the fully automated implementation, the system was calibrated using surface water samples, which were verified with laboratory water quality analyses. The trailer was then left to collect data for nearly four days, before a 1% average exceedance probability (AEP) rain event halted further monitoring due to disruptions from a subsequent flash flood. Despite this, the data were collected through Eagle.IO, where no on-site monitoring was needed after the initial calibration of the equipment.

3. Results

3.1. Phase I. Laboratory Calibration

Figure 3 illustrates a comparison between the nitrate readings obtained with the NiCaVis 705 IQ sensor and a laboratory spectrophotometer, for concentrations ranging between 0.05 mg L^{-1} and 500 mg L^{-1} without interference sources applied.

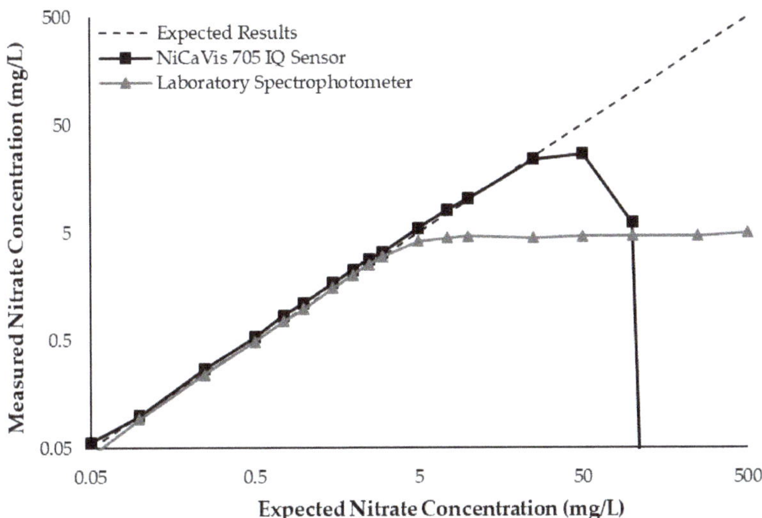

Figure 3. Expected concentration from known solutions vs. measured nitrate concentrations, NiCaVis 705 IQ sensor and laboratory spectrophotometer. Initial calibration data.

For the NiCaVis 705 IQ sensor, good accuracy was achieved up to a concentration of 25 mg L^{-1}, with a rapid drop in accuracy thereafter. The sensor was unable to measure nitrate concentrations beyond 250 mg L^{-1}. The spectrophotometer was shown to be accurate for nitrate concentrations only up to 3 mg L^{-1}; readings beyond this showed a constant concentration of 5 mg L^{-1}. This was believed to have resulted from an Inner Filter Effect (IFE), which occurred as a result of the analyte concentration being too high, causing signal loss in the spectrophotometric analysis [23]. Additional figures to emphasise the performance of the sensor and spectrophotometer when measuring interference sources such as turbidity, pH, temperature, bromide, humic acid and salinity are presented in Appendix D. The results of these experiments display a similar trend to the one observed in Figure 3, emphasising the higher accuracy of the sensor when compared with traditional water quality monitoring methods.

Overall, NiCaVis was able to determine concentrations within 0.1 mg L^{-1} of the baseline in 81% of the samples tested (n = 240), while the spectrophotometer was only able to achieve this in 63% of tests. Furthermore, the sensor read concentrations within a 0.5 mg L^{-1} range in 96% of cases, while the spectrophotometer had the same result in only 84% of samples. If accounting for the error in samples >5 mg L^{-1}, the number of samples within that range of accuracy decreases to 67%.

This study also found that the accuracy of NiCaVis can be improved through extensive compensation modelling and reading adjustments. Turbidity was found to be the largest interference source: elevated levels of turbidity with lower nitrate concentrations had the tendency to reduce the sensor's ability to accurately detect the level of nitrate. Figure 4 displays this correlation, indicating that a smaller nitrate presence causes a larger signal loss with increasing turbidity. Figure 5 then displays the relationship between the expected and experimental nitrate concentrations, based on different turbidity levels.

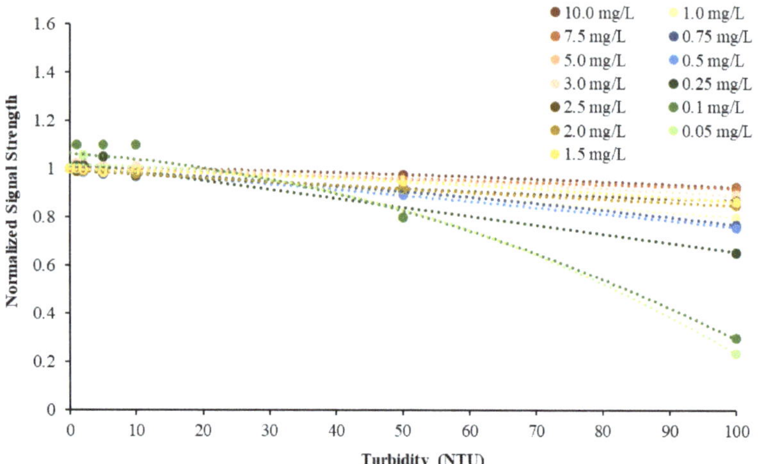

Figure 4. NiCaVis 705 IQ signal strength vs turbidity.

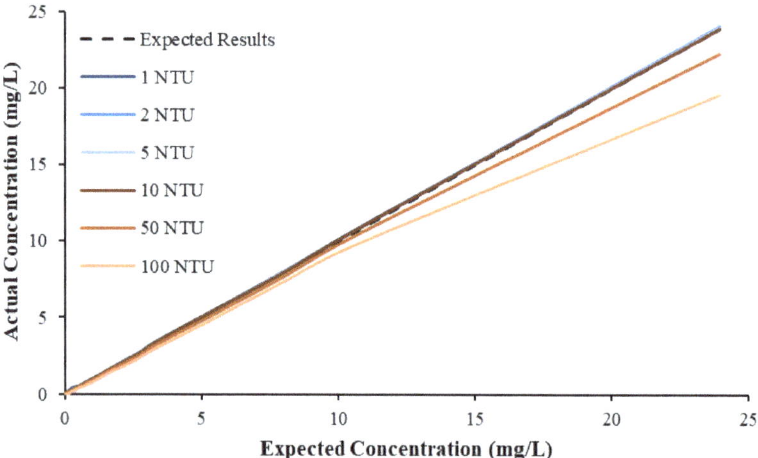

Figure 5. Expected vs. experimental nitrate concentrations for varying turbidity.

From the above data, a preliminary compensation model was developed, as shown in Equation (1) ($R^2 > 0.99$).

$$[NO_3,A] = 0.961 * [NO_3,S] + 0.003 * Tb - 0.119, \quad (1)$$

where $[NO_3,A]$ is the actual corrected nitrate concentration of the sensor, $[NO_3,S]$ is the nitrate concentration read by the sensor and Tb is the turbidity of the waterbody.

This research found that there were limitations to the model, and hence further analyses would be necessary to reinforce the validity of interference compensations, while additional studies could help to develop more complex threshold models to better determine the corrected nitrate reading based on more refined intervals. Regardless, the model was practical in correcting erroneous results for nitrate concentrations >0.1 mg L^{-1}; however, significant inaccuracies were noted for lower concentrations. Despite this, the likelihood of having concentrations <0.1 mg L^{-1} in farmland rivers or creeks susceptible to nutrient releases is low, which is evidenced by the real-time data collected during the pilot study (Figure 6).

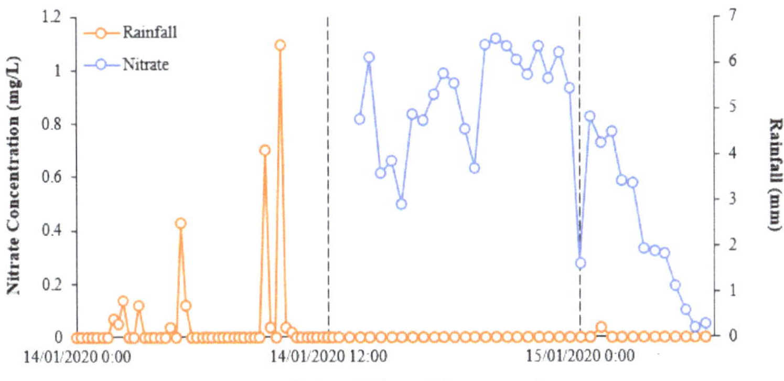

Figure 6. Nitrate–rainfall data at Hilliards Creek, 14th to 15th January 2020.

3.2. Phase II and III. Mobile Trailer Development and Pilot Field Study

Figure 6 shows the initial relationship between nitrate concentrations and rainfall over the pilot study monitoring period, based on the data extracted from Eagle.IO. Figure A20 in Appendix E displays the virtual graphing interface that the software uses to visualise readings collected by the sensors.

Figure 6 highlights the nitrate–rainfall correlation for the operational cycle of the NiCaVis sensor until it was turned off for maintenance. Furthermore, approximately 12 hours of rainfall preceding the NiCaVis' commencement of recording data is also shown.

The data seem to show that rainfall events caused delayed decreases in nitrate concentrations due to the dilution of water entering the creek. Such a delayed dilution effect was expected due to the inherent processes involved with runoff generation following rainfall, as well as flow movement downstream. In addition, given that the location of the rainfall gauge is 3.4 km upstream from the trailer deployment site, this further emphasised why the delay between rainfall and nitrate readings existed. Although the lack of nitrate data for the 14th of January does not allow us to validate such claims, the dilution effect portrayed in Figure 6 is also supported historically, with Figure 7 showing the variation of nitrate concentrations that have occurred seasonally from 2003 to 2014.

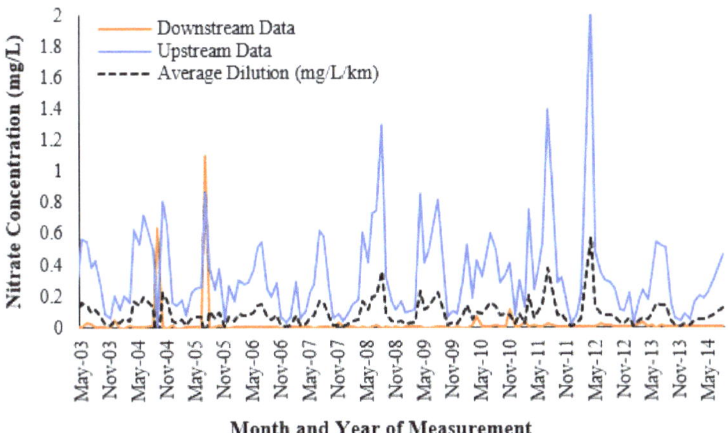

Figure 7. Historical Hilliards Creek nitrate data from May 2003 to May 2014.

Figure 7 displays an evident trend where nitrate concentrations had the tendency to lower significantly during the wet season (November to April) and rise during drier periods. Therefore, this data supports the results observed in Figure 6, displaying the inverse correlation between rainfall volume and nitrate concentration in waterways.

In addition, Figure 7 also presents a significant difference between the nitrate concentrations at Hilliards Creek (recorded 200 metres downstream from the current study) and at a location further downstream (3.61 km away). As a result, comparing the historical monthly upstream data to the results highlighted in Figure 6 indicates that an inflow source (the WWTP) was causing significant increases in nitrate concentrations, which then diluted downstream.

Furthermore, the variation of the data in Figure 6 also emphasises the necessity of real-time and high-frequency nutrient monitoring as significant variations are possible in short timespans. The data in Figure 7 show significant historical variations on a monthly basis and hence indicates that the typical sampling frequency (i.e., monthly) used for monitoring is unlikely to be truly representative of a waterway's health or of shorter-term nutrient fluctuations, which appear to occur.

Figure 8 displays the turbidity data collected by the Xylem Analytics' EXO2 Sonde, coupled with rainfall data.

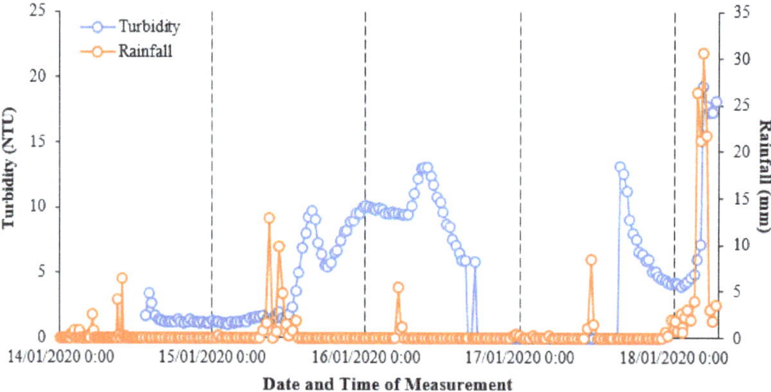

Figure 8. Turbidity–rainfall data at Hilliards Creek, 14th to 18th January 2020.

The correlation between preceding rainfall and turbidity is evident due to a lag of eight hours between the commencement of the rainfall event and the first increase in turbidity. In addition to causing an optical interference on nitrate readings, it is possible that more turbid waters may reduce the ability of chlorophyll-*a* to absorb energy from the sun, in turn reducing algal blooms and preventing nutrient uptake [24]; hence, simultaneous monitoring of turbidity and nitrates is critical for accuracy. Further analyses would be required to affirm this correlation; however, collecting more consistent nitrate, chlorophyll-*a*, turbidity and rainfall data could reveal some patterns that may be modelled and integrated into future studies to predict and manage the physicochemical conditions of waterways.

At 5:30PM on 16/01/2020, the mobile trailer system was turned off to complete operational maintenance. When the system was re-established, a significant rainfall event occurred afterwards, thus allowing further correlations to be made, as illustrated by the data presented after 12:30PM on 17/01/2020. Consequently, Figure 8 displays significant turbidity rise associated with the overnight rainfall event that occurred between 3:00AM and 5:00AM on January 18, 2020. This sudden and large burst of precipitation, noted as a one-in-100-year event, caused an increase in turbidity from 1.3 NTU (which was the base level detected with no preceding rainfall), to 19.33 NTU, an almost 15-fold increase. However, due to the severity of this event, elevated creek levels occurred such that the mobile system was inoperable for several days thereafter.

Despite this, the data presented in Figure 8 showed that rainfall events, particularly those of large magnitudes, may cause significant elevations in turbidity, and hence the data would require compensating to ensure that the readings made by the NiCaVis 705 IQ sensor and its auxiliary units are maintained at an accurate level.

4. Discussion

The completion of the pilot study consisting of laboratory calibrations and compensation modelling, as well as the trailer development and its subsequent deployment, has the potential to provide numerous benefits. The advantages of obtaining high-frequency nutrient data were explored in this study, where the 30-minute interval results provided the opportunity to observe the highly dynamic nature of water quality parameters. Consequently, the next sections briefly discuss the potential applications of the mobile trailer, as well as the implications of collecting data at a high frequency over long time-periods.

4.1. Agricultural Benefits in Rural Regions

The overproduction of crops to meet increased food demand has caused increases in fertiliser usage, leading to significant environmental damage and exemplifying the need for real-time, remote monitoring at vulnerable waterways. For lower socioeconomic nations, such as those within the tropics, routine monitoring through traditional methods is not feasible due to their diminished capability to proportionately improve their agriculture production [25]. As a result, countries in the tropics are struggling to implement monitoring regimes that can mitigate environmental harm and minimise economic losses from mismanaged fertiliser usage. Thus, one such application of a real-time, high-frequency system for water quality monitoring is to provide vulnerable nations with a way to continually monitor agricultural effluent. In doing so, fertiliser runoff can be effectively managed using real-time monitoring, thus providing an option to assist decision-makers in the reduction of economic losses associated with high-volume discharges.

4.2. Minimisation of Nutrient Pollution from Aquaculture Practices

Aquaculture industries pose a significant threat to waterway health. In recent years, nutrient inputs have significantly declined, with the conversion efficiency to assist in the growth of aquatic organisms also improving [26]. However, excretion products from these organisms persist and have the potential to cause significant environmental and ecosystem changes. This therefore presents issues where excess nutrient loads can not only lead to eutrophication but can cause disruptions to marine species. One study [27] found that up to 45% of nitrogen provided as a food source could be excreted by some organisms, emphasising the significant detriment that mass aquaculture production may have on water quality. Therefore, real-time, high-frequency monitoring can be applied to better equip stakeholders to manage effluent releases into general waterways. In doing so, the timing and volume of nutrient-rich discharge can be managed to provide a lower overall threat to waterways and ecosystems.

4.3. Improvement of Environmental Health in Environmentally Sensitive Areas

As previously mentioned, traditional water quality data are usually obtained, analysed and reported at monthly intervals. This minimises labour and technological costs while also providing relatively accurate results regarding the health of waterways. However, this study has highlighted that the overall complexity of river/creek systems is high, with the dynamic nature of these systems causing large fluctuations of physicochemical and nutrient concentrations in short time-periods. For typical waterbodies, the importance of measuring the significance of these variations may not be particularly necessary; however, some regions are very susceptible to small changes in water quality. The Great Barrier Reef (GBR) is an example of this, and it is an environmentally sensitive area; coral bleaching is a well-known issue that is associated with changes in water temperature, nutrient levels (particularly nitrate) and lighting [28,29]. As a result, this presents an opportunity for high-frequency monitoring, which has already been conducted and validated as part of this study. Monthly observations of areas such as the GBR are insufficient for maintaining the health of these regions, and hence the potential to obtain real-time data that can be readily acted on has the potential to mitigate environmental damage to vulnerable areas.

4.4. Efficacy of Real-Time, High-Frequency Sensors for Routine Water Quality Monitoring

Traditional high-frequency monitoring systems are too immobile to assist in targeted water quality analyses [8]; however, the mobile system developed in this study has the capability to be deployed at different targeted locations on-demand.

The accuracy of high-frequency optical sensors has been previously questioned [8]. Efforts are being made to improve the reliability of implementing such devices for prolonged use; however, compensating readings accurately has also proven to be beneficial, as a study by de Oliveira et al. [9] has suggested. This study further emphasised this significance by obtaining data to develop a turbidity-nitrate compensation. The calibration results shown in Figures 4 and 5 are in line with the findings of de Oliveira et al. [9] since an increase in turbidity resulted in a decrease in signal strength of the real-time optical sensor. However, we also proved that such a decrease is not only proportional to turbidity concentration but also to the actual nitrate concentration, with low nitrate levels being more susceptible to turbidity interferences and thus a loss in sensor accuracy. Our compensation model extended from [9] to include this second key variable.

5. Conclusions

Due to a lack of resources and effective water monitoring methods, nations in the tropics are at a greater risk of waterway pollution from excess nutrient loads. Current monitoring methods for nitrates are outdated and fail to deliver consistent solutions to water quality issues in river/creek systems. Furthermore, accessibility and affordability limitations of monitoring systems are significant drawbacks to routine checking of agricultural waterways in rural areas. As a result, a mobile monitoring station was developed in this study to collect high-frequency nitrate and physicochemical water data. The implications of this effective and practical monitoring system, which can be reliably compensated, are significant. Throughout this study, the overall user-input requirements of the monitoring system have been greatly optimised; wireless internet streaming and a nearly complete remote functionality have provided the opportunity for efficient and constant data transmission to occur with minimal human involvement. Furthermore, by developing the streaming platform through code-based software, the necessity for human interactions with the monitoring system, aside from the initial setup and calibration, is minimal.

This study has also identified several findings relating to the modernisation of water quality monitoring. Firstly, extensive laboratory calibrations linked interference sources including pH, turbidity, temperature, bromide, salinity and organic matter, to the potentially reduced performance of a real-time optical nitrate sensor. In doing so, it was found that turbidity had the greatest effect, and hence a compensation model was developed to maintain the accuracy of reported results, even in turbid waterway conditions. The benefits of this process are potentially significant: the reliability of real-time, high-frequency sensors considerably improves through such modelling, where the simultaneous employment of a remote sensor, linked to an online cloud-based data transmitter, can provide accurate data based on compensations associated with environmental conditions. Furthermore, these calibrations have also shown that real-time sensors can be significantly more accurate than traditional laboratory spectrophotometers, thus highlighting the necessity of modernising water quality monitoring methods to minimise time wasted and improve the accuracy and reliability of reported data. Finally, this study has shown how crucial high-frequency measurements are when monitoring water quality as several parameters, particularly nitrates, can have very short-term fluctuations in concentration. Given waterway susceptibility to eutrophication, such erratic variances can have significant consequences, and hence this study was important to further highlight the necessity of high-frequency monitoring.

Future work will expand on the research described in this paper, and remote, live monitoring and advanced data analytics will be coupled to create a fertiliser management decision-support system that can be deployed to combat targeted issues such as fertiliser dosage mismanagement and aquaculture-based nitrate monitoring. Not only would this provide farmers with an on-demand method of managing the water quality of rivers surrounding their sites, but they can also optimise fertiliser dosages and nutrient feeds for a variety of crop types and aquatic organisms to mitigate waterway pollution. Optimal fertiliser usage pays off economically for farmers due to reduced annual demand, while aquaculture monitoring assists in preventing ecosystem damage to critical species and environments.

To further refine the concept, additional trailers at different points along a targeted waterway would provide additional opportunities to verify the concentrations of nitrate from farmland areas. This study has already begun to highlight different factors that affect the severity of nitrate runoff; however, more continual high-frequency data would provide significant opportunities to interrelate these parameters. Overall, more in depth studies are required to develop specific monitoring regimes with high-frequency data. Despite this, the current pilot monitoring program has shown potential, and hence the success of future projects could provide significant benefits to agricultural stakeholders and aquaculture locations, as well as environmentally sensitive areas, particularly in more vulnerable regions, which are susceptible to the effects of overpopulation and climate change.

Author Contributions: Conceptualization, M.J.L.J., E.B., R.A.S.; methodology, M.J.L.J., E.B.; software, M.J.L.J., E.B; validation, M.J.L.J., E.B.; formal analysis, M.J.L.J., E.B.; investigation, M.J.L.J., T.S.; resources, T.S.; data curation, M.J.L.J.; writing—original draft preparation, M.J.L.J.; writing—review and editing, E.B., R.A.S. All authors have read and agreed to the published version of the manuscript.

Funding: This research received no external funding.

Acknowledgments: We are thankful to Xylem Analytics Inc. and TEW for their in-kind contribution of several key instruments, materials and resources which were used in this project. Furthermore, we would like to acknowledge the Griffith Scientific Services Laboratory for allowing us to use their facilities throughout the calibration phase, in-kind. We would also like to thank Lawrence Hughes (Griffith University) for his assistance with the monitoring station's design. Finally, we would like to thank Ian Underhill and his staff at the Griffith University Engineering Laboratory for providing us with advice and facilities to assist in developing our monitoring station.

Conflicts of Interest: The authors declare no conflict of interest.

Appendix A. Summary of Typical Water Quality Measurement Methods

Table A1. Advantages and disadvantages of different nitrate measurement methods.

Analysis Method	Advantages	Disadvantages
Ion Selective Electrode [30]	- Large pH range (2.5–11). - Large range of nitrate concentrations detectable. - Portable.	- Long preparation time (30 min–1 h). - Requires soaking in a high standard nitrate solution. - Relies on samples taken from the middle of a water body, below the surface (not always practical). - Easy to contaminate if not thoroughly rinsed. - Requires time-intensive calibration. - Several ions may contaminate readings.
Nitrate Test Strips [31]	- Quick to use. - Easy to measure (no experience required). - Short preparation time. - Practical results (ppm reading).	- Inaccurate—no exact concentration readable. - Large jump in concentration band with little colour band change. - Potentially difficult to read. - Significant margin of error. - Not reusable. - Expiry date significantly reduces accuracy.

Table A1. *Cont.*

Analysis Method	Advantages	Disadvantages
Cadmium Reaction [32,33]	- Uses nitrite colour change as a proxy for nitrate data. - Minimal interference from temperature/salinity. - Useful for monitoring.	- Long preparation time. - Significantly affected by human error. - Requires some laboratory experience to prepare/carry out the procedure. - Requires a specific wavelength (543 nm) to measure the absorbance to obtain a result.
Laboratory Spectrophotometry [32,33]	- Large wavelength spectrum available. - Useful for monitoring.	- Long preparation time. - Significantly affected by human error. - Potentially subject to interference sources that may affect results. - Requires some laboratory experience to prepare/carry out the procedure.
Real-time In-situ Spectrophotometry [10]	- Fast response time. - Real-time data. - Can be monitored remotely. - Automatic ultrasound cleaning/low maintenance. - Large spectrum for UV-VIS spectrophotometry. - Several parameters measurable. - Useful for decision making.	- Need to ensure that the device is secure (to prevent theft/damage). - Potentially subject to interference sources that may affect results.

Table A1 shows that several methods have significant disadvantages, which may minimise the reliability, accuracy or affordability of regular monitoring due to their dependency on manual sampling, in-situ requirements or the time required to obtain accurate data. Laboratory spectrophotometry is commonly used in industry water quality monitoring programs, particularly by government organisations who use the data to assist in decision-making regarding targeted waterways. However, this method also has limitations where it largely relies on water quality scientists to extract samples for analysis, which may not be feasible in rural areas due to the significant distance from city centres and monitoring laboratories [32,33].

Appendix B. Piping and Instrumentation Diagram of the Mobile Trailer

A piping and instrumentation diagram is shown in Figure A1 to reiterate the functionality of the monitoring trailer, emphasising how the design was developed to obtain water quality data.

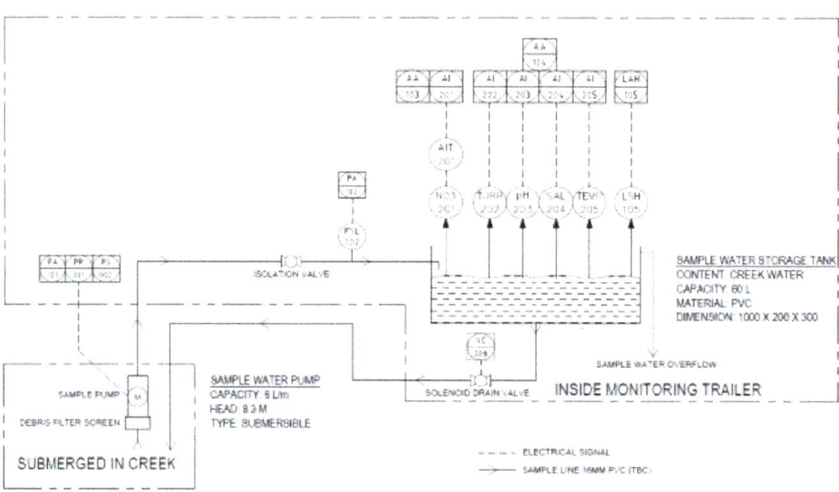

Figure A1. Piping and instrumentation diagram of the monitoring trailer's functionality.

Appendix C. Map of the Study Site in Ormiston, QLD

Figure A2. Preliminary field site location at Hilliards Creek, Ormiston [34].

Appendix D. Additional Figures Comparing the NiCaVis 705 IQ to a Spectrophotometer under Different Interference Conditions

Generally, the data shown in this section highlights the efficacy of the NiCaVis 705 IQ sensor and indicates that it is much more consistent in the results it provides under a range of conditions when compared with the spectrophotometer results. Overall, the spectrophotometer was unable to accurately identify nitrates whose concentrations exceed 5 mg/L due to a saturated absorbance and the production of nitrites. For the NiCaVis 705 IQ sensor, the accuracy dropped after 25 mg/L due to the Inner Filter Effect (IFE) causing the signal to decrease.

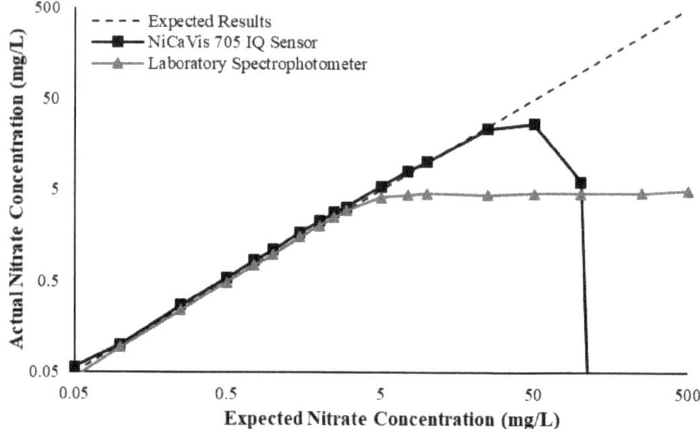

Figure A3. Comparison between the expected results and the spectral/sensor data under standard conditions.

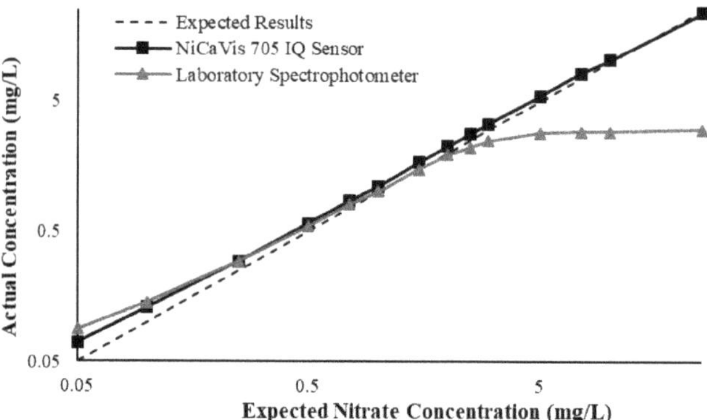

Figure A4. Comparison between the expected results and the spectral/sensor data with 10 mg L^{-1} of bromide added.

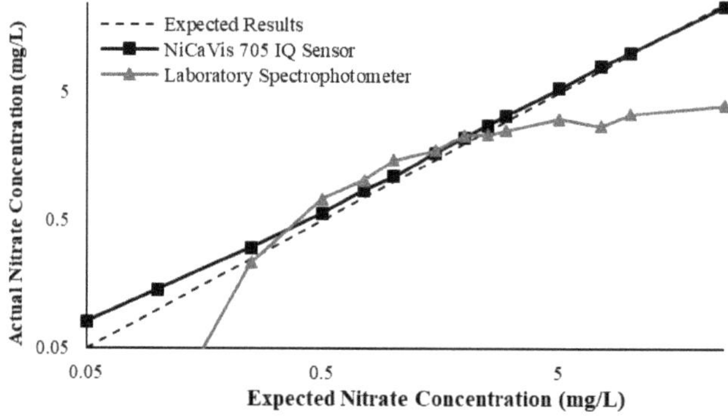

Figure A5. Comparison between the expected results and the spectral/sensor data with 20 mg L^{-1} of bromide added.

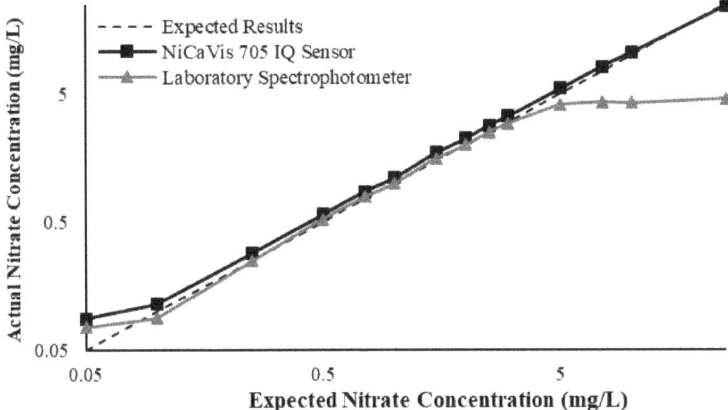

Figure A6. Comparison between the expected results and the spectral/sensor data with 5 μL per 50 mL potassium chloride (58,670 μs cm^{-1}) salinity added to the samples.

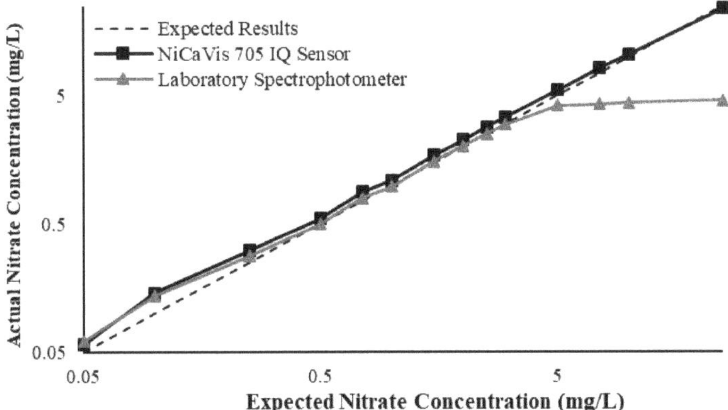

Figure A7. Comparison between the expected results and the spectral/sensor data with 10 μL per 50 mL potassium chloride (58,670 μs cm^{-1}) salinity added to the samples.

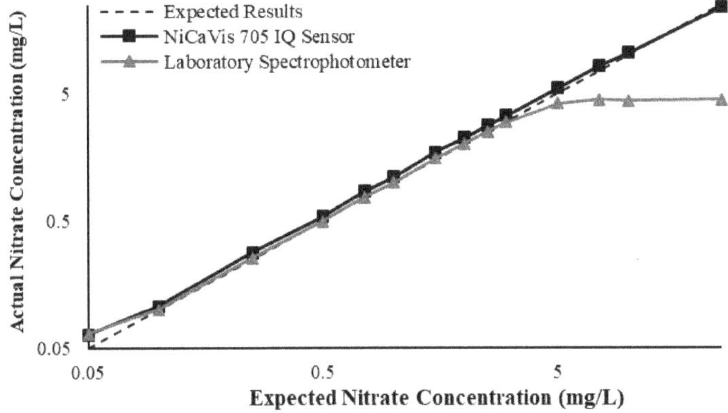

Figure A8. Comparison between the expected results and the spectral/sensor data with samples at pH 4.5.

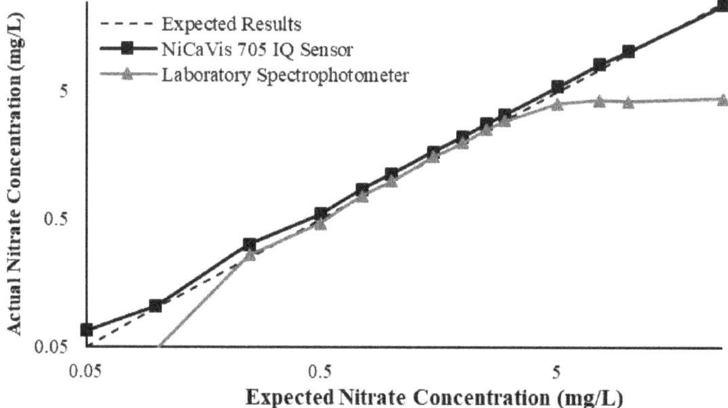

Figure A9. Comparison between the expected results and the spectral/sensor data with samples at pH 10.

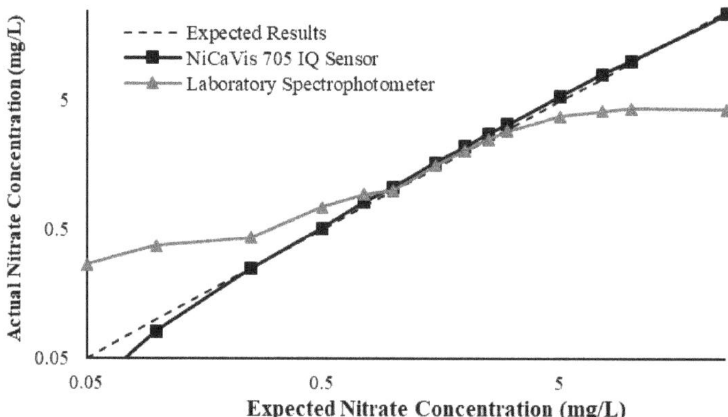

Figure A10. Comparison between the expected results and the spectral/sensor data with samples at 4.3 °C.

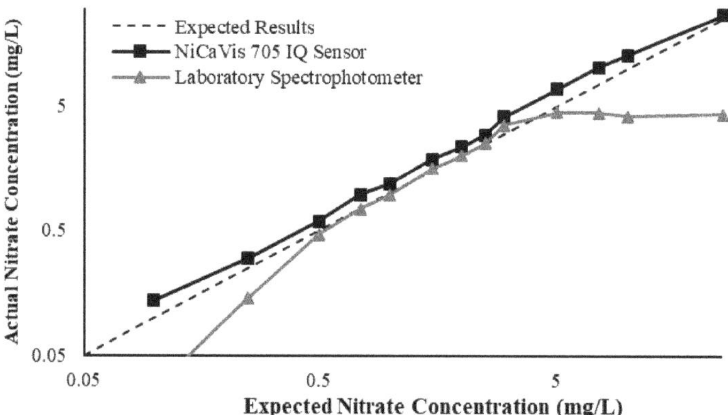

Figure A11. Comparison between the expected results and the spectral/sensor data with samples between 35–45 °C.

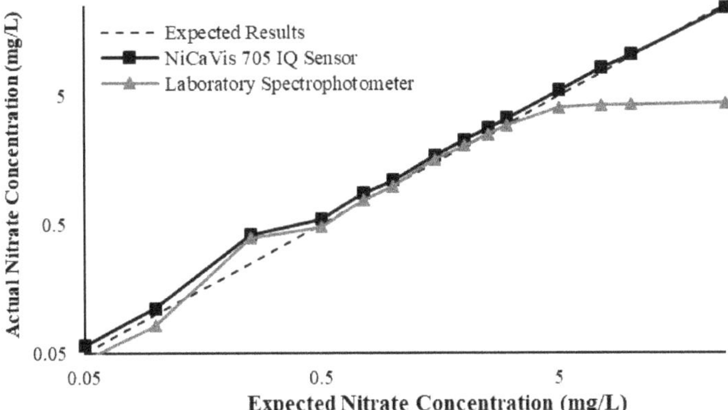

Figure A12. Comparison between the expected results and the spectral/sensor data with 1 NTU turbidity added.

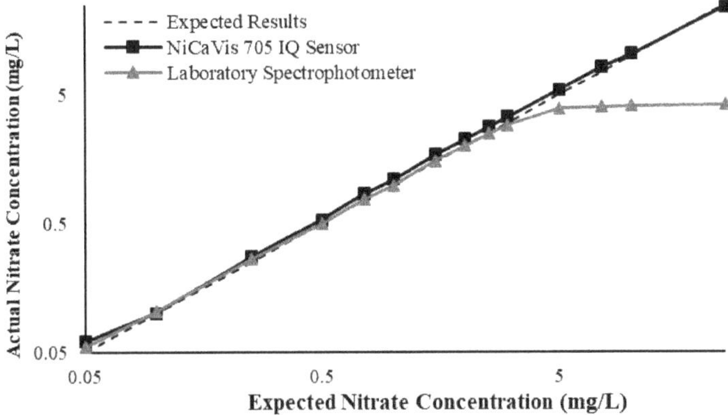

Figure A13. Comparison between the expected results and the spectral/sensor data with 2 NTU turbidity added.

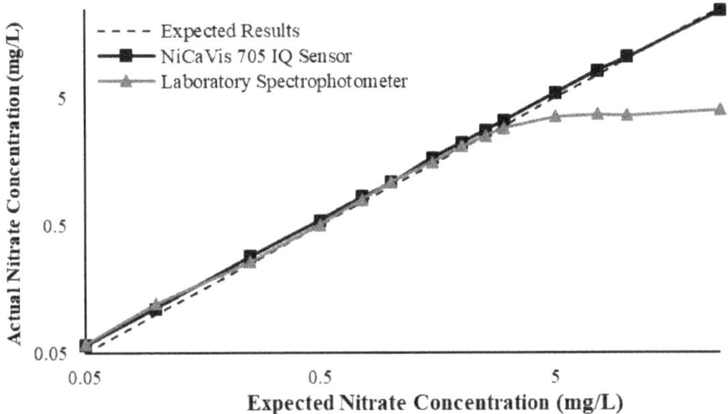

Figure A14. Comparison between the expected results and the spectral/sensor data with 5 NTU turbidity added.

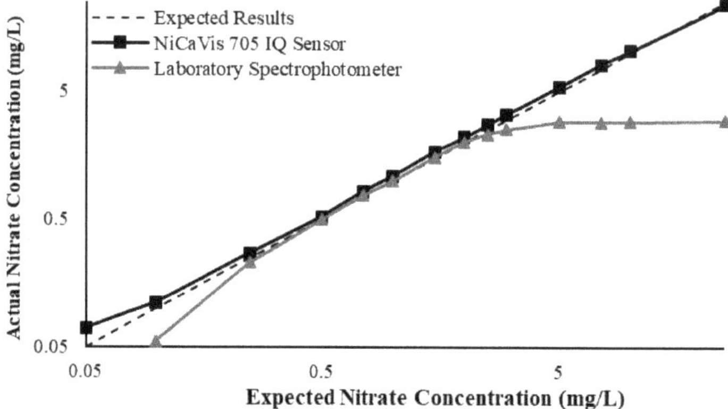

Figure A15. Comparison between the expected results and the spectral/sensor data with 10 NTU turbidity added.

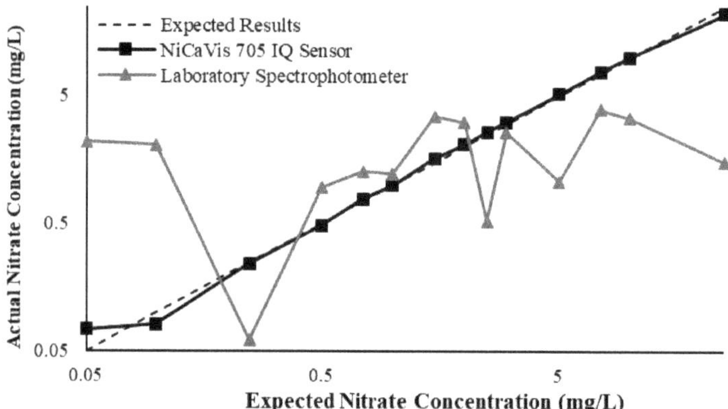

Figure A16. Comparison between the expected results and the spectral/sensor data with 50 NTU turbidity added.

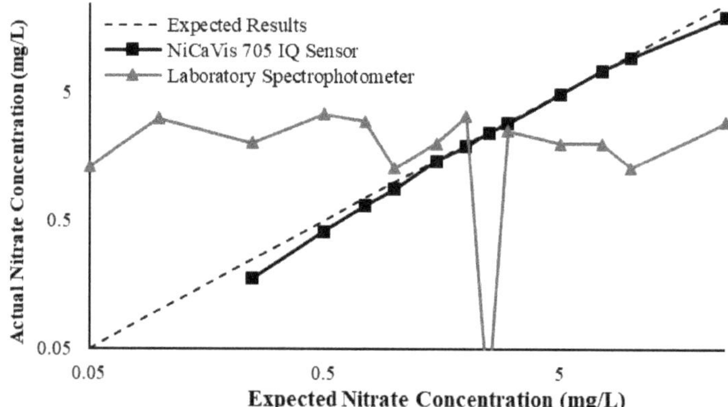

Figure A17. Comparison between the expected results and the spectral/sensor data with 100 NTU turbidity added.

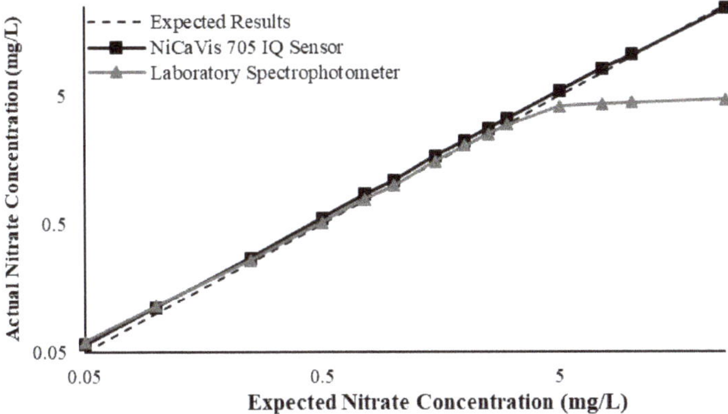

Figure A18. Comparison between the expected results and the spectral/sensor data with 10 mg L^{-1} of humic acid added.

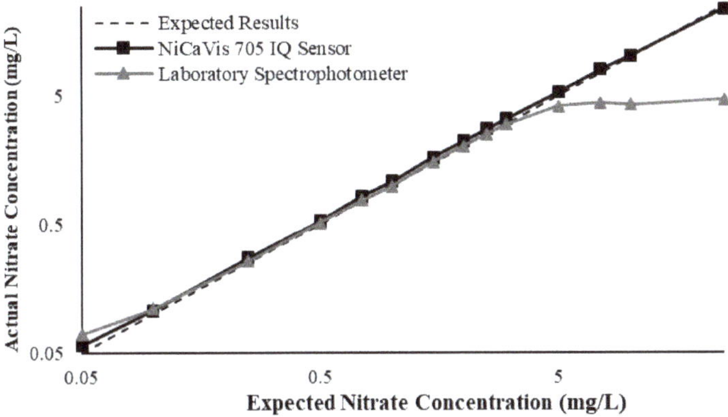

Figure A19. Comparison between the expected results and the spectral/sensor data with 20 mg L^{-1} of humic acid added.

Appendix E. Eagle.IO Online Interface

Figure A20 illustrates the online Eagle.IO interface, which was used to view data trends and identify any potential errors with the system's operation.

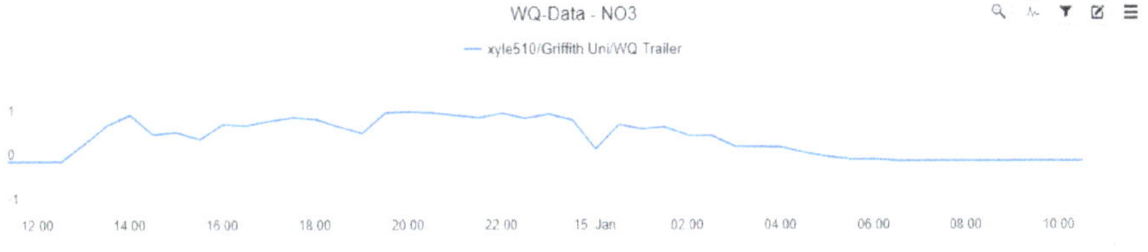

Figure A20. Eagle.IO online interface.

The interface provides the data on a fixed scale, with Figure A20 highlighting initial nitrate concentration trends on a −1 to 1 scale. From a data analysis perspective, the Eagle.IO interface was not useful in assisting to identify how consistent the reporting and results were. The data were downloaded in a.csv format for further and more detailed analyses.

References

1. Omarova, A.; Tussopova, K.; Hjorth, P.; Kalishev, M.; Dosmagambetova, R. Water Supply Challenges in Rural Areas: A Case Study from Central Kazakhstan. *Int. J. Environ. Res. Public Health* **2019**, *16*, 688. [CrossRef] [PubMed]
2. Matiatos, I. Nitrate source identification in groundwater of multiple land-use areas by combining isotopes and multivariate statistical analysis: A case study of Asopos basin (Central Greece). *Sci. Total Environ.* **2016**, *541*, 802–814. [CrossRef]
3. Mamun, K.A.; Islam, F.R.; Haque, R.; Khan, M.G.M.; Prasad, A.N.; Haqva, H.; Mudliar, R.R.; Mani, F.S. Smart Water Quality Monitoring System Design and KPIs Analysis: Case Sites of Fiji Surface Water. *Sustainability* **2019**, *11*, 7110. [CrossRef]
4. Parvez Mahmud, M.A.; Ejeian, F.; Azadi, S.; Myers, M.; Pejcic, B.; Abbassi, R.; Razmjou, A.; Asadnia, M. Recent progress in sensing nitrate, nitrite, phosphate, and ammonium in aquatic environment. *Chemosphere* **2020**, *259*, 127492. [CrossRef]
5. Christensen, V.G.; Rasmussen, P.P.; Ziegler, A.C. Real-time water quality monitoring and regression analysis to estimate nutrient and bacteria concentrations in Kansas streams. *Water Sci. Technol.* **2002**, *45*, 205–219. [CrossRef] [PubMed]
6. Kim, G.; Lee, Y.-W.; Joung, D.-J.; Kim, K.-R.; Kim, K. Real-time monitoring of nutrient concentrations and red-tide outbreaks in the southern sea of Korea. *Geophys. Res. Lett.* **2006**, 33. [CrossRef]
7. EcoLAB 2. Available online: https://www.labtech.com.mx/files/ecolab_2.pdf (accessed on 23 May 2020).
8. Pellerin, B.A.; Stauffer, B.A.; Young, D.A.; Sullivan, D.J.; Bricker, S.B.; Walbridge, M.R.; Clyde, G.A.; Shaw, D.M. Emerging Tools for Continuous Nutrient Monitoring Networks: Sensors Advancing Science and Water Resources Protection. *JAWRA* **2016**, *52*, 993–1008. [CrossRef]
9. de Oliveira, G.F.; Bertone, E.; Stewart, R.A.; Awad, J.; Holland, A.; O'Halloran, K.; Bird, S. Multi-Parameter Compensation Method for Accurate In Situ Fluorescent Dissolved Organic Matter Monitoring and Properties Characterization. *Water* **2018**, *10*, 1146. [CrossRef]
10. WTW Carbovis Sensor. Available online: https://www.xylem-analytics.in/media/pdfs/wtw-carbon-carbovis-brochure.pdf (accessed on 18 May 2020).
11. What Makes a River? Available online: https://www.americanrivers.org/rivers/discover-your-river/river-anatomy/ (accessed on 19 May 2020).
12. Lindeburg, K.S.; Drohan, P.J. Geochemical and mineralogical characteristics of loess along northern Appalachian, USA major river systems appear driven by differences in meltwater source lithology. *Catena* **2019**, *172*, 461–468. [CrossRef]
13. United States Geological Survey. Nutrients in the Nation's Waters: Identifying Problems and Progress. Available online: https://pubs.usgs.gov/fs/fs218-96/ (accessed on 20 June 2020).
14. Ayers, R.S.; Westcot, D.W. Water Quality for Agriculture. *FAO Irrig. Drain.* **1989**, 29.
15. Nitrates in Groundwater Beneath Horticultural Properties. Available online: https://www.agric.wa.gov.au/water-management/nitrates-groundwater-beneath-horticultural-properties (accessed on 19 May 2020).
16. Hansen, B.; Thorling, L.; Schullehner, J.; Termansen, M.; Dalgaard, T. Groundwater nitrate response to sustainable nitrogen management. *Sci. Rep.* **2017**, *7*, 8566. [CrossRef] [PubMed]
17. McLay, C.D.A.; Dragten, R.; Sparling, G.; Selvarajah, N. Predicting groundwater nitrate concentrations in a region of mixed agricultural land use: A comparison of three approaches. *Environ. Pollut.* **2001**, *115*, 191–204. [CrossRef]
18. Neill, M. Nitrate concentrations in river waters in the south-east of Ireland and their relationship with agricultural practice. *Water Resour.* **1989**, *23*, 1339–1355. [CrossRef]
19. Jiao, L.Z.; Dong, D.M.; Zheng, W.G.; Wu, W.B.; Feng, H.K.; Shen, C.J.; Yan, H. Determination of Nitrite Using UV Absorption Spectra Based on Multiple Linear Regression. *Asian J. Chem.* **2013**, *25*, 2273–2277. [CrossRef]
20. Inagaki, T.; Watanabe, T.; Tsuchikawa, S. The effect of path length, light intensity and co-added time on the detection limit associated with NIR spectroscopy of potassium hydrogen phthalate in aqueous solution. *PLoS ONE* **2017**, *12*, e0176920. [CrossRef]
21. Vidal Salgado, L.E.; Vargas-Hernandez, C. Spectrophotometric Determination of the pKa, Isosbestic Point and Equation of Absorbance vs. pH for a Universal pH Indicator. *Am. J. Anal. Chem.* **2014**, *5*, 1290–1301. [CrossRef]

22. Kim, B.Y.; Yun, J.-I. Temperature effect on fluorescence and UV–vis absorption spectroscopic properties of Dy(III) in molten LiCl–KCl eutectic salt. *J. Lumin.* **2012**, *132*, 3066–3071. [CrossRef]
23. Panigrahi, S.K.; Mishra, A.K. Inner filter effect in fluorescence spectroscopy: As a problem and as a solution. *J. Photochem. Photobiol. C Photochem. Rev.* **2019**, *41*, 100318. [CrossRef]
24. Balali, S.; Hoseini, S.A.; Ghorbani, R.; Kordi, H. Relationships between Nutrients and Chlorophyll a Concentration in the International Alma Gol Wetland. Iran. *J. Aquac. Res. Dev.* **2013**, *4*, 173. [CrossRef]
25. World Economic Outlook Database. Available online: https://www.imf.org/external/pubs/ft/weo/2019/02/weodata/index.aspx (accessed on 21 May 2020).
26. Ackefors, H.; Enell, M. The release of nutrients and organic matter from aquaculture systems in Nordic countries. *J. Appl. Ichthyol.* **1994**, *10*, 225–241. [CrossRef]
27. Wang, X.; Olsen, L.M.; Reitan, K.I.; Olsen, Y. Discharge of nutrient wastes from salmon farms: Environmental effects, and potential for integrated multi-trophic aquaculture. *Aquac. Environ. Interact.* **2012**, *2*, 267–283. [CrossRef]
28. Marine Park Authority. Great Barrier Reef Water Quality: Current Issues. 2001. Available online: http://elibrary.gbrmpa.gov.au/jspui/bitstream/11017/358/1/GBR-water-quality-current-issues.pdf (accessed on 20 June 2020).
29. Coral Bleaching and the Great Barrier Reef. Available online: https://www.coralcoe.org.au/for-managers/coral-bleaching-and-the-great-barrier-reef (accessed on 20 June 2020).
30. Determining the Amount of Nitrate in Water. Available online: https://www.westminster.edu/about/community/sim/documents/SDeterminingtheAmountofNitrateinWater.pdf (accessed on 18 May 2020).
31. Sterling, I. The Beginner's Guide to Aquarium Test Strips. Available online: https://fishlab.com/aquarium-test-strips/ (accessed on 18 May 2020).
32. Nitrates. Water: Monitoring and Assessment. Available online: https://archive.epa.gov/water/archive/web/html/vms57.html (accessed on 18 May 2020).
33. HRWC. Nitrate: Its Importance in Water and Its Measurement. 2015. Available online: https://www.hrwc.org/wp-content/uploads/2015/03/Nitrates3-2015.pdf (accessed on 18 May 2020).
34. Queensland Globe. Available online: https://qldglobe.information.qld.gov.au/ (accessed on 19 May 2020).

Article

A Vulnerability Assessment in Scant Data Context: The Case of North Horr Sub-County

Velia Bigi [1,*], **Alessandro Pezzoli** [1], **Elena Comino** [2] **and Maurizio Rosso** [2]

[1] Interuniversity Department of Regional and Urban Studies and Planning (DIST), Politecnico di Torino & Universita di Torino, 10125 Torino, Italy; alessandro.pezzoli@polito.it

[2] Department of Environment, Land and Infrastructure Engineering (DIATI), Politecnico di Torino, 10129 Torino, Italy; elena.comino@polito.it (E.C.); maurizio.rosso@polito.it (M.R.)

* Correspondence: velia.bigi@polito.it

Received: 28 May 2020; Accepted: 14 July 2020; Published: 27 July 2020

Abstract: In Kenyan rural areas belonging to the Arid and Semi-Arid Lands (ASALs), water quantity and water quality are major issues for the local population. In North Horr Sub-County water quality is threatened by nitrate contamination due to fecal matter pollution. This research, hence, aims at assessing the vulnerability of open shallow water sources to nitrate contamination due to fecal intrusion following flooding events and nitrate percolation in groundwater. The present research, indeed, provides, on one hand, new insights into the analysis of the vulnerability in a scant data context; on the other hand, it assesses the adaptation measures contained in the local development plan. Applying the reference definition of the Intergovernmental Panel on Climate Change (IPCC), the results demonstrate that the open shallow water sources in the northern part of the sub-county are more vulnerable to nitrate contamination. Furthermore, the consistency of the results proves the suitability of the methodology selected. Understanding the vulnerability at the local scale is key to planning risk-reduction strategies as well to increasing the local population's knowledge about flood-related risks and water quality.

Keywords: vulnerability; rural area; scant data; nitrate contamination; water; flood

---

1. Introduction

In Kenyan rural areas belonging to the Arid and Semi-Arid Lands (ASALs), water quantity and water quality are major issues for the local population [1–3]. Although access to water is the main livelihood concern, water quality has an important impact on health. Therefore, the issue of access to clean and safe water has garnered increased attention.

In North Horr Sub-County, in northern Kenya, water quality is threatened, among others, by nitrate contamination after flooding events.

Flood-related hazards, in fact, are posing serious threats to the local population and are worthy of further study. However, flood-related hazards in Kenya are poorly addressed in the literature. In particular, there is a lack of quantitative flood-related risk and vulnerability assessment, mainly due to scant data issues [4], namely the absence or the difficulty of getting access to data in African countries [5–7] as well as problems with data quality (sustainable, continuous, credible, publicly accessible, quality assured dataset) [8]. The most relevant studies are conducted in the framework of vector-borne infectious diseases [9–12] or water-borne diseases, especially fecal-oral diseases [13,14].

This study focuses on nitrate contamination of water in open shallow water sources due to fecal intrusion following flooding events and nitrate percolation in groundwater. This work evaluates the vulnerability of nitrate (NO_3^-) contamination in open shallow water sources for human and livestock consumption in a scant data context. Understanding the vulnerability to nitrate contamination is pivotal since the consumption of contaminated water can have severe outcomes in humans, such as methemoglobinemia, hypertension, increased infant mortality, central nervous system birth defects, diabetes, spontaneous abortions, respiratory tract infections and changes to the immune system [15–19], as well as methemoglobinemia, spontaneous abortion and even death in livestock [20–22].

River and groundwater contamination is generally caused by chemical fertilizer, manure and nitrogenous waste products, all containing nitrogen, used for both agricultural and industrial purposes [23]. Therefore, contamination can affect rural, semi-urban and industrialized areas. In the case here analyzed, the source of nitrates is the deposit of fecal matters. As a consequence of dry climate conditions and pastoralist-related livelihoods, northern Kenya mostly features the availability of open shallow water sources, sometimes found exactly in the riverbed (e.g., open shallow wells) [24]. The seasonal rainfall pattern causes the activation of nitrate contamination hazard. In the case of shallow wells, seasonal streams have an intermittent runoff and pastoralists can let their livestock graze near the well while waiting for watering, allowing the deposit of fecal matter directly into the watercourse. For this reason, there may be direct or indirect contamination of the water source depending on whether "fecal matter is deposited directly into waterways or so close that the potential for wash-in is very high, or via surface runoff and subsurface seepage or drainage" [25]. Although there is certainly a direct pathway of fecal matter in the well during the wet season, it is probable that another means of contamination is nitrate percolation through the permeable surface to the subsurface water reservoir. Nitrate percolation activates during rain events without adequate soil contact time for efficient denitrification and retention [26]. Therefore, soil characteristics modelling is crucial for nitrate vulnerability assessment, but its role in contamination dynamics is not known due to the uniqueness of this means of contamination. Studies in New Zealand focus on nitrate movements along a shallow groundwater flow path in a riparian wetland [26] and on estimating fecal yields from agricultural catchments for water quality purposes [27–29]. Another study in Minnesota assesses private wells vulnerability to nitrate leaching, focusing on future extreme rainfall estimates and flood- and nitrate-sensitive well identification [30]. Optimal tools for shallow groundwater risk assessment are the DRASTIC and GOD methods, widely applied in arid contexts [31–35]. The DRASTIC method is considered as one that does not need extensive, site-specific pollution data, but able to provide a solution for evaluating the vulnerability to pollution of groundwater based on known hydrogeological parameters (depth to groundwater, net recharge, aquifer media, soil media, general topography or slope, vadose zone and hydraulic conductivity of the aquifer). The GOD model, indeed, is based on only three parameters (groundwater confinement, overlying strata, and depth to groundwater. However, data availability in a developed context, where the research on nitrate pollution of shallow groundwater is well-developed, is different from a Global South context. In Kenya, these types of data are not available or partially available from project output (difficult to find and obtain) or from a global dataset (based on estimations and not fully reliable). Alternative methods [36,37] require the characteristics of the hazard, i.e., nitrate source yields which are also unknown. In general, therefore, in North Horr Sub-County high spatial and high temporal resolution data, as well as point data, are scarce. Thus, in this paper, a flexible methodology was applied focused on the vulnerability assessment in a scant data context. Since the vulnerability is multidimensional, it is almost impossible to define a universal measurement methodology [38] as well as a finite set of potential indicators [39]. Therefore, using the indications of the Intergovernmental Panel on Climate Change (IPCC) [40,41], nitrate contamination specific sub-indicators of exposure, sensitivity and adaptive capacity were identified and combined in order to obtain the indicator of vulnerability for nitrate contamination hazard in the area. The quantitative approach used in this research to evaluate the vulnerability in the North Horr area is based on a tested methodology [42] applied, for the first time, in a scant data context. Although both qualitative and quantitative methods for the evaluation of the vulnerability from nitrate in open shallow water sources are possible, few studies have tried to tackle the issue with a quantitative approach. The flexible methodology [42] applied aims at linking a quantitative approach with the socio-economic aspect evaluation [40]. This innovative approach introduces new results about the analysis of the water sources' vulnerability in North Horr Sub-County.

The result of this research may be relevant for other researchers, since this methodology applies to different types of scant data contexts and risks, as well as to decision makers in Marsabit County as an evaluation tool for the measures undertaken to face water issues contained in the development plan.

The remaining part of the paper proceeds as follows. Section 2 gives an overview of the study area and specific key information on the context on which the analysis is based, as well as information on the materials used; Section 3, separated for clarity reasons from Section 2, addresses the methodology of the analysis and the specific issues regarding the construction of sub-indicators of exposure, sensitivity and adaptive capacity. In Section 4, the main results are analyzed and discussed. Section 6 presents the conclusion of this research.

2. Materials

In a scant data context, like in the northern part of Kenya, availability of data is a pivotal issue. The materials here used have two different types of source: public structured data, i.e., climatic data, demographic data, spatial data, and planning strategies and data collected in the framework of projects. In particular, data on water sources are retrieved from the One Health platform, developed by the Italian start-up TriM (at the moment only for internal use) in the framework of the One Health project (http://www.ccm-italia.org/one-health-uomo-animale-ambiente-north-horr-2) and from the international organization Concern Worldwide (https://www.concern.net/) which shared data collected in a joint campaign with governmental institutions. Except for soil-related data, the data collected were post-processed to construct geo-referenced data in the form of a point vector defined at village or water-source level.

In addition, the limiting factors regarding nitrate behavior, mainly due to scant data context, are identified and assessed. Nitrate behavior in soils, indeed, is dependent on soil characteristics of drainage and texture to consider its transport to groundwater. The role of nitrate degradation was intentionally left out. Attenuation factors like denitrification processes and dilution can occur in the aquifer and influence nitrate concentration [43]. However, the kinetic of nitrate reduction is likely a zero-order reaction and estimations are very low (μM/L per year) [44]. For this reason and for a precautionary hypothesis, denitrification processes are not taken into consideration. Dilution processes are already assessed through the sub-indicators of physical drought exposure and catchment area.

2.1. Study Area: Geographic Positioning and Climatic Context

The area covered by this study (Figure 1) is North Horr Sub-County (Marsabit County) in the northeastern region with a particular focus on the surrounding area of eight main villages.

This region is considered as part of the ASALs since the area is mostly desert and partially covered by a shrub savannah. Local communities are mostly nomadic pastoralists breeding camels, sheep, and goats with traditional extensive livestock practice. This practice relies only on naturally available resources of pastures and water and requires energy-intensive movements of herds. From the onset of the dry season, pastoralists move to the so-called *fora* pastures and can walk distances up to 40 km to reach the watering points.

Rainfalls have a bimodal pattern (two rainy seasons alternated with two dry seasons) and a high variability of rainfall due to cycles of wet periods and droughts, although the variations of these events are not well known [45–47].

For clarity reasons, we will here refer to the example of shallow wells. However, the existence of other shallow water sources is based on similar characteristics. In this region, due to the bimodal rainfall pattern, watercourses have a seasonal pattern and runoff can appear once or twice a year [24]. During the rainy season, the seasonal stream's flow is restored and the wells in the riverbed are completely inundated. During the dry season, the water supply is provided by wells dug exactly in the streambed. However, the water extracted is not withdrawn from subsurface runoff, but from the porosity of sand substrates or cones of depression trapped between upward dykes formed of clay soil [24,48]. Those water sources are used both for human and livestock consumption and water can be extracted by hand or by means of motor pumps. For watering purposes, herds of livestock descend the riverbank and wait for their turn, also for hours, close to the well. Therefore, fecal matter can be left directly in the riverbed while waiting for watering. When the stream's flow is restored, nitrate contamination occurs through direct invasion of the fecal matter in the well, added to an indirect component of nitrate percolation in the ground. Thereby, settling water is infected when the water level decreases under the ground level. Then, after the onset of the dry season, water withdrawal and evapotranspiration cause water reduction in the well. Therefore, the concentration of nitrates increases. This process recurs at each season change from wet to dry season.

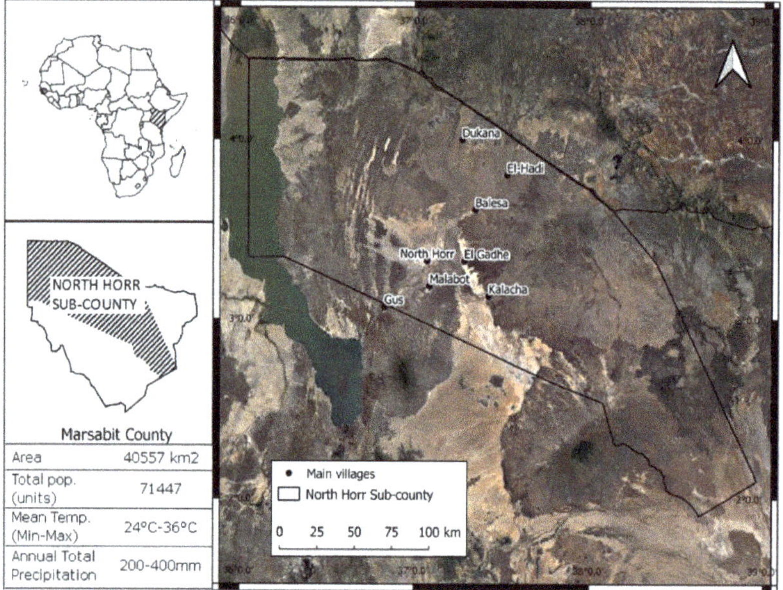

Figure 1. Study area framework. The main figure represents the study area focused on the surroundings of the main eight villages (Dukana, El Hadi, Balesa, El Gadhe, North Horr, Gus, Malabot, Kalacha). The study area is comprised in North Horr Sub-County (Marsabit County) in northern Kenya.

2.2. Climatic Data

Precipitation data are the result of a quantile mapping bias correction applied to the Kenya Meteorological Department precipitation gridded dataset [49]. Precipitation data cover the study area with high resolution (0.0375 × 0.0375 degrees) for the period 1983–2014 with a decadal temporal resolution.

2.3. Demographic Data

Population data are derived from the National Census 2019 [50]. However, there may have been underestimation bias due to severe difficulties in taking a census in an area where people are mainly nomadic with very low population density of about 4 persons per square kilometer. The Kenya National Bureau of Statistic itself declared that the enumeration of nomadic pastoralist population is complex and therefore may lead to population underestimation in ASALs.

2.4. Water Sources

Water sources in the area are mainly represented by boreholes, shallow wells and pans [51], while piped coverage remains limited [52]. Moreover, 66% of sources have contaminated water, which must be treated before drinking [53].

Water resources and allocation data of all kinds are theoretically available for purchase from Kenya governmental agencies, but in reality these data are often difficult or impossible to obtain [54].

Therefore, water source data, the categories of shallow wells, earth pans and rock catchments are retrieved by the One Health platform and through the Concern Worldwide survey. The Second County Integrated Development Plan for Marsabit establishes the presence of 220 shallow wells, 50 pans and 10 rock catchments in the whole North Horr Sub-County. Through the One Health platform and through Concern Worldwide survey, 54 shallow wells, 6 earth pans/dams and 2 rock catchments are identified and analyzed for nitrate contamination vulnerability in the study area (Table 1). Even if shallow wells are the target for the analysis of nitrate contamination vulnerability, earth pans/dams and rock catchments are also taken into consideration as they are subjected to the same conditions of openness, lack of area protection and confinement.

Table 1. Water Source Data in the Study Area.

	Shallow Wells	Earth Pans/Dams	Rock Catchments	Total Water Points
One Health platform	11	3	0	14
Concern Worldwide	43	3	2	48
	54	6	2	62

2.5. Hydraulic Conductivity

The factors that influence the transport and accumulation of nitrate from the land surface to ground water include sediments, rock type and landscape characteristics [43]. However, land characteristics (land use and slope) are here considered negligible factors. The transport capacity of an aquifer is introduced and defined by the fundamental property of hydraulic conductivity (HC) [55]. In fact, the greater the hydraulic conductivity related to the permeability and porosity of the soil, the higher the resulting nitrate infiltration to shallow underground reservoir. The variables taken into consideration are:

- Vertical HC or texture: Texture class (USDA system [56]) at 7 standard depths predicted using the global compilation of soil ground observations (available at www.isric.org/explore/isric-soil-data-hub).
- Lateral HC or drainage: Drainage classes are defined according to the Guidelines for Soil Description [56] predicted using the Africa Soil Profiles Database: Very poor, Poor, Imperfect, Moderate, Well, Somewhat Excessive, Excessive (available at www.isric.org/explore/isric-soil-data-hub).

2.6. Civil and Hydraulic Structures

The proposed analysis investigates the vulnerability at water-source level. In order to understand future changes in water source density, attention is drawn to the actions contained in the Marsabit Second County Integrated Development Plan 2018–2022 (SCIDP) [53]. Among the measures that will be undertaken, only new water sources or rehabilitation of existing water sources contained in the SCIDP are taken into consideration.

3. Methods

This study on spatial vulnerability assessment for nitrate contamination of shallow wells in riverbed is based on the identification of the IPCC approach [40,41]. In particular, the concept of vulnerability refers to "the propensity of exposed elements such as human beings, their livelihoods, and assets to suffer adverse effects when impacted by hazard events" and can be seen as a situation-specific determinant of risk [41].

Therefore, considering the context and the hazard, the model used is [42]:

$$V = \frac{ExS}{AC} \quad (1)$$

Both the Exposure (*E*) and the Sensitivity (*S*) represent the negative effects of the changing conditions, while the indicator of Adaptive Capacity (*AC*) is the parameter which may counteract the negative effect of the impact and therefore improves the vulnerability.

Figure 2 presents an overview of the vulnerability assessment model describing the sub-indicators taken into consideration for the construction of this quantitative vulnerability analysis. As previously mentioned, nitrate contamination of open shallow water sources is a climate-driven issue since the wet periods activate contamination pathways and dry periods reduce the contaminant-to-solution ratio (1). The maximum number of settlements, potentially using the water source, is introduces as a two-way sub-indicator considering the exposure of these settlements and the pressure they bring on water sources (2). Moreover, as highlighted by the literature on the topic, nitrate contamination can occur through direct or indirect pathways (3), therefore sub-indicators of lateral and vertical hydraulic conductivity are taken into consideration. The location of the open shallow water sources (inside, outside or near the stream) (4) is introduced as a sensitivity factor while the adaptation strategies are evaluated through the water sources density (5).

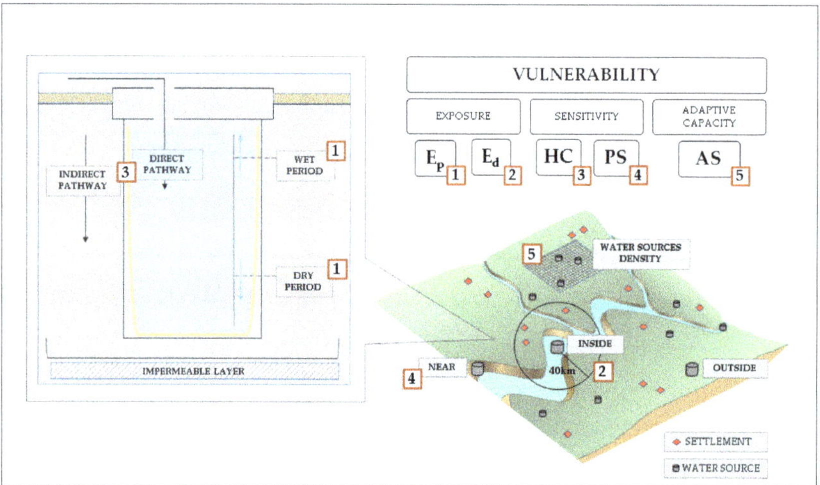

Figure 2. Vulnerability assessment model. The reference numbers in the red boxes are used in the text to refer to nitrate contamination dynamics in association with the related sub-indicators.

Technical details on the sub-indicators of Exposure, Sensitivity and Adaptive Capacity are provided in Table 2.

Table 2. Indicators and Sub-Indicators of the Vulnerability to Nitrate Contamination.

Indicator	Sub-Indicators		Data Source	Original Type of Source	Original Dimension/Resolution	Projection
Exposure (E)	Physical drought and flood exposure (E_p)	Weighted SPI indices	BCKMD [49]	Point vector	Village level	WGS 84/UTM zone 37N
	Demographic Exposure (E_d)	Weighted maximum number of settlements potentially using the water source	Kenya Population Census 2019 (KNBS)	Point vector	Water-source level	WGS 84/UTM zone 37N
Sensitivity (S)	Hydraulic conductivity (HC)	Drainage and texture	ISRIC http://www.isric.org/explore/isric-soil-data-hub	Raster	0.002° × 0.002°	WGS 84
	Presence of the stream (PS)	Water source distance from the stream		Point vector	Water-source level	WGS 84/UTM zone 37N
Adaptive Capacity (AC)	Adaptive strategies (AS)	Density of water sources	Second County Integrated Development Plan 2018–2022	Point vector	Water-source level	WGS 84/UTM zone 37N

The indicators of Exposure, Sensitivity, Adaptive Capacity and Vulnerability are represented with a pixel-based visualization with a resolution of 230 m (based on the highest resolution represented by the Drainage and Texture sub-indicators' raster resolution), depending on the location of water points, with the aim of providing a local scale vulnerability analysis.

3.1. Exposure

3.1.1. Physical Exposure: Standardized Precipitation Index (SPI) for Drought and Flood Assessment

The *SPI* [57] is recommended as the main meteorological drought index from the World Meteorological Organization (WMO) [58]. The *SPI* is used to measure the degree of drought and flood stress through a quantitative description method [42,59,60]. The contamination process, indeed, is influenced both by wet and dry periods. The nitrates contained in fecal matter deposited by livestock contaminate the water through direct and indirect pathways that activate in the wet season. The degree of intensity of the dry period influences the water level in the water source, thus, the contaminants-to-solution ratio. Using the SPI, it is possible to detect the susceptibility of the area to wet and dry conditions.

The single parameter required for its calculation is the precipitation (Equation (2)) and it is generally used as an effective and simple tool for drought assessment. The *SPI* index can be calculated for different time scales (from 1 to 24 months) in order to assess different drought types and wetness conditions [61,62]. Short time scales (1 to 3 months) are mainly related to soil water content and river discharge in headwater areas; medium time scales (3 to 12 months) are related to reservoir storages and discharge in the medium course of the rivers; long time scales (12 to 24 months) are related to variations in groundwater storage.

$$SPI = \frac{P - P^*}{\sigma_p} \tag{2}$$

where P is the monthly precipitation (in mm), P^* is mean monthly precipitation (in mm) and σ_p is the standard deviation of precipitation.

The *SPI* index was calculated at village level and for the three existing stations present in a 250 km radius (Lodwar, Marsabit and Moyale town) using the *SPI* program (available online at www.drought.unl.edu/droughtmonitoring/SPI/SPIProgram.aspx). The *SPI* index is calculated using the BCKMD dataset, a bias-corrected satellite-derived precipitation dataset based on the KMD dataset (issued by the official national meteorological service dataset, available at http://kmddl.meteo.go.ke:8081/SOURCES/.KMD/) corrected with the GPCC [49].

The *SPI* index was computed for different time scales depending on the target. For drought detection the 3-month, 6-month and 12-month time scales were computed, while for flood detection the 1-month and 3-month time scales were detected. Specific weight and ranking values were calculated for drought or flood severity and drought or flood frequencies following the method proposed by [63] for drought risk assessment and already applied in risk assessment studies [42]. Using this methodology, the value of the weighted *SPI* can range from a minimum of 0 to a maximum value of 24.

A known limit of the proposed approach is the presence of many zero rainfall accumulations due to the arid climate. However, if there are many historical zero rainfall accumulations, the estimated gamma distribution may not adequately fit the frequency distribution of the historical rainfall. Therefore, in arid regions, the *SPI* indicator should be interpreted with care [64]. Despite this, the use of monthly cumulative precipitation in this research significantly reduces the number of zero rainfall accumulations and the analysis of *SPI* is reliable.

3.1.2. Demographic Exposure

Nitrate contamination affects human and animal populations. However, statistics on animal population are only partially or not at all available. For this reason, since the local population is mainly part-devoted to pastoralism, we can estimate a proportion between human and animal population and we could assume that the greater the human population is, the greater the number of animals would be. However, due to the low population density of the area, the number of settlements was used instead. To understand the sensitivity of the water sources, the demographic exposure was quantified as the potential maximum number of settlements relying on a single water source. In this area, the pastoralists can walk up to 40 km to water their animals. The people in each settlement in a radius of maximum 40 km can potentially reach the water source. However, they would prefer the nearest sources and then gradually move to farther sources. Therefore, the counting of settlements relying on the water source ($N_{settlements}$) is weighted using the inverse of the distance between each settlement and the water source (d_x).

$$N_{settlements} = \sum_{x=1}^{n} \frac{1}{d_x} \tag{3}$$

Thus, settlements that are closer to the water source have higher weights, while the farther ones have lower weights. The values of the weighted maximum number of settlements potentially using the water source range from 1.2 (minimum value) to 480 (maximum value).

3.2. Sensitivity

3.2.1. Presence of the Stream

The presence of the stream constitutes a sensitivity sub-indicator due to the already explained contamination dynamics. In fact, the water points situated in the riverbed are highly sensitive, the water points near the stream are moderately sensitive, the water points and springs far from the riverbed show a low sensitivity (classified as "outside"). The inside/near/outside classification of the different water sources is made through the direct comparison of satellite images and the geo-referenced water sources since this operation would not have been possible with automatic processes. The sensitivity is then expressed through quantitative weights (adapted from [65]) (Table 3).

Table 3. Weights Assigned for the Presence of the Stream.

Presence of the Stream	Weight
Inside	1
Near	0.6
Outside	0.2

3.2.2. Hydraulic Conductivity

Soil texture and soil drainage classes were reclassified according to coefficients of vertical [66] and lateral [67] hydraulic conductivity respectively (Tables 4 and 5). The soil texture sub-indicator was then averaged across the seven layers to obtain a single vertical HC layer.

Table 4. Texture Classes Classifies According to the Coefficients of Vertical Hydraulic Conductivity.

Soil Texture	
Soil Texture Class	**Vertical HC (m/Day)**
Sand	15.206
Loamy Sand	13.504
Sandy Loam	2.998
Silt Loam	0.622
Loam	0.605
Sandy Clay Loam	0.544
Silty Clay Loam	0.121
Clay Loam	0.216
Sandy Clay	0.190
Silty Clay	0.086
Clay	0.112
Peat	0.691

Table 5. Drainage Classes Classified According to the Coefficients of Lateral Hydraulic Conductivity.

Soil Drainage	
Drainage Class	**Lateral HC (m/Day)**
Very poor	10
Poor	15
Imperfect	20
Moderately well	25
Well	30
Rapid	40
Excessive	50

3.3. *Adaptive Capacity*

The evaluation of the adaptation strategies focused on adaptation actions rather than the development or improvement of the institutional framework since they are long-period approaches or other strategies like risk-transfer methods since they are not problem-solving oriented. Other strategies to avoid vulnerability issue like local and conventional actions were discarded. In fact, local strategies in the ASALs against water scarcity mostly rely on traditional adaptation mechanisms, above all, the adoption of nomadic life or increase of watering distance as closer waterpoints are depleted [68]. Conventional adaptation strategies, such as restriction of water use in pans and boreholes during rainy season until surface runoff has been exhausted, controlling the number of livestock that access the pans and boreholes, and paying infrastructure maintenance fees to help increasing water availability for longer periods, are sparsely adopted. However, a more sedentary life and demographic increase are threatening water availability in inhabited areas [69]. For these reasons, the limits of local adaptation capacity must be overcome through the adaptive actions set in place by local governmental institutions. Based on the SCIDP [53], there are eight actions that contribute to the improvement of water sources: installation of gensets and solar panels, construction of shallow wells, supply of fresh and clean piped water, drilling of boreholes, installation of water towers, construction of a dam, installation of underground tanks piped water filled, water trough (Table 6).

Table 6. Adaptation Actions (Installation of Gensets and Solar Panels, Construction of Shallow Wells, Supply of Fresh and Clean Piped Water, Drilling of Boreholes, Installation of Water Towers, Construction of a Dams, Installation of Underground Tanks Piped Water Filled, Water Trough) Divided According to the Intervention Areas.

Place Name	Gensets and Solar Panel	Construction of Shallow Wells	Piped Water	Drilling of Borehole	Water Towers	Dam	Underground Tanks	Water Trough
Durte	•		•					
Barambate			•					
Khob dertu		•						
Elbeso	•		•	•				•
Konon Gos							•	
Qorqa Gudha	•							
El-Isacko Mala			•	•				
El-Gufu		•						
Barambate								
El-Buka		•						
Boji			•					
Goricha village	•		•				•	
Malabot	•		•	•				
North Horr			•	•				
Galas	•		•	•				
El Boru Magado	•			•				
Eredheri					•			
Lag Balal						•		
Tiniqo	•				•			
Qorqa Diqa	•							•
Hurri Diqa					•			
Elmuda		•						
Wanno								•
Bara	•							
Ruso	•	•						
Horri Gudha			•		•			

Compared to the current adaptation scenario (AC_0), the implementation of the actions stated in the SCIDP will move to a future adaptation scenario (AC_1). Therefore, two adaptation scenarios were constructed (AC_0 and AC_1) where the water sources density in the project area is used as proxy of the sub-indicator of the walking distance to water sources as suggested by [70]. The AC_0 scenario, thus, represents water source density ante-SCIDP and the AC_1 scenario represents water source density (i.e., ante-SCIDP water sources plus new water sources planned) post-SCIDP.

3.4. *Vulnerability*

The vulnerability is the summary indicator obtained from the combination, based on Equation (1), of the previously analyzed indicators and sub-indicators. Each indicator (Exposure, Sensitivity and Adaptive Capacity) was rasterized (WGS 84/UTM zone 37 N, 230 m × 230 m) and normalized in the range 0–10 to be comparable. Then, Exposure and Sensitivity were multiplied together obtaining the potential impacts and finally divided by the Adaptive Capacity counteracting the negative effects of the impacts. As per the scale interpretation, high values correspond to high values of each indicator and of vulnerability.

4. Results

4.1. *Exposure*

4.1.1. Demographic Exposure

The water sources in the surroundings of Dukana show a higher weighted number of settlements and therefore these are the most exposed areas in North Horr Sub-County (see Figure 3a). Some water sources in the western part of the sub-county have a medium demographic pressure. The rest of the water sources are less stressed.

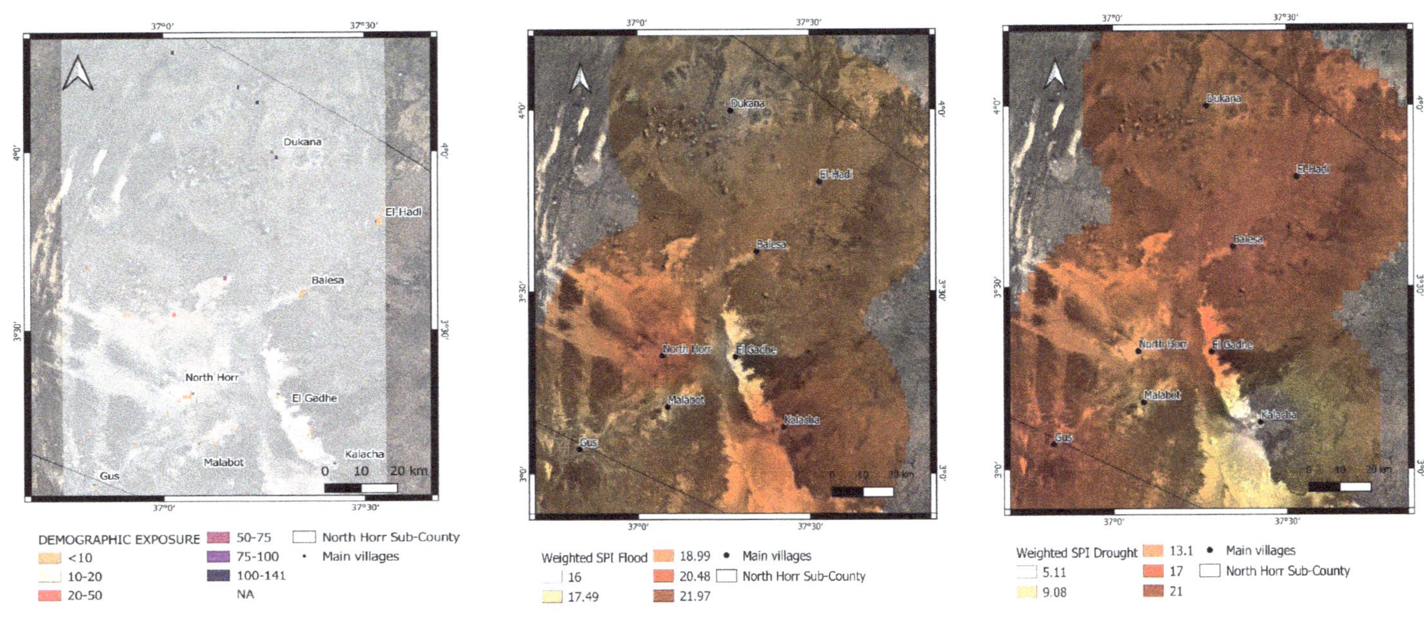

Figure 3. Exposure sub-indicators: (**a**) demographic exposure, (**b**) physical flood exposure, and (**c**) physical drought exposure.

4.1.2. Weighted SPI Index

Although there is a suitable timescale for each specific context, it may be helpful to present results for alternative timescales [71] to observe the variability of the index. SPI index weighted for each main village and station shows different performance according to the time scale (Table 7).

Table 7. SPI Index at 3-, 6- and 12-Month Timescales for All the Facilities.

	Physical Exposure (E_p)				
	Flood		Drought		
	SPI1	SPI3	SPI3	SPI6	SPI12
Facilities					
Balesa	18	13	9	21	20
Dukana	16	15	11	15	17
El Gadhe	16	13	6	19	21
El Hadi	18	11	12	19	20
Gus	17	19	6	21	15
Kalacha	21	14	3	5	15
Malabot	17	15	3	13	19
North Horr	22	15	6	15	17
Stations					
Lodwar	16	12	6	18	15
Marsabit	17	15	6	12	18
Moyale	15	17	6	18	15

The following considerations on the choice of the timescale for detecting flood and drought spells are since, at this latitude, rainfalls are concentrated into two seasons of three months each. Rain events that trigger fecal matter intrusion in shallow water are extreme events. The 1-month SPI can assess deviation from normal monthly cumulated precipitation better than the 3-month SPI, which can better assess seasonal deviations. Therefore, the 1-month SPI was used as a sub-indicator for physical flood exposure, as it can detect areal susceptibility to wetness conditions (see Figure 3b). The 6-months SPI was preferred to the 3- and 12-month timescale as it can detect two consecutive dry seasons and potentially the onset of a drought period. Therefore, we used it as the sub-indicator for physical drought exposure as it can be very effective in showing reduced streamflow and reservoir levels (see Figure 3c).

4.2. *Sensitivity*

4.2.1. Presence of the Stream

The greatest number of water sources that are situated exactly in the riverbed is found in the North and in the East of the sub-county. Therefore, this is where it is possible to find the open water sources that are more sensitive to nitrate contamination (see Figure 4a).

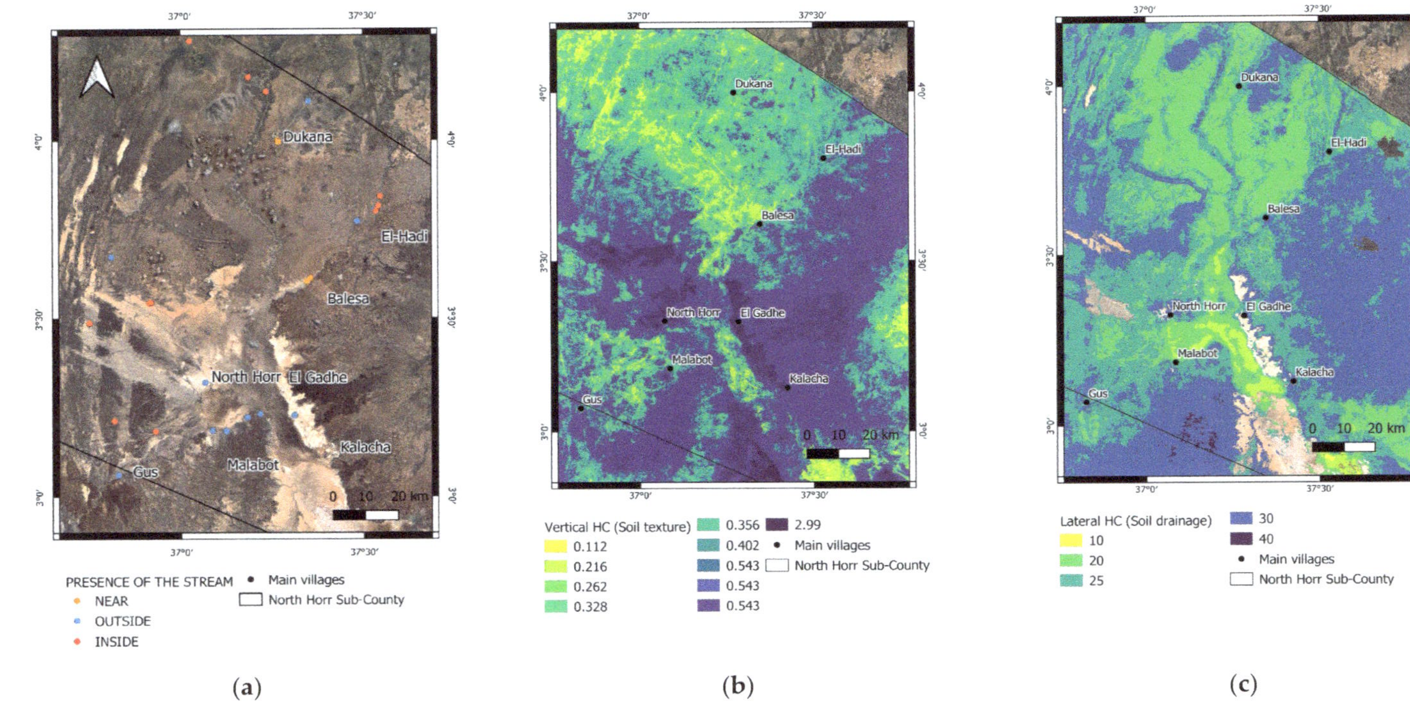

(a) (b) (c)

Figure 4. Sensitivity sub-indicators: (**a**) presence of the stream, (**b**) soil texture, and (**c**) soil drainage.

4.2.2. Hydraulic Conductivity

The two sub-indicators of hydraulic conductivity (vertical and lateral hydraulic conductivity) describe the permeability to contaminants in the area. In the lowlands in the central-southern part of the North Horr Sub-County there are the greatest rate of vertical hydraulic conductivity due to the presence of the northern part of the Chalbi desert. In the northern part of the sub-county, the vertical hydraulic conductivity decreases with the increase of the altitude (see Figure 4b).

Regarding the lateral hydraulic conductivity, higher values of drainage are found along the watercourses and in the highlands at the starting points of the watersheds (see Figure 4c).

4.3. Adaptive Capacity

Among the eight actions that contribute to the improvement of water sources, only four contribute significantly to the reduction of the walking distance to safe water sources. These are construction of shallow wells, supply of fresh and clean piped water, drilling of boreholes, installation of underground tanks filled with piped water. Therefore only the following water sources were taken into consideration for the water sources density (Table 8).

Table 8. Number and Distribution According to the Type of Water Sources Taken into Consideration for the Water Sources Density in AC_0 and AC_1.

	Boreholes	Earth Pans/Dams	Hand-Dug Wells	Shallow Wells	Piped Water	Rock Catchments	Springs	Others (Underground Tanks, Tanks)
AC_0	10	6	13	54	0	2	11	14
AC_1	16	6	13	59	10	2	11	16

The increase of the number of safe water sources in the study areas is considered as an improvement of the adaptation capacity. It lowers the animal gatherings at water sources, it reduces the exploitation of water sources themselves and improves access to water.

The scenario AC_0, as said, is the current state of density of water sources, while the scenario AC_1 is the possible future scenario if all the planned measures are set in place by the end of 2022. In both scenarios, North Horr and El Hadi are the inhabited centers with the higher concentration of water sources.

4.4. Vulnerability

The vulnerability is the resulting combination of normalised indicators: Normalised Exposure (see Figure 5), Normalised Sensitivity (see Figure 6) combined with the Scenarios AC_0 and AC_1 (see Figure 7) after their normalization in Normalised AC_0 (Figure 8a) and Normalised AC_1 (Figure 8b). The vulnerability analysis has a major outcome: both in Scenario 0 and Scenario 1 (Figure 9) the northern part of the sub-county is more vulnerable compared to the rest of area. After the implementation of adaption measures there is a timid improvement in vulnerability values in the southern sub-county (whitish spots). In Scenario 1 (Figure 9b), where there are no new adaptation measures (northern part), the values of vulnerability show no change.

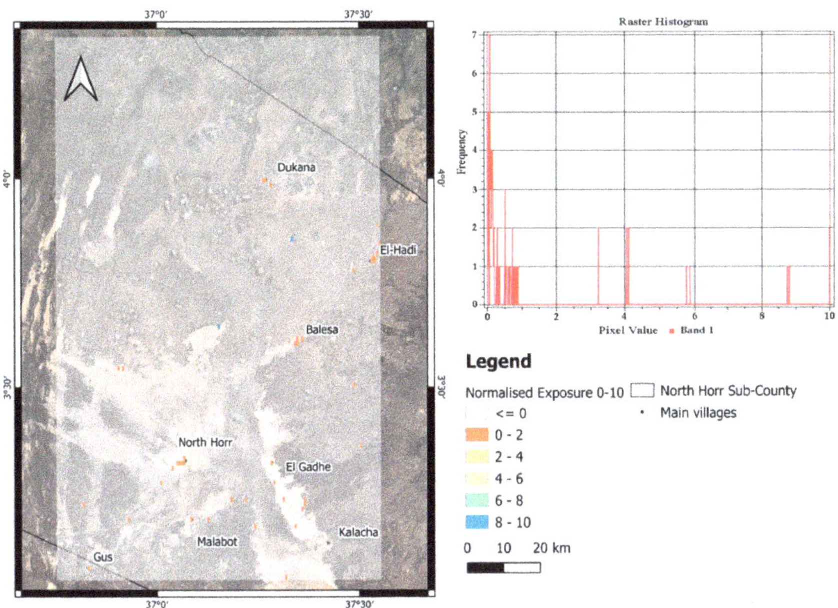

Figure 5. Normalised Exposure.

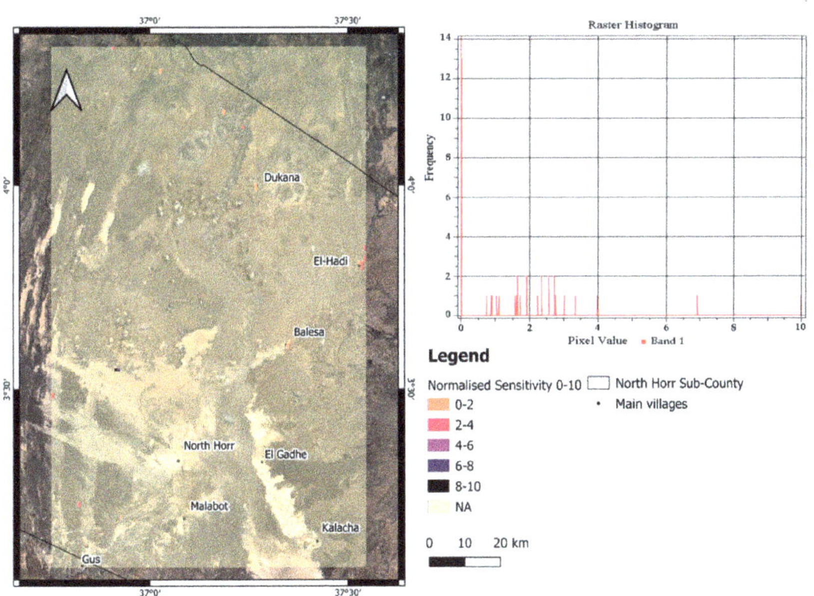

Figure 6. Normalised Sensitivity.

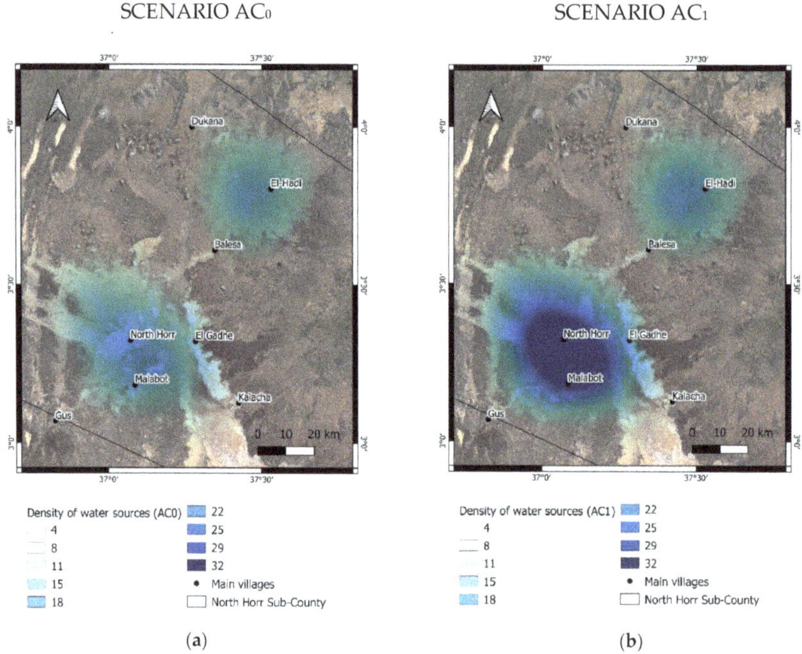

Figure 7. Adaptation sub-indicator: (**a**) density of water sources (AC_0) and (**b**) density of water sources (AC_1).

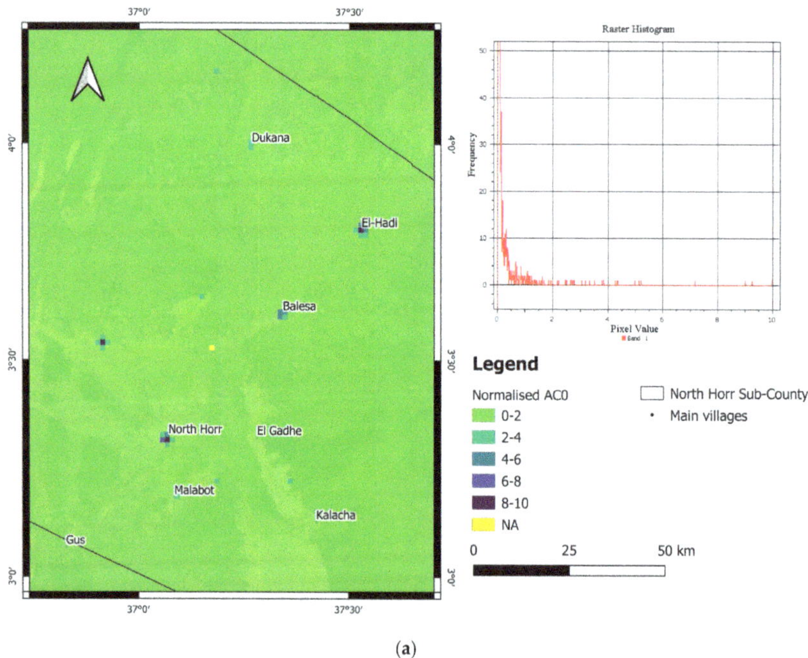

Figure 8. *Cont.*

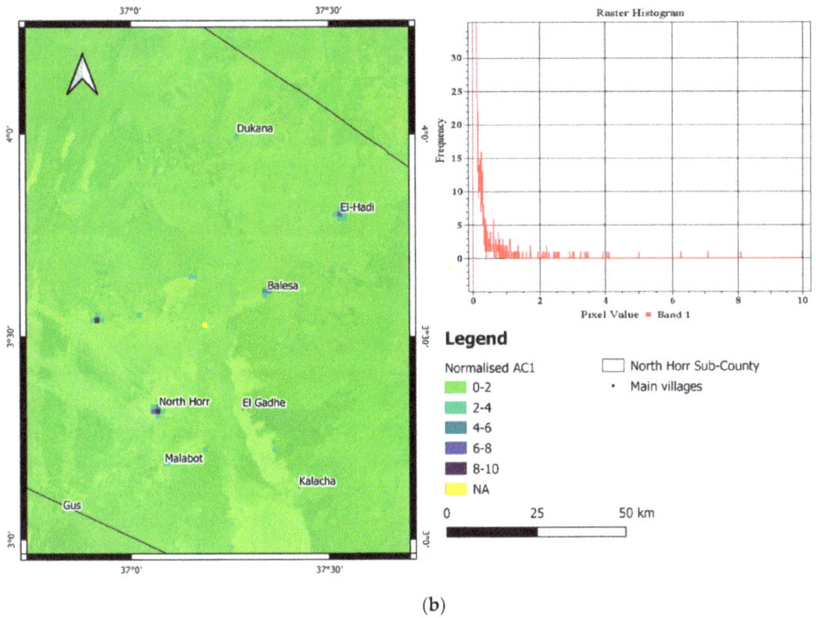

(b)

Figure 8. Adaptation indicators: (a) Normalised AC_0 and (b) Normalised AC_1.

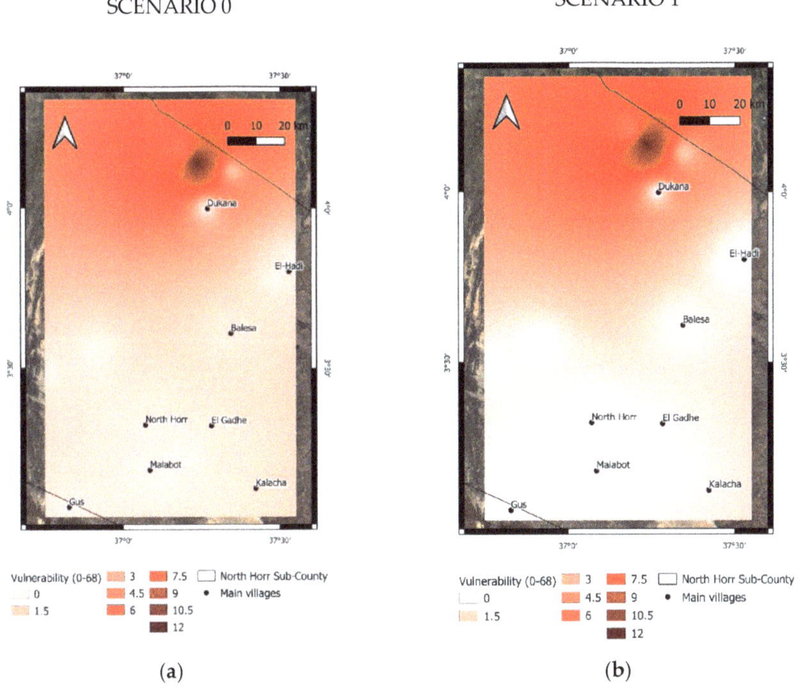

(a) (b)

Figure 9. Vulnerability indicators in two adaptation scenarios (density of water sources): (a) Scenario 0 with the current state of adaptation (b) Scenario 1 with implementation of the measures contained in the SCIDP.

5. Discussion

The result showed that the values of vulnerability in Scenarios 0 and 1 are greater in the northern part of the sub-county (dark red areas) with respect to the southern part (white to light red areas). Very few changes occurred from Scenario 0 to Scenario 1.

In fact, compared to Scenario 0, with the implementation of the adaptation measures—concentrated in the southern and central part of the sub-county—only the areas surrounding Gus, Malabot, North Horr, Kalacha, El Gadhe, Balesa and El-Hadi reduces their vulnerability. The area in the North do not benefit from the adaptation actions. Indeed, the areas in the North that were more vulnerable in Scenario 0 remain the most affected also in Scenario 1, while the areas in the South improve their capacity to adapt to nitrate contamination risk.

The measures contained in the SCIDP that aim to improve water access are concentrated in the southern-central part of the sub-county around the main town of North Horr (Figure 10a). Indeed, it seems possible to observe a radial pattern distribution of the areas of intervention starting from North Horr. As a matter of fact, North Horr is the chief town of the sub-county. Although it is difficult to reach North Horr from Marsabit town passing through the Chalbi desert, the areas in the North are way more difficult to get to and very close to the border with Ethiopia. The organization of similar interventions in the northern area could be challenging from a logistic point of view. Therefore, North Horr seems to be prioritized in the decision-making process. Moreover, the availability of funds is hardly employed in remote areas where the population density is one of the lowest in all of Kenya (about four inhabitants per km^2) [50]. In fact, at the moment, funds of sustainable growth are preferably diverted in the areas along the Great North Road (the road going from El Cairo to Cape Town passing through Marsabit County), where a rapid population influx is expected [53].

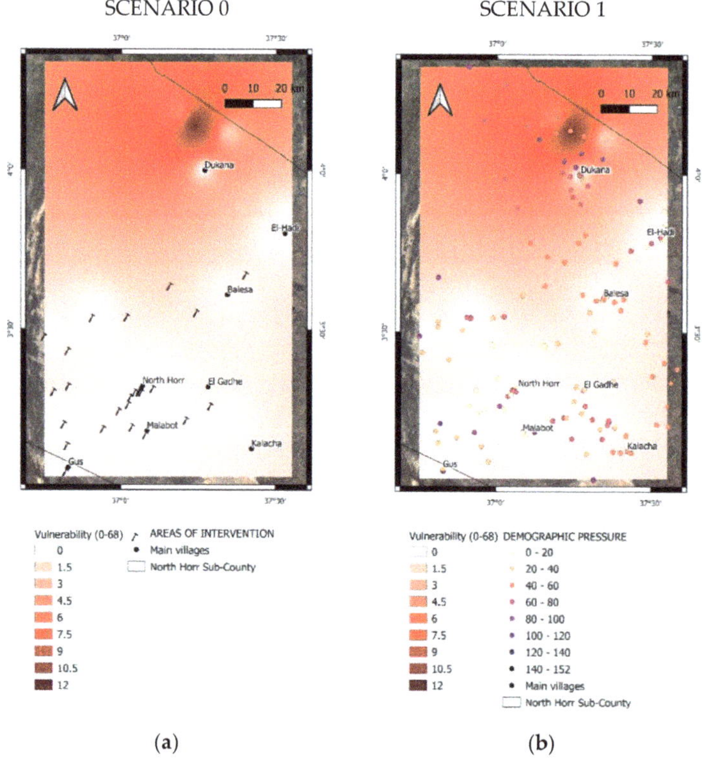

Figure 10. (**a**) Scenario 0 and the planned areas of intervention; (**b**) Scenario 1 and the demographic exposure.

Besides the low population density in the northern part of the sub-county, the demographic pressure on few water sources is an issue. In fact, a great number of settlements rely on few water sources creating significant stress for water supply dynamics (Figure 10b).

Therefore, remoteness of the northern part of the sub-county, security problems, scarce interest, and high demographic pressure on the scarce number of water source create suitable conditions for water scarcity and low water quality. Despite the SCIDP's intention to promote the development of the area for the period 2018–2022, the plan will not be able to rectify the asymmetries and enhance the easiest access to quality water. If not addressed soon, the associated consequences could be an increase in migration sometimes identified as urbanization processes and conflict escalation in the area over water resources. Moreover, worsening consequences could be deteriorated by climate changes and, in particular, by more hydro-climatic variability in space and time [72].

To prevent social inequalities in access to clean and safe water, a broader understanding of the existing water quality in Kenya is necessary. However, poor knowledge and application of the Groundwater Vulnerability Assessment are found [73], even if contamination are common problems that affect groundwater quality [74].

These results and the related discussion confirm that the adopted methodology solves the issue of vulnerability assessment in a scant data context. In fact, the methodology proposed can identify the most vulnerable areas of the sub-county, together with most vulnerable shallow water sources, considering a selection of specific available data. Thus, the scant data limit is overcome and addressed. A limitation could be a lack of groundwater NO_3 concentration data; however, an in-situ extensive data collection campaign would be a major effort for the abovementioned logistics issues. Thanks to the methodology proposed, more vulnerable areas are identified and a more narrow group of water sources to be analyzed could be selected, as done by [75]. Alternative applications of this methodology could be useful to assess the vulnerability in other scant data contexts as support for decision making or as preliminary assessment for more in-depth analysis of the water sources regarding nitrate contamination processes.

6. Conclusions

This research aimed at assessing the vulnerability of the area to nitrate contamination of water in open shallow water sources due to fecal intrusion following flooding events and nitrate percolation in groundwater. The vulnerability indicator was constructed according to two different scenarios pre- and post-interventions contained in the SCIDP, respectively Scenario 0 and Scenario 1. The result obtained shows the partial inefficacity of the interventions since they are concentrated in the central part of the sub-county, where it is already possible to find most of the water sources of the area. The northern part of the sub-county is remote, insecure, and scarcely populated, therefore, there is limited ability and low interest in improving water access in these areas.

Since the adaptation measures planned do not intervene in the most vulnerable areas, at least until 2022, the reduction of nitrate levels in the water supplied may rely on different control methods based on the acceptable level of human health risk, nitrate-control cost and technical feasibility. These methods include nitrate source control, blending of two or more water supplies and direct treatment of nitrates [76]. These methods of nitrates control should be coupled with an awareness campaign on nitrate contamination control.

Moreover, the consistency of the results confirms the goodness of the proposed methodology for scant data context based on the selection of specific quantitative sub-indicators, combining socio-economic aspects with physical features.

In conclusion, the present study contributed to the understanding of the present and near-future vulnerability of the area, as well as of the limitations of the SCIDP in contrasting the risk of nitrate contamination of shallow water sources. Policymakers should be aware of the existing imbalances between the North and South of the sub-county and try to incorporate these findings, albeit achieved in a scant data context, for future policy planning. At the same time, it is demonstrated that a quantitative approach can be easily used to evaluate the vulnerability to climate changes also in a scant data situation and in remote areas.

Author Contributions: Conceptualization, V.B. and. A.P.; methodology, V.B., A.P., E.C. and M.R.; formal analysis, V.B.; writing—original draft preparation, V.B.; writing—review and editing A.P., V.B., E.C. and M.R.; supervision, A.P., E.C. and M.R. All authors have read and agreed to the published version of the manuscript.

Funding: This research received no external funding.

Acknowledgments: This study is conducted in the framework of the International Cooperation Project "ONE HEALTH: Multidisciplinary approach to promote the health and resilience of shepherds' communities in North Kenya" funded by the Italian Agency for Development Cooperation (AICS). The authors would like to thank the project coordinator (CCM) and project partners (TriM and VSF-Germany) and the Kenyan Meteorological Department. Finally, a due thank to Concern Worldwide which shared data on water sources collected in a joint campaign with the Government of Kenya.

Conflicts of Interest: The authors declare no conflict of interest.

References

1. Marshall, S. The water crisis in Kenya: Causes, effects and solutions. *Glob. Major. E J.* **2011**, *2*, 31–45.
2. Mulei Kithiia, S. Water quality degradation trends in kenya over the last decade. In *Water Quality Monitoring and Assessment*; Voudouris, K., Voutsa, D., Eds.; InTech: Rijeka, Croatia, 2012; pp. 509–526. ISBN 9789535104865.
3. Moraa, H.; Otieno, A.; Salim, A. *Water Governance in Kenya: Ensuring Accessibility, Service Delivery and Citizen Participation*; iHubResearch: Nairobi, Kenya, 2012.
4. UNEP. *A Snapshot of the World's Water Quality: Towards a Global Assessment*; United Nations Environment Programme: Nairobi, Kenya, 2016; ISBN 9789280735550.
5. Borel-Saladin, J. Data dilemmas: Availability, access and applicability for analysis in sub-saharan african cities. *Urban Forum* **2017**, *28*, 333–343. [CrossRef]
6. Osuteye, E.; Johnson, C.; Brown, D. The data gap: An analysis of data availability on disaster losses in sub-Saharan African cities. *Int. J. Disaster Risk Reduct.* **2017**, *26*, 24–33. [CrossRef]
7. Carter, T.R.; Jones, R.; Lu, X.; Bhadwal, S.; Conde, C.; Mearns, L.; Rounsevell, M.; Zurek, M.; Parry, M.; Canziani, O.; et al. New assessment methods and the characterisation of future conditions. Climate Change 2007: Impacts, Adaptation and Vulnerability. In *Contribution of Working Group II to the Fourth Assessment Report of the Intergovernmental Panel on Climate Change*; Parry, M.L., Canziani, O.F., Palutikof, J.P., van der Linden, P.J., Hanson, C.E., Eds.; Cambridge University Press: Cambridge, UK, 2007; pp. 133–171.
8. Grasso, V.F.; Dilley, M. *A Comparative Review Of Country-Level and Regional Disaster Loss and Damage Databases*; United Nations Development Programme: New York, NY, USA, 2013.
9. Grady, S.C.; Messina, J.P.; McCord, P.F. Population vulnerability and disability in Kenya's tsetse fly habitats. *PLoS Negl. Trop. Dis.* **2011**, *5*. [CrossRef] [PubMed]
10. Mulefu, F.O.; Mutua, F.N.; Boitt, M. Malaria risk and vulnerability assessment GIS approach. Case study of busia county, Kenya. *IOSR J. Environ. Sci.* **2016**, *10*, 104–112.
11. Githeko, A.K.; Lindsay, S.W.; Confalonieri, U.E.; Patz, J.A. Climate change and vector-borne diseases: A regional analysis. *Bull. World Health Organ.* **2000**, *78*, 1136–1147.
12. Onyango, E.A.; Sahin, O.; Awiti, A.; Chu, C.; Mackey, B. An integrated risk and vulnerability assessment framework for climate change and malaria transmission in East Africa. *Malar. J.* **2016**, *15*, 1–12. [CrossRef] [PubMed]
13. Stoltzfus, J.D.; Carter, J.Y.; Akpinar-Elci, M.; Matu, M.; Kimotho, V.; Giganti, M.J.; Langat, D.; Elci, O.C. Interaction between climatic, environmental, and demographic factors on cholera outbreaks in Kenya. *Infect. Dis. Poverty* **2014**, *3*, 37. [CrossRef]
14. Olago, D.; Marshall, M.; Wandiga, S.O.; Opondo, M.; Yanda, P.Z.; Kangalawe, R.; Githeko, A.; Downs, T.; Opere, A.; Kabumbuli, R.; et al. Climatic, socio-economic, and health factors affecting human vulnerability to cholera in the Lake Victoria Basin, East Africa. *Ambio* **2007**, *36*, 350–358. [CrossRef]
15. Fewtrell, L. Drinking-water nitrate, methemoglobinemia, and global burden of disease: A discussion. *Environ. Health Perspect.* **2004**, *112*, 1371–1374. [CrossRef]

16. Malberg, J.W.; Savage, E.P.; Osteryoung, J. Nitrates in drinking water and the early onset of hypertension. *Environ. Pollut.* **1978**, *15*, 155–160. [CrossRef]
17. Fan, A.M.; Steinberg, V.E. Health implications of nitrate and nitrite in drinking water: An update on methemoglobinemia occurrence and reproductive and developmental toxicity. *Regul. Toxicol. Pharmacol.* **1996**, *23*, 35–43. [CrossRef] [PubMed]
18. Ward, M.H.; Jones, R.R.; Brender, J.D.; de Kok, T.M.; Weyer, P.J.; Nolan, B.T.; Villanueva, C.M.; van Breda, S.G. Drinking water nitrate and human health: An updated review. *Int. J. Environ. Res. Public Health* **2018**, *15*, 1557. [CrossRef]
19. Johnson, C.J.; Kross, B.C. Continuing importance of nitrate contamination of groundwater and wells in rural areas. *Am. J. Ind. Med.* **1990**, *18*, 449–456. [CrossRef]
20. MG, B. Environmental factors associated with nitrate poisoning in livestock in Botswana. *J. Pet. Environ. Biotechnol.* **2012**, *3*, 131. [CrossRef]
21. Ozmen, O.; Mor, F.; Sahinduran, S.; Unsal, A. Pathological and toxicological investigations of chronic nitrate poisoning in cattle. *Toxicol. Environ. Chem.* **2005**, *87*, 99–106. [CrossRef]
22. Campbell, J.B.; Davis, A.N.; Myhr, P.J. Methaemoglonaemia of livestock caused by high nitrate contents of well water. *Can. J. Comp. Med. Vet. Sci.* **1954**, *18*, 93–101. [PubMed]
23. WHO. Nitrate and Nitrite in Drinking-Water: Background Document for Development of WHO Guidelines for Drinking Water Quality. Available online: http://www.who.int/water_sanitation_health/dwq/chemicals/nitratenitrite2ndadd.pdfS (accessed on 18 June 2020).
24. Nissen-Petersen, E. *Water from Dry Riverbeds*; Danish International Development Agency in Kenya: Nairobi, Kenya, 2006.
25. Collins, R.; Mcleod, M.; Hedley, M.; Donnison, A.; Close, M.; Hanly, J.; Horne, D.; Ross, C.; Davies-Colley, R.; Bagshaw, C.; et al. Best management practices to mitigate faecal contamination by livestock of New Zealand waters. *N. Z. J. Agric. Res.* **2007**, *50*, 267–278. [CrossRef]
26. Burns, D.A.; Nguyen, L. Nitrate movement and removal along a shallow groundwater flow path in a riparian wetland within a sheep-grazed pastoral catchment: Results of a tracer study. *N. Z. J. Mar. Freshw. Res.* **2010**, 371–385. [CrossRef]
27. Davies-Colley, R.J.; Nagels, J.W.; Smith, R.A.; Young, R.G.; Phillips, C.J. Water quality impact of a dairy cow herd crossing a stream. *N. Z. J. Mar. Freshw. Res.* **2004**, *38*, 569–576. [CrossRef]
28. Collins, R.; Rutherford, K. Modelling bacterial water quality in streams draining pastoral land. *Water Res.* **2004**, *38*, 700–712. [CrossRef] [PubMed]
29. Nagels, J.W.; Davies-Colley, R.J.; Donnison, A.M.; Muirhead, R.W. Faecal contamination over flood events in a pastoral agricultural stream in New Zealand. *Water Sci. Technol.* **2002**, *45*, 45–52. [CrossRef] [PubMed]
30. Hoppe, B.O.; Raab, K.K.; Blumenfeld, K.A.; Lundy, J. Vulnerability assessment of future flood impacts for populations on private wells: Utilizing climate projection data for public health adaptation planning. *Clim. Change* **2018**, *148*, 533–546. [CrossRef]
31. Djoudi, S.; Boulabiez, F.; Pistre, S.; Houha, B. Assessing groundwater vulnerability to contamination in a semi-arid environment using DRASTIC and GOD models, Case of F'kirina Plain, North of Algeria. *IOSR J. Environ. Sci.* **2019**, *13*, 39–44.
32. Boulabeiz, M.; Klebingat, S.; Agaguenia, S. A GIS-Based GOD model and hazard index analysis: The quaternary coastal collo aquifer (NE-Algeria). *Groundwater* **2019**, *57*, 166–176. [CrossRef] [PubMed]
33. Ghazavi, R.; Ebrahimi, Z. Assessing groundwater vulnerability to contamination in an arid environment using DRASTIC and GOD models. *Int. J. Environ. Sci. Technol.* **2015**, *12*, 2909–2918. [CrossRef]
34. Bataineh, S.; Curtis, C.; In, M.'; Alghwazi, Z. *Groundwater Resources, the DRASTIC Method and Applications in Jordan*. Available online: http://courses.washington.edu/cejordan/SbCcMa_Presentation.pdf (accessed on 7 February 2020).
35. Zhou, Z.; Ansems, N.; Torfs, P. *A Global Assessment of Nitrate Contamination in Groundwater Internship Report*; International Groundwater Resources Assessment Center: Delft, The Netherlands, 2015.
36. Martínez-Salvador, C.; Moreno-Gómez, M.; Liedl, R. Estimating pollutant residence time and NO3 concentrations in the Yucatan karst aquifer; considerations for an integrated karst aquifer vulnerability methodology. *Water* **2019**, *11*, 1431. [CrossRef]
37. Panagopoulos, Y.; Makropoulos, C.; Baltas, E.; Mimikou, M. SWAT parameterization for the identification of critical diffuse pollution source areas under data limitations. *Ecol. Modell.* **2011**, *222*, 3500–3512. [CrossRef]

38. Birkmann, J. Measuring Vulnerability to promote disaster-resilient societies: Conceptual framewors and definitions. In *Measuring Vulnerability to Natural Hazards: Towards Disaster Resilient Societies*; United Nation University Press: Tokyo, Japan, 2006; pp. 9–54.
39. Science in support of adaptation to climate change. In Proceedings of the Conference of the Parties to the United Nations Framework Convention on Climate Change, Buenos Aires, Argentina, 7 December 2004.
40. Field, C.B.; Barros, V.R.; Dokken, D.J.; Mach, K.J.; Mastrandrea, M.D.; Bilir, T.E.; Chatterjee, M.; Yuka, K.L.E.; Estrada, O.; Genova, R.C.; et al. *Climate Change 2014 Impacts, Adaptation, and Vulnerability Part A: Global and Sectoral Aspects Working Group II Contribution to the Fifth Assessment Report of the Intergovernmental Panel on Climate Change*; Cambridge University Press: New York, NY, USA, 2014; ISBN 978-1-107-05807-1.
41. Cardona, O.-D.; van Aalst, M.K.; Birkmann, J.; Fordham, M.; McGregor, G.; Perez, R.; Pulwarty, R.S.; Lisa Schipper, E.F.; Tan Sinh, B.; Décamps, H.; et al. Determinants of risk: Exposure and vulnerability. In *Managing the Risks of Extreme Events and Disasters to Advance Climate Change Adaptation*; Cambridge University Press: Cambridge, UK; New York, NY, USA, 2012.
42. Belcore, E.; Pezzoli, A.; Calvo, A. Analysis of gender vulnerability to climate-related hazards in a rural area of Ethiopia. *Geogr. J.* **2019**, 1–15. [CrossRef]
43. Nolan, B.T.; Hitt, K.J. Vulnerability of shallow groundwater and drinking-water wells to nitrate in the United States. *Environ. Sci. Technol.* **2006**, *40*, 7834–7840. [CrossRef]
44. Tesoriero, A.J.; Puckett, L.J. O^2 reduction and denitrification rates in shallow aquifers. *Water Resour. Res.* **2011**, *47*. [CrossRef]
45. Orindi, V.A.; Ochieng, A. Case study 5: Kenya seed fairs as a drought recovery strategy in Kenya. *IDS Bull.* **2005**, *36*, 87–102. [CrossRef]
46. Karanja, F.; Mutua Nairobi, F. *Reducing the Impact of Environmental Emergencies through Early Warning and Preparedness-the Case of el Niño-Southern Oscillation (ENSO)*; UNFIP/UNEP/NCAR/WMO/DNDR/UNU: Nairobi, Kenya, 2000; Available online: https://profiles.uonbi.ac.ke/coludhe/publications/reducing-impacts-environmental-emergencies-through-early-warning-and-preparedne (accessed on 26 March 2020).
47. Ogalo, L.; Owiti, Z.; Mutemi, J. Linkages between the Indian Ocean Dipole and East African Rainfall Anomalies. *J. Kenya Meteorol. Soc.* **2008**, *2*, 3–17.
48. Kuria, Z. Groundwater distribution and aquifer characteristics in Kenya. In *Developments in Earth Surface Processes*; Elsevier: Amsterdam, The Netherlands, 2013; Volume 16, pp. 83–107.
49. Vigna, I.; Bigi, V.; Pezzoli, A.; Besana, A. Comparison and bias-correction of satellite-derived precipitation datasets at local level in Northern Kenya. *Sustainability* **2020**, *12*, 2896. [CrossRef]
50. *Kenya National Bureau of Statistics 2019 Kenya Population and Housing Census Volume 1: Population by County and Sub-County*; KNBS: Nairobi, Kenya, 2019; Volume I, ISBN 9789966102096.
51. Rutten, M. Shallow Wells: A Sustainable and Inexpensive Alternative to Boreholes in Kenya. In Proceedings of the EU Conference Support to Marginal Rural Areas in Somalia, Nairobi, Kenya, 23–26 November 2004.
52. *Water Supply and Sanitation in Kenya Turning Finance into Services for 2015 and Beyond An AMCOW Country Status Overview*; Water and Sanitation Program: Nairobi, Kenya, 2011.
53. County Government of Marsabit Second County Integrated Development Plan 2018–2022. Available online: http://marsabit.go.ke/wp-content/uploads/2019/10/Marsabit-CIDP-2018-2022.pdf (accessed on 26 March 2020).
54. Mumma, A.; Lane, M.; Kairu, E.; Tuinhof, A.; Hirji, R. *Kenya Groundwater Governance Case Study*; Water Papers; Worldbank: Washington, DC, USA, 2011; Available online: http://water.worldbank.org/water/sites/worldbank.org.water/files/GWGovernanceKenya.pdf (accessed on 26 March 2020).
55. Di Molfetta, A.; Sethi, R. *Ingegneria Degli Acquiferi*; Springer Science & Business Media: Milano, Italy, 2012; ISBN 9788847018501.
56. IUSS Working Group WRB. *World Reference Base for Soil Resources 2014: International Soil Classification System for Naming Soils and Creating Legends for Soil Maps*; Food and Agriculture Organization of the United Nations: Rome, Italy, 2014; ISBN 9789251083697.
57. Mckee, T.B.; Doesken, N.J.; Kleist, J. The relationshio of drought frequency and duration to time scales. In Proceedings of the Eighth Conference on Applied Climatology, Anaheim, CA, USA, 17 January 1993.
58. World Meteorological Organization (WMO). *Standardized Precipitation Index User Guide (WMO-No. 1090)*; World Meteorological Organization: Geneva, Switzerland, 2012.

59. Liu, X.; Wang, Y.; Peng, J.; Braimoh, A.K.; Yin, H. Assessing vulnerability to drought based on exposure, sensitivity and adaptive capacity: A case study in middle Inner Mongolia of China. *Chinese Geogr. Sci.* **2013**, *23*, 13–25. [CrossRef]
60. Tahmasebi, A. *Pastoral Vulnerability to Socio-Political and Climate Stresses: The Shahsevan of North Iran*; LIT Verlag: Münster, Germany, 2013.
61. Liu, D.; You, J.; Xie, Q.; Huang, Y.; Tong, H.; Xie, J.F.; Huang, Q.J.; Tong, Y.Y.; Liu, D. Spatial and temporal characteris-tics of drought and flood in Quanzhou based on Standardized Precipitation Index (SPI) in recent 55 years. *J. Geosci. Environ. Prot.* **1960**, *6*, 25–37.
62. Seiler, R.A.; Hayes, M.; Bressan, L. Using the standardized precipitation index for flood risk monitoring. *Int. J. Climatol.* **2002**, *22*, 1365–1376. [CrossRef]
63. Shahid, S.; Behrawan, H. Drought risk assessment in the western part of Bangladesh. *Nat. Hazards* **2008**, *46*, 391–413. [CrossRef]
64. Copernicus European Drought Observatory. *Standardized Precipitation Index (SPI)*. Available online: https://edo.jrc.ec.europa.eu/edov2/php/index.php?id=1101 (accessed on 24 February 2020).
65. Huizinga, J.; De Moel, H.; Szewczyk, W. *JRC Technical Reports Global Flood Depth-Damage Functions*; Publications Office of the European Union: Brussels, Belgium, 2017.
66. Fan, Y.; Miguez-Macho, G. A simple hydrologic framework for simulating wetlands in climate and earth system models. *Clim. Dyn.* **2011**, *37*, 253–278. [CrossRef]
67. Ju, W.; Chen, J.M.; Black, T.A.; Barr, A.G.; McCaughey, H.; Roulet, N.T. Hydrological effects on carbon cycles of Canada's forests and wetlands. *Tellus B* **2006**, *58*. [CrossRef]
68. Elema, S.U. *Effects of Climate Variability on Water and Pasture Availability in Turbi Division of Marsabit County, Kenya, 2018*; Kenyatta University: Nairobi, Kenya, 2018.
69. Hazard, B.; Adongo, C.; Wario, A.; Ledant, M. *Comprehensive Study of Pastoral Livelihoods, WASH and Natural Resource Managment in Northern Marsabit*; IFRA: Nairobi, Kenya, 2012.
70. WHO. *Guidelines for Drinking-Water Quality: Fourth Edition Incorporating the First Addendum*; World Health Organization: Geneva, Switzerland, 2014; ISBN 9789241549950.
71. Vicente-Serrano, S.M.; Beguería, S.; Lorenzo-Lacruz, J.; Camarero, J.J.; López-Moreno, J.I.; Azorin-Molina, C.; Revuelto, J.; Morán-Tejeda, E.; Sanchez-Lorenzo, A. Performance of drought indices for ecological, agricultural, and hydrological applications. *Earth Interact.* **2012**, *16*, 1–27. [CrossRef]
72. Niang, I.; Ruppel, O.C.; Abdrabo, M.A.; Dube, P.; Leary, N.; Schulte-Uebbing, L.; Field, C.; Dokken, D.; Mach, K.; Bilir, T.; et al. *Climate Change 2014: Impacts, Adaptation, and Vulnerability. Part B: Regional Aspects. Contribution of Working Group II to the Fifth Assessment Report of the Intergovernmental Panel on Climate Change*; Barros, V.R., Field, C.B., Dokken, D.J., Mastrandrea, M.D., Mach, K.J., Bilir, T.E., Chatterjee, M., Ebi, K.L., Estrada, Y.O., Genova, R.C., et al., Eds.; Cambridge University Press: Cambridge, UK; New York, NY, USA, 2014; pp. 1199–1265.
73. Rendilicha, H.G. A review of groundwater vulnerability assessment in Kenya. *Acque Sotter. Ital. J. Groundw.* **2018**, 7. [CrossRef]
74. REACH. *Country Diagnostic Report, Kenya. REACH Working Paper 3*; University of Oxford: Oxford, UK, 2015; ISBN 9781874370611.
75. Nyilitya, B.; Mureithi, S.; Boeck, P. Tracking sources and fate of groundwater nitrate in Kisumu City and Kano Plains, Kenya. *Water* **2020**, *12*, 401. [CrossRef]
76. Yong Lee, B.W.; Member, S.; Dahab, M.F.; Bogardi, I. Nitrate risk management under uncertainty. *J. Water Resour. Plann. Manag.* **1992**, *118*, 151–165.

Article

Mitigation of the Water Crisis in Sub-Saharan Africa: Construction of Delocalized Water Collection and Retention Systems

Adolfo F. L. Baratta [1,*], Laura Calcagnini [1], Abdoulaye Deyoko [2], Fabrizio Finucci [1], Antonio Magarò [3] and Massimo Mariani [3]

1 Department of Architecture, Roma Tre University, 00153 Rome, Italy; laura.calcagnini@uniroma3.it (L.C.); fabrizio.finucci@uniroma3.it (F.F.)
2 École Supérieure D'ingénierie, D'architecture et D'urbanisme, 3228 Bamako, Mali; abdoulaye.deyoko@wanadoo.fr
3 Department of Architecture, University of Florence, 50121 Florence, Italy; antonio.magaro@unifi.it (A.M.); massimo.mariani@unifi.it (M.M.)
* Correspondence: adolfo.baratta@uniroma3.it

Citation: Baratta, A.F.L.; Calcagnini, L.; Deyoko, A.; Finucci, F.; Magarò, A.; Mariani, M. Mitigation of the Water Crisis in Sub-Saharan Africa: Construction of Delocalized Water Collection and Retention Systems. *Sustainability* **2021**, *13*, 1673. https://doi.org/10.3390/su13041673

Academic Editors: Maurizio Tiepolo, Vieri Tarchiani and Alessandro Pezzoli

Received: 28 December 2020
Accepted: 31 January 2021
Published: 4 February 2021

Abstract: This paper presents the results of a three-year research project aimed at addressing the issue of water shortage and retention/collection in drought-affected rural areas of Sub-Saharan Africa. The project consisted in the design, construction, and the upgrade of existing *barrages* near Kita, the regional capital of Kayes in Mali. The effort was led by the Department of Architecture of Roma Tre University in partnership with the Onlus Gente d'Africa (who handled the on-the-ground logistics), the Department of Architecture of the University of Florence and the École Supérieure d'Ingénierie, d'Architecture et d'Urbanisme of Bamako, Mali. The practical realization of the project was made possible by Romagna Acque Società delle Fonti Ltd., a water utility supplying drinking water in the Emilia-Romagna region (Italy) that provided the financing as well as the operational contribution of AES Architettura Emergenza Sviluppo, a nonprofit association operating in the depressed areas of the world. The completion of the research project resulted in the replenishment of reservoirs and renewed presence of water in the subsoil of the surrounding areas. Several economic activities such as fishing and rice cultivation have spawned from the availability of water. The monitoring of these results is still ongoing; however, it is already possible to assess some critical issues highlighted, especially with the progress of the COVID-19 pandemic in the research areas.

Keywords: water crisis in Africa; water collection and retention systems; sand dam; migration; climate change

1. Introduction

Mali is a landlocked country located in West Africa and 51% of its land is occupied by desert. The cultivated area is 4.7 million hectares, approximately 4% of the entire territory [1]. The country is characterized by the following major types of land:

- Mildly ferralitic soils [2], about 2 million hectares in the extreme south of Mali;
- Tropical iron-rich soils [3], over 17 million hectares in the southern area of the Sahel and south of Sudan. These are highly fertile soils;
- Arid soils in the same areas;
- Semi-arid soils, with a very dry climate, about 43 million hectares, corresponding to 35% of the territory;
- Hydromorphic soils [4] and vertosols [2], characterized by an excess of water due to a temporary or permanent clogging of the soil. This type of soil is dominant in depressions and basins, especially in the inner Niger delta, and it contributes to the alluvial behavior of seasonal streams.

About 47% of the Malian territory is made up of the Niger basin, while the basin of the smaller watercourse called the Senegal River covers 11% of the territory. The Volta basin, the third largest river, corresponds to 1% of the country's surface area, while the remaining 41% is covered by the Sahara Desert. Of note, 1700 km of the 4200 km of the Niger River run through Mali. The Niger and Senegal rivers, and the intricate network of tributaries, provide most of the permanent sources of surface waters. The total average surface water volume is estimated at around 50 km^3/year. Niger alone contributes 35 km^3/year, a third of which is wasted by evaporation. Renewable water resources, present in the subsoil, can be estimated at around 20 km^3/year, half of which is water in common between surface and subsoil. Therefore, in the whole country, the total volume of renewable water is equal to 60 km^3/year. The surface water resources entering the country amount to 40 km^3/year, mostly from New Guinea (33 km^3/year) and from the Ivory Coast (7 km^3/year) [5]. For the exploitation of these water resources, the country counts on the presence of five dams (Table 1), for a total water capacity of 13.8 km^3 (Figure 1) [6].

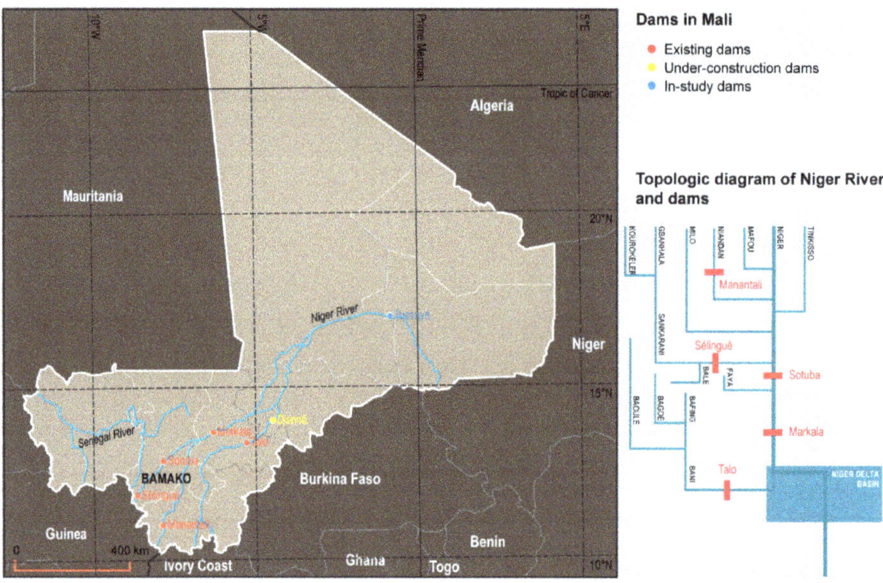

Figure 1. Water flows of the Niger River system and dams' position (Source: Authors').

Table 1. Description of the main dams on Mali's rivers [6].

Dam	River	Capacity	Description
Sélingué	Sankarani	2.17 km^3	Produces hydroelectric power, controls the flow rate (reduced to 75 m^3/s at the level of Markala) and irrigates approximately 2000 hectares of land
Sotuba	Niger	0.005 km^3	Feeds a small hydroelectric power station and the Baguineda artificial canal, which irrigates 3000 hectares of land
Markala	Niger	0.175 km^3	Feeds a series of irrigation canals
Talo	Bani	0.18 km^3	Irrigates 20,000 hectares of land through controlled flooding
Manantali	Bafing	11.27 km^3	Prevents water shortages in the Senegal River during the dry season. Provides the reserve water for the irrigation of 15,000 hectares in Mali, 240,000 hectares in Senegal and 120,000 hectares in Mauritania and the production of electricity.

Only 5% of renewable water resources available to Mali are exploited. Almost all of the water comes from seasonal surface sources, available only in the period from June to December. Only the withdrawals for the communities come from underground water resources, except for the city of Bamako, whose water is drawn from the Niger River [7]. The reasons why groundwater resources are so poorly exploited can be found in:

- Irregular feeding of the aquifers;
- Difficulties in locating subsoil water;
- High water withdrawal costs.

Furthermore, the complexity of the country's resource management organization hinders the implementation of any water policy (Table 2).

Table 2. Organization chart of the Ministries, Departments and Offices that deal directly or indirectly with water resource management [6].

Ministry	Department/Office
Ministère de l'Agriculture (Ministry of Agriculture)	DNGR—Direction Nationale du Génie Rural (National Directorate of Rural Engineering), in charge of developing policies and strategies on hydro-agricultural management and services for rural communities;
	DNA—Direction Nationale de l'Agriculture (National Directorate of Agriculture), responsible for looking after green areas development and plant protection
	IER—Institut d'économie rurale (Institute of Rural Economy), in charge of economic studies and research on the agricultural sector
	CPS—Cellule de Planification et de Statistique (Planning and Statistics Unit), which deals with the collection and processing of data relating to agriculture
	Rural development offices, in charge of the management and development of medium and large-sized state-run irrigation grids
Ministère de l'Energie et de l'Eau (Ministry of Energy and Water)	DNH—Direction Nationale de l'Hydraulique (National Directorate of Hydraulics), responsible for national water policy, as well as for coordinating and monitoring its implementation, with powers over the inventory and management of resources
Ministère de l'Environnement et l'Assainissement (Ministry of Environment and Sanitation)	DNACPN—Direction Nationale de l'Assainissement, du Contrôle des Pollutions et des Nuisances (National Directorate of Sanitation, Pollution and Nuisance Control)
	AEDD—Direction Nationale des Eaux et Forêts, de l'Agence de l'Environnement et du Développement Durable (National Directorate of Water and Forests, of the Environment and Sustainable Development Agency), which intervenes in the event of environmental damage

In addition, a new agency was created in 2002. The *Agence du Bassin du Fleuve Niger* (ABFN) (Niger River Basin Agency) was mandated with safeguarding the Niger basin, as well as the management and integration of water resources in coordination with the corresponding cross-border agencies. The intricate context of governing bodies has given rise to an equally complex regulatory framework through which the management of water resources enjoys little optimization. Ultimately, Mali is characterized by the presence of important water resources, but these are ill distributed over the territory and their strong seasonality is poorly managed.

The access rate to drinking water in Mali is 61% in rural areas and 69.2% in urbanized areas [8]. Bad distribution is followed by bad management, made up of a plethora of bodies with overlapping competencies, scarce powers and a complex regulatory system detached from the local realities (Table 3).

Besides its health implications, access to clean water is considered a human right and is a prerequisite for the realization of other human rights [9]. Therefore, the United Nations, with resolution 64/292 [10], has asked countries and international organizations "to provide financial resources, help capacity-building and technology transfer to help developing countries to provide safe, clean, accessible and affordable drinking water and sanitation for all".

Table 3. Main regulations on water management in Mali [1].

Year	Description
1998	National Environmental Protection Policy that aims to ensure a healthy environment and sustainable development, considering the environmental issue as central to every regulatory area
2002	Law number 02-006 of 31 January 2002, which updates the previous regulations on the subject and, in addition to declaring the public ownership of water, specifies the methods of management and protection, determining the rights and obligations of the State, local bodies and users
2003	National Wetland Policy that defines the long-term vision for the management of wetland ecosystems
2006	Strategic Framework for Growth and Poverty Reduction: a single legislative text that covers all medium-term development policies and strategies. It is the main document on which the negotiation with technical and financial partners is based.
2006	National Water Policy: provides strategic guidelines for the sustainable management of the country's water resources, respecting the balance between the land and aquatic ecosystems
2006	Law on Agricultural Orientation, which constitutes a unifying regulatory framework for all interventions concerning the agricultural sector
2007	National Health Policy and related sectoral strategies addressing solid waste from households and industry, the management of wastewater, special waste, and rainwater
2007–2011	Economic and Social Development Program that mainly concerns the agricultural sector, but, although indirectly, also the question of water resources
2008	National Irrigation Development Strategy which aims to standardize current approaches, identifying and highlighting the priority actions to be taken to make the most of the human and financial resources available. The strategy is based on the participatory and inclusive principle of the beneficiaries, who are involved in the definition, implementation, and management of irrigation projects. The strategy sets an increase rate of 9000 ha/year of additional irrigated areas: this goal was achieved every year from 2000 to 2010.
2009	Rice Cultivation Development Strategy, which aims to satisfy the internal consumption of cereals with the aim of transforming Mali into a net exporter of rice
2010	Code of State Property and Land, which includes groundwater and surface water, which are considered public property owned by the State

The paper focuses on the research, design, and implementation process aimed at the construction and upgrade of existing *barrages* in rural areas of Sub-Saharan Africa.

Water Crisis, Climate Change, Internal Conflict, and Migration

Climate change is among the factors with the greatest impact on the water crisis throughout Sub-Saharan Africa. It will affect those countries, such as Mali, that heavily depend on more less diversified and strongly seasonal agriculture [11].

The climatic conditions of the country (Figure 2) are characterized by average temperatures between 27 and 30 °C annually, with large temperature variations occurring mainly in the desertic areas of the north [12]. In 2015, the maximum recorded temperature was 51 °C and the minimum was 10 °C [13]. The rainy season varies according to latitude: in the south of the country, it lasts up to 6 months, with a marked increase in rainfall between June and October, while in the north it is reduced to just three months between July and September.

The rainfall in the areas close to Sahara is only 50 mm/year, in the Sahel area it is between 100 and 1100 mm/year, while in southern Mali it exceeds 1100 mm/year [12]. Furthermore, Mali is in the so-called Intertropical Convergence Zone where the typical monsoons of West Africa occur. Due to climate change, between 1960 and 2015, the average temperatures increased by 1.2 °C, with a future expectation of linear growth: it is estimated that, by 2050, temperatures could increase between 0.9 and 1.5 °C, with the largest increase in the Kayes region. These changes could have an impact on the amount and patterns of rainfall, the main source of hydrological supply. The rain cycle in Mali is decades-long and has undergone a decrease of 4.4 mm/year from 1950 to 1983 and an increase of only 2.6 mm/year between 1983 and 2015 (Figure 3). Mathematical models predict a slight average increase in precipitation (between 1% and 3%) together with a major decrease in the northern driest regions. Furthermore, the variation in the seasonal distribution could shift the wettest period towards the early part, between June and July with a subsequent reduction (between 6% and 10%) for the rest of the period (Figure 3). In addition, rain-related destructive phenomena are expected to increase [14].

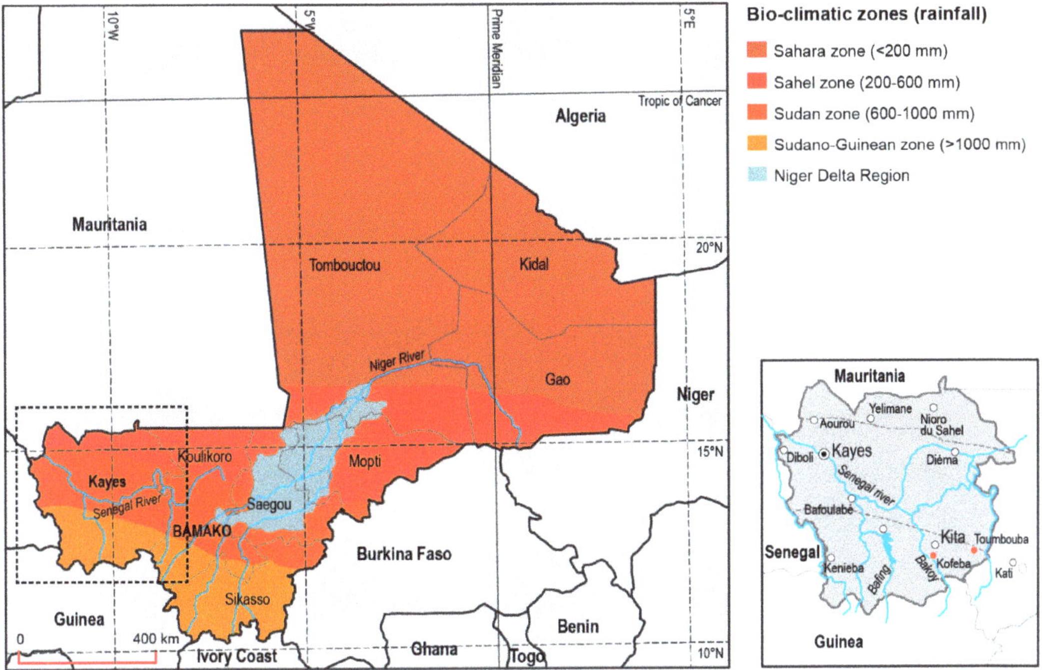

Figure 2. Mali bioclimatic zones and geographic classification of Kayes region (Source: Authors').

The impact on water resources could be substantial: there would be a reduction in the rate of subsoil resources and, at the same time, an increase in their need because the surface resources would tend to be less available due to intrinsic phenomena (such as the increase in evaporation because of the increase in temperature) and extrinsic (such as the growth in demand for water as a consequence of population growth) [15]. The result of climate change could be a decrease in food security (already under acceptable levels in the regions of Gao, Segou, Tomboctou, and Mopti) and malnutrition (acute childhood malnutrition already affects 13% of children under 5), resulting in increased mortality and reduced life expectancy [16].

The difficult political situation linked to internal conflict, which in June 2019 caused the forced migration of almost 148,000 Malians [17], could exacerbate this perspective. Political instability in Mali has its roots in the period following independence from France, obtained in 1960. The first independent government pursued real-socialism policies and nationalized all the industries in the country except for cotton. In the following five years, the country's economy nearly collapsed and Mali was forced to ask for financial support for currency to its former colonizer. A series of periods of crisis followed, leading to five coup attempts between 1970 and 1990. From the early nineties onwards, multiparty democracy began to consolidate, not without violence. Since 1990, the situation has become complicated due to the revolts of the Tuareg people, who have settled in the north of the country. They gave life to the revolutionary movement *Mouvement Populaire de l'Azaouad* (MPA) with the aim of "liberating" the territory to the north by force [18]. The conflict officially ended in 1995 and re-exploded in 2007 due to the dissatisfaction of the Tuareg soldiers integrated into the Malian army. In 2012, the MPA was transformed into the National Movement for Liberation of Azawad (MNLA) [19], which found support in Islamic terrorist groups (Al Qaeda and Ansar). Moreover, Libya, until the fall of Gaddafi, supplied arms to the Tuareg that were superior to those available to the Malian army [20]. The declaration of independence of northern Mali, by the MNLA, saw the emergence of conflicting views between the Tuareg and Ansar. The MNLA and the Jihadist group clashed and brought the Tuareg closer to the national government [21]. However, in 2015 the Tuareg accused the Malian government of not respecting the agreements and the terrorist attacks began again.

Figure 3. Anomaly of total precipitation [m] compared with anomaly of 2 m temperature (°C), historical series, in Sahel (12° N–17° N, 18° W–42° E). (Data source: Climate Change Institute, University of Maine, 2020).

Climate change, food insecurity, and social, economic, and political instability cause migration. Migration flows abroad come mainly from rural areas (73%), are characterized by a male majority (66%) and have as their primary destination the Ivory Coast (70%). As for long-distance destinations, due to the tightening of entry conditions in some European countries of traditional destinations such as France, the choices of Malian migrants fall on southern European countries such as Spain and Italy.

Rationalizing and improving the management of water resources, especially at the local level, would promote food safety, with the consequent reduction in mortality. In addition, improving the country's agropastoral economy would mean reducing political instability, and, over time, popular revolts. Therefore, a stable water supply would improve the living conditions in the country and reduce migratory phenomena.

2. Materials and Methods

This three-year research project (2017–2020) started with the cultural and scientific cooperation agreement between the nonprofit organization *Gente d'Africa* and the Department of Architecture of Roma Tre University (P.I., Prof. Adolfo F. L. Baratta, Research unit coordinator Prof. Fabrizio Finucci), with the aim of providing guidelines for the self-construction of infrastructure dedicated to health, food, and water in the depressed areas of Sub-Saharan Africa. Since 2015, the research group of Roma Tre University has been involved with addressing housing, social, and health problems in marginal areas of the world. With the aim of intervening specifically to address the water crisis in Mali, the two partners obtained the operational collaboration of the University of Florence and *Architettura Emergenza Sviluppo* (AES), a nonprofit organization founded in 2016. *Romagna Acque Società delle Fonti* Ltd. provided most of the financing. Furthermore, since 2019, an international cooperation program has been in place between the Department of Architecture of the University of Roma Tre and the *Ecole Superieure d'Ingénierie d'Architecture et d'Urbanisme* (ESIAU) of Bamako (P.I.: Prof. Abdoulaye Deyoko). ESIAU represents the only point of reference for architecture teaching in the entire country. This collaboration produces the planning, design, and construction of small and medium-sized infrastructures, with prevalent hydraulic characteristics in urban and rural areas. These water infrastructures include the distribution network, terminals, and water capture systems aimed at creating artificial basins, the so-called *barrages*. The *barrages* are barriers with a dam function, equipped with locks for the control of flooding.

The project was carried out in the following phases:

1. Analysis: identification of the architectural, technological, structural, and environmental issues influencing the design choices; adoption of appropriate survey methodologies with the aim of identifying morphological, functional, and technological solutions for the selection of suitable construction techniques; cost-benefit analysis of the solutions identified.
2. Design proposal and execution of the works: the university teams dealt with the final projects design, from the elements and components design up to the complete construction. In this phase, students, researchers, and professors were involved, with the aim of theoretical and practical cultural exchange. The nonprofit organizations took care of the institutional relations and logistical aspects, while the financing partner supported the proposition activity through its know-how.
3. Collection and dissemination of results. The nonprofit organizations, in particular AES, took care of the promotion and communication of the initiative which was presented at an International Conference held on 15 September 2018 at the Department of Architecture of the Roma Tre University and exhibited at EXCO 2019, an event held in Rome and dedicated to cooperation and development of distressed areas.
4. Monitoring of results, currently in progress.

The Italian and Malian universities have focused on the opening of the Erasmus Program beyond the European borders, implemented by the European Commission through the Key Action 107, International Credit Mobility, responding to the Call 2019 and being recently (August 2020) awarded the funding for carrying out the activities of cultural cooperation and international educational exchange.

Although the French term *barrages* is mainly used to indicate medium and large dams, in the context of rural settlements it is used to allude to small structures, able to stop or channel the waters coming from the swelling of the streams during the rainy season. In Africa, such structures, also known as sand dams, help the replenishment of the aquifers [22]. *Barrages* are containment structures that can be classified according to the ability to convey subsoil water or surface water into two types (Figure 4):

- Underground *barrages*;
- Surface *barrages*.

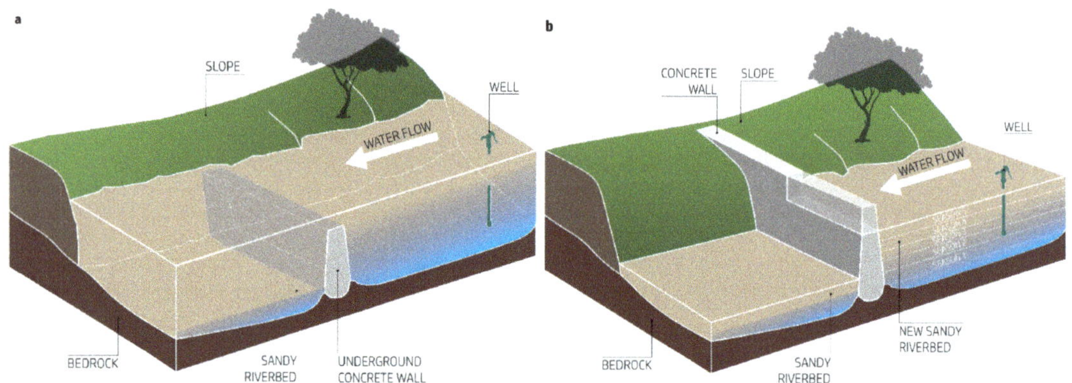

Figure 4. Concept diagram of *barrage* (sand dam): (**a**) Underground *barrages*. (**b**) Surface *barrage* (Source: Authors').

The underground *barrages* are made by digging a deep excavation down to a rocky layer or sufficiently compact soil, allowing for an ordinary foundation of limited size and preventing water from penetrating to greater depths, determining the depth of the aquifer.

The surface *barrages* follow the same principle; however, they are set on a shallow foundation and are designed to stop the watercourse when it swells on the surface because of abundant rainfall on geologically impermeable and compact clay soils.

The *barrage* structure works in a simple but effective way with minimum environmental impact. Because of its small size, it allows overtops water to flow over the *barrage* and avoids draining the aquifer downstream, slowing down the flow of water and facilitating the filling of wells [23].

The surface *barrages* generate an artificial basin which, fed during the rainy season, forms a water reservoir ahead of the dry season. Both *barrage* types, when locally built, use simple construction techniques and intuitive technical principles based on locally available materials. As a result, most of the existing surface *barrages* in Mali are weak structures, often without any foundation. They are incapable of resisting the water pressure as they exploit their weight and shape instead of the characteristics of the material. The poor culture of maintenance means that they are destined to become partially buried due to the accumulation of debris on the upstream-facing side. Furthermore, the lack of a system of local know-how transfer prevents the diffusion of the construction techniques, especially when not based on traditional methods like the use of raw earth, and the correct propagation of a "rule of the art" that could consolidate over the time and could spur any innovations.

However, in many cases the surface and underground *barrages* have been combined in a single structure capable of slowing down the surface flow and increasing the infiltration of water into the subsoil, ensuring greater availability of water resources.

3. Results

One of the main results of this project was the reconstruction and monitoring of two *barrages*, respectively near the village of Toumbouba and the village of Kofeba, close to Kita, one of the most important towns of the Kayes region, 200 km from Bamako. Both are surface *barrages* although they are different in size and required a completely different approach.

3.1. The Barrage in Toumbouba

Located at the geographical coordinates 13°01′07.58″ North Latitude and 9°21′01.04″ West Longitude, the Toumbouba *barrage* (locally known as Tumumba) is grafted onto a preexisting structure built a few decades ago. The original *barrage* was a stone masonry structure, 60 m long, grafted to the north in a retaining wall of an embankment, orthogonal to the position of the main structure. The containing wall resisted the horizontal thrust due to its shape, having a triangular section with the hypotenuse inclined by about 45°. The structure was built on ordinary foundation made by an excavation of a few tens of centimeters below the ground level, filled with large aggregate material bound using cement. The elevation structure had a larger size than the foundation, laying at least two thirds directly on the riverbed. The historical memory of the village traces its construction back to the mid-eighties, but the oral tradition did not confirm any certain source.

During the first on-site survey, in April 2017, it was determined that the power of the watercourse in 2007 caused the overturning of the structure and its breaking into three sections. The first section remained firmly buried for its entire length, while the most stressed section, about 15 m long, divided into two parts, laying down on the bed of the watercourse showing the weakness of the construction technique. The torsion due to the overturning caused damage in need of repair to the orthogonal retaining wall, and the erosion of the water caused the dissolution of a large part of the embankment it contained, jeopardizing the necessary shape of the reservoir. During the survey, the research team acquired the preliminary knowledge of the availability of materials and the skills of the local workforce. It learned of the almost absent theoretical and practical knowledge of the use of steel for construction of flexible structures, and, on the other hand, of the recent use of concrete blocks. These could be easily made in large quantities with handmade block molds as demonstrated in the large urbanizations of Bamako and Kita, where they have replaced the use of the traditional Malian raw earth adobe techniques. The survey ended with the inspection of the damaged structure, to proceed with the design phase. The mold for the construction of the concrete blocks for the vertical elevation was designed first, while the two overturned sections were demolished, and the rubble removed. Those operations were carried out with the help of nearby villagers. The design of an armed and buttressed structure was prepared next, one able to offer the greatest resistance to horizontal thrust, combining shape and flexibility. The construction costs were summarized in a bill of quantities, using the prices recorded in the initial inspection phase as a reference and cross-checking them with the international price list integrated in the CYPE software package [24]. The second mission, aimed at the construction, was organized into two phases. During the first phase, in February 2018, the base of the *barrage* was built. While waiting for the setting and curing of the concrete, the concrete blocks were manufactured. During the second phase, in April 2018, the *barrage* was built (Figure 5), and the gaps in the retaining wall of the embankment and on the embankment itself filled.

The result is a wall structure with variable thickness, strongly reinforced in both directions, buttressed, 15 m long and 2 m high, connected to the preexisting structure through the insertion of steel bars Φ16 and concrete injections every 40 cm in elevation for a depth of at least 60 cm. The structure was finished with a 3 cm thick cement-based plaster, to account for the expected lack of maintenance in the future (Figure 6).

During the survey it was decided to intervene with the consolidation of the structure and the reinforcement of the embankment adjacent to the *barrage*. Furthermore, the correct hydraulic functioning of the structure was tested and proved capable of creating a water reservoir of 6.8 hectares with a depth varying between 30 cm and 2 m, corresponding to a water reservoir of about 45 thousand m^3 in the period of maximum capacity. The basin serves, directly or indirectly, a very vast territory comprising about 20 villages, or 40 thousand inhabitants.

3.2. The Barrage in Kofeba

The work for the Kofeba *barrage* benefited from the experience gained by the group during the work in Toumbouba. The introduction of several practical innovations in the workflow resulted in a shortened construction time without compromising full functionality.

Identified by the geographic coordinates 13°01′01.77″ North Latitude and 9°35′43.60″ West Longitude, the *barrage* is located near a village which in the maps takes the name of Forongo, locally known as Kofeba, located 15 km west of Kita. Like the *barrage* in Toumbouba, the *barrage* in Kofeba was a preexisting structure in stone laid with cement,

215 m long and 80 cm high, with a particular shape whose triangular section provides an acute angle of just 30° near the shutter plane. The regulation of the capacity of the basin was carried out through the manual handling of two rudimentary iron closures, hinged as doors without the presence of a linear counter frame or fixed frame.

Figure 5. The rebuilding process in Toumbouba: (**a**) the old *barrage* down; (**b**) the building phase; (**c**) the wall complete; (**d**) the *barrage* working phase (Source: Authors').

Over time, this simple moving mechanism has yielded, bending from the rotation plane of the hinges due to the constant horizontal thrust. Because of both the design weakness of this system and the difficulty in finding suitable metal artifacts, it was decided to fill the openings with a wall. Like for the *barrage* in Toumbouba, some demolition works of part of the structure became necessary, because between the two doors there was a wall fragment about 1 m wide.

The demolition left a gap of about 3 m to be filled with the new wall. Given the small size, steps were taken with the aim of making the construction process more efficient and avoid an additional trip. Following a first mission carried out in December 2019, it was possible to ascertain the presence of a solid foundation structure, as wide as the thickness of the base of the wall, or about 2 m, and about 50 cm deep. Therefore, it was decided to proceed with the drilling of holes for the insertion of steel bars ahead of the reinforcement of the structure in elevation, filling the holes with concrete injections. The new wall was built with two imbedded water outlets to control water overflow.

While the work on the Toumbouba *barrage* was carried out without the aid of machinery or construction equipment, for the Kofeba *barrage* it was necessary to use a hammer drill capable of realizing the appropriate depth holes, powered by an electric power generator. In the lapse of time between the two missions, concrete blocks were built with the same procedure already used in Toumbouba, and the material necessary for construction was set aside.

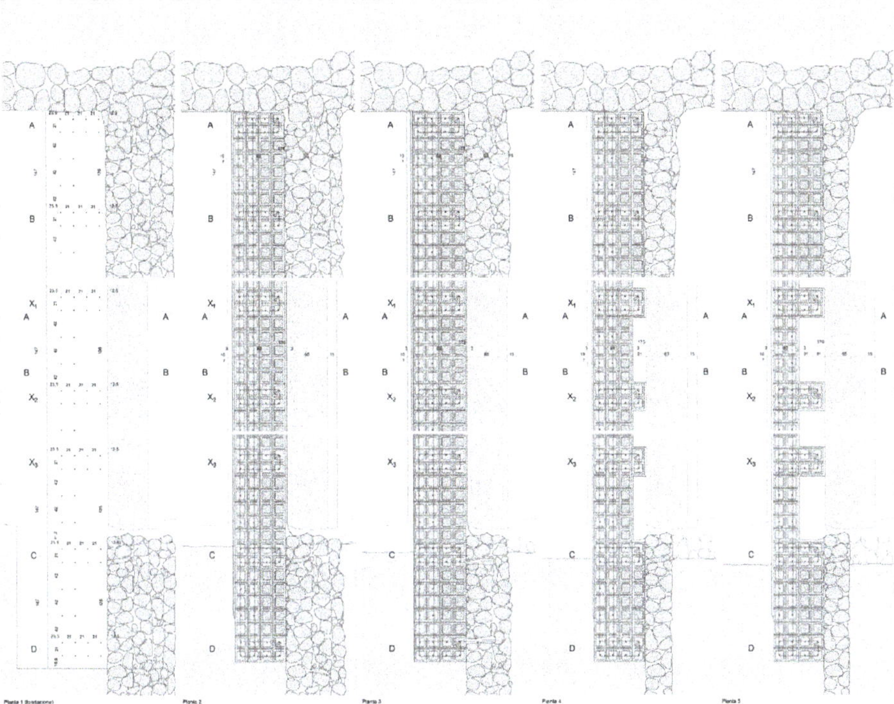

Figure 6. The executive design of the *barrage* in Toumbouba (Source: Authors').

The experience previously gained allowed the completion of the construction phase in only 4 working days (Figure 7). The monitoring phase of the Kofeba *barrage* has just started, since, at the time of writing, the rainy season had just begun, but the first photographic surveys to verify the strength of the structure appear satisfactory. At full capacity, in the absence of further settling of the reservoir grounds and with the help of small additional earthworks, the Kofeba basin could cover an area between 9.5 and 20 hectares.

Although belonging to the same typology, the two *barrages* have very different dimensional characteristics (Table 4).

Table 4. Comparison between the main data of the two *barrages*.

	Barrage of Toumbouba	*Barrage* of Kofeba
Dimensional characteristics of the entire structure [m]	Length: 60 Height: 2	Length: 215 Height: 0.8
Extension of the body of water [ha]	6.5–8.5	9.5–20
Depth of the basin [m]	Min 0.3–Max 2	Min 0–Max 0.7
Amount of water [m^3]	Min 45,000–Max 60,000	Min 90,000–Max 140,000
Number of beneficiary villages [units]	20	35
Number of beneficiary inhabitants [units]	40,000	70,000
Construction time [days]	20	7

The Toumbouba *barrage*, although smaller in size, required considerable effort to restore the reinforced concrete foundation along the entire collapsed section. In addition, the water on the Toumbouba *barrage* reaches a very high speed, generating problems of resistance to dynamic thrust. For this reason, both the foundations and the wall in elevations required a greater mass and a longer construction time.

Figure 7. The rebuilding process in Kofeba: (**a**) the old *barrage*; (**b**) the building phase; (**c**) the wall complete; (**d**) the *barrage* working phase (Source: Authors').

Kofeba's intervention (although it is a larger *barrage*) is more contained in terms of size, moreover, it was possible to use the existing foundation section. The Kofeba basin tends to fill gradually with respect to that of Toumbouba and with a lower height of the water level. These aspects have allowed a smaller wall thickness and, therefore, a much shorter period of work. At the same time, however, the Kofeba area required numerous smaller restoration interventions of the natural embankment, carried out with boulders and concrete by the local labor force.

Both interventions on the *barrage*, for the number of villages and beneficiaries' inhabitants, are characterized by an excellent cost–benefit ratio.

4. Discussion

The reconstruction of the two small *barrages* has yielded several benefits that can be classified into two categories: direct benefits and indirect benefits.

Regarding direct benefits, the following can be mentioned:

- Reduction of water shortages during the dry season in a vast territorial area including dozens of villages and thousands of people. Replenishing the aquifer had an impact on the availability of water in wells even tens of kilometers away;
- Improvement of food supply, thanks to the possibility of growing rice crops in excess to the amounts needed for local consumption and the renewed presence of fish proteins. Furthermore, the nutritional diversity has been increased by the higher productivity of the gardens that can enjoy irrigation for more than one season and by the improved quality of the meat of the animals raised, especially poultry and sheep that can be adequately fed;
- Birth of small private business activities such as, for example, a small fish farm;
- Improvement of environmental well-being conditions, due to the mitigation of temperatures in the microclimatic environment surrounding the Toumbouba *barrage*, thanks to the presence of a mass of water capable of activating thermal inertia mechanisms.

About indirect benefits, the following should be noted:

- Improvement of internal social cohesion, due to the principle of sharing food resources derived from the creation of the water basin;
- Improvement of relations with neighboring communities, through the consolidation of existing exchanges, the activation of new relationships, and the presence of the conditions for a better territorial cohabitation, thanks to the reduction of conflicts;
- Improvement of economic prospects, through the possibility of diversifying income-generating activities due to the emergence of new needs. This mechanism can produce collateral micro-entrepreneurial activities to those related to the presence of water, improving the internal economy, and providing the conditions for improving living conditions;
- Transfer of knowledge to the local parties, both scientific partners and the general community;
- Improvement of the social function of the *barrages*. Thanks to the new shape of the *barrages* with terraced seats, people walk and stay on them.

Given the quality of the results achieved when compared to the very low costs of both interventions (less than 25 thousand euros), the benefits far exceeded the costs incurred. In particular, the benefits can be classified as:

- Environmental benefits, such as replenishment of groundwater reserves and retention of surface water reserves for the dry season; presence of water in the wells of numerous villages near the *barrage*; regulation of the microclimate in the hottest period thanks to the presence of water; improvement of the conditions of farmed animals;
- Economic benefits, such as the flourishing of various common business activities, directly related to water (such as rice cultivation, fish farming) or indirectly induced by the additional disposable income (tailoring, sale of meat and eggs, etc.);
- Social benefits such as improvement of the living conditions of the inhabitants; improvement of relations between inhabitants and social cohesion; improvement of food quality; activation of periodic parties; reduction of migrations of young people from the village to the capital; improvement of the quality of food, such as meat and locally produced vegetables.

The monitoring phase of the project is still underway to assess its full environmental and economic benefits. A feasibility study is currently looking at the construction of a control station for the acquisition and monitoring of environmental variables, such as temperature, humidity, and pressure, with the aim of demonstrating the micro-climatic variations due to the presence of the water basin.

Data collection relating to the economic benefits indirectly related to the presence of the water basins is ongoing. Several micro-entrepreneurial activities are being observed, such as commercial fish farming, rice cultivation for the sustenance of the community and for retail sale, the appearance of several vegetable gardens, and the introduction of the seasonal rotation of crops.

In particular, in the fish farm in Toumbouba, an artificial pond of about 150 m^3 in capacity was built. This pond was built by excavating; the walls were covered with reinforced concrete, while on the bottom a layer of about 20 cm of compacted earth was created, containing no less than 10% clay, measured by empirical sedimentation tests. This waterproof system, often used for depths not exceeding 3 m, is not recommended for greater depths, where it is advisable to proportionally increase the thickness of the compacted earth. In addition, depths that exceed the height of the aquifer should not be waterproofed in this way to facilitate filtrating of water. The tank requires pumping for the loading and drainage of water, which is periodically necessary for the removal of organic sediments. This periodic maintenance avoids regulating the concentration of oxygen in the water by means of additional mechanical systems. The water of the farm is derived from the *barrage*, by means of mechanical pumping. Therefore, a discontinuous flow was preferred, necessary to replace the evaporated water. To begin breeding, fifty pairs of a local catfish species weighing between 300 and 400 g were placed in the tank. To facilitate reproduction, artificial nests were inserted into the tank made on the bottom with ballasted polyethylene pipes, 50 cm long with a diameter of 20 cm. Fishes were fed on alternate day manual shedding commercial feed. Despite an insignificant mortality, the fish are sold when they reach a weight between 2 and 2.5 kg, allowing to control the density in the tank. A high density can cause that bigger fish become predator of the others. The small entrepreneurial reality of the fish farm has allowed to create three new jobs in the village.

Favored by a rural economy still largely based on the bartering of goods and services, a series of artisanal activities were born in the villages with the aim of having fish in exchange. Three tailoring and fabric processing activities were born.

The production of concrete blocks for the construction of the *barrage* has allowed the specialization of some inhabitants. Two of them, in particular, have started production (through manual molding) and sale of the blocks to other villages for a total of 10,000–12,000 blocks per year.

Some of these activities are still very fragile and highly dependent on the stability of the retail market. Unfortunately, the recent unpredictable pandemic events have led to a significant reduction in the demand for fish farm products in neighboring markets, leading to the loss of fish stock already in the tanks. The pandemic has highlighted the fragility of entrepreneurial activities undertaken in these contexts that have to face a much higher entrepreneurial risk than elsewhere.

5. Conclusions

The water crisis in Sub-Saharan African countries, and specifically in Mali, is not a new phenomenon but in recent years has taken on characteristics with geopolitical and international implications. The ongoing climate changes highlight the dramatic consequences of water scarcity: absence of health and hygiene, malnutrition especially in children, with increased mortality rates in under-fives, and the constant decrease in arable land. These factors increase political instability and determine the incessant escalation of revolts, and popular uprisings often linked to international terrorist groups. In these conditions, migratory flows can only intensify, expanding the scope of the phenomenon and establishing the need for international intervention.

To solve the water issue it is necessary to promote the self-construction of small local infrastructures such as *barrages* and to simplify the unnecessarily complex and articulated state management and regulatory systems.

The three-year research project led by the Department of Architecture of Roma Tre University, in collaboration with the University of Florence and in partnership with the Onlus *Gente d'Africa*, *Romagna Acque Società delle Fonti* Ltd. and AES *Architettura Emergenza e Sviluppo*, has led to the reconstruction of two damaged and no longer operating *barrages*, in the Kayes region, near Kita.

The operational phase of the project saw the participation of local communities and the involvement of municipal authorities brought together to restore the functionality and increase the performance of two small and damaged infrastructures.

The research project encompassed a series of on-site missions involving information gathering, surveying, prospecting, design, manufacturing, construction and the collection and dissemination of results. The reconstruction of the basins produced several indirect benefits. These range from improved cohesion of local communities to the increase in micro-business activities related to the presence of water. In the current phase, the monitoring of the results is mainly aimed at assessing the environmental impact of the *barrages*, through the direct measurement of a series of basic environmental parameters and the indirect detection of the improvement of the living conditions of humans and animals.

Author Contributions: All authors wrote the paper. All authors have read and agreed to the published version of the manuscript.

Funding: The research was supported by Romagna Acque Società delle Fonti Ltd., Gente d'Africa onlus, Roma Tre University, AES Architettura Emergenza Sviluppo.

Institutional Review Board Statement: Not applicable.

Informed Consent Statement: Not applicable.

Data Availability Statement: Not applicable.

Acknowledgments: A heartfelt thanks to Sara Negroni, Mauro Foli, and Massimo Mantuano from *Gente d'Africa onlus,* And to Tonino Bernabè for *Romagna Acque Società delle Fonti.* A further heartfelt thanks goes to Diawoye Tounkara, local reference for *Gente D'Africa* but, above all, irreplaceable guide, and support in Malian territory. Finally, a special thanks to the students who took part in the various missions in Mali: Giovanni Baratta, Iacopo D'Orazi, Francesca Limongelli and Pietro Marinari.

Conflicts of Interest: The authors declare no conflict of interest.

References

1. FAO. *Profil de Pays—Mali, FAO Aquastat Rapports, Organisation des Nations Unies pour L'alimentation et L'agriculture, 2015*; FAO: Rome, Italy, 2015.
2. Giordano, A. *Pedologia*; Edizioni UTET: Torino, Italy, 1999.
3. Desio, A. *Geologia Applicate All'ingegneria*; Ulrico Hoepli: Milano, Italy, 2003.
4. Magaldi, D. *Conoscere il Suolo. Introduzione Alla Pedologia*; ETAS Libri Edizioni: Milano, Italy, 1984.
5. DNH, UNESCO-WWAP. *Rapport National sur la Mise en Valeur des Ressources en Eaux. Etude de cas Préparé e Pour le 2ème Rapport Mondial des Nations Unies sur la \ en Valeur des Ressources en eau, Direction Nationale de L'hydraulique, Programme Mondial Pour L'évaluation des Ressources en Eau, Organisation des Nations Unies pour L'éducation, la Science et la Culture, 2006*; UNESCO: Paris, France, 2006.
6. MEME. *Plan d'Action National de Gestion Intégrée des Ressources en Eau (PAGIRE). 1ère Partie: Etat des Lieux des Ressources en eau et de leur Cadre de Gestion*; Ministère de l'Energie, des Mines et de l'Eau: Bamako, Mali, 2007.
7. MEA. *Rapport national sur l'Etat de l'environnement 2009*; Ministère de l'Environnement et de l'Assainissement: Bamako, Mali, 2010.
8. Dao, A. Mali: Problématique de L'accès à L'eau Potable et Assainissement: Au Cœur d'un Forum ce Mardi. 2014. Available online: https://maliactu.net/mali-problematique-de-lacces-a-leau-potable-et-assainissement-au-coeur-dun-forum-ce-mardi/ (accessed on 30 July 2020).
9. UN. General Comment No. 15: The Right to Water. UN Committee on Economic, Social and Cultural Rights. 2017. Available online: http://www.refworld.org/docid/4538838d11.html (accessed on 30 July 2020).
10. UN. Resolution A/RES/64/292. United Nations General Assembly. 2017. Available online: http://www.un.org/en/ga/search/view_doc.asp?symbol=A/RES/64/292 (accessed on 30 July 2020).
11. Rigaud, K.A.; de Sherbinin, A.; Jones, B.; Bergmann, J.; Clement, V.; Ober, K.; Schewe, J.; Adamo, S.; McCusker, B.; Heuser, S.; et al. *Groundswell: Preparing for Internal Climate Migration*; World Bank: Washington, DC, USA, 2018.
12. USAID. Climate Risk Profile: Mali. 2018. Available online: www.climatelinks.org/sites/default/files/asset/document/Mali_CRP_Final.pdf (accessed on 2 August 2020).
13. NCEA Netherlands Commission for Environmental Assessment. Dutch Sustainability Unit. 2015. Available online: ees.kuleuven.be/klimos/toolkit/documents/690_CC_mali.pdf (accessed on 2 August 2020).
14. MMES—Mali Ministry for Environment and Sanitation. Elements of National Policy for Adaptation to Climate Change: Final Report. 2018. Available online: www.adaptation-undp.org/sites/default/files/downloads/mali_-_national_policy_2008.pdf (accessed on 2 August 2020).
15. USAID. *Climate Change in Mali: Key Issues in Water Resources. African and Latin American Resilience to Climate Change Project, 2013*; USAID: Washington, DC, USA, 2013.
16. USAID. Mali: Nutrition Profile. 2018. Available online: www.usaid.gov/sites/default/files/documents/1864/Mali-Nutrition-Profile-Mar2018-508.pdf (accessed on 2 August 2020).
17. USAID. Food Assistance Fact Sheet: Mali. 2019. Available online: www.usaid.gov/sites/default/files/documents/1866/FFP_Fact_Sheet_Mali.pdf (accessed on 2 August 2020).
18. Caspian Reports. Origin of Mali's Tuareg Conflict. 2013. Available online: www.youtube.com/watch?v=IlCpaUt7DfY (accessed on 2 August 2020).
19. Morgan, A. The Causes of the Uprising in Northern Mali. 2012. Available online: thinkafricapress.com/causes-uprising-northern-mali-tuareg/ (accessed on 2 August 2020).
20. Dyxon, R.; Labous, J. Gains of Mali's Tuareg Rebels Appear Permanent, Analysts Say. 2014. Available online: www.latimes.com/world/la-xpm-2012-apr-04-la-fg-mali-tuaregs-20120404-story.html (accessed on 2 August 2020).
21. Bate, F.; Diarra, A. New North Mali Arab Force Seeks to Defend Timbuktu. 2012. Available online: www.reuters.com/article/us-mali-north/new-north-mali-arab-force-seeks-to-defend-timbuktuidUSBRE8380MB20120409 (accessed on 2 August 2020).
22. Robaux, A. Les barrages souterrains. *Terres et Eaux* **1954**, *6*, 23–37.

23. Nilsson, A. *Groundwater Dams for Small-Scale Water Supply*; Intermediate Technology Publications Ltd.: New York, UK, USA, 1988.
24. CYPE. *Software pour l'Architecture et l'Ingénierie de la Construction*. 2020. Available online: produits.cype.fr/ (accessed on 2 August 2020).

Article

Groundwater Resources Assessment for Sustainable Development in South Sudan

Manuela Lasagna [1,2], Sabrina Maria Rita Bonetto [1,2,*], Laura Debernardi [1], Domenico Antonio De Luca [1,2], Carlo Semita [1,2] and Chiara Caselle [1]

1 Earth Sciences Department, Turin University, 10125 Torino, Italy; manuela.lasagna@unito.it (M.L.); laura.debernardi@unito.it (L.D.); domenico.deluca@unito.it (D.A.D.L.); carlo.semita@unito.it (C.S.); chiara.caselle@unito.it (C.C.)

2 CISAO, Interdepartmental Centre of Research and Technical and Scientific Cooperation with Africa, Turin University, 10125 Torino, Italy

* Correspondence: sabrina.bonetto@unito.it

Received: 29 May 2020; Accepted: 8 July 2020; Published: 10 July 2020

Abstract: The economic activities of South Sudan (East-Central Africa) are predominantly agricultural. However, food insecurity due to low agricultural production, connected with weather conditions and lack of water infrastructure and knowledge, is a huge problem. This study reports the results of a qualitative and quantitative investigation of underground and surface water in the area of Gumbo (east of Juba town) that aims to assure sustainable water management, reducing diseases and mortality and guaranteeing access to irrigation and drinking water. The results of the study demonstrate the peculiarity of surface and groundwater and the critical aspects to take into account for the water use, particularly due to the exceeding of limits suggested by the WHO and national regulation. The outcomes provide a contribution to the scientific overview on lithostratigraphic, hydrochemical and hydrogeological setting of a less-studied area, characterized by sociopolitical instability and water scarcity. This represents a first step for the improvement of water knowledge and management, for sustainable economic development and for social progress in this African region.

Keywords: water resources; sustainable management; local development; water for food security

1. Introduction

Environmental sustainability plays a major role in the United Nations (UN) Agenda 2030 for Sustainable Development. In this framework, the balance between protecting environmental resources and satisfying social and economic needs has become a key development issue.

The global variety of ecosystems and landscapes and the different accessibility to natural resources contribute to shaping social and economic factors, which affect the development of local territories and communities, providing special inputs to their economic growth and resilience. The critical relationship between economic development, growth and environmental protection, if not reconciled in sustainable ways, risks causing the depletion of natural resources.

In South Sudan, as in several Sub-Saharan African countries, environmental conditions are often vulnerable to several shocks (climate, population growth, human activities impact, conflicts and security), which may produce heavy effects on their development.

Many development projects, such as the project "Women Empowerment and Sustainable Agricultural Development to Achieve Food Security in South Sudan (WOSA) AID 10915" funded by the Italian Agency for Cooperation and Development (AICS), support the sustainable management and use of natural resources through actions for protecting water, air and soil, and for preserving biodiversity and combating desertification, including the elaboration and implementation of measures for mitigating the effects of climate change and for fostering resilience. Other important activities concern the energy and infrastructure sector, which includes transportation and water networks, the distribution of electricity generated by renewable resources and the broad sector of territorial planning.

In developing these approaches, the environmental characterization is promoted to identify landscape character types, geological stratification, biogeographical regions, etc. The recognized spatial frameworks would be useful in assessing interventions and impacts of policies to ensure local food security, social development and sustainable natural resources management. This information should be shared with local administrations, with the scientific community and with any organization that operates in the area.

The economy of South Sudan (East-Central Africa) is predominantly agrarian, with almost 60% of the total workforce engaged either directly or indirectly in agriculture. However, this sector remains underdeveloped due to the political instability of the country and the primitive method of farming systems. Food insecurity is a major problem due to low agricultural production connected with weather conditions and lack of water infrastructure and knowledge [1].

For these reasons, the present study aims to propose an analysis of the local hydrogeological framework in order to answer to the main needs linked to the deficit in the food production and to guarantee a sustainable access to irrigation and drinking water, reducing diseases and mortality.

The study focuses on the city of Gumbo, located at the east side of Juba town. Juba is divided into three sub-districts named Sherikat (Central Gumbo), Jebel Lemon (east of Gumbo) and Adodi (North-West Gumbo). The population of Gumbo is roughly 5000 households, and more than 70% of the population depends on agriculture. About 80% of the population comes from different parts of the country, particularly from east side of Juba. They are internally displaced persons or refugees as a result of the civil war, which broke out in December 2013, and its impact.

The aim of this work is to provide a hydrogeological, lithostratigraphic and hydrochemical reconstruction of the area. These activities are useful for programming drilling campaigns of new wells and in defining sustainable uses or interventions for water according to the quality [2–10]. The shortage of previous studies in the area, also due to the long civil war that occurred in the region in recent years, makes particularly important the collection and the organization of available scientific knowledge on the territory and the execution of new studies to fill the gap of data at the local scale. This may help to create a database for the necessary future interventions that have to be planned in the area.

The study is organized as a first collection of geologic and hydrogeological data of South Sudan, starting from the scarce literature that was completed with a more detailed lithostratigraphic and hydrogeological analysis, specifically focused on the area of Juba.

2. Material and Methods

2.1. Collection of Information on the Study Area from the Scientific Literature

The collection of geological and hydrogeological information on South Sudan started from the scarce available scientific literature and brought to the reconstruction of a general framework of the climate, the vegetation and the regional geological and hydrogeological setting. The absence of more detailed data at the local scale was filled with a lithostratigraphic, hydrogeological and hydrochemical reconstruction in the area of Juba, the capital of South Sudan. Specifically, the proposed study focuses on an area surrounding Gumbo (Central Equatoria—Payam Rajaf) (Figure 1), which is located in a flat plain at a distance of about 1.5 km eastwards from the Nile River, referred to here as the "White Nile".

Figure 1. Location of the study area (modified from [11]).

2.2. On-Site Survey

2.2.1. Lithostratigraphic Reconstruction

The lithostratigraphic reconstruction was conducted combining data from a geophysical survey, performed in August 2017 by the United Drilling Company, and from the execution of four new wells drilled in 2018.

The geophysical survey was implemented in an area suffering from lack of water (Figure 2). It consisted of two vertical electrical soundings (VESs), which provided an insight into both thicknesses and types of the overburden strata and of the presence of fractured bedrock, based on resistivity contrasts. VESs were conducted using potential electrode spacing (MN/2) of 0.5 and 5 m, reaching a depth of about 50 m.

2.2.2. Hydrogeological Reconstruction

The new wells were drilled in areas where people lived without free access to water for human, agriculture and livelihood consumption (WPC, W1, W2, W3 in Figure 2). The depth of wells ranges between 60 and 90 m. Stratigraphic logs and water table levels (meters below ground level, m b.g.l.) contribute to the stratigraphic and hydrogeological reconstruction. Water table levels of spring–summer 2018 were interpolated to create a preliminary water table map, indicating the main flow direction of groundwater.

2.2.3. Hydrochemical Analyses

The reconstruction of the hydrochemical setting was performed through two sampling campaigns, conducted in August 2017 (1 water sample, called WPC) and July 2018 (6 water samples). This second campaign includes both groundwater samples (W1, W2 and W3) and surface water samples, taken from seasonal streams (S1, S2 and S3) (Figure 2). Due to the political instability, military checkpoints conditioned the mobility of persons and vehicles, limiting the access to many areas, particularly along the main streams. Hence, the sampling was only performed for the new drilled wells (groundwater) and areas of free access (seasonal streams samples).

The chemical analyses, performed by the Nuovi Servizi Ambientali (NSA) laboratory in Robassomero (Turin, Italy), included a first step of assessment of data quality, which was accomplished by calculating the balance of positive and negative ions. Water fulfils the principle of electroneutrality and is therefore always uncharged. The level of error was calculated by using the following formula [12]:

- Error of ion balance = ((Σcations − Σ anions) / (Σcations + Σ anions)) * 100

An error of up to ±5% was considered as tolerable. The results of chemical analyses were then displayed through a Piper diagram.

Since water from wells and seasonal streams is currently used for human consumption, the chemical results were compared with the limits established by the regulations in force in South Sudan and the WHO regulations to provide information on the quality of water in the study area. More specifically, the following regulations were considered:

- World Health Organization (WHO) guidelines [13];
- Maximum permissible limit for drinking water quality in South Sudan.

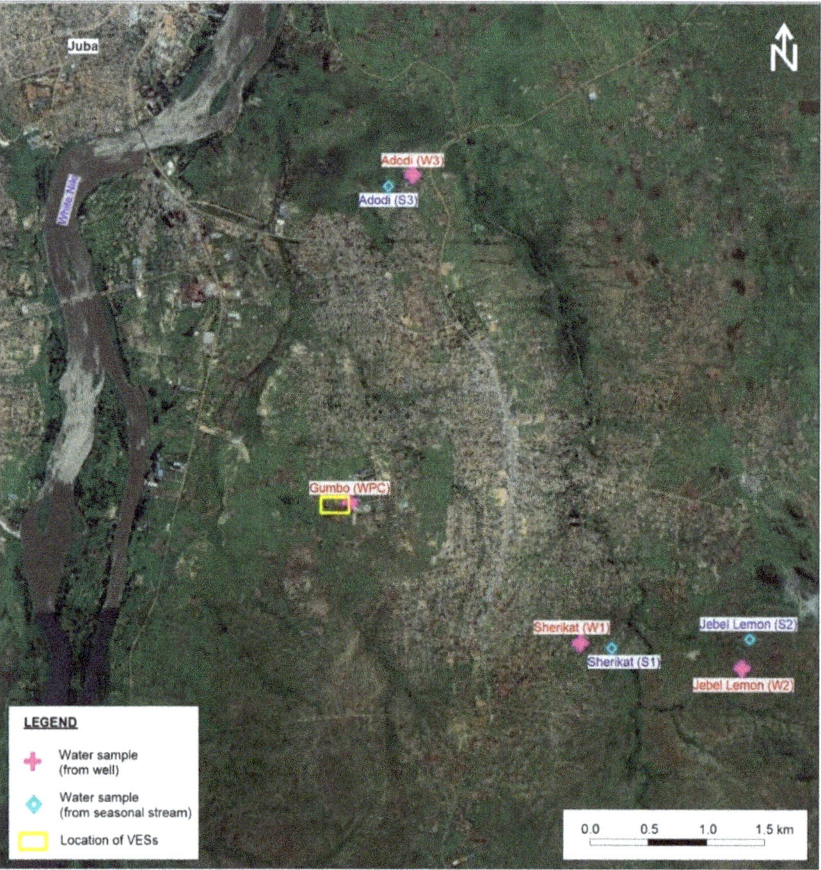

Figure 2. Location of the water samples taken for hydrochemical analyses (wells and seasonal streams) and vertical electrical soundings (VESs).

3. State-of-the-Art Knowledge in the Study Area

The state-of-the-art geological and hydrogeological knowledge in South Sudan only includes general studies, at the regional or even national scale. The purpose of this review is to collect of all the available information, creating a solid framework on which setting up a more detailed local-scale dataset proposed in this study.

3.1. Climate and Vegetation

The average altitude of South Sudan is 470 m above sea level, with a gentle slope to the north. In general, the terrain is mainly plain with thick equatorial vegetation and savannah grasslands, and the climate is equatorial. As regards to the annual distribution of precipitation, the rainfall over the River Nile basin is characterized by highly uneven seasonal and spatial distribution. Most of the basins show only one rainy season, typically in the summer months. Only the equatorial zone has two distinct rainy periods. The probability of occurrence and volume of precipitation generally declines moving northwards, with the arid regions in Egypt and the northern region of Sudan receiving insignificant annual rainfall [14]. The area of Juba (specific object of this study) is in a sector characterized by a total rainfall (average annual for the period 1960–1990) between 900 and 1200 millimeters (Figure 3).

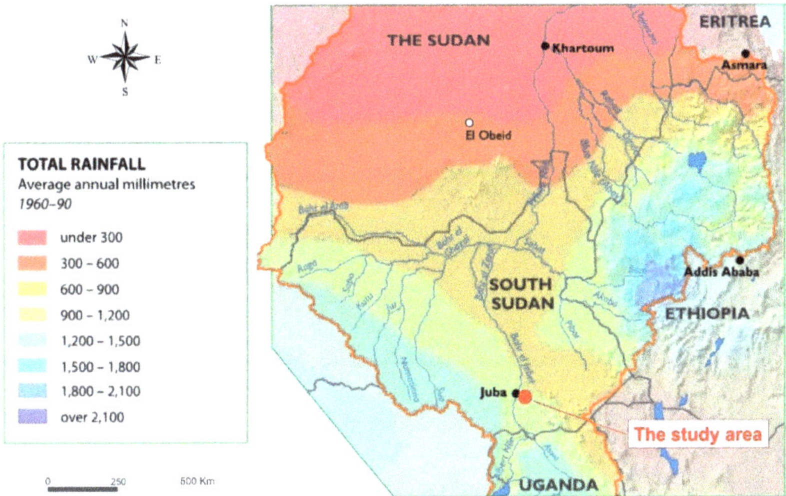

Figure 3. Average annual rainfall for the period 1960–1990 in South Sudan [14] and location of the study area. The orange line borders the River Nile Basin.

For a detailed description of the study area, data of the weather station in Juba [15] were collected and analyzed. The climate of Juba is tropical with average yearly temperature between 33.8 and 20.7 °C. Maximum temperature is during February–March, and minimum temperature is during December–January (Figure 4). The rainy season is from May to October. The dry season starts from November and ends in March and occurs particularly in January and February. Average yearly total rainfall is 974 mm. (Figure 5). The highest number of rainy days occurs from April to October, whereas the lowest number of rainy days are reported in January. The average annual number of rainy days is 104 (Figure 6).

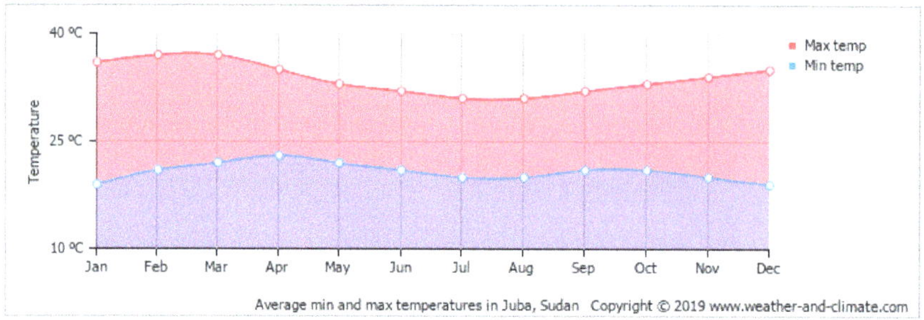

Figure 4. Monthly average minimum and maximum temperature of Juba [15].

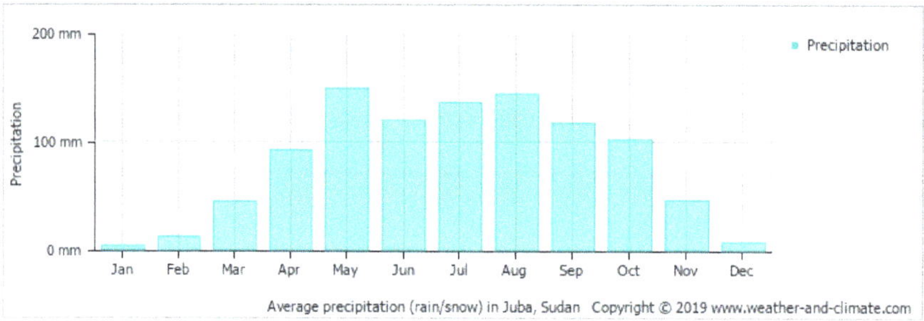

Figure 5. Monthly precipitation of Juba [15].

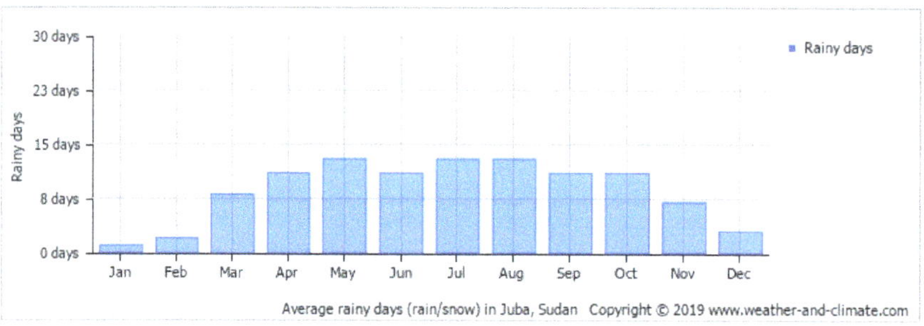

Figure 6. Average monthly rainy days over the year of Juba [15].

Across the Nile region, actual and potential evapotranspiration vary markedly [14]. Seasonal/monthly variability of evapotranspiration is indeed a function of temperature, wind speed, relative humidity, solar radiation and biomass production. More specifically, the study area is located in a sector characterized by a potential evapotranspiration (average annual for the period 1960–1990) between 1600 and 1800 mL per year (Figure 7).

In the study area, the type of vegetation is biologically diverse, and it is not generally very dense. Examples of common trees are teak, shrubs and mango trees.

The rate of weathering is very high due to the combination of both high temperature and rainfall. This condition facilitates hydrolysis, oxidation and reduction and physical types of weathering such as the action of the plant's roots.

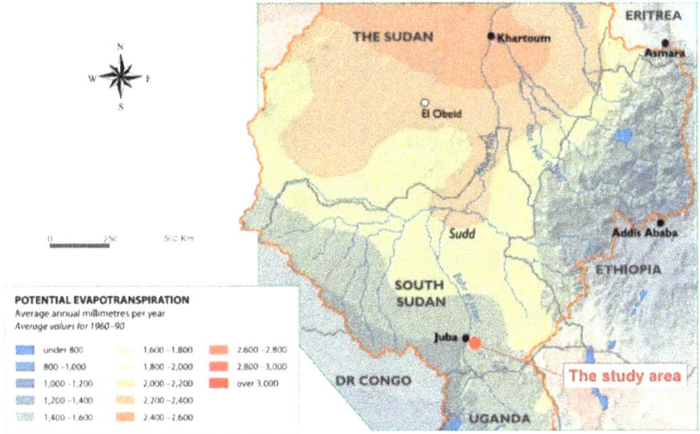

Figure 7. Average annual potential evapotranspiration for the period 1960–1990 [14] and location of the study area. The orange line borders the River Nile Basin.

3.2. Geological Setting

The geological setting of South Sudan comprises three main geological frameworks [16–18]:

- an extensive metamorphic basement complex and spatially associated granitic and basaltic intrusions;
- volcanic rifts;
- a widespread Tertiary–Quaternary cover sequence, mainly comprising poorly consolidated sands, clays and lateritic soils.

The basement complex is dominated by Proterozoic rocks of medium to high metamorphic grade, with isolated areas of probably older higher-grade granulitic rocks. In broad terms, the Proterozoic metamorphic basement comprises three main units: (i) variably banded and foliated granitic gneiss and migmatites (basement complex, dominantly banded magmatic gneiss in Figure 8), (ii) biotite-amphibole schist/amphibolites and (iii) calcareous meta-sediments and quartz meta-sediments (basement complex, dominantly schists and metasediments in Figure 8).

The crystalline basement complex hosts the majority of mineral occurrences. However, the recorded mineral presence is limited, because the country lacks a significant mineral resource evaluation.

Effusive basic volcanic rocks, mainly basaltic lava and tuffs, are found in the south-eastern part of the country (Rocks volcanics, mainly basalts in Figure 8) and are related to the activity of the East African Rift system.

The Neogene sequence comprises unconsolidated sands, gravels, clay sands and clays (Umm Rwaba Formation in Figure 8) characterized by rapid facies changes. Conditions of deposition of the Umm Rwaba formation are fluvial and lacustrine, with sediments laid down in a series of land deltas similar to those existing in present day South Sudan. The age of sediments is considered to range from Tertiary to Quaternary [19]. This formation hosts most of the ongoing oil exploration activities, and numerous holes/wells were drilled through the cover into the underlying basins. According to oil exploration data from Central and Southern Sudan, the maximum drilled thickness of the Umm Rwaba Formation is higher than 4600 m. According to geophysical data, the maximum recorded thickness is higher than 8200 m in some places [16].

Extensive alluvial deposits underlie the flood plain of the Nile and many of its tributaries within the study area and toward the north of Juba.

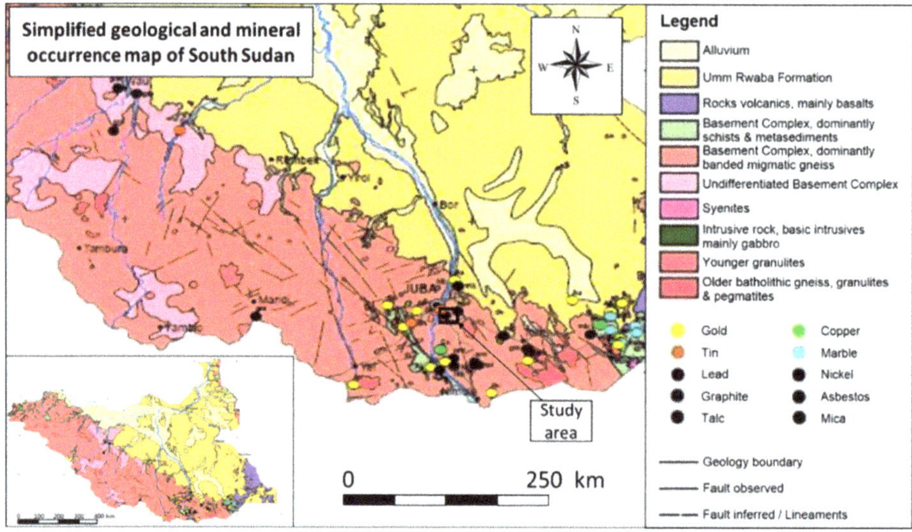

Figure 8. Map of the geology and mineral deposits of South Sudan [20] and location of the study area.

The Juba area is generally underlain by metamorphic rocks of the basement complex. The basement complex is overlain unconformably by alluvial and surficial deposits that vary in thickness from one location to another. In detail, the basement complex of Juba is composed of gneiss that ranges from medium to high grade metamorphism. Both banded and unbanded gneiss occurs in the area; different types such as augen gneiss, grey gneiss, etc., are present. The gneiss is intruded by several doleritic, gabbroic and granitic intrusions that occur as plutons, or doleritic, gabbroic dikes with east–west trending. The rocks in the study area are mildly deformed. The geological structures consist of both ductile structures, which include folds, foliation and crenulation cleavage, and structures with joints and faults.

Effects of weathering, due to the tropical climate of the area, can be seen in the change of surface color (usually dark or reddish brown) of most rocks, exfoliations and other types of physical weathering.

Close to the Nile River and its tributaries, the geology of the area is dominated by recent alluvium, terraces, deltas and swamp deposits.

3.3. Hydrogeological Setting

As shown in Figure 9, three different hydrogeological sectors may be identified in South Sudan, based on the structure of aquifers and on the recharge rate [14,21]:

- Major aquifers: The northern part of the state, on the border with Sudan, and the eastern part, on the border with Ethiopia. In these sectors, large and rather uniform groundwater basins are present, usually in large sedimentary basins, that may offer good conditions for groundwater exploitation;
- Local and shallow aquifers: Southern and western parts of the state, bordering Congo, Uganda and Central African Republic. These areas are characterized by limited groundwater resources in local and shallow aquifers;
- Complex hydrogeological structure: A small area in the south-eastern part of the state. Complex hydrogeological structures are found in heterogeneous folded or faulted regions, where highly productive aquifers are found in close proximity to areas without significant aquifers.

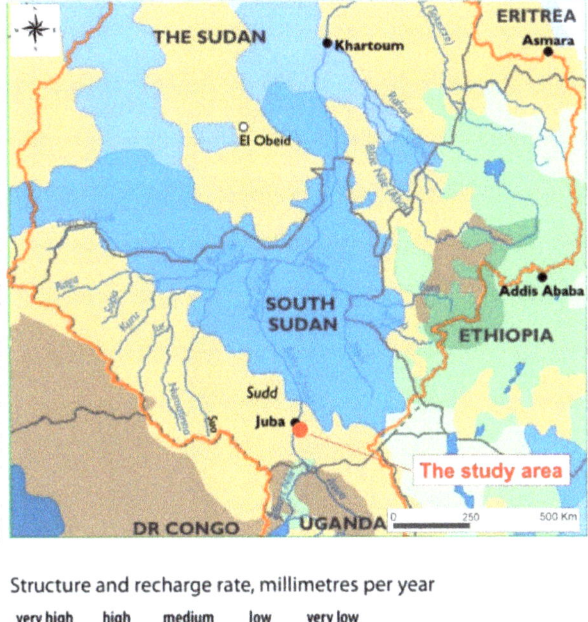

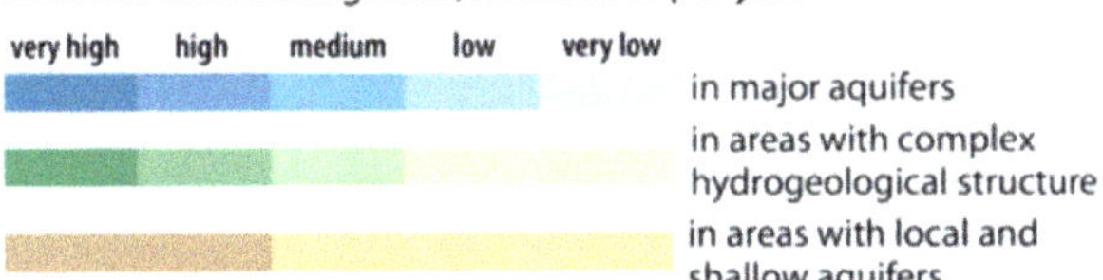

Figure 9. Sketch of the structure and recharge rate of the aquifers [14]. The orange line borders the River Nile Basin. The red circle indicates the study area.

The study area is located in a sector with local and shallow aquifers; the groundwater recharge rate is between medium and very low.

The shallow aquifers are usually located in the overburden or in a fractured upper part of the bedrock. The recharge of shallow aquifers is generally dependent on the size of the catchment area and the lithological character of the overburden.

In the study area, the cover sequence may contain a phreatic aquifer if constituted by coarse deposits (gravel, sand, pebble) of a large thickness. Groundwater in the basement formations generally occurs in the weathered (overburden) and fractured rocks. Weathered rocks, indeed, may have a good transmissivity and storage. However, the best aquifers are generally found at the contact zone between the overburden and the rocks. This zone has fewer secondary clay minerals, resulting in a higher transmissivity. Lastly, the highest yielding aquifers can be expected in the fractured bedrock. Boreholes are usually drilled into the fractured bedrock where the permeability is high and where the storage can be provided by the overburden. Fractured aquifers may be recharged through a connected system of fractured zones.

The hydrography of the study area consists of the Nile River and several other streams connected to the Nile River.

4. Results of the On-Site Survey

4.1. Lithostratigraphic Reconstruction

The analysis of the stratigraphic logs of the four new drilled wells (Figure 10) highlighted the presence of a clayey and sandy soil on the surface, with a thickness between 1 and 8 m, followed in depth by an alternation of non-fractured granite layers and altered wet granite or fractured granite.

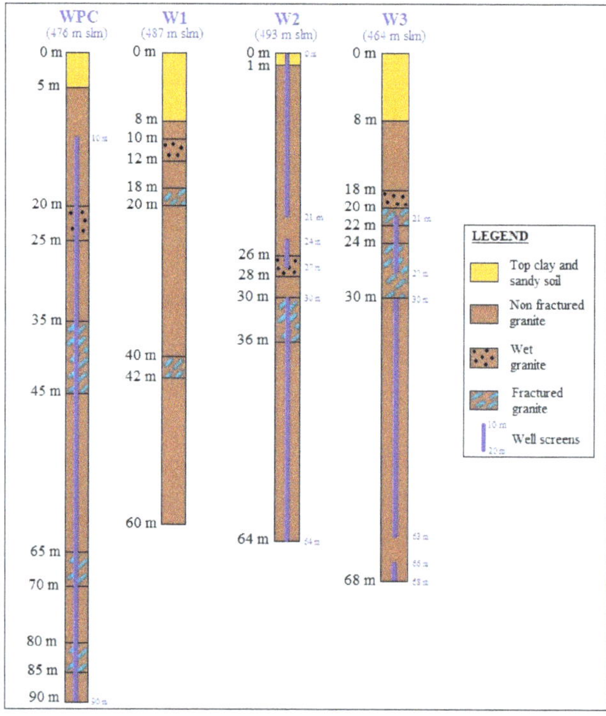

Figure 10. Stratigraphic logs of four new wells (not to scale).

The wet granite has a thickness of 2 to 5 m and is located at a depth ranging between 10 and 20 m. The first level of fractured granite has a variable thickness of 2 to 10 m and depth ranging between 18 and 35 m. The second level of fractured granite has thickness of 2 to 6 m and depth ranging between 24 and 65 m.

The screens are mainly positioned in the granitic basement along the entire depth of the wells, with the exception of well W2, which also catches water from the overburden, suggesting the role of both the overburden and bedrock as local aquifers.

The two VESs performed near to the WPC well enabled the identification of layers with homogeneous resistivity (Figure 11). VES1 has slightly lower apparent resistivity values than VES2, but their results are coherent. More specifically, they show the presence in the subsoil of three main layers with different thicknesses and apparent resistivity. In general, the data show the existence of a first layer, with a thickness of about 2 m, characterized by intermediate resistivity that is followed in depth by a second layer with low resistivity, up to about 20–25 m. The resistivity profile ends with a third layer with high resistivity and up to 50 m of depth.

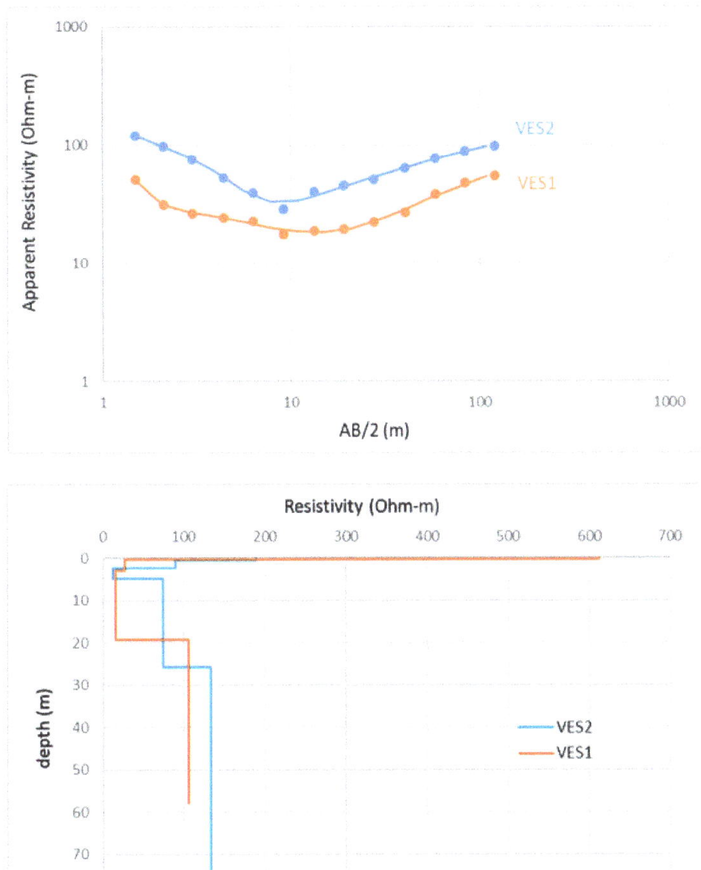

Figure 11. Graph with the results of the two VESs and their elaboration.

4.2. Hydrogeological Reconstruction

Table 1 reports the position, depth and diameter of the wells and the respective water level measurements. The depth of water table ranges between 8.0 and 24.3 m b.g.l. (i.e., piezometric level between 456.0 and 468.7 m a.s.l). The direction of groundwater flow is predominantly SSE–NNW, in the direction of the White Nile, which serves as a drainage axis of the entire surrounding plain sector.

Table 1. Data used for the piezometric reconstruction (spring–summer 2018).

Location	Id. Code	Latitude (N)	Longitude (E)	Well Depth (m)	Well Diameter (m)	Depth to Water Table from Ground Level (m b.g.l.)	Water Table Level (m a.s.l.)
Sherikat	W1	4.7849722	31.6433055	60	0.1016	12.0	475.0
Jebel Lemon	W2	4.783132	31.6562968	64	0.1016	24.3	468.7
Adodi	W3	4.8202146	31.6301272	68	0.1016	8.0	456.0
Gumbo	WPC	4.7953400	31.6249200	90	0.1524	8.0	468.0

Figure 12 reports the resulting water table map. Since the piezometric reconstruction is based on only four wells, the map only indicates the main direction of groundwater flow.

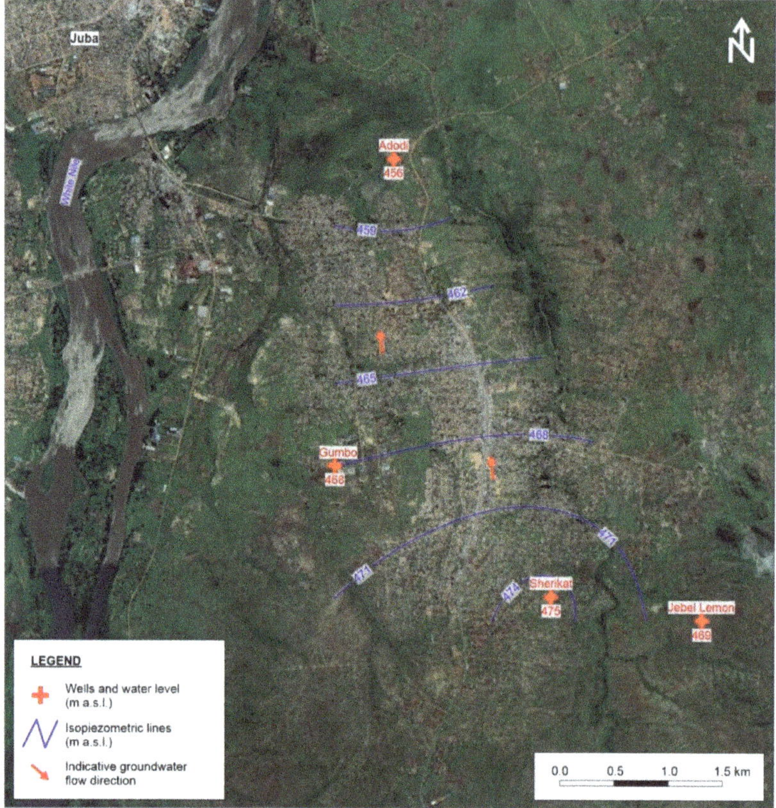

Figure 12. Water table map for spring–summer 2018. Due to the scarce data, the map gives indicative information about the hydraulic gradient and groundwater flow direction.

4.3. Hydrochemical Analyses

Table 2 reports the results of the water chemical analyses. The error of ion balance was always less than ±5%, and consequently it was considered tolerable. Table 2 also reports the comparison with the limits established by the regulations in force in South Sudan and the WHO regulations, providing information on the quality of water for human consumption in the study area.

Results show that both the content in individual elements and the total mineralization are very variable, depending on the analyzed sample. In general, however, surface water from seasonal streams has lower electrolytic conductivity and chloride levels.

The WPC groundwater sample, taken in 2017, shows different chemical features when compared to groundwater samples taken in W1, W2 and W3 wells in 2018. The latter, indeed, have higher heavy metal content. This difference is probably due to the different positions of screens in the wells and thus to the different sampled aquifers.

The limit value of total hardness of the South Sudan regulation (200 mg/L) is only respected in samples W3, W2 and S2, whereas the W1, S1 and S3 exceed it.

The single ions exceeding the suggested limits by the WHO and national regulation that may assume negative effects for human health in the long term and negative consequences for the soil in case of agricultural use are chromium (according to WHO limits) and sodium, fluorides and manganese (with regards to the South Sudan legislation limits).

Table 2. Results of chemical analyses and comparison with permissible limits for the South Sudan legislation and WHO guidelines for drinking water quality. The parameters exceeding the permissible limits for the South Sudan legislation are in light blue; those that exceed the WHO guidelines for drinking water quality are in red (N.M. = not measured).

Sample Date				WPC "Gumbo" August 2017	W1 "Sherikat" July 2018	W2 "Jebel Lemon" July 2018	W3 "Adodi" July 2018	S1 "Sherikat" July 2018	S2 "Jebel Lemon" July 2018	S3 "Adodi" July 2018
Parameter	U.M.	Limit Values (South Sudan)	Limit Values (WHO 2011)	Value	Value	Value	Value	Value	Value	Value
pH	pH	6.5–9.5		N.M.	7.71	7.81	8.09	7.30	7.43	7.71
Total suspended solids	mg/L			N.M.	25.0	29.0	30.0	23.0	27.0	25.0
Electrolytic Conductivity (20 °C)	µS/cm	1500		N.M.	1138	553	1248	221	397	788
Aluminum	µg/L			1.1	<1.0	<1.0	<1.0	<1.0	<1.0	<1.0
Antimony	µg/L		20	<0.5	<0.5	<0.5	<0.5	3.6	<0.5	<0.5
Silver	µg/L			<1.0	<1.0	<1.0	<1.0	<1.0	<1.0	<1.0
Arsenic	µg/L	50	10	4.8	<1.0	<1.0	<1.0	<1.0	<1.0	<1.0
Beryllium	µg/L			<0.1	<0.1	<0.1	<0.1	<0.1	<0.1	<0.1
Cadmium	µg/L	3–5	3	<0.2	<0.2	<0.2	<0.2	<0.2	<0.2	<0.2
Cobalt	µg/L			<1.0	<1.0	<1.0	<1.0	<1.0	<1.0	<1.0
Total Chromium	µg/L		50	<1.0	55.1	59.8	38.5	11.1	8.8	63.5
Chromium VI	µg/L			<3.0	<3.0	<3.0	<3.0	<3.0	<3.0	<3.0
Iron	µg/L	500		<1.0	192	206	136	43.7	35.1	219
Mercury	µg/L		6		0.2	0.2	0.5	0.1	1.3	<0.1
Nickel	µg/L		70	<1.0	58.6	64.6	42.5	14.2	11.0	60.8
Lead	µg/L		10	<1.0	3.1	1.2	1.5	<1.0	<1.0	<1.0
Copper	µg/L	1500	2000	<1.0	2.1	2.1	2.9	1.3	1.0	2.0
Selenium	µg/L		40	<1.0	<1.0	<1.0	<1.0	<1.0	<1.0	<1.0
Manganese	µg/L	400		<1.0	55.9	469	6.3	2.0	1.6	8.9
Thallium	µg/L			<0.2	<0.2	<0.2	<0.2	<0.2	<0.2	<0.2
Zinc	µg/L	3000		31.6	1183	418	622	37.7	46.1	182
Potassium	mg/L	25–50		14.7	12.5	11.2	4.46	5.01	3.91	10.9
Calcium	mg/L	80–150		109	85.9	52.5	10.7	24.9	46.7	68.2
Sodium	mg/L	100		115	102	98.8	103	29.2	62.8	103
Magnesium	mg/L			29.8	45.7	15.1	6.03	6.75	9.32	26.1
Boron	µg/L		2400	<15.0	53.6	40.6	43.3	42.5	37.0	50.3
Titanium	µg/L			<1.0	N.M.	N.M.	N.M.	N.M.	N.M.	N.M.
Total phosphorus (P)	µg/L			32.9	N.M.	N.M.	N.M.	N.M.	N.M.	N.M.
Molybdenum	µg/L			<1.0	N.M.	N.M.	N.M.	N.M.	N.M.	N.M.
Tin	µg/L			<1.0	N.M.	N.M.	N.M.	N.M.	N.M.	N.M.
Fluorides	µg/L	1000	1500	822	1139	1342	1047	431	1300	1430
Nitrite (NO2)	µg/L	500	3000		<50	<50	<50	<50	<50	<50
Sulphates	mg/L	200		9.62	99.9	0.68	36.5	1.29	2.64	36.1
Chlorides	mg/L	200		17.1	69.8	2.32	54.0	2.73	10.9	38.0
Phosphates (PO4)	mg/L			N.M.	<0.10	<0.10	0.39	0.15	<0.10	<0.10
Bromides	mg/L			N.M.	<0.10	<0.10	<0.10	<0.10	<0.10	<0.10
Nitrates (NO3)	mg/L	30	50	4.9	0.4	0.4	0.4	0.4	1.4	25.1
Total Cyanides	µg/L			N.M.	<10	<10	<10	<10	<10	<10
Total Hardness	F°			N.M.	40.3	19.3	5.1	26.5	15.5	27.8
Total Hardness	mg/L CaCO3	200		394	403	193	51	265	155	278
CO3– + HCO3-	mg/L			701.5	585.6	561.2	195.2	201.3	372.1	439.2

The levels of sodium are usually higher in groundwater than in surface water. WPC, W1, W3 and S3 have sodium concentrations slightly higher than South Sudan limits.

Greater attention must be paid to fluoride, which presents medium–high concentrations (between 800 and 1400 μg/L) in water samples W1, W2, W3, S2 and S3. High fluoride levels may cause damage to bones, kidneys and teeth.

Manganese is usually found in concentrations less than 10 μg/L, with the exception of wells W1 (55.9 μg/L) and W2 (469 μg/L, greater than the South Sudan legislation limit).

The hexavalent form of Chromium, which is very toxic and carcinogenic to humans, is always less than 3 μg/L. On the contrary, the trivalent form, that is not carcinogenic, is present in concentrations higher than the WHO limits for the samples W1, W2 and S3.

To clarify the distribution of parameters exceeding the permissible limits, Figure 13 reports the samples with a content higher than South Sudan national legislation and WHO guidelines, specifying the water concentration.

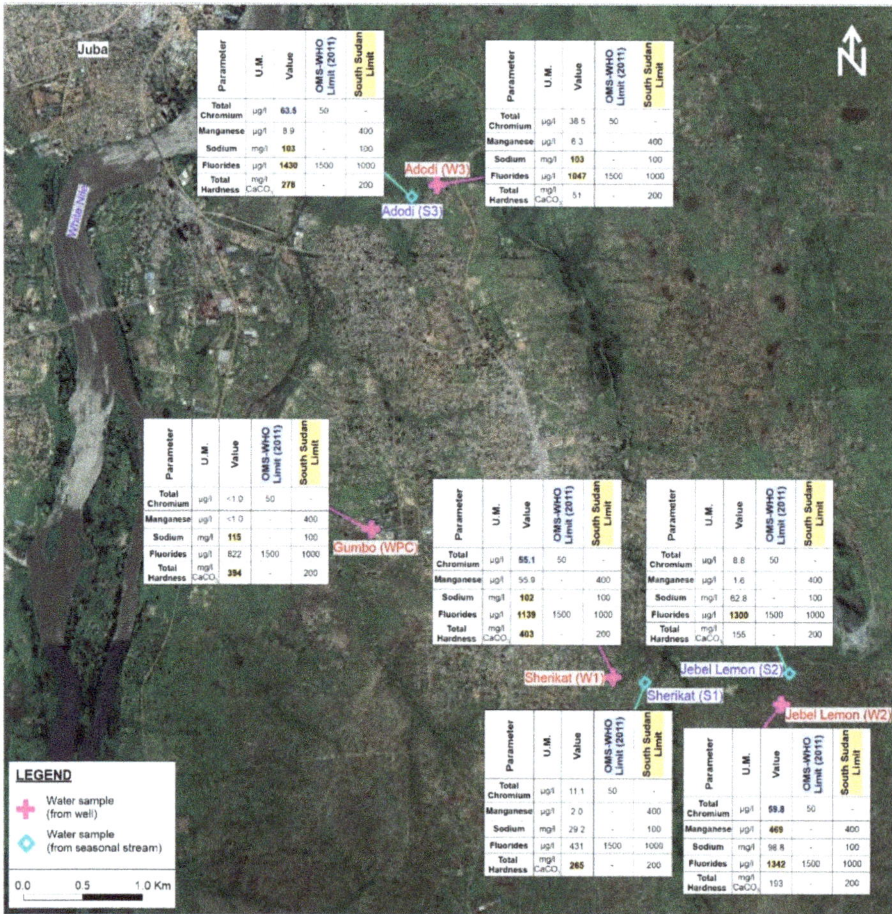

Figure 13. Map of the distribution of the chemical parameters exceeding the permissible limits for the South Sudan national legislation and WHO guidelines for drinking water quality.

5. Discussions

The set of lithostratigraphic, hydrogeological and hydrochemical data collected with the on-site survey allowed a local framework for the distribution and the quality of the water resource to be defined.

The analysis of VES results and borehole logs coherently suggest the presence of a few meters of overburden (intermediate resistivity) followed by altered/fractured granite (low resistivity) up to 20 m depth and no fractured granite (higher resistivity) at a lower depth. Considering the depth of the wells and the reliability of information coming from direct geological investigation, the borehole logs suggest the alternation of fractured and non-fractured granite up to the end of the wells (60 to 90 m depth).

Based on this lithostratigraphic setting (overburden and igneous bedrock), Figure 14 proposes four possible hydrogeological scenarios. Data suggest the presence of local and shallow aquifers with medium to very low groundwater recharge rate. Phreatic aquifers are hosted in the coarse unconsolidated deposits or in the altered basement, whereas the local aquifers are associated with the fractured basement. The best aquifers are generally located at the contact zone between the overburden and the rocks and in the fractured rocks of the bedrock, which hosts the highest yielding aquifers.

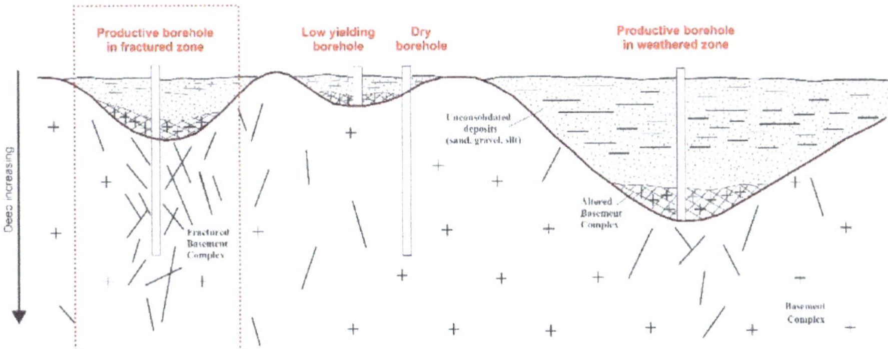

Figure 14. Possible scenarios for wells drilled in locations similar to the Gumbo area from a lithostratigraphic–hydrogeological point of view.

This conceptual model is coherent with the hydrogeological framework at the national scale retrieved from the available literature [14,21], which identifies the major aquifers in the northern and in the eastern parts of the state (i.e., on the borders with Sudan and Ethiopia), where the geological setting describes the presence of thick fluvial and lacustrine deposits. The southern and western parts of the region, where the Gumbo area is located, mainly consists of granite and other crystalline rocks, only creating the conditions for local and shallow aquifers. As a consequence of this general assessment, the report [22] describes a difference in the potential between the northern and eastern aquifers and the ones in the south and west wide areas of the country. The former are described as excellent aquifers, with high potential mainly depending on the depth and the good permeability properties of the deposits; the latter, being located in areas characterized by the presence of the basement complex, are described as low potential aquifers, of which yield of groundwater is only enough for rural or urban water supply.

In this hydrogeological framework, despite a general low expected productivity, the local conditions may allow water production to be optimized if the drilling is performed in accordance with the conditions defined as "Productive borehole in fractured zone" (Figure 14), where the boreholes are drilled both in the overburden and fractured bedrock and the latter is characterized by layers of high permeability due to alteration (altered basement complex) or fractures (fractured basement complex). The storage can be provided by the overburden.

The hydrochemical characterization of surface and groundwater is a complete novelty in the panorama of available data on South Sudan. The results furnish the main chemical features of water of the area, normally used for human consumption, irrigation and livelihood.

The data registered a local exceeding of the suggested limits, with assumed possible negative effects for the human health in the long term and negative consequences for the soil in case of agricultural use, for chromium, sodium, fluorides, manganese and total hardness. In general, surface water, coming from seasonal streams, highlighted a lower water quality and scarcer and more ephemeral discharge. Indeed, even if it represents the most direct, cheapest and most easily obtained water resource for humans and animals, it shows higher vulnerability to biological and chemical pollution and is easy to contaminate by anthropic input. Hence, the best supply is represented by groundwater.

Even if the number of samples is very low, some considerations of the relationship between single parameters were made. Chlorides vs. sodium, sodium vs. electrolytic conductivity and chlorides vs. electrolytic conductivity graphs show in general good correlations for surface water (Figure 15).

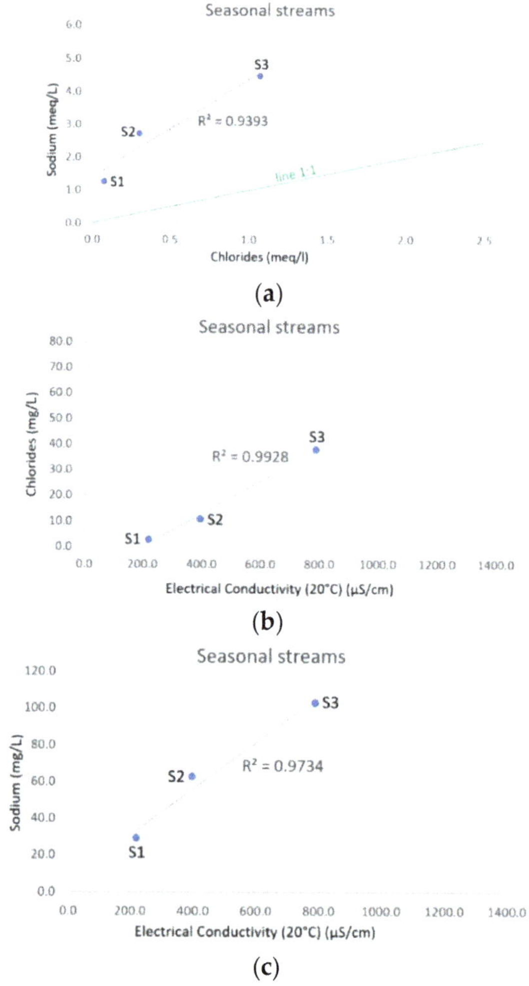

Figure 15. (**a**) Chloride vs. sodium graph (meq/L); (**b**) electrolytic conductivity vs. chlorides graph; (**c**) electrolytic conductivity vs. sodium graph for the surface analyzed waters.

New data have to be added in order to verify the indicative trends obtained.

Chemical data were also plotted in the Piper diagram, allowing for a classification of groundwater and surface water in chemical facies and for a simultaneous comparison of different water samples. The Piper diagram (Figure 16) shows that the samples belong to the Na-HCO3 facies. These waters are typical of deep groundwater environments, influenced by ion exchange.

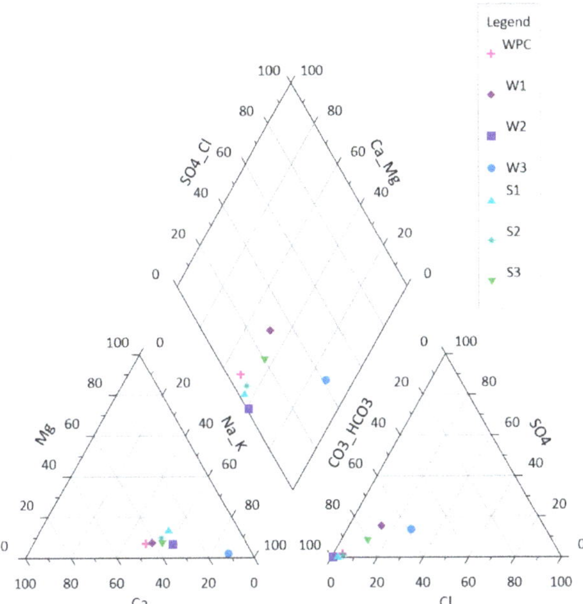

Figure 16. Piper diagram, showing hydrochemical facies of groundwater and surface water samples.

The Na-rich facies resulting from the Piper classification is coherent with the high concentration of sodium that slightly exceeds the national regulation in three groundwater samples out of four. The remaining sample (W2) has, however, an Na concentration of 98.8 mg/L (the limit is 100 mg/L).

Nevertheless, according to the experimental results, this high sodium concentration is not related to an equivalent abundance of chlorides (i.e., the correlation between sodium and chlorides is unbalanced in favor of sodium). The hypothesis of an anthropic contamination that could be considered in presence of salty waters (i.e., similar concentrations of Na and Cl) seems therefore to be unsuitable in this case, suggesting, rather, a natural origin. The described geological framework, reporting a regional bedrock mainly consisting of granite rocks rich in Na minerals (e.g., albite and Na feldspars) supports this hypothesis. Similar results and conclusions were obtained by the author of [23], who proposed an analysis of the water pollution in Sudan (i.e., in the northern geographical position with respect to our study area) and observed a prevalence of sodium bicarbonate water facies where the aquifer was in contact with the rock basement.

A similar natural origin of water pollution may be suggested for the fluorides that exceed the national regulation limits in the majority of the samples. Fluoride was, indeed, described as one of the most important natural pollutants of water in Africa [24] as a consequence of volcanic activities, presence of thermal waters, gases emissions and granitic and gneissic rocks. These natural conditions are often enhanced by low pumping rates, which increase the contact time and favor water–rock interactions, resulting in higher fluoride releases in groundwater [24,25].

As already stated, the concentration of total chromium, which exceeds the WHO limits in three of the samples, is related to a low percentage (<5%) of hexavalent chromium (i.e., the chromium form usually linked to anthropic contamination and more dangerous for the human health). This suggests, again, a natural origin of the contamination connected with lithological setting (i.e., presence of granite or gneiss rocks)

The last anomalous record in the results involves sample W2, which has a content of manganese that slightly exceeds the national regulations and is significantly higher than in all the other samples. Since the depth of the W2 well is not significantly different from the others, this anomalous concentration should be attributed to local conditions. The absence of nitrates and sulphates and the low electrolytic conductivity that characterize this water sample suggest the existence of a reducing environment, possibly connected to an organic contamination that progressively introduces a dissolution of manganese (and iron) already present in the surrounding rocks.

6. Conclusions

The groundwater resources of a country are crucial for the development of several economic sectors. In semi-arid and arid environments, groundwater enables millions of human beings and animals to exist in terrains that would otherwise be uninhabitable.

The General Assembly of the United Nations published the document "Transforming our world: the 2030 Agenda for Sustainable Development" as a result of the UN summit for the adoption of the post-2015 development agenda. In terms of the Sustainable Development Goal target for drinking water, Sustainable Development Goal (SDG) target 6.1 is to "achieve universal and equitable access to safe and affordable drinking water for all by 2030".

The availability and quality of water, besides representing a fundamental resource as drinking water, is necessary for a good development of agricultural and breeding activities and, consequently, for a sustainable development of the local economy. The water shortage, together with the great frequency and severity of drought periods and the excessive heat conditions, often causes agricultural vulnerability in arid or semi-arid areas of Sub-Saharan Africa. These conditions of vulnerability are enhanced by the global climate changes of recent years, the effects of which are expected to be large and far-reaching predominantly in the developing world [26,27].

For these reasons, the present study proposes an investigation of the lithostratigraphic and hydrogeological setting of the Gumbo village (SE of Juba) to make water extraction systems available and ensure their sustainable management to guarantee food security and economic independence for the populations in the area. Due to the political and social instability, the possibility to move and work in the field during the project was negatively affected, implying the difficulty of collecting data and water samples. Nevertheless, considering the scarce scientific and technical literature available for South Sudan, the collected data and proposed interpretations represent a first fundamental step for the characterization of the Gumbo area. The results contribute to defining a conceptual hydrogeological model and the critical aspects to take into account for the water use in the sampled sites.

In any context, the knowledge of the hydrogeological and hydrochemical setting is an important prerequisite for a risk-informed management of water resources. The planning of hydrogeological-based interventions may guarantee a wide access to good quality water for the local population for drinking, irrigation and livestock purposes. This study represents, therefore, a fundamental step to ensuring food security as well as sustainable economic development and social progress in this African region.

The collected data can be very useful for Administrative Local Authorities, recently established in South Sudan, as well as researchers and the local population, in order to improve water knowledge and management in this little-known area. Moreover, the implementation of a scientific dataset represents the first step toward sustainable economic development and social progress, helping to improve the livelihood of the rural communities and enhancing sustainable and environmentally friendly agriculture, soil conservation and remunerable livestock.

According to the Local Authorities, a survey program increasing the number of water points to monitor the quantitative and qualitative distribution of the water resource should help in better defining the local knowledge gained in the present study and consolidating the scenario that is defined (higher quality of groundwater than surface water, "productive borehole in fractured zone" hydrogeological model and natural contaminations due to a geological setting with local increasing of exceeding parameters).

Future developments of the study may also include the long-term monitoring of the groundwater. The repetition of the analyses proposed in this study (i.e., measure of groundwater static level, hydrochemical analyses of surface and groundwater samples) at different times throughout the year, together with an increase of the number of sampling points, may allow for the assessment of the real potential of the water resource, with an evaluation of the influence of the seasonal variations on the resource availability. The temporal and spatial increment of collected samples would also contribute to validating the hydrogeological model, with an improved understanding of the flow directions and of the relations between groundwater and surface waters.

Further information should moreover include socioeconomic and demographic data (population, land use, cultivated crops, livestock activities, etc.) of the study area to better understand the various human needs for water to neither exhaust the water sources and the local economy nor have long term negative impact on the environment.

In addition, the monitoring of the chemical contaminations over time, possibly enriched with the local sampling of waters at different depths, may quantify the real danger related to each anomalous concentration and explain the distribution of the contaminations in the aquifer [28,29]. As a function of the specific use (e.g., drinking water, agriculture), the results of these analyses may help to assess the necessity and the features of interventions that may guarantee the safe use of the water resource.

In relation to the intended use of water for agricultural purposes, additional analyses may be specifically directed to the evaluation of the sodium adsorption ratio (SAR), which is a good indicator of the suitability of water for irrigation.

Due to the dry climate conditions of South Sudan, and of the Sub-Saharan African regions in general, another important development of the research could be the assessment of water resources' vulnerability to climate change [30]. An interesting approach to investigate this topic is the analysis of the water resource potential; after the selection of a number of target stations, the study should provide an estimation of rainfall amount on average over the last 30 years and a comparison of rainfall data and survey campaign outputs. The results would show whether the climate change affected the groundwater, with a quantitative evaluation of the entity of this influence on both availability and quality of the water resource [31,32].

A similar development of the research should represent the first step to defining Local Development Territorial Plans that the government needs in order to identify sustainable social and economic assessment (economic activities to encourage, territorial distribution of the refugees, stabilization or development of nomadism). It would be recommended for all decision-makers and politicians but also for rural development program managers to employ a similar approach, which emphasizes the knowledge of the characteristics of the territory and its specificities to deal with the management of natural resources, water in particular, in a sustainable way.

Author Contributions: Conceptualization, M.L. and S.M.R.B.; data curation, L.D., C.C. and C.S.; formal analysis, M.L. and S.M.R.B.; funding acquisition, S.M.R.B.; Investigation, L.D.; methodology, M.L. and S.M.R.B.; supervision, D.A.D.L.; visualization, C.C. and C.S.; writing—original draft, M.L., S.M.R.B., C.C. and C.S.; writing—review & editing, M.L. and S.M.R.B. All authors have read and agreed to the published version of the manuscript.

Funding: This research was funded by AICS (Agenzia Italiana per la Cooperazione allo Sviluppo), OSC-2016 competition, grant number AID 10915/VIDES/SSD, Title of the project "Women Empowerment and Sustainable Agriculture Development to Achieve Food Security in South Sudan (WOSA)".

Acknowledgments: The authors thank Comina Cesare for his help in the interpretation of geophysical data.

Conflicts of Interest: The authors declare no conflict of interest.

References

1. Bonetto, S.; Facello, A.; Camaro, W.; Cristofori, E.I.; Demarchi, A. An approach to integrate spatial and climatological data as support to drought monitoring and agricultural management problems in South Sudan. In Proceedings of the EGU General Assembly, Vienna, Austria, 17–22 April 2016.
2. Anudu, G.K.; Essien, B.I.; Onuba, L.N.; Ikpokonte, A.E. Lineament analysis and interpretation for assessment of groundwater potential of Wamba and adjoining areas, Nasarawa State, north-central Nigeria. *J. Appl. Technol. Environ. Sanit.* **2011**, *1*, 185–192.
3. Ayenew, T.; Demlie, M.; Wohnlich, S. Hydrogeological framework and occurrence of groundwater in the Ethiopian aquifers. *J. Afr. Earth Sci.* **2008**, *52*, 97–113. [CrossRef]
4. Bonetto, S.; De Luca, D.; Lasagna, M.; Lodi, R. Groundwater distribution and fluoride content in the West Arsi Zone of the Oromia Region (Ethiopia). In *Engineering Geology for Society and Territory, Volume 3: River Basins, Reservoir Sedimentation and Water Resources*; Lollino, G., Arattano, M., Rinaldi, M., Giustolisi, O., Marechal, J.-C., Grant, G.E., Eds.; Springer: Berlin/Heidelberg, Germany, 2015; pp. 579–582.
5. Bonetto, S.; Facello, A.; Cristofori, E.I.; Camaro, W.; Demarchi, A. An Approach to Use Earth Observation Data as Support to Water Management Issues in the Ethiopian Rift. In *Climate Change Adaptation in Africa. Climate Change Management*; Springer: Berlin/Heidelberg, Germany, 2017; pp. 357–374.
6. Bonetto, S.; Facello, A.; Umili, G. The contribution of Curva Tool semi-automatic approach in structural and groundwater investigations. A case study in the Main Ethiopian Rift Valley. *Egypt. J. Remote Sens. Space Sci.* **2018**, *23*, 97–111. [CrossRef]
7. Rango, T.; Bianchini, G.; Beccaluva, L.; Tassinari, R. Geochemistry and water quality assessment of central Main Ethiopian Rift natural waters with emphasis on source and occurrence of fluoride and arsenic. *J. Afr. Earth Sci.* **2010**, *57*, 479–491. [CrossRef]
8. Yenne, E.Y.; Anifowose, A.Y.B.; Dibal, H.U.; Nimchak, R.N. An assessment of the relationship between lineament and groundwater productivity in a part of the basement complex, Southwestern Nigeria. *IOSR J. Environ. Sci. Toxicol. Food Technol.* **2015**, *9*, 23–35.
9. Lasagna, M.; Dino, G.A.; Perotti, L.; Spadafora, F.; De Luca, D.A.; Yadji, G.; Tankari, D.-B.A.; Moussa, I.; Harouna, M.; Moussa, K. Georesources and environmental problems in Niamey city (Niger): A preliminary sketch. *Energy Procedia* **2015**, *76*, 67–76. [CrossRef]
10. Bucci, A.; Franchino, E.; De Luca, D.A.; Lasagna, M.; Malandrino, M.; Bianco, P.A.; Hernandez, S.H.O.; Macario, C.I.; Sac, E.E.; Hernandez, A. Ground water chemistry characterization using multi-criteria approach: The upperSamala River basin (SW Guatemala). *J. South Am. Earth Sci.* **2017**, *78*, 150–163. [CrossRef]
11. UNITED NATIONS. Available online: https://www.un.org/Depts/Cartographic/map/profile/southsudan.pdf (accessed on 3 December 2019).
12. Appelo, C.A.J.; Postma, D. *Geochemistry, Groundwater and Pollution*; Balkema: Rotterdam, The Netherlands, 1996.
13. WHO. *Guidelines for Drinking-Water Quality*, 4th ed.; WHO Library Cataloguing-in-Publication Data: Geneva, Switzerland, 2011.
14. Nile Basin Initiative (NBI), Initiative du Bassin du Nil. State of the River Nile basin. 2012. Available online: https://www.nilebasin.org/documents-publications (accessed on 7 December 2019).
15. Climate and Average Weather in South Sudan. Available online: http://www.weather-and-climate.com (accessed on 4 June 2019).
16. EISOURCEBOOK. Good-Fit Practice Activities in the International Oil, Gas & Mining Industries—South Sudan. Available online: http://www.eisourcebook.org/1250_SouthSudan.html (accessed on 6 June 2019).
17. Vail, J.R. Outline of the Geochronology and tectonic units of the basement complex of Northeast Africa. *Proc. R. Soc. Lond. Ser. A* **1976**, *350*, 127–141.
18. Vail, J. Research in Sudan, Somalia, Egypt and Kenya. *J. Afr. Earth Sci.* **1990**, *11*, 231–232. [CrossRef]
19. Vail, J.R.; Kuron, J. High level igneous emplacements in the Red Sea Hills, Sudan. *Geol. Rundsch.* **1978**, *67*, 521–530. [CrossRef]
20. Adak, S.M. *The Geology of Juba and the Surrounding Area, Central Equatoria*; University of Juba: Juba, South Sudan, 2015.
21. Vrba, J.; Richts, A. *The Global Map of Groundwater Vulnerability to Floods and Droughts. Explanatory Notes*; UNESCO: Paris, France, 2015.

22. Ministry of Electricity, Dams, Irrigation &Water Resources of the Republic of South Sudan. *Irrigation Development Master Plan (Final Report)—Annex 3: Irrigation Development Potential Assessment*; Ministry of Electricity, Dams, Irrigation &Water Resources of the Republic of South Sudan: Juba, South Sudan, 2005.
23. Abdo, G. *Status of Groundwater Quality and Pollution Risk in Sudan. Groundwater Protection Network in the Arab Region*; IHP Publication: No 14; UNESCO CAIRO Office: Paris, France, 2003.
24. Malago, J.; Makoba, E.; Muzuka, A.N.N. Fluoride Levels in Surface and Groundwater in Africa: A Review. American. *J. Water Sci. Eng.* **2017**, *3*, 1–17. [CrossRef]
25. Schmoll, O.; Howard, G.; Chilton, J.; Chorus, I. *Protecting Groundwater for Health—Managing the Quality of Drinking-Water Sources*; World Health Organization (WHO): London, UK, 2006.
26. Altieri, M.A.; Nicholls, C.I. The adaptation and mitigation potential of traditional agriculture in a changing climate. *Clim. Chang.* **2017**, *140*, 33–45. [CrossRef]
27. Meijer, S.S.; Catacutan, D.; Ajayi, O.C.; Sileshi, G.W.; Nieuwenhuis, M. The role of knowledge, attitudes and perceptions in the uptake of agricultural and agroforestry innovations among smallholder farmers in sub-Saharan Africa. *Int. J. Agric. Sustain.* **2015**, *13*, 40–54. [CrossRef]
28. Lasagna, M.; De Luca, D.A. The use of multilevel sampling techniques for determining shallow aquifer nitrate profiles. *Environ. Sci. Pollut. Res.* **2016**, *23*, 20431–20448. [CrossRef]
29. Chapman, S.; Parker, B.; Cherry, J.; Munn, J.; Malenica, A.; Ingleton, R.; Jiang, Y.; Padusenko, G.; Piersol, J. Hybrid multilevel system for monitoring groundwater flow and agricultural impacts in fractured sedimentary bedrock. *Groundw. Water Monit. Remediat.* **2015**, *35*, 55–67. [CrossRef]
30. Caselle, C.; Bonetto, S.; De Luca, D.A.; Lasagna, M.; Perotti, L.; Bucci, A.; Bechis, S. An interdisciplinary approach to the sustainable management of territorial resources in Hodh el Chargui, Mauritania. *Sustainability* **2020**, *12*, 5114. [CrossRef]
31. Lasagna, M.; Ducci, D.; Sellerino, M.; Mancini, S.; De Luca, D.A. Meteorological variability and groundwater quality: Examples in different hydrogeological settings. *Water* **2015**, *12*, 1297. [CrossRef]
32. Fallahati, A.; Soleimani, H.; Alimohammadi, M.; Dehghanifard, E.; Askari, M.; Eslami, F.; Karami, L. Impacts of drought phenomenon on the chemical quality of groundwater resources in the central part of Iran—Application of GIS technique. *Environ. Monit. Assess.* **2020**, *192*, 64. [CrossRef] [PubMed]

Article

Water Resource Management and Sustainability: A Case Study in Faafu Atoll in the Republic of Maldives

Maurizio Filippo Acciarri [1,*], Silvia Checola [2], Paolo Galli [3], Giacomo Magatti [4] and Silvana Stefani [5]

[1] Department of Materials Science, Università degli Studi di Milano Bicocca, 20126 Milan, Italy
[2] REF-E S.R.L., 20123 Milan, Italy; silvia.checola@ref-e.com
[3] MaRHE Center, Università degli Studi di Milano-Bicocca, 20126 Milan, Italy; paolo.galli@unimib.it
[4] Department of Earth and Environmental Sciences, Università degli Studi di Milano Bicocca, 20126 Milan, Italy; giacomo.magatti@unimib.it
[5] Department of Statistics and Quantitative Methods, Università degli Studi di Milano Bicocca, 20126 Milan, Italy; silvana.stefani@unimib.it
* Correspondence: maurizio.acciarri@unimib.it; Tel.: +39-02-6448-5136

Abstract: This paper contributes to the existing literature in proposing an integrated approach to water management and energy renewable production in a fragile environment. After the 2004 tsunami, in many outer islands in The Republic of Maldives, the lens freshwater natural reservoir was deeply damaged. Currently, the populations of rural atolls use rainwater and water in plastic bottles imported from the mainland for drinking. To provide safe and sustainable drinking water, we analyze the feasibility of two different actions: a desalination system fed by a diesel plant or by a photovoltaic (PV) plant with batteries. The current situation (business as usual, (BAU)) is also evaluated and taken as a benchmark. After illustrating the technical and economic features of desalination and PV plants, a financial and environmental analysis is conducted on the two alternatives plus BAU, showing that the desalination fed by the PV plant results in optimization both on the financial and the environmental side. The levelized cost of water (LCOW) and the CO_2 levelized emissions of water (LEOW) are calculated for each alternative. The case study is developed in Magoodhoo Island, Faafu Atoll and can be extended to other islands in The Republic of Maldives and in general to small island developing states (SIDS).

Keywords: photovoltaic energy; desalination system; SIDS; CO_2 emissions; LCOW; LEOW

Citation: Acciarri, M.F.; Checola, S.; Galli, P.; Magatti, G.; Stefani, S. Water Resource Management and Sustainability: A Case Study in Faafu Atoll in the Republic of Maldives. *Sustainability* **2021**, *13*, 3484. https://doi.org/10.3390/su13063484

Academic Editor: Maurizio Tiepolo

Received: 11 February 2021
Accepted: 15 March 2021
Published: 22 March 2021

1. Introduction

Most of the small island developing states (SIDS), to which The Republic of Maldives (in short, The Maldives) belongs, share similar sustainable development challenges: a small but increasing population, vulnerability to external shocks, strong dependence on import trade, and a fragile environment [1]. Development is a need and many SIDS now recognize, as a primary objective, the move towards low carbon sustainable economies [1], while at the same time improving the standard of living of the local population.

Moreover, islands enjoy generally beautiful sceneries, fishing resources and unique natural settings that must be preserved and guaranteed for future generations.

As reported by the Asian Development Bank [2] in November 2020, the tourism industry comprises around 25% of the national gross domestic product (GDP) in SIDS. In [3], a road map establishes the guidelines to transition from a fossil-fuel-based energy sector to a cost-effective, business-competitive, affordable, and sustainable renewable energy. This scenario forecasts a continuous and sustained moderate transformation of the energy matrix that will result in 29% of fossil fuels being reduced compared to a business-as-usual situation. As recently reported by IEA [4], the share of renewables in global world electricity generation is nearly 28% in 2020, but there are important differences among countries.

Even though it is almost universally recognized that renewable energies must gradually substitute fossil fuels [5], the use of renewables on small islands is still quite low based on high transaction costs and poor knowledge of the market potential [6].

Electricity is also dependent on imported fuel, and the level of CO_2 emissions is quite high [7]. Taking the Magoodhoo island in The Maldives as a case study, our aim is to analyze an important issue: the coupled nexus water/energy, and to show that feasible solutions are possible, from the financial and environmental point of view.

Scarcity of energy and water are among the major drawbacks for SIDS [8]. This is particularly the case of many outer islands in The Maldives, where, since the 2004 tsunami, the lens freshwater natural reservoir has been damaged, leading to a significant increase in costs as well as an increase in the scarcity of drinking water.

In [9], an extensive review of the recent situation of SIDS is illustrated. Like many SIDS all over the world, The Maldives depends upon imported conventional sources of energy, and scarcity of water severely restricts economic and social development. As far as energy is concerned, the dependence on fossil fuels makes SIDS vulnerable to oil price rises with a significant impact on the local economy. Energy systems and water supply are closely coupled, and the nexus is particularly relevant in SIDS. The ultimate goal is to make SIDS independent from the mainland in energy and water. A desalination system, combined with renewable energy, can definitely be the solution for solving the scarcity of water and contributing to the reduction of pollution from fossil fuels.

In [10], 1087 islands are classified with the use of a cluster analysis according to climatic as well as physical and socio-economic parameters. The conjecture is that an energy supply system proven to be successful on one island could also be successfully implemented on another island.

From the literature analysis, it is quite evident that in SIDS, an energy transition to more sustainable energy sources is desirable and different studies state that the transition should take place in the next few years [11–25].

In The Maldives, energy sources are traditionally based on imported conventional supply, of which the most common fuel is diesel which fuels small power plants, often with low efficiency and heavy polluting. Due to the diesel production, emissions of CO_2 into the environment are large and in the medium term may cause damages to the environment. Renewable resources are scarcely used. While solar water heaters are quite widespread, photovoltaic (PV) panels and wind plants are rare. In 2016, the total power generation from renewable energy was 6 MW. In the medium term, the government has planned to install renewable energy systems for up to 30% of the daytime peak load in all inhabited islands [26]. Thus, the efforts of the Maldivian government and local authorities are on implementing renewable systems. The aim is to transit from energy dependence to independence from the mainland.

Studies on energy, applied or theoretical, for the assessment and feasibility of renewable projects in SIDS are quite numerous; less numerous are studies and experiences in desalination combined with renewables.

An extensive study is carried on a reverse osmosis (RO) desalination plant, located in the Sinai Peninsula in Egypt, fed by conventional and unconventional energy sources in [27].

A project for a desalination plant of 1000 m^3/day is proposed in the Tyra Island in Greece [28]. The report illustrates two case scenarios: solar and wind power, compared to diesel generators, feed a desalination plant. Diesel generators are used as back up in both cases.

In [29] the feasibility of desalination systems fed by PV or high concentration solar plants of medium and large size in Saudi Arabia is discussed as opposed to diesel as the source of energy. The cost of solar is still high given the heavy subsidizing of diesel in Saudi Arabia.

In Kaya et al. (2019) [30] an analysis based on the levelized cost of water indicates that the current thermal desalination methods used in Abu Dhabi should be substituted by solar PV systems and reverse osmosis (RO) technology for desalination.

Helal [31] explores the economic feasibility of three alternative configurations of an autonomous seawater reverse osmosis (SWRO) unit in remote areas of the UAE. Three different scenarios are proposed here for the powering system where the unit is driven either by a diesel generator, a PV-diesel hybrid system, or solely driven by solar panels without battery backup. This paper can be considered an interesting benchmark, as the low desalination plant dimension [20 m^3/day] is comparable with our scenario.

The global average levelized cost of water (LCOW) from desalination plants could decline from around \$2.86/m^3 in 2015 to \$1.25 by 2050 if solar, storage systems and other renewable energies are used to decarbonize the sector [32]. The researchers forecast that in specific regions of China, India, Australia, and the U.S., drinking water may cost less than a dollar.

An extensive report from IRENA [33] gives useful information on systems already functioning. Different renewable energy sources are proposed to reduce the environmental impact.

Mainly, studies are based on the use of wind turbines or solar plants, but there are also proposals where other renewable sources, such as sea currents, are used [34]. This application could be interesting particularly for some islands where the scarcity of drinking water is more severe.

A recent paper published in 2020 [35] reports a study about the economic, technical, and environmental impacts of different desalination system configurations (centralized or decentralized, components, and technologies) on transition plans to achieve a higher share of renewable energy and desalination supplies for regions facing water scarcity. The analysis forecasts a reduction of LCOW from the current value of \$2.19–2.46/m^3 to \$0.79–1.01/m^3 in 2040.

In The Maldives, scarcity and quality of water are relevant issues. Traditionally, Maldivians used lens freshwater from household wells for both potable and non-potable purposes. The elevation of the highest point in most islands is less than 2 m above sea level, so the freshwater lens thickness of each island is low. The use of rainwater for drinking began in 1978 due to a cholera outbreak. In 1995, the first municipal water supply project began its services. It started from the capital Malé and extended to other population centers, including all touristic resorts. A public water system became available to 49% of the local population and was fully operative by the end of 2017 [36].

However, many outer island populations still depend on collecting rainwater for domestic use, as in the study case illustrated in the present paper. Furthermore, recently and with increased frequency, the collected rainwater stores run out during the dry season, causing expensive emergency water supply from the mainland. Moreover, after the 2004 tsunami, the lens freshwater deposit was deeply damaged, and freshwater is now brackish. While tourist resorts and the city of Malé are all equipped with (thermal) desalination plants, in many outer islands, people use rainwater and a supply of bottled water imported from the mainland for drinking, rainwater for cooking and a small amount of lens water for washing, flushing and other purposes [26].

However, even though The Maldives shows a wide deficit in energy and freshwater resources, they have unique advantages in term of renewable energy (sun) with excellent radiation and seawater potential. An integrated approach to water management and renewable energy production can definitely improve the living in rural atolls and substantially reduce CO_2 emissions.

In this paper, we also carried out a life cycle assessment (LCA) for the different solutions proposed. In the literature, different studies on LCA for water systems of different dimensions can be found [37,38]: in integrated urban water systems or for the specific analysis of plastic bottle impact [39].

A synthesis of the literature discussed here, for water desalination and energy source solution adopted, is reported in the discussion section. The data will be used as a comparison with our results.

This paper contributes to the gradual road to energy/water independence from the mainland by proposing a financial and environmental evaluation for the installation and maintenance of a desalination plant. The off-grid energy to the desalination system is provided by diesel or by a PV plant. Results in this paper show that a desalination system plus a PV plant with batteries is the optimal choice, as opposed to the use of diesel, and more feasible than the current scenario, both from the financial and the environmental point of view.

The project is carried out in Magoodhoo Island, in Faafu Atoll in The Maldives. The project is jointly financed by the Marine Research and High Education Center (MaRHE Center). MaRHE Center, officially inaugurated on January 28, 2009, is in Magoodhoo Island. Purpose of the Center is to carry out research and teaching activities in the fields of environmental sciences and marine biology, science of tourism and human geography, to teach how to protect a fragile environment and its biodiversity, how to use and manage its resources in a responsible way. https://marhe.unimib.it/ accessed on 1 March 2021) and by the University Milano Bicocca (Italy) and the Italian Ministry of Environment.

2. Materials and Methods

The Faafu Atoll is located 137 km from Malé and covers an area of 30 km long and 27 km wide, in the Indian Ocean (Figure 1). The Faafu Atoll is made of 23 islands, covering 1.6 km^2 of land, of which five are inhabited by locals and one is an island-resort in Filitheyo. The capital is Nilandhoo and the population of the atoll is about 4200 inhabitants who live mainly from tourism and fishing.

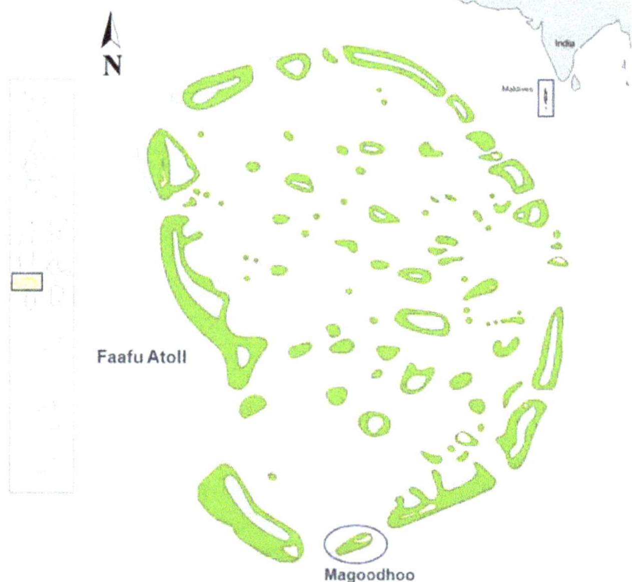

Figure 1. The Faafu Atoll in The Maldives.

Magoodhoo Island is in the Faafu Atoll (3°4′26″ N, 72°57′15″ E), with a population of 800 inhabitants including the MaRHE Center personnel (35 researchers approx.). In total, there are 133 families of 6 members each on average.

In this project, we have sized a desalination plant on the water needs of the local population. Data collection has been brought forward by two co-authors of the present paper with interviews of local people, inspections of water tanks, and delivery checks of bottled water to the island.

The project involves four main steps: Section 2.1. Understanding the community water demand. Section 2.2. Choice of a desalination technology. Section 2.3. Technological and economic analysis of two alternative actions. Section 2.4. Choice of the best alternative among ACTION 1, ACTION 2 and the current situation (business as usual) based on financial and environmental criteria.

2.1. Understanding the Community Water Demand

In every family house, including the MaRHE Center, rainwater is used for cooking and drinking, while ground water is used for washing and flushing (see also [36] for a recent survey on the quality of water in The Maldives). In every family house, in the municipality compound and in the MaRHE Center, the ground water is taken from household and public wells, while the rainwater is collected in big plastic water tanks, each containing 2500 L of water. The rainwater pours into the tank through gutters on the roof. The roof is cleaned occasionally, no more than once a year, and the only filters used are pieces of clothes stuffed into the pipe hole. The water tank, once filled up, is closed, and reopened when necessary. The water flows from the tank through a faucet. The tank fills up during the rainy season and occasionally when it rains out of the rainy season. The tanks, provided by the Maldivian government more than 20 years ago and partly by the Red Cross after the 2004 tsunami, are substituted occasionally (Figure 2).

Figure 2. Public water tank used by the inhabitants of Magoodhoo to collect rainfall water for daily use.

A reliable estimate puts consumption at 23.81 L per day per family, thereby using up a full tank in approximately three/four months. This amounts to the full use of 3–4 tanks per year (no data about diseases (potential presence of micro-bacteria and plastic microparticles from the tanks) caused by non-filtered water are available). People in the MaRHE Center drink the same water as the village, taken from public tanks, but filter it out by their own filters through a water dispenser located in the MaRHE Center canteen.

The water from the tanks is felt by local people as not safe for drinking, but it is extensively used for cooking. As an alternative to water tanks, water in plastic bottles is used. The bottled water, in plastic usually of 1.5 L, is given mainly to children and sick people to drink and consumption is large. From Table 1, the consumption of water (rainwater and bottled water) per person per day amounts to 4.35 L. The World Health Organization sets the minimum quantity of water needed at 7.5 L per person per day [40]. Statistics on water consumption in touristic resorts and hotels are rare. An estimate from a big hotel chain in Europe puts water consumption in the range 380–1100 L/guest/night. Swimming pool accounts for an equivalent of 60 L/guest/night [1].

Table 1. Water consumption (liters) and cost ($) per person, per family, of the whole island.

		(person)	(family)	(island)
PLASTIC BOTTLES	Daily	0.38 L 0.09 $	2.30 L 0.54 $	307 L 71.44 $
	Monthly	11.51 L 2.68 $	69 L 16.11 $	9200 L 2143 $
	Yearly	138 L 32.17 $	830 L 193 $	110,400 L 25,721 $
RAINWATER	Daily	3.97 L	23.81 L	3166 L
	Monthly	119.05 L	714 L	94,962 L
	Yearly	1429 L	8568 L	1,139,544 L

Data in Table 1 correspond to the quantity for drinking and cooking, not considering water for washing and flushing, taken from underground wells.

The bottled water is transported to the village by a supply boat (dhoni) (Figure 3) that carries, on average, 2300 L of water per week. The water is sold in the little shops at an average cost of approximately 3.6 MVR per liter ($1 = 15.45 MVR, as of March 2019. the water sold in little shops is bought by the local population only. The price of water varies according to supply conditions and demand. People of the MaRHE Center, practicing an anti-plastic policy, drink the water from the dispenser available in the Center. The total consumption of plastic bottled water sums up to 830 L of water per family, per year. This corresponds to an average cost of $193 per year per resident family the median yearly income of a family in the Faafu Atoll is $17,087 (264,000 MVR) [http://statisticsmaldives.gov.mv/nbs/wp-content/uploads/2016/03/Presentation-Income-HIES2016.pdf] (accessed on 1 March 2021). The consumption of plastic bottles corresponds to 1.1% of yearly income (Table 1).

Figure 3. Weekly supply via dhoni of water bottles for the island of Magoodhoo.

The current individual consumption of water amounts to an average of 4.35 L per day. This water is rainwater and/or bought in the market in plastic bottles.

The aim of the project is to provide 6 L of safe drinking water per day to all inhabitants of the island (Table 2). The supply of water will be provided by a desalination plant fed by an off-grid source of energy that can be renewable or conventional.

Table 2. Current consumption and target per person, per family, of the whole island (liters per day).

CURRENT	4.35 L/day	26.11 L/day	3471 L/day
DESALINATED WATER	6 L/day	36 L/day	4800 L/day

2.2. Choice of a Desalination Technology

Given the size of the island, a small scale decentralized desalination unit can be sufficient. The plant will be fed by off-grid electricity.

The strength and weakness of the available desalination technologies were analyzed to choose the appropriate technology. The advantages and disadvantages of some well-known commercial desalination technologies are reported in the literature [41–44].

When choosing a technology, an important factor is the required energy demand for the desalination technology as well as the use of a mature technology when designing a decentralized small-scale desalination unit. In choosing desalination technologies to be coupled with a renewable energy source, it is also important to know the amount of the conventional energy required by the desalination processes (Table 3). The reverse osmosis plant cost has been determined according to an offer by a local installer in Magoodhoo (Table 4). As can be seen from Table 4, the cost of the desalination plant is $16,500 but the complete installation, including complementary tools, amounts to $50,000. If working 12 h/day, the plant is designed to satisfy the local demand, guaranteeing 5 m^3/day (5000 L/day), which is 40% more of the current consumption.

Table 3. Technical specifications of the reverse osmosis (RO) plant.

Characteristic	Unit	Quantity
Capacity	m^3/day	5
Feed temperature	°C	25
Water quality	ppm Total Dissolved Solids g/mL	<500
Power consumption	kW	3.7
Daily operating	h/day	12
Total energy required	kWh/day	44.4

Table 4. Cost of the RO plant.

Components	Cost ($)
Reverse osmosis desalination plant-10 PTD	16,500
Drinking water storage tank	1500
Water storage for toilet service	3000
Sea water intake and reject line	15,000
Sea water intake pump system with control board	1850
Registration	5000
Installation materials	3000
Masonry (to arrive a 50,000)	1319
GST 6%	2830
Total	**50,000**

We recall that a minimum of 7.5 L per capita per day will meet the requirements of most people under most conditions [40]. With this project, the desalination plant will provide 6 L of safe drinkable water per day per person. Rainwater and ground water are still available for other purposes.

The RO plant design includes a pre-treatment process in which fouling elements would be removed. Conventional pre-treatment was selected due to the low incurred costs.

For our requirements, a RO desalination plant of 3.7 kW is selected with a production capacity of 417 L/h. The water produced is stored in two different tanks. As the renewable energy source, photovoltaic is the source commonly coupled with the desalination plants [23,26]. We compare this scenario with a conventional one where the electricity required by the desalination plant is delivered by a diesel generator. Another alternative we consider is the current scenario (business as usual).

2.3. Technological and Economic Analysis of Two Alternative Actions

RETScreen is a software tool for energy system modelling which analyzes the energy scenario among the energy mix suggested by the user and provides a detailed financial analysis and emission analysis [45,46] software was used to estimate the energy production using solar PV and economic assessment of the plant. The main components of the proposed RO system are shown in Figure 4. The system consists of PV panels, a battery bank, an inverter, a high-pressure pump (HP), a membrane module and a storage tank.

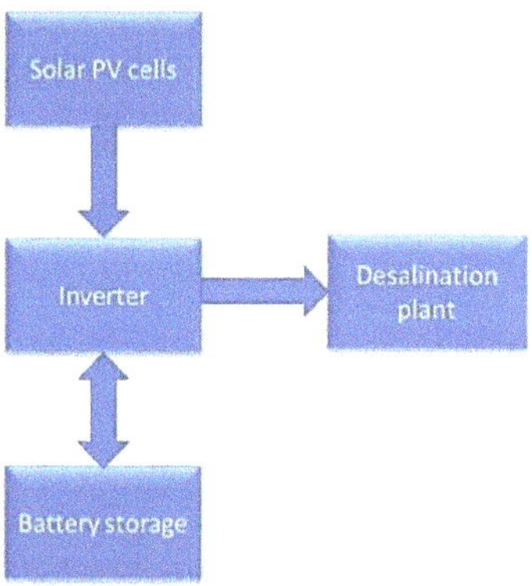

Figure 4. Main components of the RO + photovoltaic (PV) configuration.

The proposed integrated system for the RO load with 417 kW peak demand and the combined renewable energy sources were modelled using RETScreen software.

First, an energy production analysis involving inputs and costs was carried out for the PV plant. Input values and energy production are illustrated in Table 5. The power capacity of the PV system is 10.8 kW. The output of the inverter in AC kW is 10 kW. Magoodhoo Island, and The Maldives in general, are in a very good position to benefit from the sun's energy through solar PV technology, due to excellent solar radiation all year round. The solar radiation in Magoodhoo varies from 5.9 to 6.9 kWh/m^2/day (Figure 5).

Table 5. Input values for the RETScreen energy model sheet.

Item	Input Value
Solar tracking mode	One-axis
Slope	2
Azimuth	0°
PV module type	Si-monocrystalline-Jkm315M-72
Capacity of one PV unit	16.23%
Number of PV units	32
Annual solar horizontal radiation (MWh/m^2)	NASA 22-year monthly average
Miscellaneous losses	5% (dust and sandstorms)
Inverter efficiency (DC to AC)	96%
Inverter capacity	10 kW
Inverter losses	1%
Total annual PV production (kWh/y)	17,600
Daily PV energy production (kWh/day)	48

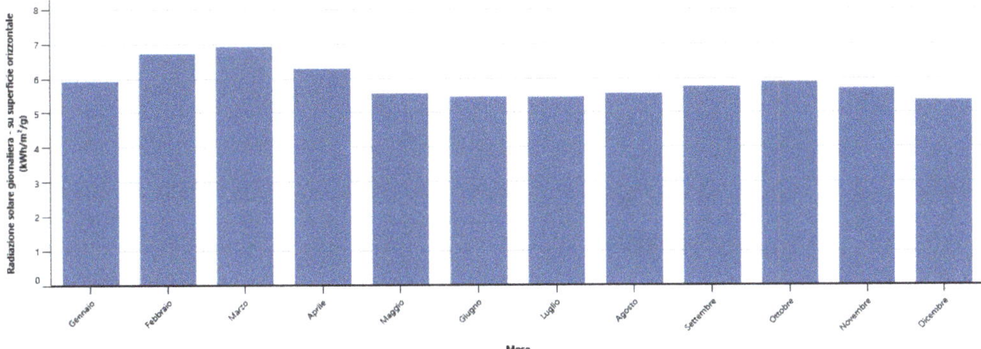

Figure 5. Daily average solar radiation in the site [RETScreen data].

The annual electricity produced is about 17,600 kWh, which means a production of about 48 kWh/day. The total solar collector area is 62 m^2 and can be located on a roof (see also [24] for a feasibility analysis on roof-mounted solar PV systems in Hulhumale Island in The Maldives for energy production) (Figure 6).

Figure 6. Photovoltaic plant under construction on the rooftop of Magodhoo Municipality.

The total average energy consumption is 48 kWh/day to produce 5 m^3 of desalinated water per day (H12 water production).

The cost of the PV plant with all components is $30,000 (tax included) and has been determined according to an offer by a local installer (Table 6). Prices in the mainland are probably lower, but the island is remote and local installers set the price.

Table 6. PV plant cost (in $).

	Components	Cost $/Wp
Module cost ($)	8427	0.84
Inverter	2520	0.25
Balance of System	4435	0.44
Installation labor	3226	0.32
Installer margin and overhead	11,392	1.15
Total	**30,000**	**3.00**

Due to the intermittent nature of solar energy, in a PV based power system, battery storage facilities are needed to ensure a constant power supply. The cost of battery storage is around $1500. The lifetime is estimated at 10 years.

2.4. Choice of the Best Alternative among ACTION 1, ACTION 2 and the Current Situation (Business as Usual) Based on Financial and Environmental Criteria

To proceed with the choice of the most feasible solution, three alternatives have been considered:

1. The current scenario, business as usual (BAU);
2. ACTION 2, desalination system fed by conventional fuel (diesel);
3. ACTION 2, desalination system fed by a PV plant.

For all alternatives, the net present value (NPV), the payback period (PBP) and the CO_2 emissions are calculated and compared. NPV is the difference between the present value of cash inflows and the present value of cash outflows over the lifetime of the project. NPV is used in capital budgeting to analyze the profitability of a projected investment. The best alternative has the highest NPV.

PBP refers to the amount of time it takes to recover the cost of the projected investment; or in other words, PBP is the length of time it takes for a projected investment to reach its breakeven point. The best alternative has the lowest PBP.

The current scenario (BAU)

In the current scenario, approximately 6133 bottles of 1.5 L capacity are brought from Malé on the island every month; that is 73,600 plastic bottles per year.

A range between 75 and 180 g of CO_2 per bottle of 1.5 L emitted in the life cycle can be estimated [47–52]. This is also confirmed by an LCA evaluation carried out by the Polaris Research Center of the University of Milano–Bicocca. The results state that a 1.5 L plastic bottle is responsible for the emission of 108 grams of CO2 equivalent over its life cycle (data not published). Here, we consider for the LCA of the bottles from the extraction of raw materials to waste disposal.

This means 662.36 kg of CO_2 per month and almost 7.95 tons of CO_2 per year. We must add to it the emissions produced during the travel by boat. The 1 ton cargo supply dhoni (typology of the boat is referred to the LCA database software Gabi at this link http://gabi-documentation-2018.gabi-software.com/xml-data/processes/3a819ab9-1979-45b7-a8e9-b4f633d5a662.xml accessed on 1 March 2021) issues in the atmosphere 0.02324 kg CO_2 eq per km. The distance from Malé to Magoodhoo Island is almost 137 km, and it is traveled once a week for water transport. It amounts to 25.47 kg of CO_2 per month and almost 305.65 kg of CO_2 per year.

The average cost of water is 3.6 MVR per liter ($1 = 15.45 MVR) for a total expenditure of $25,721 per year (almost $193 per family per year) (Table 1). This is a cost sensible market change and is expected to grow according to the increasing number of inhabitants. Costs are summarized in Table 7.

Table 7. Business as usual (BAU).

Total plastic water consumption per year: 110,400 L
Annual cost: $25,721
Annual CO_2 consumption: 8.255 tons
Pros: no need of investments for water supply
Cons: Consumption of plastic water bottles and large production of CO_2, low predictability of the rainwater amount (there have been recent phenomena of water shortage, solved by municipality intervention), potential damage to health from nonfiltered water for drinking and cooking.

In Table 7, the current scenario is shown. Advantages (pros) are that no investment costs are required. However, disadvantages (cons) are big in damages for the environment and potential dangers for health from the non-filtered water kept in the tanks for months.

ACTION 1 Desalination and Conventional Fuel.

The objective of ACTION 1 is to provide safe water for the community for drinking and cooking.

In this scenario, a RO desalination plant is installed with a working period of H12, 3.7 kW and a capacity of 417 L/h, for a total of 5000 L/day. The market price is $50,000. In addition, the desalination plant requires an estimated O&M cost of $1000 per year.

A fuel generator (diesel) powers the plant. The total energy required is 16,200 kWh/y. The fuel generator is sold at a market price of $5000.

In order to feed the generator (working H12) we need 4524 L of fuel per year. The local price for fuel is 0.91 $/L which amounts to $4116.84 per year. This is a yearly cost necessary to power the system, but subjected to market conditions. The emission of CO_2/L of diesel is 2.66 kg [21] which amounts to 12,033.84 kg of CO_2 issued in the atmosphere per year.

In ACTION 1, the whole demand of water for drinking and cooking is fulfilled. Rainwater is available for other domestic uses, like washing and for personal use. Flushing is guaranteed by the ground water.

From Table 8, plastic bottles (with a saving of CO_2 emissions) are not used but the fuel causes high pollution. In case of disruption, rainwater can still be used.

Table 8. ACTION 1 Desalination and conventional fuel ($).

Desalination plant capital expenditure: 50,000
Fixed Annual cost (O&M desalination plant): 1000
Fuel generator initial cost: 5000
Fixed Annual cost (fuel alimentation): 4116.84
Annual CO_2 consumption: 12.033 tons
Pros: no use of plastic bottles, stability of production during operating hours, possible use of rainwater for drinking and cooking in case of failure. In case of major need, the system may work until 24 h a day at a higher cost.
Cons: CO_2 emissions are almost 50% more than those in BAU, cost of fuel variable, depending on market conditions.

ACTION 2 Desalination and PV Plant with Battery Storage.

The objective of ACTION 2 is to provide safe water for the community for drinking and cooking.

As in ACTION 1, the market price of the RO desalination plant is $50,000.

The total energy required is 16,200 kWh/y, provided by the PV plant at 10.8 kW with battery packs of a max capacity of 260 Ah.

In ACTION 2, the initial investment cost is high. The market cost is $30,000. The price of batteries is $405 per kWh of installed capacity, which amounts to $1500 [24] with a life expectancy of 10 years. O&M PV costs amount approximately to $100 per year.

In ACTION 2, the whole demand of water for drinking and cooking is fulfilled. Rainwater is available for other domestic uses, like washing and for personal use. Flushing is guaranteed by the ground water.

From Table 9, the advantage in terms of the environment is very clear; however, initial capital expenditure is high.

The benefit of investing in ACTION 1 and ACTION 2 comes from the elimination of the "perpetua" costs in BAU by the purchase of water in plastic, which amounts to $25,721 per year, paid by the whole community.

In ACTION 1, the initial cost of the investment is $55,000 and the yearly cash flow in the following years is $20,604.16. This revenue comes from saving $25,721, obtained by the elimination of plastic bottles, less the fixed cost of fuel to feed the generator ($4116.84) and the annual desalination plant O&M costs ($1000).

In ACTION 2, the initial cost of investment is $80,000 and the future annual benefit is $24,621, given by the "revenue" $25,747, which comes from the elimination of plastic bottles, less the fixed costs of maintenance of the PV plant ($100) and desalination plant ($1000). The cost of the batteries ($1500) is added to the cost of the PV plant.

Table 9. ACTION 2 Desalination and PV plant with battery storage ($).

Desalination plant capital expenditure: 50,000
PV plant capital expenditure: 30,000
Battery cost: 1500
Fixed Annual cost (O&M desalination plant): 1000
Fixed Annual cost (O&M PV plant): 100
Annual CO_2 consumption: 0 kg
Pros: no use of plastic bottles, possible use of rainwater for drinking and cooking in case of failure, totally CO_2 free, energy free of charge and stable due to batteries, total independence from local energy provider.
Cons: higher initial capital expenditure. In case of extreme bad weather, the PV production may decrease but the storage allows a stable production. In case of major need, the system should be equipped with additional PV modules and battery storage at a higher cost.

Cash flows of the three alternatives, the NPV and the PBP on a 10 year time horizon are calculated. The discount rate (7%) is taken from official 2017 data [53]. Cash flows, NPVs and PBPs are shown in Figure 7 and Table 10 respectively.

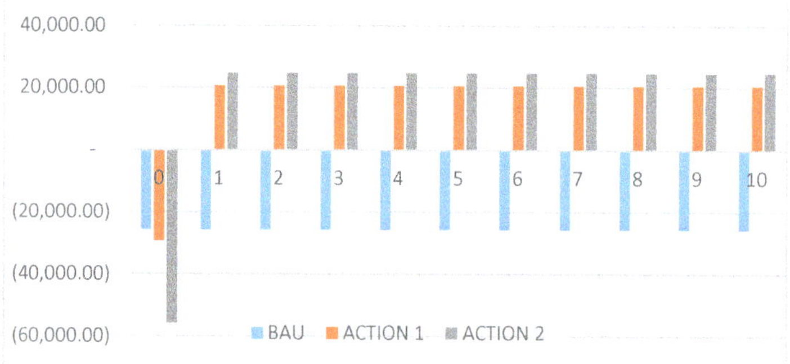

Figure 7. BAU, ACTION 1, ACTION 2; cash flows.

Table 10. Net Present Value (NPV), Payback period (PBP).

	NPV ($)	PBP (Years)
BAU	−206,374	Not applicable
ACTION 1	115,436	1.42
ACTION 2	117,149	2.27

From Table 10, it is obvious that action must be taken. BAU at this stage is very expensive for the whole community, let alone the health risks of drinking rainwater kept in plastic tanks for months.

ACTION 1 and ACTION 2 require initial investments that are compensated by the savings from the use of plastic bottles. In a 10 year time horizon, the NPV for both alternatives is high and the PBP is less than three years for both actions. This means that the breakeven point for both actions is reached in less than three years and the initial high investments are compensated by the returns in the next years for the whole community. ACTION 2 shows the best NPV with a slightly higher PBP.

Finally, we calculate the levelized cost of water (LCOW) for all actions, including BAU. LCW allows the collapse of the entire analysis into a unique indicator, which is useful for making the final choice among all available options. LCOW is the cost per m^3 of water generated during the whole lifetime. LCW is defined as

$$LCOW\left(\frac{\$}{\mathrm{m}^3}\right) = \frac{\sum_{k=0}^{n} \frac{C_k}{(1+i)^k}}{\sum_{k=1}^{n} \frac{W_k}{(1+i)^k}}$$

where $C_{k,}$ $k = 0, \ldots, n$ are the yearly cash flows including the initial capital expenditure and C_0 . $W_{k,}$ $k = 1, \ldots, n$ is the amount of water produced each year. The best alternative shows the lowest LCOW. Using an interest rate of 7% for a lifetime of 10 years, the results in Table 11. Levelized cost of water (LCOW) and levelized emission of water (LEOW) of each alternative for the whole island. For BAU, we have included rainwater and plastic water, for a total of 1249.944 m^3/y. For ACTION 1 and ACTION 2 we have not considered the savings from not purchasing plastic water. For ACTION 1 and ACTION 2, the quantity of water produced has been taken as the maximum capacity of the RO plant (5 $\mathrm{m}^3/\mathrm{day}$ and 1825 m^3/y).

Table 11. Levelized cost of water (LCOW) and levelized emission of water (LEOW) of each alternative for the whole island.

		LCOW ($/m³)	**LEOW (kg/m³)**
BAU	1249.944 m^3/y	22.58	6.6
ACTION 1	1825 m^3/y	7.095	6.59
ACTION 2	1825 m^3/y	6.961	0

We have expanded this concept to the levelized emission of water (LEOW) to redirect the impact on the environment in term of CO_2 per m^3 of water produced in the different alternatives. LEOW is defined as

$$LEOW\left(\frac{\$}{\mathrm{m}^3}\right) = \frac{\sum_{k=0}^{n} \frac{E_k}{(1+i)^k}}{\sum_{k=1}^{n} \frac{W_k}{(1+i)^k}}$$

where E_k , $k = 0, \ldots, n$ co are the yearly CO_2 emissions and W_k , $k = 1, \ldots, n$ is the amount of water produced each year.

Results are shown in Table 11. All emissions refer to the production phase: for BAU, that is the use and disposal of plastic bottles; for ACTION 1, the use of diesel after the plant installation; for ACTION 2, there are no CO_2 emissions from PV production after the plant installation.

Tables 10 and 11 show clearly that ACTION 2 turns out to be optimal, both in terms of financial and environmental evaluation: ACTION 2 shows the lowest LCOW and no CO_2 emissions. On the contrary, ACTION 1 is not sustainable and the impact on the environment does not change practically from BAU.

3. Discussion

A summary of the literature on desalination plants fed by different energy sources is reported in Table 12. For the sake of comparison, our results are included in the table.

As can be seen from Table 12, the LCOW calculated for actions 1 and 2 (Table 11) is in line with small scale RO plants powered by PV or diesel. Economies of scale apply in desalination plants and LCOW varies according to dimension and location. The production of water on islands, especially if remote, is in general more expensive. Apparently, using only LCOW as an indicator to evaluate the whole project, there is no difference between diesel or PV, as in [30].

Table 12. Summary of the literature data; our data are included.

Reference	Desalination Technology	Power Technology	Location	Dimension m³/Day	LCOW (USD/m³)
[26]	Reverse Osmosis	PV	Sinai Peninsula Egypt	100	0.59
[27]	Multi-stage flash	PV	Thira, Greece	1000	0.49
	Reverse Osmosis	Wind			1.97
[29]	Multi-stage flash	PV	The Emirate of Abu Dhabi	90,000	0.8–1.5
	Multi-effect distillation			90,000	0.7–1.2
	Reverse Osmosis			90,000	0.2–0.4
[28]	Reverse Osmosis	PV	Saudi Arabia	6550	1.21–1.42
				190,000	0.85–0.89
[30]	Reverse Osmosis	Motor diesel	United Arab Emirates	20	7.64
		PV			7.34
[33]	Reverse Osmosis	Wave energy converter	USA	2000–5800	1.79–2.2
[34]	Reverse Osmosis	PV	Middle East	1–400	1.5–33.0
		Wind		50–3360	0.7–9.0
[47]	Reverse Osmosis	PV	United Arab Emirates	1344	0.83
Our solution	Reverse Osmosis	BAU	Faafu Atoll The Maldives	5	22.6
		Motor diesel			7.09
		PV +battery			6.96

With both ACTIONs 1 and 2 the yearly cost of water per family lowers: while in the current situation it is $193 (Table 1), with ACTIONs 1 and 2, the cost lowers to $126 (including a markup of 30% for management of the plants by the municipality). The population pays less for more water, as the desalination plant provides 40% more of the current available water.

However, when considering environmental aspects, the choice is driven strongly towards PV.

Finally, it must be noted that the environmental analysis we have performed is in fact underestimating the effect of plastic disposal since microplastic and charred microplastic may have strongly damaging effect on seawater life. Even though Magoodhoo is remote, the population is relatively scarce and touristic annual afflux is practically null, microplastics and especially charred microplastics are found abundantly nearby the island, related to local practices of burning plastic waste at the shoreline [54]. No indicators exist that can quantify the phenomenon.

4. Conclusions

Magoodhoo Island, like many rural atolls in The Maldives, suffers from scarcity of energy and water and has a strong dependence on imports from the mainland.

While rainfall is obviously carbon free, the use of plastic bottles has a negative impact on the environment and high costs for the local population.

In order to provide a proper amount of safe drinkable water to the population, installation of a desalination plant is planned, fed by diesel or by a PV plant equipped with battery storage.

Results are in favor of the combination RO + PV plant (ACTION 2), both from a financial and environmental point of view. In less than three years (PBP indicator), the initial investment can be offset by the savings in the use of water imported from mainland. RO + PV is also environmentally friendly (LEOW indicator).

This is one of the first projects in The Maldives for a sustainable water supply system and we strongly believe that if all islands undertake and strive to change the way they manage water and energy significantly, in the next years The Maldives will appreciate a substantial reduction of CO_2 emissions, while guaranteeing good water supply.

The local population living conditions must improve, but not at the expense of a fragile environment. Therefore, any choice involving substantial investments must be accompanied by a financial and environmental analysis. Our project considers both aspects.

Author Contributions: Conceptualization, S.S. and M.F.A.; methodology, S.S., S.C. and M.F.A.; software, M.A. and G.M.; validation, P.G. and S.C.; resources, S.S. and. S.C.; data curation, G.M.; writing—original draft preparation, S.S. and S.C.; writing—review and editing M.F.A., S.S. and M.F.A.; supervision, P.G. All authors have read and agreed to the published version of the manuscript.

Funding: This research received no external funding.

Institutional Review Board Statement: Not applicable.

Informed Consent Statement: Not applicable.

Data Availability Statement: Not applicable.

Acknowledgments: Research developed within the MaRHE Center and B.A.S.E. Center of University Milano Bicocca. The authors are grateful to reviewers for their constructive suggestions.

Conflicts of Interest: The authors declare no conflict of interest.

References

1. Styles, D.; Schonberger, H.; Galvez Martos, J.L. *Best Environmental Management Practice in the Tourism Sector*; Publications Office of the European Union: Luxembourg, 2013. [CrossRef]
2. Asian Development Bank. *A Brighter Future for Maldives Pwered by Renewables Road Mapfor the Energysector 2020–2030*; Asian Development Bank: Mandaluyong, Philippines, 2020; ISBN 978-92-9262-514-6.
3. UN. Available online: https://www.un.org/ohrlls/content/about-small-island-developing-states (accessed on 1 March 2021).
4. IEA. *Global Energy Review*; IEA: Paris, France, 2020. Available online: https://www.iea.org/reports/global-energy-review-2020 (accessed on 1 March 2021).
5. Michalenaa, E.; Hills, J.M. Paths of renewable energy development in small island developing states of the South Pacific. *Renew. Sustain. Energy Rev.* **2018**, *82*, 343–352. [CrossRef]
6. Blechinger, P.; Cader, C.; Bertheau, P.; Huyskens, H.; Seguin, R.; Breyer, C. Global analysis of the techno-economic potential of renewable energy hybrid systems on small islands. *Energy Policy* **2016**, *98*, 674–687. [CrossRef]
7. IEA. *Global CO2 Emissions by Sector*; IEA: Paris, France, 2020. Available online: https://www.iea.org/data-and-statistics/charts/global-co2-emissions-by-sector-2018 (accessed on 1 March 2021).
8. Notaro, V.; Puleo, V.; Fontanazza, C.M.; Sambito, M.; La Loggia, G. A decision support tool for water and energy saving in the integrated water system. *Procedia Eng.* **2015**, *119*, 1109–1118. [CrossRef]
9. Liu, J.; Mei, C.; Wang, H.; Shao, W.; Xiang, C. Powering an island system by renewable energy—A feasibility analysis in the Maldives. *Appl. Energy* **2018**, *227*, 18–27. [CrossRef]
10. Meschede, H.; Holzapfel, P.; Kadelbach, F.; Hesselbach, J. Classification of global island regarding the opportunity of using RES. *Appl. Energy* **2016**, *175*, 251–258. [CrossRef]
11. Kuang, Y.; Zhang, Y.; Zhou, B.; Li, C.; Cao, Y.; Li, L.; Zeng, L. A review of renewable energy utilization in islands. *Renew. Sustain. Energy Rev.* **2016**, *59*, 504–513. [CrossRef]
12. Mendoza-Vizcaino, J.; Raza, M.; Samper, A.; Diaz Gonzales, F.; Galceran-Arellano, S. Integral approach to energy planning and electric grid assessment in a renewable energy technology integration for a 50/50 target applied to a small island. *Appl. Energy* **2019**, *233–234*, 524–543. [CrossRef]
13. Meschede, H.; Esparcia, E.A., Jr.; Holapfel, P.; Berthaud, P.; Ang, R.C.; Blanco, A.C.; Ocon, J.D. On the transferability of smart energy systems on off-grid islands using cluster analysis—A case study for the Philippine archipelago. *Appl. Energy* **2019**, *251*, 113290–113296. [CrossRef]
14. Segurado, R.; Krajacic, G.; Duic, N.; Alves, L. Increasing the penetration of renewable energy resources in S. Vicente, Cape Verde. *Appl. Energy* **2011**, *2*, 466–472. [CrossRef]
15. Berthaud, P.; Cader, C. Electricity sector planning for the Philippine islands: Considering centralized and decentralized supply options. *Appl. Energy* **2019**, *251*, 113393–113398. [CrossRef]
16. Surroop, D.; Raghoo, P. Energy landscape in Mauritius. *Renew. Sustain. Energy Rev.* **2017**, *73*, 688–694. [CrossRef]
17. Shea, R.P.; Ramgolam, Y.K. Applied levelized cost of electricity for energy technologies in a small island developing state: A case study in Mauritius. *Renew. Energy* **2019**, *132*, 1415–1424. [CrossRef]
18. Aguirre-Mendoza, A.M.; Diaz-Mendoza, C.; Pasqualino, J. Renewable energy potential analysis in non interconnected islands. Case study: Isla Grande, Corales del Rosario Archipelago, Colombia. *Ecol. Eng.* **2017**. [CrossRef]
19. Handayani, K.; Krozer, Y.; Filatova, T. From fossil fuels to renewables: An analysis of long-term scenarios considering technological learning. *Energy Policy* **2019**, *127*, 134–146. [CrossRef]

20. Dornan, M.; Jotzo, F. Renewable technologies and risk mitigation in small island developing states: Fiji's electricitysector. *Renew. Sustain. Energy Rev.* **2015**, *48*, 35–48. [CrossRef]
21. Soomauroo, Z.; Blechinger, P.; Creutzig, F. Unique Opportunities of Island States to Transition to a Low-Carbon Mobility System. *Sustainability* **2020**, *12*, 1435. [CrossRef]
22. Masrur, H.; Howlader, H.O.R.; Elsayed Lofty, M.; Khan, K.R.; Guerrero, J.M.; Seniyu, T. Analysis of Techno-Economic-Environmental Suitability of an Isolated Microgrid System Located in a remote island of Bangladesh. *Sustainability* **2020**, *12*, 7. [CrossRef]
23. Eras-Almeida, A.A.; Egido Aguilera, M.A.; Blechinger, P.; Berendes, S.; Caamaño, E.; García-Alcalde, F. Decarbonizing the Galapagos Islands: Techno-Economic Perspectives for the Hybrid Renewable Mini-Grid Baltra-Santa Cruz. *Sustainability* **2020**, *12*, 2282. [CrossRef]
24. Ali, I.; Shafiullah, G.M.; Urmee, T. A preliminary feasibility of roof-mounted solar PV systems in the Maldives. *Renew. Sustain. Energy Rev.* **2018**, *83*, 18–32. [CrossRef]
25. Wijayatunga, P.; George, L.; Lopez, A.; Aguado, J.A. Integrating Clean Energy in Small Island Power Systems: Maldives Experience. *Energy Procedia* **2016**, *103*, 274–279. [CrossRef]
26. Ministry of Environment. Island Electricity Data Book. 2017. Available online: https://www.environment.gov.mv/v2/en/download/6796 (accessed on 1 March 2021).
27. Ibrahim, S.A.; Bari, M.R.; Miles, L. Water Resources Management in Maldives with an Emphasis on Desalination. Available online: http://citeseerx.ist.psu.edu/viewdoc/download?doi=10.1.1.113.913&rep=rep1&type=pdf (accessed on 1 March 2021).
28. Alanis-Noyola, A.E.; Prasanna, A.; Rannu, T.; Roja Solorzano, L. Pre-Feasibility Analysis of a Desalination Plant Powered by Renewable Energy in Thira, Greece. In Proceedings of the International Conference on Renewable Energies and Power Quality (ICREPQ'12), Santiago de Compostela, Spain, 28–30 March 2012. [CrossRef]
29. Fthenakis, V. *New Prospects for PV Powered Water Desalination Plants: Case Studies in Saudi Arabia*; Brookhaven National Laboratory: Upton, NY, USA, 2015.
30. Kaya, A.; Tok, M.E.; Koc, M. A Levelized Cost Analysis for Solar-Energy-Powered Sea Water Desalination in The Emirate of Abu Dhabi. *Sustainability* **2019**, *11*, 1691. [CrossRef]
31. Helal, A.M.; Al-Malek, S.A.; Al-Katheeri, E.S. Economic feasibility of alternative designs of a PV-RO desalination unit for remote areas in the United Arab Emirates. *Desalination* **2008**, *221*, 1–16. [CrossRef]
32. Caldera, U.; Breyer, C. Strengthening the global water supply through a decarbonised global desalination sector and improved irrigation systems. *Energy* **2020**, *200*, 17507–117554. [CrossRef]
33. IRENA. Renewable Desalination: Technology Options for Islands. 2015. Available online: https://www.irena.org/publications/2015/Dec/Renewable-Desalination-Technology-Options-for-Islands (accessed on 1 March 2021).
34. Yu, Y.; Dale, J. Analysis of a Wave-Powered, Reverse-Osmosis System and Its Economic Availability in the United States. In Proceedings of the 36th International Conference on Ocean, Offshore and Arctic Engineering OMAE, Trondheim, Norway, 25–30 June 2017.
35. Ahmadi, E.; McLellan, B.; Ogata, S.; Ivatloo, B.M.; Tezuka, T. An Integrated Planning Framework for Sustainable Water and Energy Supply. *Sustainability* **2020**, *12*, 4295. [CrossRef]
36. Jaleel, M.I.; Shaheeda, A.I.; Afsal, H.; Mustafa, M.; Pathirana, A. A Screening Approach for Assessing Groundwater Quality for Consumption in Small Islands: Case Study of 45 Inhabited Islands in the Maldives. *Water* **2020**, *12*, 2209. [CrossRef]
37. Sambito, M.; Freni, G. LCA Methodology for the Quantification of the Carbon Footprint of the Integrated Urban Water System. *Water* **2017**, *9*, 395. [CrossRef]
38. Jeong, H.; Minne, E.; Crittenden, J.C. Life cycle assessment of the City of Atlanta, Georgia's centralized water system. *Int. J. Life Cycle Assess.* **2015**, *20*, 880–891. [CrossRef]
39. Dettore, C. Comparative Life-Cycle Assessment of Bottled vs. Tap Water System. Master's Thesis, University of Michigan, Ann Arbor, MI, USA, 2009.
40. World Health Organization. Available online: https://www.who.int/water_sanitation_health/emergencies/qa/emergencies_qa5/en/ (accessed on 1 March 2021).
41. Alkaisi, A.; Mossad, R.; Sharifian-Barforoush, A. A review of the water desalination systems integrated with renewable energy. *Energy Procedia* **2017**, *110*, 268–274. [CrossRef]
42. Monnot, M.; Martínez Carvajal, G.D.; Laborie, S.; Cabassud, C.; Lebrun, R. Integrated approach in eco-design strategy for small RO desalination plants powered by photovoltaic. *Desalination* **2017**, *435*, 246–258. [CrossRef]
43. Nagaraj, R.; Thirugnanamurthy, D.; Murthy Rajput, M.; Panigrahi, B.K. Techno-Economic Analysis of Hybrid Power System siZing Applied to Small Desalination Plants for Sustainable Operation. *Int. J. Sustain. Built Environ.* **2016**, *5*, 269–276. [CrossRef]
44. Atallah, M.O.; Farahat, M.A.; Lotfy, M.E.; Senjyu, T. Operation of conventional and unconventional energy sources to drive a reverse osmosis desalination plant in Sinai Peninsula, Egypt. *Renew. Energy* **2020**, *145*, 141–152. [CrossRef]
45. *RETScreen International (2001–2004), Clean Energy Project Analysis: RETScreen Engineering and Cases Textbook*; Minister of Natural Resource: Verennes, QC, Canada, 2005. Available online: http://msessd.ioe.edu.np/wp-content/uploads/2017/04/Textbook-clean-energy-project-analysis.pdf (accessed on 1 March 2021).
46. Alsheghri, A.; Sharief, S.A.; Rabbani, S.; Aitzhan, N.Z. Design and Cost Analysis of a Solar Photovoltaic Powered Reverse Osmosis Plant for Masdar Institute. *Energy Procedia* **2015**, *75*, 319–324. [CrossRef]

47. Lagioia, G.; Calabró, G.; Amicarelli, V. Empirical study of the environmental management of Italy's drinking water supply. *Resour. Conserv. Recycl.* **2012**, *60*, 119–130. [CrossRef]
48. Nessi, S.; Rigamonti, L.; Grosso, M. LCA of waste prevention activities: A case study for drinking water in Italy. *J. Environ. Manag.* **2012**, *108*, 73–83. [CrossRef]
49. Perugini, F.; Mastellone, M.L.; Arena, U. A life cycle assessment of mechanical and feedstock recycling options for management of plastic packaging wastes. *Environ. Prog.* **2005**, *24*, 137–154. [CrossRef]
50. Bez, J.; Heyde, M.; Goldhan, G. Waste treatment in product specific life cycle inventories. *Int. J. LCA* **1998**, *3*, 100–105. [CrossRef]
51. Shen, L.; Worrell, E.; Patel, M.K. Open-loop recycling: A LCA case study of PET bottle-to-fibre recycling. *Resour. Conserv. Recycl.* **2010**, *55*, 34–52. [CrossRef]
52. Papong, S.; Malakul, P.; Trungkavashirakun, R.; Pechda, W.; Chom-in, T.; Nithitanakul, M.; Sarobol, E. Comparative assessment of the environmental profile of PLA and PET drinking water bottles from a life cycle perspective. *J. Clean. Prod.* **2014**, *65*, 539–550. [CrossRef]
53. Maldives Central Bank Discount Rate 31 December 2017, Index Mundi. Available online: https://www.indexmundi.com/maldives/central_bank_discount_rate.html (accessed on 1 March 2021).
54. Saliua, F.; Montano, S.; Garavaglia, M.G.; Lasagnia, M.; Seveso, D.; Galli, P. Microplastic and charred microplastic in the Faafu Atoll, Maldives. *Mar. Pollut. Bull.* **2018**, *136*, 464–471. [CrossRef]

Article

Effects of Land Use Change from Natural Forest to Livestock on Soil C, N and P Dynamics along a Rainfall Gradient in Mexico

Daniela Figueroa [1,2], Patricia Ortega-Fernández [1,3], Thalita F. Abbruzzini [1,4], Anaitzi Rivero-Villlar [1], Francisco Galindo [5], Bruno Chavez-Vergara [4], Jorge D. Etchevers [6] and Julio Campo [1,*]

1 Instituto de Ecología, Universidad Nacional Autónoma de México, Mexico City 04510, Mexico; daniielafigueroa@gmail.com (D.F.); orteg.patricia92@gmail.com (P.O.F.); thalita@geologia.unam.mx (T.F.A.); anaitzirv@iecologia.unam.mx (A.R.V.)
2 Instituto de Geografía, Universidad Nacional Autónoma de México, Mexico City 04510, Mexico
3 Centro de Investigación en Ciencias de Información Geoespacial, Consejo Nacional de Ciencia y Tecnología, Mexico City 14240, Mexico
4 Instituto de Geología, Universidad Nacional Autónoma de México, Mexico City 04510, Mexico; bruno.chavez@gmail.com
5 Facultad de Medicina Veterinaria y Zootecnia, Universidad Nacional Autónoma de México, Mexico City 04510, Mexico; fgalindomaldonado@gmail.com
6 Colegio de Posgraduados, Campus Montecillos, Mexico State 56230, Mexico; jetchev@colpos.mx
* Correspondence: jcampo@ecologia.unam.mx

Received: 9 September 2020; Accepted: 9 October 2020; Published: 19 October 2020

Abstract: The effects of converting native forests to livestock systems on soil C, N and P contents across various climatic zones are not well understood for the tropical region. The goal of this study was to test how soil C, N and P dynamics are affected by the land-use change from natural forests to livestock production systems (extensive pasture and intensive silvopastoral systems) across a rainfall gradient of 1611–711 mm per year in the Mexican tropics. A total of 15 soil-based biogeochemical metrics were measured in samples collected during the dry and rainy seasons in livestock systems and mature forests for land-use and intersite comparisons of the nutrient status. Our results show that land-use change from natural forests to livestock production systems had a negative effect on soil C, N and P contents. In general, soil basal respiration and C-acquiring enzyme activities increased under livestock production systems. Additionally, reduction in mean annual rainfall affected moisture-sensitive biogeochemical processes affecting the C, N and P dynamics. Our findings imply that land-use changes alter soil C, N and P dynamics and contents, with potential negative consequences for the sustainability of livestock production systems in the tropical regions of Mexico investigated.

Keywords: climate change; drought; rainfall regime; soil biogeochemistry

---

1. Introduction

Globally, soils store at least three times as much carbon (C) as is found in either the atmosphere or living plants [1–3], with the largest portion of it found in tropical forests [4,5]. A global examination of the terrestrial C stocks highlighted that the largest soil organic C stocks in tropical lands are subject to the greatest risks [2], and land-use change continues to pose a threat to tropical forests [6–8]. Moreover, earth-system models have shown that these C stocks in soils will become increasingly vulnerable during the twenty-first century, since reduced rainfall and large-scale agricultural transformation of forests, primarily to pastureland [9,10], are predicted for large areas of the tropics [11]. Tropical ecosystems dominate the exchange of CO_2 between the atmosphere and terrestrial biosphere, yet our understanding of how nutrients control the tropical C dynamics remains far from complete. A better understanding of nitrogen (N) and phosphorus (P) balances in soils can help guide the implementation of mitigation policies and land-use management, since the supplies of one or both nutrients constrain C uptake in the tropics [12]. In contrast to the clear inventory-based assessments of aboveground C on global scales [4], C cycling in soils remains less well understood, due to high biogeochemical variation in the tropics [13].

Environmental factors such as rainfall decline, together with agricultural expansion, generate abrupt, large-scale changes in forestlands and alter soil C, N and P dynamics [14–16]. The impacts of converting natural forests to systems for livestock production on soil C, N and P dynamics have been well examined at the stand scales, with studies relating variations in above- and below-ground C input through litterfall and root exudation to these inherently different management practices [17]. However, the influence of wider regional or global-scale conversion of natural forests to livestock on soil C, N and P dynamics is not yet well understood. This is perceived as a key bottleneck in improving the prediction and evaluation of the results of soil C mitigation efforts related to land-use change; it is also an impediment to better understanding the degradation of soil quality. Previous studies have suggested that converting native forests to livestock systems significantly impacts the quantity and quality of C and nutrient inputs [18,19], and that these changes could be sensitive to rainfall regime. Converting annual cropland to perennial pastureland enhanced microbial biomass and enzyme activities involved in C, N and P cycling due to lower fluctuations in soil water content in pastureland [20]. Campo et al. [21] reported that changes in the land cover can increase surface soil C pools in the dry tropics, while converting native forests to pasturelands in the humid tropics significantly decreases C contents in soils. The changes in land use and management may decrease soil organic matter content and significantly alter soil water dynamics, which would have effects on soil organic carbon [22].

Mexico, with a land area of 1.96 million km^2, covers climatic gradients, from humid to dry climate zones, and has experienced diverse and intensive land-use change for animal production, mainly in the tropical regions of the country [23], which are responsible for 50% of national livestock production [24]. Today, beef production in the tropics of Mexico is undergoing major changes with the introduction of management practices for long-term sustainable intensification (intensive silvopastoral systems) through more efficient use of water, materials, and energy [25,26]. Intensive silvopastoral systems are an emerging response to the increasing consumption of animal protein in the country [27] and an alternative to extensive pasture systems, which clear the natural cover to open space for cattle grazing [28,29], reducing the biodiversity and ecological services [30–34], and this intensification of livestock production systems could become a key climate change mitigation technology [35]. For example, intensive silvopastoral systems can produce 12 times more meat than extensive pasture systems [36], and their annual methane emissions per tonne of meat produced are 1.8 times lower than those of extensive pastures systems [37]. With its variety of tropical climates and livestock production systems, Mexico can be viewed as a unique laboratory containing complex interactions between climatic zones and human activities, and thus providing an excellent opportunity to examine simultaneous climate and livestock production impacts on soil fertility, particularly on soil C sequestration and background nutrient status.

To better understand the complex interactions among rainfall, livestock strategies, soil, and C, N and P dynamics, we analyzed C and nutrient dynamics across a gradient of sites varying in rainfall amount (from 1661 to 711 mm $year^{-1}$), using a robust experimental design. The gradient contains 12 sites, including mature tropical forests (as natural forests), extensive pasture and intensive silvopastoral systems located in humid, subhumid and semiarid climates in the southeast of Mexico. In particular, we studied how C, N and P dynamics in the topsoil layer (upper 10 cm of soil) differed between adjacent native forests and livestock systems, and if those differences were related to rainfall amount or livestock production system (i.e., pastures and silvopastoral). We examined soil bulk organic

C, and total N and P concentrations. We also measured labile concentrations of soil organic C, total N and total P, since their contents cycle more rapidly and should show a greater proportional response to changes in chemistry due to land use, compared to the response of bulk content. For the study, we also include measures of soil basal respiration, net N transformations, and C, N and P enzyme activity, as indicators of soil C and nutrient biogeochemistry.

2. Material and Methods

2.1. Study Areas

The study was carried out at four sites with different mean annual rainfall (MAR) in Mexico; two located in a dry tropical region, where MAR decreases from 917 to 711 mm, and two in a humid tropical region, where MAR decreases from 1661 to 1232 mm (Table 1). At each location, we evaluated two types of livestock system: extensive pastures systems and intensive silvopastoral systems. In addition, at each location, a natural forest site (tropical forest in all cases) was used as a reference for original soil fertility conditions. The criteria used for the selection of sites were: (i) sites that shared the same mean annual temperature, but located across a gradient of MAR; (ii) the presence of natural forests, pastures and silvopastures within the same edaphic conditions within each site; (iii) extensive pastures with around 25 years of land use; (iv) silvopastures with around 5 years of land use after around 20 years of extensive pasture systems; (v) both livestock production systems located at a distance of no more than 5 km from the reference (i.e., natural forests). To ensure that none of the sites had been subjected to other human activity, sites were selected following consultations with local owners and a review of the regional government's land-use database.

Long-term climate data from weather stations show that all sites are characterized by a distinct period of low precipitation (four to seven months with rainfall below 100 mm; Table 1). The four sites differ strongly in aridity index (i.e., mean annual rainfall divided by mean annual evapotranspiration). Across the sites, variation in mean annual temperature is less than 1.5 °C, and the climate, semiarid to subhumid, would support from tropical dry to humid forests in the Holdridge Life Zone System [38]. Soils (Inceptisols at the wettest end of the rainfall gradient, and Entisols at all three drier sites) have bulk density and clay content that decrease across sites with decreased rainfall amount, and pH that increases from the wettest site to the driest site of the gradient.

Table 1. Location and characteristics of four study sites across the southeast of Mexico.

Location	**20°51′ N, 89°37′ W**	**21°08′ N, 88°09′ W**	**19°24′ N, 96°22′ W**	**20°01′ N, 97°06′ W**
Altitude (m)	10	20	14	121
Mean annual temperature (°C) [2]	25.7	25.3	25.4	24.4
Mean annual rainfall (mm year^{-1}) [2]	711	917	1232	1661
Potential evapotranspiration (mm year^{-1}) [1]	1719	1677	1561	1215
Aridity index [2]	0.41 (semiarid)	0.55 (dry subhumid)	0.79 (humid)	1.37 (humid)
Water stress months	October–May	November–May	October–May	January–April
Mean month rainfall in water stress months				
(mm month^{-1}) [3]	27	53	29	80
Soil bulk density (g cm^{-3})	0.8	0.8	0.9	1.1
Soil clay content (%)	12.1	12.5	15.4	15.5
Soil pH (H_2O)	8.0	7.6	7.4	5.1

[1] Long-term climatic data (1975–2019 period; *Comisión Nacional de Aguas* personal communication). [2] Aridity Index [39,40]. [3] Water stress months are the number of months per year with rainfall < 100 mm.

Three mature forest stands at each site were selected. The predominant vegetation at the three drier sites is the seasonally dry tropical forest (i.e., forests subjected to prolonged dry season; [41]), while humid tropical forests dominate at the wettest end of the rainfall gradient. Floristically, *Leguminosae* is the most important family across studied sites [42–44], and the abundance of *Leguminosae* trees increases with the decrease in MAR.

The average size of extensive pasture system generally increases with the rainfall amount (87.3 ha in the 1232 mm site, 100 ha in the 711 mm site, 123 ha in the 917 mm and 170 ha in the 1661 mm of MAR site). The extensive pasture system at all sites is dual-purpose, for dairy and meat productions. The average age of pastures of this type was 25 years at the four study sites. At the drier sites (711 and 917 mm of MAR), in pasture systems, one head of cattle per hectare graze for ~2 days and rotate every ~35 days, while an average of 1.3 animals per hectare graze for ~7 days and rotate every ~45 days at the wetter locations (1232 and 1661 mm of MAR). Livestock are usually supplemented with poultry litter and mineral salts at both drier sites, and the pastures do not receive applications of inorganic fertilizers. The most common species used as forage are *Penniseptum purpureum*—Heinrich C.F. Schumacher, *Pannicum maximum*—Jacq., *Cynodon nlemfuensis*—McVaugh, and *Brachiaria brizantha*—Hochst. ex A. Rich. Stapf [45]. Pastures do not receive fertilization at the wetter sites of the rainfall gradient, and the species used as forage are *Brachiaria brizantha*—Hochst. ex A. Rich. Stapf, *Paspalum vaginatum* Swartz, *Cynodon plectostachyus* (K. Schum.) Pilg., *Digitaria decumbens* Stent., *Pennisetum clandestinum* Hochst. ex Chiov, and *Panicum maximum* Jacq. [45]. The establishment of pastures at all study sites was carried out via traditional slash-and-burning of natural forests, followed by ploughing.

The implementation of silvopastoral systems along the rainfall gradient is relatively recent, since the age of the plots range from 4 to 6 years. The establishment of the silvopastoral systems was carried out at sites previously dedicated to extensive pastures for around 20 years, followed by ploughing and the planting of a monoculture of native trees (*Leucaena leucocephala* (Lam.) de Wit.) in high density (~10,000 per hectare). The average animal load in silvopastoral systems is 3 animals per hectare at both drier sites, which graze for ~1 day, while a larger number of livestock (4 to 6 animals per hectare) graze for ~4 days in the wetter sites; cattle are rotated every ~30 days at all four study sites.

2.2. Soil Sampling and Analysis

Sampling of soils was carried out in the dry (April) and rainy (October) seasons of 2017. One plot (10 × 50 m) was established at each forest stand, pasture and silvopastoral system, and five equidistant (10 m) samples were taken for each type of land use. The topography of all selected plots was even to minimize the effects of local terrain on soil fertility [46]. Samples were taken from the topsoil (0–10 cm in depth), combined into one composite sample per plot and stored at 4 °C prior to analysis. A total of three composite samples (from three independent landowners) per site for each type of land use were taken. The upper 10 cm of the soil profile concentrates organic C in tropical dry and tropical humid forests of Mexico [21]. Soil samples were air-dried and sieved (2 mm mesh), and gravimetric water content in fresh soil samples was determined prior to analysis to correct the soil weight used in each determination. Although water content in fresh soils did not vary among land uses or across sites, samples taken in the dry season had less moisture than those taken in the rainy season (23.7 ± 2.1 and 43.9 ± 4.1%, respectively; mean ± 1 standard error).

Soil texture [47] and pH (in water) were determined prior to C, N and P analyses. Total and inorganic C (carbonates) concentrations were analyzed in an automated C analyzer (SCHIMADZU 5005A), by grinding a 5-g air-dried subsample (100-mesh screen). Organic C was estimated from the difference between total and inorganic C concentrations. The concentration of total N and total P was determined by acid digestion [48] using an NP analyzer (Technicon Autoanalyzer III). Soil available P was extracted using the Bray and Kurtz [49] method for acidic soil (pH ≤ 7), and the Olsen method was used for alkaline soils (pH > 7). After extraction, available soil P concentrations were determined with the colorimetrical method [48].

Carbon and N concentrations in soil microbial biomass were determined by chloroform fumigation–extraction methods [50] using replicated samples of fresh soil. Fumigated and non-fumigated samples were incubated at 24 °C for 24 h. Microbial biomass C was extracted with 0.5 MK_2SO_4, filtered (Wathman No. 42 paper), and the concentration of C was measured using an automated C analyzer. Microbial C was estimated from the difference between C concentrations in the fumigated and the non-fumigated extracts, and a conversion factor k_C equal to 0.45 [50] was used. Microbial biomass N was extracted in a similar way, after filtering through Wathman No. 1 paper, the filtrate was digested in acid and the total concentration of N was determined using an automated NP analyzer. Microbial N concentration was calculated in a similar way to microbial C, using a conversion factor k_N of 0.57 [51].

Soil basal respiration was determined from duplicated fresh subsamples (20 g) incubated at 25 °C. Soil subsamples were moistened to 50% water-filled pore space following light tamping in a jar (700 mL) containing vials of water to maintain humidity and 10 mL of 1.0 M NaOH to absorb CO_2. Alkali traps were replaced at 1, 2, 3, 5, 7, 14, 21 and 28 days and were removed at 35 days. Evolved CO_2 was determined by titration of alkali with 0.5 M HCl [52]. Soil basal respiration was calculated using measurements from days 3 to 35 to avoid the majority of flush due to drying and rewetting. The CO_2 flux produced was standardized per gram of soil organic C.

We measured mineral N concentrations (NO_3 plus NH_4) and net N mineralization and nitrification rates using 2 M KCl extraction and aerobic incubation methods. Mineral N concentrations were measured by extracting a 15-g sub-sample in 100 mL 2 M KCl [53]. The soil KCl solution was shaken for 1 h and allowed to settle overnight. A 20 mL aliquot supernatant was transferred to vials and frozen for analysis (initial mineral N concentration). Nitrogen mineralization and nitrification rates were measured during 15-day aerobic incubations [53]. A second sub-sample was wetted to field water holding capacity with distilled water, maintained at field capacity moisture and incubated at 25 °C for 15 days before extraction, using KCl (final mineral N concentration). Analysis of both initial and final mineral N concentrations was done on an Autoanalyzer system using procedures to determine NO_3–N plus NO_2–N, which were reported as NO_3–N, and using the salycilate–hypochlorite procedure for NH_4–N. Nitrogen mineralization rate was determined from the difference between mineral N concentrations at the start and end of the incubation, and results were expressed on a basis of mean daily mineral N production. Likewise, nitrification rate was determined from the difference in NO_3–N concentration at the beginning and end of the incubation, and results were expressed in similar units.

We measured the indicator enzymes most commonly used to infer growth-limiting C sources and nutrients: ß-1, 4-glucosidase (BG; enzyme commission number: EC 3.2.1.21) and polyphenol oxidase (POX; EC 1.10.3.1) to infer C-acquiring enzymes, ß-1, 4-N-acetylglucosaminidase (NAG; EC 3.2.1.14) to infer N acquiring enzymes, and acid phosphatase (AP; EC 3.1.3.1) to infer P acquiring enzymes [54], following the method proposed by Jackson et al. [55]. The enzyme assays were incubated at 25 °C for 2 h and their absorbance was recorded using a microplate reader at 410 nm for ß-1, 4-glucosidase, ß-1, 4-N-acetylglucosaminidase and acid phosphatase activities, and at 460 nm for polyphenol oxidase activity. The concentration of pNP (or tyrosine, for polyphenol oxidase) detected in the soil sample was corrected by subtracting the sum of absorption from the sample and substrate control wells, and enzyme activities were calculated as follows:

$$EA = (\text{final absorbance})/(C \times \text{incubation time} \times \text{soil dry mass})$$

where EA is the enzyme activity expressed in µmol of pNP (or tyrosine, for polyphenol oxidase) released per gram of soil per hour (µmol g^{-1} h^{-1}), and C is the conversion factor that relates absorbance to µmol of pNP (or tyrosine, for polyphenol oxidase) for each enzyme activity.

2.3. Statistical Analyses

For each metric, a one-way ANOVA was performed, testing the effects of the rainfall regimen and the effect of the land-use change. The ANOVA residuals were explored for normality using the

Shapiro–Wilk's test. Data were transformed logarithmically when the assumptions of normality did not occur; the following soil metrics violated the normality assumptions: microbial biomass C and N concentrations, net N transformation rates and the activity of the ß-1, 4-glucosidase. The honest significant difference (HSD) test was used when statistical differences ($p < 0.05$) were observed across sites, or among land uses in each site. The interactions between soil metrics, land uses (natural forests, pastures and silvopastoral systems) and rainfall regime (MAR) were explored using a principal component analysis (PCA). In addition, correlation matrices were used as a way to depict the relationships between soil metrics within each site in both sampling seasons (i.e., dry and rainy seasons). All statistical analyses were performed using R statistical software [56].

3. Results

3.1. Carbon, Nitrogen and Phosphorus Concentrations

Soil organic C, total N, total P and, microbial biomass C and N concentrations were the highest at the driest end of the rainfall gradient and decrease with rainfall increase, meanwhile NH_4 concentration was higher in the wettest end of the rainfall gradient (Figures 1–3, Table 2). Rainfall regime did not have a significant effect on soil NO_3 and available P concentrations (Figure 2). Rainfall seasonality did not affect the organic C, total N, total P, and microbial biomass C and N concentrations in soils. However, NH_4 and NO_3 concentrations were higher in the rainy season than in the dry season. In contrast to the seasonal variation in soil inorganic N concentrations, available P concentration was consistently higher in soils taken in the dry season than those taken in the rainy season.

Land-use change from natural forests to both livestock production systems (i.e., extensive pasture and intensive silvopastoral systems) decreased soil organic C, total N and P available concentrations (Figures 1 and 2, Table 2). In contrast, no significant changes in NO_3, NH_4, total P and microbial biomass C and N concentrations were detected with land-use change (Figures 1–3). Finally, no significant changes in soil C, N and P concentrations were detected between livestock production systems.

Table 2. Site, season and land-use effects on soil carbon, nitrogen and phosphorus metrics along a rainfall gradient in Southeast Mexico.

Metric	Source of Variation		
	Site	**Season**	**Land Use**
		F	
Organic C (OC)	34.6 ***	0.595 NS	8.45 **
Total N (TN)	50.5 ***	1.33 NS	6.47 **
NH_4	23.0 ***	13.9 ***	0.626 NS
NO_3	4.36 NS	59.8 ***	1.55 NS
Total P (TP)	43.0 ***	0.676 NS	2.61 NS
Available P (AP)	0.565 NS	29.9 ***	6.48 **
Microbial biomass C (MBC)	32.0 ***	1.14 NS	3.44 NS
Microbial biomass N (MBN)	16.4 ***	0.860 NS	0.042 NS
Soil basal respiration (SBR)	2.95 *	0.124 NS	6.76 **
Net N mineralization (MIN)	17.4 ***	4.20 *	1.13 NS
Net nitrification (NIT)	32.9 ***	4.15 *	1.05 NS
ß-1, 4-glucosidase (BG)	8.06 *	8.94 **	3.36 NS
Polyphenol oxidase (POX)	37.0 ***	2.85 NS	2.21 NS
ß-1, 4-N-acetylglucosaminidase (BNA)	10.1 **	0.851 NS	4.69 *
Acid phosphatase (PHO)	70.1 ***	0.305 NS	6.84 *

Significance main effect: NS, $p > 0.05$; *, $p < 0.05$; **, $p < 0.01$; ***, $p < 0.001$.

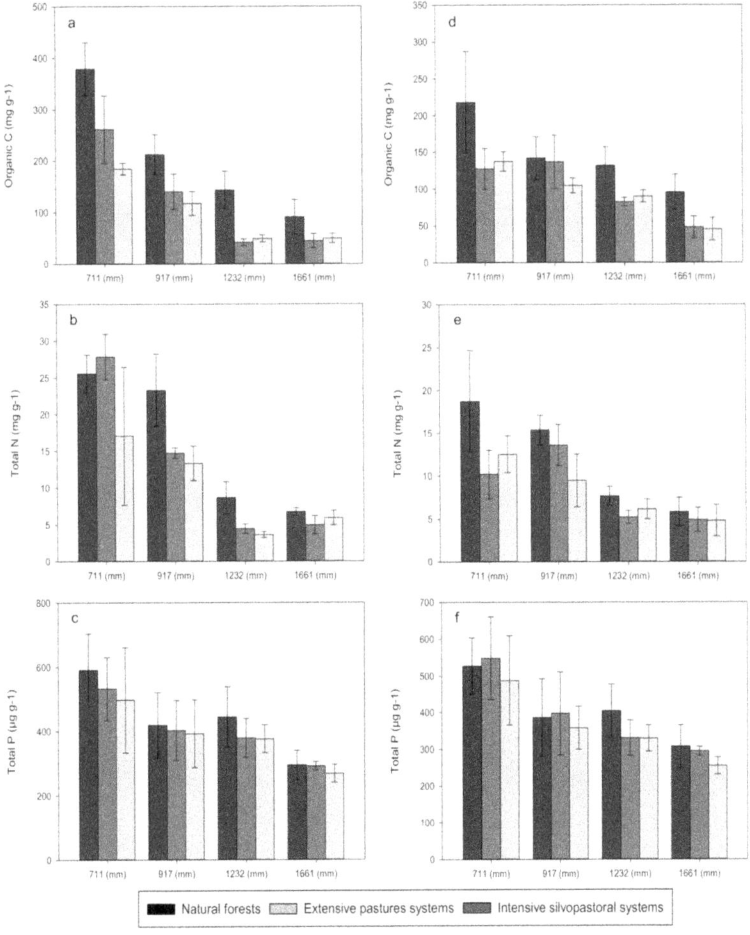

Figure 1. Soil organic carbon, total nitrogen and total phosphorus concentrations in dry (**a**–**c**) and rainy (**d**–**f**) seasons under natural forests, pastures and silvopastoral systems along a rainfall gradient. Data are means and confidence intervals.

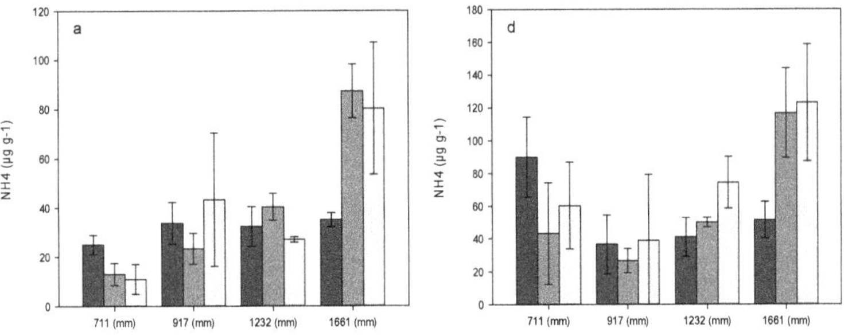

Figure 2. *Cont.*

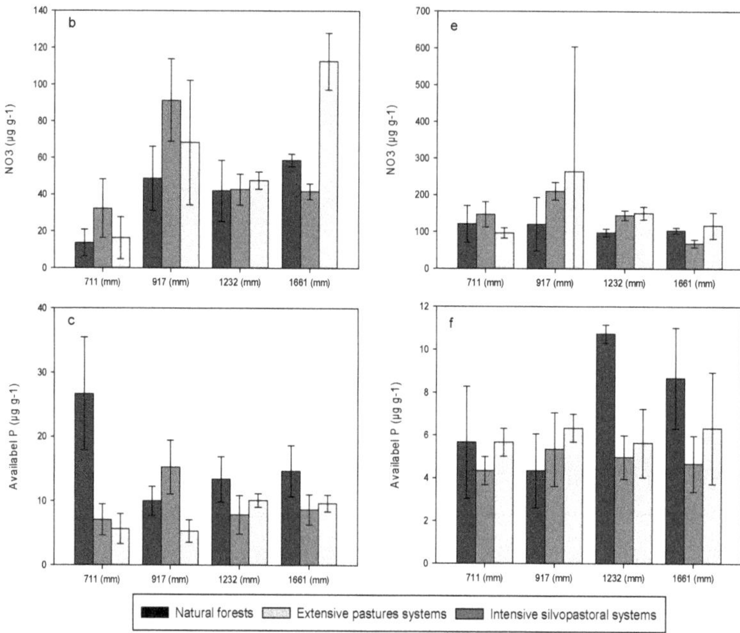

Figure 2. Soil ammonium, nitrate and available phosphorus concentrations in dry (**a–c**) and rainy (**d–f**) seasons under natural forests, pastures and silvopastoral systems along a rainfall gradient. Data are means and confidence intervals.

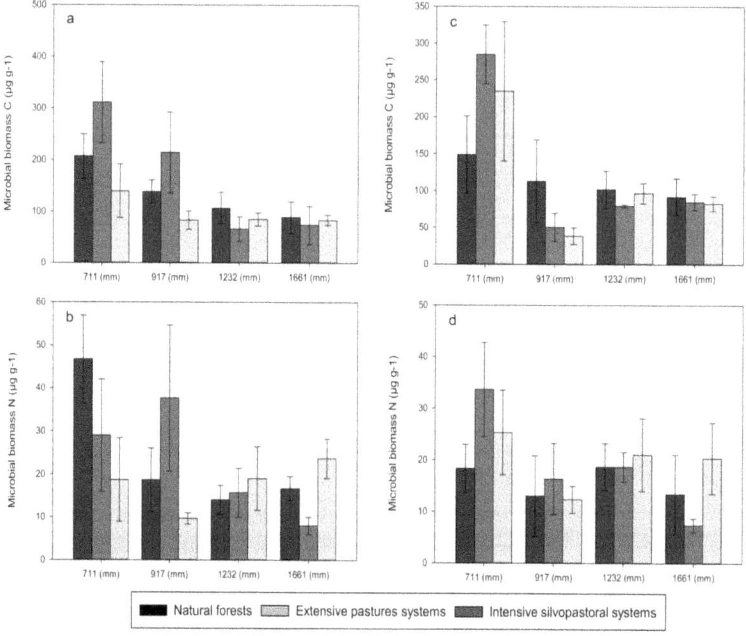

Figure 3. Soil microbial biomass carbon and microbial biomass nitrogen concentrations in dry (**a**,**b**) and rainy (**c**,**d**) seasons under natural forests, pastures and silvopastoral systems along a rainfall gradient. Data are means and confidence intervals.

3.2. Soil Basal Respiration and Net Nitrogen Transformations

The soil basal respiration was greater in wetter sites (i.e., sites with 1232 and 1661 mm of MAR) than in the drier counterparts (i.e., sites that receive 711 and 917 mm of MAR) (Figure 4, Table 2). In addition, net N transformations differed considerably among sites reflecting changes in rainfall amount. However, the paired comparisons using the Tukey–Kramer HSD test show that soils from the driest end of the rainfall gradient had the highest N transformations, and net N mineralization and net nitrification rates decreased with increase in rainfall amount. The season did not have a significant effect on soil basal respiration. However, large variation in net N transformations between seasons were observed, with higher net N mineralization and net nitrification rates in the rainy season than in the dry season.

Land-use change from natural forests to extensive pasture and intensive silvopastoral systems increased soil basal respiration (Figure 4, Table 2). However, net N transformations did not change with land-use change irrespective of the livestock production system. No significant changes in these soil C and N fluxes were observed between pasture and silvopastoral systems.

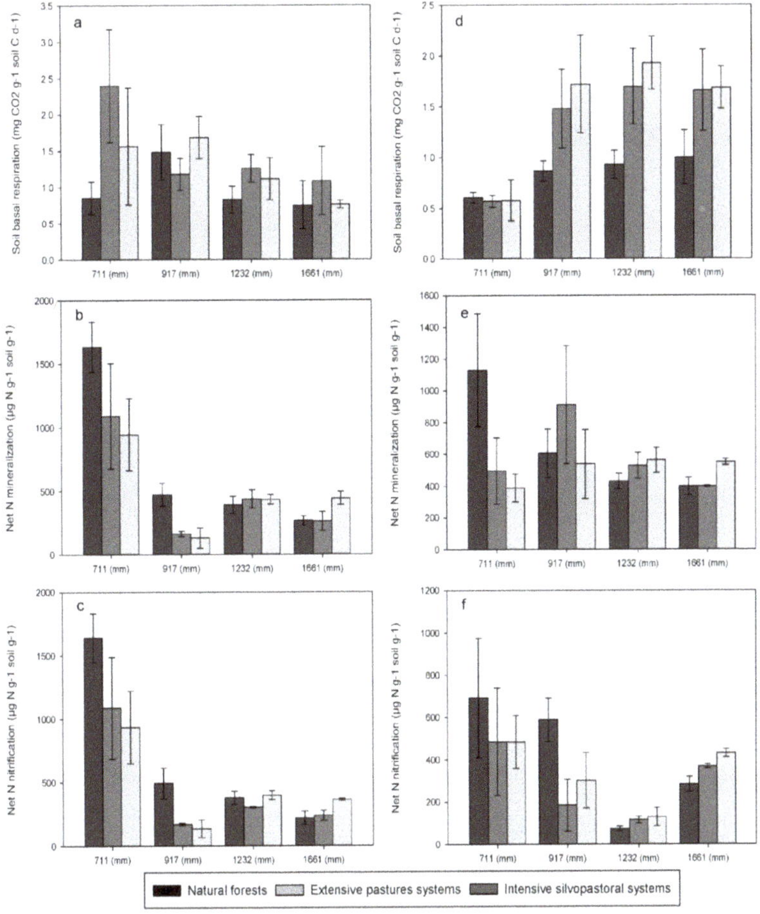

Figure 4. Soil basal respiration, net nitrogen mineralization and net nitrification in dry (**a**–**c**) and rainy (**d**–**f**) seasons under natural forests, pastures and silvopastoral systems along a rainfall gradient. Data are means and confidence intervals.

3.3. Enzyme Activities

The activity of C- and N-acquiring enzymes varied significantly among sites; but changes were not consistently related to rainfall regime (Figure 5, Table 2). Intermediate sites across the gradient (i.e., sites that receive 911 and 1232 mm of MAR) had higher polyphenol oxidase and 4-N-acetylglucosaminidase, or ß-1, 4-glucosidase activities, respectively, whereas the wettest site showed the lowest activity for the three enzymes. In contrast, soils from the wettest end of rainfall gradient had the highest acid phosphatase activity, reflecting the low concentration of available P. Rainfall seasonality did not have consistent effect on the activity of C-, N- and P-acquiring enzymes (e.g., on polyphenol oxidase, ß-1, 4-N-acetylglucosaminidase, and acid phosphatase) across sites. However, generally the activity of ß-1, 4-glucosidase was greater in soils collected in the dry season than in soils collected in the rainy season.

The land use did not affect the activity of C-acquiring enzymes in soils (Figure 5, Table 2). However, there were significant differences in the activity of N- and P-acquiring enzymes between natural forests and intensive silvopastoral systems across sites, with an increase in the ß-1, 4-N-acetylglucosaminidase and acid phosphatase activities under the legume-silvopastoral system. No significant changes in N- and P-acquiring enzyme activities were detected in between extensive pastures and natural forests, or with intensive silvopastoral systems.

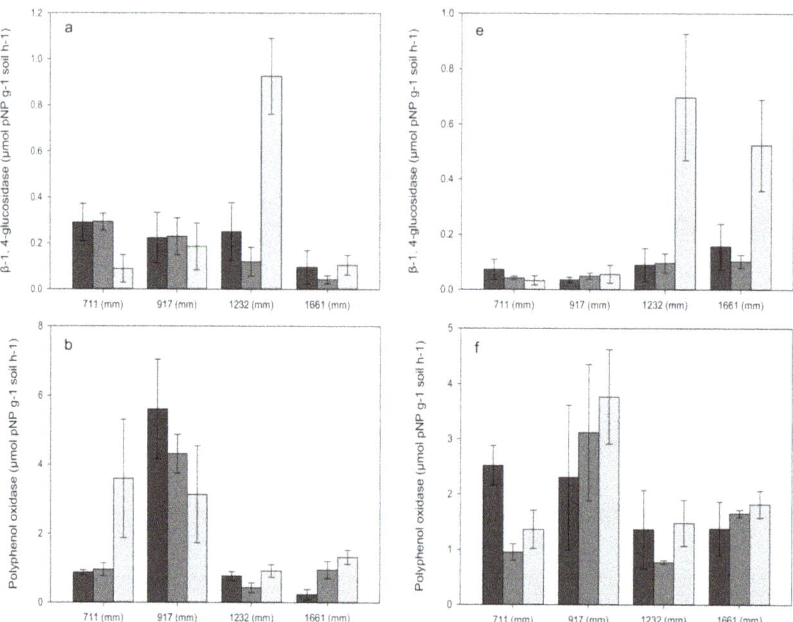

Figure 5. *Cont.*

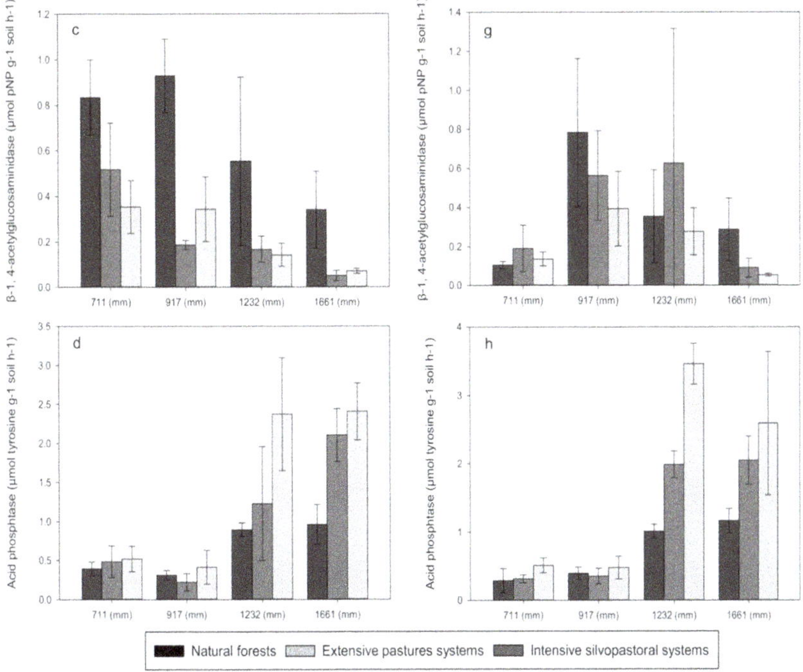

Figure 5. Activities of ß-1, 4-glucosidase, polyphenol oxidase, ß-1, 4-N-acetylglucosaminidase and acid phosphatase in soils in dry (**a**–**d**) and rainy (**e**–**h**) seasons from natural forests, pastures and silvopastoral systems along a rainfall gradient. Data are means and confidence intervals.

3.4. Multivariate Analysis and Soil Metric Relationships

The PCA allowed us to observe the effects of rainfall on soil metrics in each land-use system in the dry and rainy seasons (Figure 6). The two firsts principal components explained 61 and 52 percent of data variation for the dry and rainy seasons, respectively (Tables 3 and 4). Across the entire data set, the first principal axis summarized 47 percent of the variation in soil metrics in the dry season (Figure 6a, Table 3). Organic C, total N, NH_4 and acid phosphatase had the highest correlation scores on this axis. Sites found at the negative end of PC1 exhibited the lowest activity of acid phosphatase (pastures and silvopastoral systems at 1232 and 1661 mm of MAR) and the highest NH_4 concentrations (pastures and silvopastoral systems at the wettest end of the rainfall gradient) (Figures 2 and 5). By contrast, sites found at the positive end (natural forests and pastures at the driest end of the rainfall gradient) exhibited the highest organic C and total N concentrations (Figure 1). The second axis accounted for 14 percent of the variation. Available P and polyphenol oxidase explained the second-largest fraction of the explained variation in soil characteristics. Sites found at the negative end of PC2 exhibited the highest polyphenol oxidase activities irrespective of land use (sites at 917 mm of MAR). By contrast, sites found at the positive end (all land uses at the driest end of the gradient) exhibited low phosphatase activity.

On the other hand, in the rainy season, the first principal axis summarized 34 percent of the variation (Figure 6b, Table 4). The importance of organic C, total N and acid phosphatase detected in the soils of the dry season was also observed in the first PC of the rainy season soil samples. Sites found at the negative end of PC1 exhibited high values in acid phosphatase activities under livestock production systems (the wetter sites; i.e., extensive pastures and silvopastoral systems at 1232 and 1661 mm of MAR). Furthermore, the positive end of the PC1, reflect the highest organic C and total N concentrations in soils under the natural forests at the 711 mm of MAR. The second axis accounted for 17 percent of the variation. Polyphenol oxidase and microbial biomass C and N explained the second-largest fraction of the explained variation in soils. Sites found at the negative end of PC2 exhibited the highest concentrations of microbial biomass C and N under extensive pastures and intensive silvopastoral systems at the driest end of the rainfall gradient. By contrast, sites found at the positive end of this spectrum exhibited the highest polyphenol oxidase activity under extensive pastures and intensive silvopastoral systems, but in sites at 917 mm of MAR.

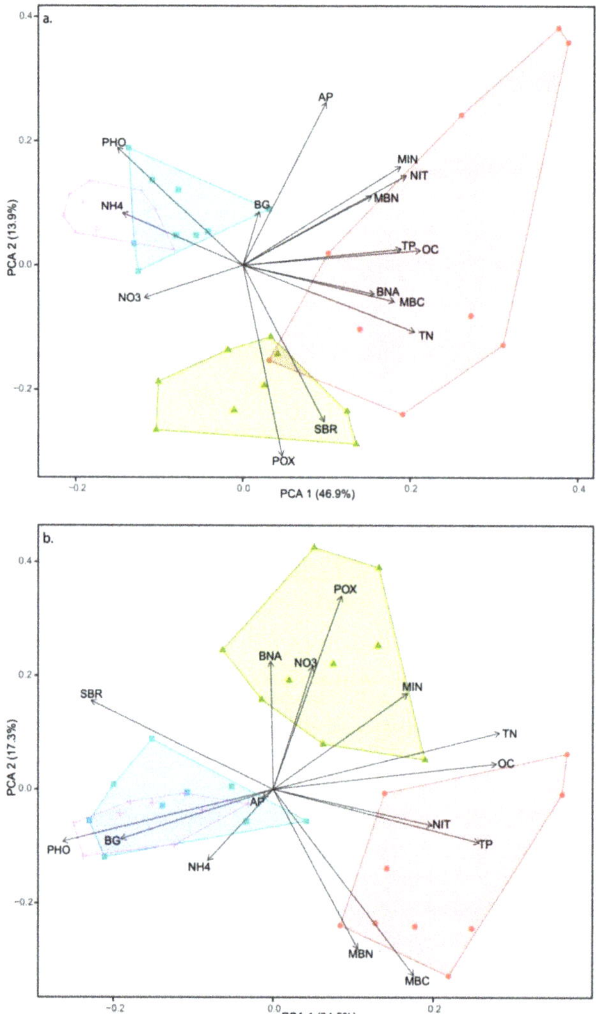

Figure 6. Principal components (PC) analyses biplots of the soil metric data for sites along a rainfall gradient in dry (**a**) and rainy (**b**) seasons. Red circles 711-mm of mean annual rainfall sites; green triangles 917 mm of mean annual rainfall sites; blue squares 1232 mm of mean annual rainfall sites; lilac symbols 1611 mm of mean annual rainfall sites.

Table 3. Eigenvalues, cumulative percent variation, and eigenvectors of the first three principal components (PCs) for the soil metrics in the dry season.

	PC1	PC2	PC3
Eigenvalue	7.045	2.091	1.440
Cumulative percent variation	46.9	60.8	70.4
Eigenvectors			
Organic C (OC)	0.94	0.06	0.14
Total N (TN)	0.91	−0.26	0.12
NH_4	−0.64	0.20	0.12
NO_3	−0.52	−0.13	0.58
Total P (TP)	0.84	0.06	−0.14
Available P (AP)	0.44	0.63	0.49
Microbial biomass C (MBC)	0.80	−0.14	0.001
Microbial biomass N (MBN)	0.68	0.27	0.31
Soil basal respiration (SBR)	0.43	−0.61	−0.41
Net N mineralization (MIN)	0.84	0.39	−0.15
Net nitrification (NIT)	0.87	0.35	−0.13
ß-1, 4-glucosidase (BG)	0.08	0.21	−0.43
Polyphenol oxidase (POX)	0.21	−0.74	0.34
ß-1, 4-N-acetylglucosaminidase (BNA)	0.70	−0.11	0.17
Acid phosphatase (PHO)	−0.67	0.46	−0.29

Table 4. Eigenvalues, cumulative percent variation, and eigenvectors of the first three principal components (PCs) for the soil metrics in the rainy season.

	PC1	PC2	PC3
Eigenvalue	5.174	2.601	2.025
Cumulative percent variation	34.4	51.7	65.2
Eigenvectors			
Organic C (OC)	0.89	0.10	0.11
Total N (TN)	0.90	0.23	0.22
NH_4	−0.26	−0.28	0.74
NO_3	0.16	0.49	0.22
Total P (TP)	0.82	−0.21	−0.04
Available P (AP)	−0.04	−0.03	−0.14
Microbial biomass C (MBC)	0.56	−0.74	0.01
Microbial biomass N (MBN)	0.34	−0.63	0.03
Soil basal respiration (SBR)	−0.72	0.35	0.30
Net N mineralization (MIN)	0.53	0.38	0.58
Net nitrification (NIT)	0.63	−0.14	0.46
ß-1, 4-glucosidase (BG)	−0.61	−0.20	0.51
Polyphenol oxidase (POX)	0.27	0.77	0.29
ß-1, 4-N-acetylglucosaminidase (BNA)	−0.01	0.51	−0.50
Acid phosphatase (PHO)	−0.84	−0.21	0.34

The correlation matrix (Pearson's correlation test) across the rainfall gradient show a relevant pattern among soil C, N and P metrics impacted by the rainfall regime (Tables S1–S4). Consistently, two groups of interacting soil metrics appeared. The first group was composed by the sites at the driest and the wettest ends of the gradient with more significant associations between soil metrics (44 in the case of the site that receive 711 mm of rainfall, and 48 in the case of the site that receive 1661 mm of rainfall). The second group was composed by sites at the intermediate range of the rainfall gradient, that show a lower number of significant associations among soil metrics (20 and 35, for sites at 917 mm and 1232 mm of MAR, respectively).

4. Discussion

4.1. Effects of Forest Conversion on Soil Carbon, Nitrogen and Phosphorus Dynamics

The present study provides a quantitative overview of C, N and P concentrations, and metrics of their fluxes in the upper mineral soil layer under natural forests and converted livestock lands across the southeast of Mexico. First, we found a loss of organic C, total N and available P with land-use change. Although this organic C trend is consistent with the meta-analysis of Guo and Gifford [57], who reported that the accumulation of organic C in the mineral soil is altered with the conversion of forest cover to pastures, the significant variation in C values across sites following the conversion from natural forests to livestock production systems was probably due to differences in soil type and vegetation, and effects of land-use disturbance on soil parameters. Nonetheless, our study still suggests that land-use change alters the C-holding capacity of soil as regards short C retention capacity, and that intensive management practices (e.g., clear-cutting and slash burning for site preparation and pruning) and rainfall conditions could favor increased soil C losses [58]. On the other hand, the loss observed in the total N is unexpected, considering the inputs of N with the animal excreta in both livestock production systems. Moreover, although biological N fixation is the primary source of N input in tropical lands [59], the observed loss in soil total N at study sites after the conversion to intensive silvopastoral system indicates that land-use change did alter significantly the balance between N input and loss, with consequences for N retention in this agricultural ecosystem.

The significant differences in C and N observed at study sites due to forest conversion could be related to differences in both quantity and quality of C and N inputs through litter and different management practices between native forests and livestock production systems. Plant species differ in their C sequestration potential, and land-use change by changing woody tree species covers with recalcitrant C compounds [60] to non-woody pasture species or leguminous trees, would be expected to alter the sequestration potential of C. Besides the expected large differences in litter inputs to the soil between natural forests and N-rich litter in silvopastoral systems, forest floor mass and nutrients under the natural forests species had a large quantity of organic materials poorly decomposed that would be incorporated into the mineral soils, as has been observed in these forest ecosystems [61], while a high rate of litter decomposition is expected in silvopastoral systems due to high N concentration and low content of recalcitrant compounds in litter from leguminous trees [62].

Soil basal respiration has been used as a powerful index to evaluate the alterations of soil C cycle [63]. In the present study, land-use change increased the soil basal respiration suggesting a higher ratio of labile soil organic C (i.e., soil organic C pool with "fast" decomposition [64,65]) to the total soil organic C under pastures and *Leucaena* plantations than under natural forests, probably reflecting the expected differences across land uses in the quantity and/or quality of the C inputs. It is known that it is the ratio of labile soil organic C to total soil organic C, rather than the total soil organic C content itself, that influences soil quality [66], due to its key role for metabolically active microbes [67] and nutrient supply to growing plants [68]. These expected differences in the contribution of labile soil organic C to soil total organic C among land uses could be related to the chemical recalcitrance of root tissue in the studied native forests [60]. On the other hand, management involving no-till in livestock production systems allows the accumulation of labile forms of organic C in soil surface layers [69,70], while more stable organic C accumulates under native forest soils [71]. In addition, inputs of dung-derived C and N often induce microbial priming effects, promoting soil organic matter decomposition [72]. Aside from these possible drivers of changes in soil organic C mineralization, the observed differences in CO_2 fluxes between natural forests and livestock systems could persist for an extended period considering that organic C stabilization in soil is relatively long-lasting (between 20 and 60 years; [73]).

Additionally, we observed a generalized loss in soil available P values after land-use change, despite the fact that dung is rich in P [74]. Moreover, the activity of the acid phosphatase under silvopastoral systems increased relative to its activity under natural forests, with largest changes in the wettest end of the rainfall gradient. We cannot determine whether such increase in P-acquiring enzymes under intensive silvopastoral systems resulted from an increase in P-demand by *Leucaena* plantations (thus exacerbating the possible direct nutrient limitation to soil microbes). Generally, P supply limits ecosystem function in tropical zones [12] and in the study region [74]. The land-cover change of native forests to extensive pasture systems maybe does not change this situation. However, the intensive silvopastoral systems maybe makes P-limitation worse, as indicated also by the low availability of P in the soils. Although this hypothesis of pervasive P limitation in silvopastoral systems could be verified with a study of primary production at the ecosystem level, because microbial biomass C:N:P ratios are relatively invariant across ecosystems [75], differences in microorganism efforts to acquire P should be expected between ecosystems that differ in nutrient supply in order to maintain microorganism homeostasis [76], (i.e., soils poor in P have high levels of activity of P-acquiring enzymes, as was observed in the intensive silvopastoral systems).

4.2. Effects of Rainfall Regime on Soil Carbon, Nitrogen and Phosphorus Dynamics

Climatic variables have strong impacts on soil organic C, N and P contents and fluxes in tropical and extratropical soils [5,77], and previous studies suggested that rainfall is the primary factor controlling soil C, N and P dynamics in water-limited tropical ecosystems [15,78]. Our study indicates that organic C, total N and available P, and microbial biomass C and N concentrations increased with decreased rainfall in both natural forests and livestock production systems. This result is in line with those obtained by Burke et al. [79] who reported that rainfall clearly has a direct role regionally and globally in the amount of soil C and nutrients stored. These patterns indicate that C and nutrient accumulation reflect a strong decrease in decomposition rate with the decrease in MAR reflecting also enrichment in organic matter recalcitrance with decrease in rainfall amount [60,80]. Moreover, the soil basal respiration rates decreased with the decrease in rainfall amount, suggesting a moisture limitation for soil C mineralization. In addition, the net N transformations differed considerably among sites with changes in rainfall amount. However, in contrast with the pattern observed for the C mineralization across sites, the observed highest N transformations in soils from the driest end of the gradient, with the largest values in samples taken in the dry season, suggested that there is a large potential of N losses when the rainy season starts. Thus, the expected reduction in rainfall in the southeast of Mexico [81] could open the N cycle irrespective of land use. Thus, our study also suggests that climate change could affect moisture-sensitive biogeochemical processes, and a N limitation could occur in native forests and livestock production systems if drought intensity increases.

Overall, our study demonstrated that soils (both under native forests and livestock systems) at the driest end of the rainfall gradient had C, N and P accumulation during the dry season, probably due to the decrease in decomposition [62] and soil leaching during the rainless period. These large pools of resources in the dry season could be a large input of C and nutrients to microbial biomass and vegetation demands when the rainy season starts. Correlation analysis allowed us to identify variations in the interaction among soil C and nutrient dynamics across sites affected by both the amount and seasonality of rainfalls. In line with Finzi et al. [82], and Delgado-Baquerizo et al. [83] who propose that drought increase would uncouple the C, N and P cycles, our study indicates that the strength of relationships among soil metrics for C, N and P cycles decreased with the decreases in MAR from the 1661 mm rainfall site to the 917 mm rainfall site, indicating that the cycles of C, N and P in the soil are less coupled with drought. A recent study in Texas found that greater soil water content favored soil bacterial communities [84]. Contrary to this trend, the association between soil variables increase below 917 mm of MAR. This could be due to more intense and prolonged seasonal drought that limits soil leaching and favors the microbial control of soil C and nutrient dynamics [85–87].

Regarding the sustainability of the livestock sector in the southeast of Mexico, our study on the consequences of the land-use change and climatic variation in soil nutrient dynamics allows us to conclude that the regulatory mechanisms of soil fertility in these two grazing livestock systems will vary, depending on the details of the site's nutrient limitations and the expected drought intensification.

Aside from these short- and long-term scenarios, the large extension of degraded land in the region [88] creates the opportunity for scaling up tropical forest restoration plans [89] or even the use of intensive leguminous tree plantations, which could be a positive contribution to reduce the massive impact of land use on tropical ecosystems, increase the landscape ecosystem services [8], and limit the contribution of the livestock sector to national N emission [90].

5. Conclusions

The analysis derived from our soil biogeochemical measurements provides an estimate of the response of tropical forest ecosystem and the sensitivity of land-use-change effects to increase in drought in the tropical region of Mexico. Our study indicates that expected climate change could impact moisture-sensitive biogeochemical processes, altering future carbon, nitrogen and phosphorus balances in soils under natural forests, and also in agricultural ecosystems, with potential negative consequences for the sustainability of livestock production systems in these tropical regions. Achieving the agricultural sustainability in these regions depends on limiting drought effects on soil carbon and nutrient losses with a better understanding of their biogeochemical dynamics as well as of the interactions among their cycles.

Supplementary Materials: The following are available online at http://www.mdpi.com/2071-1050/12/20/8656/s1, Table S1: Pearson correlation coefficients for soil metrics in sites that receive 711 mm of total annual rainfall, Table S2: Pearson correlation coefficients for soil metrics in sites that receive 917 mm of total annual rainfall, Table S3: Pearson correlation coefficients for soil metrics in sites that receive 1232 mm of total annual rainfall, Table S4: Pearson correlation coefficients for soil metrics in sites that receive 1661 mm of total annual rainfall

Author Contributions: Conceptualization—J.C., F.G.; formal analysis and data management—D.F., P.O.-F., T.F.A., A.R.-V., B.C.-V., J.D.E., J.C.; writing—original draft preparation—T.F.A., J.C.; writing—review and editing—J.C. All authors have read and agreed to the published version of the manuscript.

Funding: This research was made possible through support from the Mexican Consejo Nacional de Ciencia y Tecnología to D.F. and P.O.-F.; and by the Universidad Nacional Autónoma de México (PAPIIT grant 2015 RV200715).

Acknowledgments: We would like to thank Enrique Solís and Ofelia Beltrán-Paz for their valuable contribution on the soil analyses. We thank the landowners and managers for allowing access to their land and logistical help in the field.

Conflicts of Interest: The authors declare no conflict of interest.

References

1. Cias, P.; Sabine, C.; Bala, G.; Bopp, L.; Brovkin, V.; Canadell, J.; Chhabra, A.; DeFries, R.; Galloway, J.; Heimann, M.; et al. Carbon and Other Biogeochemical Cycles. In *Climate Change 2013: The Physical Science Basis. Contribution of Working Group I to the Fifth Assessment Report of the Intergovernmental Panel on Climate Change*; Stocker, T.F., Qin, D., Plattner, G.-K., Tignor, M., Allen, S.K., Boschung, J., Nauels, A., Xia, Y., Bex, V., Midgley, P.M., Eds.; Cambridge University Press: Cambridge, UK, 2013; pp. 465–570.
2. Scharlemann, J.P.W.; Tanner, E.V.J.; Hiederer, R.; Kapos, V. Global soil carbon: Understanding and managing the largest terrestrial carbon pool. *Carbon Manag.* **2015**, *5*, 81–91. [CrossRef]
3. Quéré, C.L.; Andrew, R.J.; Canadell, J.G.; Sitch, S.; Korsbakken, J.I.; Peters, G.P.; Manning, A.C.; Boden, T.A.; Tans, P.P.; Houghton, R.A.; et al. Global carbon budget 2016. *Earth Syst. Sci. Data* **2016**, *8*, 605–649. [CrossRef]
4. Pan, Y.; Birdsey, R.A.; Fang, J.; Houghton, R.; Kauppi, P.E.; Kurz, W.A.; Phillips, O.L.; Shvidenko, A.; Lewis, S.L.; Canadell, J.G.; et al. A large and persistent carbon sink in the World's forests. *Science* **2011**, *333*, 988–993. [CrossRef] [PubMed]
5. Carvalhais, N.; Forkel, M.; Khomik, M.; Bellarby, J.; Jung, M.; Migliavacca, M.; Mu, M.; Saatchi, S.; Santoro, M.; Thurner, M.; et al. Global covariation of carbon turnover times with climate in terrestrial ecosystems. *Nature* **2014**, *514*, 213–217. [CrossRef] [PubMed]
6. Friedlingstein, P.; Houghton, R.A.; Marland, G.; Hackler, J.; Boden, T.A.; Conway, T.J.; Canadell, J.G.; Raupach, M.R.; Ciais, P.; Le Quéré, C. Update on CO_2 emissions. *Nat. Geosci.* **2010**, *3*, 811–812. [CrossRef]
7. Houghton, R.A.; Nassikas, A.A. Global and regional fluxes of carbon from land use and land cover change 1850–2015. *Glob. Biogeochem. Cycles* **2017**, *31*, 1–17. [CrossRef]

8. Veldkamp, E.; Schmidt, M.; Powers, J.S.; Corre, M.D. Deforestation and reforestation impacts on soils in the tropics. *Nat. Rev. Earth Environ.* **2020**. [CrossRef]
9. Graesser, J.; Aide, T.M.; Grau, H.R.; Ramankutty, N. Cropland/pastureland dynamics and the slowdown of deforestation in Latin America. *Environ. Res. Lett.* **2015**, *10*, 034017. [CrossRef]
10. DeSy, V.; Herold, M.; Achard, F.; Avitabile, V.; Baccini, A.; Carter, S.; Clevers, J.G.; Lindquist, E.; Pereira, M.; Verchot, L. Tropical deforestation drivers and associated carbon emission factors derived from remote sensing data. *Environ. Res. Lett.* **2019**, *14*, 094022. [CrossRef]
11. Chadwick, R.; Good, P.; Martin, G.; Rowell, D.P. Large rainfall changes consistently projected over substantial areas of tropical land. *Nat. Clim. Chang.* **2016**, *6*, 177–181. [CrossRef]
12. Townsend, A.R.; Cleveland, C.C.; Houlton, B.Z.; Leck, C.; White, J.W. Multi-element regulation of the tropical forest carbon cycle. *Front. Ecol. Environ.* **2011**, *9*, 9–17. [CrossRef]
13. Townsend, A.R.; Asner, G.P.; Cleveland, C.C. The biogeochemical heterogeneity of tropical forests. *Trends Ecol. Evol.* **2008**, *23*, 424–431. [CrossRef] [PubMed]
14. Kauffman, J.B.; Sanford, R.L., Jr.; Cummings, D.L.; Salcedo, I.; Sampaio, E.V.S.B. Biomass and nutrient dynamics associated with slash fires in Neotropical dry forests. *Ecology* **1993**, *74*, 140–151. [CrossRef]
15. Allen, K.; Dupuy, J.M.; Gei, M.G.; Hulshof, C.; Medvigy, D.; Pizano, C.; Salgado-Negret, B.; Smith, C.M.; Trierweiler, A.; van Bloem, S.J. Will seasonally dry tropical forests be sensitive or resistant to future changes in rainfall regimes? *Environ. Res. Lett.* **2017**, *12*, 094002. [CrossRef]
16. Singh, J.S.; Chaturvedi, R.K. *Tropical Dry Deciduous Forest: Research Trends and Emerging Features*; Springer: New York, NY, USA, 2017.
17. Post, W.M.; Kwon, K.C. Soil carbon sequestration and land-use change: Processes and potential. *Glob. Chang. Biol.* **2000**, *6*, 317–327. [CrossRef]
18. Asner, G.P.; Archer, S.R. Livestock and the Global Carbon Cycle. In *Livestock in a Changing Landscape: Drivers, Consequences and Responses*; Steinfeld, H., Mooney, H.A., Schneider, F., Neville, L.E., Eds.; Island Press: Washington, DC, USA, 2010; pp. 69–82.
19. Galloway, J.; Dentener, F.; Burke, M.; Dumont, E.; Bouwman, A.F.; Kohn, R.A.; Mooney, H.A.; Seitzinger, S.; Kroeze, C. The Impact of Animal Production Systems on the Nitrogen Cycle. In *Livestock in a Changing Landscape: Drivers, Consequences and Responses*; Steinfeld, H., Mooney, H.A., Schneider, F., Neville, L.E., Eds.; Island Press: Washington, DC, USA, 2010; pp. 83–95.
20. Bhandari, K.B.; West, C.P.; Acosta-Martínez, V.; Cotton, J.; Cano, A. Soil health indicators as affected by diverse forage species and mixtures in semi-arid pastures. *Appl. Soil Ecol.* **2018**, *132*, 179–186. [CrossRef]
21. Campo, J.; García-Oliva, F.; Navarrete-Segueda, A.; Siebe, C. Stocks and dynamics of organic carbon in tropical forest ecosystems of Mexico. *Terra Latinoamericana* **2016**, *34*, 31–38.
22. Suseela, V.; Conant, R.T.; Wallenstein, M.D.; Dukes, J.S. Effects of soil moisture on the temperature sensitivity of heterotrophic respiration vary seasonally in an old-field climate change experiment. *Glob. Change Biol.* **2012**, *18*, 336–348. [CrossRef]
23. Mendoza-Ponce, A.; Corona-Núñez, R.; Kraxner, F.; LeDuc, S.; Patrizio, P. Identifying effects of land use cover changes and climate change on terrestrial ecosystems and carbon stocks in Mexico. *Glob. Environ. Change* **2018**, *56*, 12–23. [CrossRef]
24. Galeana-Pizaña, J.M.; Couturier, S.; Monsivais-Huertero, A. Assessing food security and environmental protection in Mexico with a GIS-based Food Environmental Efficiency index. *Land Use Pol.* **2018**, *76*, 442–454. [CrossRef]
25. Rivera Huerta, A.; Güereca, L.P.; Rubio Lozano, M.S. Environmental impact of beef production in Mexico through life cycle assessment. *Resour. Conserv. Recycl.* **2016**, *109*, 44–53. [CrossRef]
26. Rivera Huerta, A.; Rubio Lozano, M.S.; Padilla-Rivera, A.; Güereca, L.P. Social sustainability assessment in livestock production: A social life cycle assessment approach. *Sustainability* **2019**, *11*, 4419. [CrossRef]
27. Tello, J.; Garcillán, P.P.; Ezcurra, E. How dietary transition changed land use in Mexico. *Ambio* **2020**. [CrossRef] [PubMed]
28. Dirzo, R.; García, M.C. Rates of deforestation in Los Tuxtlas, a neotropical area in Southeast Mexico. *Conserv. Biol.* **1992**, *6*, 84–90. [CrossRef]
29. Bray, D.B.; Klepeis, P. Deforestation, forest transitions, and institutions for sustainability in Southeastern Mexico, 1900–2000. *Environ. Hist.* **2005**, *11*, 195–223. [CrossRef]

30. Balvanera, P.; Cotler, H.; Oropeza, O.A.; Aguilar Contreras, A.; Aguilera Peña, M.; Aluja, M.; Andrade Cetto, A.; Arroyo Quiroz, I.; Ashworth, L.; Astier, M.; et al. State and Trends of Ecosystem Services. In *Natural Capital of Mexico, vol II.*; Sarukhán, J., Dirzo, R., González, R., March, I.J., Eds.; CONABIO: Mexico City, Mexico, 2009; pp. 185–245. (In Spanish)
31. Reid, R.S.; Bedelian, C.; Said, M.Y.; Kruska, R.L.; Mauricio, R.M.; Castel, V.; Olson, J.; Thornton, P.K. Global Livestock Impacts on Biodiversity. In *Livestock in a Changing Landscape: Drivers, Consequences and Responses*; Steinfeld, H., Mooney, H.A., Schneider, F., Neville, L.E., Eds.; Island Press: Washington, DC, USA, 2010; pp. 111–137.
32. Broom, D.M.; Galindo, F.A.; Murgueitio, E. Sustainable, efficient livestock production with high biodiversity and good welfare for animals. *Proc. R. Soc. Biol. Sci.* **2013**, *280*, 2013–2025. [CrossRef]
33. Nahed-Toral, J.; Valdivieso-Pérez, A.; Aguilar-Jiménez, R.; Cámara-Cordova, J.; Grande-Cano, D. Silvopastoral systems with traditional management in southeastern Mexico: A prototype of livestock agroforestry for cleaner production. *J. Clean. Prod.* **2013**, *57*, 266–279. [CrossRef]
34. Williams, D.R.; Alvarado, F.; Green, R.E.; Manica, A.; Phalan, B.; Balmford, A. Land-use strategies to balance livestockproduction, biodiversity conservation and carbon storage in Yucatan, Mexico. *Glob. Chang. Biol.* **2017**, *23*, 5260–5272. [CrossRef]
35. Havlik, P.; Valin, H.; Herrero, M.; Obersteiner, M.; Schmid, E.; Rufino, M.C.; Mosnier, A.; Thornton, P.K.; Böttcher, H.; Conant, R.T.; et al. Climate change mitigation through livestock system transitions. *Proc. Nat. Acad. Sci. USA* **2014**, *111*, 3709–3714. [CrossRef]
36. Murgueitio, E.; Chará, J.; Barahona, R.; Cuartas, C.; Naranjo, J. Los sistemas silvopastoriles intensivos (SSPi), herramienta de mitigación y adaptación al cambio climático. *Trop. Subtrop. Agroecosys.* **2014**, *17*, 501–507.
37. Murgueitio, E.; Calle, Z.; Uribe, F.; Calle, A.; Solorio, B. Native trees and shrubs for the productive rehabilitation of tropical cattle ranching lands. *For. Ecol. Manag.* **2011**, *261*, 1654–1663. [CrossRef]
38. Holdridge, L.R.; Grenke, W.C.; Hatheway, W.H.; Liang, T.; Tosi, J.A. *Forest Environment in Tropical Life Zones: A Pilot Study*; Pergamon Press: New York, NY, USA, 1971.
39. García, E. *Modificación al Sistema de Clasificación Climática de Koeppen*; Universidad Nacional Autónoma de México: Mexico City, Mexico, 2004.
40. Middleton, N.; Thomas, D. *World Atlas of Desertification*; Edward Arnold: London, UK, 1997.
41. Mooney, H.A.; Bullock, S.H.; Medina, E. Introduction. In *Seasonally Dry Tropical Forests*; Bullock, S.H., Mooney, H.A., Medina, E., Eds.; Cambridge University Press: Cambridge, UK, 1995; pp. 1–8.
42. Flores-Guido, J.S.; Durán, R.; Ortíz, J.J. Comunidades Vegetales Terrestres. In *Biodiversidad y Desarrollo Humano en Yucatán*; Durán, R., Méndez, M., Eds.; CICY, PPD-FMAM, CONABIO, SEDUMA: Merida, Mexico, 2010; pp. 125–129.
43. Gómez-Pompa, A.; Krömer, T.; Castro-Cortéz, R. *Atlas de la Flora de Veracruz. Un Patrimonio Natural en Peligro*; Gobierno del Estado de Veracruz: Veracruz, Mexico, 2010.
44. Burgos-Hernández, M.; Castillo-Campos, G.; Vergara-Tenorio, M. Potentially useful flora from the tropical rainforest in central Veracruz, Mexico: Considerations for their conservation. *Acta Botánica Mexicana* **2014**, *109*, 55–77. [CrossRef]
45. INEGI. Actualización del Marco Censal Agropecuario. 2016. Available online: http://www.beta.inegi.org.mx/proyectos/agro/amca/ (accessed on 24 February 2020).
46. Porder, S.; Johnson, A.H.; Xing, H.X.; Brocard, G.; Goldsmith, S.; Pett-Ridge, J. Linking geomorphology, weathering and cation availability in the Luquillo Mountains of Puerto Rico. *Geoderma* **2015**, *249*, 100–110. [CrossRef]
47. Gee, G.W.; Bauder, J.W. Particle-size Analysis. In *Methods of Soil Analysis. Part 1*; Klute, A., Ed.; Soil Science Society of America: Madison, WI, USA, 1986; pp. 383–411.
48. Anderson, J.M.; Ingram, J.S.I. *Tropical Soil Biology and Fertility: A Handbook of Methods*; CAB International: Wallingford, UK, 1993.
49. Bray, R.H.; Kurtz, L.T. Determination of total, organic and available forms of phosphorus in soils. *Soil Sci.* **1945**, *59*, 39–46. [CrossRef]
50. Vance, E.D.; Brookes, P.C.; Jenkinson, D.S. An extraction method for measuring soil microbial biomass. *Soil Biol. Biochem.* **1987**, *19*, 703–707. [CrossRef]
51. Jenkinson, D.S. The Determination of Microbial Biomass Carbon and Nitrogen in Soil. In *Advances in Nitrogen Cycling in Agricultural Ecosystems*; Wilson, J.R., Ed.; CAB International: Wallingford, UK, 1988; pp. 368–386.

52. Anderson, J.P.E. Soil Respiration. In *Methods of Soil Analysis: Chemical and Microbiological Properties, Part II*; Page, A.L., Miller, R.M., Kenny, D.R., Eds.; American Society of Agronomy: Madison, WI, USA, 1982; pp. 831–871.
53. Robertson, G.; Wedin, D.; Groffman, P.; Blair, J.; Holland, E.; Nadelhoffer, K.; Harris, D.F. Soil Carbon and Nitrogen Availability: Nitrogen Mineralization, Nitrification, and Soil Respiration Potentials. In *Standard Soil Methods for Long-Term Ecological Research*; Robertson, G.P., Coleman, D.C., Bledsoe, C.S., Sollins, P., Eds.; Oxford University Press: New York, NY, USA, 1999; pp. 258–271.
54. Sinsabaugh, R.L.; Hill, B.H.; Follstad Shah, J.J. Ecoenzymatic stoichiometry of microbial organic nutrient acquisition in soil and sediment. *Nature* **2009**, *468*, 795–798. [CrossRef] [PubMed]
55. Jackson, C.R.; Tyler, H.L.; Millar, J.J. Determination of microbial extracellular enzyme activity in waters, soils and sediments using high throughput microplate assays. *J. Vis. Exp.* **2013**, *80*, e50399. [CrossRef]
56. R Core Team. *R: A Language and Environment for Statistical Computing*; R Foundation for Statistical Computing: Vienna, Austria, 2013; Available online: https://www.R-project.org/ (accessed on 28 July 2020).
57. Guo, L.B.; Gifford, M. Soil carbon stocks and land use change: A meta-analysis. *Glob. Chang. Biol.* **2002**, *8*, 345–360. [CrossRef]
58. Yang, Y.S.; Guo, J.; Chen, G.; Xie, J.; Gao, R.; Li, Z.; Jin, Z. Carbon and nitrogen pools in Chinese fir and evergreen broadleaved forests and changes associated with felling and burning in mid-subtropical China. *Forest Ecol. Manag.* **2005**, *216*, 216–226. [CrossRef]
59. Wang, Y.P.; Houlton, B.Z. Nitrogen constraints on terrestrial carbon uptake: Implications for the global carbon-climate feedback. *Geophys. Res. Lett.* **2009**, *36*, L24403. [CrossRef]
60. Campo, J.; Merino, A. Variations in soil carbon sequestration and their determinants along a precipitation gradient in seasonally dry tropical forest ecosystems. *Glob. Chang. Biol.* **2016**, *22*, 1942–1956. [CrossRef] [PubMed]
61. Cuevas, R.M.; Hidalgo, C.; Payán, F.; Etchevers, J.D.; Campo, J. Precipitation influences on active fractions of soil organic matter in seasonally dry tropical forests of the Yucatan: Regional and seasonal patterns. *Eur. J. For. Res.* **2013**, *132*, 667–677. [CrossRef]
62. Bejarano, M.; Crosby, M.M.; Parra, V.; Etchevers, J.D.; Campo, J. Precipitation regime and nitrogen addition effects on leaf litter decomposition in tropical dry forests. *Biotropica* **2014**, *46*, 415–424. [CrossRef]
63. Hertel, D.; Harteveld, M.A.; Leuschner, C. Conversion of a tropical forest into agroforest alters the fine root-related carbon flux to the soil. *Soil Biol. Biochem.* **2009**, *41*, 481–490. [CrossRef]
64. Jenkinson, D.S.; Ladd, J.N. Microbial Biomass in Soil: Measurement and Turnover. In *Soil Biochemistry*; Paul, E.A., Ladd, J.N., Eds.; Marcel Dekker: New York, NY, USA, 1981; pp. 415–471.
65. Falloon, P.D.; Smith, P. Modelling refractory soil organic matter. *Biol. Fertil. Soils* **2000**, *30*, 388–398.
66. Loveland, P.; Webb, J. Is there a critical level of organic matter in the agricultural soils of temperate regions: A review. *Soil Tillage Res.* **2003**, *70*, 1–18. [CrossRef]
67. Fontaine, S.; Bardoux, G.; Abbadie, L.; Mariotti, A. Carbon input to soil may decrease soil carbon content. *Ecol. Lett.* **2004**, *7*, 314–320. [CrossRef]
68. Lagomarsino, A.; Moscatelli, M.C.; De Angelis, P.; Grego, S. Labile substrates quality as the main driving force of microbial mineralization activity in a poplar plantation soil under elevated CO_2 and nitrogen fertilization. *Sci. Total Environ.* **2006**, *372*, 256–265. [CrossRef]
69. Angers, D.A.; Eriksen-Hamel, N.S. Full-inversion tillage and organic carbon distribution in soil profiles: A meta-analysis. *Soil Sci. Soc. Am. J.* **2008**, *72*, 1370–1374. [CrossRef]
70. Powlson, D.S.; Stirling, C.M.; Jat, M.L.; Gerard, B.G.; Palm, C.A.; Sanchez, P.A.; Cassman, K.G. Limited potential of no-till agriculture for climate change mitigation. *Nat. Clim. Chang.* **2014**, *4*, 678–683. [CrossRef]
71. Ludwig, M.; Achtenhagen, J.; Miltner, A.; Eckhardt, K.-U.; Leinweber, P.; Emmerling, C.; Thiele-Bruhn, S. Microbial contribution to SOM quantity and quality in density fractions of temperate arable soils. *Soil Biol. Biochem.* **2015**, *81*, 311–322. [CrossRef]
72. Bloor, J.M.G. Additive effect of dung amendment and plant species identity on soil processes and soil inorganic nitrogen in grass monocultures. *Plant Soil* **2015**, *396*, 189–200. [CrossRef]
73. West, T.O.; Post, W.M. Soil organic carbon sequestration rates by tillage and crop rotation: A global data analysis. *Soil Sci. Soc. Am. J.* **2002**, *66*, 1930–1946. [CrossRef]
74. Campo, J.; Vázquez-Yanes, C. Effects of nutrient limitation on aboveground carbon dynamics during tropical dry forest regeneration in Yucatán, Mexico. *Ecosystems* **2004**, *7*, 311–319. [CrossRef]

75. Cleveland, C.C.; Liptzin, D. C:N:P stoichiometry in soil: Is there a "Redfield ratio" for the microbial biomass? *Biogeochemistry* **2007**, *85*, 235–252. [CrossRef]
76. Sinsabaugh, R.L.; Belnap, J.; Findlay, S.G.; Shah, J.J.F.; Hill, B.H.; Kuehn, K.A.; Kuske, C.R.; Litvak, M.E.; Martinez, N.G.; Moorhead, D.L.; et al. Extracellular enzyme kinetics scale with resource availability. *Biogeochemistry* **2014**, *121*, 287–304. [CrossRef]
77. Vitousek, P. *Nutrient Cycling and Limitation: Hawai'i as a Model. System*; Princeton University Press: Princeton, NJ, USA, 2004.
78. Ochoa-Hueso, R.; Arca, V.; Delgado-Baquerizo, M.; Hamonts, K.; Piñeiro, J.; Serrano-Grijalva, L.; Shawyer, J.; Power, S.A. Links between soil microbial communities, functioning, and plant nutrition under altered rainfall in Australian grassland. *Ecol. Monogr.* **2020**. [CrossRef]
79. Burke, I.C.; Yonker, C.M.; Parton, W.J.; Cole, C.V.; Flach, K.; Schimel, D.S. Texture, climate, and cultivation effects on soil organic matter content in U.S. grassland soils. *Soil Sci. Soc. Am. J.* **1989**, *53*, 800–805. [CrossRef]
80. Jasso-Flores, I.; Galicia, L.; Chávez-Vergara, B.; Merino, A.; Tapia-Torres, Y.; García-Oliva, F. Soil organic matter dynamics and microbial metabolism along an altitudinal gradient in Highland tropical forests. *Sci. Total Environ.* **2020**, *741*, 140143. [CrossRef]
81. Conde, C.; Estrada, F.; Martínez, B.; Sánchez, O.; Gay, C. Regional climate change scenarios for Mexico. *Atmósfera* **2011**, *24*, 125–140.
82. Finzi, A.C.; Austin, A.T.; Cleland, E.E.; Frey, S.D.; Houlton, B.Z.; Wallenstein, M.D. Responses and feedbacks of coupled biogeochemical cycles to climate change: Examples from terrestrial ecosystems. *Front. Ecol. Environ.* **2011**, *9*, 61–67. [CrossRef]
83. Delgado-Baquerizo, M.; Maestre, F.T.; Gallardo, A.; Bowker, M.A.; Wallenstein, M.D.; Quero, J.L.; Ochoa, V.; Gozalo, B.; García-Gómez, M.; Soliveres, S.; et al. Decoupling of soil nutrient cycles as a function of aridity in global drylands. *Nature* **2013**, *502*, 672–676. [CrossRef] [PubMed]
84. Bhandari, K.B.; Longing, S.D.; West, C.P. Soil microbial communities in corn fields treated with atoxigenic Aspergillus flavus. *Soil Syst.* **2020**, *4*, 35. [CrossRef]
85. Singh, J.S.; Raghubanshi, A.S.; Singh, R.S.; Srivastava, S.C. Microbial biomass acts as a source of plant nutrients in dry tropical forests and savanna. *Nature* **1989**, *338*, 499–500. [CrossRef]
86. Campo, J.; Jaramillo, V.J.; Maass, J.M. Pulses of soil phosphorus availability in a tropical dry forest: Effects of seasonality and level of wetting. *Oecologia* **1998**, *115*, 167–172. [CrossRef]
87. Gei, M.G.; Powers, J.S. Nutrient Cycling in Tropical Dry Forests. In *Tropical Dry Forests in the Americas*; Sánchez-Asofeifa, A., Powers, J.S., Fernandes, G.W., Quesada, M., Eds.; CRC Press: Boca Raton, FL, USA, 2014; pp. 141–156.
88. Krasilnikov, P.; Gutiérrez-Castorena, M.C.; Ahrens, R.J.; Cruz-Gaistardo, C.O.; Sedov, S.; Solleiro-Rebolledo, E. *The Soils of Mexico*; Springer: New York, NY, USA, 2013.
89. Tobón, W.; Urquiza-Haass, T.; Koleff, P.; Schröter, M.; Ortega-Álvarez, R.; Campo, J.; Lindig-Cisneros, R.; Sarukhán, J.; Bonn, A. Restoration planning using a multi-criteria approach to guide Aichi targets in a megadiverse country. *Conserv. Biol.* **2017**, *31*, 1086–1097. [CrossRef]
90. Instituto Nacional de Ecologia y Cambio Climático. Inventario Nacional de Gases y Compuestos de Efecto Invernadero. Available online: https://www.gob.mx/inecc/acciones-y-programas/inventario-nacional-de-emisiones-de-gases-y-compuestos-de-efecto-invernadero (accessed on 5 October 2020).

Publisher's Note: MDPI stays neutral with regard to jurisdictional claims in published maps and institutional affiliations.

Article

Water Erosion Risk Assessment in the Kenya Great Rift Valley Region

George Watene [1,2,3,4], Lijun Yu [1,*], Yueping Nie [1], Jianfeng Zhu [1,2], Thomas Ngigi [3,4], Jean de Dieu Nambajimana [2,5] and Benson Kenduiywo [3]

1 Aerospace Information Research Institute, Chinese Academy of Sciences, Beijing 100101, China; gchege@jkuat.ac.ke (G.W.); nieyp@radi.ac.cn (Y.N.); zhujf@radi.ac.cn (J.Z.)
2 University of Chinese Academy of Sciences, Beijing 100049, China; njado52@gmail.com
3 Department of Geomatic Engineering and Geospatial Information Systems, Jomo Kenyatta University of Agriculture and Technology, P.O. Box 62000-00200 Nairobi, Kenya; tgngigi@jkuat.ac.ke (T.N.); bkenduiywo@jkuat.ac.ke (B.K.)
4 Sino-Africa Joint Research Centre, P.O. Box 62000-00200 Nairobi, Kenya
5 Key Laboratory of Mountain Surface Processes and Ecological Regulation, Institute of Mountain Hazards and Environment, Chinese Academy of Sciences, Chengdu 610041, China
* Correspondence: yulj@radi.ac.cn; Tel.: +86-010-6485-8122

Abstract: The Kenya Great Rift Valley (KGRV) region unique landscape comprises of mountainous terrain, large valley-floor lakes, and agricultural lands bordered by extensive Arid and Semi-Arid Lands (ASALs). The East Africa (EA) region has received high amounts of rainfall in the recent past as evidenced by the rising lake levels in the GRV lakes. In Kenya, few studies have quantified soil loss at national scales and erosion rates information on these GRV lakes' regional basins within the ASALs is lacking. This study used the Revised Universal Soil Loss Equation (RUSLE) model to estimate soil erosion rates between 1990 and 2015 in the Great Rift Valley region of Kenya which is approximately 84.5% ASAL. The mean erosion rates for both periods was estimated to be tolerable (6.26 t ha^{-1} yr^{-1} and 7.14 t ha^{-1} yr^{-1} in 1990 and 2015 respectively) resulting in total soil loss of 116 Mt yr^{-1} and 132 Mt yr^{-1} in 1990 and 2015 respectively. Approximately 83% and 81% of the erosive lands in KGRV fell under the low risk category (<10 t ha^{-1} yr^{-1}) in 1990 and 2015 respectively while about 10% were classified under the top three conservation priority levels in 2015. Lake Nakuru basin had the highest erosion rate net change (4.19 t ha^{-1} yr^{-1}) among the GRV lake basins with Lake Bogoria-Baringo recording annual soil loss rates >10 t ha^{-1} yr^{-1} in both years. The mountainous central parts of the KGRV with Andosol/Nitisols soils and high rainfall experienced a large change of land uses to croplands thus had highest soil loss net change (4.34 t ha^{-1} yr^{-1}). In both years, forests recorded the lowest annual soil loss rates (<3.0 t ha^{-1} yr^{-1}) while most of the ASAL districts presented erosion rates (<8 t ha^{-1} yr^{-1}). Only 34% of all the protected areas were found to have erosion rates <10 t ha^{-1} yr^{-1} highlighting the need for effective anti-erosive measures.

Keywords: soil erosion; Great Rift Valley Lakes; ASAL; Kenya; desertification

Citation: Watene, G.; Yu, L.; Nie, Y.; Zhu, J.; Ngigi, T.; Nambajimana, J.d.D.; Kenduiywo, B. Water Erosion Risk Assessment in the Kenya Great Rift Valley Region. *Sustainability* **2021**, *13*, 844. https://doi.org/10.3390/su13020844

Received: 26 November 2020
Accepted: 11 January 2021
Published: 16 January 2021

1. Introduction

Soil erosion poses a serious threat to global agricultural production [1] with worldwide mean soil erosion rates and total annual soil loss estimated to be between 12 to 15 t ha^{-1} yr^{-1} and 2.5 to 4 billion tons [2], respectively. In East Africa (EA), particularly for countries within the east side of the Sudano-Sahelian region, rapid economic expansions resulting to unsustainable use of natural resources coupled with recent climatic changes have exacerbated on-site and off-site effects of soil erosion including flooding, environmental degradation and loss of agricultural land productivity [3–5]. Loss of productive soil by erosion in turn negatively impacts food security [1] as more than 50% of the population in most sub-Saharan countries depends on agriculture for their livelihood [6]. A multiple of factors make tropical developing countries more vulnerable to the processes of soil erosion including high soil erodibilities, deforestation, desertification, agricultural intensification, poor soil conservation methods, and convergence of intense climatic regimes [7].

Desertification contributes nearly 80% of soil and land degradation in Kenya which has 88% of its land mass classified as arid and semi-arid lands (ASAL) [5]. Low vegetation cover, population increases combined with climatic changes has enhanced water erosion which further increases the risk of dryland degradation. These ASALs have been identified as most "vulnerable regions to climatic change" as well as the main cause of most socio-economic problems in the country [8]. With a total contribution of about 34.2% to the Gross Domestic Product (GDP-2019) [9], agriculture forms the mainstay of Kenya's economy. Mulinge et al. [10] estimated the total economic value of land degradation at 1.3 billion USD annually between 2001 and 2009 in Kenya which has over 12 million people living in degraded areas. Food crop productivity in Kenya's highly erodible soils [11] has in the recent past been hampered by soil erosion, increasing population and dynamic weather changes. Forest logging on mountain ranges and unchecked land-use activities along with intensive tropical precipitation increases soil erosion rates in Kenya's highland areas [12]. Water-induced soil erosion in the country's croplands has been estimated at 26 t ha^{-1} yr^{-1} [4] and has resulted to a permanent reduction of soil productivity in about 20% of its total area [10]. Reference [11] which is a recent Land and Degradation Assessment (LADA 2016) report in Kenya shows only about 2.2% of the total surface area has minimal risk of degradation while 61.4% suffers from severe degradation through soil erosion. In addition, the total forestland in Kenya decreased by 1% while croplands and bare lands (mostly in ASAL areas) increased by 7.3% and 2.6% respectively between 1990 and 2010 [11]. Land degradation thus threatens the source of the livelihoods of millions of Kenyans who predominantly depend on small scale farming [13]. In the central highlands of Kenya as with other most rural catchments, the total annual soil loss predictions vary from 134 t ha^{-1} yr^{-1} to 549 t ha^{-1} yr^{-1} which surpasses the estimated soil tolerance for tropical highland areas of 2.2–10 t ha^{-1} yr^{-1} [14,15]. The Great Rift Valley (GRV) region of Kenya which has 84.5% of its coverage classified as ASAL [16], in particular is prone to erosion due to steep topography, severe droughts, continued population pressure within its highland areas and dynamic land use and land cover changes over the last decade [17,18]. These changes include; overgrazing due to inadequate animal husbandry by pastoralists that leads to soil desertification [19], increased deforestation in the highland forest areas, e.g., Mau Forest and Cherangani Hills [17], massive shift from shrub lands to grasslands, human migration from lowlands areas (ASAL) to highland areas, and reduction in cropland areas. Land misuse coupled with climatic changes have made rills, gullies and sheet erosion are prevalent occurrence in the region, e.g., in the Baringo-Kerio valley [20]. The region is important as it forms the largest part of Kenya's grain basket zone that produces maize, wheat, beans, and tea [21]. The area has recorded high amounts of rainfall during the East African Short Rains (EASR) (October–December) periods in the recent past due to the Indian Nino resulting in rising lake levels in the GRV lakes [22]. Flooding of these lakes has also been attributed to a 50-year cyclic climatic event [23]. This has potentially heightened the risks of water erosion, pollution, siltation, flooding, and landslides [7,12].

A series of programs have been instigated in Kenya to address land degradation problems [10,21]. For instance, the forest policy has boosted reforestation measures which enabled the country to attain 6.99% national forest cover in 2014 [10]. Establishment of national parks and protected areas has reduced overgrazing and other adverse human-related activities in large rangelands although this significantly increases the pressure on the carrying capacity of the abutting lands. Some districts have adopted some Soil and Water Conservation (SWC) measures, e.g., conservation tillage practices and terracing, commonly referred to as *Fanya Juu* terraces, have shown considerable results [24]. With the current population density, the magnitude of soil erosion rates have continued to rise due to extreme weather changes, over-cultivation, desertification, and relative scarcity of productive farming land resulting to unsustainable sub division of land [3]. Since about

75 percent of the soils in Kenya are environmentally fragile and soil erosion is a long-term process involving complex combination of physical and hydrological factors [10,25], there is a need for methodical monitoring in order to establish critical areas and plan or devise for targeted soil conservation measures. Additionally, agriculture in Kenya strongly depends on irrigation water [21], and therefore it is important to monitor soil erosion spatial processes so as arm policy makers with efficient solutions to reduce soil transported into reservoirs.

Various studies have applied different models to compute soil loss by water erosion and sediment yields [26]. Some of the main hydrological models include the Universal Soil Loss Equation (USLE) [27], and its revised version (Revised Universal Soil Loss Equation (RUSLE)), Agricultural Non-Point Source Model (AGNPS), Morgan–Morgan–Finney (MMF), Soil and Water Assessment Tool (SWAT), Water Erosion Prediction Project (WEPP), Erosion Productivity Impact Calculator (EPIC), European Soil Erosion Model (EUROSEM), and The Limberg soil erosion mode (LISEM) [26]. These are broadly classified into physical and experiential models [28]. The applicability of a particular model generally depends on watershed spatial scale or characteristics, data accessibility and efficiency. Despite their complexity and high data requirements, physical models have "inbuilt process-based sub models" [26] that represent erosion processes more realistically. At regional scales and large catchments with limited data, empirical models such as USLE and RUSLE are most commonly applied [29] to estimate potential water erosion rates. The RUSLE is an updated version of the USLE model that has been extensively applied in many areas with different terrain characteristics and climatic zones to estimate long-term potential annual soil erosion rate, mainly because of its easy integration with geospatial technologies and low data requirements. Recent advancements in Geographic Information Systems (GIS) have enhanced RUSLE to allow for erosion monitoring at varied spatial and temporal scales [30].

Many of the past soil erosion studies in Kenya focus on the catchment scale or are at the local level [12,24,25,31]. Defersha et al. [32] applied WEPP and EROSION 3D physical models to estimate erosion rates and sediment yields at the Mara River Basin and revealed that the mean annual erosion rates for cultivated lands (120 t ha^{-1} yr^{-1}) was higher than that of bush land (7 t ha^{-1} yr^{-1}) or grasslands (3 t ha^{-1} yr^{-1}). Mati et al. [33] assessed the applicability of EUROSEM in two small catchments and found it inadequate for dry rangeland areas. Baker et al. [34] used the SWAT model in River Njoro watershed on the floor of Kenya's Rift Valley and showed that surface runoff increased proportionately with changes in land use. Similarly, Hunink et al. [35] indicated that coffee and maize growing areas presented mean erosion rates of 50 t ha^{-1} yr^{-1} and 10 t ha^{-1} yr^{-1} in the Upper Tana basin, respectively. In the USLE study by Mati et al. [24], 36% of the Upper Ewaso Ng'iro basin was predicted to suffer from mean erosion rates above the tolerable rate of 10 t ha^{-1} yr^{-1} mostly in the overgrazed rangelands. However, despite the presence of soil erosion in the physiographical regions of Kenya, few studies have applied the RUSLE model for spatial temporal evaluations particularly at the regional or national level [25,28,31,36]. The present study targets the GRV region of Kenya which covers about 33% of the country's total surface area with an aim to quantify (i) estimate the magnitude of potential soil loss rates in 1990 and 2015; (ii) assess the spatial changes among soil erosion risk classes between the two periods; (iii) identify priority areas for SWC; and (iv) quantify annual soil loss rates in Kenya Great Lakes ASAL basins, topography and protected areas.

2. Materials and Methods

2.1. Study Area

The Kenya Great Rift Valley (KGRV) region is located in the tropical zone of East African Rift System (EARS) and geographically lies between latitudes 4°12′ N and 3°15′ S and longitudes 34°00′ E and 38°05′ E (Figure 1). The region shares its northern border with Sudan–Ethiopia border, southern border with Tanzania and about half of its western border with Uganda. It has a total area of 194,291.73 Km^2, corresponding to approximately

33% of the country's total area The area is characterized by undulating volcanic and tectonic terrain with altitudes ranging from 360 to 4170 m.a.s.l with a mean altitude of approximately 1200 m.a.s.l. The Eastern Rift Valley traverses north south across the region approximately 720 km and 110 km wide. The Kenya Lake System UNESCO World Heritage Site is located in this region [37] and Lake Turkana which is located in the north is both the world's largest permanent desert lake and largest alkaline lake. The area encompasses Kenya's four main water towers: the Mau Forest Complex which supports an important ecosystem—including an equivalent market value of 229 million USD for the tea and tourism economic sectors only [38]—Cherangani Hills and sections of Mt. Elgon and Aberdare Ranges. These vital ecosystems face constant threats from both anthropogenic forces (encroachment and deforestation) and natural hazards that have resulted to drying up of some rivers and streams within the region. Figure 2 shows Landsat time-series images indicating rising water levels in Lake Baringo due increased rainfall in the East Africa (EA) region (EASR) [22] and large-scale deforestation within the Mau forest Complex highlands [17].

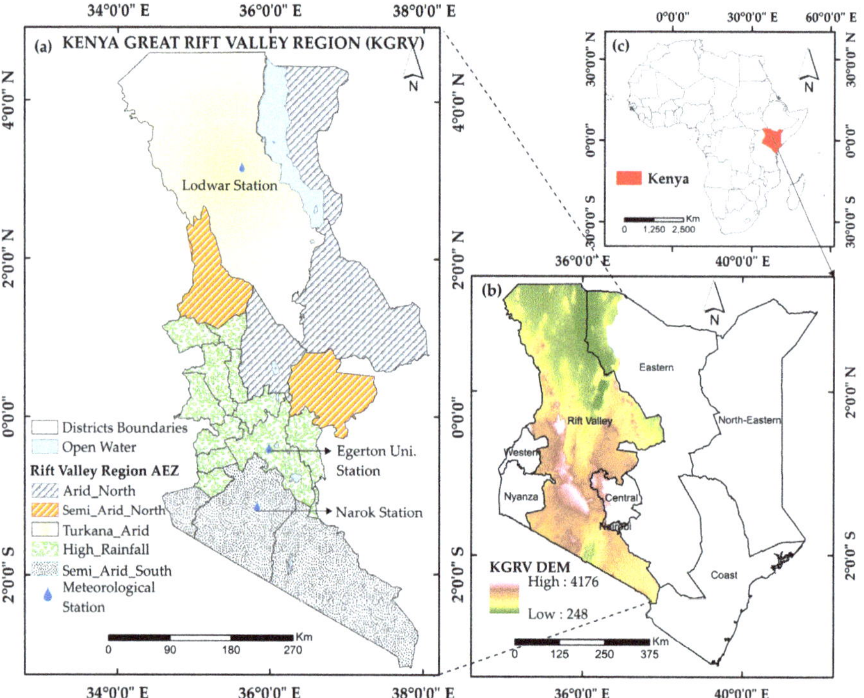

Figure 1. Map of the study area: (a) administrative districts and Agro-Ecological Zones (AEZ) [16,21] within the Great Rift Valley region of Kenya and (b) the location of Great Rift Valley region in Kenya (c) location of Kenya in the Africa continent.

Figure 2. Landsat time-series images on selected areas in the Kenya GRV region: (**a**,**b**) indicating rising water levels in Lake Baringo between 30th May, 2013 and 1st May, 2020 respectively; (**c**,**d**) indicating large-scale deforestation in the Mau Forest Complex between 1st February, 2002 and 16th February, 2016 respectively.

The area has a tropical climate with a mean annual precipitation of about 614 mm and two wet seasons (March–May (*Masika*) and September–December (*Vuli*)). The highland areas including Mau Forest (2000 to 2800 m above sea level) enjoy high intensity rainfall ranging between 1000 and 2000 mm and mean annual temperatures of 10 and 22 °C. The long-term (1970–2015) mean annual precipitation of Lodwar (8635000), Egerton University (KE0863), and Narok (9135001) meteorological stations representing Upper, Central, and Lower climatic zones of the KGRV, respectively, ranges from about 5 mm to 150 mm as shown in Figure 3. Daily rainfall data from the Kenya Meteorological Department (KMD) showed that Lodwar, Egerton University, and Narok stations recorded 2, 26, and 14 days with rainfall measurement greater than 10 mm, respectively, in the year 1990. Similarly and following the same order, the stations recorded 6, 28, and 19 days in the year 2015. The maximum daily rainfalls for Lodwar, Egerton University, and Narok stations in 1990 are 30.1, 72.2, and 46.5 mm, respectively, and 30.8, 41.4, and 36.2 mm, respectively, in 2015. The northern Lotikipi plains in Turkana district experience low amounts of rainfall ranging from 3 to 55 mm yearly with mean annual temperatures varying from 28 to 31 °C. The tropical highlands of the KGRV are mostly associated with Andosols and Nitisols soils that are developed from volcanic material. Cambisols are within areas with medium elevation while Lithosols, Solonchaks, and Regosols are prevalent in the ASAL regions. The dominant soil categories (Figure 4) include Lithosol (29.8%), Regosols (15.0%), Nitosols (10.4%), Cambisols (7%), and Ferrasols (6.8%) [39]. The area has five agro-ecological zones (AEZ) (Figure 1) include Arid North (with a mean annual rainfall of 506 mm), Semi-Arid North (759 mm), Semi-Arid South (762 mm), High Rainfall (1188 mm), and Turkana (258 mm) with a proportion of each zone contributing 24.8%, 9.5%, 20.5%, 15.5%, 29.7%, and 15.5%, respectively, of the total region area [16,21,40]. The region presents ideal conditions to conduct soil loss analysis for the country due to its unique environmental diversity (i.e., combination of ASAL and High Rainfall AEZ) that covers four of the five drainage basins in Kenya [40]. Topography and landforms largely shapes the region's drainage pattern (Figure 4). Several rivers branch from the central Kenya highlands into the endorheic Great Rift Valley basin, rivers in the western areas flow westward into Lake Victoria while streams from the Sudan-Ethiopia border drain into Lake Turkana (Figure 4). The lithology of the region is mainly dominated by igneous rocks around the mountainous landforms with sedimentary and metamorphic rocks mainly occupying the northern and western parts respectively. The four dominant landform types include Plains (29.2%), Plateaus (17.8%), Hills and Mountain foot ridges (12.6%), and Mountains (10.4%). Based on the 2019 national census the area has an estimated population of 13.8 million [9] which derive its livelihood mainly from agriculture in the High Rainfall AEZ and animal husbandry in the ASALs.

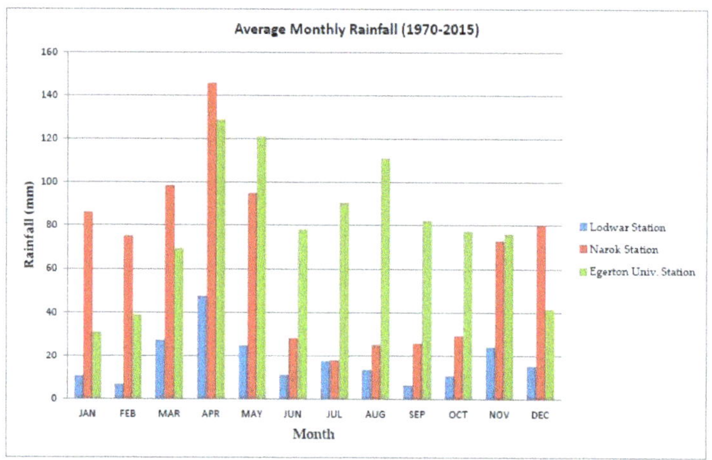

Figure 3. Average monthly rainfall (mm) for the Kenya Great Valley Region.

2.2. Data Collection

Table 1 shows the key datasets used in this study obtained from different sources. All the data was re-projected to the World Geodetic System (WGS) 1984_Universal Traverse Mercator (UTM) and resampled to match the data with coarse spatial resolution (250 m) using the SDMtoolbox in ArcGIS 10.5 software.

2.3. Land Use and Land Cover (LULC) Maps

For this study, 1990 and 2015 land use and land cover (LULC) maps for the GRV in Kenya (Figure 5) were acquired from the Department of Resource Surveys and Remote Sensing (DRSRS), Kenya in order to mask out non-erodible areas, estimate erosion rates for different LULC categories [41] and the impact of land use land cover changes (LULCC) on soil los rates. Erosion-prone areas covered a total area of 185,884.3 Km^2 that included Dense Forest (4.9%), Open Forest (3.3%), shrub land (48.6%), grassland (21.6%), cropland (4.9%), and bare lands (16.7%), in 1990. Following a similar order, the proportion of land uses were 4.2%, 3.0%, 45%, 20.7%, 11.8%, and 15.3% in the year 2015.

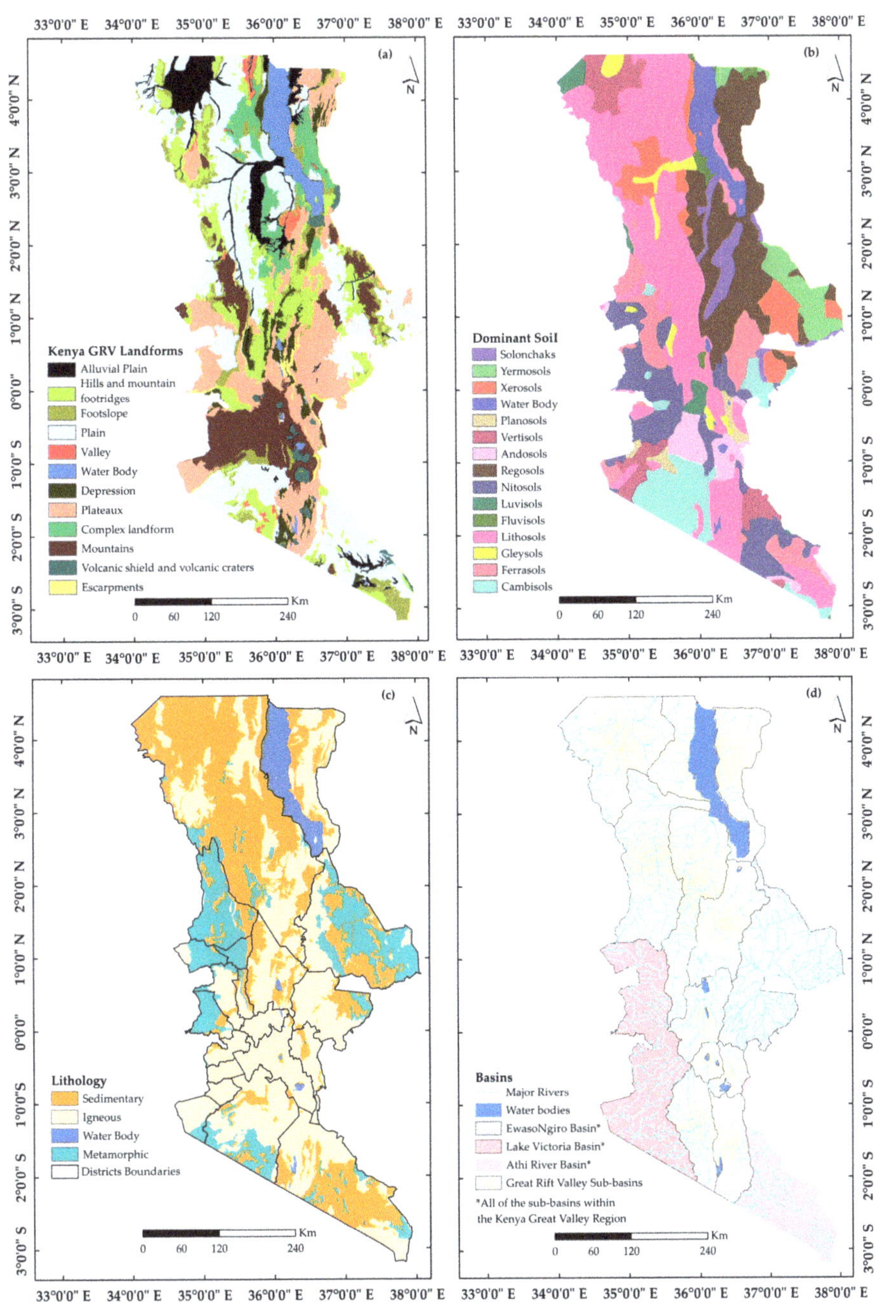

Figure 4. Maps of the Kenya Great Rift Valley (KGRV) region: (**a**) Landform types (**b**) Dominant Soil types region in Kenya (**c**) Lithological map (**d**) Major river basins.

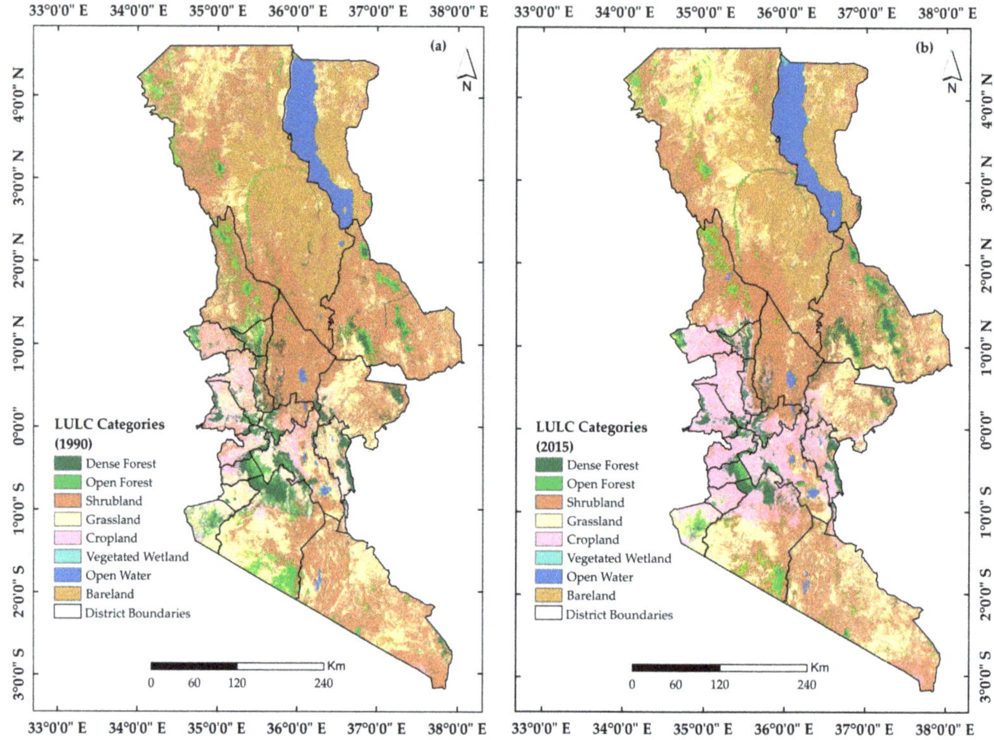

Figure 5. Land use/land cover distribution maps for the Kenya Great Rift Valley Region: (**a**) 1990 (**b**) 2015.

Table 1. The key datasets used for the Revised Universal Soil Loss Equation (RUSLE) model in the KGRV.

Variables	Datasets	Data Source
Terrain	DEM	SRTM GL1 version (30 m) [42]
		http://earthexplorer.usgs.gov/ (NASA)
Soil	Soil Properties	Africa Soil Information Service (AfSIS) (250 m) [43]
		https://files.isric.org/public/afsis250m/ (ISRIC-WSI)
Land	LULC Map	Land Use Land Cover (LULC) Maps of 1990 and 2015 (30 m)
		Department of Resource Surveys and Remote Sensing (DRSRS), MoE&F(K)
Climate	Rainfall Map	CHIRP Global Rainfall Data (5 arcsec) [44] between 1981 to 2015 periods
		ftp://ftp.chg.ucsb.edu/pub/org/chg/products/ (USGS-EROS)

DEM—Digital Elevation Model; SRTM GL1—Shuttle Radar Topography Mission Global 1 arc second; NASA—National Aeronautics and Space Administration, US; AfSIS—Africa Soil Information Service; ISRIC-WSI—International Soil Reference and Information Centre, World Soil Information; MoE&F(K)—Ministry of Environment and Forestry (Kenya); CHIRPS—Climate Hazards Group InfraRed Precipitation with Station data; USGS-EROS—United States Geological Survey, Earth Resource Observation Center.

2.4. RUSLE Model Application

Due to its unique merits that include easy integration with geospatial technologies [30], applicability in areas with limited data and adaptability at different spatial scales the Revised Universal Soil Loss Equation (RUSLE) empirical model has broadly been applied to estimate soil erosion rates worldwide.

The RUSLE model was chosen in this study since it has been tested in different landscapes, its application expediency and low data requirements [45]. The RUSLE equation [45] (Equation (1)) incorporates five different environmental variables using geo-informatics techniques to estimate the characteristics of soil erosion (Figure 6). These factors are rainfall erositivity (R), soil erodibility (K), slope length and steepness (LS), cover management (C), and support practice (P) [27].

$$A = R * K * LS * C * P \quad (1)$$

where A = annual average soil loss (t ha^{-1} yr^{-1}); R = rainfall erosivity factor (MJ mm ha^{-1} h^{-1}y^{-1}); K = soil erodibility factor (t ha h ha^{-1} MJ^{-1}mm^{-1}); LS = slope length and slope steepness factor (dimensionless); C = cover management factor (dimensionless); P = support practice factor (dimensionless).

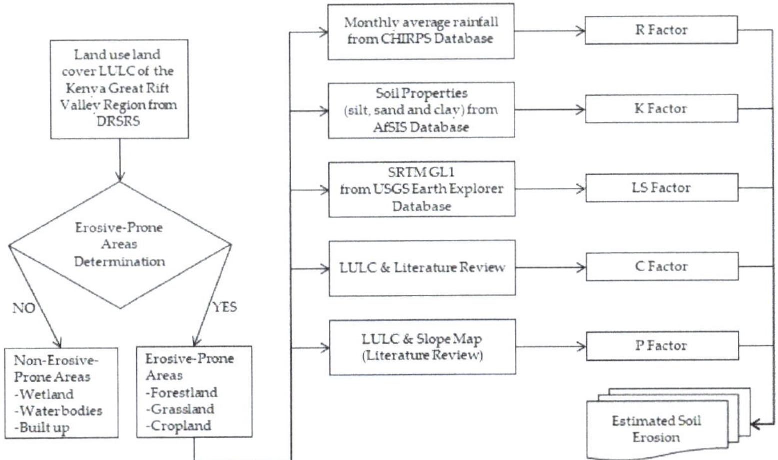

Figure 6. The methodological framework for estimating water-induced soil erosion rates using RUSLE model in the Great Rift Valley region of Kenya. LCLU: Land Use and Land Cover, DRSRS: Department of Resource Surveys and Remote Sensing, CHIRPS: Climate Hazards Group InfraRed Precipitation with Station data, SRTM: Shuttle Radar Topography Mission, AfSIS: Africa Soil Information Service.

2.5. Determination of RUSLE Factors

2.5.1. Rainfall Erosivity (R) Factor

Due to the direct correlation between rainfall intensity and erosion, the R factor represents the main driving factor of soil erosion [30] and contributes almost 80% of total soil loss [46]. The classical Wischmeier and Smith (1978) calculation method for R factor requires the use of storm erosivity index (EI) values of at least 20 years to account for seasonal variabilities and rainfall intensities [14,27]. Globally, not many areas have such gauged data readily available especially in developing countries [28]. The Lo et al. [47] method (Equation (2)) was adapted to estimate the rainfall erosivity factor since it has been applied by several studies in the East Africa region with significant results [29,48].

$$R = 38.46 + 3.48 \times P \quad (2)$$

where R = Rainfall Erosivity in MJ mm ha^{-1} h^{-1} yr^{-1}, P is the mean annual precipitation in mm. The mean annual precipitation for the time periods 1981–1999 and 1999–2015 were computed from the monthly average precipitation downloaded from the Climate Hazards Group InfraRed Precipitation with Station data (CHIRPS) [44] database and used to calculate rainfall erosivity factors for the periods 1990 and 2015, respectively [31]. CHIRPS data is readily available at high spatial-temporal resolutions and has been shown to have significant results in determining long-term rainfall trends when compared to rain gauge station datasets for Kenya [49,50] as well as the entire East Africa region [7,51].

2.5.2. Soil Erodibility (K) Factor

The *K* factor indicates the ability or resistance of soil particles to disintegrate and be transported by surface water runoff. This is dependent on the inherent soil properties including soil texture, organic matter, soil structure, and permeability [52]. To determine the *K* factor, the EPIC (erosion-productivity impact calculator) [53] model was applied to the sand, organic, silt and sand soil fractions of the area as compiled by the Africa Soil Information Service (AfSIS) [43].

$$K = F_{csand} \times F_{si-cl} \times F_{orgc} \times F_{hisand} \times 0.1317, \tag{3}$$

where

$$F_{csand} = \left[0.2 + 0.3exp\left(-0.0256SAN\left(1 - \frac{SIL}{100}\right)\right)\right], \tag{4}$$

$$F_{si-cl} = \left[\frac{SIL}{CLA + SIL}\right]^{0.3}, \tag{5}$$

$$F_{orgc} = \left[1.0 - \frac{0.0256C}{C + exp(3.72 - 2.95C)}\right], \tag{6}$$

$$F_{hisand} = \left[1.0 - \frac{0.70\ SN1}{SN1 + exp(-5.51 + 22.9\ SN1)}\right], \tag{7}$$

where *SAN*, *SIL*, and *CLA* are percent sand, silt and clay content, respectively; *C* is the organic carbon content; and *SN1* is sand content subtracted from 1 and divided by 100. F_{csand} (Equation (4)) = gives a low soil erodibility factor for soil with coarse sand and a high value for soil with little sand content. $F_{si\text{-}cl}$ (Equation (5)) = gives a low soil erodibility factor with high clay to silt ration F_{orgc} (Equation (6)) = is the factor that reduces soil erodibility for soil with high organic content. F_{hisand} (Equation (7)) = is the factor that reduces soil erodibility for soil with extremely high sand content.

2.5.3. Slope Length and Slope Steepness (LS) Factor

The *LS* factor represents the impact of topography on soil erosion [54]. It expresses the effects of local landscape on soil loss and is taken as the product of two terrain attributes: slope length (*L*) and slope steepness factor (*S*). Increasing the slope length and slope steepness values leads to higher overland flow speed and accelerates erosion rates [48]. The (*LS*) can be defined as the ratio of soil loss on a given slope length and steepness to soil loss from a seedbed with 22.13 m slope length and a steepness of 9% where all other conditions are held constant [54]. SRTM GL1 version 3 (30 m resolution) dataset provided by the United States Geological Survey (U.S.G.S.) [42] for the region was acquired to derive terrain attributes using the Raster Calculator tool from the Spatial Analyst extension of ArcMap 10.5 (Environment Systems Research Institute (Esri) Inc., Redlands, CA, USA). The *L* factor was calculated following algorithm (Equation (8)) proposed by Desmet and Govers (1996) while the S factor was estimated using the McCool et al. (1987) method (Equation (11)) [55,56]

$$L_{i.j} = \frac{\left(A_{i.j-in} + D^2\right)^{m+1} - A_{i.j-in}^{m+1}}{D^{m+2} \cdot x_{i.j}^{m} \cdot (22.13)^{m}}, \tag{8}$$

$$m = \frac{\beta}{1 + \beta} \tag{9}$$

$$\beta = \frac{\sin\theta / 0.0896}{3(\sin\theta)^{0.8} + 0.56} \tag{10}$$

$$S_{i.j} = \begin{cases} 10.8\sin\theta_{i.j} + 0.03, & \tan\theta_{i.j} < 9\% \\ 16.8\sin\theta_{i.j} - 0.50, & \tan\theta_{i.j} \geq 9\% \end{cases} \tag{11}$$

where $L_{i.j}$ = slope length factor for the grid cell with coordinates *(i.j)*; D = the grid cell size (*m*); $X_{i.j}$ =; $a_{i.j}$ = aspect direction for the grid cell with coordinates (*i.j*); $A_{i.j\text{-}in}$ = Flow accumulation or contributing area at the inlet of a grid cell with coordinates (*i.j*) (m^2), β = the ratio of inter-rill erosion, θ = the slope in degrees [56].

2.5.4. Cover Management Factor (C) Factor

The *C* factor corresponds to impact of vegetation canopy and land management practices on soil loss [27]. Reference [27] defined the C factor as the proportion of soil loss from land cropped under specific conditions to the corresponding soil loss under clean-tilled, continuous fallow land. The cover management factor ranges from 0 for non-erosive areas with thick vegetation cover to 1 which indicates very high susceptibility to erosion due to intensive tillage or exposed smooth surfaces. Nyssen et al. [57] emphasized on the importance of *C* factor in soil erosion assessments thus misrepresentations of *C* factor coefficients for different land covers can result in high over or under estimations of erosion rates. For this study, *C* value coefficients were adopted from different literatures mainly focusing on the East Africa region (Table 2) and assigned to the corresponding thematic LULC raster maps.

Table 2. Adopted *C* values for different land use patterns (after past literatures in East Africa (EA)).

Land Use	C Factor	Sources
Dense Forest	0.001	[58,59]
Open Forest	0.01	[58–60]
Shrubland	0.08	[36,61]
Grassland	0.05	[58,60,62]
Cropland	0.15	[58,60,61,63]
Bareland	0.40	[59,63]

2.5.5. Support Practice (P) Factor

The *P* factor represents land management control practices aimed at decreasing the rate of surface water runoff which in turn reduces soil erosion [27,56]. Conventional conservation measures include contouring, strip-cropping and terracing. Determining *P* factor values at large regional scales is nontrivial due to scarcity of data regarding conservation practices as well as complexities presented by different land uses [28]. A maximum value of "1" indicating poorest conservation practices was set to the P-factor [27] for the entire region since precautionary measures are often overlooked in regional soil erosion investigations [4]. In addition, Reference [64] revealed that the impact of soil and water conservation measures rapidly diminish in East African semi-arid areas. This study also separately estimated mean erosion rates in agricultural areas within the central and southern parts of the Kenya GRV region (Figure 1) in the year 2015 taking into account the three traditional soil erosion control practices proposed by Shin (1999) [65] (Table 3) to better understand their effect on croplands.

Table 3. *P* factor estimates for the common erosion control practices focusing on slope (%) [65].

Slope (%)	Conservation Support Practices (*P* Factor)		
	Contouring	Strip Cropping	Terracing
0.0–7.0	0.55	0.27	0.10
7.0–11.3	0.60	0.30	0.12
11.3–17.6	0.80	0.40	0.16
17.6–26.8	0.90	0.45	0.18
>26.8	1.00	0.50	0.20

3. Results

3.1. Estimated Soil Erosion Rates in the Great Rift Valley Region of Kenya

Figures 7 and 8 show the RUSLE results for the two periods while their statistical details are provided in Table 4. Rainfall Erosivity Factor (*R*) value varied between 359 and 8241 MJ mm ha^{-1} h^{-1} yr^{-1} (Figure 7a) with a mean of 2626.7 MJ mm ha^{-1} h^{-1} yr^{-1} in the year 1990. For the year 2015, the value varied from 340 and 7974.9 MJ mm ha^{-1} h^{-1} yr^{-1} with an average of 2162 MJ mm ha^{-1} h^{-1} yr^{-1}. In both years the highland areas recorded rainfall erosivity values >5000 MJ mm ha^{-1} h^{-1} yr^{-1} as with the work of Reference [7]. The central areas of the Kenya Great Rift Valley region is largely dominated by high *R* value while at the upper and lower parts (ASAL zones), *R* values are in the low range. These *R* values are consistent with the spatial distribution of the average annual rainfall across the region. Soil erodibility Factor (*K*) values ranged from 0.014 to 0.026 t ha h ha^{-1} MJ^{-1} mm^{-1} (Figure 8a). The topographic factor (*LS*) values were classified into five categories (Figure 8b) while the cover management Factor (C) ranged between 0 and 0.4 (Figure 7d).

The mean erosion rate for the year 1990 was estimated at 6.26 t ha^{-1} yr^{-1} with a standard deviation of 50.71 while the year 2015 presented a rate of 7.14 with a standard deviation of 40.38. In both years, the estimated mean rate of annual soil loss fell within the normal soil loss tolerances (from 5 to 11 t ha^{-1} yr^{-1}) [14,27,46]. The amount of total annual soil loss in the KGRV region was 116 Mt yr^{-1} in 1990 and 132 Mt yr^{-1} in 2015. To show the spatial distribution of water erosion and their areal extents in 1990 and 2015, the study area was classified into six erosion risk categories ranging from "very low" to "extremely high". In both years, the very low and low erosion classes when combined constitute greater sections of the total study area. In 1990, the two classes totaled to 154,822.1 Km^2 (83.3% of the total study area) while in 2015, the total was 150,695.2 Km^2 (81.1% of the total study area). The areal extent covered by medium, high medium, high, very high, and extremely high increased from 16,249.6 Km^2, 10,783.2 Km^2, 3 564.3 Km^2, 257.9 Km^2, and 207.0 Km^2 in 1990 to 16,906.1 Km^2, 11,838.3 Km^2, 5430.8 Km^2, 561.4 Km^2, and 452.4 Km^2, respectively, in 2015. However, the low erosion class reduced from 99,265.1 Km^2 to 93,733.1 Km^2 over the study period.

Table 4. The erosion risk classes and their corresponding erosion rate net changes between the 1990 and 2015 periods.

Erosion Risk Class	Erosion Rate Class (t/ha/y)	Year 1990			Year 2015			Net Change (t/ha/y)
		Area (10^4 ha)	Extent%	Erosion Rate (t/ha/y)	Area (10^4 ha)	Extent%	Erosion Rate (t/ha/y)	
Very Low	0–1	555.57	29.88	0.45	569.62	30.67	0.44	−0.01
Low	1–10	992.65	53.4	3.42	937.33	50.37	3.49	0.07
Medium	10–20	162.50	8.75	14.19	169.06	9.09	14.2	0.01
High medium	20–40	107.83	5.8	27.79	118.38	6.41	27.98	0.19
High	40–80	35.64	1.91	52.28	54.31	2.92	53.68	1.40
Very High	80–100	2.58	0.13	88.41	5.61	0.3	88.57	0.16
Extremely High	>100	2.07	0.13	284.41	4.52	0.24	168.85	−79.56

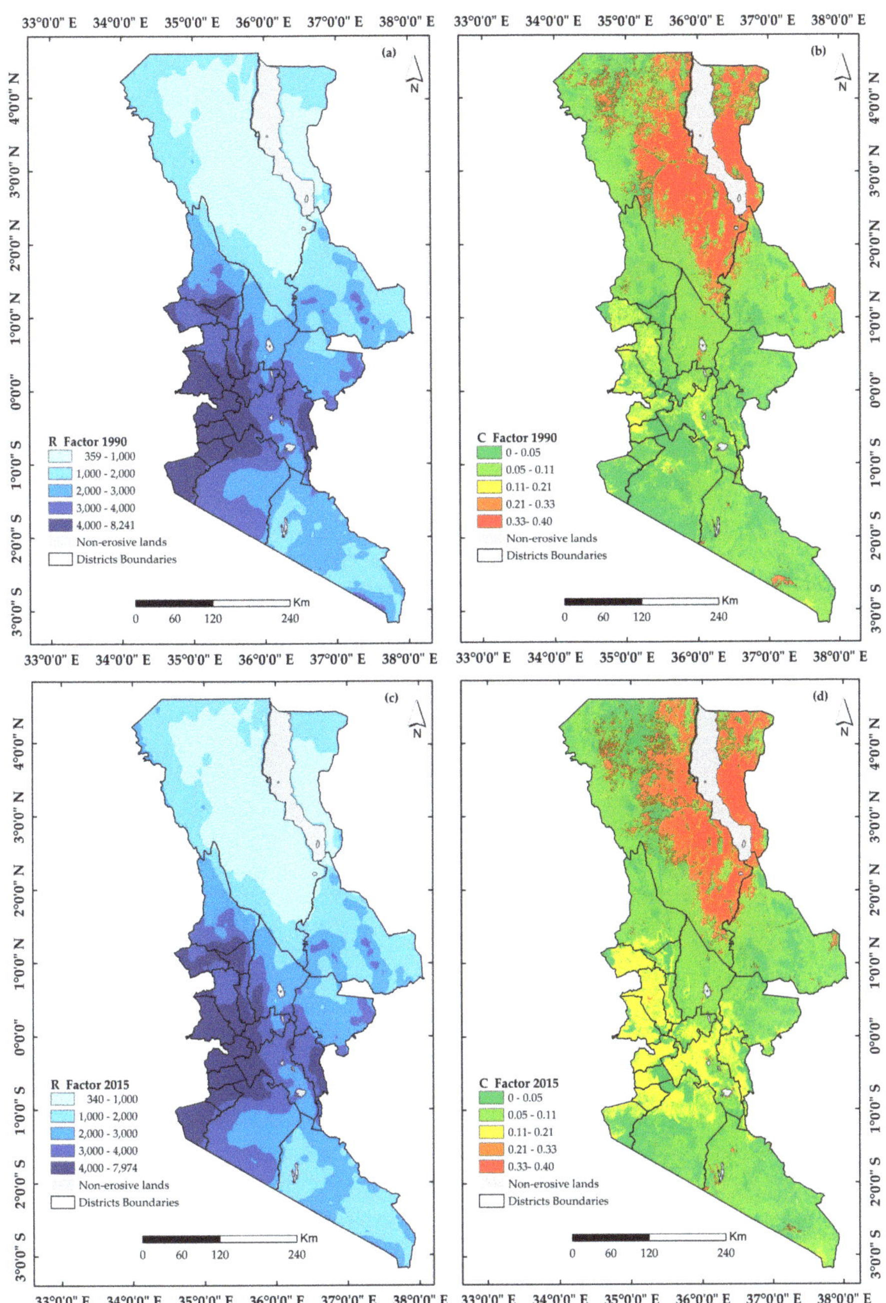

Figure 7. The Revised Universal Soil Loss Equation (RUSLE) factor maps of the KGRV: (**a**) rainfall erosivity factor, 1990 (supplementary materials); (**b**) cover management factor, 1990 (supplementary materials); (**c**) rainfall erosivity factor, 2015 (supplementary materials); and (**d**) cover management factor, 2015 (supplementary materials).

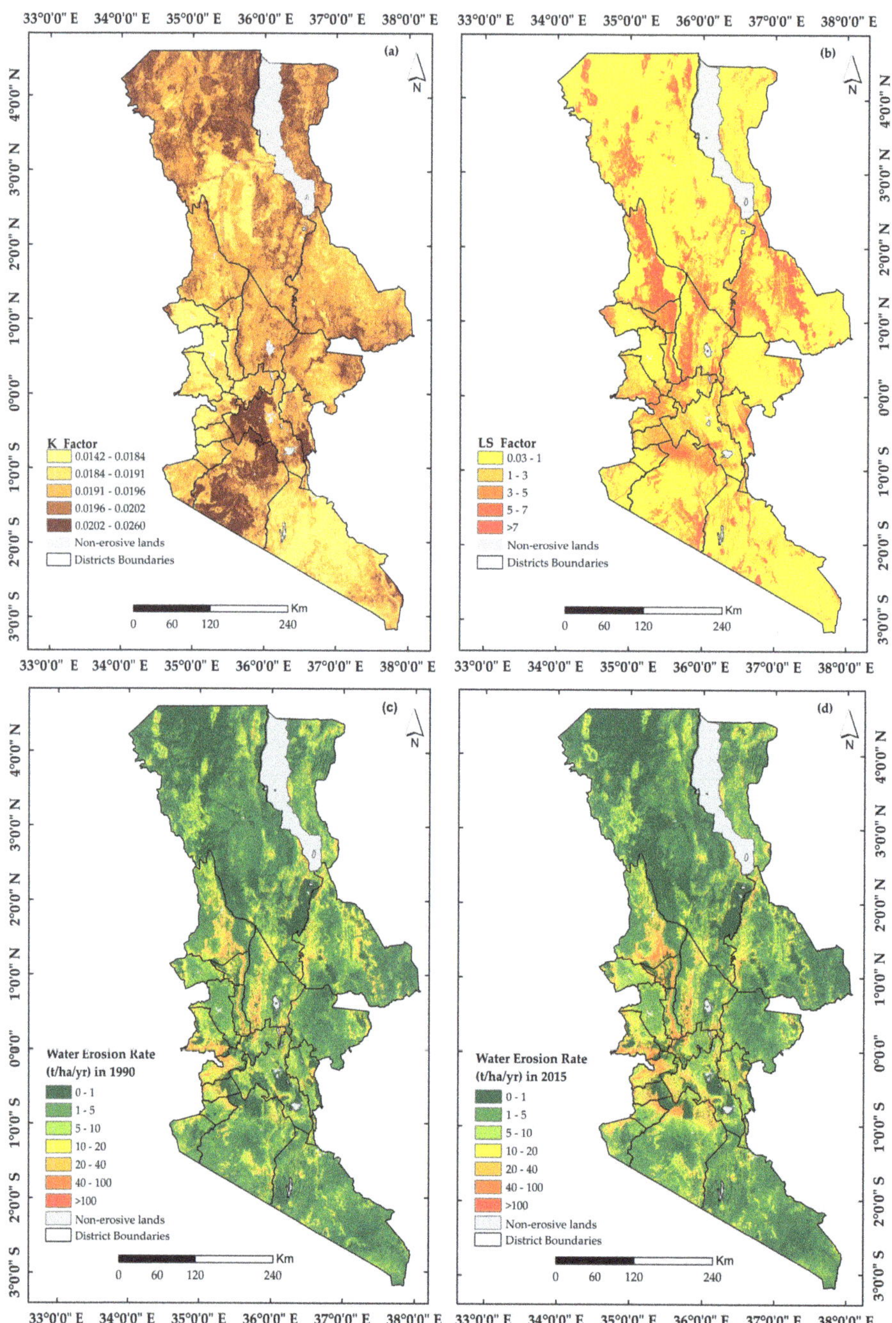

Figure 8. The Revised Universal Soil Loss Equation (RUSLE) factor and results maps of the KGRV: (**a**) soil erodibility factor (supplementary materials); (**b**) slope length and slope steepness factor (supplementary materials); (**c**) spatial distribution of man annual soil loss in 1990 (supplementary m aterials); and (**d**) spatial distribution of man annual soil loss in 1990 (supplementary materials).

3.2. Land Use/Land Cover Changes (LULCC) and Soil Erosion in the Great Rift Valley Region of Kenya

Table 5 shows the comparison of soil loss estimates between changes in LULC types within the investigated area. The results suggest a significant incline of mean rate of soil loss in cropland (from 15.8 t ha^{-1} yr^{-1} in 1990 to 20.6 t ha^{-1} yr^{-1} in 2015) and dismal reduction across all the other LULC types. LULCC occurred in the area of about 71,044.6 Km^2 (38.2%), while 114,839.7 Km^2 (61.8%) remained unchanged over the study period. Notable LULC types experiencing conversions (Table A1) include shrubland to grassland 15,077.3 Km^2, grassland to shrubland 12,289.9 Km^2, grassland to cropland 7530.7 Km^2, bareland to grassland 4686.5 Km^2, and dense forest to cropland 1643.6 Km^2. Among LULCC that contributed to increased soil losses are dense forest to bareland (59.0 t ha^{-1} yr^{-1}), dense forest to cropland (31.0 t ha^{-1} yr^{-1}), open forest to cropland (31.0 t ha^{-1} yr^{-1}), shrubland to cropland (10.7 t ha^{-1} yr^{-1}), grassland to cropland (10.2 t ha^{-1} yr^{-1}), and open forest to grassland (5.7 t ha^{-1} yr^{-1}). Conversely, LULCC that contributed to reduced soil losses include cropland to dense forest (−26.4 t ha^{-1} yr^{-1}), shrubland to open forest (−10.1 t ha^{-1} yr^{-1}), and shrubland to dense forest (−19.5 t ha^{-1} yr^{-1}) (Table 5).

Table 5. Estimated mean erosion rates per LULCC (Land use/land cover change) category (1990–2015) for Dense Forest, Open Forest, Shrubland, Grassland, Cropland and Bareland.

LULC Types	Dense Forest	Open Forest	Shrub Land	Grass Land	Crop Land	Bare Land	Erosion Rate 2015
Dense Forest	-	1.6	15.5	9.2	31.0	59.0	1.2
Open Forest	−3.8	-	7.6	5.7	26.3	10.8	2.9
Shrubland	−19.5	−10.1	-	−0.9	10.7	6.5	7.1
Grassland	−9.6	−3.8	1.0	-	10.2	3.7	2.0
Cropland	−26.4	−14.5	−5.3	−4.7	-	12.2	20.6
Bareland	−23.6	−4.6	−6.3	−3.0	−11.2	-	6.3
Erosion Rate 1990	1.5	4.0	7.2	3.5	15.8	6.4	-
Net Change (1990–2015)	−0.3	−1.1	−0.1	−1.5	4.8	−0.1	0.9

Spatial analysis of erosion risk conducted at the district level revealed that most of the districts within the mountainous landform part of the KGRV recorded soil loss rates >10 t ha^{-1} yr^{-1} (Table A2). Kericho district was consistent in presenting mean rates greater than 20 t ha^{-1} yr^{-1} in both study periods while most ASAL (e.g., Turkana, Laikipia and Kajiado) districts had rates <5 t ha^{-1} yr^{-1}. Keiyo and Marakwet districts had the highest soil loss increment of about 10 t ha^{-1} yr^{-1} over the study period. This is agreement with work of Reference [31] who reported an average increase of about 6 t ha^{-1} yr^{-1} for the western parts of the KGRV. This can be attributed to the high altitude and high mean rainfall which favored intensive farming on highly erosive soils in the region [31]. Table A3 shows the distribution of LULCC in relation to soil erosion at a district level. It represents the area coverage of the LULCC that occurred per district and the corresponding mean soil loss rates. The results suggest that though Keiyo district experienced the least conversions, it had highest erosion rates due to its high slope (19.8%) and mean rainfall (about 1200 mm). On the contrary, the driest districts characterized by semiarid desert plateaus recorded low erosion rates despite their high LULC conversion exchanges (e.g., Turkana and Samburu).

3.3. Estimated Soil Erosion Rates in the Protected Areas within the Great Rift Valley Region of Kenya

The United Nations Environment Program (UNEP) and the World Conservation Monitoring Center (WCMC) puts the total number of protected areas of Kenya at 411 with coverage of approximately 72,545 Km^2 [28,66] The number of the protected areas listed in the Great Rift Valley region of Kenya (Table A4) with a soil rate of <10 t ha^{-1} yr^{-1} went down from 57% in 1990 to 34% in 2015. Most of the protected areas that recorded an inclined of mean erosion rate are located in the areas occupied by characterized steep gradients topography and high rainfall intensity. Some of the endangered areas with mean erosion rates >35 t ha^{-1} yr^{-1} in year 2015 that need intervention planning include Kessop

(51.86 t ha^{-1} yr^{-1}), Sogotio (40.7 t ha^{-1} yr^{-1}), Kaisungor (47.04 t ha^{-1} yr^{-1}), Chemurokoi (40.87 t ha^{-1} yr^{-1}), and Kimojoch (41.45 t ha^{-1} yr^{-1}). Internationally acclaimed areas like Masai Mara, Lake Nakuru, and Amboseli national parks had consistent low soil loss averages of 1.5 to 4 t ha^{-1} yr^{-1} in both years of study and dismal changes. Nevertheless, areas that experienced high cases of deforestation between the two periods presented sharp incline erosion rates, e.g., Eastern Mau (from 5.94 to 18.18 t ha^{-1} yr^{-1}), South Western Mau (from 4.56 to 13.45 t ha^{-1} yr^{-1}), Southern Mau (from 13.13 to 19.22 t ha^{-1} yr^{-1}), Kipkabus (from 17.77 to 34.15 t ha^{-1} yr^{-1}), and Timboroa (from 14.17 to 24.55 t ha^{-1} yr^{-1}).

3.4. Classification Estimated Mean Erosion Rates by Severity and Conservation Priority

To prioritize for conservation planning, the quantitative soil erosion loss map loss of the Great Rift Valley region of Kenya were classified into 6 erosion classes following the methodology by Koirala et al. [67] in order to identify conservation priority areas (Table 6). The erosion severity ordinal classes are namely: slight (0–5 t ha^{-1} yr^{-1}), moderate (5–10 t ha^{-1} yr^{-1}), high (10–20 t ha^{-1} yr^{-1}), very high (20–40 t ha^{-1} yr^{-1}), severe (40–80 t ha^{-1} yr^{-1}), and very severe (>80 t ha^{-1} yr^{-1}). Areas with very severe erosion levels have been categorized as first priority whereas slight erosion values allocated 6th conservation priority. The results show that areas under slight erosion decreased from 71.73% of the total erosive lands in 1990 to 69.46% in 2015. However, the extent of total erosive lands for moderate, high, very high, severe and very severe classes increased from 11.57%, 8.74%, 5.8%, 1.91%, and 0.25% in 1990 to 11.63%, 9.09%, 6.36%, 2.92%, and 0.54%, respectively, in 2015.

Table 6. The distribution of estimated soil erosion rates per different severity classes.

Erosion Class (t/ha/y)	Severity Class	Year 1990			Year 2015			Net Change (t/ha/y)	Priority Level
		Area (10^4 ha)	Extent (%)	Erosion Rate (t/ha/y)	Area (10^4 ha)	Extent (%)	Erosion Rate (t/ha/y)		
0–5	Slight	1333.01	71.73	1.6	1290.70	69.46	1.54	−0.06	6th
5–10	Moderate	215.21	11.57	7.07	216.25	11.63	7.11	0.04	5th
10–20	High	162.50	8.74	14.19	169.06	9.09	14.2	0.01	4th
20–40	Very High	107.83	5.8	27.79	118.38	6.36	27.98	0.19	3rd
40–80	Severe	35.64	1.91	52.28	54.31	2.92	53.68	1.40	2nd
>80	Very Severe	4.65	0.25	175.67	10.14	0.54	124.39	−51.28	1st

3.5. Estimated Soil Erosion Rates by Slope and Elevation

The mean annual soil loss for areas with high slopes (β = 17.6–26.8%) was 14.99 t ha^{-1} yr^{-1} resulting in a total loss of approximately 19.2 Mt yr^{-1} in the year 1990 (Table 7). The erosive lands within this slope category had the largest erosion net change in the year 2015 which presented a rate of 18.53 t ha^{-1}yr^{-1}. Total soil loss for gentle slopes (β < 7%) increased slightly from 20.1 Mt yr^{-1} to 21.1 Mt yr^{-1}. The elevation raster map of the investigated area was reclassified into five different categories and the corresponding mean soil loss rates extracted using the ArcGIS Spatial Analyst tool set. The mean erosion rate for elevation of <500 m.a.s.l were 3.42 t ha^{-1} yr^{-1} in 1990 and 2.32 t ha^{-1} yr^{-1} in 2015 (Table 8). Areas with elevation greater than 2000 m.a.s.l characterized by high mean rainfall had the highest net change in soil loss (5.87 t ha^{-1} yr^{-1}) while regions with an elevation less than 1500 m.a.s.l having slightly reduced erosion rates.

Table 7. The distribution of estimated soil erosion rates by slope.

Slope (%)	Area (10^4 ha)	Extent %	Year 1990		Year 2015		Net Change (t/ha/y)	Average Rainfall (mm/y)
			Erosion Rate (t/ha/y)	Soil Loss (Mt/y)	Erosion Rate (t/ha/y)	Soil Loss (Mt/y)		
0–7	1142.27	61.45	1.76	20.1	1.85	21.1	0.09	522
7–11.3	238.24	12.81	5.75	13.7	6.59	15.7	0.84	724
11.3–17.6	163.36	8.78	10.23	16.7	12.48	20.4	2.25	807
17.6–26.8	128.00	6.88	14.99	19.2	18.53	23.7	3.54	809
>26.8	186.97	10.08	24.95	46.6	27.54	51.4	2.59	756

Table 8. The distribution of estimated soil erosion rates by elevation.

Elevation (m.a.s.l)	Area (10^4 ha)	Extent %	Year 1990		Year 2015		Net Change (t/ha/y)	Average Rainfall (mm/y)
			Erosion Rate (t/ha/y)	Soil Loss (Mt/y)	Erosion Rate (t/ha/y)	Soil Loss (Mt/y)		
<500	137.99	7.42	3.42	4.7	2.32	3.2	−1.1	298
500–1000	654.75	35.22	3.85	25.2	3.63	23.7	−0.22	313
1000–1500	371.19	19.96	7.02	26.1	6.66	24.7	−0.36	612
1500–2000	425.87	22.93	8.05	34.3	9.14	38.9	1.09	900
>2000	269.05	14.47	9.67	26.0	15.54	41.8	5.87	1128

3.6. Soil Erosion in the Major River Basins

The Great Rift Valley region of Kenya covers four of the five major basins that drain in Kenya [40,68]: the entire Great Rift Valley Area basin (GRVA) and sections of Ewaso Ngiro, Lake Victoria and Athi River basins (Figure 4) that had mean annual soil rates of 6.16 t ha^{-1} yr^{-1}, 5.09 t ha^{-1} yr^{-1}, 9.42 t ha^{-1} yr^{-1}, and 3.0 t ha^{-1} yr^{-1} in 1990 and 6.79 t ha^{-1} yr^{-1}, 4.37 t ha^{-1} yr^{-1}, 13.7 t ha^{-1} yr^{-1}, and 3.6 t ha^{-1} yr^{-1} in 2015, respectively (Table 9). Of all the major sub basins within the KGRV region, only Lake Bogoria-Baringo had a mean annual soil loss rate >10 t ha^{-1} yr^{-1} in 1990 while Lake Victoria, Lake Nakuru, Lake Naivasha, and also Lake Bogoria-Baringo recorded soil loss rates >10 t ha^{-1} yr^{-1} in the year 2015.

Table 9. The estimated soil erosion rates and the corresponding net changes in the major river basins of the KGRV.

Sub-Basin Name	Area (10^4 ha)	Year 1990		Year 2015		Net Change (t/ha/y)	Average Slope (%)
		Erosion Rate (t/ha/y)	Soil Loss (Mt/y)	Erosion Rate (t/ha/y)	Soil Loss (Mt/y)		
Lake Victoria	263.10	9.42	24.7	13.70	36.0	4.28	12.6
Lake Nakuru	23.20	6.74	1.5	10.93	2.5	4.19	12.0
Ewaso Ngiro South-Narok	88.22	5.94	5.2	9.21	8.1	3.27	13.4
Lake Naivasha	32.39	8.70	2.8	10.80	3.5	2.10	13.5
Lake Bogoria-Baringo	77.40	10.13	7.8	12.11	9.3	1.98	14.6
Turkwel River	203.46	5.93	12.1	6.74	13.7	0.81	14.1
Kerio Valley	174.42	7.05	12.3	7.75	13.5	0.70	13.5
Suguta	130.62	8.10	10.6	8.14	10.6	0.04	16.3
Lotikipi Plains	202.19	2.55	5.1	2.45	4.9	−0.10	11.1
Ewaso Ngiro South-Kajiado	82.76	6.03	5.0	5.89	4.9	−0.14	11.9
Athi River	137.06	3.00	4.1	2.69	3.6	−0.31	9.40
Lake Turkana	188.40	5.97	11.2	5.37	10.1	−0.60	10.9
Ewaso Ngiro	255.60	5.09	13	4.37	11.1	−0.72	12.9

3.7. Estimated Soil Erosion Rates by Major Landform and Soil Types within the KGRV

Tables 10 and A4 present the statistical details of the soil loss rates within the major landform types and dominant soils within the Kenya Great Rift Valley region respectively. In both years of study, areas around escarpment had the highest mean erosion rates of >25 t ha^{-1} yr^{-1} with plains recorded lowest rates of about 2.5 t ha^{-1} yr^{-1}. Mountainous parts around the central KGRV had the highest incline in mean erosion rates (4.34 t ha^{-1} yr^{-1}) while depressions had decreased soil rates (−1.43 t ha^{-1} yr^{-1}) over the study period. The results show that all of the dominant soil groups had mean soil loss rates <10 t ha^{-1} yr^{-1} in the year 1990 while Andosols and Nitisols (soils associated with volcanic material) recorded 11.53 t ha^{-1} yr^{-1} and 13.9 t ha^{-1} yr^{-1} in 2015, respectively. In addition, the two soil groups presented high soil loss rate net changes over the study period with Andosols (with high agricultural production) increasing by 5.52 t ha^{-1} yr^{-1}. Despite their resistance to water erosion due to good aggregate stability and high water permeability, Andosols can become susceptible to erosion particularly in intensively cultivated and deforested areas [69]. The heavy clayey Vertisols and the mostly water logged Gleysols had consistent low soil loss rates (about 3.0 t ha^{-1} yr^{-1}) in both years.

Table 10. The estimated soil erosion rates per landforms categories in the Great Rift Valley region of Kenya.

Landform Types	Area (10^4 ha)	Extent (%)	Year 1990		Year 2015		Net Change (t/ha/y)	Average Slope (%)
			Erosion Rate (t/ha/y)	Soil Loss (Mt/y)	Erosion Rate (t/ha/y)	Soil Loss (Mt/y)		
Mountains	201.98	10.86	10.57	21.3	14.91	30.1	4.34	19.35
Escarpment	15.13	0.81	28.09	4.2	30.89	4.6	2.8	32.86
Plateaus	342.70	18.43	6.95	23.8	8.47	29.0	1.52	9.67
Mountain Foot ridges	241.16	12.97	12.55	30.2	13.62	32.8	1.07	21.19
Foot Slope	147.56	7.93	4.41	6.5	5.25	7.7	0.84	7.03
Plain	564.82	30.38	2.4	13.5	2.34	13.2	−0.06	4.98
Complex Landforms	93.72	5.1	6.46	6.0	6.39	5.9	−0.07	9.11
Alluvial Plain	125.04	6.72	1.35	1.6	1.16	1.4	−0.19	3.13
Valley	21.46	1.15	6.01	1.2	5.35	1.1	−0.66	8.24
Volcanic Shield/Craters	26.56	1.42	9.34	2.4	8.66	2.3	−0.68	17.29
Depression	78.69	4.23	6.17	4.8	4.74	3.7	−1.43	5.21

3.8. Sensitivity Analysis of the RUSLE Model Factors used in the KGRV

Each of the five RUSLE parameters has a different role or impact on the total magnitude of the mean erosion rate [52]. The descriptive statistics in the model shown in Table 11 revealed that rainfall erosivity parameter (R) and soil erodibility parameter (K) are the two strongest controlling parameters for soil erosion in the Kenya GRV region. Regions with severe erosion rates increased significantly after removing parameter K.

Table 11. Sensitivity analysis of parameters used in the KGRV RUSLE model.

Description	Minimum Erosion Rate (t ha^{-1}y^{-1})	Maximum Erosion Rate (t ha^{-1}y^{-1})	Average Erosion Rate (t ha^{-1}y^{-1})	Standard Deviation
Removal of R	0	142.78	0.010	0.049
Removal of K	1.88	8,138,827	1127.43	4034.52
Removal of LS	1.24	38.23	11.25	4.35
Removal of C	0.19	545,818.88	93.19	295.96

3.9. Estimated Soil Erosion Rates in the Agricultural areas within the Central and Southern Rift Valley Region of Kenya (in 2015)

The central and southern region that includes the High Rainfall AEZ where agriculture is mostly practiced within the study area [21] (Figure 8), was separately analyzed in 2015 to understand the effect of common conservation practices on cropland areas (i.e., contouring, strip-cropping, and terracing). This can assist in making risk-informed decisions to conserve croplands that substantially contributed to increased soil loss rates over the period of study. Under the baseline conditions with P factor values set to "one", the total cropland area (21,864.2 Km^2) Figure 9a had a moderate mean soil erosion rate of 18.0 t ha^{-1} yr^{-1}; only 4.5% of the croplands had a sustainable mean soil loss <1 t ha^{-1} yr^{-1} while 28.6% of the croplands had severe soil loss rates >20 t ha^{-1} yr^{-1} mostly in central highland areas. This estimated mean annual soil loss rate is higher than the normal soil tolerances (from 5 to 11 t ha^{-1} yr^{-1}) [27] and the highland threshold for agro-ecological zones in tropical areas (from

0.2 to 11 t ha^{-1} yr^{-1}) [48]. The study predicted that, compared to the baseline scenario, terraces Figure 9b would decrease mean soil loss by 84.4% (from 18.0 to 2.8 t ha^{-1} yr^{-1}) while strip-cropping Figure 9c and contouring Figure 9d would slow soil erosion rate by approximately 2.5 and 1.2 times (from 18.0 to 7.2 t ha^{-1} yr^{-1}) and (from 18.0 to 14.4 t ha^{-1} yr^{-1}), respectively.

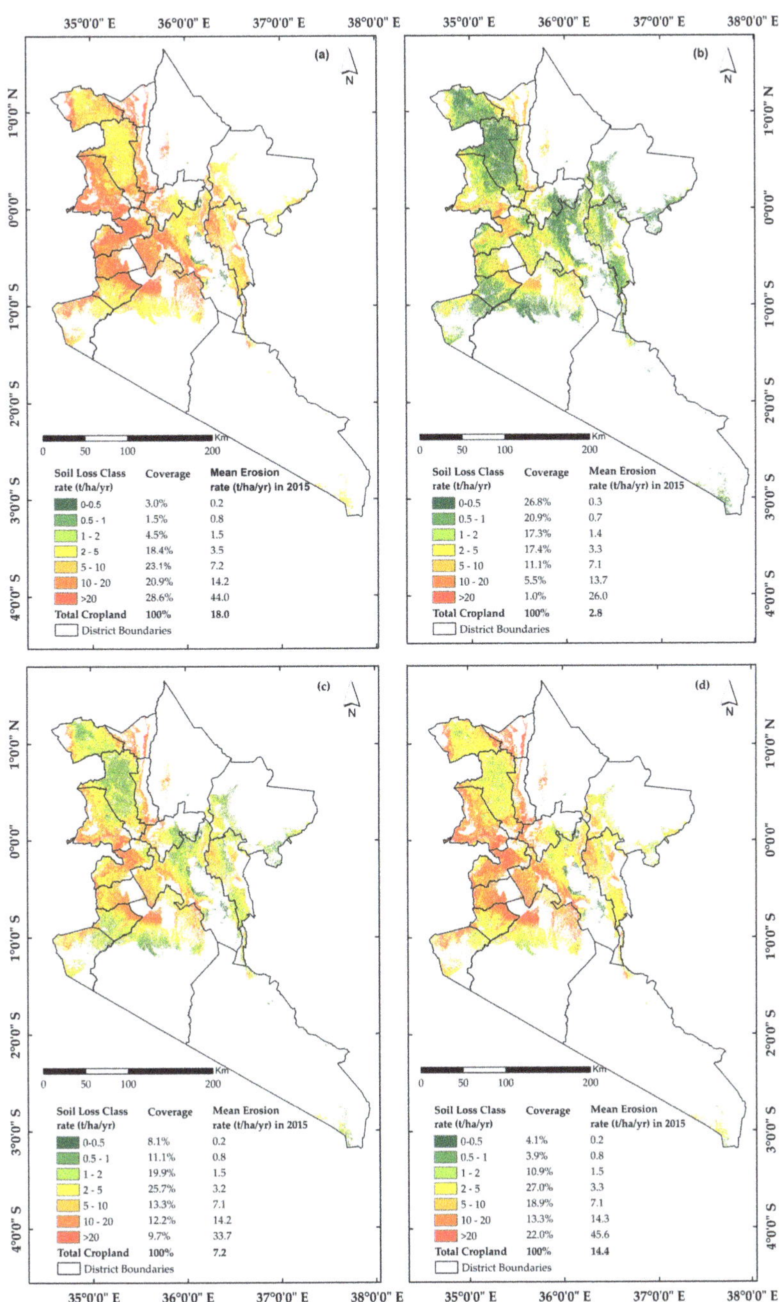

Figure 9. The estimated mean soil erosion rates for croplands (21,864.2 Km^2) within the central and southern regionsof the KGRV under different conservation practices in 2015 period: (**a**) baseline scenario; (**b**) terracing; (**c**) strip-cropping; and (**d**) contouring.

4. Discussion

Overview of Estimated Soil Erosion Risk in the Great Rift Valley Region of Kenya

The present study found that the mean erosion rate for the entire area was estimated at 6.26 t ha^{-1} yr^{-1} with a total soil loss of 116 $Mtyr^{-1}$ in 1990 (Figure 6) and 7.14 t ha^{-1} yr^{-1} with a total soil loss of 132 $Mtyr^{-1}$ in 2015. This estimated rate is within the range of erosion rate for Africa (10.8–146 t ha^{-1} yr^{-1}) [70] and slightly below the tolerable limits for mountainous environments (below 25 t ha^{-1} yr^{-1}) [67]. The mean erosion rate is also within range of the normal soil loss tolerances (ranging between 5 and 11 t ha^{-1} yr^{-1}) [27,48]. For both years, a greater proportional of the investigated region fell under the tolerable category (Table 4) as per the recommended maximum threshold of soil loss tolerance of 10 t ha^{-1} yr^{-1} for tropical areas [71]. Lake Bogoria-Baringo basin presented mean annual erosion rates >10 t ha^{-1} yr^{-1} for both periods while Lake Naivasha which recorded 10.8 t ha^{-1} yr^{-1} (Table 9) in year 2015. These values are in agreement with study by Mati et al. [24] which reported >10 t ha^{-1} yr^{-1} mean erosion rates in the Ewaso Ngi'ro basin. Flooding and the adverse effects of soil erosion within the Kenya GRV lakes has been attributed to the geomorphology of the lakes' environment and climatic factors [23]. Research is ongoing to find more definitive explanations for the recent rising levels of these Great Lakes [72,73]. Lakes Baringo and Naivasha are bordered mostly by flats lands while Lakes Bogoria and Nakuru are located in valleys enclosed by eastern and western rift escarpments [17]. Mubea and Menz [74,75] works revealed increasing urbanization and land degradation patterns in Nakuru district. This compounded with recent climatic changes [50] can offer some explanations why Lake Nakuru basin recorded the highest rise in soil loss rates among the Great Lakes. Rapid increment in floriculture and horticulture farming, over cultivation near river banks, and population growth are some of anthropogenic factors contributing to land degradation in the Lake Naivasha basin [76]. Willy et al. [77] survey study in Lake Naivasha basin showed low implementation of soil conservation practices with only 16% of sample households employing a combination of terracing, contouring, and grass strips. Lakes Baringo and Bogoria are located in Baringo County whose economy relies heavily on livestock which contributes 70% of its total income and supports 90% of its population [9]. Baringo County experiences frequent droughts; therefore, overgrazing can put pressure on land resulting to desert-like conditions [78]. Despite the highest mean annual precipitation and slope gradients for tree cover areas, forests presented lowest mean erosion rates for both periods (Table 5) emphasizing on the values of trees in soil erosion control. However, croplands had highest mean erosion rates indicating that intensive agricultural activities in areas with steep slopes significantly increase soil erosion threat within the Kenya GRV region. Fenta et al. [7] noted the high susceptibility to soil erosion of Kenyan highlands especially in disturbed forests or under sparse vegetation. Previous studies including [14,41] have also reported high erosion rates in highland areas with forestlands or poor vegetation due to deforestation, overgrazing, wildfires, and land cover changes. Koirala et al. [67] is in agreement with the concept that soil erosion increased proportionately with slope in mountainous regions while Schürz et al. [36] also reported high erosion rates in forested districts located on highlands e.g., West Pokot and Marakwet. In addition, recent study by Kogo et al. [31] in western Kenya (a subset of the KGRV region) revealed high rates around highlands e.g., the Mt. Elgon. Large bare areas in the ASAL which might potentially have high erosion rates recorded relatively low actual mean rates due to low rainfall and erosivity values. Significant bareland to grassland land cover conversions resulted in slight reduction of soil loss rates in ASAL areas, e.g., Turkana and Marsabit districts. The top three priority regions (Table 6) with erosion rates > 20 t ha^{-1} yr^{-1} contribute approximately 10% of the total erosive prone areas in the year 2015 and include highland districts located across the steep escarpment and ranges. Such areas are in urgent need of soil water conservation measures to mitigate heavy soil losses. Most of the protected areas are within forests or highland areas and recorded high erosion rates (Table A5) which is comparable with other estimated erosion rates for protected sites in other tropical lands as shown in References [28,29].

To assess the validity of RUSLE method for this region, the findings of the study were compared with areas of similar geo-environment and climatic conditions in the Eastern Rift Valley (EAR) region and seen to be analogous. For instance, our results coincide with those of previous studies by: Aneseyee et al. [79] evaluated the mean soil erosion rate in the neighboring Omo-Gibe Basin in the Ethiopian Rift Valley to be 17.65 t ha^{-1} yr^{-1}, Tamene et al. [71] found the mean soil loss rate of Laelaywkro catchment in Northern Ethiopia to be 20 t ha^{-1} yr^{-1}, Gizaw et al. [80] revealed that mean annual soil loss of Somodo watershed in South West Ethiopia is 18.69 t ha^{-1} yr^{-1} and Ligonja and Shrestha [81] reported a mean erosion rate of 15.7 t ha^{-1} yr^{-1} in Kondoa, Tanzania (Table 12). The districts' mean erosion rate values within the study area were consistent though not equal to the median and mean of soil losses values that resulted from the USLE model ensemble calculated by Schürz et al. [36]. In line with other studies in the East Africa region, terracing was found to be a highly effective soil erosion control measure especially for croplands located on the Kenya GRV highlands and the High Rainfall AEZ. Terracing and use of stone bunds was in practice across the Eastern Rift Valley (ERV) in small scale since the prehistoric periods as evidenced by ancient agricultural landscapes situated in the ASAL areas, i.e., Marakwet (Kenya), Engaruka (Tanzania) (Figure 10a), and Konso (Ethiopia) [82]. In Kenya, many of these terraces that had shown significant results were demolished or abandoned in retaliation to the colonial authority [20]. A sizeable number of smallholder farmers currently employ terracing and contouring within the Kenya GRV region (Figure 10b).

Table 12. Previous studies in and around the Great Rift Valley region of Kenya.

Studies	Case Study	Mean Erosion Rates (t ha^{-1} y^{-1})	Method
Fenta et al. [4]	Kenya & All croplands	6.95 and 26.0	RUSLE
Haregeweyn et al. [63]	Cultivated lands, Upper Blue Nile	28.8	RUSLE
Kogo et al. [31]	Lake Victoria Basin, Western Kenya	7.5–12.3	RUSLE
Defersha et al. [32]	Bush land, Mara River Basin	7	WEPP & EROSION 3D
Aneseyee et al. [79]	Omo-Gibe Basin (ERV)	17.65	RUSLE
Hategekimana et al. [28]	Kenyan Coast	10–27.9	RUSLE
Ligonja and Shrestha [81]	Kondoa, Tanzania (ERV)	15.7	USLE
Sutherland and Bryan [83]	Lake Baringo sub-basin	16–96	Plot study
Mati et al. [24]	Upper EwasoNg'irosub-basin	0–51.3	Plot study
Kiepe [84]	Machakos, Kenya	16–36	Plot study
Tiffen et al. [85]	Athi basin area	15.0	Plot study

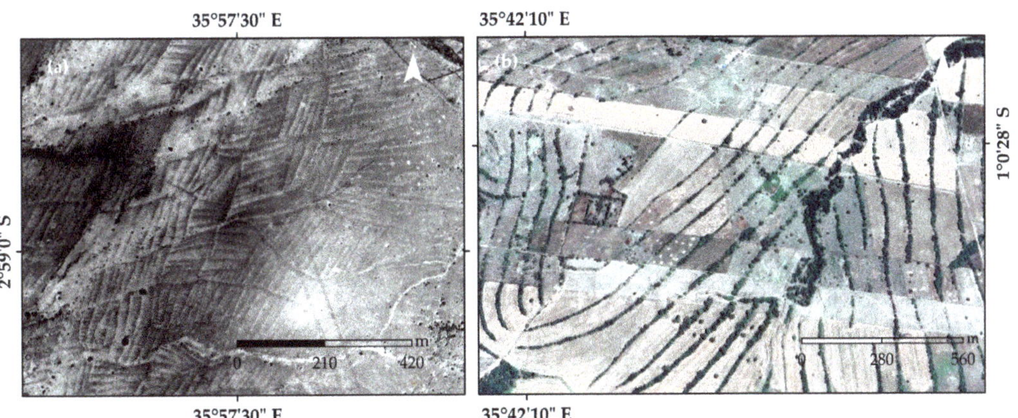

Figure 10. Soil conservation practices in the Eastern Rift Valley (ERV) region: (**a**) an archived aerial image (1960s) showing abandoned stone bunds and terraces of the semi-arid, historical agricultural landscape on the foot of the ERV escarpment at Engaruka, Tanzania; (**b**) Google Earth image taken on 26 July 2019 depicting soil conservation measures in Narok County, Kenya GRV.

Ruto et al. [86] revealed that terracing reduced soil erosion activity in Narok County and significantly increased maize and beans yields. Despite their effectiveness in controlling runoff, terraces, and stone bunds can be the source of erosion if poorly maintained or abandoned over time [64]. The high mean water erosion rates in the river basins can be attributed to negative land use land changes (Figure 2b) as well as neglect in adopting effective soil water conservation measures (Figure 10). Zhunusovaet al. [87] indicated that the single use of terraces had negative impact on crop yield in the Lake Naivasha basin while Reference [88] found that mulching and ground cover can be ineffective in controlling runoff flow on croplands with steep slopes as those in Kenya GRV region. In addition, Willy et al. [77] reported combined control measures (multiple soil conservation practices) can be adequate within the Lake Naivasha basin. Ten out of the thirteen water basins in the Kenya GRV region are located in ASAL regions characterized by lowland pasture, desert shrubs, exposed barren areas with sparse vegetation and poor land, and animal husbandry; thus, susceptible to water erosion (Figure 11). In Kerio Valley basin, increase in barelands, degraded forests coupled with recent high rainfall intensity increased soil loss resulting in heavy sedimentation as evidenced by very low water levels of Lake Kamnarok in the downstream areas [12]. Although the RUSLE method has been applied widely in different landscapes with significant results, its accuracy largely depends on the type of dataset (resolution, up-to-date, preference of primary over secondary data) and data manipulation methods [30]. The method suffers some limitations and its applicability in mountainous terrain remains doubtful [89]; possibly explaining the very high mean erosion rates recorded on escarpments and ranges within the Kenya GRV region.

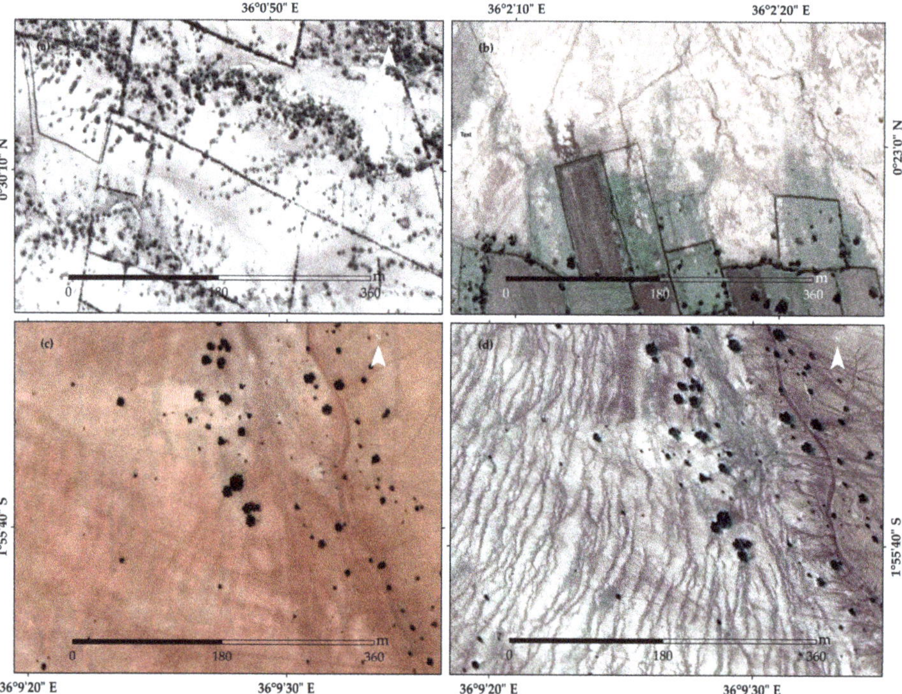

Figure 11. Google Earth images showing spatial occurrence of water erosion in the Kenya GRV region: (**a**) image taken 15th May, 2017 showing exposed soil areas in Marigat District within the Lake Baringo basin; (**b**) image taken 15th May 2017 indicating poor soil conservation measures in Lorwak in Baringo County; (**c**,**d**) a time series images taken 9t October 2017 and 18th December 2017 respectively showing water erosion severity on a bare area in Narok County during the short rain period (October–December, 2017).

The soil erodibility factor for this study did not include pertinent factors, e.g., level of soil weathering, resistance against dispersion and crusting [90]. The mean annual erosion estimated in the current study also does not account for rainfall erositvity and vegetation seasonal variability. The Van der Knijff algorithm $C = exp\left(-a * \frac{NDVI}{\beta - NDVI}\right)$ [91] where a and β are the parameters that determine the shape of the (Normalized Difference Vegetation Index) NDVI-C curve have been employed by several studies [28,29] in the East Africa (EA) region to estimate vegetation cover factor although it can lead to overestimation of C values in tropical regions with high rainfalls [30,41,92] (e.g., the forested areas of the KGRV). This equation gave high range of values for the KGRV (0.0–1.6) when compared to the values generated by the Durigon equation [92] $C = (-NDVI + 1)/2$ (0.05–0.65) recommended for tropical areas. This is after applying these equations on eleven MODIS (Moderate Resolution Imaging Spectroradiometer) NDVI datasets taken during the KGRV's wet seasons in the year 2015 for comparison purposes. Methods that assign C values to different LULC classes based on on-field determinations or combination techniques (e.g., involving image transformations and geostatistical analysis) are recommended for better estimation of C factor [4,30,36,71]. Table 13 shows the zonal statistics details between C factor map given by the Durigon equation on MODIS NDVI and the 2015 LULC map. The mean C factor values for forests are high (approximately 150 times higher than those used in this study) resulting to overestimation of the corresponding mean soil loss rates. This can give a false interpretation that croplands are better than forests in controlling water erosion in the KGRV.

Table 13. The mean C factor values for different land uses in KGRV and their corresponding soil loss rates for 2015.

LULC	Mean C Factor Value (Using Durigton Equation with MODIS NDVI Data)	Mean Erosion Rate (2015) (t ha^{-1} y^{-1})
Dense Forest	0.16	52.9
Open Forest	0.22	40.5
Shrubland	0.30	22.2
Grassland	0.33	9.8
Cropland	0.21	26.9
Bareland	0.42	6.59
Overall	0.32	19.0

Few works have been undertaken to estimate P factor values for the EA region [93] thus a maximum value of "one" was assigned to the Kenya GRV study area to indicate none conservation measures. The P factor method previously applied by other researchers in the region in which agricultural lands are assigned P values in relation to percent slope [41,46] (0–5%, P = 0.10; 5–10%, P = 0.12; 10–20%, P = 0.14; 20–30%, P = 0.19; 30–50%, P = 0.25; and 50–100%, P = 0.33) was noted to generate a comparatively low mean erosion rate (6.3 t ha^{-1} yr^{-1}) for croplands within the Kenya GRV after including it in the 2015 RUSLE model. Use of high resolution imagery with field data can be more sufficient [94]. In addition, generation of RUSLE parameters from published datasets with medium or coarse resolutions (data interpolations) may produce spatially variable erosion rates that can exceed acceptable tolerance levels [95]. These limitations notwithstanding, the RUSLE method was seen as a fast and practical approach in pinpointing potential erosion hotspots for such a vast region using limited data and the study results will be valuable in the management of the Kenya GRV region as well as provide useful guidelines in soil erosion investigations in tropical areas.

5. Conclusions

Water erosion is major source of soil degradation in Kenya whose land mass is dominated by arid and semi-arid lands. These ASALs are susceptible to natural hazards including soil erosion that can destroy vegetation cover resulting to land degradation hence increase desertification risk. This poses a significant threat to agricultural production and food security in Kenya. The present study examined the magnitude of soil erosion rates using the RUSLE model in the Great Rift Valley region of Kenya due to its environmental diversity (i.e., combination of ASALs and agricultural lands) and important ecosystem services it provides in the country. The study is among the first attempts to quantify multi-temporal soil loss rates at the national scale for Kenya (with about 33% of the total land mass). Annual soil loss was found to be severe in the central and southern parts of region particularly along mountain fringes with high rainfall intensity for both years. The overall mean soil loss rate for the entire area fell under the tolerable erosion rate of 10 t ha^{-1} yr^{-1} in 1990 and 2015 although the substantial net changes in erosion rates for croplands underscores the need to devise effective anti-erosive interventions. Areas that require prioritized in soil conservation measures include more than half of all the protected areas as well as croplands in the central and southern regions of the Kenya GRV that presented mean erosion rates higher than 15 t ha^{-1} yr^{-1}. This also includes water basins: e.g., Lake Nakuru which has high urbanization trends as well ASAL basins, and Lake Bogoria-Baringo basin which has been adversely affected by human activities, e.g., agriculture with poor SWC measures and over reliance to pastoralism. Outcomes of this study can inform watershed managers on ways to reduce soil erosion rates, e.g., value of integrated conservation practices, curbing unregulated land use, overgrazing, limiting mass migration and deforestation as well as encouraging conservation tillage. The findings can further help policy makers plan for sustainable soil management strategies as the country gears towards achieving land degradation-neutrality.

Supplementary Materials: The RUSLE model factor maps for KGRV are available online at https://www.mdpi.com/2071-1050/13/2/844/s1.

Author Contributions: Formal analysis, methodology and investigation and writing; G.W., L.Y. and J.Z. Project supervision and administration; L.Y. and Y.N. Reviewing, statistical analysis and discussion; T.N., B.K., J.d.D.N. and Y.N. All authors have read and agreed to the published version of the manuscript.

Funding: This research was funded by the National Key Research and Development Project of China through grant agreement No 2020YFC1521900 and 2020YFC1521901. This work has also been supported by National Key Research and Development Program (No. 2014A8007007020) and China Scholarship Council 2014GXYB33 through Sino-Africa Joint Research Centre.

Institutional Review Board Statement: Not applicable.

Informed Consent Statement: Not applicable.

Data Availability Statement: Data available in a publicly accessible repository that that does not issue DOIs.

Acknowledgments: The authors would like to thank the editor and anonymous reviewers for their valuable comments and suggestions to improve the quality of this paper. The authors thank Yves Hategekimana, Fidele Kamarage and Felix Mutua for their valuable assistance.

Conflicts of Interest: The authors declare no conflict of interest.

Appendix A

Table A1. Estimated Water-induced Mean soil loss rates under different Land Use and Land Cover (LULC) Conversions within the Great Rift Valley region of Kenya.

LULC Conversions	Area (ha)	Average Erosion Rate (t/ha/y)	Average Slope (%)
Bareland to Cropland	2810	−11.2	7.5
Bareland to Dense Forest	2724	−23.6	15.0
Bareland to Grassland	468,651	−3.0	4.0
Bareland to Open Forest	12,554	−4.6	4.3
Bareland to Shrubland	450,977	−6.3	8.1
Cropland to Bareland	3683	12.2	4.7
Cropland to Dense Forest	13,471	−26.4	14.6
Cropland to Grassland	75,764	−4.7	6.3
Cropland to Open Forest	3015	−14.5	11.5
Cropland to Shrubland	68,429	−5.3	12.1
Dense Forest to Bareland	1441	59.0	9.8
Dense Forest to Cropland	164,363	31.0	15.7
Dense Forest to Grassland	45,609	9.2	14.2
Dense Forest to Open Forest	73,087	1.6	20.7
Dense Forest to Shrubland	132,667	15.5	19.9
Grassland to Bareland	240,603	3.7	4.3
Grassland to Cropland	753,068	10.2	9.3
Grassland to Dense Forest	37,950	−9.6	17.6
Grassland to Open Forest	50,646	−3.8	10.5
Grassland to Shrubland	1,228,990	1.0	7.1
Open Forest to Bareland	4492	10.8	6.8
Open Forest to Cropland	35,447	26.3	15.0
Open Forest to Dense Forest	98,021	−3.8	34.6
Open Forest to Grassland	24,458	5.7	13.8
Open Forest to Shrubland	272,772	7.6	18.1
Shrubland to Bareland	429,738	6.5	8.6
Shrubland to Cropland	532,070	10.7	12.4
Shrubland to Dense Forest	142,017	−19.5	29.7
Shrubland to Grassland	1,507,730	−0.9	5.8
Shrubland to Open Forest	227,208	−10.1	23.2

Table A2. Estimated Water-induced Mean soil loss rates per district within the Great Rift Valley region of Kenya.

Districts Names	Area (10^4 ha)	Year 1990		Year 2015		Net Change (t/ha/y)	Average Slope (%)
		Erosion Rate (t/ha/y)	Soil Loss (Mt/y)	Erosion Rate (t/ha/y)	Soil Loss (Mt/y)		
Keiyo	14.38	14.54	2.1	24.77	3.6	10.23	19.8
Marakwet	15.85	13.97	2.2	23.81	3.8	9.84	25.1
Bomet	14.28	10.09	14	16.20	2.3	6.11	10.2
Kericho	20.83	20.05	4.2	26.06	5.4	6.01	16.0
Nandi	28.36	18.61	5.3	24.11	6.8	5.50	15.3
Buret	13.90	13.83	1.9	18.43	2.6	4.60	15.1
Trans Nzoia	24.42	7.37	1.8	11.57	2.8	4.20	11.5
Nakuru	74.82	7.48	5.4	11.64	8.4	4.16	11.4
Uasin Gishu	33.29	5.34	1.8	8.47	2.8	3.13	9.1
Narok	150.23	5.77	8.7	8.62	13	2.85	13.2
Trans Mara	27.84	6.31	1.8	9.10	2.5	2.79	9.5
Nyandarua	32.68	8.44	2.8	11.20	3.7	2.76	13.6
West Pokot	90.01	11.15	10	13.84	12	2.69	20.7
Koibatek	23.09	10.49	2.4	13.17	3	2.68	15.8
Laikipia	94.17	4.12	3.9	4.77	4.5	0.65	8.9
Baringo	86.45	11.23	9.5	11.43	9.7	0.20	16.3
Kiambu	3.23	12.75	0.4	12.63	0.4	−0.12	15.9
Turkana	499.8	3.46	17.3	3.21	19	−0.25	10.9
Kajiado	216.83	3.72	8.1	3.43	7.4	−0.29	9.9
Marsabit	184.66	7.28	8.3	6.2	7	−1.08	6.7
Samburu	209.71	6.96	14	5.75	12	−1.21	16.5

Table A3. Estimated mean soil loss and mean slope per district of erosion-prone areas experiencing Land Use/Land Cover Changes (LULCC) between 1990 and 2015.

Districts Names	Extent of LULCC (%)	Average Erosion Rate (t/ha/y)	Average Rainfall (mm/y)	Average Slope (%)
Keiyo	0.37	10.24	1195	19.8
Marakwet	0.44	9.83	1143	25.1
Bomet	0.55	6.16	1315	10.2
Kericho	0.49	6.01	1458	16
Nandi	0.7	5.49	1514	15.3
Buret	0.33	4.6	1661	15.1
Trans Nzoia	0.83	4.17	1172	11.5
Nakuru	2.03	4.13	988	11.4
Uasin Gishu	1.0	3.14	1055	9.1
Narok	3.23	2.85	921	13.2
Nyandarua	0.99	2.78	1140	13.6
West Pokot	1.6	2.74	772	20.7
Koibatek	0.48	2.73	1055	15.8
Trans Mara	0.77	2.7	1365	9.5
Laikipia	1.96	0.64	746	8.9
Baringo	0.79	0.14	801	16.3
Marsabit	1.88	−0.12	344	6.7
Turkana	12.06	−0.23	258	10.9
Kajiado	4.11	−0.31	574	9.9
Kiambu	0.08	−0.5	923	15.9
Samburu	3.52	−1.21	528	16.5

Table A4. Distribution of estimated soil erosion rates under different soil types in the KGRV.

Dominant Soil Type	Area (10^4 ha)	Extent (%)	Year 1990		Year 2015		Net Change (t/ha/y)	Average Slope (%)
			Erosion Rate (t/ha/y)	Soil Loss (Mt/y)	Erosion Rate (t/ha/y)	Soil Loss (Mt/y)		
Andosols	66.82	3.59	6.01	4.0	11.53	7.7	5.52	12.75
Nitosols	199.03	10.7	9.46	18.8	13.9	27.6	4.44	11.87
Planosols	21.17	1.13	7.31	1.5	9.91	2.0	2.6	8.26
Ferrasols	132.62	7.13	4.88	6.4	6.89	9.1	2.01	7.92
Cambisols	136.24	7.32	5.56	7.5	6.34	8.6	0.78	8.65
Lithosols	576.78	31.02	6.66	38.4	7.27	41.9	0.61	11.52
Vertisols	101.79	5.47	2.83	2.9	3.26	3.3	0.43	5.52
Luvisols	29.22	1.57	5.12	1.4	5.48	1.6	0.36	7.92
Gleysols	41.28	2.22	2.32	0.95	2.47	1.0	0.15	3.9
Solonchaks	56.60	3.12	3.89	2.2	3.66	2.0	−0.23	7.14
Fluvisols	15.07	0.81	3.02	0.45	2.78	0.41	−0.24	3.89
Regosols	283.58	15.25	6.98	19.7	6.71	19.0	−0.27	11.32
Xerosols	123.39	6.63	4.62	5.7	3.02	3.7	−1.6	8.29
Yermosols	75.24	4.04	6.15	4.6	4.21	3.1	−1.94	13.08

Table A5. Estimated Water-induced Mean soil loss rates per Protected Areas within the Great Rift Valley region of Kenya.

Protected Area Names	Area (10^4 ha)	Year 1990		Year 2015		Net Change (t/ha/y)	Average Slope (%)
		Erosion Rate (t/ha/y)	Soil Loss(Mt/y)	Erosion Rate (t/ha/y)	Soil Loss (Mt/y)		
Chemurokoi	0.39	12.91	0.05	40.87	0.16	27.96	27.4
Kaisungor	0.11	22.47	0.02	47.04	0.05	24.56	31.1
Sogotio	0.33	17.71	0.06	40.70	0.13	22.99	29.9
Metkei	0.17	9.42	0.02	26.52	0.05	17.1	45.6
Kipkabus (Elg-Marak)	0.06	17.77	0.01	34.15	0.02	16.37	43.7
Kessop	0.25	39.43	0.10	51.86	0.13	12.43	38.6
Eastern Mau	6.59	5.94	0.39	18.18	1.20	12.24	15.7
Kerrer	0.32	15.35	0.05	27.35	0.09	12	26.4
Kipkabus (Uasin/Gishu)	0.70	7.16	0.05	17.79	0.12	10.63	12.2
Bahati	1.12	8.90	0.10	19.32	0.22	10.42	15.1
Timboroa	0.53	14.17	0.07	24.55	0.13	10.37	17.8
Kapchemutwa	0.92	28.55	0.26	38.79	0.36	10.23	37.5
Lelan	1.10	21.65	0.24	31.60	0.35	9.94	37.1
Embobut	2.00	17.98	0.36	27.76	0.56	9.78	34.2
Mau Narok	0.07	3.39	0.00	12.99	0.01	9.6	23.4
Marmanet	2.08	5.92	0.12	15.41	0.32	9.49	16.1
Kapolet	0.13	8.91	0.01	18.18	0.02	9.26	26.3
South-western Mau	8.32	4.56	0.38	13.45	1.12	8.88	16.5
Kaptagat	1.18	8.39	0.10	16.97	0.20	8.57	13.4
Transmara	3.77	5.51	0.21	13.91	0.52	8.39	26.4
Maji Mazuri	0.80	5.43	0.04	13.77	0.11	8.33	15.6
Kiptaberr	1.07	8.80	0.09	16.87	0.18	8.06	22.4
Londiani	2.15	7.74	0.17	15.00	0.32	7.26	15.6
Taressia	0.04	14.18	0.01	20.66	0.01	6.48	10.6
Katimok	0.19	23.72	0.05	30.17	0.06	6.45	29.3
Kipkunurr	1.53	10.05	0.15	16.40	0.25	6.35	28.8
Southern Mau	0.01	13.13	0.00	19.22	0.00	6.09	12.6
Kakamega	0.70	20.78	0.15	26.74	0.19	5.96	15
Kilombe Hill	0.32	13.07	0.04	19.01	0.06	5.94	19.3
Ol-arabel	0.99	7.82	0.08	13.72	0.14	5.9	14
Molo	0.09	13.02	0.01	18.78	0.02	5.75	
North Nandi	1.13	12.00	0.13	17.67	0.20	5.67	11.6
Ol-pusimoru	3.57	4.49	0.16	10.14	0.36	5.65	15.7
Mount Londiani	2.22	5.39	0.12	10.38	0.23	4.98	18.7
Northern Tinderet	2.94	6.45	0.19	11.35	0.33	4.9	17.3
Kapsaret	0.13	4.77	0.01	9.19	0.01	4.42	8.7
Tinderet	3.40	7.13	0.24	11.33	0.38	4.2	17.1
Chemorogok	1.12	9.84	0.11	13.87	0.15	4.02	19.6
Ngong Hills	0.33	8.86	0.03	11.78	0.04	2.92	11.3
Ol-bolossat	0.36	3.24	0.01	6.03	0.02	2.79	7.1
South Nandi	1.99	11.15	0.22	13.90	0.28	2.75	14.4
Menengai	0.55	10.61	0.06	13.28	0.07	2.67	13.5
Leshau	0.02	3.64	0.00	6.27	0.00	2.63	6.4
Kitalale	0.20	5.04	0.01	7.62	0.01	2.57	7.3
Sekhendu	0.06	5.43	0.00	7.99	0.01	2.55	7.5
Mount Elgon	0.99	9.36	0.09	11.68	0.12	2.32	27.1
Kipipiri	0.43	12.86	0.06	15.12	0.07	2.26	32.3
Mukogodo	2.94	10.54	0.31	12.69	0.37	2.14	24.1
Uaso Narok	0.12	4.82	0.01	6.96	0.01	2.14	8.2
Chepalungu	0.50	2.25	0.01	4.36	0.02	2.11	5.6
Rumuruti	0.62	2.95	0.02	5.03	0.03	2.08	6.8
Kapkanyar	0.70	2.72	0.02	3.83	0.03	1.11	20.2
Eburu	0.82	8.09	0.07	9.17	0.07	1.07	27.2
Lariak	0.52	3.81	0.02	4.87	0.03	1.05	6.7
Sibiloi	14.68	2.57	0.38	3.06	0.45	0.48	9.1
Kerio Valley	0.49	1.27	0.01	1.45	0.01	0.18	5.7
Nakuru	0.05	0.95	0.00	1.06	0.00	0.1	8.1
Nasolot	0.76	11.78	0.09	11.78	0.09	0	21.1
Perkerra Catchment	0.53	13.23	0.07	13.22	0.07	0	14.1

Table A5. *Cont.*

Protected Area Names	Area (10^4 ha)	Year 1990		Year 2015		Net Change (t/ha/y)	Average Slope (%)
		Erosion Rate (t/ha/y)	Soil Loss(Mt/y)	Erosion Rate (t/ha/y)	Soil Loss (Mt/y)		
Masai Mara	14.99	3.71	0.56	3.70	0.55	−0.01	9
South Turkana	10.45	5.14	0.54	5.11	0.53	−0.02	15.4
Lake Nakuru	1.87	2.82	0.05	2.78	0.05	−0.04	8
Longonot	0.28	10.43	0.03	10.07	0.03	−0.36	17.2
Mount Nyiru	3.83	6.44	0.25	6.03	0.23	−0.41	37.6
Maralai	1.72	6.47	0.11	5.87	0.10	−0.6	16.2
Mukobe	0.08	7.62	0.01	6.99	0.01	−0.62	11.7
Samburu	1.59	3.96	0.06	3.17	0.05	−0.79	9.5
South Island	0.86	20.25	0.17	19.08	0.16	−1.16	9.7
Amboseli	4.01	2.69	0.11	1.50	0.06	−1.18	7
Hell's Gate	1.19	13.63	0.16	12.17	0.15	−1.45	18.6
Lake Bogoria	0.85	19.26	0.16	17.10	0.14	−2.16	22.5
Loitokitok	0.22	7.62	0.02	5.24	0.01	−2.38	9.2
Kabarak	0.15	27.43	0.04	24.58	0.04	−2.85	27.5
Kimojoch	0.08	44.33	0.03	41.45	0.03	−2.87	38.8
Saimo	0.10	33.20	0.03	29.46	0.03	−3.74	34.1
Namanga Hill	1.06	10.74	0.11	6.58	0.07	−4.15	32.6
Ndotos Range	9.51	12.42	1.18	6.98	0.66	−5.43	34.1
Matthews Range	9.41	10.70	1.01	5.23	0.49	−5.46	31.2
Central Island	0.11	18.06	0.02	9.90	0.01	−8.16	10.5

References

1. Gomiero, T. Soil Degradation, Land Scarcity and Food Security: Reviewing a Complex Challenge. *Sustainability* **2016**, *8*, 281. [CrossRef]
2. Chuenchum, P.; Xu, M.; Tang, W. Estimation of Soil Erosion and Sediment Yield in the Lancang–Mekong River Using the Modified Revised Universal Soil Loss Equation and GIS Techniques. *Water* **2020**, *12*, 135. [CrossRef]
3. Blake, W.H.; Rabinovich, A.; Wynants, M.; Kelly, C.; Nasseri, M.; Ngondya, I.; Patrick, A.; Mtei, K.; Munishi, L.; Boeckx, P.; et al. Soil erosion in East Africa: An interdisciplinary approach to realising pastoral land management change. *Environ. Res. Lett.* **2018**, *13*, 124014. [CrossRef]
4. Fenta, A.A.; Tsunekawa, A.; Haregeweyn, N.; Poesen, J.; Tsubo, M.; Borrelli, P.; Panagos, P.; Vanmaercke, M.; Broeckx, J.; Yasuda, H.; et al. Land susceptibility to water and wind erosion risks in the East Africa region. *Sci. Total Environ.* **2020**, *703*, 135016. [CrossRef] [PubMed]
5. Nguru, P.M.; Rono, D.K. Combating Desertification in Kenya. In *Combating Desertification in Asia, Africa and the Middle East*; Heshmati, G.A., Squires, V., Eds.; Springer Science & Business: New York, NY, USA; London, UK, 2013; pp. 139–151. [CrossRef]
6. Van Straaten, P. *Rocks for Crops: Agrominerals of Sub-Saharan Africa*; ICRAF: Nairobi, Kenya, 2002.
7. Fenta, A.A.; Yasuda, H.; Shimizu, K.; Haregeweyn, N.; Kawai, T.; Sultan, D.; Ebabu, K.; Belay, A.S. Spatial distribution and temporal trends of rainfall and erosivity in the Eastern Africa region. *Hydrol. Process.* **2017**, *31*, 4555–4567. [CrossRef]
8. Ameso, E.A.; Bukachi, S.A.; Olungah, C.O.; Haller, T.; Wandibba, S.; Nangendo, S. Pastrol Resilience among the Maasai Pastrolists of Laikipia County, Kenya. *Land* **2018**, *7*, 78. [CrossRef]
9. Government of Kenya. Statistical Abstract. 2019. Available online: https://www.knbs.or.ke/?wpdmpro=statistical-abstract-2019 (accessed on 3 September 2020).
10. Mulinge, W.; Gicheru, P.; Murithi, F.; Maingi, P.; Kihiu, E.; Kirui, O.K.; Mirzabaev, A. Economics of Land Degradation and Improvement in Kenya. In *Economics of Land Degradation and Improvement—A Global Assessment for Sustainable Development*; Nkonya, E., Mirzabaev, A., von Braun, J., Eds.; Springer International Publishing: Cham, Switzerland, 2016; pp. 471–498. [CrossRef]
11. Ministry of Environment and Forestry. Land Degradation Assessment in Kenya-March. 2016. Available online: http://www.environment.go.ke/wp-content/uploads/2018/08/LADA-Land-Degradation-Assessment-in-Kenya-March-2016.pdf (accessed on 12 July 2020).
12. Boitt, M.; Albright, O.; Kipkulei, H. Assessment of Soil Erosion and Climate Variability on Kerio Valley Basin, Kenya. *J. Geosci. Environ. Prot.* **2020**, *8*, 97–114. [CrossRef]
13. Cohen, M.J.; Brown, M.T.; Shepherd, K.D. Estimating the environmental costs of soil erosion at multiple scales in Kenya using emergy synthesis. *Agric. Ecosyst. Environ.* **2006**, *114*, 249–269. [CrossRef]
14. Angima, S.; Stott, D.; O'Neill, M.; Ong, C.K.; Weesies, G.A. Soil Erosion Prediction Using RUSLE for Central Kenya Highland Conditions. *Agric. Ecosyst. Environ.* **2003**, *97*, 295–308. [CrossRef]
15. Gachene, C.K.K.; Mbuvi, J.P.; Jarvis, N.J.; Linner, H. Soil Erosion Effects on Soil Properties in a Highland Area of Central Kenya. *Soil Sci. Soc. Am. J.* **1997**, *61*, 559–564. [CrossRef]

16. Vigani, M.; Dudu, H.; Ferrari, E.; Mainar, A. *Estimation of Food Demand Parameters in Kenya A Quadratic Almost Ideal Demand System (QUAIDS) Approach*; Publications Office of the European Union: Luxembourg, 2019. [CrossRef]
17. Were, K.O.; Dick, Ø.B.; Singh, B.R. Remotely sensing the spatial and temporal land cover changes in Eastern Mau forest reserve and Lake Nakuru drainage basin, Kenya. *Appl. Geogr.* **2013**, *41*, 75–86. [CrossRef]
18. Muriithi, F.K. Land use and land cover (LULC) changes in semi-arid sub-watersheds of Laikipia and Athi River basins, Kenya, as influenced by expanding intensive commercial horticulture. *Remote Sens. Appl. Soc. Environ.* **2016**, *3*, 73–88. [CrossRef]
19. Muchena, F.N. *Indicators for Sustainable Land Management in Kenya's Context. GEF Land Degradation Focal Area Indicators*; ETC-East Africa: Nairobi, Kenya, 2008.
20. Dregne, H.E. Land Degradation in the Drylands. *Arid Land Res. Manag.* **2002**, *16*, 99–132. [CrossRef]
21. Ministry of Agriculture Livestock Fisheries and Irrigation. *Towards Sustainable Agricultral Transformation and Food Security in Kenya 2019–2029*; Government Press: Nairobi, Kenya, 2019. Available online: http://www.kilimo.go.ke/wp-content/uploads/2019/05/ASTGS-Long-version.pdf (accessed on 3 September 2020).
22. Hirons, L.; Turner, A. The Impact of Indian Ocean Mean-State Biases in Climate Models on the Representation of the East African Short Rains. *J. Clim.* **2018**, *31*, 6611–6631. [CrossRef]
23. Obando, J.A.; Onywere, S.; Shisanya, C.; Ndubi, A.; Masiga, D.; Irura, Z.; Mariita, N.; Maragia, H. Impact of Short-Term Flooding on Livelihoods in the Kenya Rift Valley Lakes. In *Geomorphology and Society, Advances in Geographical and Environmental Sciences*; Meadows, M.E., Lin, J.-C., Eds.; Springer: Tokyo, Japan, 2016; pp. 193–215. [CrossRef]
24. Mati, B.; Morgan, R.; Quinton, J.; Brewer, T.; Liniger, H. Assessment of erosion hazard with the USLE and GIS: A case study of the Upper Ewaso Ng'iro North basin of Kenya. *Int. J. Appl. Earth Obs. Geoinf.* **2000**, *2*, 78–86. [CrossRef]
25. Luvai, A.; Obiero, J.P.; Omuto, C. Methods for Erosion Estimates in Assessment of Soil Degradation: A Review for Catchments in Kenya. *Int. J. Eng. Res.* **2020**. [CrossRef]
26. Avwunudiogba, A.; Franklin Hudson, P. A Review of Soil Erosion Models with Special Reference to the needs of Humid Tropical Mountainous Environments. *Eur. J. Sustain. Dev.* **2014**, *3*, 299. [CrossRef]
27. Wischmeier, W.H.S.; Smith, D.D. *Predicting Rainfall Erosion Losses—A Guide to Conservation Planning*; US Department of Agriculture Science and Education Administration: Washington, DC, USA, 1978.
28. Hategekimana, Y.; Allam, M.; Meng, Q.; Nie, Y.; Mohamed, E. Quantification of Soil Losses along the Coastal Protected Areas in Kenya. *Land* **2020**, *9*, 137. [CrossRef]
29. Karamage, F.; Zhang, C.; Liu, T.; Maganda, A.; Isabwe, A. Soil Erosion Risk Assessment in Uganda. *Forests* **2017**, *8*, 52. [CrossRef]
30. Phinzi, K.; Ngetar, N.S. The assessment of water-borne erosion at catchment level using GIS-based RUSLE and remote sensing: A review. *Int. Soil Water Conserv. Res.* **2019**, *7*, 27–46. [CrossRef]
31. Kogo, B.K.; Kumar, L.; Koech, R. Impact of Land Use/Cover Changes on Soil Erosion in Western Kenya. *Sustainability* **2020**, *12*, 9740. [CrossRef]
32. Defersha, M.B.; Melesse, A.M.; McClain, M.E. Watershed scale application of WEPP and EROSION 3D models for assessment of potential sediment source areas and runoff flux in the Mara River Basin, Kenya. *CATENA* **2012**, *95*, 63–72. [CrossRef]
33. Mati, B.; Morgan, R.; Quinton, J. Soil erosion modelling with EUROSEM at Embori and Mukogodo catchments, Kenya. *Earth Surf. Process. Landf.* **2006**, *31*, 579–588. [CrossRef]
34. Baker, T.J.; Miller, S.N. Using the Soil and Water Assessment Tool (SWAT) to assess land use impact on water resources in an East African watershed. *J. Hydrol.* **2013**, *486*, 100–111. [CrossRef]
35. Hunink, J.E.; Niadas, I.A.; Antonaropoulos, P.; Droogers, P.; de Vente, J. Targeting of intervention areas to reduce reservoir sedimentation in the Tana catchment (Kenya) using SWAT. *Hydrol. Sci. J.* **2013**, *58*, 600–614. [CrossRef]
36. Schürz, C.; Mehdi, B.; Kiesel, J.; Schulz, K.; Herrnegger, M. A systematic assessment of uncertainties in large scale soil loss estimation from different representations of USLE input factors—A case study for Kenya and Uganda. *Hydrol. Earth Syst. Sci. Discuss.* **2019**, *2019*, 1–35. [CrossRef]
37. UNESCO. World Heritage List-Kenya Lake System in the Great Rift Valley. Available online: https://whc.unesco.org/en/list/1060 (accessed on 12 July 2020).
38. Eshiamwata, G.W. *Monitoring Habitat at Key Biodiversity Sites in Africa Using Remote Sensing: Land Cover Change at important Bird Areas in Eastern Africa*; University of Nairobi: Nairobi, Kenya, 2012; Available online: http://erepository.uonbi.ac.ke/handle/11295/6777 (accessed on 7 June 2020).
39. FAO. Digital Soil Map of the World. Available online: http://data.fao.org/maps/wms?styles=geonetwork_DSMW_14116_style (accessed on 3 September 2020).
40. World Resources Institute. Kenya GIS Data-Annual Projected Water Balance by Subdrainage Area in Kenya, 2000 and 2010. In *World Resources Institute*; 2007; Available online: https://www.wri.org/resources/data-sets/kenya-gis-data#water (accessed on 3 September 2020).
41. Nambajimana, J.d.D.; He, X.; Zhou, J.; Justine, M.F.; Li, J.; Khurram, D.; Mind'je, R.; Nsabimana, G. Land Use Change Impacts on Water Erosion in Rwanda. *Sustainability* **2019**, *12*, 50. [CrossRef]
42. United States Geological Survey. U.S. Geological Survey Earthexplorer. Available online: http://earthexplorer.usgs.gov/ (accessed on 3 September 2020).

43. Hengl, T.; Heuvelink, G.B.M.; Kempen, B.; Leenaars, J.G.B.; Walsh, M.G.; Shepherd, K.D.; Sila, A.; MacMillan, R.A.; Mendes de Jesus, J.; Tamene, L.; et al. Mapping Soil Properties of Africa at 250 m Resolution: Random Forests Significantly Improve Current Predictions. *PLoS ONE* **2015**, *10*, e0125814. [CrossRef]
44. Funk, C.; Peterson, P.; Landsfeld, M.; Pedreros, D.; Verdin, J.; Shukla, S.; Husak, G.; Rowland, J.; Harrison, L.; Hoell, A.; et al. The climate hazards infrared precipitation with stations—A new environmental record for monitoring extremes. *Sci. Data* **2015**, *2*, 150066. [CrossRef]
45. Renard, K.G.; Foster, G.R.; Weesies, G.; McCool, D.; Yoder, D. *Predicting Soil Erosion by Water: A Guide to Conservation Planning with the Revised Universal Soil Loss Equation (Rusle)*; United States Department of Agriculture: Washington, DC, USA, 1997; Volume 703.
46. Weldu Woldemariam, G.; Edo Harka, A. Effect of Land Use and Land Cover Change on Soil Erosion in Erer Sub-Basin, Northeast Wabi Shebelle Basin, Ethiopia. *Land* **2020**, *9*, 111. [CrossRef]
47. Lo, A.; El-Swaify, S.A.; Dangler, E.W.; Shinshiro, L. *Effectiveness of EI30 as An Erosivity Index in Hawaii*; El-Swaify, S.A., Moldenhauer, W.C., Lo, A., Eds.; Soil Conservation Society of America: Ankeny, IA, USA, 1985.
48. Woldemariam, G.W.; Iguala, A.D.; Tekalign, S.; Reddy, R.U. Spatial Modeling of Soil Erosion Risk and Its Implication for Conservation Planning: The Case of the Gobele Watershed, East Hararghe Zone, Ethiopia. *Land* **2018**, *7*, 25. [CrossRef]
49. Ayugi, B.; Tan, G.; Ullah, W.; Boiyo, R.; Ongoma, V. Inter-comparison of remotely sensed precipitation datasets over Kenya during 1998–2016. *Atmos. Res.* **2019**, *225*, 96–109. [CrossRef]
50. Kimaru, A.N.; Gathenya, J.M.; Cheruiyot, C.K. The Temporal Variability of Rainfall and Streamflow into Lake Nakuru, Kenya, Assessed Using SWAT and Hydrometeorological Indices. *Hydrology* **2019**, *6*, 88. [CrossRef]
51. Dinku, T.; Funk, C.; Peterson, P.; Maidment, R.; Tadesse, T.; Gadain, H.; Ceccato, P. Validation of the CHIRPS Satellite Rainfall Estimates over Eastern of Africa: Validation of the CHIRPS Satellite Rainfall Estimates. *Q. J. R. Meteorol. Soc.* **2018**, *144*. [CrossRef]
52. Sujatha, E.R.; Sridhar, V. Spatial Prediction of Erosion Risk of a Small Mountainous Watershed Using RUSLE: A Case-Study of the Palar Sub-Watershed in Kodaikanal, South India. *Water* **2018**, *10*, 1608. [CrossRef]
53. Williams, J.R. The EPIC Model. In *Computer Models of Watershed Hydrology*; Water Resources Publications: Highlands Ranch, CO, USA, 1995.
54. Morgan, R.P.C.; Morgan, D.D.V.; Finney, H.J. A predictive model for the assessment of soil erosion risk. *J. Agric. Eng. Res.* **1984**, *30*, 245–253. [CrossRef]
55. Desmet, P.J.J.; Govers, G. A GIS procedure for automatically calculating the USLE LS factor on topographically complex landscape units. *J. Soil Water Conserv.* **1996**, *5*, 427–433.
56. McCool, D.K.; Brown, L.C.; Foster, G.R.; Mutchler, C.K.; Meyer, L.D. Revised slope steepness factor for the universal soil loss equation. *Trans. ASAE* **1987**, *30*, 1387–1396. [CrossRef]
57. Nyssen, J.; Simegn, G.; Taha, N. An upland farming system under transformation: Proximate causes of land use change in Bela-Welleh catchment (Wag, Northern Ethiopian Highlands). *Soil Tillage Res.* **2009**, *103*, 231–238. [CrossRef]
58. Hurni, H. Erosion-Productivity-Conservation Systems in Ethiopia. In Proceedings of the IV International Conference on Soil Conservation on Soil Conservation, Maracey, Venezuela, 3–9 November 1985; pp. 654–674.
59. Yang, D.; Kanae, S.; Oki, T.; Koike, T.; Musiake, K.; Yang, D.W.; Shinjiro, K.; Taikan, O.; Toshio, K.; Katumi, M. Global potential soil erosion with reference to land use and climate changes. Hydrological Process. *Hydrol. Process.* **2003**, *17*, 2913–2928. [CrossRef]
60. Girma, R.; Gebre, E. Spatial modeling of erosion hotspots using GIS-RUSLE interface in Omo-Gibe river basin, Southern Ethiopia: Implication for soil and water conservation planning. *Environ. Syst. Res.* **2020**, *9*, 19. [CrossRef]
61. Bewket, W.; Teferi, E. Assessment of soil erosion hazard and prioritization for treatment at the watershed level: Case study in the chemoga watershed, blue nile basin, Ethiopia. *Land Degrad. Dev.* **2009**, *20*, 609–622. [CrossRef]
62. Gelagay, H.S.; Minale, A.S. Soil loss estimation using GIS and Remote sensing techniques: A case of Koga watershed, Northwestern Ethiopia. *Int. Soil Water Conserv. Res.* **2016**, *4*, 126–136. [CrossRef]
63. Haregeweyn, N.; Tsunekawa, A.; Poesen, J.; Tsubo, M.; Meshesha, D.T.; Fenta, A.A.; Nyssen, J.; Adgo, E. Comprehensive assessment of soil erosion risk for better land use planning in river basins: Case study of the Upper Blue Nile River. *Sci. Total Environ.* **2017**, *574*, 95–108. [CrossRef]
64. Taye, G.; Poesen, J.; Vanmaercke, M.; Wesemael, B.V.; Martens, L.; Teka, D.; Nyssen, J.; Deckers, J.; Vanacker, V.; Haregeweyn, N.; et al. Evolution of the effectiveness of stone bunds and trenches in reducing runoff and soil loss in the semi-arid Ethiopian highlands. *Z. Geomorphol.* **2015**, *59*, 477–493. [CrossRef]
65. Shin, G. *The Analysis of Soil Erosion Analysis in Watershed Using GIS*; Gang-Won National University: Chuncheon, Korea, 1999.
66. UNEP-WCMC. Protected Areas Delimited by the United Nations Environment Programme (UNEP) and the World Conservation Monitoring Centre (WCMC). Available online: https://www.protectedplanet.net/ (accessed on 3 September 2020).
67. Koirala, P.; Thakuri, S.; Joshi, S.; Chauhan, R. Estimation of Soil Erosion in Nepal Using a RUSLE Modeling and Geospatial Tool. *Geosciences* **2019**, *9*, 147. [CrossRef]
68. GoK. National Water Master Plan 2030, Volume I—Executive Summary. 2013. Available online: https://wasreb.go.ke/downloads/National%20Water%20Master%20Plan%202030%20Exec.%20Summary%20Vol.%201%20Main%201.pdf (accessed on 12 July 2020).
69. FAO. World Reference Base for Soil Resources 2014. In *International Soil Classification System for Naming Soils and Creating Legends for Soil Maps*; FAO: Rome, Italy, 2015; Volume 106.

70. Stocking, M. Rates of erosion and sediment yield in the African environment. In Proceedings of the Challenges in African Hydrology and Water Resources, Harare, Zimbabwe, 11–16 July 1984; pp. 285–295.
71. Tamene, L.; Adimassu, Z.; Aynekulu, E.; Yaekob, T. Estimating landscape susceptibility to soil erosion using a GIS-based approach in Northern Ethiopia. *Int. Soil Water Conserv. Res.* **2017**, *5*, 221–230. [CrossRef]
72. Wambui, M.; Opere, A.; Githaiga, J.; Karanja, F. Assessing the impacts of climate variability and climate change on biodiversity in Lake Nakuru, Kenya. *Bonorowo Wetl.* **2018**, *8*, 13–24. [CrossRef]
73. Olago, D.; Opere, A.; Barongo, J. Holocene palaeohydrology, groundwater and climate change in the lake basins of the Central Kenya Rift. *Hydrol. Sci. J. J. Des. Sci. Hydrol.* **2009**, *54*. [CrossRef]
74. Mubea, K.; Menz, G. Monitoring Land-Use Change in Nakuru (Kenya) Using Multi-Sensor Satellite Data. *Adv. Remote. Sens.* **2012**, *1*, 74–84. [CrossRef]
75. Mubea, K.; Goetzke, R.; Menz, G. Simulating Urban Growth in Nakuru (Kenya) Using Java-Based Modelling Platform XULU. In Proceedings of the 2013 European Modelling Symposium, Manchester, UK, 20–22 November 2013; pp. 103–108.
76. Harper, D.M.; Mavuti, K.M. Lake Naivasha, Kenya: Ecohydrology to guide the management of a tropical protected area. *Ecohydrol. Hydrobiol.* **2004**, *4*, 287–305.
77. Willy, D.K.; Zhunusova, E.; Holm-Müller, K. Estimating the joint effect of multiple soil conservation practices: A case study of smallholder farmers in the Lake Naivasha basin, Kenya. *Land Use Policy* **2014**, *39*, 177–187. [CrossRef]
78. Eric, O.O.; Onyango, J.; Obudho, P.A. *Lake Baringo; Experince and Lessons Learnt Brief*; University of Nairobi: Nairobi, Kenya, 2006.
79. Aneseyee, A.B.; Elias, E.; Soromessa, T.; Feyisa, G.L. Land use/land cover change effect on soil erosion and sediment delivery in the Winike watershed, Omo Gibe Basin, Ethiopia. *Sci. Total Environ.* **2020**, *728*, 138776. [CrossRef] [PubMed]
80. Gizaw, T.; Yalemtsehay, D.; Kalkidan, F. Soil Erosion Risk Assessment Using GIS Based USLE Model for Soil and Water Conservation Planning in Somodo Watershed, South West Ethiopia. *Int. J. Environ. Agric. Res.* **2018**, *4*, 35–43.
81. Ligonja, P.; Shrestha, R. Soil Erosion Assessment in Kondoa Eroded Area in Tanzania using Universal Soil Loss Equation, Geographic Information Systems and Socioeconomic Approach. *Land Degrad. Dev.* **2013**, *26*. [CrossRef]
82. Widgren, M.S. *Islands of Intensive Agriculture in Eastern Africa*; Ohio University Press: Athens, OH, USA, 2004.
83. Sutherland, R.A.; Bryan, K.B. Runoff and erosion from a small semiarid catchment, Baringo district, Kenya. *Appl. Geogr.* **1990**, *10*, 91–109. [CrossRef]
84. Kiepe, P. Cover and Barrier Effect of Cassia Siamea Hedgerows on Soil Conservation in Semi-Arid Kenya. *Soil Technol.* **1996**, *9*, 161–171. [CrossRef]
85. Tiffen, M.; Mortimore, M.; Gichuki, F. *More People Less Erosion*; Environmental Recovery in Kenya; ACTS Press: Nairobi, Kenya, 1994.
86. Ruto, A.; Gachene, C.; Gicheru, P.; Mburu, D.; Khalif, Z. Crop yields along the toposequence of terraced andosols in Narok, Kenya. *Trop. Subtrop. Agroecosyst.* **2017**, *20*, 35–47.
87. Zhunusova, E.; Kyalo Willy, D.; Holm-Müller, K. An Analysis of Returns to Intergrated Soil Conservation Practices in the Lake Naivasha Basin, Kenya. In Proceedings of the 4th International Conference of the African Association of Agricultural Economists, ICAAAE, Hammamet, Tunisia, 22–25 September 2013.
88. Baumhardt, R.L.B.-C. Soil: Conservation Practices. In *Encyclopedia of Agriculture and Food Systems*; Alfen, N.V., Ed.; Elsevier: San Diego, CA, USA, 2014; Volume 5, pp. 153–165.
89. Thapa, P. Spatial estimation of soil erosion using RUSLE modeling: A case study of Dolakha district, Nepal. *Environ. Syst. Res.* **2020**, *9*, 15. [CrossRef]
90. Le Roux, J.J.; Morgenthal, T.L.; Malherbe, J.; Pretorius, D.J.; Sumner, P.D. Water erosion prediction at a national scale for South Africa. *Water SA* **2008**, *34*, 305–314. [CrossRef]
91. Van der Knijff, J.J.R.; Montanarella, L. *Soil Erosion Risk Assessment in Italy*; European Soil Bureau, Joint Research Centre of the European Commission: Brussels, Belgium, 2000.
92. Durigon, V.L.; Carvalho, D.F.; Antunes, M.A.H.; Oliveira, P.T.S.; Fernandes, M.M. NDVI time series for monitoring RUSLE cover management factor in a tropical watershed. *Int. J. Remote Sens.* **2014**, *35*, 441–453. [CrossRef]
93. Xiong, M.-Q.; Sun, R.; Chen, L.-D. Global analysis of support practices in USLE-based soil erosion modeling. *Prog. Phys. Geogr. Earth Environ.* **2019**, *43*. [CrossRef]
94. Wang, R.; Zhang, S.; Yang, J.; Pu, L.; Yang, C.; Yu, L.; Chang, L.; Bu, K. Intergrated Use of GCM, RS, and GIS for the Assessment of Hillslope and Gully Erosion in the Mushi River Sub-Catchment, Northeast China. *Sustainability* **2016**, *8*, 317. [CrossRef]
95. Molnár, D.K.; Julien, P.Y. Estimation of upland erosion using GIS. *Comput. Geosci.* **1998**, *24*, 183–192. [CrossRef]

Article

Climate-Smart Adaptations and Government Extension Partnerships for Sustainable Milpa Farming Systems in Mayan Communities of Southern Belize

Kristin Drexler

American Public University System, Charles Town, WV 25414, USA; kristin.drexler@mycampus.apus.edu

Abstract: There are disproportionate adverse impacts related to climate change on rural subsistence farmers in southern Belize, Central America who depend directly on natural resources for their food and livelihood security. Promoting a more resilient farming system with key climate-smart agriculture (CSA) adaptations can improve productivity, sustainability, and food security for Mayan milpa farming communities. Once a sustainable system, the milpa has become less reliable in the last half century due to hydroclimatic changes (i.e., droughts, flooding, hurricanes), forest loss, soil degradation, and other factors. Using interviews with both milpa farmers and Extension officers in southern Belize. This qualitative study finds several socio-ecological system linkages of environmental, economic, socio-cultural, and adaptive technology factors, which influence the capacity for increasing CSA practices. Agriculture Extension, a government service of Belize, can facilitate effective CSA adaptations, specifically, an increase in mulching, soil nutrient enrichment, and soil cover, while working as partners within Maya farming traditions. These CSA practices can facilitate more equitable increases in crop production, milpa farm system sustainability, and resilience to climate change. However, there are several institutional and operational barriers in Extension which challenge their efficacy. Recommendations are presented in this study to reduce Extension barriers and promote an increase in CSA practices to positively influence food and livelihood security for milpa communities in southern Belize.

Keywords: climate-smart agriculture; socio-ecological systems; extension; Belize; milpa; food security; sustainability

Citation: Drexler, K. Climate-Smart Adaptations and Government Extension Partnerships for Sustainable Milpa Farming Systems in Mayan Communities of Southern Belize. *Sustainability* **2021**, *13*, 3040. https://doi.org/10.3390/su13063040

Academic Editors: Maurizio Tiepolo and Marc Rosen

Received: 25 January 2021
Accepted: 7 March 2021
Published: 10 March 2021

1. Introduction

For centuries, the traditional practice of milpa farming has been sustainable and reliable as the major food and livelihood source for Mayan milpa communities in southern Belize [1–4] as farmers allow areas to regenerate to a mosaic of forest succession stages and crop diversity [5–8]. In the last 50 years, however, the slash-and-burn aspect of the milpa has become less reliable and less sustainable due to environmental factors, such as hydroclimatic changes (i.e., droughts, flooding, hurricanes), forest and biodiversity loss, pests and crop disease, soil degradation and other factors in combination with socio-economic and governance factors such as poverty, population growth, land tenure, and marginalization [8–16]. These factors have multiple systemic impacts to the resilience of milpa communities.

Government response and action is needed to promote climate-smart agriculture (CSA) practices and positively influence food and livelihood security in Belize. CSA practices can "increase productivity in an environmentally and socially sustainable way, to strengthen farmers' resilience to climate change, and to reduce agriculture's contribution to climate change" [17] p. 14. Government agricultural Extension service in Belize is in an effective position to promote CSA practices in Maya milpa communities because Extension works within the cultural traditions of the milpa system as partners in the process [10]. However, there are multiple barriers for Extension which challenge its efficacy, including milpa farmers' land tenure, taxation, and poverty, and Extension's lack of operational budget, lack of technical training in CSA technologies, and a lack of staff [15]. Without a more effective Extension service facilitating CSA adaptations, there are implications for unsustainable crop production and food and livelihood insecurity in milpa communities of rural southern Belize [10,16–18]. The purpose of this study is to both examine three promising CSA practices—mulching, soil nutrient enrichment, and cover plants—and make policy recommendations to reduce Extension barriers to promote CSA practices for positive influences on food and livelihood security in southern Belize.

2. Background

Food and livelihood security for milpa farmers in southern Belize depends largely on Government response and promotion of climate-smart practices. Food security is the ability to provide present and future generations with a reliable food supply; it considers multiple factors and depends upon reliable crop production while sustaining a healthy ecological balance in a farming system [19–22]. Food security is dependent upon sustainable agriculture—the enhancement of crop production while sustaining a healthy ecological balance within agro-ecosystems [22]. Sustainable agriculture involves economic, environmental, social and other factors to promote food and livelihood security for communities [19,21].

2.1. Forest Loss and Climate Change Vulnerability in Belize

Rural communities in Belize are vulnerable to resource loss and degradation due to climate change, forest and biodiversity loss, and other factors [10,16–18]; these impacts, due, in part, to agro-industry and slash-and-burn agriculture, are exacerbated by rapid population growth, increased input of fertilizers, and farming on degraded soils [18,23]. There are implications for unsustainable agricultural systems and with that, the loss of forest, water availability, erosion control, and other needed natural resources and ecosystem services unless there is a strong government response in Belize. These ecosystem changes are "expected to threaten the sustainability of social, economic, and ecological systems" [24] p. 8.

Intact forests regulate climate, protect soils and water, and contain over 75 percent of global terrestrial biodiversity [25]. In Belize, the forest cover is roughly 60% but declining [26]. The growing rate of forest and biodiversity loss in Belize has compounding ecosystem pressures related to climate change, pollution, environmental degradation, and continual expansion of farms into forests [18,27,28]. Agriculture is the most significant anthropogenic driver of deforestation globally [11,25,29] and in Belize [12,26]; agriculture in the tropics directly impacts forest loss and is "responsible for nearly 85% of deforestation [and] 45% of deforestation in the humid tropics [is] due to shifting cultivation" [30], para. 17.

Large scale climate and ecosystem changes in southern Belize have distinct impacts on the environment, crop production and economy, food security, public health, culture, and other factors in Belizean milpa communities [24,31–33]. Climate change impacts perceived by farmers in Belize include a lack of rain, increased heat and sun exposure, offset rainy seasons, increased storm intensity, and an increase in pests and crop diseases [32]. In addition, climate change accelerates soil erosion and land degradation. These factors negatively impact crop reliability, which is linked to livelihoods and resource security, community health, cultural traditions, and other factors [15,34]. There are disproportionate adverse impacts related to climate and ecosystem change on the rural poor, who depend directly on natural resources for their food and livelihood security [12,15,17,35,36]; these impacts perpetuate a cycle of environmental degradation, poverty, and vulnerability to climate and ecosystem changes [37].

2.2. The Milpa Farming System in Belize

A milpa is a small-scale shifting cultivation system of subsistence farming [4,38] traditionally involving slash-and-burn and/or slash-and-mulch practices [39,40]. Mulching and nutrient enrichment have also been a part of the traditional Maya milpa farming practice for centuries [40–42]. The milpa is a significant aspect of Maya culture and tradition as Maya identity, ceremony, community, and livelihood are all rooted in the milpa [9,43]. Milpa crop production is used for subsistence and selling at local markets [4,38]; milpas provide most of a family's need for food, wood, and income [11,44].

Milpa practices include clearing small areas of forest to plant a diversity of crops—primarily corn, beans, and squash—on nutrient-rich soil [7,44]. Through crop diversity, the milpa can sustainably increase milpa productivity and promote food security and food sovereignty, the "right to healthy and culturally appropriate foods" [43], p. 396. However, the milpa system is "not indefinitely resilient, particularly in an era of global economic and environmental change" [12], p. 75. Specifically, the slash-and-burn aspect of traditional milpa farming (clearing and burning of small areas of forests for crop rotation) is no longer sustainable with changing climate conditions, increasing human population, and natural resource competition [38]. Milpa farmers who exclusively practice slash-and-burn agriculture are more vulnerable to livelihood and food insecurity [10,12,17]. Burning reduces carbon stocks and the intense heat during burning can destroy critical root and seed banks [45]. Moreover, water-holding and nutrient status declines which dramatically increases "risks of accelerated erosion, water runoff, and crop failure in times of below normal rainfall" [46], p. 112.

2.3. *Managing Climate-Smart Agriculture (CSA) in Belize*

There is a need to manage resource loss and climate change vulnerability in Belize. Climate variability and extreme events (e.g., droughts, storms) are expected to become more frequent and damaging to water resources and agro-ecological systems in Belize in the coming decades [31]. The agriculture sector in Belize is especially vulnerable to climate change "not only due to its geo-physical location and hydro-meteorological hazards, but it is also due to the shortcomings of the current disaster risk reduction and response mechanisms to effectively mitigate the impacts" [31]. There is also a lack of institutional expertise to handle foreseeable climate change impacts [31].

The aim of climate-smart agriculture (CSA) is to "increase productivity in an environmentally and socially sustainable way, to strengthen farmers' resilience to climate change, and to reduce agriculture's contribution to climate change" [17], p. 14. CSA practices such as mulching and nutrient enrichment, along with improved land and water management, can result in higher and more stable yields, less production risk, increased system resilience to climate change, and lower greenhouse gas emissions. Therefore, CSA practices contribute to better food and livelihoods security for farming communities [34,47,48].

In Belize, the government vision for agriculture—the main engine of economic growth in Belize—incorporates several pillars including sustainable production and innovative technologies (Pillar 1), nutrition and food security, especially for rural populations (Pillar 3), and "climate change adaptation, environmentally sound production practices, conservation of natural resources, and risk management mechanisms" (Pillar 4) [49], p. 12. Continuing challenges to sustainable agriculture in Belize include: High and increasing poverty and unemployment, the rising inequality and access to food vulnerable populations, the steady decline in agriculture competitiveness, and increased scarcity of natural resources worsened by natural disasters and climate change [49]. These challenges are exacerbated by global financial instabilities which have negatively impacted employment, food and nutrition security, poverty, and inequity in Belize [49].

To improve the agricultural sustainability and reduce impacts of natural disasters and climate change, government action is needed, including the "promotion of more resilient farming systems and practices [e.g., climate-smart practices], as well as sound coordination, exchange of information, methodologies, and tools between experts and institutions" (31, para. 12). Increasing CSA practices can sustainably mitigate climate change impacts and support food security under a changing climate [15,50–52], while maintaining the health of ecosystems [20] and potential equitable increases in production in Belize. Otherwise, "marginal areas may become less suited for arable farming" [17], p. 15, resulting in less food and livelihood security for milpa farming communities in Belize.

2.3.1. Mulching and Soil Cover for Water and Nutrient Holding

In Belize, climate-smart agriculture practices include mulching, soil nutrient enrichment, and soil cover; mulching avoids burning of debris and allows farmers to let organic matter decay on site. Mulching improves water holding capacity, soil organic matter (SOM), fertility, and stability, as well as reducing runoff and weed growth [12,42,47,51]. Further, mulching can improve soil water-holding by adding crop residues and manure to soil which effects soil properties and nutrient cycling, as well as lowering emissions [15,53]. Mulching has also been found to regulate surface temperatures, thus improving moisture and germination as well as other benefits for crop productivity [39,40,54].

Practiced by about half the milpa farmers in southern Belize' Toledo District [32], mulching has similar planting and harvesting timing and is beneficial because it restores degraded soils, provides shorter fallow periods, and stabilizes crop yields [39,40,54,55]. Also, mulching in addition to soil cover, such as mucuna beans can lead to higher yields due to decreased on-farm erosion and nutrient leaching, lower grain losses due to pests, and reduced labor for weeding and fertilizer application [47,56,57].

2.3.2. Soil Nutrient Enrichment

Soil nutrient enrichment involves farming inputs that improve the soil conditions for production [39,40,54]. Soil enrichment practices can include adding chemical or non-chemical fertilizers and integrating effective microorganisms (EM) to break down slashed debris faster and build soil fertility [10,15]. There may be a need for farmers to purchase and use fertilization inputs in mulch systems, although that is debated in the literature. Two studies state fertilization inputs are essential to achieve good yields under fire-free conditions, although this cost may be offset from increased yields [15,58]. Other studies find that external fertilizer inputs were avoided with mulching where there was an increase in soil organic matter and water holding capacity [12,59].

There are some disadvantages to mulching and nutrient enrichment for farmers. Aside from the potential need to purchase fertilization inputs to enrich soil, mulching might also have a potential to increase snakes or animal vector encounters. Additionally, although mulching benefits are largely agreed to increase sustainability of yields, the increase in yield amount is debated due to slower nutrient release from the decomposing vegetation compared to burning [51]. Overall, mulching was found to have important farm system sustainability benefits, such as improving soil nutrients, regulating surface temperatures, improving moisture and germination, and increased crop productivity and sustainability [39,40,54].

2.4. Government Agriculture Extension in Belize

CSA practices and sustainable crop production depend upon government policies and action to shape agriculture adaptation response at the local level [17,32,49]. Government Extension services provide scientific knowledge and promote climate-smart agriculture practices, technologies, and innovations through farmer education and demonstrations [49,60,61]. Globally, Extension has a strong institutional expectation to inform, educate, and facilitate best practices for farmers [60] and to improve agriculture sustainability and promote "more resilient farming systems and practices" [31] para. 12 by working within the local socio-cultural traditions. Smallholder farmers are able to adapt to environmental changes using their traditional knowledge and experience and by adopting climate-smart agriculture adaptions [62].

In Belize, agriculture Extension is in an effective position to promote climate-smart practices in Maya milpa communities because they can work within the cultural traditions of the milpa system as partners in the process [10]. In doing so, Extension can influence adaptive capacity by transferring and promoting CSA technologies and site-specific technologies, such as water management, cover plants, enrichment, and mulching [10,12]. These practices facilitate a more productive and resilient agriculture system [31] and build resilience in milpa communities [2,10,63,64].

With Extension support, farmer capacity can be improved to innovate, solve problems, and adopt CSA practices; however, this will require continuous facilitation, capacity-building and support over time [56,65]. There are multiple barriers for Extension efficacy in promoting CSA practices in southern Belize, including governance, land tenure, taxation, poverty, and other challenges [15,49]. Further, there are institutional barriers within Extension including a lack of operational budget, lack of technical training in CSA technologies, and a lack of staff allocated from the national Extension office [49]; presently, there are four Extension Officers in southern Belize' Toledo District who are responsible for a large rural district of 52 communities [32].

2.5. Socio-Ecological Systems (SES) Framework

As part of the socio-ecological system (SES), milpa communities experience system impacts and can be more vulnerable to ecosystem changes such as climate change. Socio-ecological systems (SES) is an effective framework to study climate-smart agricultural adaptation in milpa communities in Belize [10,17,66] as SES is complex, systemic, cumulative, and intertwined with human systems [67]. This study examines perceptions of climate-smart practices from milpa farmers and agricultural Extension officers in southern Belize using a SES framework. SES is a flexible framework which considers the interrelationships, linkages, and synergies between multiple trans-disciplinary factors—social, economic, environmental, cultural, governance, justice and other factors; SES also involves inclusion and community-based partnerships and adaptive management [68–70]. Milpa farmers and Extension Officers in southern Belize can become more enabled partners in climate-smart solution finding. The socio-ecological system of milpa communities is a linked network where an impact on one part of the system—the loss or degradation of soil due to storm erosion, for example—can affect the human system, such as food security and farmer livelihoods [11,67,71].

3. Methods

This qualitative study uses Phenomenology and face-to-face interviews to examine common-lived experiences of climate-smart practices from the perceptions of milpa farmers and Extension officers. Phenomenology is well-suited for this study as it is both a philosophy and an inquiry strategy used to "develop an understanding of complex issues that may not be immediately implicit in surface responses" [72], p. 301. Phenomenology is useful to "investigate the relationship between participatory Extension methods and farmers changing to more sustainable practices" [73], p. 22.

The semi-structured interviews were both "purposive and prescribed from the start" [72], p. 302 and allow flexibility to ask participants deeper follow-up questions [32], in order to hear their stories, and to see the emergence of common experiences or phenomena through the participant's own words and descriptions [74–76]. During interviews, participants were asked both demographic questions and open-ended questions on topics such as milpa farming practices, socio-ecological system linkages to ecological changes (i.e., forest loss, climate change), barriers and conduits of sustainable agriculture practices, and other topics. Using Phenomenology, specific patterns, categories, and themes emerged from the interview data collected and analyzed [77–79]. New Mexico State University Institutional Review Board (IRB) approved all study protocols and interview questions; all interviews followed a voluntary and informed consent procedure.

3.1. Setting of the Study

In the southern Toledo District of Belize, three Extension officers and five milpa farmers from Pueblo Viejo and Indian Creek villages were interviewed for this study. Toledo District is the southernmost district in Belize; its population is nearly 50% Q'eqchi'

(Kekchi) Maya, 20% Mestizo, and 17% Mopan Maya. There are also Garifuna, Creole, East Indian, and Mennonite populations [80]. Milpa households in each village were selected using a stratified random design. The sample subpopulation of 'primary (head) milpa farmer' for each selected household was intentional to elicit the perspective of farmers who have the most direct knowledge of local forests, soils, and agriculture systems. Two participants selected were Maya cultural and political leaders in their villages who spoke to the importance of the milpa as part of their cultural practice. Interviews of Extension officers were conducted in both office and field settings; three (of the four total) Extension officers in the Toledo District in southern Belize were interviewed.

3.2. Data Analysis

Using the multi-perspectival framework of Socio-ecological Systems (SES), a combination of processes was used in the data analysis, including 1. Open (analytical), 2. Axial (reduction and clustering of categories), and 3. Selective coding [77,81–86]. These processes involved creating categories of like taxonomies and assembling structures or groups of themes into conceptual diagrams to show relationships and linkages [82]. Selective coding is a form of data synthesis where the intersection or integration of emergent thematic categories are first "crystallized" [10]. Crystallization uses multiple perspectives to blend data to produce thick description and knowledge of a phenomenon as well as a deepened, inclusive, multi-perspectival, and complex interpretation of it [85,86]. Selective coding results as the intersection of the main categories and themes [77] where categories are systematically related or conceptually linked in a multi-perspectival and holistic way [86]. From this data analysis and synthesis, dominate themes emerged and were categorized in the following Results section.

4. Results

Through socio-ecological system (SES) examination of interview data from milpa farmers and Extension officers, this study finds direct and indirect influences of climate-smart agriculture (CSA) practices—specifically mulching, soil enrichment practices, and ground and soil cover methods—on milpa farming sustainability. Table 1 summarizes results from interviews with milpa farmers and Extension Officers. Participants in this study perceived direct and indirect, (a) environmental, (b) economic, (c) socio-cultural influences, and (d) adaptive technology potential from CSA practices on the sustainability and resiliency of Belizean milpa communities.

Table 1. Summary of Results: Perceived Environmental, Economic, and Socio-cultural influences, and Adaptive Technology potential from climate-smart agriculture (CSA) practices (mulching, soil enrichment, and ground cover) on milpa sustainability in Belize.

Perceptions (Influences)	Environmental (Air, Soil, Water, Forests, Erosion, Pests)	Economic (Farm Income, Expenses, Tourism; Extension Budget)	Socio-Cultural (Culture, Traditions, Adaptation, Resilience)	Adaptive Technology (Potential)
Milpa Farmers	Positive. Mulching/ground cover increase soil fertility and decrease erosion (compared to burning). Negative: More snakes. More time for decomposition. Clearing forests allows rotating on nutrient-rich soil- organic/less fertilizer use.	Positive. Stabilizes and increases production (subsistence and sale at small markets); non-chemical enrichment/cover plants = low/no cost. Shorter fallow periods. Intact forests increase tourism income. Mixed: perceived higher and lower fertilizer expenses (Overall: more yield = more income). Negative: More pests/more need for costly pesticides.	Positive. Working directly with Extension to manage and sustain crop production with changing climate/seasons (temps/rainfall), uncertainty, insecurity. Slash-and-burn (only) not culturally sustainable (degrades); mulching adds nutrients.	Positive: Soil enrichment and intercropping: Need more information and financial support, technology/innovation (to mimic nutrient-rich forest soil) and keep forests intact. Use more organics (chicken manure). Pesticides/Integrated Pest Management: Need more information and assistance.
Extension Officers	Positive. Mulching and ground cover have erosion control, increases soil moisture (germination) and fertility, regulates soil temperature, reduces nutrient loss; reduces air emissions (no burning); better water management; sharing solutions for less pesticide use.	Positive. Sustaining production (proves Extension efficacy). Negative. Institutional and operational barriers (lack of budget, CSA training, staff); need to share resources (vehicles, fuel, staff) with other local government and non-government entities. Large district with 52 communities and only 4 Extension staff.	Positive. Working withing milpa traditions for proven adaptations (i.e., pest and water management, soil enrichment) increases sustainable farming = security = community resilience, and Extension efficacy. Negative. Cultural barriers: Adaptive practices are slow-moving (but faster with young farmers); land tenure, poverty	Positive. Promoting non-synthetic soil cover and enrichment with integrated pest management, effective microorganisms (EM); and nitrogen-fixing and cover plants: Arachis, mucuna beans used by Amish/Mennonites).
Overall	Positive. Increased soil nutrients, ground cover and moisture, erosion control, managing crop pests and disease.	Positive. Sustained production). Mixed: Both higher/lower fertilizer expenses. Negative: Lack of information; higher costs for pests/pesticides. Extension barriers (lack of budget, training and staff)	Positive. Adaptation = sustainable farming. Adapting cultural practices maintains cultural traditions. Negative: Socio-cultural barriers adopting new CSA technologies.	Government vision and priority for sustainable agriculture and community resilience to climate change impacts (food insecurity, livelihoods, storm and disaster resilience).

4.1. Environmental Influences

Milpa farmers and Extension officers perceived positive environmental influences from CSA practices on milpa sustainability. For the purposes of this article, environmental influences, include impacts to air quality, soil nutrients, soil water holding, land erosion, and pests and disease. This study finds CSA practices of mulching, soil nutrient enrichment, and soil cover have positive environmental influences for increasing soil nutrients, ground cover and moisture, controlling erosion, and managing crop pests and disease. Extension officers interviewed for this study perceived both mulching and nutrient enrichment to be climate-smart and beneficial for milpa famers.

4.1.1. Benefits from Mulching (vs. Burning)

Milpa farmers and Extension officers perceived less impact from mulching over burning. All milpa farmers interviewed for this study practice slash-and-burn; one farmer explained his process: "I will soon start to chop bush, and then it dries, and then [I] burn it, and then plant it. You chop more bush to plant more [crops]." Some also practice mulching; one milpa farmer described his preference for mulching over burning due to the benefit of less erosion: "[We] just leave [debris] there and it'll get rotten, right? Leave the stump right there, because the stump—it holds a lot of soil [and] when it's raining, it won't flush off. So, just leave the stump right there until it gets rotten".

Extension Officers interviewed explained mulching provides effective ground cover and erosion control; also mulching keeps more moisture and fertility in the soil. Whereas, burning exposes and heats up the soil, causing nutrient loss. An Extension officer explained the benefits of leaving the vegetation to rot in the mulching process: "[The grass] covers the soil [and] ... there's a little moisture by the roots of the plant [and] it will keep the soil cool instead of in the hot sun ... so it does work. It does work." One Officer explained burning and not leaving debris on the soil causes erosion: " ... then you have a long drought [and then] how do you keep moisture? And, those are the things that we have to make farmers aware of—it's a chain of reaction." There is a disadvantage to mulching in "that it's too bushy and people don't want to go in there ... because it attracts maybe snakes and other things" [32]. However, there was an overall positive environmental influence of mulching, nutrient enrichment, and soil cover on milpa sustainability.

Extension Officers see other CSA benefits with mulching, including reduced air emissions, better water management, and the use of non-chemical inputs for crop pests and disease. Climate change creates the condition for unreliable water and a higher incidence of crop pests and disease; one Extension Officer described how this affects crop production: "A high incidence of pests (are) noticeable now ... so, all of these things—and a limited water supply—all of these things are affecting agriculture, in general ... and those things limits our work as well." He stated that years ago, they did not have to think about the climate, but now they have to "be [climate] smart." An Extension officer stated they "need to do a little bit more public awareness in terms of the negative effects [of burning]" due to air pollution, global warming, and other effects.

4.1.2. Benefits of Nutrient Enrichment and Soil Cover

Milpa farmers traditionally rotate crops on nutrient-rich "black" soil due to the nutrient depletion in farmed soil over time [10]. One farmer stated if there were soil enrichment assistance for farmers—specifically information and financial assistance for inputs—he would not need to "chop" forest to use the enriched soil. Extension Officers want to demonstrate to farmers that climate-smart alternatives (i.e., mulching and enrichment) work. Increasing nutrient enrichment such as effective microorganisms (EM) and using nitrogen-fixing cover plants can benefit milpa farmers. One Officer educates farmers and promotes EM. He stated, "A lot of farmers, they are starting to use organic material—meaning chicken manure. They are using a lot of EM agriculture to build up the soil fertility." The same officer also explained the benefits of mucuna beans for nutrient enrichment:

> We have some farmers that benefit from the training as well, because, at some point, we introduce some types of fertilizer that you incorporate in the soil ... [for example] mucuna beans: The Mennonites [presuming he means the less mechanized Amish community] use it a lot, you know; they don't use a lot of synthetic fertilizer, they only use these types of mucuna beans.

Another Extension officer promotes arachis (*Arachis glabrata*), a wild peanut perennial. Arachis is useful for milpa farmers as an effective ground and soil cover and as a nitrogen-fixing plant. These climate-smart practices mimic or replicate the nutrient cycling in forest ecosystems while allowing for sustainable production of agriculture [46].

4.2. Economic Influences

Milpa farmers and Extension officers perceived mostly positive and some negative economic influences from CSA practices on milpa sustainability. For the purposes of this article, economic influences include impacts to farmer income, farmer expenses, farmer time, Extension expenses, and impacts to alternative income (tourism). This study finds climate-smart practices, particularly mulching, soil nutrient enrichment, and soil cover, have overall positive economic influences for increasing farmer income and reducing farmer expenses. However, Extension's limited budget to promote CSA practices is a barrier.

From an economic perspective, milpa farmers are concerned about maintaining production (for subsistence and selling at local markets) and reducing costs and expenses—related primarily to controlling pests and disease and adding fertilizer inputs—in their daily farming practice. Some farmers interviewed worried about increasing pests resulting from climate warming; their lack of knowledge of pests and pesticides management was a top concern [32]. Most often in milpa communities, pest management includes the use of chemical fertilizers and pesticides, which are an additional cost for farmers [32].

4.2.1. Soil Enrichment Cost-Benefit

Soil enrichment involving fertilizer inputs (chemical or nonchemical) is a cost to farmers; all farmers interviewed for this study stated they buy and/or use fertilizer inputs. Although an upfront cost, one study concluded that using fertilizers increases farmer yields enough to compensate for fertilizer costs [58]. Moreover, adding nonchemical enrichment or soil cover (i.e., arachis) can be low to no cost. Milpa farmers who rotate crops in cleared forest areas avoid fertilizer cost by using nutrient-rich black soil. One farmer explained that otherwise, the soil gets too dry and hard; "but, if we change every year, it doesn't need fertilizer. Yah, just normal planting—organic . . . That's why we maintain for we [sic] forest." Conversely, another farmer explained that keeping forests intact is important for his village's economic development and tourism industry:

> We understand the slash and burn is [bad]—sometimes for humans, for us and also for a wildlife—and, so, we are trying to avoid that now. We are working very closely with the village leaders [to develop potential tourism in the village] . . . because we need to take care of our forest, including creeks, rivers, and streams, and so forth.

4.2.2. Extension Barriers

Climate change negatively impacts farmer food security and livelihood which can impact Extension efficacy in promoting and facilitating sustainable agriculture practices in milpa farming communities [32]. However, there are institutional and operational barriers in Extension services, including a lack of government funding for daily operating, a lack of technical training in CSA technologies, and a lack of staff. One Extension officer noted: "We need support from [the national office] because we cannot do it alone . . . We need to prioritize [climate-smart] topics because everything now is climate change . . . everything is focused around climate change and resilience." To cope with the low numbers of staff, Extension officers stated they have to collaborate and share resources with other government agencies and nongovernmental organizations to carry out some aspects of their Extension duties (i.e., sharing vehicles to reach farmers in remote communities).

4.3. Socio-Cultural Influences

Milpa farmers and Extension officers perceived both positive and negative socio-cultural influences from CSA practices on milpa sustainability. For the purposes of this article, socio-cultural influences are defined as impacts to farmer traditions and heritage, adaptation responses, adopting new technologies (i.e., effective microorganisms), and community resilience. This study finds climate-smart practices, particularly mulching, soil nutrient enrichment, and soil cover have overall positive socio-cultural influences for maintaining milpa socio-cultural practices and traditions. However, there are barriers for some farmers in adopting newer CSA technologies and practices.

4.3.1. Climate Change Impacts to Milpa Culture

Milpa farmers and Extension staff perceived direct impacts from climate change—more intense sun/heat, lack of rain, offset rainy season, increase pests and disease [32]; in turn, these impacts effect the cultural traditions of milpa communities and their ability to provide corn and beans for subsistence. One farmer stated: "I am waiting for that rain because I have some corn that is (small) [demonstrates small size] that haven't gotten water for a good while—so, I'm hoping and wishing that this week it will soon rain." Another stated: "I tried to plant vegetables but he [sic] no grow—he dead. I planted, but he neva (never) grow good. I don't know why, because maybe he got sun too much;" and "the rains are different than what they were because, it doesn't rain too much … I am worried about crops—worried about the weather." A third farmer described the offset rainy season: When they expect the rainy season, there is strong sun and heat; they don't know "which time is for the right time" to plant. He explained: "We need to study the climate changes and the temperature (so we can) try to manage." This farmer also explained the exclusive practice of slash-and-burn, a traditional form of milpa agriculture, may not be culturally sustainable:

> The only way we could damage [the milpa farming culture] for us is if we continue to slash and burn, and burn, and slash and burn—and, we believe that one day our crop will never come out good again because the fertile[ity] of the ground is washed off, so everything goes in the creeks, in the river; and, the land becomes poor and poor and poor and poor—and, so, now, we don't want to practice that because we understand the situation there. So, we believe that to maintain the soil, to treat the soil in a proper way … not to cut down the trees or not to burn it—even though if you want to fall something—but, leave it there—just, leave it there, and it'll get rotten.

4.3.2. Extension Barriers

Extension Officers interviewed for this study stated that climate change impacts to milpa farmers creates a barrier for Extension service efficacy. One Officer stated climate change is a real factor now: "A high incidence of pests (are) noticeable now … so, all of these things—and a limited water limited water supply—all of these things are affecting agriculture, in general … and those things limit our work as well." Another Extension Officer added: "Climate change is very, very important because all these pests and disease … in a hot climate or little water available—always climate change issue is very critical now and there … and we make [farmers] aware of that".

4.3.3. Cultural Adaptations and Adaptive Technologies

Increasing CSA practices such as soil nutrient enrichment can help milpa farmers plant on soil which mimics rich black soil (i.e., converted from cleared forest). However, some farmers perceive this socio-cultural adaptation is not needed; one farmer stated he prefers to clear forest because "black soil is better [to farm]" and that "works for us." Two farmers interviewed were interested in learning new technologies and adapting their practice; one stated an interest in intercropping and effective microorganisms (EM) for soil enrichment:

> It would be interesting to bring something with the soil and mix it up—and put plants there like tomatoes. You could plant when you mix up the soil … the [plants] come very good. And, with corn too … Yes, yes—that would be interesting … interesting. You bring some soil, you just mix it up, and plant some there.

Extension officers are promoting milpa farmers increase climate-smart practices, including slash-and-mulch farming and soil nutrient enrichment, as they are beneficial and sustainable milpa practices [32] which can increase farmer production and income. One Officer stated about half of the farmers in the district practice mulching; however, there is a need for more public awareness of the negative effects of slash-and-burn practices by making it a priority to show farmers proof of workable CSA practices:

> We need to continue to educate the farmers, right(?) ... because whenever you do your field visit and so on, you can see that some farmers, yes, they do adaptive technology very slowly; others, they go a little bit faster ... but some of them it's very hard. So, what I have found is that (we should) continue capacity-building, education ... (and) at some point, of course, especially the young farmers that are coming up—try to teach them the right way of doing agriculture—sustainably.

Another Extension officer is trying to educate and promote nutrient enrichment technologies such as effective microorganisms (EM), mucuna beans, and arachis: "A lot of farmers, they are starting to use organic material (such as) chicken manure, right? They are using a lot of EM agriculture to build up the soil fertility." He explained the benefits of mucuna beans:

> We have some farmers that benefit from our training ... we introduce some types of fertilizer that you incorporate in the soil ... (for example) mucuna beans: the Mennonites [presuming meaning the less mechanized Amish community] use it a lot, you know; they don't use a lot of synthetic fertilizer, they only use these types of mucuna beans.

He also explained the benefits of arachis, a wild peanut perennial; arachis can be used for chicken feed in a mobile chicken coop. He explained arachis is an excellent ground and soil cover and as a nitrogen-fixing plant.

4.3.4. Extension Partnership

Extension Officers perceived milpa farming practices are sustainable due to cultural traditions and knowledge passed down through generations. However, the Officers interviewed stated milpa traditions will only stay sustainable if farmers adapt to include CSA practices. An Extension supervisor in the national office stated district Extension officers can work within the cultural traditions of the milpa system to promote sustainable practices:

> [We need] a way to demonstrate to [the farmer] a way to adequately compensate for what they are moving ... we need to look at injecting proportionate technology in the milpa system, and then look at how the farmers react to that injection. It's a learning process, not to challenge traditional [farming methods, but try to promote] a few [effective] agricultural practices like soil conservation, irrigation systems, and integrated pest management [87].

With time and resources, Extension Officers are in an effective position to promote sustainable agriculture production because they are partners with milpa farmers in both maintaining traditional milpa practices and adopting more sustainable CSA practices [15].

5. Discussion

The purpose of this study is to both (1) examine three promising climate-smart agriculture (CSA) practices—mulching, soil nutrient enrichment, and cover plants—and (2) make policy recommendations to reduce Extension barriers to promote the CSA practices for positive influences on food and livelihood security in southern Belize. From interviews with milpa farmers and Extension officers in Belize, this study finds CSA practices were perceived to have overall positive socio-ecological system influences on Maya milpa farming communities, including: (a) Economic, (b) Environmental, and (c) Socio-cultural influences, as well as (d) Adaptive technology potential; these influences were perceived as conduits for sustainable milpa agriculture. A flow diagram or conceptual impact model (Figure 1) of the influence categories was constructed (via PowerPoint software) using this study's perception data gathered during interviews. The model shows relationships and linkages between the SES influences as they relate to CSA practices.

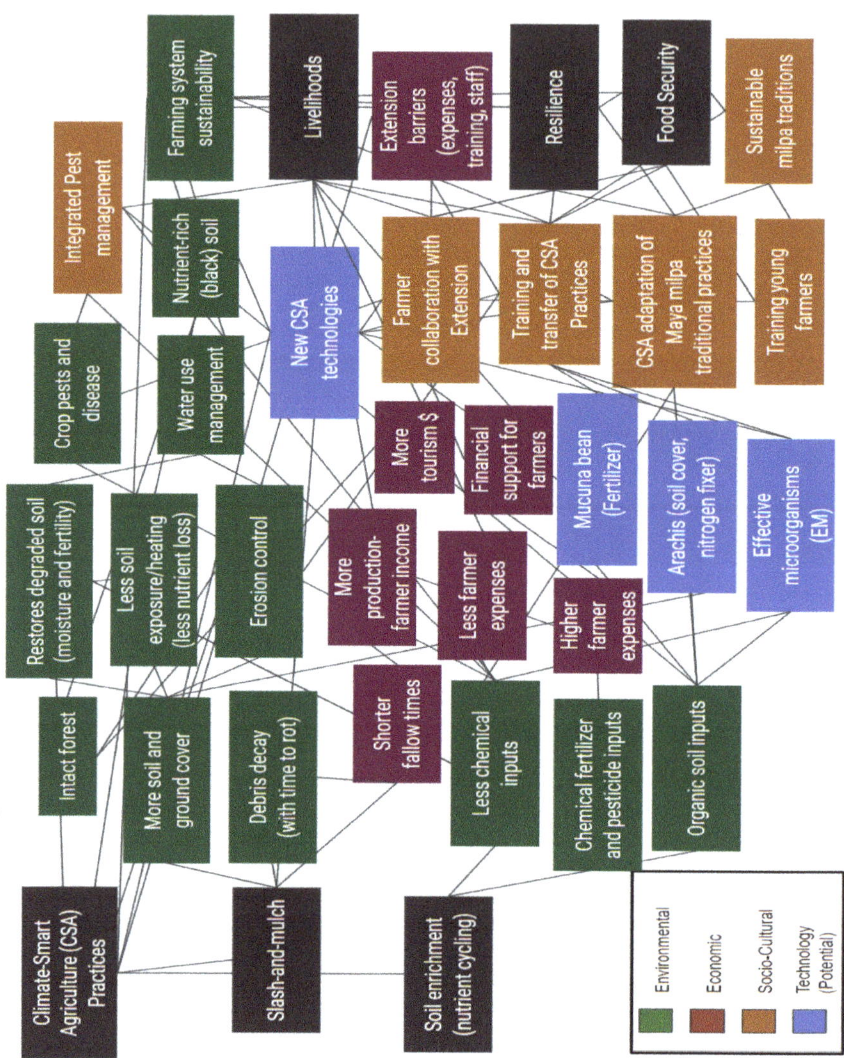

Figure 1. A Socio-ecological Systems impact model from this study's climate-smart agriculture (CSA) influences—environmental, economic, socio-cultural, and adaptive technology (potential)—on milpa farming sustainability and resilience in southern Belize.

Figure 1 is based on the conceptual framework by SES architect Dr. Elinor Ostrom, widely considered to be the foremost researcher on SES; her model shows a multilevel and multi-perspectival examination of SES factors, drivers, interactions, and outcomes with implications for adaptive management, multiple stakeholder coordination, collective action, and community-level application [68–70]. Other similar models informing Figure 1 which demonstrate complex, multi-perspectival linkages in a socio-ecological system include Parrot, et al. [70], Flora and Flora [88] community capitals, Cote and Nightingale [89], Gonzales, et al. [90], and Tenza, et al. [91]; these works show an emphasis on feedback dynamics where human systems can shape ecological components and vice versa.

The data analysis processes of selective coding and crystallization described in Section 2 created a complex and multi-perspectival interpretation where dominant themes emerged. The major influence categories were systematically linked [77,86] and show multiple intersections which directly and indirectly relate to CSA practices in milpa communities. In Figure 1, Environmental influences are represented by green boxes, Economic influences are burgundy, Socio-cultural influences are gold, and Adaptive technology potential are blue.

Figure 1 demonstrates linkages that can be useful to inform government policy and action. Intersections of linkages can be used to help identify priority areas which may have implications on the larger socio-ecological system. For example, the model demonstrates how soil enrichment practices are linked to potential adaptive technologies such as mucuna beans, arachis, and effective microorganisms; those are linked to less chemical inputs, a reduced need for costly fertilizer inputs, more soil stability by keeping forests intact (i.e., not needing to cut forests for black soil), maintaining Maya socio-cultural traditions, and so on—altogether demonstrating a positive influence on the larger milpa socio-ecological system. Therefore, one Extension intervention (i.e., facilitating increased use of mucuna beans) can have positive impacts to other parts of the milpa socio-ecological system. Overall, the model shows increasing CSA practices of mulching, soil enrichment, and cover plants can foster higher crop production with more resource stability; the implications for this can mitigate disproportionate climate change impacts related to poverty, climate justice, resilience, and food and livelihood insecurity in milpa communities.

Figure 1 is intended to be a small picture of an otherwise larger and more complex milpa agroecological system with multi-dimensional, dynamic, non-linear (circular) feedbacks and flows [32]. Understanding these system relationships—and how each factor functions in the complex whole of the SES—is important as each decision a farmer makes to adopt CSA practices can advance the entire milpa agriculture system further [92,93]. Elements of Figure 1, specifically related to Extension and government action, were used to inform the policy recommendations in this study.

6. Conclusions and Policy Recommendations

From interviews with milpa farmers and Extension officers in Belize, this study finds climate-smart agriculture (CSA) practices of mulching, soil nutrient enrichment, and soil cover can have overall positive socio-ecological system (SES) influences on milpa sustainability in southern Maya Belize communities. Although milpa farming has been sustainable for centuries, global climate change and other factors such as poverty, population growth, forest loss, and land degradation have made the practice less so over the last 50 years. Promoting the increase of CSA practices on milpa farms has overall positive (a) environmental, (b) economic, and (c) socio-cultural influences and (d) adaptive technology potential on milpa sustainability, food and livelihood security, and resilience. By examining the perceived SES influences of CSA practices, this study makes policy recommendations to reduce government Extension barriers to promote the CSA practices for positive influences on food and livelihood security in milpa communities of southern Belize.

Promoting CSA practices necessitates Government involvement and action. Agriculture Extension in Belize is in an effective position to facilitate an increase in CSA practices because it has a strong institutional expectation to inform, educate, and demonstrate best practices to the public. Working within milpa cultural traditions, Extension Officers can promote an increase in climate-smart practices, while including milpa farmers as partners in the process. Specifically, promoting the practices of mulching, soil nutrient enrichment, and soil cover can have positive socio-ecological system influences and potential equitable increases in crop productivity. Government agriculture Extension services are needed to promote CSA practices in the milpa communities they serve. Recommendations here are targeted to Government Extension services in Belize at the national and district levels:

1. Increase District operational funds (i.e., vehicles, fuel), technical training, and the number of trained officers to enable Extension to promote farmer adaptation such as mulching, soil nutrient enrichment, and other nonchemical technologies (e.g., effective microorganisms, mucuna beans);

2. Use socio-ecological system and agroecological research and analysis to inform policy in adaptive management for climate and ecosystem changes; producing food in a more sustainable way involves multidisciplinary factors and a participatory and action-oriented approach to sustainable and just food systems [66,94–96]; and
3. Develop appropriate scaling [97] to enable local village and farmer leadership in CSA technologies [98] through capacity-building, community participation and collective action, and sustainable-focused programs, such as farmer field schools and youth involvement programs. Working within the cultural traditions of milpa farmers and including farmers as partners in the process, Extension services can promote CSA practices for a more sustainable milpa farming system for food and livelihood security in southern Belize.

Incorporating these recommendations while continuing to work within the cultural traditions of milpa farmers as farmers as partners in the process, Extension services can promote CSA practices for a more sustainable milpa farming system for food and livelihood security in southern Belize.

Funding: This research received no external funding; it was part of Drexler's doctoral dissertation research (Drexler, 2019).

Institutional Review Board Statement: The study was conducted according to the guidelines of the Declaration of Helsinki, and approved by the Institutional Review Board of New Mexico State University; submission 17466 | Extension Leadership and Sustainable Agriculture in Belize Forest Farming Communities: A Socio-ecological Systems Approach (dissertation); approved 2018-12-06 11:36:23.

Informed Consent Statement: Informed consent was obtained from all subjects involved in the study.

Data Availability Statement: The data presented in this study are available on request from the author. The data are not publicly available due to confidentiality and consent protocols related to audio recording of interviews; coding and categorizing of some transcribed responses is available.

Acknowledgments: We thank the Director of Extension and the Toledo District milpa farmers and Extension Officers for their time contributing interviews to this study.

Conflicts of Interest: The author declares no conflict of interest. Having no external funding source, there was no outside role in the design of the study; in the collection, analyses, or interpretation of data; in the writing of the manuscript, nor in the decision to publish the results.

References

1. Altieri, M.A.; Toledo, V.M. The agroecological revolution in Latin America: Rescuing nature, ensuring food sovereignty and empowering peasants. *J. Peasant Stud.* **2011**, *38*, 587–612. [CrossRef]
2. Benitez, M.; Fornoni, J.; Garcıa-Barrios, L.; López, R. Dynamical networks in agroecology: The milpa as a model system. In *Frontiers in Ecology, Evolution and Complexity*; Benítez, M., Miramontes, O., Valiente-Banuet, A., Eds.; CopIt-arXives: Mexico City, Mexico, 2014.
3. Ford, A.; Nigh, R. *The Maya Forest Garden: Eight Millennia of Sustainable Cultivation of the Tropical Woodlands*; Routledge: London, UK, 2016. [CrossRef]
4. Nigh, R.; Diemont, S.A. The Maya milpa: Fire and the legacy of living soil. *Front. Ecol. Environ.* **2013**, *11*, 45–54. [CrossRef]
5. Daniels, A.E.; Painter, K.; Southworth, J. Milpa imprint on the tropical dry forest landscape in Yucatan, Mexico: Remote sensing & field measurement of edge vegetation. *Agric. Ecosyst. Environ.* **2008**, *123*, 293–304. [CrossRef]
6. Isakson, S.R. Between the Market and the Milpa: Market Engagements, Peasant Livelihood Strategies, and the On-Farm Conservation of Crop Genetic Diversity in the Guatemalan Highlands. Ph.D. Thesis, University of Massachusetts Amherst, Amherst, MA, USA, 2007.
7. Pleasant, J. Food yields and nutrient analyses of the Three Sisters: A Haudenosaunee cropping system. *Ethnobiol. Lett.* **2016**, *7*, 87–98. [CrossRef]
8. Shal, V. The Mayas and Their Land (Report No. 489). *Belize Development Trust.* 2002. Available online: https://ambergriscaye.com/BzLibrary/trust489.html (accessed on 9 January 2020).
9. De Frece, A.; Poole, N. Constructing livelihoods in rural Mexico: Milpa in Mayan culture. *J. Peasant Stud.* **2008**, *35*, 335–352. [CrossRef]
10. Drexler, K. Government extension, agroecology, and sustainable food systems in Belize milpa farming communities: A socio-ecological systems approach. *J. Agric. Food Syst. Community Dev.* **2020**, *9*, 85–97. [CrossRef]

11. Levasseur, V.; Olivier, A. The farming system and traditional agroforestry systems in the Maya community of San Jose, Belize. *Agrofor. Syst.* **2000**, *49*, 275–288. [CrossRef]
12. Lozada, S.B. Securing food and livelihoods: Opportunities and Constraints to Sustainably Enhancing Household Food Production in Santa Familia Village, Belize. Master's Thesis, University of Montana, Missoula, MT, USA, 2014. Available online: https://scholarworks.umt.edu/etd/4202/ (accessed on 9 January 2020).
13. New Agriculturist. Country Profile—Belize. 2005. Available online: http://www.new-ag.info/en/country/profile.php?a=847 (accessed on 15 January 2018).
14. Steinberg, M.K. Political ecology and cultural change: Impacts on swidden-fallow agroforestry practices among the Mopan Maya in southern Belize. *Prof. Geogr.* **1998**, *50*, 407–417. [CrossRef]
15. Kongsager, R. Barriers to the adoption of alley cropping as a climate-smart agriculture practice: Lessons from maize cultivation among the Maya in southern Belize. *Forests* **2017**, *8*, 260. [CrossRef]
16. Flint, R.W. Belize Pollution Problems. Sustainability Now for Belize, C.A. 2015. Available online: https://web.archive.org/web/20160420132354/http://sustainability-now.org/belize-pollution-problems/ (accessed on 9 January 2020).
17. Oremo, F.O. Small-Scale Farmers' Perceptions and Adaptation Measures to Climate Change in Kitui County, Kenya. Master's Thesis, University of Nairobi, Nairobi, Kenya, 2013.
18. Young, C.A. Belize Ecosystems: Threats and Challenges to Conservation in Belize. Conservation Letter. *Trop. Conserv. Sci.* **2008**, *1*, 18–33. Available online: https://tropicalconservationscience.org (accessed on 9 January 2020). [CrossRef]
19. Esri. GIS for Sustainable Agriculture. 2008. Available online: http://www.esri.com/library/bestpractices/sustainable-agriculture.pdf (accessed on 9 January 2020).
20. FAO (Food and Agricultural Organization of the United Nations). The 10 Elements of Agroecology Guiding the Transition to Sustainable Food and Agriculture Systems. N.d. Available online: http://www.fao.org/3/i9037en/i9037en.pdf (accessed on 9 January 2020).
21. Mazumdar, S. Geographic Information Systems in the Application of Precision Agriculture for Sustainable Sugarcane Production in the Republic of Panama. Master's Thesis, McGill University, Montreal, QC, Canada, 2008. Available online: https://escholarship.mcgill.ca/concern/theses/6q182n86q (accessed on 9 January 2020).
22. Rao, M.; Waits, D.; Neilsen, M. A GIS-based modeling approach for implementation of sustainable farm management practices. *Environ. Model. Softw.* **2000**, *15*, 745–753. [CrossRef]
23. Meerman, I.; Cherrington, E. (2005) as cited in Chicas, S.D.; Omine, K; Ford, J.B. Identifying erosion hotspots and assessing communities' perspectives on the drivers, underlying causes and impacts of soil erosion in Toledo's Rio Grande Watershed: Belize. *Appl. Geogr.* **2016**, *68*, 57–67. Available online: https://doi.org/10.1016/j.apgeog.2015.11.010 (accessed on 9 January 2020).
24. Richardson, R. Belize and Climate Change: The Costs of Inaction. Human Development Issues Paper. *United Nations Development Programme*. 2009. Available online: https://www.undp.org/content/dam/belize/docs/UNDP%20BZ%20Publications/Belize-and-Climate-Change-The-Costs-of-Inaction.pdf (accessed on 9 January 2020).
25. United Nations Food and Agricultural Organization (FAO). *The State of the World's Forests. Forest and Agriculture: Land-Use Challenges and Opportunities*; United Nations Food and Agricultural Organization (FAO): Rome, Italy, 2016. Available online: http://www.fao.org/3/a-i5588e.pdf (accessed on 9 January 2020).
26. Cherrington, E.; Cho, P.; Waight, I.; Santos, T.; Escalante, A.; Nabet, J. *Executive Summary: Forest Cover and Deforestation in Belize, 2010–2012*; CATHALAC: Panama City, Panama, 2012.
27. Ruscalleda, J. Deforestation in Belize: Why Does the Agriculture Sector Need Standing Forests? The Belize Agriculture Report. 10 May 2016. Available online: http://ufdc.ufl.edu/UF00094064/00039 (accessed on 9 January 2020).
28. Young, O.R.; Berkhout, F.; Gallopin, G.C.; Janssen, M.A.; Ostrom, E.; Van der Leeuw, S. The globalization of socio-ecological systems: An agenda for scientific research. *Glob. Environ. Chang.* **2006**, *16*, 304–316. [CrossRef]
29. Fischer, A.A. Integrating Rural Development and Conservation: The Impacts of Agroforestry Projects on Small Farmers in Panama. Ph.D. Thesis, Dalhousie University, Halifax, NS, Cananda, 1998.
30. Lanly, J.P. Deforestation and Forest Degradation Factors. *XII World Forestry Congress*. 2003. Available online: http://www.fao.org/docrep/article/wfc/xii/ms12a-e.htm#P14_73 (accessed on 9 January 2020).
31. FAO. Brief Profile of the Agriculture Sector (Belize). 2010. Available online: http://www.fao.org/climatechange/download/28928-0e0bc49fd6e72690e82c4c3b9d4dc6f52.pdf (accessed on 9 January 2020).
32. Drexler, K. Extension Leadership and Sustainable Agriculture in Belize Forest Farming Communities: A Socio-Ecological Systems Approach. Ph.D. Thesis, New Mexico State University, Las Cruces, NM, USA, 2019.
33. Chicas, S.D.; Omine, K.; Ford, J.B. Identifying erosion hotspots and assessing communities' perspectives on the drivers, underlying causes and impacts of soil erosion in Toledo's Rio Grande Watershed: Belize. *Appl. Geogr.* **2016**, *68*, 57–67. [CrossRef]
34. Sikka, A.K.; Islam, A.; Rao, K.V. Climate-smart land and water management for sustainable agriculture. *Irrig. Drain.* **2018**, *67*, 72–81. [CrossRef]
35. Adger, W.N. Social capital, collective action and adaptation to climate change. *Econ. Geogr.* **2003**, *79*, 387–404. [CrossRef]
36. John, L.; Firth, D. Water, watersheds, forests and poverty reduction: A Caribbean perspective. *Int. For. Rev.* **2005**, *7*, 311–319. [CrossRef]
37. Wildcat, D.R. Introduction: Climate change and indigenous peoples of the USA. In *Climate Change and Indigenous Peoples in the United States*; Springer: Berlin/Heidelberg, Germany, 2013; pp. 1–7.

38. Downey, S.S. Resilient Networks and the Historical Ecology of Q'eqchi' Maya Swidden Agriculture. Ph.D. Thesis, University of Arizona, Tucson, AZ, USA, 2009. Available online: http://hdl.handle.net/10150/195686 (accessed on 9 January 2020).
39. Johnston, K.J. The intensification of pre-industrial cereal agriculture in the tropics: Boserup, cultivation lengthening, and the Classic Maya. *J. Anthropol. Archaeol.* **2003**, *22*, 126–161. [CrossRef]
40. Thurston, H.D. *Slash/Mulch Systems: Sustainable Methods for Tropical Agriculture*; Westview: Boulder, CO, USA, 1997.
41. Ong, C.K.; Kho, R.M. A framework for quantifying the various effects of tree-crop interactions. In *Tree–Crop Interactions, Agroforestry in a Changing Climate*, 2nd ed.; Ong, C.K., Black, C.R., Wilson, J., Eds.; CABI International: Wallingford, UK, 2015; pp. 1–23. [CrossRef]
42. Altieri, M.A.; Nicholls, C.I.; Montalba, R. Technological approaches to sustainable agriculture at a crossroads: An agroecological perspective. *Sustainability* **2017**, *9*, 349. [CrossRef]
43. Falkowski, T.B.; Chankin, A.; Diemont, S.A.; Pedian, R.W. More than just corn and calories: A comprehensive assessment of the yield and nutritional content of a traditional Lacandon Maya milpa. *Food Secur.* **2019**, *11*, 389–404. [CrossRef]
44. Emch, M. The human ecology of Mayan cacao farming in Belize. *Hum. Ecol.* **2003**, *31*, 111–131. [CrossRef]
45. Uhl, C. Factors controlling succession following slash-and-burn agriculture in Amazonia. *J. Ecol.* **1987**, *75*, 377–407. [CrossRef]
46. Kidd, C.V.; Pimental, D. (Eds.) *Integrated Resource Management: Agroforestry for Development*; Elsevier: San Diego, CA, USA, 2012.
47. Branca, G.; McCarthy, N.; Lipper, L.; Jolejole, M.C. Climate-smart agriculture: A synthesis of empirical evidence of food security and mitigation benefits from improved cropland management. *Mitig. Clim. Chang. Agric. Ser.* **2011**, *3*, 1–42.
48. Teklewold, H.; Mekonnen, A.; Kohlin, G. Climate change adaptation: A study of multiple climate-smart practices in the Nile Basin of Ethiopia. *Clim. Dev.* **2019**, *11*, 180–192. [CrossRef]
49. Government of Belize (GOB). The National Agriculture and Food Policy of Belize (2015–2030). Ministry of Agriculture and Fisheries, 2015. Available online: https://www.agriculture.gov.bz/wp-content/uploads/2017/05/National-Agriculture-and-Food-Policy-of-Belize-2015-2030.pdf (accessed on 9 January 2020).
50. FAO. Climate-Smart Agriculture Sourcebook. Food and Agriculture Organization of the United Nations Report. 2013. Available online: http://www.fao.org/3/i3325e/i3325e.pdf (accessed on 9 January 2020).
51. Spurney, J.A. Effects of milpa and conventional agriculture on soil organic matter, structure and mycorrhizal activity in Belize. *Caribb. Geogr.* **2000**, *11*, 20–33.
52. Hellin, J.; Fisher, E. Climate-smart agriculture and non-agricultural livelihood transformation. *Climate* **2019**, *7*, 48. [CrossRef]
53. Wickel, A.J.; van de Giesen, N.C.; SÁ, T.D.A.; Denich, M.; Vlek, P.L.; Vielhauer, K. Effects of Slash & Burn vs. Slash & Mulch on Water, Solute and Sediment Dynamics at the Watershed Level. In Proceedings of the Embrapa Amazônia Oriental-Resumo em Anais de Congresso (ALICE), German-Brazil Workshop, Hamburg, Germany, 3–8 September 2000.
54. Mkhize, T. Exploring Farming Systems and the Role of Agroecology in Improving Food Security, Productivity and Market access for Smallholder Farmers. Master's Thesis, University of KwaZulu-Natal, Glenwood, South Africa, 2016. Available online: http://hdl.handle.net/10413/14569 (accessed on 9 January 2020).
55. Erenstein, O. Smallholder conservation farming in the tropics and sub-tropics: A guide to the development and dissemination of mulching with crop residues and cover crops. *Agric. Ecosyst. Environ.* **2003**, *100*, 17–37. [CrossRef]
56. Ayarza, M.A.; Welchez, L.A. Drivers affecting the development and sustainability of the Quesungual slash and mulch agroforestry system (QSMAS) on Hillsides of Honduras. *Compr. Assess. Bright Spots Proj. Final Rep. Noble A* **2004**, 187–201.
57. CIAT. *Quesungual Slash and Mulch Agroforestry System (QSMAS): Improving Crop Water Productivity, Food Security and Resource Quality in the Sub-Humid Tropics*; CPWF Project Report; Foreign, Commonwealth & Development Office (FCDO): Cali, Colombia, 2009.
58. Denich, M.; Vlek, P.L.; de Abreu Sá, T.D.; Vielhauer, K.; Lücke, W. A concept for the development of fire-free fallow management in the Eastern Amazon, Brazil. *Agric. Ecosyst. Environ.* **2005**, *110*, 43–58. [CrossRef]
59. Singh, B.K.; Bardgett, R.D.; Smith, P.; Reay, D.S. Microorganisms and climate change: Terrestrial feedbacks and mitigation options. *Nat. Rev. Microbiol.* **2010**, *8*, 779–790. [CrossRef] [PubMed]
60. Seevers, B.; Graham, D. *Education Through Cooperative Extension*, 3rd ed.; University of Arkansas Press: Fayetteville, AR, USA, 2012.
61. USDA NIFA (U.S. Department of Agriculture, National Institute of Food and Agriculture). *Joint Cooperative Extension Programs at 1862 Land-Grant Institutions and University of the District of Columbia Public Postsecondary Education Reorganization Act Program*; U.S. Department of Agriculture: Washington, DC, USA, 2019. Available online: https://nifa.usda.gov/sites/default/files/program/2020-Smith-Lever-UDC-Joint-2aug19.pdf (accessed on 9 January 2020).
62. Singh, R.; Singh, G.S. Traditional agriculture: A climate-smart approach for sustainable food production. *Energy Ecol. Environ.* **2017**, *2*, 296–316. [CrossRef]
63. MAFFESDI (Ministry of Agriculture, Fisheries, Forestry, the Environment, Sustainable Development and Immigration, Government of Belize). *Extension* 20 February 2019. Available online: http://www.agriculture.gov.bz/extension-2/ (accessed on 9 January 2020).
64. Tandon, H. *Strengthening Sustainable Agriculture in the Caribbean*. A Report for the FAO. 2014. Available online: https://competecaribbean.org/wp-content/uploads/2015/02/Strengthening_Sustainable_Agriculture_in-the-Caribbean_web.pdf (accessed on 9 January 2020).
65. Khatri-Chhetri, A.; Aggarwal, P.K.; Joshi, P.K.; Vyas, S. Farmers' prioritization of climate-smart agriculture (CSA) technologies. *Agric. Syst.* **2017**, *151*, 184–191. [CrossRef]

66. Steenwerth, K.L.; Hodson, A.K.; Bloom, A.J.; Carter, M.R.; Cattaneo, A.; Chartres, C.J.; Hatfield, J.L.; Henry, K.; Hopmans, J.W.; Horwath, W.R.; et al. Climate-smart agriculture global research agenda: Scientific basis for action. *Agric. Food Secur.* **2014**, *3*, 1–39. [CrossRef]
67. Molnar, S.; Molnar, I.M. *Environmental Change and Human Survival*; Prentice Hall: Upper Saddle River, NJ, USA, 2000.
68. Olsson, P.; Folke, C.; Berkes, F. Adaptive co-management for building resilience in social–ecological systems. *Environ. Manag.* **2004**, *34*, 75–90. [CrossRef]
69. Ostrom, E. A general framework for analyzing sustainability of social-ecological systems. *Science* **2009**, *325*, 419–422. [CrossRef]
70. Parrott, L.; Chion, C.; Gonzalès, R.; Latombe, G. Agents, individuals, and networks: Modeling methods to inform natural resource management in regional landscapes. *Ecol. Soc.* **2012**, *17*. [CrossRef]
71. Lal, R. Soils and sustainable agriculture: A review. *Agron. Sustain. Dev.* **2008**, *28*, 57–64. [CrossRef]
72. Goulding, C. Grounded theory, ethnography and phenomenology: A comparative analysis of three qualitative strategies for marketing research. *Eur. J. Mark.* **2005**, *39*, 294–308. [CrossRef]
73. Dunn, T.; Gray, I.; Phillips, E. From personal barriers to community plans: A farm and community planning approach to the extension of sustainable agriculture. In *Case Studies in Increasing the Adoption of Sustainable Resource Management Practices*; Land & Water Resources Research & Development Corporation: Canberra, Australia, 2000; p. 15. [CrossRef]
74. Myers, M.D.; Newman, M. The qualitative interview in IS research: Examining the craft. *Inf. Organ.* **2007**, *17*, 2–26. [CrossRef]
75. Rabionet, S.E. How I learned to design and conduct semi-structured interviews: An ongoing and continuous journey. *Qual. Rep.* **2011**, *16*, 563.
76. Creswell, J.W. *Qualitative Inquiry and Research Design: Choosing among Five Approaches*; SAGE Publications: Thousand Oaks, CA, USA, 2013.
77. Gall, M.D.; Gall, J.P.; Borg, W.R. *Educational Research: An Introduction*, 8th ed.; Pearson: Boston, MA, USA, 2007.
78. Ravitch, S.M.; Carl, N.M. *Qualitative Research: Bridging the Conceptual, Theoretical, and Methodological*; SAGE Publications: Thousand Oaks, CA, USA, 2016.
79. SIB (Statistical Institute of Belize). Postcensal Estimates by Administrative Area and Sex (2010–2018). 2018. Available online: http://www.sib.org.bz/statistics/population (accessed on 9 January 2020).
80. Guide to Belize. Sights by District. Available online: www.guidetobelize.info/en_sights.html (accessed on 19 April 2019).
81. LeCompte, M.D. Analyzing qualitative data. *Theory Pract.* **2000**, *39*, 146–154. [CrossRef]
82. Stuckey, H.L. The second step in data analysis: Coding qualitative research data. *J. Soc. Health Diabetes* **2015**, *3*, 7. [CrossRef]
83. Charmaz, K. *Constructing Grounded Theory: A Practical Guide through Qualitative Analysis*; Sage: Thousand Oaks, CA, USA, 2006.
84. Pereira, L. Developing perspectival understanding. In *Contemporary Qualitative Research: Exemplars for Science and Mathematics Educators*; Taylor, P.C., Wallace, J., Eds.; Springer: Dordrecht, The Netherlands, 2007; pp. 189–203. [CrossRef]
85. Strauss, A.; Corbin, J. Grounded theory methodology. In *Handbook of Qualitative Research*; Denzin, N.K., Lincoln, Y.S., Eds.; Sage: Thousand Oaks, CA, USA, 1994; pp. 273–285.
86. Ellingson, L.L. *Engaging Crystallization in Qualitative Research: An Introduction*; Sage: Thousand Oaks, CA, USA, 2009.
87. Ramirez, G.; Extension, Belmopan, Belize. Personal Communication, 26 September 2007.
88. Flora, C.; Flora, J. *Rural Communities' Legacy and Change*, 4th ed.; Westview Press: Boulder, CO, USA, 2013.
89. Cote, M.; Nightingale, A.J. Resilience thinking meets social theory: Situating social change in socio-ecological systems (SES) research. *Prog. Hum. Geogr.* **2012**, *36*, 475–489. [CrossRef]
90. González, J.A.; Montes, C.; Rodríguez, J.; Tapia, W. Rethinking the Galapagos Islands as a Complex Social-Ecological System: Implications for Conservation and Management. *Ecol. Soc.* **2008**, *13*, 13. [CrossRef]
91. Tenza, A.; Pérez, I.; Martínez-Fernández, J.; Giménez, A. Understanding the decline and resilience loss of a long-lived social-ecological system: Insights from system dynamics. *Ecol. Soc.* **2017**, *22*, 15. [CrossRef]
92. Koutsouris, A. Higher education facing sustainability: The case of agronomy. *Int. J. Learn. Annu. Rev.* **2008**, *15*, 269–276. [CrossRef]
93. University of California-Davis (UC-Davis). (n.d.) *What Is Sustainable Agriculture?* Agriculture Sustainability Institute: Davis, CA, USA. Available online: http://asi.ucdavis.edu/programs/sarep/about/what-is-sustainable-agriculture (accessed on 9 January 2020).
94. Méndez, V.E.; Bacon, C.M.; Cohen, R. Agroecology as a transdisciplinary, participatory, and action-oriented approach. *Agroecol. Sustain. Food Syst.* **2013**, *37*, 3–18.
95. Rivera-Ferre, M.G. The resignification process of agroecology: Competing narratives from governments, civil society and intergovernmental organizations. *Agroecol. Sustain. Food Syst.* **2018**, *42*, 666–685. [CrossRef]
96. Wezel, A.; Bellon, S.; Doré, T.; Francis, C.; Vallod, D.; David, C. Agroecology as a science, a movement and a practice. A review. *Agron. Sustain. Dev.* **2009**, *29*, 503–515. [CrossRef]
97. Totin, E.; Segnon, A.C.; Schut, M.; Affognon, H.; Zougmoré, R.B.; Rosenstock, T.; Thornton, P.K. Institutional perspectives of climate-smart agriculture: A systematic literature review. *Sustainability* **2018**, *10*, 1990. [CrossRef]
98. Arif, M.; Jan, T.; Munir, H.; Rasul, F.; Riaz, M.; Fahad, S.; Adnan, M. Climate. In *Global Climate Change and Environmental Policy*; Springer: Singapore, 2020; pp. 351–377.

Article

Impact of Climate Change Adaptation on Household Food Security in Nigeria—A Difference-in-Difference Approach

Oyinlola Rafiat Ogunpaimo [1,2,*], Zainab Oyetunde-Usman [3] and Jolaosho Surajudeen [2]

1 Rural Economy & Development Programme, Teagasc, Mellows Campus, Athenry, Co., Galway H65 R718, Ireland
2 Department of Agricultural Economics and Farm Management, Federal University of Agriculture, Abeokuta 112251, Nigeria; surajudeenjolaosho@yahoo.com
3 National Resources Institute, University of Greenwich, Chatham Maritime ME4 4TB, UK; zainabus23@gmail.com
* Correspondence: oyinlolaadams@gmail.com

Abstract: Studies have shown that climate change adaptation options (CCA) are implemented to buffer the unfavorable climatic changes in Nigeria causing a decline in food security. Against the background of measuring the impact of CCA options using cross-sectional data, this study assessed how CCA had affected food security using panel data on farming households from 2010–2016 obtained from Nigerian General Household Survey (GHS). Data were analyzed using the Panel probit model (PPM), Propensity Score Matching (PSM), and Difference-in-Difference (DID) regression. PPM showed that the probability of adopting CCA options increased with farm size ($p < 0.01$), extension contact ($p < 0.01$), and marital status ($p < 0.01$), but decreased with the age of the household head ($p < 0.01$). Credit facilities ($p < 0.05$), ownership of farmland ($p < 0.01$), household size ($p < 0.01$), years of schooling ($p < 0.01$), household asset ($p < 0.01$), and location ($p < 0.05$) also had a significant but mixed effect on CCA choices. PSM revealed that farming households that adopted CCA strategies had 9% higher food security levels than non-adopters. Furthermore, the result of the DID model revealed a significant positive effect of CCA on household food security ($\beta = 5.93$, $p < 0.01$). It was recommended that education and provision of quality advisory services to farmers is crucial to foster the implementation of CCA options.

Keywords: developing countries; welfare; panel probit model; adoption; propensity score matching

Citation: Ogunpaimo, O.R.; Oyetunde-Usman, Z.; Surajudeen, J. Impact of Climate Change Adaptation on Household Food Security in Nigeria—A Difference-in-Difference Approach. *Sustainability* **2021**, *13*, 1444. https://doi.org/10.3390/su13031444

Academic Editor: Maurizio Tiepolo
Received: 29 December 2020
Accepted: 26 January 2021
Published: 29 January 2021

Publisher's Note: MDPI stays neutral with regard to jurisdictional claims in published maps and institutional affiliations.

1. Introduction

The agricultural practices in African nations especially Nigeria largely rely on the natural weather conditions of the locality. Changes in the climatic condition of the country are evident in increased desert encroachment and extreme droughts in the Northern region [1,2], likewise the problem of persistent flood and erosion occurrence in the Southern region. Climatic variability and changes have been linked to erosion, increased flooding, environmental degradation [3], and a decrease in agricultural productivity [4,5].

Frequent and intense weather events as a result of climate change are likely to impact the welfare and food security status of both the rural and urban populace through poor food production, poor land availability, and reduced opportunities [6]. The optimal usage of land for crop and animal production, biodiversity restoration, health, and well-being can also be negatively impacted by increased temperature and precipitation changes, and increased weather fluctuations [7–9].

In accordance with the 2020 global food security index, Nigeria's food insecurity status is considered serious in the severity chart [10]. The Federal Ministry of Agriculture of Nigeria in 2014 estimated that 65% of the population is food insecure despite having more than half of all employments dependent on agriculture [5]. Among several other factors, heightened food insecurity among farm households is caused by limited access to credit, poor storage and improved agricultural facilities, and negative environmental influences such as erosions and floods [5]. Other reasons include the lower household income necessary for food purchases needed to attain food security [11], increased population growth [12,13], and a huge reliance on imported food items [14]. The 2030 United Nations Sustainable Development Goals are new global policies with the objective to restructure regional and national development plans over the next 10 years. The global policy aims to put an end to poverty and hunger, food insecurity, sustaining natural resources and the environment, and promote food and agriculture sustainability [15].

The International Symposium on Climate and Food Security (ISCFS) and Intergovernmental Panel on Climate Change (IPCC) Fourth Assessment Report (AR4) also recognized three critical global problems of poverty and hunger, increased population growth, and unfavorable weather and climate [16]. Agricultural production and climate change exhibit a feedback relationship, while agricultural activities may result in increased emissions and pollutions leading to climatic changes, climate change also influences agricultural output. Research has indicated that by 2030, the negative consequence of climate change on agriculture will be more severe across all the countries of the world [17]. Climatic changes already have an antagonistic influence on food security with the number of chronically undernourished people in the world estimated to have increased by 38 million in 2016, in addition to 777 million recorded in 2015 [18], thus, without proper implementation of adaptation and mitigation measures, climate variability and change will threaten the achievements of the SDG goals in eradicating poverty and hunger [17].

As stated by the United Nations Framework Convention on Climate Change (UNFCCC, 2007) AR4 and buttressed by Field [19] and Steynor and Pasquini [20], Africa is predicted to be the continent most susceptible to climate change and variability; adverse climatic impacts are worsening the livelihoods, welfare, and regional and household food security in tropical developing regions [17]. Consensus exists that climate change and variability will have a significantly negative impact on all these aspects of food security in Africa [21]. Like other African countries, studies such as Ebele and Emodi [22] reported that the projected impact of climate change on West Africa's agricultural productivity could lead to a 4% reduction in the region's GDP; in Nigeria, the adverse influence of climate change on food security is evidenced by the changes in plant duration and output of cereal crops, reductions in aquatic life [23], and livestock failure [24]. In tackling climate change and variability, studies [25,26] indicate that CCA can promote household food security. While research had been conducted on factors influencing the choice of CCA by farming households and the impact of CCA options on household food security, very few studies if any had applied a panel data analysis approach to investigate the influence of CCA options on household food security in Nigeria.

Morland [27] stated that small scale farmers in Africa experience weather variability and other climate change related events. Among farm households, varieties of climate-adaptation methods abound and this includes diversification and crop rotation, engaging in non-agricultural income-generating activities, practicing soil and water conservation techniques, adjusting the times they sow their lots, use of irrigation and creation of flood barriers, adopting improved seed varieties and fertilizers, tree-planting, and integrating crop production with livestock.

Adaptation to climate change and variability means anticipating the adverse effects of climate variability and taking appropriate action to prevent and or minimise the damage they cause or taking advantage of opportunities that may arise [28]. It involves the use of climate sustainable practices to reduce the negative consequences of climate change.

In many African countries, especially Nigeria, access to food will be severely affected by climate change. Africa is the region where climate change and variability have had the biggest impact on acute food insecurity and malnutrition, affecting 59 million people in 24 countries [7]. Given these discouraging prospects, it is no surprise that the adaptation strategies are vital to support the climate effects on food security. These strategies can indeed buffer against climate variability and play a crucial role in promoting the food security of farm households, thereby reducing the negative effect of climate variability on food security.

Thus, this research sought to investigate the impact of climate change adaptation on farm households' food security status by providing answers to the following questions: (1) What are the factors affecting climate adaptation strategies employed by farm households in Nigeria? (2) What is the impact of climate change adaptation on household food security?

2. Materials and Methods

This section describes the study area, the data source, a summary of farm household characteristics, the description of variables, and the method of data analysis.

2.1. Study Area

The study area is Nigeria situated 4° and 14° N, and longitudes 2° and 15° E. Nigeria is located in West Africa with its capital at Abuja created in 1976 having a total area of 923,769 km^2 (356,669 sq mi) [29], making it the world's 32nd largest country. Nigeria is bordered to the north by Niger, to the east by Chad and Cameroon, to the south by the Gulf of Guinea of the Atlantic Ocean, and the west by Benin [29].

Like most countries in sub-Saharan Africa, the climatic condition in Nigeria is tropical with varying rainy and dry seasons. The Nigeria vegetation belts, also known as the agroecological zones range from the mangrove, freshwater swamps, and tropical rainforest which extends from the South-South, South-East to South-West regions. The tropical savanna grasslands zone is dominant in the middle belts with Sudan and Sahel Savanna in the Northern regions. Agricultural production is dominant in the northern agroecological zones, however, climatic changes and human activities such as continuous cropping, overgrazing, and bush burning especially in densely populated areas have highly impacted the agricultural vegetation.

Specifically, in the far northern areas, the nearly total disappearance of plant life has facilitated a gradual southward advance of the Sahara [29]. The map of Nigeria showing the different regions and states is illustrated as shown in Figure 1.

2.2. Data Source

Panel data on relevant socioeconomic, demographic, consumption, and production data of farming households from 2010–2016 were obtained from the World Bank and the Nigeria Bureau of Statistics (NBS) General Household Survey (GHS). The GHS is a nationally representative survey with respondents obtained from the 36 states of Nigeria including the Federal Capital Territory. Three waves of the Living Standard Measurement Survey (First wave—2010/2011 Ref. NGA_2010_GHSP-W1_v03_M [30], Second wave—2012/2012 Ref. NGA_2012_GHSP-W2_v02_M [31] and the Third-wave—2015/2016 Ref. NGA_2015_GHSP-W3_v02_M [32]) was employed such that there are six-period panel data for the farm households from which 3500 farm households were used for this study.

2.3. Summary of Characteristics of Farm Households in the Study Area

The summary of farm households' characteristics is shown in Table 1. Age is an important determinant of farm activities. It is believed that younger people commit more energy to production activities, while older farmers are likely to be more experienced. The mean age of respondents was 50.4 and 53.2 years in 2010 and 2016 respectively. In relation to the gender of the household head, the majority of family households were male-headed. Male household heads accounted for 88.36% in 2010 to a slight reduction of 84.14% in 2016. This report buttresses the dominance of males in farming activities in Nigeria as reported by [33]. However, the decline in the proportion of male-headed households in the study area may be due to increasing awareness of female empowerment and capacity building in the area.

Table 1. Summary of Farm Households' Characteristics in the Study Area.

Characteristics		Year			
		2010		2016	
		Percentage	Mean	Percentage	Mean
Gender of HH	Female	11.64		15.86	
	Male	88.36		84.14	
Marital status	Single	15.00		20.43	
	Married	85.00		79.57	
Accessed credit	Yes	2.94	-	24.01	
	No	97.06		75.99	
Extension contact	Yes	14.38		86.79	
	No	85.62		13.21	
Ownership of land	Yes	29.07		1.24	
	No	70.93		98.76	
Age of HH			50.4		53.2
HH SIZE			6		8
Years of schooling			6		7
Household asset			11		41
Plot size			11,372.38		10,264
Secondary income			13,465.25		5672.59
Total food expenditure			16,382.38		21,867.58
Total non-food expenditure			7137.23		8350.23
Total expenditure			23,519.61		30,217.83

Author's Computation from LSMS Data 2010:2016.

As illustrated in Table 1, the majority (85%) of farming households in 2010 were married, this came down to 79.57% in 2016. In terms of educational level, the average years of schooling were 6 years with no significant change in 2016 implying that although farm households have access to formal education, the level of education among farming households in Nigeria is still low. However, by 2016 the average years of schooling had increased by a year in the country due to various campaigns and probable enlightenment on the need for female education in the region. Following apriori expectation education is likely to increase the probability of farming households adapting to climate changes because it can be assumed that education will increase farm household's awareness of CCA.

The size of households is also shown in Table 1. It is evident that the average household size for the sampled farm households' was approximately six persons in 2010 to eight persons in 2016. Previous literature [35–37] argued that the probability of adopting labour-intensive adaptation measures increases with family size due to the availability of free or inexpensive man-power. Also, large families divert part of their labour force into non-farm activities to generate more income [38,39].

From Table 1, the majority of respondents (85.62%) had no access to extension contact in 2010, however, in 2016 the majority (86.79%) had access to extension contact. This showed the intensification of extension services contact increased over the years within the study area. For access to formal credit, the majority (97.5%) of the respondents did not have access; this may limit the ability of the farmers to expand their scale of production. This result is buttressed by the findings of [40] who assessed the trend of formal credit allocation to food crop production in Nigeria; their results showed that there was a decreasing trend in the credit allocation to the food crop production since 2011. It can be argued that agriculture production is constrained in Nigeria by poor credit delivery; the delivery of credit facilities in the country is largely in favour of the wealthy farmers as opposed to poor farmers; the wealthy farmers may utilize the loan acquired for ulterior motives rather than the initial function of agricultural production [40,41]. This limited access to credit facilities may be as a result of high-interest rates credit facilities provided by financial institutions among other bureaucratic delays inherent in loan assessment, acquisition, and disbursement in Nigeria.

Ownership of land can serve as an indicator of the wealth status of farming households [25], and thus it can be expected that the increased wealth status of farming households leads to increased food security of farm households. The distribution of the ownership of land is illustrated in Table 1; the result obtained showed that the majority (98.76%) of farm-households owned land compared to 2010 where the majority (69.96%) of farm households had no access to land.

The summary of average food expenditures and, by extension, food security of farm households, is presented in Table 1. A household is categorized to be very vulnerable to food insecurity if more than 75% of its total expenditure is spent on food items whereas people spending 65–75% are considered to have high food insecurity [11,42]. Following Engel's law, the higher the food expenditure share, the lower the food security of farm households. Thus, the food security status of farm households measured as household food expenditure share slightly reduced in 2016 with average farm households spending approximately 72% of their expenditure on food compared to 2010 when an average farm household spent about 69.6% of their expenditures on food.

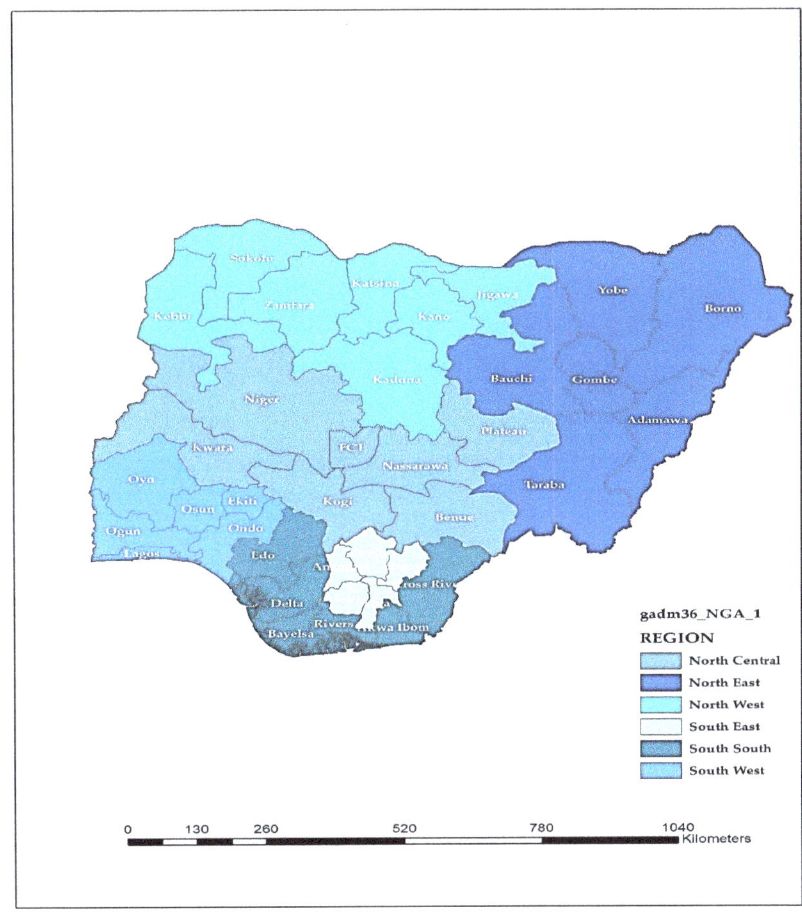

Figure 1. Geographical subdivisions of Nigeria showing the 36 states, Federal Capital Territory, and the six geo-political regions of the country obtained by computation of GADM [34] geo package Data.

2.4. Analytical Techniques

The method of data analysis adopted in this study to achieve the stated objectives includes:

2.4.1. Panel Probit Model

The pooled probit model specification of the panel data model was employed to evaluate the factors affecting CCA strategies employed by farm households for this study [43,44]. Following Akerele et al. [43], the panel probit model is expressed as follows:

$$y^*_{jit} = a + X_{jt}\beta + e_{jit} \tag{1}$$

where

y^*_{jit} = the latent (underlying) variable that determines whether farm household j would be classified as an adopter of CCA measure i at time t;
β = a vector coefficient;
X_{jt} = a matrix of explanatory variables;
a = the constant term; and
e_{jit} = the idiosyncratic errors assumed to have zero mean and unit variance. The relationship between the latent variable y^*_{jit} and the observed outcome y_{jit} is represented as

$$y_{jit} = \begin{cases} 0 \ if \ y^*_{jit}<0 \\ j \ if \ y^*_{jit}>0 \end{cases} \quad for\ i = 1, \ldots\ldots\ldots, n \quad and\ j = 1, \ldots\ldots p-1 \tag{2}$$

where $y_{jit} = 1$ if a farm household adopts a CCA strategy for each adaptation strategy.

The selection of the variables was motivated by previous literature [4,25,45–47], availability of data [48], and economic theories [4] on factors influencing the choice of CCA and are presented in Table 2. Farm and household attributes were included as explanatory variables for assessing determinants of CCA options. Ownership of land or tenancy status was used as a determinant mainly because these can act as proxies for the wealth status of the farm households [25].

Table 2. The Description Measurement and A Priori Expectation of the Variables.

Variable	Description	Measurement	A Priori Expectation
X1	Age of household head	Years	±
X2	Accessed credit	Dummy: 1 for yes; 0 no	+
X3	Tenancy status	Dummy: 1 for farm owner; 0 otherwise	+
X4	Farm size	Hectares	+
X5	Accessed extension contact	Dummy: 1 for yes; 0 no	+
X6	Household size	Number	±
X7	Gender of the HH	Dummy: 1 for male; 0 otherwise	±
X8	Marital status	Dummy: 1 for married; 0 otherwise	⊥
X9	Educational level of HH	Years	+
X10	Secondary occupation income	Naira	+
X11	Quantity of household asset	Number	+
X12	North Central	Dummy: 1 for yes; 0 otherwise	+
X13	North East	Dummy: 1 for yes; 0 otherwise	+
X14	North West	Dummy: 1 for yes; 0 otherwise	+
X15	South East	Dummy: 1 for yes; 0 otherwise	+
X16	South West	Dummy: 1 for yes; 0 otherwise	+
Reference category	South-South		

Following Teklewold et al. [4], socio-demographic characteristics of farm households important in implementing CCA options were controlled for, these factors include age, gender, household size, gender, and educational level of the household head. Resource constraint was also considered while accounting for the factors influencing CCA options. Quantity of household assets can act as a proxy in measuring the wealth status of farm households [4]; access to credit facilities extension services was included as one of the explanatory factors since extension services provide crucial education and information needed to adopt CCA options [25,46].

Following from this, the four categories of adaptation strategies were considered in this study, these measures were selected based on the popularity of these measures amongst farm households across all the geo-political zones in Nigeria considered in this study:

Diversify more into other crops
Used Irrigation facilities
Diversify into off-farm activities
Implement soil conservation techniques

Adoption of crop diversification is a CCA method that may involve the planting of high yield variety and drought-resistant crops or intercropping [48], which has been extensively identified from previous studies as an option that can help farmers and farming households buffer the negative effects of climate change. Planting of crops such as cereals that are highly affected with sporadic fluctuations in weather patterns along with turgid crops such as cassava will minimize crop losses due to weather events [33,49].

Implementation of soil conservation techniques such as fallowing and practicing alley cropping can aid in the restoration of soil nutrients, minimize nutrient loss, protects the vegetation cover, and also reduces organic matter oxidation in the soil [49]. Alley cropping can aid in reducing soil erosion while also serving as windbreaks.

Water as a resource is crucial for optimal crops and livestock cultivations [48]. Several studies such as [48–51] documented the importance of implementation of irrigation facilities especially in regions prone to drought or low rainfall occurrences. The adoption of irrigation is encouraged to augment the rainfall amount required for optimal crop cultivation.

Off-farm diversification has been extensively used as a CCA strategy as evidenced by previous literature [51,52]. Farming households may undergo off-farm activities or other occupation during the dry season, unfavourable climate conditions [51] or mainly to complement income sources in order to meet household food security status.

Estimating the impact of climate change adaptation on household food security in Nigeria was achieved in two stages using a combination of two analytical tools, which were described as 2.4.2 (Propensity Score Matching) and 2.4.3 (Difference-In-Difference).

2.4.2. Propensity Score Matching

This study employed PSM to estimate the impact of the adaptation strategies on farm household food security status. The PSM is defined as the conditional probability that a farm household adopts the new adaptation strategies, given pre-adoption characteristics [53]. To mimic a typical randomized controlled experiment, the PSM assumes the unconfoundedness assumption, also known as conditional independence assumption, which implies that once Z is controlled for, technology adoption is random and uncorrelated with the outcome variables. The PSM also accounts for this sample selection bias [25,54].

Following the framework of Ali and Erenstein [25], the PSM was used to estimate the impact of climate change adaptation on farm households' food security. After investigating the choice determinants of CCA practices using PPM, a propensity score matching approach was employed to analyse the impact of adaptation practices on food security.

The farm households were classified as food secure or food insecure based on their share of total household expenditure spent on food. Following Ali and Erenstein [25,42], Smith, et al. [55], households spending more than 75% of their expenditures on food were categorized as food insecure households and were assigned a dummy value of zero; while farm households were categorized as food secure and assigned a value of 1 when the food expenditure is below the threshold level (75%) of total expenditure.

A risk-averse farm F_i opts for a few strategies (S_j). It is assumed that households that have opted for adaptation strategies have higher utility levels compared to those that have not: [25].

$$U[F(S_1)] > U[F(S_0)] \tag{3}$$

The PSM can be expressed as according to Ali and Erenstein [25]:

$$\mathrm{P}\,(Z) = \Pr\{I = 1|Z\} = E\{1|Z\} \tag{4}$$

where I = is the indicator for adoption and

Z = the vector of pre-adoption characteristics.

The conditional distribution of Z, given $p(Z)$, is similar in both groups of adopters and non-adopters.

The expected treatment effect for the treated population is of primary significance and is given as

$$\tau|_{i=1} = E(\tau|I = 1) = E(R_1|I = 1) - E(R_0|I = 1) \tag{5}$$

where τ = the average treatment effect for the treated (ATT),

R_1 = denotes the value of the outcome for adopters of the adaptation, and

R_0 is the value of the same variable for non-adopters.

As noted above, the major problem is that we do not observe $E(R_0|I = 1)$, in other words, it is potentially a biased estimator.

After estimating the propensity scores, the average treatment effect for the treated (ATT) can then be estimated as [25,56]

$$\begin{aligned} \tau = E(R_1 - R_0|I = 1) &= E\{E\{R_1 - R_0|I = 1,\ p(Z)\}\} \\ &= E\{E\{R_1|I = 1,\ p(Z)\} - E\{R_0|I = 0,\ p(Z)\}\} \end{aligned} \tag{6}$$

PSM is based on two underlying assumptions, that is: the common support and the conditional independence assumption [25]. A diagnostic test of matching quality must be carried out after matching to estimate the standard errors and treatment effects. Some balancing tests were to be carried out to access the matching quality, mean absolute bias, t-statics, and the bias reduction before and after matching [57,58].

2.4.3. Difference-in-Difference

DID was also used to assess the impact of CCA on household food security; unlike the PSM which estimates the impact of CCA on household food security between adopters of CCA and non-adopters, DID evaluates the impact of CCA over time, that is from 2010 to 2016.

Difference-in-difference (DID) methods, compared with PSM, assume that unobserved heterogeneity in adoption is present but that such factors are time-invariant. With data on project and control observations before and after the CCA adoption, therefore, this fixed component can be differenced out. Some variants of the DID approach have been introduced to account for potential sources of selection bias. Combining PSM with DID methods can help resolve the problem of selection bias, by matching units in the common support [56]. The propensity score can be used to match participant/adopters and control/non-adopters units in the base year, and the CCA impact is calculated across adopters and matched control units within the common support. For two time periods t = {1,2}, the DID estimate for each adoption area i will be calculated as

$$DID_i = \left(Y_{i2}^T - Y_{i1}^T\right) - \sum_{j \in C} \omega(i,j)\left(Y_{i2}^C - Y_{i1}^C\right) \tag{7}$$

where

$\omega(i,j)$ is the weight (using a PSM approach) given to the jth control area matched to adoption area i.

Y_{i2}^{T} = Farm household food security status of CCA adopters in 2016.
Y_{i1}^{T} = Farm household food security status of CCA adopters in 2010
Y_{i2}^{C} = Farm household food security status of non-adopters of CCA in 2016
Y_{i1}^{C} = Farm Household food security status of non-adopters of CCA in 2010.

3. Results and Discussion

The findings of this research work, interpretations, and also discussion of the result are presented in this section.

3.1. Determinants of Farm Households Climate Change Adaptation Options

A panel probit model was used in this study to estimate the factors affecting adaptation strategies employed by farming households. Adaptation options identified include

Irrigation
Soil conservation
Crop diversification
Diversification into non-farm activities

The likelihood ratio test from the Panel probit model showed the overall significance of the models at ($p < 0.01$) probability level, which signified that the model is useful in explaining factors influencing decisions of farming households to adapt to climate change.

Age of Household Head: As shown in Table 3, the age of the household head is an important determinant in the decision of farming households to use irrigation ($p < 0.01$), and diversify into non-farm activities ($p < 0.01$). The sign of the parameter is negative, implying that the older the household head, the less likely their probability to adopt irrigation and diversify into non-farm activities. It can be deduced from the result that with a year increase in the age of farmers the probability of implementing irrigation facilities and practicing non-farm diversification decreases by 1% and 0.4% respectively. These findings suggest that younger farmers are more likely to adopt these CCA strategies compared to their older counterparts, possibly because they are innovative and keen to try new technology and methods to improve agriculture, whereas older farmers through years of experience may understand the negative economic implications of practicing such strategies. These findings are in support of Ali and Erenstein [25]; where the age of the household head had a negative relationship with CCA adoption, they claimed that older farmers may be conservative about trying new and innovative agricultural practices despite increased awareness. However, the result was against the findings of [59–61] who found that age had a positive association with CCA adoption among farming households.

Access to Credit Facilities: Access to credit facilities was positively significant ($p < 0.01$) for practicing soil conservation and off-farm diversification (shown in Table 3), which is in support of Hassan and Nhemachena [62] and Ojo and Baiyegunhi [63]. In their study, they opined variations to farmers' adaptation options, which are largely dependent on their access to credit and information on credit. On the other hand, access to credit facilities has a negative but significant ($p < 0.10$) effect on the probability of using irrigation facilities. This may be due to the cost implication attached to the use of irrigation and the predictive risk of being unable to refund the credit owed when due. The effect on irrigation facilities is in line with [59] but opposed to findings in [64]. Hisali et al. [59] reported that households without credit have a greater likelihood of implementing CCA options, the suggested situations like this may occur where repayment of credit leads to resource constraint needed for CCA adoption or that credit can be used for other purposes other than climate change adaptation.

Table 3. Parameter Estimates of Panel Probit Model of Determinants of Farming Households CCA Strategies.

	Irrigation	Soil Conservation	Crop Diversification	Diversify into Other Occupation
Credit	−0.14 * (−1.82)	0.13 *** (2.61)	−0.20 (−0.38)	0.23 *** (5.30)
Tenancy Status	0.05 (1.19)	0.10 *** (39.49)	0.26 *** (7.97)	−0.08 *** (−2.99)
Farm size	0.02 *** (3.55)	0.03 * (1.86)	0.03 *** (4.10)	0.08 (1.62)
Extension contact	0.17 *** (2.89)	−0.09 * (−1.61)	0.24 *** (4.07)	−0.07 (−1.64)
Household size	0.02 * (1.90)	0.06 *** (9.17)	0.01 (0.68)	0.04 *** (6.27)
Age of HH	−0.01 *** (−3.03)	0.01 (0.69)	−0.01 (−0.11)	−0.004 *** (−4.74)
Gender of HH	−0.07 (−0.75)	0.30 *** (4.69)	−0.02 (−0.38)	−0.11 ** (−2.38)
Marital status	0.24 ** (2.19)	−0.16 ** (−2.55)	0.28 *** (5.17)	0.26 *** (5.13)
Years of schooling	−0.01 (−0.04)	−0.04 *** (−12.35)	−0.02 *** (−6.29)	0.02 *** (7.27)
HH asset	−0.01 *** (−3.16)	−0.09 *** (−15.66)	−0.01 *** (−7.12)	0.003 *** (3.79)
North Central	0.35 ** (2.23)	0.49 *** (7.65)	0.28 *** (3.80)	−0.54 *** (−7.51)
North East	−0.03 (−0.17)	0.56 *** (8.68)	0.15 ** (2.10)	−0.42 *** (−5.88)
North West	0.56 *** (3.71)	0.48 *** (7.37)	1.12 *** (12.35)	−0.18 *** (−2.52)
South East	−0.01 (−0.04)	0.55 *** (8.48)	1.43 *** (15.11)	−0.25 *** (−3.52)
South West	−0.35 (−1.62)	0.50 *** (6.38)	0.06 (0.99)	0.47 *** (5.37)
Constant	−2.12 (−9.79)	−1.99 (−19.02)	0.38 (3.50)	−0.16 (−1.70)
Number of Observations	**18,873**	**18,873**	**18,873**	**18,873**
Log-Likelihood	**−0.07 *****	**−4262.18 *****	**−6173.30 *****	**−9332.21 *****
Wald Chi2(14)	**0.21 *****	**2371.78 *****	**745.85 *****	**431.58 *****
***p*-Value**	**−0.05 ****	**0.000 *****	**0.000 *****	**0.000 *****

*, ** and *** represents statistical significance at 10%, 5% and 1% respectively. Authors computation of LSMS data 2010–2016.

Tenancy Status: The influence of ownership of farmland is reported in Table 3. As indicated ownership of farmland has a mixed effect on adaptation options, it has a direct and significant relationship with soil conservation ($p < 0.01$) and crop diversification ($p < 0.01$). With ownership of land, the decision on the usage of land rests solely on the farmer, due to the availability of lands, it is easier for the farmer to leave some portion of his land for fallowing and also utilize the farm for the cultivation of crops with varying lifecycles since he does not have to fear he may lose his tenancy status. The cost of incurring land is null, therefore, there are more funds available to go into planting various crops. Quan [65] and Kokoye, et al. [66] concluded that land ownership availability can be an incentive for farmers to invest in resources for farming because farmers can pass their land on to the next generation; therefore, they are more willing to care for the land by adopting practices that can aid to maintain its productivity and food security in the context of climate change. Conversely, ownership of farmland has a negative and significant relationship in practicing alley cropping and diversifying into non-farm activities. The relationship between ownership of farmland and diversifying into nonfarm activities is expected because farmers may have invested so much in their farming business; another reason is that owning land may increase the profitability of the business. Previous studies, however, showed mixed results for the relationship between tenancy status and adoption of CCA options. While some studies [67,68] posit a direct relationship between land ownership and adoption of CCA options, other studies such as [25,69–73] reported a negative correlation. The latter are variously associated with the need for farmers in this category to have more agriculturally reliant livelihoods.

Farm size: Farm size has a significant influence on CCA options. An increase in farm size increases the probability of farmers adopting irrigation ($p < 0.01$), implementing soil conservation techniques ($p < 0.01$), and practicing crop diversification. From Table 3, a 1 hectare increase in farm size increases the likelihood of farm households implementing irrigation, soil conservation techniques, and crop diversification by 2%, 3%, and 3% respectively. Findings are in support of several studies that generally reported a positive association between CCA adoption and farm size [25,70,74,75]. Farmers with large land possessions are likely to have more capacity to try out and invest in climate risk-coping strategies. As reported by Arunrat et al. [64], an increase in farm size and land ownership reduces bureaucratic delays with regards to decisions about CCA adoption, mainly because of their ability to procure the high capital and landholdings, and the freedom required to implement innovative practices on their land.

Extension Contact: Studies such as Adams [39], Tambo [51], Boansi, Tambo and Müller [61], and Gbetibouo [76] have shown significant effects of access to extension contact on adopting CCA options. The result of the PPM confirmed that access to the extension has a positive and significant ($p < 0.01$) impact on irrigation use and crop diversification; from Table 3, it can be inferred that a unit increase in farm households' access to extension contact increases the likelihood of implementing irrigation and crop diversification by 17% and 24% respectively. The reason behind it is that extension services help disseminate innovations likely denoting the role of advisory services, and access to information among other resources may motivate the farm household to implement such CCA strategies [70,71]. These findings support those of Tambo [51], Boansi, Tambo and Müller [61], and Gbetibouo [76], which showed that extension services enhanced the availability of information on CCA options.

Household Size: It is positive and significant ($p < 0.01$) for the probability of households to diversify into non-farm activities and implement soil conservation techniques. Increasing household size results in an increase in food expenditure and the compulsion to meet this need comes from non-agricultural income sources. Ali and Erenstein [25], Deressa, Hassan and Ringler [45], and Arshad, et al. [77] revealed similar results of the increase in household size, which increases the probability of adopting a strategy. This is likely due to the prevalence of family labour, which makes task achievement more effective, especially during peak periods. Adams [39], Temesgen, Hassan, Tekie, Mahmud and Ringler [47], and Le Dang, et al. [78] contradict the positive influence of household size; they opined that household size has a negative and significant impact on the probability of choosing adaptation strategies.

Gender of Household Head: Results obtained in Table 3 are partially in tandem with previous findings [46,50,79,80] that male-headed households often have a higher likelihood of adopting agricultural innovations and thus are better adapted to climate change. Being a male-headed household increases the chances of practicing soil conservation compared to their female counterparts. However, the likelihood to diversify into other occupations increases with being a female-headed household because females in the household especially in Nigeria are found to play supportive roles (such as processors and traders) in the households by diversifying the household income, thus easing the financial burden of the family. Females in households also tend to make financial plans for unforeseen circumstances. Adams [39] and Ogunpaimo, et al. [81] shared a similar view on females tilting towards the adoption of occupation diversification compared to the males.

Marital status: Table 3 showed that married farmers have the likelihood to use adaptation strategies such as irrigation facilities, crop diversification, and nonfarm diversification compared to singles. This is likely because more efforts come into making decisions when being married compared to being single. On the other hand, being married has a negative but significant influence on implementing soil conservation techniques.

Years of schooling: From Table 3, it is shown that the years of schooling of the farm household head have mixed effects on the choice of CCA. This variable significantly and positively affected practicing diversification into non-farm activities ($p < 0.01$); this result shown in Table 3 supported the work of Ali and Erenstein [25], Alam [48], Gebrehiwot and Van Der Veen [49], and Alam, Alam and Mushtaq [60]. The papers all agreed that educated farmers may be more aware and perceive climate change, as they can easily understand and interpret information compared to farmers with a lower level of education. However, this philosophy did not work for the adoption of some strategies; years of schooling negatively influenced the probability of practicing soil conservation ($p < 0.01$) and crop diversification ($p < 0.01$).

Quantity of Household Asset: The quantity of household assets, which is a proxy of the wealth status of farming households, is an important variable that reflects farmers' choice of climate change adaptation options. Results shown in Table 3 contradict Ali and Erenstein [25] in that quantity of household assets enacted a negative influence on climate change adaptation options except for non-farm diversification ($p < 0.01$). It can be implied that income from the use and sale of household size is diverted mainly into non-farm diversification with non-farm diversification serving as secondary income to the farming households.

Location: Location typically plays an important role in CCA adoption [25,82–85]. In this study, we included dummies for agroecological zones to control for the location effect on adaptation strategies, with South-South being the base for the model. The result indicated a significant positive and significant probability of farm households in North-Central and North-West to implement irrigation facilities compared to farm households in the South-South region. The likelihood of adopting soil conservation techniques increases with residing in all other regions of the country in relation to the South-South zone. The findings in Table 3 also highlighted that all the zones, except for the South-West zone, negatively affect the probability of farming households to diversify into non-farm activities concerning those in the South-South region.

3.2. *Impact of Climate Change Adaptation on Household Food Security*

A combination of PSM and DID was used to evaluate the impact of CCA adoption on household food security. It is therefore imperative to discuss the result of the impact of CCA options on household food security between adopters and non-adopters from 2010 to 2016.

3.2.1. PSM Result of Impact of CCA on Household Food Security

The with and without effect of climate change adaptation options is explained by PSM. Table 4 presents the impacts of adaptation methods used on household food security based on propensity score matching. The impact of climate change adaptation on household food security was significant with adopters having 9% higher food security than non-adopters in 2010. This result is in support of Ali and Erenstein [25], but against Weldegebriel and Prowse [86] who found that the adaptation strategy reduced farm income and, with that, food security due to the exclusion of important variables. Ali and Erenstein [25] stated that CCA practices help to enhance the food security and welfare of rural households. Thus, farm households should be encouraged to adopt a few CCA practices to improve welfare outcomes. Farm households not adopting CCA practices are more likely to be food insecure. Shiferaw, et al. [87] also opined that the average treatment effect on the treated (ATT) of adaptation on household food security was positive and significant, which implied that CCA options foster household food security.

Table 4. PSM Showing the Impact of CCA Adoption on Household Food Security.

Outcome Variable	ATT	t-Values	Mean Bias	Median Bias	Bias Reduction
Food security	0.09 **	2.15	4.2	4.3	83%

Authors computation of LSMS data 2010–2016. ** represents statistical significance at 5% respectively.

Table 5 shows the covariate balancing tests before and after matching. As indicated in the table, the balancing test revealed that the bias was relatively higher before matching.

Table 5. Covariate Balancing test for the difference between CCA Adopters and non-adopters.

Variable	Before Matching-Mean Absolute Bias	After Matching Mean Absolute Bias	t-Vals of Covariates before Matching	t-Vals of Covariates After Matching	% Reduction Bias
Credit	0.001	0.01	0.15	1.63	75.8
Tenancy Status	0.06	0.01	1.90	1.13	74.7
Farm size	0.07	0.01	0.11	1.44	70.4
Extension contact	0.06	0.002	2.82	0.13	97.2
Age of HH	1.62	0.28	1.49	0.41	82.6
Gender of HH	0.07	0.002	3.94	1.39	97.2
Years of Schooling	0.49	0.09	2.07	0.25	82
Marital Status	0.03	0.01	1.19	0.69	65
HH SIZE	0.08	0.01	0.41	1.33	88.7
HH ASSET	0.516	0.112	0.87	0.41	78.1

Authors computation of LSMS data 2010–2016.

For instance, before matching tenancy status ($p < 0.10$), extension contact ($p < 0.05$) and gender of the household head ($p < 0.01$) could cause selection bias when assessing the influence of CCA options on household food security status. The percentage bias reduction is between 65–97.2%. These indicators of covariates balancing showed the results obtained satisfied the balancing of covariates following matching and the application of the common support condition. The result implied no selection bias when matching adopters and non-adopters, thus differences in food security levels are mainly due to the adoption of CCA measures.

3.2.2. DID Result of Impact of CCA on Household Food Security

The true impact of CCA on household food security over time can be measured by looking at the effects of adaptation options between adopters and non-adopters, which was illustrated by the result of the PSM and then measuring the impact of the adaptation measures over the period of adoption using the DID. The adopter and the non-adopter groups within the same common support in the PSM analysis for the base period were appended, after which the DID analysis was carried out.

The difference in difference (DID) estimation combined with propensity score matching (PSM) was used to evaluate the average impact of the CCA options on household food security. The average treatment effects of CCA options were evaluated, which compares food security in the adoption state (Y_1) with the outcomes in the control or the counterfactual (Y_0) conditional on receiving treatment.

Contrary to previous studies [25,88], which used cross-sectional data to assess the impact of CCA options, this study used panel data for six time periods to assess the impact of CCA options on household food security. Similar to Kangmennaang et al. [26] and Kabunga, et al. [89], this approach allowed for the combination of propensity score matching with DID estimation to control for selection bias and temporal impact variability. The estimated results showed that adopting CCA options intervention positively influenced household food security.

As shown in Table 6, the F-value is significant at ($p < 0.01$), which indicated that the model was useful in assessing the impact of CCA on household food security over time. The result in Table 6 showed that the coefficient of time trend (y16) was significant ($\beta = -4.02$, $p < 0.01$); this implied that household food security was trending down with time. The result of the DID is positive and significant ($\beta = 5.93$, $p < 0.01$), which reveals that the impact of the CCA options increases household food security between adopters and non-adopters. This finding confirms that CCA options had a significant positive impact on farm households' food security status. This finding shared similar results with Noltze, et al. [90], and Kangmennaang et al. [26] who found that agroecological practices in the form of CCA promote food security after 2 years. However, while this study adopted the use of HFES as a measure of food security, Kangmennaang et al. [26] used the Household Food Insecurity Access Scale (HFIAS).

Table 6. DID Showing the Impact of CCA Options on Household Food Security Without Covariates.

Household Food Expenditure	Coefficient	Robust Standard Error	t-Values
Adapt	4.86 ***	1.09	4.45
y16	−4.02 ***	0.50	−8.06
DID	5.93 ***	0.68	8.73
Constant	76.00	0.91	83.29
F-value	**33.85 *****		
Prob > F	**0.000**		

Authors computation of LSMS data 2010–2016. *** represent statistical significance at 1% respectively.

It must be noted that CCA adoption may be implemented by farming households before the year 2010, however, 2010 was used as the baseline due to data availability. The findings in this study also support the result of other studies that confirmed the direct effects of CCA options on household food security. Becerril and Abdulai [91] reported that increased farm output can lead to higher consumption, off-farm diversification, and increased farm incomes. Surpluses from farm yield may also be used to increase the household quantity of assets increasing the adaptive capacity of households to climate change, thus promoting households' food security status [92,93]. Khonje, et al. [94] also reported that sustainable practices such as crop diversification and other CCA options can lead to improved welfare and food security outcomes. Other effects of adopting CCA options reported may include promoting women empowerment, capacity-building, and knowledge exchange within the community, which may further lead to increased food production at the community level, increased consumption, and better living standard conditions [26]. Adopting CCA options can also foster collective relationships among farming households within the communities, allowing for the sharing of risks and burdens associated with farm activities.

To control for any selection bias between the adopters and non-adopters of CCA, the results obtained in Table 6 were controlled for covariates influences on the impact of CCA on household food security status, as shown in Table 7. The result obtained indicated that even after controlling for potential covariate influence, DID had a positive and significant ($\beta = 4.15$, $p < 0.01$) effect on household food security status. This result was corroborated by [26] who found that covariates control does not influence the outcome of the DID result or the influence of CCA on the household food security status.

Table 7. DID Showing the Impact of Adaptation Options on Household Food Security with Covariates.

Food Security Outcome	Coefficient	Robust Standard Error	t-Values
Adapt	4.34 ***	1.22	3.55
Time	−3.13	3.17	−0.99
DID	4.15 ***	1.01	4.11
Constant	79.90	2.90	36.36
F-value	**13.55 *** **		
Prob > F	**0.000**		

Authors computation of LSMS data 2010–2016. *** represent statistical significance at 1% respectively.

4. Conclusions

This study assessed the impact of climate change adaptation (CCA) on household food security among farm households in Nigeria. Against previous works of literature that adopted cross-sectional approaches to investigate CCA impacts on welfare outcomes, this research work adopted a panel data analysis, thus measuring the impacts of CCA on household food security across space and time. We recognized that there are other CCA options not considered in this study mainly due to lack or limited data of such CCA strategies in the LSMS data. However, this study has provided useful insights and information on the relationship between CCA options and household food security in Nigeria. Based on the aforementioned findings, this study confirmed the need for adaptation to climate change by farming households, which increase with an increase in farm size, extension contact, and marital status, with access to credit, ownership of farmland, household size, the gender of household size, years of schooling, household asset and location having mixed effects on the choice of adaptation strategies. From the study, it was shown that climate change adaptations have helped farming households improve their food security status in the face of prevalent climatic conditions. Therefore, the study recommends that farming households should practice continual implementation of CCA options to foster improvement in household food security status. Also, access to credit facilities and extension contacts remains a catalyst for implementing adaptation measures, thus constant and quality extension contacts and credit facilities with low-interest rates should be provided to farming households to enable them to adapt better to climate changes and improve household food security status.

Author Contributions: Conceptualization, O.R.O.; Formal analysis, O.R.O.; Methodology, O.R.O.; Software, O.R.O.; Validation, Z.O.-U., and J.S.; Writing—original draft, O.R.O.; Writing—review & editing, Z.O.-U., and J.S. All authors have read and agreed to the published version of the manuscript.

Funding: This research received no external funding.

Institutional Review Board Statement: Not applicable.

Informed Consent Statement: Not applicable.

Data Availability Statement: Publicly available datasets were analyzed in this study. This data can be found here: [https://microdata.worldbank.org/index.php/catlog/lsms/] at reference number [Ref: NGA_2010_GHSP-W1_v03_M, Ref: NGA_2012_GHSP-W2_v02_M, Ref: NGA_2015_GHSP-W3_v02_M].

Conflicts of Interest: No conflict of interest was declared by the authors of this paper.

References

1. Medugu, I.N.; Majid, M.R.; Choji, I. A comprehensive approach to drought and desertification in Nigeria. *Manag. Environ. Qual. Int. J.* **2008**. [CrossRef]
2. Abaje, I.; Ati, O.; Iguisi, E.; Jidauna, G. Droughts in the sudano-sahelian ecological zone of Nigeria: Implications for agriculture and water resources development. *Glob. J. Hum. Soc. Sci. B Geogr. Geo Sci. Environ.* **2013**, *13*, 1–10.
3. Echendu, A.J. The impact of flooding on Nigeria's sustainable development goals (SDGs). *Ecosyst. Health Sustain.* **2020**, *6*, 1791735. [CrossRef]

4. Teklewold, H.; Kassie, M.; Shiferaw, B. Adoption of multiple sustainable agricultural practices in rural Ethiopia. *J. Agric. Econ.* **2013**, *64*, 597–623. [CrossRef]
5. El-ladan, I. Climate change and food security in Nigeria. In Proceedings of the International Conference on Possible Impacts of Climate Change on Africa, Institute of African Research and Studies, Cairo University, Cairo, Egypt, 18–20 May 2014; p. 16.
6. Firdaus, R.R.; Gunaratne, M.S.; Rahmat, S.R.; Kamsi, N.S. Does climate change only affect food availability? What else matters? *Cogent Food Agric.* **2019**, *5*, 1707607. [CrossRef]
7. FAO; IFAD; UNICEF; WFP; WHO. *The State of Food Security and Nutrition in the World 2018: Building Climate Resilience for Food Security and Nutrition*; Food & Agriculture Organization: Rome, Italy, 2018.
8. Edame, G.E.; Ekpenyong, A.; Fonta, W.M.; Duru, E. Climate change, food security and agricultural productivity in Africa: Issues and policy directions. *Int. J. Humanit. Soc. Sci.* **2011**, *1*, 205–223.
9. FAO. *Climate Change and Food Security: A Framework Document*; Food and Agriculture Organization of the United Nations: Rome, Italy, 2008.
10. von Grebmer, K.; Bernstein, J.; Alders, R.; Dar, O.; Kock, R.; Rampa, F.; Wiemers, M.; Acheampong, K.; Hanano, A.; Higgins, B.; et al. *2020 Global Hunger Index: One Decade to Zero Hunger: Linking Health and Sustainable Food Systems*; Welthungerhilfe: Bonn, Germany; Concern Worldwide: Dublin, Iraland, 2020.
11. Adebayo, O.; Olagunju, K.; Kabir, S.K.; Adeyemi, O. Social crisis, terrorism and food poverty dynamics: Evidence from Northern Nigeria. *Int. J. Econ. Financ. Issues* **2016**, 6. [CrossRef]
12. Korir, L.; Rizov, M.; Ruto, E. Analysis of Household Food Demand and Its Implications on Food Security in Kenya: An Application of QUAIDS Model. In Proceedings of the Agricultural Economics Society, 92nd Annual Conference 2018, Coventry, UK, 16–18 April 2018.
13. Ikudayisi, A.; Okoruwa, V.; Omonona, B. From the lens of food accessibility and dietary quality: Gaining insights from urban food security in Nigeria. *Outlook Agric.* **2019**, *48*, 336–343. [CrossRef]
14. Adeniyi, D.A.; Dinbabo, M.F. Factors Influencing Household Food Security Among Irrigation Smallholders in North West Nigeria. *J. Rev. Glob. Econ.* **2019**, *8*, 291–304. [CrossRef]
15. *The Sustainable Development Goals Report 2018*; UN General Assembly: New York, NY, USA, 2018; p. 36.
16. Chambwera, M.; Zou, Y.; Boughlala, M. *Better Economics: Supporting Adaptation with Stakeholder Analysis*; International Institute for Environment and Development (IIED): London, UK, 2011.
17. FAO. Climate Change, Agriculture and Food Security. Available online: https://reliefweb.int/sites/reliefweb.int/files/resources/a-i6030e.pdf (accessed on 18 September 2020).
18. FAO; IFAD; UNICEF; WFP; WHO. The state of food security and nutrition in the world 2017. In *Building Resilience for Peace and Food Security*; FAO: Rome, Italy, 2017.
19. Field, C.B. *Climate Change 2014–Impacts, Adaptation and Vulnerability: Regional Aspects*; Cambridge University Press: Cambridge, UK, 2014.
20. Steynor, A.; Pasquini, L. Informing climate services in Africa through climate change risk perceptions. *Clim. Serv.* **2019**, *15*, 100112. [CrossRef]
21. Field, C.B.; Barros, V.R.; Mastrandrea, M.D.; Mach, K.J.; Abdrabo, M.-K.; Adger, N.; Anokhin, Y.A.; Anisimov, O.A.; Arent, D.J.; Barnett, J. Summary for policymakers. In *Climate Change 2014: Impacts, Adaptation, and Vulnerability. Part A: Global and Sectoral Aspects. Contribution of Working Group II to the Fifth Assessment Report of the Intergovernmental Panel on Climate Change*; Cambridge University Press: Cambridge, UK, 2014; pp. 1–32.
22. Ebele, N.E.; Emodi, N.V. Climate change and its impact in Nigerian economy. *J. Sci. Res. Rep.* **2016**, 1–13. [CrossRef] [PubMed]
23. Abdulkadir, T.; Salami, A.; Aremu, A.; Ayanshola, A.; Oyejobi, D. Assessment of neural networks performance in modeling rainfall amounts. *J. Res. For. Wildl. Environ.* **2017**, *9*, 12–22.
24. Elum, Z.A.; Modise, D.M.; Marr, A. Farmer's perception of climate change and responsive strategies in three selected provinces of South Africa. *Clim. Risk Manag.* **2017**, *16*, 246–257. [CrossRef]
25. Ali, A.; Erenstein, O. Assessing farmer use of climate change adaptation practices and impacts on food security and poverty in Pakistan. *Clim. Risk Manag.* **2017**, *16*, 183–194. [CrossRef]
26. Kangmennaang, J.; Kerr, R.B.; Lupafya, E.; Dakishoni, L.; Katundu, M.; Luginaah, I. Impact of a participatory agroecological development project on household wealth and food security in Malawi. *Food Secur.* **2017**, *9*, 561–576. [CrossRef]
27. Morland, A. Climate Change, Food Security and Adaptation. Available online: https://www.thenewhumanitarian.org/analysis/2017/06/14/fact-file-climate-change-food-security-and-adaptation (accessed on 21 September 2020).
28. European Commission. Adaptation to Climate Change. Available online: https://ec.europa.eu/clima/policies/adaptation_en (accessed on 18 September 2020).
29. Ade Ajayi, J.F.; Kirk-Greene, A.H.M.; Udo, R.K.; Falola, T.O. Nigeria. Available online: https://www.britannica.com/place/Nigeria (accessed on 15 December 2020).
30. National Bureau of Statistics, Federal Republic of Nigeria. *Nigeria General Household Survey (GHS), Panel 2010, Wave 1 Ref. NGA_2010_GHSP-W1_v03_M*; National Bureau of Statistics: Abuja, Nigeria, 2010.
31. National Bureau of Statistics, Federal Republic of Nigeria. *Nigeria General Household Survey, Panel 2012-2013, Wave 2. Ref. NGA_2012_GHSP-W2_v02_M*; National Bureau of Statistics: Abuja, Nigeria, 2012.

32. National Bureau of Statistics, Federal Republic of Nigeria. *General Household Survey, Panel (GHS-Panel) 2015–2016.Ref. NGA_2015_GHSP-W3_v02_M*; National Bureau of Statistics: Abuja, Nigeria, 2015.
33. Ogunpaimo, O.R.; Dipeolu, A.O.; Ogunpaimo, O.J.; Akinbode, S.O. Determinants of choice of climate change adaptation options among cassava farmers, in southwest Nigeria. *Futo J. Ser.* **2020**, *6*, 25–39.
34. GADM. Country Profile. Available online: https://gadm.org/download_country_v3.html (accessed on 25 January 2021).
35. Birungi, P.B. *The Linkages between Land Degradation, Poverty and Social Capital in Uganda*; University of Pretoria: Pretoria, South Africa, 2008.
36. Mugi-Ngenga, E.; Mucheru-Muna, M.; Mugwe, J.; Ngetich, F.; Mairura, F.; Mugendi, D. Household's socio-economic factors influencing the level of adaptation to climate variability in the dry zones of Eastern Kenya. *J. Rural Stud.* **2016**, *43*, 49–60. [CrossRef]
37. Dolisca, F.; Carter, D.R.; McDaniel, J.M.; Shannon, D.A.; Jolly, C.M. Factors influencing farmers' participation in forestry management programs: A case study from Haiti. *For. Ecol. Manag.* **2006**, *236*, 324–331. [CrossRef]
38. Tizale, C.Y. *The Dynamics of Soil Degradation and Incentives for Optimal Management in the Central Highlands of Ethiopia*; University of Pretoria: Pretoria, South Africa, 2007.
39. Adams, O.R. *Climate Variability and Adaptation Strategies in Cassava Production in Ogun State, Nigeria*; Federal University of Agriculture: Abeokuta, Nigeria, 2015.
40. Ijaiya, M.; Abdulraheem, A.; Abdullahi, I. Agricultural credit guarantee scheme and food security in Nigeria. *Ethiop. J. Environ. Stud. Manag.* **2017**, *10*, 208–218. [CrossRef]
41. Aku, P. Comparative analysis of NACB and ACGSF Loan disbursement to agriculture in Nigeria. *J. Soc. Manag. Stud.* **1995**, *2*, 99–108.
42. Smith, L.C.; Subandoro, A. *Measuring Food Security Using Household Expenditure Surveys*; Food Security in Practice Technical Guide Series; International Food Policy Research Institute: Washington, DC, USA, 2007; p. 146.
43. Akerele, D.; Sanusi, R.; Fadare, O.; Ashaolu, O. Factors influencing nutritional adequacy among rural households in Nigeria: How does dietary diversity stand among influencers? *Ecol. Food Nutr.* **2017**, *56*, 187–203. [CrossRef] [PubMed]
44. Cameron, A.C.; Trivedi, P.K. *Microeconometrics Using Stata*; Stata Press College Station: College Station, TX, USA, 2009; Volume 5.
45. Deressa, T.T.; Hassan, R.M.; Ringler, C. Perception of and adaptation to climate change by farmers in the Nile basin of Ethiopia. *J. Agric. Sci.* **2011**, *149*, 23–31. [CrossRef]
46. Oyetunde-Usman, Z.; Olagunju, K.O.; Ogunpaimo, O.R. Determinants of adoption of multiple sustainable agricultural practices among smallholder farmers in Nigeria. *Int. Soil Water Conserv. Res.* **2020**. [CrossRef]
47. Temesgen, D.; Hassan, R.; Tekie, A.; Mahmud, Y.; Ringler, C. *Analyzing the Determinants of Farmers' Choice of Adaptation Methods and Perceptions of Climate Change in the Nile Basin of Ethiopia*; International Food Policy Research Institute (IFPRI): Washington, DC, USA, 2008.
48. Alam, K. Farmers' adaptation to water scarcity in drought-prone environments: A case study of Rajshahi District, Bangladesh. *Agric. Water Manag.* **2015**, *148*, 196–206. [CrossRef]
49. Gebrehiwot, T.; Van Der Veen, A. Farm level adaptation to climate change: The case of farmer's in the Ethiopian Highlands. *Environ. Manag.* **2013**, *52*, 29–44. [CrossRef]
50. Vo, H.H.; Mizunoya, T.; Nguyen, C.D. Determinants of farmers' adaptation decisions to climate change in the central coastal region of Vietnam. *Asia Pac. J. Reg. Sci.* **2021**, 1–23. [CrossRef]
51. Tambo, J.A. Adaptation and resilience to climate change and variability in north-east Ghana. *Int. J. Disaster Risk Reduct.* **2016**, *17*, 85–94. [CrossRef]
52. Mekuyie, M.; Mulu, D. Perception of impacts of climate variability on pastoralists and their adaptation/coping strategies in fentale district of Oromia Region, Ethiopia. *Environ. Syst. Res.* **2021**, *10*, 1–10. [CrossRef]
53. Rosenbaum, P.R.; Rubin, D.B. The central role of the propensity score in observational studies for causal effects. *Biometrika* **1983**, *70*, 41–55. [CrossRef]
54. Dehejia, R.H.; Wahba, S. Propensity score-matching methods for nonexperimental causal studies. *Rev. Econ. Stat.* **2002**, *84*, 151–161. [CrossRef]
55. Smith, L.C.; Dupriez, O.; Troubat, N. *Assessment of the Reliability and Relevance of the Food Data Collected in National Household Consumption and Expenditure Surveys*; International Household Survey Network: Oxford, UK, 2014; p. 69.
56. Khandker, S.B.; Koolwal, G.; Samad, H. *Handbook on Impact Evaluation: Quantitative Methods and Practices*; The World Bank: Washington, DC, USA, 2009. [CrossRef]
57. Becker, S.O.; Ichino, A. Estimation of average treatment effects based on propensity scores. *Stata J.* **2002**, *2*, 358–377. [CrossRef]
58. Caliendo, M.; Kopeinig, S. Some practical guidance for the implementation of propensity score matching. *J. Econ. Surv.* **2008**, *22*, 31–72. [CrossRef]
59. Hisali, E.; Birungi, P.; Buyinza, F. Adaptation to climate change in Uganda: Evidence from micro level data. *Glob. Environ. Chang.* **2011**, *21*, 1245–1261. [CrossRef]
60. Alam, G.M.; Alam, K.; Mushtaq, S. Influence of institutional access and social capital on adaptation decision: Empirical evidence from hazard-prone rural households in Bangladesh. *Ecol. Econ.* **2016**, *130*, 243–251. [CrossRef]
61. Boansi, D.; Tambo, J.A.; Müller, M. Analysis of farmers' adaptation to weather extremes in West African Sudan Savanna. *Weather Clim. Extrem.* **2017**, *16*, 1–13. [CrossRef]

62. Hassan, R.M.; Nhemachena, C. Determinants of African farmers' strategies for adapting to climate change: Multinomial choice analysis. *Afr. J. Agric. Resour. Econ.* **2008**, *2*, 83–104. [CrossRef]
63. Ojo, T.; Baiyegunhi, L. Determinants of credit constraints and its impact on the adoption of climate change adaptation strategies among rice farmers in South-West Nigeria. *J. Econ. Struct.* **2020**, *9*, 1–15. [CrossRef]
64. Arunrat, N.; Wang, C.; Pumijumnong, N.; Sereenonchai, S.; Cai, W. Farmers' intention and decision to adapt to climate change: A case study in the Yom and Nan basins, Phichit province of Thailand. *J. Clean. Prod.* **2017**, *143*, 672–685. [CrossRef]
65. Quan, J. *Land Access in the Early 21st Century: Issues, Trends, Linkages and Policy Options*; FAO Livelihoods Support Programme Working Paper 24; FAO: Rome, Italy, 2006.
66. Kokoye, S.E.H.; Yabi, J.A.; Tovignan, S.D.; Yegbemey, R.N.; Nuppenau, E.-A. Simultaneous modelling of the determinants of the partial inputs productivity in the municipality of Banikoara, Northern Benin. *Agric. Syst.* **2013**, *122*, 53–59. [CrossRef]
67. Fosu-Mensah, B.Y.; Vlek, P.L.; MacCarthy, D.S. Farmers' perception and adaptation to climate change: A case study of Sekyedumase district in Ghana. *Environ. Dev. Sustain.* **2012**, *14*, 495–505. [CrossRef]
68. Iheke, O.R.; Agodike, W.C. Analysis of factors influencing the adoption of climate change mitigating measures by smallholder farmers in IMO state, Nigeria. *Sci. Pap. Ser. Manag. Econ. Eng. Agric. Rural. Dev.* **2016**, *16*, 213–220.
69. Nabikolo, D.; Bashaasha, B.; Mangheni, M.; Majaliwa, J. Determinants of climate change adaptation among male and female headed farm households in eastern Uganda. *Afr. Crop Sci. J.* **2012**, *20*, 203–212.
70. Abid, M.; Scheffran, J.; Schneider, U.A.; Ashfaq, M. Farmers' perceptions of and adaptation strategies to climate change and their determinants: The case of Punjab province, Pakistan. *Earth Syst. Dyn.* **2015**, *6*, 225–243. [CrossRef]
71. Abid, M.; Schneider, U.A.; Scheffran, J. Adaptation to climate change and its impacts on food productivity and crop income: Perspectives of farmers in rural Pakistan. *J. Rural Stud.* **2016**, *47*, 254–266. [CrossRef]
72. Iqbal, M.; Ahmad, M.; Mustafa, G. Impact of Farm Households' Adaptation on Agricultural Productivity: Evidence from Different Agro-Ecologies of Pakistan. *Clim. Chang. Work. Pap.* **2015**, *5*, 17.
73. Javed, S.; Kishwar, S.; Iqbal, M. *From Perceptions to Adaptation to Climate Change: Farm Level Evidence from Pakistan*; Pakistan Institute of Development Economics: Islamabad, Pakistan, 2015.
74. Tiwari, V.; Wahr, J.; Swenson, S. Dwindling groundwater resources in northern India, from satellite gravity observations. *Geophys. Res. Lett.* **2009**, 36. [CrossRef]
75. Bryan, E.; Ringler, C.; Okoba, B.; Roncoli, C.; Silvestri, S.; Herrero, M. Adapting agriculture to climate change in Kenya: Household strategies and determinants. *J. Environ. Manag.* **2013**, *114*, 26–35. [CrossRef]
76. Gbetibouo, G.A. *Understanding Farmers Perceptions and Adaptations to Climate Change and Variability: The Case of the Limpopo Basin Farmers South Africa*; IFPRI Research Brief: Washington, DC, USA, 2009.
77. Arshad, M.; Amjath-Babu, T.; Kächele, H.; Müller, K. What drives the willingness to pay for crop insurance against extreme weather events (flood and drought) in Pakistan? A hypothetical market approach. *Clim. Dev.* **2016**, *8*, 234–244. [CrossRef]
78. Le Dang, H.; Li, E.; Nuberg, I.; Bruwer, J. Farmers' assessments of private adaptive measures to climate change and influential factors: A study in the Mekong Delta, Vietnam. *Nat. Hazards* **2014**, *71*, 385–401. [CrossRef]
79. Trinh, T.Q.; Rañola, R.F., Jr.; Camacho, L.D.; Simelton, E. Determinants of farmers' adaptation to climate change in agricultural production in the central region of Vietnam. *Land Use Policy* **2018**, *70*, 224–231. [CrossRef]
80. Asrat, P.; Simane, B. Farmers' perception of climate change and adaptation strategies in the Dabus watershed, North-West Ethiopia. *Ecol. Process.* **2018**, *7*, 7. [CrossRef]
81. Ogunpaimo, O.R.; Ogbe, A.; Edewor, S. Determinants of factors affecting adaptation strategies to climate change in cassava processing in South West, Nigeria. In Proceedings of the 6th African Conference of Agricultural Economists, Abuja, Nigeria, 23–25 September 2019; p. 17.
82. Vincent, K. Uncertainty in adaptive capacity and the importance of scale. *Glob. Environ. Chang.* **2007**, *17*, 12–24. [CrossRef]
83. Tiwari, K.R.; Sitaula, B.K.; Nyborg, I.L.; Paudel, G.S. Determinants of farmers' adoption of improved soil conservation technology in a middle mountain watershed of central Nepal. *Environ. Manag.* **2008**, *42*, 210–222. [CrossRef] [PubMed]
84. Hinkel, J. "Indicators of vulnerability and adaptive capacity": Towards a clarification of the science–policy interface. *Glob. Environ. Chang.* **2011**, *21*, 198–208. [CrossRef]
85. Below, T.B.; Mutabazi, K.D.; Kirschke, D.; Franke, C.; Sieber, S.; Siebert, R.; Tscherning, K. Can farmers' adaptation to climate change be explained by socio-economic household-level variables? *Glob. Environ. Chang.* **2012**, *22*, 223–235. [CrossRef]
86. Weldegebriel, Z.B.; Prowse, M. Climate-change adaptation in Ethiopia: To what extent does social protection influence livelihood diversification? *Dev. Policy Rev.* **2013**, *31*, 35–56. [CrossRef]
87. Shiferaw, B.; Kassie, M.; Jaleta, M.; Yirga, C. Adoption of improved wheat varieties and impacts on household food security in Ethiopia. *Food Policy* **2014**, *44*, 272–284. [CrossRef]
88. Ogunniyi, A.; Omonona, B.; Abioye, O.; Olagunju, K. Impact of irrigation technology use on crop yield, crop income and household food security in Nigeria: A treatment effect approach. *AIMS Agric. Food* **2018**, *3*, 154–171. [CrossRef]
89. Kabunga, N.S.; Dubois, T.; Qaim, M. Impact of tissue culture banana technology on farm household income and food security in Kenya. *Food Policy* **2014**, *45*, 25–34. [CrossRef]
90. Noltze, M.; Schwarze, S.; Qaim, M. Impacts of natural resource management technologies on agricultural yield and household income: The system of rice intensification in Timor Leste. *Ecol. Econ.* **2013**, *85*, 59–68. [CrossRef]

91. Becerril, J.; Abdulai, A. The impact of improved maize varieties on poverty in Mexico: A propensity score-matching approach. *World Dev.* **2010**, *38*, 1024–1035. [CrossRef]
92. Wanjala, B.M.; Muradian, R. Can big push interventions take small-scale farmers out of poverty? Insights from the Sauri Millennium Village in Kenya. *World Dev.* **2013**, *45*, 147–160. [CrossRef]
93. Orr, A.; Mausch, K. *How Can We Make Smallholder Agriculture in the Semi-Arid Tropics More Profitable and Resilient?* Working Paper Series No. 59; International Crops Research Institute for Semi-Arid Tropics: Nairobi, Kenya, 2014; p. 28.
94. Khonje, M.; Manda, J.; Alene, A.D.; Kassie, M. Analysis of adoption and impacts of improved maize varieties in eastern Zambia. *World Dev.* **2015**, *66*, 695–706. [CrossRef]

Article

Mainstreaming Disaster Risk Reduction into Local Development Plans for Rural Tropical Africa: A Systematic Assessment

Maurizio Tiepolo * and **Sarah Braccio**

Interuniversity Department of Regional and Urban Studies and Planning (DIST)-Politecnico and University of Turin, viale Mattioli 39, 10125 Turin, Italy; sarah.braccio@polito.it

* Correspondence: maurizio.tiepolo@polito.it; Tel.: +39-0110907491

Received: 21 January 2020; Accepted: 9 March 2020; Published: 12 March 2020

Abstract: Disaster risk reduction in rural Africa can contribute to reducing poverty and food insecurity if included in local development plans (LDPs). Five years after the Sendai Framework for Disaster Risk Reduction (DRR), we do not know how much risk reduction is practiced in rural Africa. The aim of this assessment is to ascertain the state of mainstreaming DRR in development planning in the rural jurisdictions of tropical Africa. One hundred and ninety-four plans of 21 countries are considered. Ten characteristics of the plans are examined: Climate trends, hydro-climatic hazards, vulnerability and risk assessments, alignment with Sendai Framework, vision, strategies and objectives, DRR actions, internal consistency, DRR relevance and funding sources, local and technical knowledge integration, public participation. It is found that local climatic characterization is almost always absent and risk reduction is an objective of the plans in one case out of three. Prevention actions prevail over those of preparedness. There is poor participation in the plan preparation process and this limits the implementation of the actions. A modification of the national guidelines on the preparation of LDPs, the orientation of official development assistance towards supporting climate services and the training of local planners, together with the increase of financial resources in local jurisdictions are essential for improving DRR at local scale.

Keywords: disaster risk reduction; official development assistance; public participation; risk tracking; rural development; Sendai framework; sustainable development

1. Introduction

Tropical Africa presents specific characteristics as opposed to other regions of the Global South. Firstly, 62.5% of the population is still rural [1]. Over half of jobs and 69% of income come from agriculture [2]. Despite this, 58% of the population is in conditions of food insecurity [3]. Poverty and inequalities between countries and within individual countries reach the world's highest levels [4]. In these conditions, the development of agriculture is considered better for absorbing the poor than industry and services [5–9]. However, agriculture is strongly exposed to climate change (CC), which affects rain-fed crops and livestock [10–12], casts smallholders into deeper poverty [13], and makes investments vulnerable [14] because of low adaptive capacity [15]. Finally, fluvial flooding and dust storms affecting urban areas [16–18], are formed in the surrounding rural areas and it is there that they should be treated primarily [19]. For all these reasons, rural environment remains the hot spot for disaster risk reduction (DRR) in Africa.

Smallholder farmers from tropical Africa have insufficient financial capital and willingness to change to significantly reduce disaster risk [20] and sometimes lack natural, social, and human capital [21]. DRR should first be addressed by the local governments of all those countries where the process of administrative decentralization has assigned them the task of conserving the environment and protecting the population from natural hazards. In addition, precisely for this reason Official Development Assistance (ODA) helps local governments strengthen their capacities with specific programs [15]. The substantial reduction of disaster loss and damage and the increase of local disaster risk reduction strategies by 2030 has become a target of the Sendai Framework for Disaster Risk Reduction (2015) [22,23]. Five years after the Sendai Framework we only know the number of African countries with DRR strategies in place: Just seven, according to UNDRR, which is in charge of monitoring the implementation of the Sendai Framework [24]. We do not know if this results from a lack of information or from national strategies which "often do[es] not penetrate to the local level" [25].

However, the number of strategies in place does not guarantee a reduction of risk at local scale. Peer-reviewed literature points out that DRR mainstreaming into local development plans (LDPs) is often based upon rough climatic analyses, produced with little public participation, done with few actions, mostly of little significance, and lacks sufficient resources for implementation [26–39]. Genuine public participation in the planning process is critical to reducing disaster risk. This is particularly true in rural areas, where almost all actions require the direct involvement of smallholder farmers and herders [40–45].

The problem is, therefore, to know the content of the DRR strategies in place locally and the process by which they were prepared, rather than just knowing their number. An evaluation of the local DRR strategies can therefore highlight the aspects to be improved in the DRR and inspire the ODA consequently.

The aim of this article is to ascertain the mainstreaming of DRR in local development plans for rural areas in tropical Africa.

We will proceed by identifying the local jurisdictions. To compare the LDPs, it is necessary to investigate a homogeneous climatic area. The tropical zone is the most present in Africa. Within it, the analysis is restricted to English-speaking, French-speaking, and Portuguese-speaking jurisdictions with plans in force: Municipalities, cantons, districts, counties, sometimes regions. The LDPs are identified by Google search. The plans considered are only those in force at September 2019 in jurisdictions with at least 90% of the territory in the tropical zone and rural population exceeding 50%. With these restrictions, we obtained 194 plans, covering 24% of rural jurisdictions in 21 tropical African countries.

Ten characteristics of the plans are considered: (i) Climatic trend, (ii) hazard identification, (iii) existence of vulnerability and risk assessments, (iv) alignment with the Sendai framework for DRR, (v) vision, strategies and objectives, (vi) DRR actions, (vii) internal coherence, (viii) budget size and origin, (ix) local and scientific knowledge integration, (x) public participation in the plan preparation.

Although DRR mainstreaming in the LDPs is in place in half of the countries of tropical Africa, improvements are needed in analysis and planning, as well as resources to achieve the actions, if DRR is to be consolidated at local scale.

The importance of this article is in highlighting the weaknesses of today's local DRR. However, also in proposing a simple assessment framework, repeatable over time and on a larger scale, which was still missing. These features can facilitate the transition from occasional assessments to tracking the DRR in local development plans.

2. Materials and Methods

The assessment is split into four phases. The first phase identifies the jurisdictions being studied. We begin by outlining the tropical zone according to the Köppen-Geiger classification (tropical rain forest, tropical monsoon, tropical wet and dry or savanna, hot and hot-semi-arid desert climates) based upon temperatures and precipitation, as calculated in the period 1980–2016 on a one kilometer grid [46], which updates, in greater spatial resolution and over a more recent timeframe, the work of Rubel and Kottek [47]. We then identify the government level responsible for producing local development plans. In some countries, it is the municipality, in others it is the district, the canton, the county, and sometimes the region. In Africa, these different jurisdictions do not necessarily correspond to territories of growing extension: A municipality in Cameroon, for example, may be larger and more populated than a district in Uganda. The jurisdictions of Angola, Ethiopia, Kenya, Madagascar, Malawi, Namibia, South Africa, Tanzania, Uganda, Zambia and Zimbabwe, which fall only in a small part in the tropical zone, were excluded. Finally, we considered only the jurisdictions with rural population exceeding 50% (Figure 1).

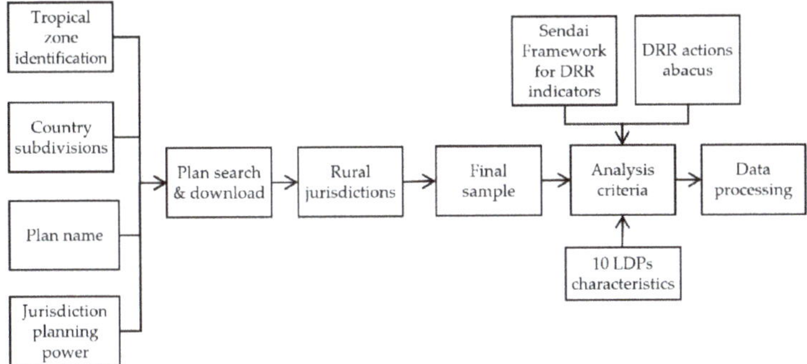

Figure 1. Flow chart of the assessment.

During the second phase, the plans are obtained. We searched Google with the name of each local jurisdiction and the integrated, local, municipal or strategic development plan and obtained 194 plans of 21 countries of which 107 were prepared after the Sendai framework for DRR, therefore likely to integrate its principles (Figure 2).

We also obtained the national strategies and guidelines for the preparation of the LDPs for each of the 21 countries.

The third phase identifies the information required to examine the 10 key characteristics of the plans: Climatic trend, hydro-climatic hazards, vulnerability and risk assessment, Sendai alignment, vision, strategy and objectives, actions, internal consistency, DRR budget and source, local and scientific knowledge integration, public participation (Table S1).

During the fourth phase, the information is collected, processed, and analyzed. The analysis of the climatic characterization ascertains if the plans are based on a series of at least thirty years of climatic observations and trace long-term local climatic scenarios. It is ascertained whether the plan is based on a CC vulnerability or risk assessment of the local communities. The alignment of the LDPs with the Sendai Framework verifies that the plan is subsequent to the National DRR Strategy and checks that at least the DRR and Sendai terms appear among the words of the plan. The analysis of the vision, strategy, and objectives verifies if DRR is present. The actions are broken down by purpose: Prevention, namely "to avoid existing and new disaster risks", and preparedness, i.e. "to recover from the impacts of likely, imminent or current disasters" [48]. Therefore, the actions are compared with those recurring in the literature on the individual hazards [49–60] to verify their completeness. The internal consistency analysis ascertains whether the plan relates to the climate trend, impacts, and possible solutions using specific frameworks. Secondly, the actions proposed by the communities are considered to verify if they subsequently appear among those retained by the plan and prioritized for the first year. The analysis of the budget reserved to DRR actions follows. In this regard, it is important to ascertain the share of the budget reserved for the DRR and to what extent it comes from the local administration's own resources, from the central government or from donors. The planning process is then examined to discover if local and technical knowledge has been integrated. Finally, the role of public participation in preparing 105 municipal development plans is ascertained: Analysis, planning, adopting, implementing, monitoring and evaluating (M&E) the plan while representing the many communities that make up a rural jurisdiction and gender constitute as many levels of participation [41].

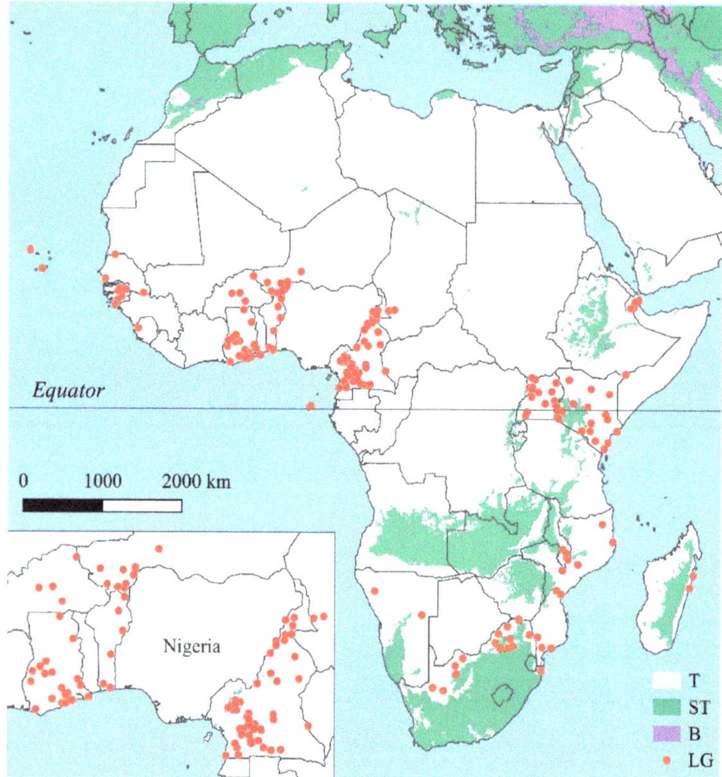

Figure 2. The 194 local jurisdictions (LG) in the tropical zone (T) with development plans in force and subtropical (ST) and boreal (B) zones.

3. Results

Monitoring the implementation of the Sendai Framework offers little information on the rural jurisdictions of tropical Africa. Considering DRR mainstreaming in 194 LDPs in force in half of the African countries gives a vision of 1.2 million km^2 populated by 37 million inhabitants. The rural jurisdictions considered, irrespective of the level (municipality, district or canton, county, region) are vast. The municipality of Kagisano Molopo (South Africa) for example, spreads across 23,800 km^2, that of Mintom (Cameroon) over 11,000 km^2. Jurisdictions contain an average of 52 settlements, as well as the municipal capital town and an average population of 193,000 inhabitants. Vast territories present several micro-climates and hazards that differ from one area to another, to be addressed with specific actions. In addition, the presence of many communities makes the representation during the plan preparation process a challenge.

3.1. Local Planning Overview

The local jurisdictions use different plans depending on the country: Municipal (district or regional) development plan, municipal (or county) integrated development plan, municipal (district or regional) strategic plan. Irrespective of the name, the plans have some common characteristics. Firstly, they are medium-term tools (3–5 years) prepared following national guidelines. Secondly, the plans are split into two parts: A factual base (with climate section) identifies the problems and available resources, then the actual planning, which usually defines the vision, strategies, objectives and actions. The latter are localized, quantified, their cost is defined, as well as the origin of the funds to be used to finance them. Some LDPs take stock of the previous plan. Other plans present the actions necessary for each community and the priority actions to be implemented during the first year. The strategies never detail the actions (quantity, costs). In the five countries with the highest number of plans, those in force and freely accessible on the web cover from 12% of the rural districts of Uganda to 94% of rural municipalities of South Africa (Table 1, Figure 3).

Table 1. Local development plans (LDPs) considered in five countries.

Country	Tropical Rural Jurisdictions	Tropical Rural Jurisdictions with Ongoing LDP	
	n.	n.	%
Cameroon	300	52	17
Ghana	46	25	54
Kenya	30	20	67
South Africa	16	15	94
Uganda	137	16	12

The Sendai Framework schedules the alignment of local DRR strategies with national ones. In fact, some countries have a national strategy (Cape Verde, Madagascar), others have a national plan or a preparation, management or response policy to the risk (Burkina Faso, Kenya, Mozambique, Namibia, South Africa, Uganda) or a national plan to strengthen DRR capacities (Chad, Niger). These heterogeneous tools at national scale define some important elements for local planning: The legal framework, the competencies of the different players, the operational procedures, the coordination between players in the case of emergency, the information path in the case of early warning, the capacities to be strengthened, the awareness-raising and participation, the methods of post-disaster recovery, and the funds to implement all of this. Today, out of the 21 investigated countries, 10 do not have a national strategy, three countries (Benin, Cameroon, and Mozambique) have LDPs already in force before the national strategy and only eight countries have a strategy following which a new generation of LDPs was formulated, thus being in line with the Sendai Framework (Figure 4). Without a national strategy, the local plans would be deprived of the aforementioned elements.

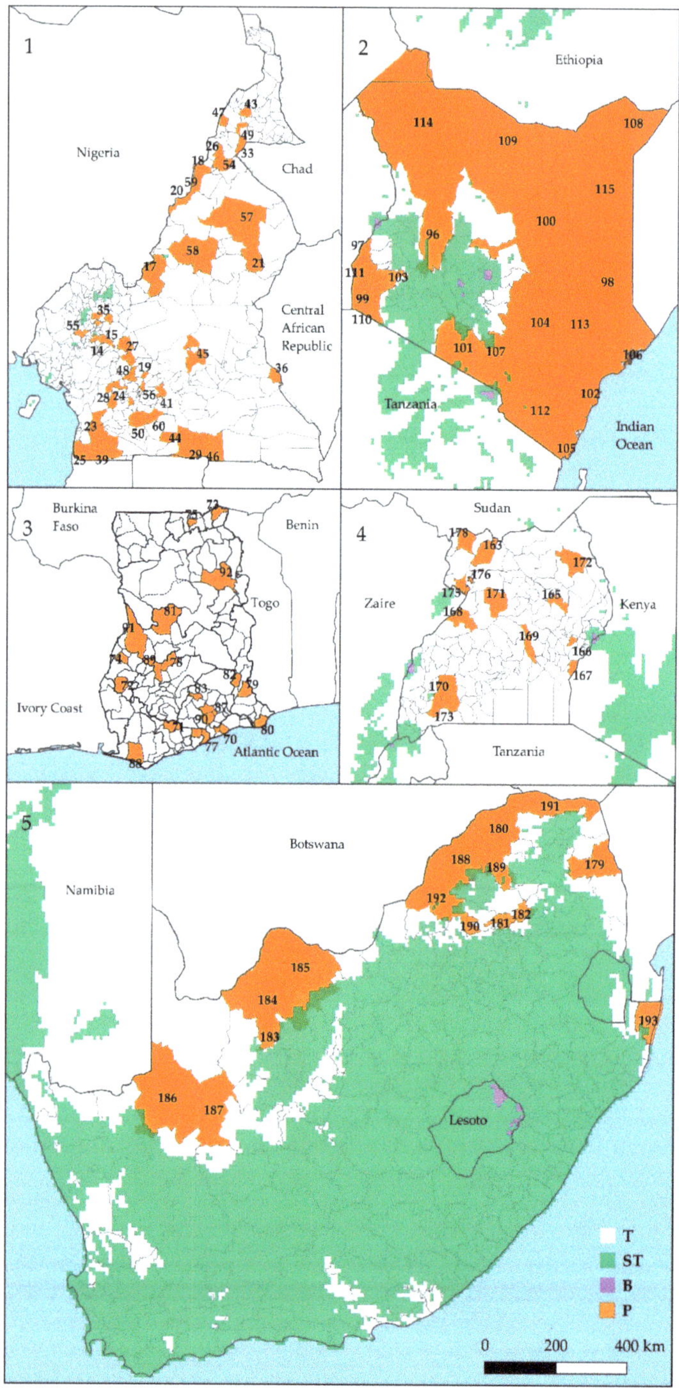

Figure 3. Cameroon (1), Ghana (3), Kenya (2), South Africa (5), Uganda (4). Ongoing local development plans in tropical zone (P) at September 2019.

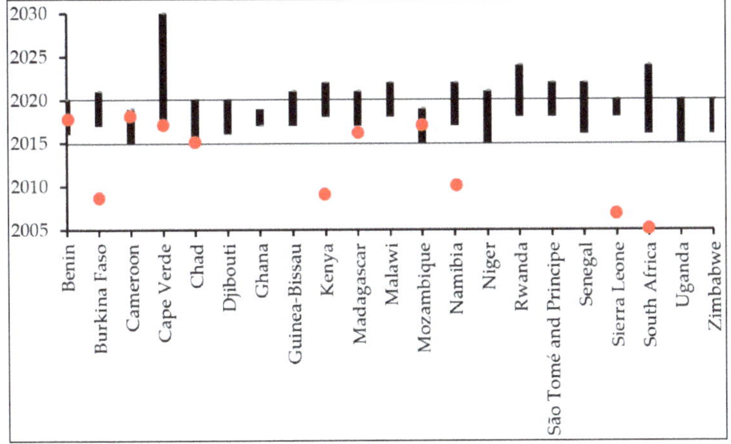

Figure 4. Ongoing LDPs (black segments), national disaster risk reduction (DRR) strategies (red dots), and the five years' time to attend targets of the Sendai framework for DRR.

3.2. Analysis Phase

3.2.1. Climate Characterization and Hazards

LDPs rarely characterize the climate as they consider too short a timeframe, or they are based only on local knowledge. In Niger, the latest generation LDPs are the exception: Precipitation and temperatures analyzed over a thirty-year series of data, identification of dry spells, sometimes of wind. Increase of temperatures, heavy rainfall, late onset of the wet season, and decline in total annual rainfall are the changes most frequently observed (Table 2). The CC local scenarios are never presented unless the national ones are reported. Conversely, the LDPs almost always identify the main hydro-climatic threats. Drought is by far the hazard most frequently reported by the plans. This is followed, with less frequency, by strong winds, bush fires, and floods. High temperatures, heavy storms, sea level rise, lightening, hailstorm, dust storms, salt-water intrusion and cyclones are reported more rarely (Table 3).

Table 2. Main local climate trends referred by 84 LDPs in rural Africa.

Climatic trend	Frequency %
Temperature increase	44
Heavy rainfall	35
Late onset of wet season	30
Yearly rainfall drop	30
Rainfall increase	21
Wet season early end	21
Air pollution	12
Dry spells increase	7
Heat wave	7
Cyclone	7

Table 3. Main hazards referred by 176 local development plans in rural Africa.

Hazard	Frequency %
Drought	56
Strong winds	38
Bush fire	29
Flood	28
Heat wave	10
Heavy storm	10
Sea level rise	8
Lightening	6
Hailstorm	5
Dust storm	3
Salt-water intrusion	3

3.2.2. Vulnerability and Risk Assessments

The considered LDPs do not have vulnerability assessments (quantification and localization of the vulnerable population to the individual hazards) and the very term "vulnerability" is found only in 43% of the plans. Even the risk assessment (probability of occurrence of the individual hazards, risk level, risk mapping) is absent. Moreover, the term hazard appears in only 46% of the plans.

3.3. *Planning Phase*

The LDPs mention the Sendai framework in 40% of cases and DRR in 46% of cases, although they include some typical DRR action without being defined as such. Many plans provide a long-term vision. The most frequent envisage a prosperous, wealthy community, which has achieved sustainable development and a high quality of life. The strategies proposed to achieve this vision involve the construction of roads, water, sanitation and hygiene (WASH), afforestation and use of green energy. The most frequent objectives remain the disaster prevention and response, the construction and maintenance of roads, increased access to drinking water, to basic services, to electricity (Table 4).

Table 4. Most common vision, strategies, and objectives in LDPs for rural Africa.

Term	Vision %	Strategy %	Objective %
Roads		42	22
WASH		29	21
Afforestation		25	
Green energy		25	
D prevention/response			24
Basic services			21
Electricity			21
Prosperous, wealthy	20		
Sustainable development	15		
High quality of life	13		
Plans n.	147	24	67

As to the actions, those of DRR prevention (119 plans) prevail over those of preparedness (49 plans). The most frequently illustrated actions of hydro-climatic DRR are afforestation (50%), the use of drought tolerant crops (23%), the construction or rehabilitation of boreholes and wells (22%) and CC sensitization (18%). The preparedness actions are much less frequent: Disaster relief (24%), early warning (24%), and disaster management plan (18%) (Table 5).

Table 5. Frequency of risk prevention and preparedness actions in 127 LDPs in tropical Africa.

Action	Prevention %	Preparedness %
Afforestation	50	
Drought tolerant crops	23	
Boreholes construction	22	
CC sensitization	18	
Durable construction materials	18	
Tree nursery	14	
Storm water drainage	12	
Windbreaks	11	
Agroforestry	11	
Cattle watering	11	
Hay cultivation	10	
Disaster relief		24
Early warning system		24
Disaster management plan		18
Volunteers groups establishment		16
Hazard mapping		10

We note the absence of some canonical actions for addressing the individual hazards. With respect to drought, there is no weather forecasting actions to encourage run-off infiltration, crop residues stocking, micro-credit, and self-help community groups [52], no crop insurance [60]. In relation to strong winds, there is no house retrofitting. There are no actions to deal with bush fires with respect to the many already trialed elsewhere in prevention (vulnerable population localization, alert protocols, access routes, lessening fuel pressure in forest areas) and in preparedness (fire break around homes) [53,54]. Half of the countries considered overlook the Ocean. Sea level rise threatens long stretches of West African coast [61]. The plans do not envisage any measure to reduce the risk of coastal flooding, such as natural buffers, beach nourishment, sea walls, and elevated houses [51]. Apart from the reduction of emissions, which has a long-term effect on the local climate, and afforestation, there are no other actions aimed at reducing the impact of heat waves. There is no early warning, no heat wave action plan. The prevention of dust storms with the stabilization of dunes with straw checkerboard and the increase of arboreal and herbaceous vegetation on denuded soils, for example, using wheatgrass, are not planned [55–57]. There are no measures to address salt-water intrusion consequent to sea level rise [58,59]: A frequent phenomenon in coastal areas that affects agriculture and access to drinking water. As to preparedness in general, there are no climate scenarios. Civil protection is mentioned by just 21% of the plans. There is no mention of simulations and drills, and even less so, actions informing the public on what to do in case of a warning (Figure 5). Resilience is a frequent word of the plan (65%).

The LDPs use different instruments to control the coherence between problems identified, objectives, and actions. The most complete are proposed by Cameroon's LDPs with the climatic threat-impact-strategies-actions framework and the logical framework (objective-outcome-indicator) (Table 6). Despite this, the DRR actions proposed by the individual communities rarely appear among the priority ones of the plan (Figure 6).

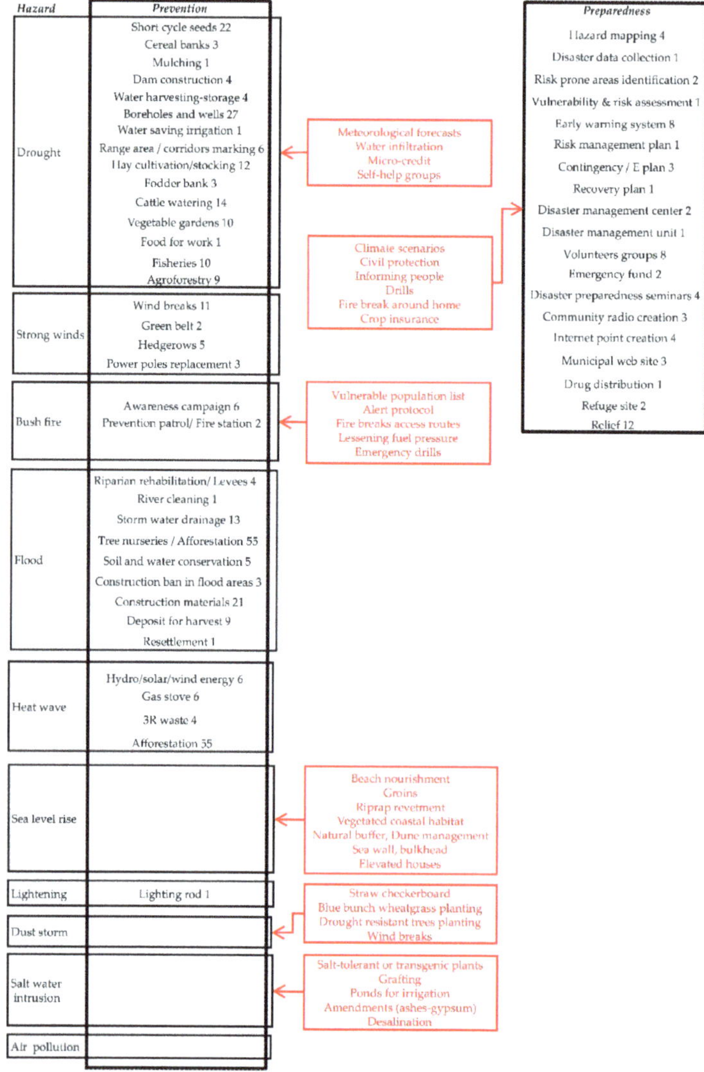

Figure 5. Fifty-seven actions explicitly addressed to DRR in 111 LDPs for rural tropical Africa (figures stand for action frequency) and missing actions (red).

Table 6. Tools for threat-actions consistency control in 100 LDPs of tropical rural Africa.

Tool	Plans %
Climate trend-impact-solution	41
Logical framework	40
SWOT	30
Strength-weakness	6
Problems and needs	5
Assets and constraints	4
Problems, impacts, needs	4

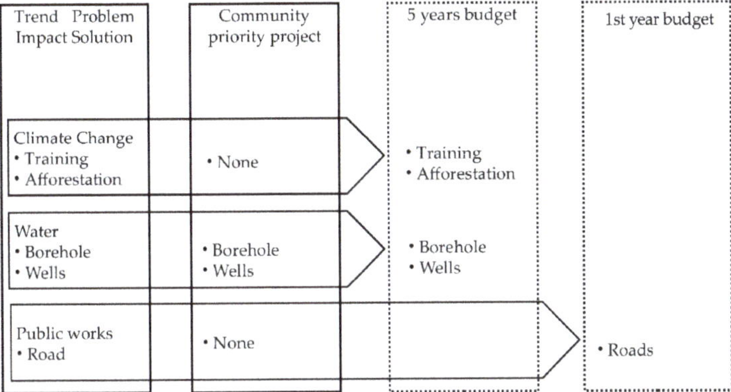

Figure 6. Consistency of climate change (CC) actions between the planning phases in Cameroun's LPDs.

The resources scheduled by the plans in the medium-term to implement DRR vary greatly from country to country, but in the majority of cases they are below 20% of overall expenditure. Benin, Niger, and South Africa have plans that reserve the highest share for DRR. Rwanda, Djibouti, and Malawi allocate the lowest share (Figure 7). Local governments never finance over 10% of DRR actions. The remaining 90% is borne by the State or by donors.

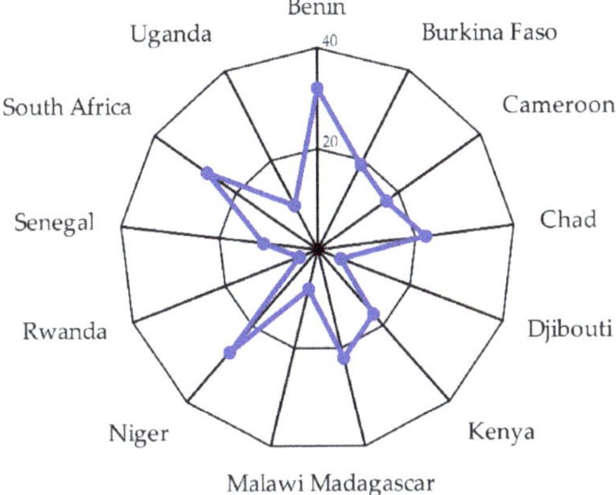

Figure 7. Budget (%) for hydro-climatic DRR in 13 African countries.

3.4. Plan Preparation Process

Three aspects of the plan preparation process are relevant for actions implementation: The integration of scientific knowledge with local knowledge, public involvement, and gender representation.

The main fields in which scientific knowledge can contribute to LDPs are climate characterization, hazard probability of occurrence, hazard prone zones identification, risk level and assessment within a back-casting exercise (expected effects of actions), definition of some preparedness actions (e.g., early warning system). In none of these fields, with the sole exception of Niger's LPDs, do the plans use scientific knowledge: The information is extracted exclusively from local knowledge using participatory rural appraisal tools in the analysis phase.

The information contained in the plans allow for two aspects of participation to be appreciated. First, the community, community-based organizations (CBOs), and individual citizens participation in analysis, planning, implementation, plan monitoring and evaluation (M&E). Second, gender representation in the analysis, plan approval, and M&E. The analysis is developed only on the municipal development plans, which are the real arena on which participation can develop.

Local jurisdictions in rural Africa contain many communities. On average, the municipalities have 52 human settlements. As a consequence, in many cases, the participatory analysis process occurs by bringing together the delegates of each community into zone centers. In the cases where the plans provide information on this phase of the preparation process, the participation attends 81% of the plans considered. In this phase, the communities provide information on their needs, sometimes solutions, expressed in the best of cases as priority micro-projects (or actions). Public participation of individual communities and individuals is explicitly required by 70% of the plans to implement afforestation and health actions (vaccinations, construction of health centers) and in financing of actions mainly. However, only one out of three plans involve communities in strategies, priority actions identification, and budgeting. Planning is reserved to the municipal council, sometimes expanded to ministerial representatives and economic operators. Representatives of CBOs and communities are excluded from the M&E committee in two plans out of three (Table 7).

Table 7. Public participation in municipal development plans preparation in rural tropical Africa.

Plan Phase	MDPs		Participation	
	Considered	Unknown	Yes	No
	n.	n.	%	%
Analysis	105	22	81	19
Planning	101	43	36	64
Implementation	44	14	70	30
M&E	102	42	34	66

Almost all of the plans call for greater participation of women and sometimes of young people and minorities in the decision-making processes. However, there is little understanding of how to achieve it. Furthermore, awareness-raising is proposed, rather than organizing activities at times that allow women to participate and to lighten the workload on their shoulders, increase the level of education. In fact, the plans contain little information on how gender involvement occurred in the preparation process. In the analysis phase, the share of women in community delegations is on average 25%. The adoption of the plans is the responsibility of the city council. In this assembly, gender representation is on average just 28%. However, the differences in gender representation from country to country are large: Senegal (44%), South Africa (43%), Cameroon (24%), Niger (16%), Burkina Faso (12%), Benin (5%) (Table S1). Monitoring and evaluation of the plan is the responsibility of the M&E committee, a body in which gender representation drops to 16% (Table 8).

Table 8. Gender participation in the municipal development plans (MDPs) preparation in rural tropical Africa.

Plan Phase	MDPs	Gender Participation		
		Unknown		Known
	n.	n.	n.	%
Analysis	18	13	5	25
Adoption	128	64	64	28
M&E	60	34	26	16

4. Discussion

In Africa, DRR is particularly important for protecting the primary sector from the impact of climate change and thus to allow agriculture, breeding, and forests to reduce poverty and food insecurity, which remains the characteristic trait of the Continent compared to other regions of the Global South. Unfortunately, the monitoring of the Sendai Framework carried out by UNDRR so far tells us little about the state of DRR in rural Africa. Through our systematic assessment, we have ascertained that almost all countries have LDPs at the scale of the municipality, district/county, canton, or region. In half of the countries, those instruments are freely accessible on the web. The analysis of 10 characteristics of the plans has allowed us to characterize the local DRR.

The plans not having climate analyses do not identify the hazards according to the probability of occurrence, and they do not outline climate scenarios. These deficiencies are serious when considering the vastness of the jurisdictions considered, which require a spatial characterization of the climate. Exceptions aside, hydro-climatic threats are identified based solely upon local knowledge and are not positioned hierarchically by severity or frequency. Vulnerability and risk assessments are lacking.

The planning process does not integrate scientific and local knowledge and presents problems already observed in other contexts. The plans prepared internally to the municipality are of higher quality than those prepared by consultants [62].

The preparation process of the LDPs at a municipal scale begins with the identification of the needs of the individual communities. When this occurs in territorial assemblies attended by the delegates of the communities of the zone, the representation of the communities and of gender is low. This is the only arena in which public participation takes place and confirms what has already been observed ten years ago with respect to the process of preparing the first generation local development plans in the Sahel [63].

Only in a third of cases do communities decide on the municipal development plan and participate in monitoring and evaluation activities. In the rest of the cases the delegates of the community do not appear in any committee, do not participate in any negotiation, do not receive any delegation of power (empowerment) [64], and do not control anything, not even the stage of progress of the plan [41].

The true planning occurs in another arena, which is accessed by the municipal councilors, technical services, representatives of the ministries, and sometimes donors. Here, gender representation is 28% only. Planning usually begins with a visioning exercise which is not, however, followed by that of back-casting: Estimating how many actions would be necessary to significantly reduce the climate risk and proceeding backwards to identify how many of them to carry out with the plan. This exercise is impeded by the lack of knowledge on the frequency of occurrence of the hazards, on the exposed zones, on the impact of DRR actions. For example, in relation to pluvial floods, we do not know how much the run-off reduces on different types of soil by virtue of infiltration works (trapezoidal bund, half-moons, stone lines, etc.). Therefore, it is impossible to estimate the reduction of risk following the risk treatment. The objectives are dominated by efforts to reduce the hydro-climatic risks, then to break isolation (roads) and meet primary needs (WASH, electricity).

When risk prevention actions are planned, the most frequent concern afforestation (as a means of conserving soil), agriculture and access to water (drought prevention). Different plans (Kenya) contain mitigation actions (renewable energy, energy saving/LED, improved stoves). However, mitigation actions are not able to contain the rise in temperatures, for which land-based mitigation (vegetation) and interventions on building materials are required [29]. Ultimately, the priority actions contain little DRR: The exposed zones are not identified, the dynamics of the settlements within them are not known, early warning systems are infrequent or not designed involving populations at risk [65], house retrofitting is not facilitated. School education as a public awareness channel [28] is not practiced, apart from tree planting.

The planned climate actions are never financed by the municipalities over 10%. The remainder is borne by donors or is simply not funded. The financial weakness of rural local administrations [33,36] is also confirmed in tropical Africa. In these conditions the implementation of the plan falls largely on the shoulders of communities and individuals, which are asked to finance the actions or to provide materials and labor but in two thirds of cases they have no voice in the planning process. Although the plan preparation process lasts an average of 10 months, genuine public participation is infrequent.

The literature is filled with examples in which the lack of participation translates into a lack of implementation [36,38,41]. Several local governments are planning to activate a municipal website.

However, uploading the LDP on the web remains the most common E-government action [66,67]. Simple information systems through smartphones are not used, with the sole exception of Senegal.

The monitoring indicators used by UNDRR do not measure the local capacity of DRR. It is not sufficient to know country by country the percentage of local jurisdictions with DRR strategies aligned with the national DRR strategy. It is necessary to know the content of those strategies, how they are formulated, what potential they have to be implemented. Widespread poverty [9] and food insecurity [3] lead many local governments to support primarily agro-pastoral production and to meet primary needs which are still unsatisfied (WASH, electricity, health, education), before dealing with DRR. By doing so, the primary sector remains exposed to hydro-climatic hazards.

The assessment has produced four unexpected findings. Firstly, the large number of LDPs in force (and more are still not freely accessible). Secondly, a very articulated palette of hazards, despite being dominated by the ubiquitous drought. Thirdly, the emergence of some actions still not widespread but important: Risk reduction local funds, considered important instruments for reducing poverty [9,13] and the use of renewable energy sources. The latter have the benefit of being present in situ (sun, wind, water), do not require transportation costs and are not subject to price increases (petrol), factors that penalize the smallholder farmers of remote rural areas. Fourthly, the lack of genuine public participation, despite this term being in many plans among the most frequently used.

The main limitation of the assessment is the number of plans examined, moreover relating to just half of the Continent's tropical countries, of which only 107 plans were formulated after the Sendai Framework for DRR. Another limitation is the fact of considering only planning and not implementation.

The major problems that a multicountry and multilanguage assessment has to face is the understanding and comparison of the plan budgets, which sometimes provide little detail (actions merged by sector, absence of total cost, etc.) and of the participation process.

The implications of this assessment concern DRR mainstreaming in the next generation of LDPs. We have observed some components of the LDPs critical for DRR. With regards to the use of the assessment results, three recommendations apply.

The first recommendation concerns the national guidelines for the preparation of LDPs. If DRR mainstreaming is to be improved in the next generation of plans, the guidelines should require climate characterization and mobilize national meteorological services to provide climate services to local governments. The plans should identify the zones (and inhabitants) exposed to hazards with greater probability of occurrence, quantify DRR, and estimate how far the risk is reduced if the actions scheduled in the plan are implemented in the 3/5 years of planning, introducing the back-casting exercise. The quantity of tables required should be reduced in favor of those strictly necessary to control the coherence between threats and priority actions. Plans that do not include DRR hydro-climatic actions among the priority ones must motivate this decision. Finally, the guidelines should require a precise description of the plan methodology and the representativeness of communities and gender in all phases of the process: Analysis, planning, adoption, and M&E.

The second recommendation concerns the ministries of those countries that have still not made the LDPs freely accessible on the internet. The portals from which it is possible to access the plans of Cameroon, Ghana, Kenya, Namibia, and South Africa are best practices to be used for inspiration.

The third recommendation concerns ODA. Donors should consider supporting the revision of the national guidelines for LDPs preparation, climate services to local administrations (including the upgrading of the local weather stations and early warning systems), and strengthening the capacity of local planning units.

These three recommendations, if implemented, would improve DRR at local scale. Monitoring the number of local risk reduction strategies does not help to understand if a substantial reduction of losses and damage and increase of disaster risk strategies is being achieved.

5. Conclusions

Five years after the Sendai Framework, we do not know whether rural municipalities in tropical Africa have truly succeeded in mainstreaming DRR in their LDPs. The problem is knowing the quality of these plans, rather than the number of local strategies in place. We, therefore, considered the mainstreaming of the DRR in 194 LDPs of rural Africa post-Sendai, observing ten characteristics of these tools.

With a few exceptions aside, we noted the absence of climate characterization and scenarios. Vision, strategies, and objectives rarely mention DRR. LDPs strengthen livelihoods (agriculture, livestock, fisheries), increase the access to basic services (WASH, electricity) and to infrastructures to break the isolation of remote communities. Some actions in these sectors are also of DRR (access to drinking water and watering) but are not enough to face sea level rise, bush fires, dust storms, salt- water intrusion. Civil protection and crop insurance are missing from preparedness actions. The DRR actions identified by the communities rarely become priority actions of the plan. Public participation is more a goal than the approach followed to prepare the plan. The plan requires the participation of the communities and individuals to be implemented but the national rules and sometimes local administrators exclude genuine public participation in the planning process which are consequently poorly representative of the communities and gender. It should be remembered that gender share among the municipal councilors stops on average at 28%. Only Senegal and South Africa approach the correct gender representation of municipal councilors. This inevitably limits the plan implementation. The planning process, therefore, needs to be reviewed if DRR is to be fully integrated at local scale and translated into solid appropriate actions by the local communities.

In developing the assessment, we considered plans that sometimes provide little information about the preparation process and budget. This restricted the sample considered to analyze these characteristics, limiting the significance of the relative results.

In many tropical African countries, it is not possible to increase significantly DRR by the action of individual smallholder farmers and herders. For this reason, DRR is an institutional task of local governments, since the first steps of the administrative decentralization process started twenty years ago in many countries of tropical Africa. The main tool used for this purpose is the LDP, as it is a medium-term planning tool, mandatory by law and with a long tradition. However, the LDP has weaknesses that should be identified to be reduced. If this is not done, disaster risk reduction will not take root and CC and variability will continue to generate food insecurity and poverty.

The problem of not knowing the content and process of the DRR strategies in tropical Africa has been addressed by considering 194 LDPs using an assessment framework that can be repeated in other countries. We recommend that any organization willing to support local disaster risk reduction should consider the assessment framework proposed in this study and use it to switch from an occasional assessment to a tracking process. It will thereby be possible to continue to have an understanding of improvements and residual fragilities in local DRR rather than merely counting how many strategies are in force.

Supplementary Materials: The following are available online at http://www.mdpi.com/2071-1050/12/6/2196/s1, Table S1. Key characters of 194 local development plans for rural tropical Africa freely accessible on the web.

Author Contributions: Conceptualization, M.T.; methodology, M.T.; investigation, M.T. and S.B.; writing—original draft preparation, M.T.; writing—review and editing, M.T. and S.B.; visualization, S.B.; funding acquisition, M.T. All authors have read and agreed to the published version of the manuscript.

Funding: This research was funded by DIST-Politecnico and University of Turin, Italy.

Conflicts of Interest: The authors declare no conflict of interest. The funders had no role in the design of the study; in the collection, analyses, or interpretation of data; in the writing of the manuscript, or in the decision to publish the results.

References

1. UNDESA-United Nations Department of Economic and Social Affairs. *World Urbanization Prospects. The 2018 Revision*; United Nations: New York, NY, USA, 2018.
2. Davis, B.; Di Giuseppe, S.; Zezza, A. Are African households (not) leaving agriculture? Patterns of households' income sources in rural Sub-Saharan Africa. *Food Policy* **2017**, *67*, 153–174. [CrossRef]
3. FAO; IFAD; UNICEF; WFP; WHO. *The State of Food Security and Nutrition in the World. Safeguarding against Economic Slowdown and Downturns*; FAO: Rome, Italy, 2019.
4. Beegle, K.; Christiaensen, L.; Dabalen, A.; Gaddis, I. *Poverty in a Rising Africa*; The World Bank: Washington, DC, USA, 2016. [CrossRef]
5. Collier, P.; Conway, G.; Venables, T. Climate change in Africa. *Oxf. Rev. Econ. Policy* **2008**, *24*, 337–353. [CrossRef]
6. Diao, X.; Hazell, P.; Thurlow, J. The role of agriculture in African development. *World Dev.* **2010**, *38*, 1375–1383. [CrossRef]
7. Imai, K.S.; Gaiha, R.; Garbero, A. Poverty reduction during the rural-urban transformation: Rural development is still more important than urbanisation? *J. Policy Modeling* **2017**, *39*, 963–982. [CrossRef]
8. Page, J.; Shimeles, A. Aid, employment and poverty in Africa. *Afr. Dev. Rev.* **2015**, *27*, 17–30. [CrossRef]
9. Barrett, C.B.; Christiaensen, L.; Sheahan, M.; Shimeles, A. On the structural transformation of rural Africa. *J. Afr. Econ.* **2017**, *26*, 11–35. [CrossRef]
10. Schlenker, W.; Lobell, D.B. Robust negative impacts of climate change on African agriculture. *Environ. Res. Lett.* **2010**, *5*, 1–8. [CrossRef]
11. Muller, C.; Cramer, W.; Hare, W.L.; Lotze-Campen, H. Climate change risks for African agriculture. *PNAS* **2011**, *108*, 4313–4315. [CrossRef]
12. Adhikan, U.; Nejadhashemi, A.P.; Woznicki, S.A. Climate change and East Africa: A review of impact on major crops. *Food Energy Secur.* **2015**, *4*, 110–132. [CrossRef]
13. Hansen, J.; Hellin, J.; Fisher, E.; Cairns, J.; Stirling, C.; Lamanna, C.; van Etten, J.; Rose, A.; Campbell, B. Climate risk management and rural poverty reduction. *Agric. Syst.* **2019**, *172*, 28–46. [CrossRef]
14. Davidson, O.; Halsnaes, K.; Huq, S.; Kok, M.; Metz, B.; Sokona, Y.; Verhagen, Y. The development and climate nexus: The case of Sub-Saharan Africa. *Clim. Policy* **2003**, *3S1*, 97–113. [CrossRef]
15. Kotir, J.H. Climate change and variability in Sub-Saharan Africa: A review of current and future trends and impacts on agriculture and food security. *Environ. Dev. Sustain.* **2011**, *13*, 587–605. [CrossRef]
16. Pelling, M.; Wisner, B. *Disaster Risk Reduction. Cases from urban Africa*; Earthscan: London, UK, 2008.
17. Adelekan, I.; Johnson, C.; Manda, M.; Matyas, D.; Mberu, B.U.; Parnell, S.; Pelling, M.; Satterthwaite, D.; Vivekananda, J. Disaster risk and its reduction: An agenda for urban Africa. *Int. Dev. Plan. Rev.* **2015**, *37*, 33–43. [CrossRef]
18. Gore, C. Climate change adaptation and African cities: Understanding the impact of government and governance on future action. In *The Urban Climate Challenge*; Johnson, C., Toly, N., Shroeder, H., Eds.; Routledge-Taylor and Francis: New York, NY, USA, 2015; pp. 205–224. [CrossRef]
19. Tiepolo, M. Flood risk reduction and climate change in large cities south of the Sahara. In *Climate Change Vulnerability in Southern African Cities. Building Knowledge for Adaptation*; Macchi, S., Tiepolo, M., Eds.; Springer: Cham, Switzerland, 2014; pp. 19–36. [CrossRef]
20. Thompson, H.E.; Berrang-Ford, L.; Ford, J.D. Climate change and food security in Sub-Saharan Africa: A systematic literature review. *Sustainability* **2010**, *2*, 2719–2733. [CrossRef]
21. Connolly-Boutin, L.; Smit, B. Climate change, food security, and livelihoods in sub-Saharan Africa. *Reg. Environ. Chang.* **2016**, *16*, 385–399. [CrossRef]
22. United Nations. *Sendai Framework for Disaster Risk Reduction 2015–2030*; UNISDR: Geneva, Switzerland, 2015.
23. UNISDR-United Nations office for Disaster Risk Reduction. *Technical Guidance for Monitoring and Reporting on Progress in Achieving the Global Targets on the Sendai Framework for DRR*; UNISDR: Geneva, Switzerland, 2017.
24. UNDRR-United Nations office for Disaster Risk Reduction. Measuring Implementation of the Sendai Framework. Available online: https://sendaimonitor.unisdr.org (accessed on 1 November 2019).
25. UNDRR. *Global Assessment Report on Disaster Risk Reduction*; United Nations Office for Disaster Risk Reduction: Geneva, Switzerland, 2019.

26. Wheeler, S.M. State and municipal climate change plans. The first generation. *J. Am. Plan. Assoc.* **2008**, *74*, 481–496. [CrossRef]
27. Bassett, E.; Shandas, V. Innovation and climate action planning. *J. Am. Plan. Assoc.* **2010**, *76*, 435–450. [CrossRef]
28. Tang, Z.; Brody, S.D.; Quinn, C.; Chang, L.; Wei, T. Moving from agenda to action: Evaluating local climate change action plans. *J. Environ. Plan. Manag.* **2010**, *53*, 41–62. [CrossRef]
29. Measham, T.G.; Preston, B.L.; Smith, T.F.; Brooke, C.; Gorddard, R.; Withycombe, G.; Morrison, C. Adapting to climate change through local municipal planning: Barriers and challenges. *Mitig. Adapt Strat. Glob Chang.* **2011**, *16*, 889–909. [CrossRef]
30. Stone, B.; Vargo, J.; Habeeb, D. Managing climate change in cities: Will climate action plans works? *Landsc. Urban Plan.* **2012**, *107*, 263–271. [CrossRef]
31. Fu, X.; Tang, Z. Planning for drought-resilient communities: An evaluation of local comprehensive plans in the fastest growing counties in the US. *Cities* **2013**, *32*, 60–69. [CrossRef]
32. Lyles, W.; Berke, P.; Smith, G. A comparison of local hazard mitigation plan quality in six states, USA. *Landsc. Urban Plan.* **2014**, *122*, 89–99. [CrossRef]
33. Reckien, D.; Flacke, J.; Dawson, R.J.; Heidrich, O.; Olazabal, M.; Foley, A.; Hamann, J.-P.; Orru, H.; Salvia, M.; De Gregorio Hurtado, S.; et al. Climate change response in Europe: What's the reality? Analysis of adaptation and mitigation plans from 200 urban areas in 11 countries. *Clim. Chang.* **2014**, *122*, 331–340. [CrossRef]
34. Araos, M.; Berrang-Ford, L.; Ford, J.D.; Austin, S.E.; Biesbroek, R.; Lesnikowski, A. Climate change adaptation planning in large cities: A systematic global assessment. *Environ. Sci. Policy* **2016**, *66*, 375–382. [CrossRef]
35. Shi, L.; Chu, E.; Carmin, J. Global patterns of adaptation planning. Results of a global survey. In *The Routledge Handbook of Urbanization and Global Environmental Change*; Seto, K.C., Solecki, W.D., Griffith, C.A., Eds.; Routledge: London, UK, 2015; pp. 336–349.
36. Tiepolo, M.; Cristofori, E. Climate change characterization and planning in large tropical and subtropical cities. In *Planning to cope with Tropical and Subtropical Climate Change*; Tiepolo, M., Ponte, E., Cristofori, E., Eds.; De Gruyter Open: Berlin, Germany, 2016; pp. 6–41. [CrossRef]
37. Tiepolo, M. Relevance and quality of climate planning for large and medium-sized cities of the tropics. In *Renewing Local Planning to face Climate Change in the Tropics*; Tiepolo, M., Pezzoli, A., Tarchiani, V., Eds.; Springer: Cham, Switzerland, 2017; pp. 199–226. [CrossRef]
38. Horney, J.; Nguyen, M.; Salvesen, D.; Dwyer, C.; Cooper, J.; Berke, P. Assessing the quality of rural hazard mitigation plans in the Southeastern United States. *J. Plan. Educ. Res.* **2017**, *37*, 56–65. [CrossRef]
39. Reckien, D.; Salvia, M.; Pietrapertosa, F.; Simoes, S.G.; Olazabal, M.; De Gregorio Hurtado, S.; Geneletti, D.; Krkoška Lorenzová, E.; D'Alonzo, V.; Krook-Riekkola, A.; et al. Dedicated versus mainstreaming approaches in local climate plans in Europe. *Renew. Sustain. Energy Rev.*. [CrossRef]
40. Chirenje, L.I.; Giliba, R.A.; Musamba, E.B. Local communities' participation in decision-making process through planning and budgeting in African countries. *Chin. J. Popul. Resour. Environ.* **2013**, *11*, 10–16. [CrossRef]
41. Arnstein, S. A ladder of citizen participation. *J. Am. Inst. Plan.* **1969**, *35*, 216–224. [CrossRef]
42. Brody, S.D.; Godschalk, D.R.; Burby, R.J. Mandating citizen participation in plan making: Six strategic planning choices. *J. Am. Inst. Plan.* **2003**, *69*, 245–264. [CrossRef]
43. Godschalk, D.R.; Brody, S.; Burby, R. Public participation in natural hazard mitigation policy formation: Challenges for comprehensive planning. *J. Environ. Plan. Manag.* **2003**, *46*, 733–754. [CrossRef]
44. Evans-Cowley, J.; Hollander, J. The new generation of public participation: Internet-based participation tools. *Plan. Pract. Res.* **2010**, *25*, 397–408. [CrossRef]
45. Finch, C. Participation in Kenya's Local Development Funds: Reviewing the Past to Inform the Future. 2015. Available online: http://documents.worldbank.org/curated/en/666021468172488909/Participation-in-Kenya (accessed on 28 February 2020).
46. Beck, H.E.; Zimmermann, N.E.; McVicar, T.R.; Vergopolan, N.; Berg, A.; Wood, E.F. Data descriptor: Present and future Köppen-Geiger climate classification maps at 1-km resolution. *Sci. Data* **2018**, *5*, 1–12. [CrossRef] [PubMed]
47. Rubel, F.; Kottek, M. Observed and projected climate shifts 1901-2100 depicted by world maps of the Köppen-Geiger climate classification. *Meteorol. Z.* **2010**, *19*, 135–141. [CrossRef]

48. UN General Assembly. *Report on the Open-Ended Intergovernmental Expert Working Group on Indicators and Terminology Relating to Disaster Risk Assessment, 71 Session*; United Nations: New York, NY, USA, 2017; Available online: Unisdr.org/we/inform/publications/51748 (accessed on 9 January 2020).
49. Government of India. *National Disaster Management Authority. Guidelines for Preparation of Action Plan-prevention and Management of Heat Wave*; NDMA: New Delhi, India, 2017.
50. CPSL-Coastal Protection and Sea Level Rise. *CSPL Third Report—The Role of Spatial Planning and Sediment in Coastal Risk Management*; Common Wedden Sea Secretariat: Wilhelm Shaven, Germany, 2010.
51. Fu, X.; Gomaa, M.; Deng, Y.; Peng, Z.-R. Adaptation planning for sea level rise: A study of US coastal cities. *Journal Environment Planning Management* **2016**, *60*, 249–265. [CrossRef]
52. UNISDR. *Drought Risk Reduction Framework and Practices. Contributing to the Implementation of the Hyogo Framework for Action*; UNISDR: Geneva, Switzerland, 2019.
53. The State of Queensland Audit Office. *Follow-up of Bushfire Prevention and Preparedness. Report 5: 2018-19*; Queensland Audit Office: Brisbane, Australia, 2018.
54. Dube, E. Improving disaster risk reduction capacity of district civil protection units in managing veld fires: A case of Mangwe district in Matabeleland South Province, Zimbabwe. *Jambá J. Disaster Risk Stud.* **2015**, *7*, 13. [CrossRef]
55. UNCDD-United Nations Convention to Combat Desertification. *Global Alarm: Dust and Sandstorms from the World's Drylands*; UNCDD: Bangkok, Thailand, 2002.
56. UNEP-United Nations Environment Program; WMO; UNCCD. *Global Assessment of Sand and Dust Storms*; UNEP: Nairobi, Kenya, 2016.
57. Middleton, N.; Kang, U. Sand and dust storms: Impact mitigation. *Sustainability* **2017**, *9*, 1053. [CrossRef]
58. Shammi, M.; Karmakar, B.; Rahman, M.M.; Islam, M.S.; Rahman, R.; Uddin, M.K. Assessment of salinity hazard of irrigation water quality in monsoon season of Batiaghata upazila, Khulna district, Bangladesh and adaptation strategies. *Pollution* **2016**, *2*, 183–197. [CrossRef]
59. Ondrasek, G.; Rengel, Z.; Veres, S. Soil salinisation and salt stress in crop production. In *Abiotic Stress in Plants. Mechanisms and Adaptation*; Shanker, A., Venkateswarlu, B., Eds.; IntechOpen: London, UK, 2011. [CrossRef]
60. Tadesse, M.A.; Bekele, A.S.; Erenstein, O. Weather index insurance for managing drought risk in smallholder agriculture: Lessons and policy implications for sub-Saharan Africa. *Agric. Food Econ.* **2015**, *3*, 3–26. [CrossRef]
61. Becker, M.; Karpytchev, M.; Papa, F. Hotspots of relative sea level rise in the Tropics. In *Tropical Extremes: Natural Variability and Trends*; Venugopal, V., Sukhatme, J., Murtugudde, R., Roca, R., Eds.; Elsevier: Amsterdam, The Netherlands, 2019; pp. 203–262.
62. Frazier, T.G.; Walker, M.H.; Kumari, A.; Thompson, C.M. Opportunities and constraints to hazard mitigation planning. *Appl. Geogr.* **2013**, *40*, 52–60. [CrossRef]
63. Tiepolo, M.; Artuso, M. Les plans de développement communal au Sahel. In *Suivi et Évaluation des Plans de Développement Communal au Sahel*; Tiepolo, M., Ed.; L'Harmattan: Turin, Italy, 2011; pp. 21–44.
64. Tshabalala, E.L.; Lombard, A. Community participation in the integrated development plan: A case study of Govan Mboki municipality. *J. Public Adm.* **2009**, *44*, 396–409.
65. Tarchiani, V.; Massazza, G.; Rosso, M.; Tierpolo, M.; Pezzoli, A.; Ibrahim, M.H.; Katiellou, G.L.; Tamagnone, P.; De Filippis, T.; Rocchi, L.; et al. Community and impact based early warning system for flood risk preparedness: The experience of the Sirba River in Niger. *Sustainability* **2020**, *12*, 1802. [CrossRef]
66. Conroy, M.M.; Evans-Cowley, J. E-participation in planning: An analysis of cities adopting on-line citizen participation tools. *Environ. Plan. C Gov. Policy* **2006**, *24*, 371–384. [CrossRef]
67. Wilson, A.; Tewdwr-Jones, M.; Comber, K. Urban planning public participation and digital technology: App development as a method of generating citizen involvement in local planning process. *Environ. Plan. B Urban Anal. City Sci.* **2019**, *46*, 286–302. [CrossRef]

MDPI
St. Alban-Anlage 66
4052 Basel
Switzerland
Tel. +41 61 683 77 34
Fax +41 61 302 89 18
www.mdpi.com

Sustainability Editorial Office
E-mail: sustainability@mdpi.com
www.mdpi.com/journal/sustainability

www.ingramcontent.com/pod-product-compliance
Lightning Source LLC
LaVergne TN
LVHW071623170726
843515LV00009B/2464

* 9 7 8 3 0 3 6 5 1 3 7 0 6 *